AF597994

# Ionic Liquids

Queen's University, Belfast,
United Kingdom

22–24 August 2011

# FARADAY DISCUSSIONS

Volume 154, 2012

RSC Publishing

**The Faraday Division of the Royal Society of Chemistry**, previously the Faraday Society, founded in 1903 to promote the study of sciences lying between Chemistry, Physics and Biology.



*Faraday Discussions* (Print ISSN 1359-6640, Electronic ISSN 1364-5498) is published 6 times a year by the Royal Society of Chemistry, Thomas Graham House, Science Park, Milton Road, Cambridge, UK CB4 0WF. Volume 154 ISBN-13: 978 1 84973 4455

2012 annual subscription price: print+electronic £709, US $1,322; electronic only £673, US $1,256. Customers in Canada will be subject to a surcharge to cover GST. Customers in the EU subscribing to the electronic version only will be charged VAT. All orders, with cheques made payable to the Royal Society of Chemistry, should be sent to RSC Distribution Services, c/o Portland Customer Services, Commerce Way, Colchester, Essex, UK CO2 8HP.
Tel +44 (0) 1206 226050;
E-mail sales@rscdistribution.org

If you take an institutional subscription to any RSC journal you are entitled to free, site-wide web access to that journal. You can arrange access *via* Internet Protocol (IP) address at www.rsc.org/ip. Customers should make payments by cheque in sterling payable on a UK clearing bank or in US dollars payable on a US clearing bank. Periodicals postage is paid at Rahway, NJ and at additional mailing offices. Airfreight and mailing in the USA by Mercury Airfreight International Ltd., 365 Blair Road, Avenel, NJ 07001, USA.

US Postmaster: send address changes to *Faraday Discussions*, c/o Mercury Airfreight International Ltd., 365 Blair Road, Avenel, NJ 07001. All despatches outside the UK by Consolidated Airfreight.

PRINTED IN THE UK

***Faraday Discussions*** documents a long-established series of *Faraday Discussion* meetings which provide a unique international forum for the exchange of views and newly acquired results in developing areas of physical chemistry, biophysical chemistry and chemical physics.

# Ionic Liquids

Faraday Discussions

**www.rsc.org/faraday_d**

A General Discussion on Ionic Liquids was held Queen's University, Belfast, United Kingdom on 22nd, 23rd and 24th August 2011.

*RSC Publishing is a not-for-profit publisher and a division of the Royal Society of Chemistry. Any surplus made is used to support charitable activities aimed at advancing the chemical sciences. Full details are available from www.rsc.org*

## CONTENTS

ISSN 1359-6640; ISBN 978-1-84973-445-5

**Cover**
See Costa Gomes *et al.*, *Faraday Discuss.*, 2012, **154**, 41–52.

The figure shows the spatial distribution functions of the alkyl side chain terminal carbon atoms of the imidazolium cations, in $[C_nC_1Im][Ntf_2]$ ionic liquids, around ethane and butane solutes. Both solutes are solvated in the non-polar domains of the ionic liquids, the first solvation shell being predominantly occupied by the alkyl chains of the cation.

Image reproduced by permission of Margarida F. Costa Gomes from *Faraday Discuss.*, 2012, **154**, 41.

### INTRODUCTORY LECTURE



### PAPERS AND DISCUSSIONS



 

## CONCLUDING REMARKS



## ADDITIONAL INFORMATION

# Ionic Liquids: Past, present and future

**C. Austen Angell,* Younes Ansari and Zuofeng Zhao**

***Received 8th November 2011, Accepted 8th November 2011***
**DOI: 10.1039/c1fd00112d**

An overview of the field of low-melting ionic liquids is given from its inception in 1886 through to the present time. The subject is divided into an introductory section that summarizes the early history of the field, and differentiates its sub-sections, before addressing matters judged of some interest in "pre-surge" and "post-surge" stages of its development, focusing on physicochemical as opposed to the prolific synthetic and industrial aspects in which the author has no competence. We give a final section specifically to protic ionic liquids, which we consider to have particular scientific potential.

## A. Introduction

This article is intended to provide some background to the subject of the present Faraday Discussion on Ionic Liquids and, to this end, will consist of three sections after this introduction. The first will be devoted to a short history of the subject intended to show how the present state of enthusiasm for low melting ionic assemblages has arisen out of a diverse background of ionic liquid studies before taking off under its own considerable force of innovation. The second section is intended to provide some understanding of the various reasons for the current state of enthusiasm for the field, and the third section will attempt to give it some additional depth and projection as it enters the stage of a maturing discipline, mostly using examples from the authors' own recent works.

Like all of electrochemistry and ionic liquid physical chemistry, the field started with Humphrey Davy's pioneering work on the electrolytic decomposition of simple molten salts under the influence of an applied dc electric field, to yield the elements that initially had been chemically combined in the salt under study. Davy's studies, collected in his major work,[1] place the investigation of ionic liquids at the very beginning of organized chemical science. They were preceded only by the alchemists' more chaotic studies of molten nitrates and ammonium salts to which no reference will be made here.

Davy worked primarily with the high-melting simple salts. The first person to consciously use an ambient temperature ionic liquid for scientific purposes seems to have been the Nobel Prize-winning physicist, Ramsay, who is reported by Schottenberger and co-workers[2] to have described "syrupy ionic liquids" that he prepared by combination of acids with picoline.[3] Although other early studies are known[2] the first that we will discuss here is the case of ethylammonium nitrate, prepared deliberately for its ionic character by Paul Walden in 1914.[4] It has probably been studied in more detail, and for more purposes, than any other ionic liquid. Walden and his works are important to us because in his earlier studies of aqueous solutions, he had formulated the Walden rule[5] that relates the equivalent conductivity $\Lambda$ of an ionically conducting liquid to its viscosity $\eta$.

$$\Lambda\eta = \text{const.} \tag{1}$$

The rule has provided the basis for a very useful classification of ionic liquids, about which more will be said below.

*Dept. of Chemistry and Biochemistry, Arizona State University, Tempe, AZ, 85287-1604, USA*

Although eqn (1) worked reliably for aqueous solutions the early German school of molten salt chemists, studying low-melting silver salts, found that the rule broke down seriously and needed to be replaced by the "fractional" Walden rule where $\gamma$ is a constant $0 < \gamma < 1$.

$$\Lambda\eta^{\gamma} = \text{const.} \tag{2}$$

The fractional Walden rule implied that the Arrhenius activation energy for conductivity was lower than that for viscosity, which was readily interpreted in terms of the smaller silver ion slipping through the quasilattice of the larger halide ions, this picture being particularly clear in the case of the silver iodide which was a good conductor even in its crystalline state at temperatures near the melting point. Such violations of the Walden rule are now referred to as "superionic" behavior, and are much sought after in the field of glassforming ionic liquids (where solid electrolytes are the prize) and particularly protonic ionic liquids (that might serve as fuel cell electrolytes). Some progress in this area will provide the substance of the last section of this article.

The silver salts of the early German studies were low-melting but were not ambient temperature liquids or even liquids with melting points below 100 °C in the sense of the "ionic liquids" that are in focus in most of the papers at this discussion. Inorganic examples of ionic liquids that satisfy this criterion do exist, however, and it would be useful at this point to briefly delineate the types of substances that together make up the sub 100 °C melting subjects of this discussion —liquids that in principle consist only of ions. Here we follow the classification given by the present author[6] in the book edited by Ohno "Electrochemical Aspects of Ionic Liquids" (Chapter 2).

## Classes of ionic liquids

**1. Aprotic ionic liquids.** The majority of ionic liquids, and certainly those responsible for the meteoric rise in the number of publications in this area since the mid-90's, are liquids in which the cations are organic molecular-ions. Examples would be the resonance stabilized alkyl pyridinium and dialklyimidazolium cations that go back to Hurley and Weir[7] in the middle of the last century. They mixed *N*-substituted alkyl and aryl pyridinium halides with various metal chlorides and nitrates and obtained low- liquids with which they performed electrochemical extractions (though they seem to be best known for their efforts on aluminum deposition on which they took the first patents). Hurley and Weir presented the first phase diagram on an aluminum chloride + organic cation halide system, which showed the existence of stable ionic liquids at temperatures of −40 °C (Fig. 1).

Much more recently developed have been the cyclic,[8] and non-cyclic[9,10] tetraalkyl ammonium salts like those with alkylpyrrolidinium cations, and particularly those with ether oxygenated sidechains.[10] Such cations are usually charge-compensated by anions of oxidic character like nitrate perchlorate or more frequently fluorinated-oxidic character like triflate. Among the most common of the latter are the $PF_6^-$, $BF_4^-$, triflate (trifluoromethane sulfonate, $CF_3SO_3$), and bis-trifluoromethanesulfonyl-imide, ($(CF_3SO_2)_2N$, or $NTf_2$) ions. The fluorinated anions are prominent because of the viscosity-lowering reduction of the van der Waals interactions (thanks to the tightly bound, hence unpolarizable, character of the fluorine electrons). The particular success of the $NTf_2^-$ anion, which is quite a large anion, may be due to the feature pointed out by Henderson,[11] namely, that this ion has two isoenergetic forms, hence can add an extra, (and therefore liquid-stabilizing) mode to the configurational entropy.

**2. Protic ionic liquids.** These are formed by the simple transfer of a proton from pure Brønsted acid to pure Brønsted base. Historically the first ILs made, as noted already, their development in the modern era was spearheaded by the Ohno[12,13] laboratory in Japan. The nature and reversibility of this process establishes a proton

 

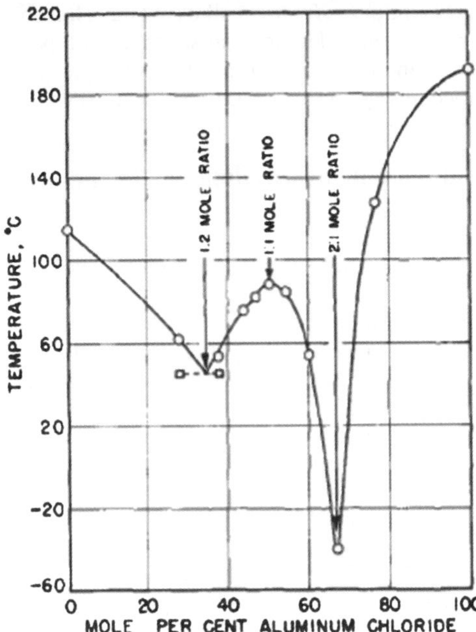


**Fig. 1** Phase diagram of the system ethylpyridinium bromide + $AlCl_3$ (from Hurley and Weir, ref. 7).

potential in the liquid product that lends this class of ionic liquids a special tunability —to which more attention will be given in the last section of this article. The $NTf_2$ anion mentioned in the previous section serve equally well to lower the cohesion, hence also the fluidity, of these liquids. But the cohesion can also be manipulated by tuning the proton transfer energy, with the result that it is possible for protic ionic liquids, according to eqn (1), to become more conductive than the aprotic cases. Indeed, by taking advantage of these features, some of the most conductive liquids ever known, have been obtained.

**3. Inorganic ionic liquids.** These may be obtained, in both aprotic and protic forms, by taking advantage of the same packing problems that lead to low-melting ILs of the organic cation type. There may be fewer of them, but there are aprotic examples like lithium chlorate (melting point 115 °C), and its glassforming eutectic with lithium perchlorate, and protic examples like hydrazinium nitrate, $T_m$ = 80 °C, not to mention low-melting mixtures of ammonium salts, and finally the largely unexplored cases of salts with inorganic molecular cations, such as $PBr_3Cl^+$, $SCl_3^+$, $ClSO_2NH_3^+$ *etc.*, with appropriate weak base anions.

**4. Solvate (chelate) ionic liquids.** These form a largely unstudied class of ILs that needs to be recognized because the class includes cases of multivalent cation salts that would not ordinarily be able to satisfy the criterion of $T_m$ < 100 °C. The first recognized members of this class were molten salt hydrates, like $Ca(NO_3)_2 \cdot 4H_2O$, whose mixtures with alkali metal salts were found to be almost ideal mixtures, most with liquidus temperatures well below ambient.[14] These were hailed as a "new class of molten salt mixtures", but there has been some question about the lifetime of the water molecules in the cation coordination shell. This should be long with respect to the diffusion time scale for the "ionic liquids" classification to be unambiguous. Recently the Watanabe laboratory has described new cases where long lifetime is guaranteed because the ligating groups all belong to the same molecule. Thus instead of 4 water molecules, Tamura *et al.*[15] use 4 alkoxy groups linked together, *e.g.* tetraglyme chelating the $Li^+$ cation, and find the salts with imide type anions to be ambient temperature liquids (with many desirable cell electrolyte properties). An anion solvate, to be identified in section C, proves to be the most conductive IL known.

## B. Post WW2 developments, preliminary to the present surge

### 1. Molten salts

There were a number of motivations for pursuing work in the field of molten salts, some of which became studies that would belong to the present sub 100 °C field, and these bear brief summarization. Much of it was motivated by the interest in molten salts as heat transfer fluids for the molten salt nuclear reactor project of the Department of Energy and centred at the Oak Ridge National laboratory. To borrow from a previous article of this type:[16]

*"At Oak Ridge, people were studying the "weak field" analogs of the "basic" $MgO$–$SiO_2$ metallurgical slags, represented by solutions of alkali fluorides and the Lewis acid $BeF_2$ (all ion charges halved relative to the $MgO$–$SiO_2$ melts). Complexation of the fluoride ions*[17] *to yield tetrahedral $BeF_4^{2-}$ and $Be_2F_7^{3-}$ anions, reduced the Coulomb energy of the liquid, lowering liquidus temperatures and also lowering viscosity*[18] *to give fluids with the excellent heat transfer characteristics needed for a successful nuclear reactor program.*

*The high degree of analogy to silicate systems presented by the $LiF$–$BeF_2$ system was not of great interest to Oak Ridge scientists, but became the center of attention in the East German laboratory of Vogel. Vogel*[19] *not only determined the phase diagram but showed that, in solutions rich in $BeF_2$, the glassforming solutions underwent a liquid-liquid phase separation*[20] *during cooling, similar to that which had been identified in silica-rich $Li_2O$–$SiO_2$ glasses.*

*A further analogy was waiting to be made, using molten chlorides closer in nature to those with which the field is currently so involved. With $ZnCl_2$ as the network glass-former and pyridinium chloride as the network breaker or "modifier", Easteal*[21] *demonstrated that the glassforming properties of the binary chloride system were highly analogous to those of the $Na_2O$–$SiO_2$ system.*

*While complex anions that could be enlarged into glassforming networks were clearly of interest, complex cations were not to be neglected. The rare earth analogs of Humphrey Davy's crimson alkali metal-alkali halide solutions were then under study at Imperial College in the group of John Bockris' former student John W. Tomlinson. The writer had the privilege of working under Tomlinson as the Imperial College Stanley Armstrong fellow, and studied the deep scarlet solution formed by the solution of metallic cadmium in $CdCl_2$.*[22] *Rather than forming the liquid equivalent of blue-purple alkali chloride crystal f-centers, (lattice vacancies containing an excess electron) which was Gruen and co-workers*[23] *interpretation of Humphrey Davy's crimson alkali halide solutions, the excess cadmium in $CdCl_2$ appeared to form metal dimers, $Cd_2^{2+}$."*

A parallel effort in molten salt chemistry was in place at Argonne National Laboratory, where some of the first applications of the chemical activity of the chloride ions compensating the positive charges on organic cations, were being made. The spectrum of Ni(II) in the tetrahedral field provided by chloride ions in the anionic complex $NiCl_4^{2-}$ was published in 1959 by Gruen and McBeth[24] and is shown in Figure 2, where it is contrasted with the equivalent spectrum obtained when the same Ni(II) species is present in an inorganic chloride melt at a somewhat higher temperature.

The striking difference in spectra between the LiCl and CsCl melts on the one hand, and the CsCl and pyridinium chloride melts on the other, illustrates an important feature of the low melting organic cation ionic liquid milieu. It is a difference that has recently been used to interpret some striking anion effects on redox potentials in ambient temperature ionic liquids.[25] It is that, because of the weaker polarizing field of the large organic cations, the anions in many of the low-melting ionic

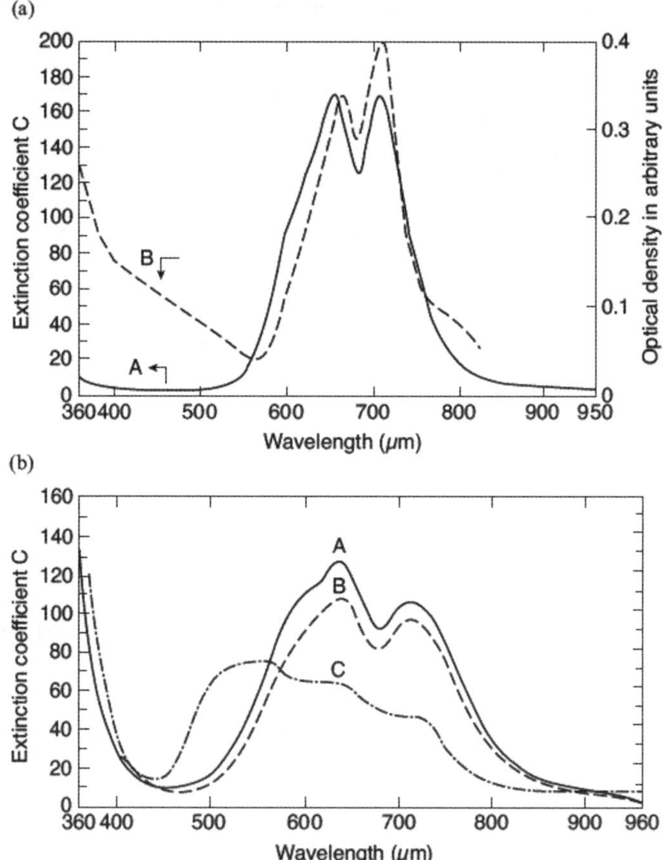


**Fig. 2** (a) The electronic (visible) spectrum of $Ni^{2+}$ dissolved in pyridinium chloride at 160 °C at a concentration of $1 \times 10^{-4}$ M (A) and in the crystalline state (solid solution of $Ni_2Cl_4$ in $Cs_2ZnCl_4$ where the Zn site is known to be tetrahedral (B). (b) Electronic spectrum of Ni(II) in liquid $Cs_2ZnCl_4$ at 650 °C (A), liquid CsCl at 700 °C (B) and liquid LiCl (C). Note the difference in intensities of the two tetrahedral-like spectra in the high temperature melt and in the organic cation melt of Fig. 2a. Redrawn in ref. 16 from figures in ref. 24, by permission of John Wiley & Sons, Inc.

liquids have a much higher chemical activity in the sense of electron density on the negative species. In consequence, the ability of these electrons to mix into the orbitals of a transition metal species, like Ni(II) in the present case —and so cause breakdown of the Laporte rule of quantum mechanics that formally forbids the transitions seen in Fig. 1— is much enhanced, and the intensity of the electronic transitions is greatly increased. This is called the *nephelauxitic* effect, an electronic phenomenon that also determines the effect of anion character on the redox potentials of redox couples dissolved in ionic liquids.

While this work was going on at Argonne, and Oak Ridge, National Laboratories some quite unrelated, but very important, studies were in progress at Purdue University. These involved proton transfer ionic liquids analogous to those of some current work in the field to be described near the end of this article. H. C. Brown, later to become Purdue's first Nobel Laureate in chemistry, and coworker Pearsall, were exploring the manner in which Friedel Crafts catalysts work, and in the process they transferred protons from HCl to toluene by simultaneously complexing the Cl to $AlCl_3$ to yield the ultra- weakly basic anion $AlCl_4^-$ and a protonated toluene species.[26] The latter must be described as the toluenium cation, which was a bright emerald green species, and undoubtedly also an extremely potent protonating agent itself. It was only stable below −45 °C, however. At that temperature it lost one mole of HCl, forming the even weaker base, the heptachlorodialuminate anion.[26]

At the same time, but with more metallurgical purpose, Hurley and Weir[7] described the ionic liquids in the system ethylpyridinium bromide + $AlCl_3$, to which we earlier made reference. They also studied other metallic halides that would dissolve in the ionic liquid, and described the cases in which electrodeposition could be achieved. However the electrodeposition of aluminum quickly became their principal interest. They cited experiments from the 1930s that showed that, even in such early times, much was understood of the potential of these systems for electrochemical manipulations. This line of research involving haloaluminates was later to become a major activity in the laboratories of first Robert Osteryoung and Gleb Mamantov, and later, John Wilkes and Charles Hussey, who together pushed the field of chloroaluminate physical chemistry and electrochemistry to sophisticated levels. A minor parallel effort in the author's laboratory, that was initiated in a study of the air and water-stable zinc chloride-pyridinium chloride system and was followed by a study of the properties of solutions of α-picolinium chloride + various inorganic cation chlorides, helped emphasize the role that the anion played in determining the fluidity (and hence conductivity) of these systems. This is explored a little further in the next section.

## 2. Role of complex anions in development of ambient temperature ionic liquids

In order to complete this introductory lecture manuscript in time for publication the following section is taken verbatim from ref. 16 in lieu of being paraphrased. The reference and figure numbers have been changed to values compatible with the present text.

*"Organic cation halides have proved very useful for the study of complex anions and their effects on molten salt properties. The dramatically different effects of complexation on the liquid viscosity are well illustrated by the 1976 study of Hodge[27] who used the lower-melting α-methyl derivative of pyridinium chloride as the source of the Lewis base chloride. Data seen in Fig. 3 show that the cohesive energy of an ionic liquid is lowered most strongly when chloride ions are complexed by $FeCl_3$ to produce the large singly charged $FeCl_4^-$ anion. This observation lead to the prediction that, amongst ionic liquids of a given cation, the tetrachloroferrate salt should be the most fluid and also most conductive, even more so than the well-studied ionic liquids containing $AlCl_4^-$ anions.[28,29] The veracity of this expectation is demonstrated in Fig. 4 for salts of the butylmethylimidazolium cation.[30] For the 25 °C conductivity, the tetrachloroferrate is half an order of magnitude above any others in the group.*

*The increase of ionic mobility by complexation seen here is no different in principle from the use of $BeF_2$ additions to alkali fluoride melts utilized in the molten salt reactor technology discussed in our introduction. There, the addition of one mole of $BeF_2$ replaced two high charge intensity fluoride anions with a doubly charged anion of much larger dimensions, for net decrease in cohesion.*

*The fluorinated anion salts (with $BF_4^-$ and $PF_6^-$) that are most frequently cited in connection with practical applications of ambient temperature ionic liquids,[31–33] are to be understood in the same terms. (Unfortunately, the tetrachloroferrate anion is slowly hydrolyzed by water, and application of ambient temperature ionic liquids usually requires the presence of water-stable anions). Fluorinated species, in addition to being water-stable, have the advantage of being unpolarizable and so their presence minimizes the contribution of van der Waals interactions to the liquid cohesion. This characteristic of fluorinated anions in the ionic liquids context was first noted in a 1983 study of binary solutions of lithium halides with organic salt halides by Cooper (see ref. 10), where the replacement of iodide anions by the equivalent amount of $BF_4^-$ anions was predicted to lead to an increase of ambient temperature conductivity, and*

 

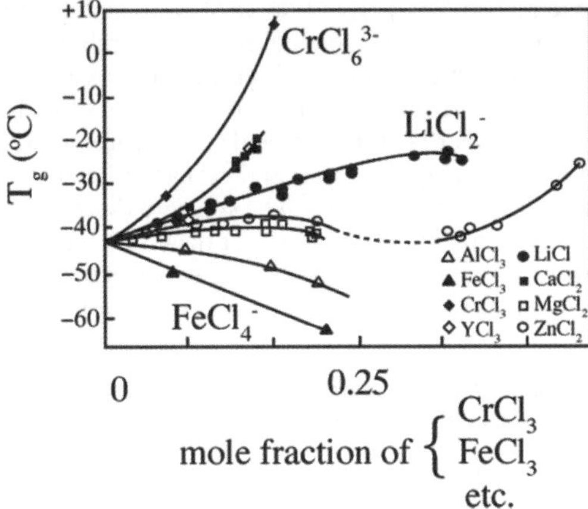


**Fig. 3** Effect of the second component on the cohesion of the liquid containing common cation α-picolinium chloride (Redrawn for ref. 37 from a figure in ref. 27, reproduced by permission of the American Chemical Society).

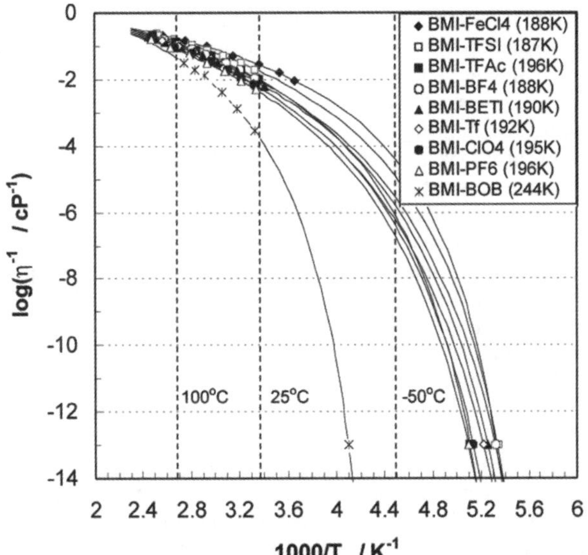


**Fig. 4** Arrhenius plot of fluidities of various salts of the common cation butylemethyl imidazolium. Note the place of the $FeCl_4^-$ salt in each case (Redrawn for ref. 37 from a figure in ref. 30, reproduced by permission of the American Chemical Society).

*then shown to do so, (in a note added in proof†)by some two orders of magnitude (due to lowering of $T_g$ by ~50 K).*[34] *Cooper's further extensive syntheses and measurements on viscosity and conductivity of a family of tetra-alkylammonium tetrafluoroborate and tetrachloroferrate salts, unfortunately remained unpublished (for over two decades) until 2003.*[30] "

† We have prepared the tetrafluoroborate analog of $(MeOEt)EtMe_2N^{+1}$, though in a slightly impure form. This room temperature molten salt has a remarkable conductivity of $1.7 \times 10^{-3}$ (ohm cm)$^{-1}$ at 25 °C and may in fact represent an important development in nonaqueous supporting electrolytes for ambient temperature applications.

After a decade had passed, Cooper, now at IBM, who had been in discussion with Wilkes about the stability of imidazole-based cations relative to pyridine-based cations and tetraalkyl ammonium cations in ionic liquids, presented basic data on imidazolium triflates (trifluoromethansulfonates) and mesylates (methanesulfonates) to the Electrochemical Society's 1991 International Conference on Molten Salts,[35] while Wilkes and coworker Zaworotko[36] submitted their parallel work to *Chem. Commun.* Both contributions appeared during 1992. The stage was set for the flowering of the present era but it took a surprisingly long time for these solvents to enter the common perspective. We take up that part of the story after considering some further applications of ionic liquid halides of the benign (non-hydrolysable) variety.

## 3. Spectroscopic applications of the weakly polarized ionic liquid anions. Coordination geometry at high pressure, and at negative pressure.

In the heyday of molten salt/ionic liquid chemistry driven by the molten salt reactor program discussed in a previous section, there was a great deal of interest in the electronic spectra of transition metals dissolved in the melts some of which was described already. The interest lay largely in the application of the ligand field energy level diagrams to the problem of coordination numbers and speciation in the solvent-free ionic melts. Specific applications of the ambient temperature melt electronic spectra were sparse.

But one of the first "task specific" applications of the low temperature ionic liquids was based on visible spectroscopy. The electronic spectra were adopted to take advantage of new ambient temperature instrumentation for reaching very high pressures— the recently invented diamond anvil cell. The exploration of coordination geometry of cobalt ions in organic cation chlorides up to 3.5 GPa, undertaken in the author's laboratory, was quite successful. The findings on the conversion of tetrahedral to octahedral cobalt coordination, and the concordance of the equilibrium $\Delta V$ values with the volume of the chloride ion, have been recalled in a recent Accounts of Chemical Research article,[37] and will not be described again here. However, there is an oppositely directed, and hence interesting, application of the visible spectroscopy of the Co(II) ion in ambient temperature ionic liquids that is worth describing. This application exploits the spectroscopic consequences of coordination geometry changes, to verify the physical reality of negative pressures.

To verify and quantify the existence of states of negative pressure in ionic liquids, a member of the fourth class of ionic liquids, *viz.*, the solvated cation class, was used. The hexahydrate of lithium chloride has chloride ions that, while not as basic as in an organic cation chloride, are nonetheless sufficiently basic to give very definitive tetrahedral $CoCl_4^{2-}$ electronic spectra when a modest amount of $CoCl_2$ is added. The solution is a bright blue color like the ambient pressure sample of the organic chloride used in the diamond anvil cell high pressure study. To demonstrate, visually, that negative pressure can be "impressed" on physical systems, two experiments are performed,[38,39] the findings of which are presented in Fig. 5. Firstly, the solution is studied at ambient temperature in a simple glass window high pressure cell up to 400 bar (40 mPa) pressure, which is enough to cause a marked change in spectral intensity at the tetrahedral coordination peak (660 nm). The solution is then degassed and transferred to a thick-walled glass optical cell which has been modified for easy conversion to a Berthelot tube, as illustrated in the inset to Fig. 5 (RH panel). Residual air is removed from the tubular cell extension and, while under vacuum, the stem is sealed off. The cell and contents are then carefully warmed until the remaining vacuum bubble has been eliminated by thermal expansion, and then spectra are taken every few degrees during cooling related to the surface energy of a nanoscopic nucleus, the bubble cannot reappear at the original temperature of elimination, and so the liquid, now cooling along an isochore, goes into tension. As the temperature approaches ambient temperature at which the spectra of the compressed sample of Fig. 5 LHS panel were taken, the tension in the liquid is

 

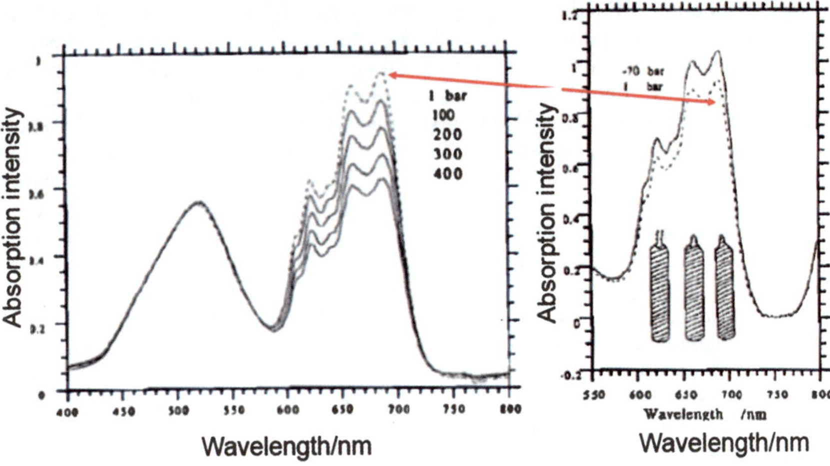


**Fig. 5** The visible spectrum Co(II) in liquid $LiCl \cdot 6H_2O$, as a function of positive pressure (LHS) and negative pressure (RHS) (Rearranged from Refs. 34, 35 by permission).

suddenly broken and the pressure returns to the vapor pressure of the solution. The spectra before and after the bubble reappears are compared in Fig. 4 RH panel.

The dashed curves are in each case the ~0 pressure spectra (1 atm and ~0.01 atm pressure respectively). The difference between the spectra before and after cavitation leave little doubt that the solution was at a pressure on the other side of zero just before it cavitated. The magnitude of the tension can be easily estimated by extrapolation of the positive pressure intensities. A value of −70 atm. was obtained. It would now be of considerable interest to repeat this simple experiment for a cobalt solution in an ionic liquid like the tetraglyme "solution" of Watanabe and coworkers,[15] in which there is no labile species to initiate a cavitation.

## C. The post Cooper–O'Sullivan, Wilkes–Zaworotko era

### 1. Sources of the surge

When the Cooper and O'Sullivan findings[35] on their study of ambient temperature triflates and mesylates of imidazolium cations were reported at the International Conference of Molten Salts, Wilkes and Zaworotko[36] quickly reported their own studies on salts of similar character. The latter authors chose the communications route and their paper appeared first, and also appeared in a location of high visibility. The Wilkes group was already well known for study of ambient temperature molten salts and so it is not surprising that they are most widely credited with the discovery of this new and practically very important class of solvent liquids, but the Cooper–O'Sullivan report is of wide perspective and should be part of everyone's reading.

It is unlikely, however, that either group's developments would have been given much attention had it not been for the vision of groups like those of Grätzel in Switzerland and MacFarlane in Australia who quickly saw the possibilities that lay ahead for these new low vapor pressure liquids in the fields of materials science and synthetic chemistry,[40] as Seddon and other synthetic organic chemists had already recognized for the chloroaluminates. The exponential rise in publications and patents in the ionic liquids field can be attributed to these early demonstrations of applied chemistry potential. Interestingly enough, Seddon's highly cited paper that foresaw the rise in applications of low vapor pressure ionic liquids[40] makes no mention of the perfluoro anion salts of the Cooper-Wilkes breakthrough (though he enquires into the factors predetermining low melting points). The first reference to tetrafluoroborates from the Seddon laboratory seems to have been the Holbrey–Seddon contribution of 1999.[41] Even Welton's enormously cited review on chemical synthesis in ionic liquids relates first to the chloroaluminate cases.[42] Applications of the fluorinated anion salts seems to have been first brought into focus by the Grätzel group, whose interest lay in involatile electrolytes for solar cells. The enormously cited 1996 paper by Bonhote *et al.*,[43] which showed awareness of Cooper and O'Sullivan's work, was quickly followed by the MacFarlane group's detailed quaternary ammonium $NTf_2$ paper,[44] though both were preceded by Fuller *et al.*s' description of the hexafluorophosphate salt of a resonance stabilized cation as "a model ambient temperature ionic liquid.[44] A vinyl polymer version of these ambient temperature ionic liquids was described by Ohno and Ito.[45]

Also in the mid-90's a variety of ionic mixtures that are liquid and highly conductive at ambient temperature and include lithium salts for battery electrolyte applications, were described by Watanabe and coworkers,[46] and by the authors' group, in search of systems that might generate "polymer-in-salt" electrolytes.[47] These have recently been improved by incorporation of the low $T_g$, mobility-conferring $NTf_2$ and chlorosulfonate anions (though the latter, like the chloroaluminate anion is easily hydrolysed)

The contribution of the MacFarlane laboratory to both the physical chemistry and chemical kinetics aspects of this field has been of enormous value, while additional impetus had been generated by the recognition by Breneke,[48] Maginn[49] and others that properly chosen ionic liquid compositions have high capacity for ingesting $CO_2$, reversibly. However the most powerful driver of the ionic liquids field must at present surely come from the recognition of Ionic Liquids as unique solvents/catalysts for chemical synthesis. The extraordinary citation counts for the overview works of authors like Welton,[42] Wasserscheid[32] and Seddon[40] are an indication of the number of people who are responding to this opportunity.

The present author is, however, not the appropriate person to assess this surge of synthetic activity, and so we will give the remainder of this article to discussion of some physico-chemical aspects, mostly of recent origin, and some new applications.

## 2. Physico-chemical aspects: The volatility of ionic liquids

**Aprotic ionic liquids and ionicity.** It is generally believed that aprotic ionic liquids are highly involatile, while protic ionic liquids are of volatility somewhere between aprotic and molecular liquids. There are some corrections to be made to this view. Firstly, relatively high vapor pressures have been observed, unexpectedly, for some aprotic cation tetrafluoroborates,[30,50] as well as other cases of expected high ionicity.[51] In the fluoroborate case, a correlation was made with conductivities that are much lower than expected from the high fluidities exhibited. The latter observation requires that there be some mechanism for coupling the cation and anion together to reduce the conductive efficiency of normal diffusive motion. The relationship is examined in Fig. 6.

Fig. 6 is an example of the use of Walden's rule to establish a classification diagram of broad utility, now commonly referred to as a "Walden plot", where the ideal line is based on the properties of classical dilute aqueous solutions for which the Walden rule was originally formulated. As reminded by MacFarlane *et al.*[52] a small correction is needed if the diffusing particles are abnormally large. Equivalent conductivities that are higher than expected from the Walden rule indicate a decoupling of one of the charge carriers from the structure that determines the viscosity, and is a very desirable characteristic. Values that are below the ideal line indicate the pairing up of the ions referred to above. The fluoroborate with data points well below the ideal line was the one that was observed to distil easily.

For the purpose of understanding how some aprotic ionic liquids can have relatively high vapor pressures, we illustrate in the figure how ion pairing can reduce the symmetric charge shielding that is characteristic of an ionic crystal lattice and

 

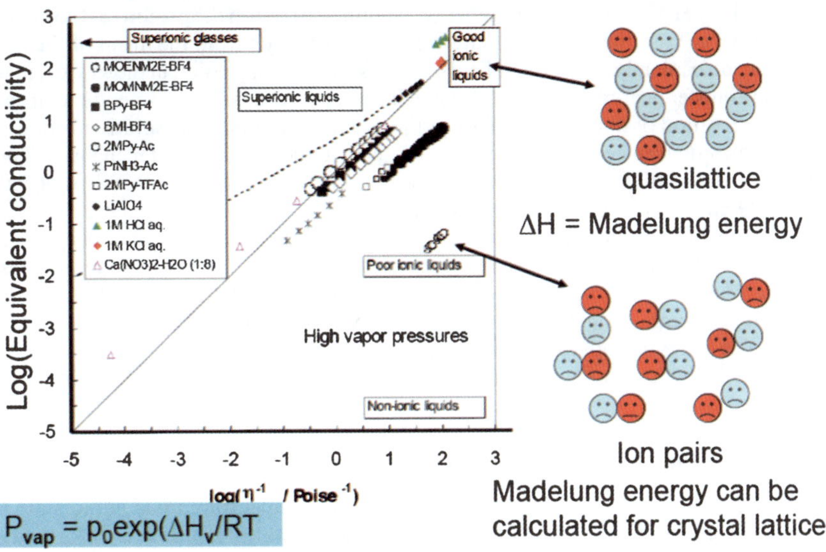


**Fig. 6** Walden plot showing the classification of ionic liquids into superionic, "good" and "poor" (subionic) and the relation between low equivalent conductivity and low ionicity (ion pairing).

determines the "Madelung energy" component of the crystal lattice energy. It is the component that makes the difference between the vapor pressure of a strongly polar molecular liquid like propylene carbonate (PC), and a (much less volatile) protic ionic liquid like ethylammonium nitrate (where the cation is comparable in size to propane and the nitrate anion is comparable in size to the carbonate unit of PC). It is also a characteristic of a "good" ionic liquid, in which each ion is surrounded by a more or less uniform shell of ions of the opposite charge. If the screen is too symmetric, however, the melting point is likely to be high, and the system will have too high a melting point to meet the IL criterion. The Madelung energy of a number of crystals forms of salts that yield low melting ion liquids have been assessed by Izgorodina *et al.*[53] If some energetic feature, like a strong hydrogen bond, interferes with the formation of the coulomb screen then the Madelung energy, will be turned off and the vapor pressure will go up. The subject of ionicity in ionic liquids has been extensively discussed by the groups of Watanabe[54,55] and MacFarlane[52] among others. A review of aprotic IL volatility has recently appeared[56], but the ionicity connection is not discussed.

In Fig. 6 it is seen that one of the tetrasubstituted ammonium tetrafluoroborates has a conductivity that is an order of magnitude below the ideal line suggesting that the liquid is heavily ion-paired, which is surprising for an aprotic ionic liquid with a weakly basic anion. That this is not a unique case is shown by Fig. 7,[50] where it is seen that three of the four examples of this sort of system behave the same way. The methoxymethyldimethylethylammonium salt of Fig. 6 was observed to be easily distilled, but similar measurements have not been made for the additional cases of Fig. 7. The origin of the ion-pairing tendency seen in Fig. 6 was presumed to originate in the enhanced acidity of the methylene protons when sandwiched between the more active N and O atoms, leading to hydrogen bonding to the fluorines of the anion,[30] but this does not explain the observations of Fig. 7. A crystal structure would be helpful but has not yet been made.

**Protic ionic liquids and the relevance of Δp$K_a$ from aqueous solutions.** For protic ionic liquids, formed by simple acid-to-base proton transfer, the origin of volatility and the connection to low ionicity is much more obvious. If the free energy of proton transfer is small then little thermal energy is required to return the proton to its original site for which case the vapor pressure will not differ much from the mean of the two original molecular liquids. Equivalently, the observed boiling point, in the case of small proton transfer energy, will not differ much from the mean of the two end members. The observation of non-linear boiling points is a mark of non-ideal solution behavior due the proton transfer and can be take as a guide to the proton transfer energy.[57] This measure of deviation from ideal behavior was found to correlate surprisingly well with the difference in aqueous $pK_a$ values for the interacting pairs which requires that the free energy of hydration for the ions of the anhydrous PIL be nearly the same. Direct evaluations of the proton transfer energies involved in PIL formation have only just begun[58,59] and are urgently needed to advance our understanding in this area. At the limit of very large proton transfer energy, as between superacid and superbase, the properties of protic and aprotic ionic liquids are not discernible.[60]

The rich phenomenology of protic ionic liquids will be given further attention in a later section.

## 3. Conductivity, viscosity and fragility

One of the most attractive features of the ambient temperature ionic liquid state is the availability of ionic conducting materials of low vapor pressure. It is unfortunate that the conductivity is not quite high enough to provide a viable alternative to the molecular solutions that are currently used in the electrolytes of portable power sources since the added safety factor would be very favorable. Since the $Li^+$ cation is so much smaller than either the cations or anions in the typical ionic liquids, it might naively have been expected that superionic behavior, like that seen for $LiAlCl_4$ in Fig. 6, would be found in mixtures of lithium salts and ionic liquids. However the opposite is found to be true (as could be anticipated from the behavior shown in Fig. 3 where the complexing of chlorides by lithium caused the $T_g$ of the ionic liquid to increase, opposite to the effect of adding a trivalent chloride like $FeCl_3$. Indeed it was this finding, for the case of the larger anion $I^-$, and the associated implication that $Li^+$ would always "win" the electrostatic tug-o'-war, and so dig itself a trap, that lead the author and Cooper to abandon the ionic liquid approach to finding superionic lithium electrolytes.[10] It also to the discovery of the early examples of low-melting tetrafluoroborate ILs.[10,61]

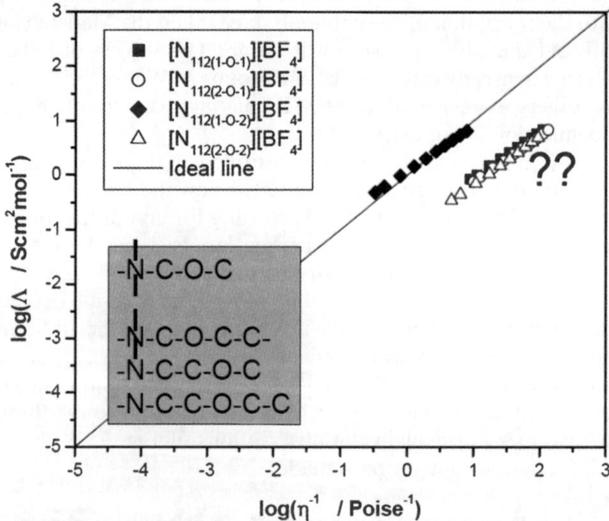


**Fig. 7** Walden plot for four oxygenated substituted ammonium tetrafluoroborates, indicating order of magnitude losses of conductivity in three cases, presumed due to ion pairing effects.

The conductivities of aprotic ionic liquids are problematic for portable energy purposes because the cohesive energies are a little too high, leading to viscosities that impede the ionic mobilities to an undesirable extent. A decrease in cohesion is necessary to overcome this factor. For some purposes this decrease can be accomplished, without affecting the ionicity, by using a protic IL PIL) in place of the aprotic substance. With the right choice of acid and base the cohesive energy can be lowered with respect to that of the nearest equivalent aprotic IL without affecting the ionicity too seriously. Certain PILs have been shown to have ionic conductivities approaching those of aqueous solutions.,[62] as for all other liquids Their consequent promise as fuel cell a electrolytes will be discussed further below.

The conductivities and viscosities of ionic liquids depend on temperature as for all other liquids, but there are qualitative differences. Where most simple liquids conform well to the Arrhenius equation, most ionic liquids do not. The reason is that they are observed in their "low temperature regimes" $T < 2T_g$, which usually means that the Arrhenius equation does not apply. Instead these transport properties follow a modified Arrhenius equation in which absolute zero is replaced by a finite divergence temperature, often called an ideal glass transition temperature, $T_0$. Thus, for transport property $P$

$$P = P_0\exp(DT_0/(T - T_0)$$

where $P_0$, $D$, and $T_0$ are constants. Before the temperature $T_0$ can be reached, however, the liquid structure becomes arrested at the glass temperature $T_g$. The sharpness of the deviation from the Arrhenius equation is determined by the value of $D$ which is infinity for Arrhenius behavior (which forces $T_0$ to 0 K). The smallest value of $D$ that has been recorded is about 3. This behavior is one of the leading puzzles of the "glassy state problem", which continues to confound theoreticians in this research area. Some examples of the deviations from Arrhenius behavior have already been seen for the case of aprotic ILs in Fig. 4. In Fig. 8 we show further cases with even higher conductivities that have been obtained with protic ionic liquids. Comparison of recently determined values for PILs of the symmetrical guanidinium cation (Gdm: $C(NH_2)_3^+$)[63], is made with the high values observed for aqueous LiCl hexahydrate, and also with some inorganic ILs of the chloroaluminate type that are used for high power energy storage systems.

Missing from Figure 8 is the extraordinary case of EMImF.2.3HF, reported by Hagiwara and coworkers.[64] This liquid has an aprotic cation but the fluoride anion is complexed by two or three additional HF molecules to make a sort of inverse solvate (class 4 IL) where it is now the anion that is solvated. The bonds in the complex anion are evidently very strong as no weight loss occurs until well above 200oC in TGA. The conductivity is almost as high as that of LiCl·6H2O in Figure 8, making it the most conductive IL known. It is also the most fluid.

The unusual temperature dependence of conductivity of the guanidinium salts in Figure 8 is to be noted. An unusually large temperature dependence is also found for the viscosities of the guanidinium salts. Using the glass transition temperature measurable for the GdmSCN-GdmFormate we can establish that the large temperature dependence is a consequence of a small value of $D$ in the VFT equation. The existence of different values of $D$ is shown in Fig. 8 in which the values of viscosity for different liquids are plotted in scaled Arrhenius form replacing the usual $1/T$ by $T_g/T$ where $T_g$ is taken as the temperature at which the viscosity reaches the value $10^{12}$ Pa.s so that all curves share the common point $T_g = 1.0$. The Gdm salt system is the most fragile ionic liquid ever studied. Eqn (1) is often interpreted in terms of the Adam Gibbs theory, according to which the temperature dependence of the transport coefficients is determined by the temperature dependence of the configurational entropy. It is suspected that in the case of Gdm salts, this quantity is enhanced by a "loose" proton contribution, as further suggested below (Fig. 9).

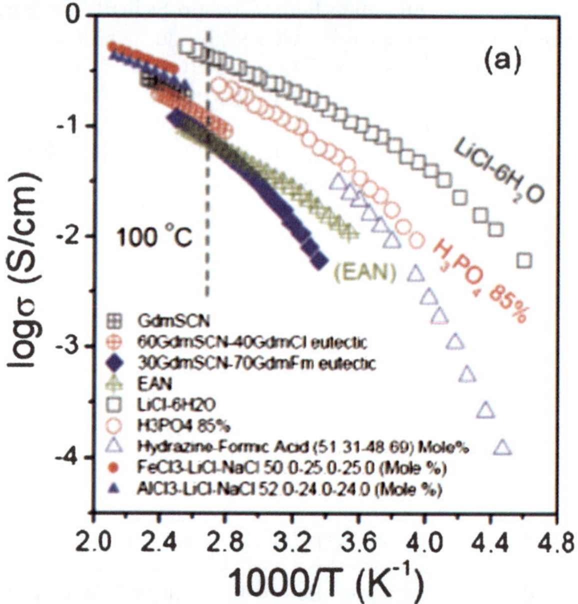


**Fig. 8** Comparison of the conductivities of protic ionic liquids of guanidinium cations with those of other high conductivity systems like aqueous LiCl, hydrated phosphoric acid, and some alkali chloroaluminate eutectics (reproduced from ref. 63, by permission of the American Chemical Society).

## D. Protic ionic liquids

We give a separate section to the case of protic ionic liquids, because of the special characteristics that the presence of a variable proton activity bestow on this class of ionic liquid. Although there is in no solvent present, hence no means by which the concentration of protonated species can be varied as in aqueous solutions, there is no doubt that the equivalent of a pH in the true thermodynamic sense of proton activity, is a feature of protic ionic liquids. Having a variable proton activity should mean having some control over proton-mediated chemical processes and, to the best of our knowledge, this additional dimension of ionic liquid science has not yet been properly exploited. Much experience with protic ionic liquids has been reviewed by Greaves and Drummond.[65]

Our interest in the special characteristics attributable to protons can be divided into two branches, *viz*., (1) those dealing with the physical consequences of the active protons, primarily proton exchange processes and the possibility of a free proton contribution to the dynamics, and (2) those dealing with the chemical and biological consequences. The latter are of particular interest because they control such vital phenomena as the folding of proteins to the native state and, under extreme departure from stabilizing conditions, the misfolding and fibrilization of proteins. The ability to control and enhance proton-catalyzed chemical reactions should be another feature of protic ionic liquid chemistry, and one that has not been deliberately explored to the best of our knowledge. We explore these two aspects briefly in the following paragraphs.

### 1. Superprotonic dynamics

It was expected, when Belieres began his wide-ranging study of PILs of different types,[66] that a number of cases with delocalized proton hopping transport would

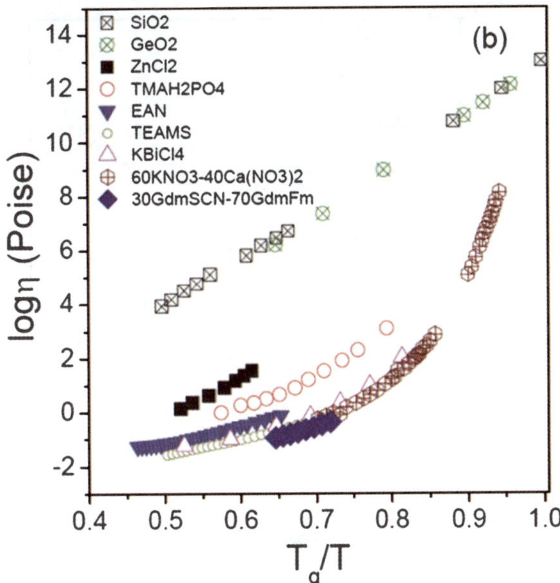


**Fig. 9** $T_g$-scaled Arrhenius plot of viscosities for aprotic and protic ionic liquids, with data comparisons with some classical "strong" and "fragile" liquids. It seems that the Gdm salts have the highest fragility observed so far (reproduced from ref. 63, by permission of the American Chemical Society).

emerge, but this proved not to be the case. Except for highly viscous and poorly conducting cases of dihydrogen phosphates, the organic cation salts all either obeyed the Walden rule expectations or fell on the subionic side of the ideal line. Even the case of ethylammonium nitrate EAN, that several authors have hailed as a water analog with a hydrogen-bonded network, fails to reach the ideal line, let alone exceed it. The only protic salt of high enough conductivity to reach the ideal line so far is the case of GdmSCN, This is marked as a red star in Fig. 10, which also includes EAN. It also the most decoupled case of all —dilute aqueous HCl under supercooled liquid conditions where the tetrahedral network of water is most highly developed, the rotational relaxation time of the molecule remains short, and the Grotthus mechanism at its most efficient. A new approach, in which acidity is varied independently of ionicity, will be described separately.[67]

## 2. Protic ionic liquids as solvents for control of protein folding and fibrilization

The solubility of proteins in aqueous solutions of several protic ionic liquids of "neutral" character, has an interesting form. It first decreases, passes through a minimum and then increases again.[68] At high EAN contents, with water : salt mole ratios of about 1 : 1, the solubility reaches extraordinary values, approaching 500 mg ml$^{-1}$ if the solubility measurements (always a problematical issue) are taken at face value. The remarkable finding is that despite the enormous concentration, the protein can be thermally unfolded and refolded many times without much loss to aggregation. The origin of such stability is unknown at this time. A refoldability index has been suggested[70] on the basis of comparison of the unfolding enthalpy in successive runs, and values of 0.98 are commonly found. The unfolding enthalpy, and the endotherm shape, is very similar to that seen in aqueous (physiological) solutions, as illustrated in Fig. 11. The unfolding retains its quasi two-state character. Most of the organic cations can be replaced by ammonium ions without changing this remarkable ability to refold.[70]

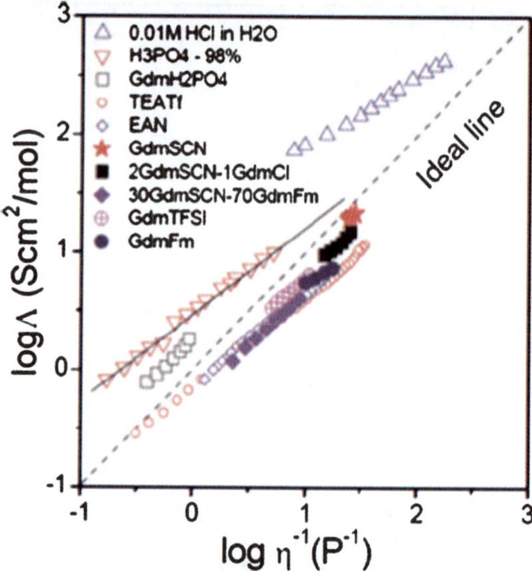


**Fig. 10** Walden plot for some Gdm salts and reference materials. Note the position of GdmSCN relative to EAN and extrapolated phosphoric acid (reproduced from ref. 63, by permission of the American Chemical Society).

If the proton activity of the solvent PIL is increased by choosing the anion of a strong acid, *e.g.* using hydrated ammonium bisulfate as solvent, then the protein is destabilized and rapidly fibrilizes. The process has been followed by beta sheet markers, and by electron microscopy.[71] In the case of proto-fibrils, the process can be reversed by redissolution in EAN and dilution to physiological conditions, when standard assays show that most of the biological activity has been restored. There is a great deal of challenging work in this field waiting to be reported and very much more waiting to be performed.

## 3. Proton activity, and the proton energy level diagram

The activity of the proton in protic ionic liquids is determined by the properties of the acid and base between which the proton exchange occurs. The effective proton activity of the PIL falls roughly half way between that of the acid and that of the base. In the absence of direct measurements, this energy gap across which the proton falls has been estimated from aqueous solution data, and theoretical estimates. An energy level diagram analogous to those proposed by Gurney for aqueous solutions and by the author for electron transfers in ionic liquids, has been presented for protic ionic liquids, initially in ref,[66] and in more advanced forms in refs. 6 and 37 and other papers in preparation.[67] Fig. 12 is a current version from which the lower (superbase section[60]) about which little is known, has been withheld. Some direct assessments have recently been made electrochemically in scrupulously dry conditions,[72] and the evidence to date is that the present version, Fig. 12, is semiquantitative. Lacking a common solvent, and a common standard state, it cannot be quantitative.[25] The application of Figure 5 to determine the conditions for forming superacidic and superprotonic ionic liquids while maintaining high ionicity, is currently being reported.[73]

## 4. New electrolytes for fuel cells.

This section would be incomplete without some mention of the role that protic ionic liquids may play as substitutes for the usual acid or base solution electrolytes in fuel

 

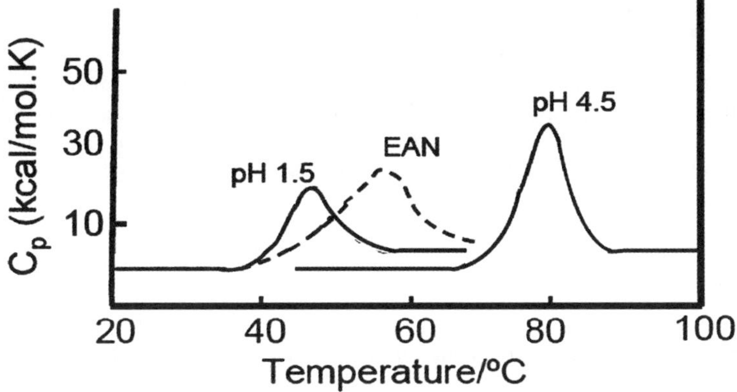


**Fig. 11** Comparison of the unfolding endotherm of lysozyme in hydrated EAN with that in physiological conditions at different values of pH by Privalov and Pfeil.[69]

cells. This possibility was only a recent recognition, first explored via the substitution of an ionic liquid for water in the Nafion membrane by the MacFarlane group,[74] closely followed by the Watanabe group,[75] who have followed it with full membrane assemblies.[76] PILs of the inorganic IL (class 3) type, can also serve in this capacity[77] and may have some stability advantages.

| | Occupied | Vacant | $pK_a$ | E (eV) |
|---|---|---|---|---|
| Super-Acid | $HSbF_6$ | $SbF_6^-$ | -25 | 1.48 |
| | ($HAlCl_4$) | $AlCl_4^-$ | | |
| | HTf | $Tf^-$ | -14 | 0.83 |
| Acid Electrolytes | $HSO_3F$ | $SO_3F^-$ | -13 | 0.77 |
| | HTFSI | $TFSI^-$ | -12.2 | 0.72 |
| | $pFPyH^+$ | pFPy | -12 | 0.71 |
| | $HClO_4$ | $ClO_4^-$ | -10 | 0.59 |
| | $H_2SO_4$ | $HSO_4^-$ | -9 | 0.53 |
| | $CH_3SO_3H$ | $CH_3SO_3^-$ (MS) | -1.9 | 0.11 |
| | $HNO_3$ | $NO_3^-$ | -1.3 | 0.08 |
| | 2-Fluoropyridine $H^+$ | 2-Fluoropyridine | -0.43 | 0.03 |
| | $CF_3COOH$ | $CF_3COO^-$ (TfAc) | -0.25 | 0.01 |
| | $H_3O^+$ | $H_2O$ | 0 | 0 |
| Neutral Electrolytes | $HPO_2F_2$ | $PO_2F_2^-$ | 0.3 | -0.02 |
| | $H_3PO_4$ | $H_2PO_4^-$ | 2.12 | -0.13 |
| | 1,2,3-1 H-triazole $H^+$ | 1,2,3-1 H-triazole | 3 | -0.18 |
| | HF | $F^-$ | 3.2 | -0.19 |
| | HCOOH | $HCOO^-$ | 3.75 | -0.22 |
| | $CH_3COOH$ | $CH_3COOH^-$ | 4.75 | -0.28 |
| | α-$MePyH^+$ | α-MePyH | 6.99 | -0.41 |
| | hydrazine $H^+$ | hydrazine | 7.96 | -0.47 |
| | $NH_4^+$ | $NH_3$ | 9.23 | -0.55 |
| | $EtNH_3^-$ | $EtNH_3$ | 10.63 | -0.63 |
| | $Et_3NH^+$ | $Et_3N$ (TEA) | 11.25 | -0.67 |
| | $C(NH_2)_3^+$ | Guanidine | 13.6 | -0.8 |
| | $H_2O$ | $OH^-$ | 14 | -0.83 |

**Fig. 12** Gurney free energy level diagram for acid/base pairs, showing superacid, acid and neutral electrolyte formation possibilities. *Proton transfer gap needs to be about 0.7 eV for a high ionicity electrolyte.

## Concluding remarks

While we have attempted to do justice to the depth of the subject of ionic liquids here, it is not possible in a work of this size to do any justice to its breadth, which continues to expand with no end in sight. It may be expected that, as new applications like those in chromatography, electron microscopy, and biochemistry, become more widely appreciated, the field will become properly recognized as one of the major accomplishments of 21$^{st}$ century science.

## Acknowledgements

The writer acknowledges support from the National Science Foundation under Experimental Chemistry under collaborative Grant No. CHE0404714. He acknowledges valuable monthly discussions of section D material with Pablo Debenedetti, Gene Stanley, and Peter Rossky. Also acknowledged is the DOD Army Research Office for support under Grant No. W911NF0710423.

## References

1 H. Davy, *Researches, Chemical and Philosophical. Biggs and Cottle, Bristol*, 1800, **1800**.
2 G. Laus, G. Bentivoglio, H. Schottenberger, V. Kahlenberg and H. Kopacka, *Lenzinger Berichte*, 2009, **84**, 71.
3 W. Ramsay, *Philos. Mag., Ser. 5*, 1876, **11**, 269.
4 P. Walden, *Bull. Acad. Imper. Sci. (St. Petersburg)*, 1914, 405.
5 P. Walden, *Z. Physik Chem.*, 1906, **55**, 207.
6 C. A. Angell; W. Xu; F. M. Yoshizawa; A. Hayashi; J.-P. Belieres; P. Lucas; M. Videa; Z.-F. Zhao; K. Ueno; Y. Ansari; J. Thomson; D. Gervasio In *Electrochemical Aspects of Ionic Liquids*; second ed.; Ohno, H., Ed.; John Wiley & sons: Hoboken, 2011.
7 F. H. Hurley and T. P. Wier, *J. Electrochem. Soc.*, 1951, **98**, 203.
8 D. R. MacFarlane, S. A. Forsyth, J. Golding and G. B. Deacon, *Green Chem.*, 2002, **4**, 444.
9 J. Sun, M. Forsyth and D. R. MacFarlane, *J. Phys. Chem. B*, 1998, **102**, 8858.
10 E. I. Cooper and C. A. Angell, *Solid State Ionics*, 1983, **9–10**, 617.
11 W. Henderson, *Invited presentation at 4th International Congress on Ionic Liquids (COIL4) (to be published)*, 2011.
12 M. Hirao, M. Yoshizawa and H. Ohno, *Electrochim. Acta*, 2000, **45**, 1291.
13 H. Ohno and M. Yoshizawa, *Solid State Ionics*, 2002, **154**, 303.
14 C. A. Angell, *J. Electrochem. Soc.*, 1965, **112**, 1224.
15 T. Tamura, T. Hachida, K. Yoshida, N. Tachikawa, K. Dokko and M. Watanabe, *J. Power Sources*, 2010, **195**, 6095.
16 C. A. Angell. In *Molten Salts and Ionic Liquids: Never the Twain?*; Seddon, M. G.-E. a. K. R., Ed.; John Wiley & Sons, Inc.: 2009, p 1.
17 B. F. Hitch and C. F. Baes, *Inorg. Chem.*, 1969, **8**, 201.
18 G. D. Robbins and J. Braunste, *Abstracts of Papers of the American Chemical Society*, 1970, 86.
19 W. Vogel, H. Reiss and J. Seifert, *Glass Technology*, 1983, **24**, 133.
20 W. Vogel, *J. Non-Cryst. Solids*, 1977, **25**, 170.
21 J. Easteala and C. a. Angell, *J. Phys. Chem.*, 1970, **74**, 3987.
22 C. A. Angell and J. W. Tomlinson, *Disc. Faraday Soc.*, 1962, **32**.
23 D. M. Gruen, *Q. Rev. Chem. Soc.*, 1965, **19**, 349.
24 D. M. Gruen and R. L. McBeth, *J. Phys. Chem.*, 1959, **63**, 393.
25 K. Ueno and C. A. Angell, *J. Chem. Phys. (in press)*, 2011.
26 H. C. Brown and H. W. Pearsall, *J. Am. Chem. Soc.*, 1951, **73**, 5347.
27 C. A. Angell; I. M. Hodge; P. A. Cheeseman. In *International conference on solten salts*; Pemsler, J. P., Ed.; The Electrochemical Society: 1976, p 138
28 J. a. Boon, R. T. Carlin, a. M. Elias and J. S. Wilkes, *J. Chem. Crystallogr.*, 1995, **25**, 57.
29 R. T. Carlin and J. S. Wilkes, *J. Mol. Catal.*, 1990, **63**, 125.
30 W. Xu, E. I. Cooper and C. A. Angell, *J. Phys. Chem. B*, 2003, **107**, 6170.
31 R. D. Rogers and K. R. Seddon, *Science*, 2003, **302**, 792.
32 P. Wasserschied and W. Keim, *Angewandte Chemie - Int. Ed.*, 2000, **39**, 3773.
33 *Ionic Liquids: Industrial Applications to Green Chemistry*; Rogers, R. D.; Seddon, K. R., ed.; Americn Chemical Society, 2002.

 

34 C. A. Angell, in *Molten Salts: From Fundamentals to Applications,* NATO Science Series II. Mathematics, Physics and Chemistry - Vol. 52, ed., M. Gaune-Escarde, Kluwer Academic Pub., Delft, 2002, p. 305–322.
35 E. I. Cooper and E. J. O'Sullivan, *Proc. 8th. Intern. Symp. Molten Salts* 92-16 1992, 92-16, 386-396.
36 J. S. Wilkes and M. J. Zaworotko, *J. Chem. Soc., Chem. Commun.*, 1992, 965.
37 C. A. Angell, N. Byrne and J.-P. Belieres, *Accounts of Chemical Research (special issue on Ionic Liquids)*, 2007, **40**, 1228.
38 Q. Zheng, J. Green, J. Kieffer, P. H. Poole, J. Shao, G. H. Wolf and C. A. Angell, *NATO-ASI (Liquids under Negative pressure)Kluwer Scientific*, 2002, 1.
39 Q. P. D. Zheng. Ph. D. Thesis, Purdue University 1991.
40 K. R. Seddon, *J. Chem. Technol. Biotechnol.*, 1997, **68**, 351.
41 J. D. Holbrey and K. R. Seddon, *J. Chem. Soc., Dalton Trans.*, 1999, 2133.
42 T. Welton, *Chem. Rev.*, 1999, **99**, 2071.
43 P. Bonhote, A. P. Dias, N. Papageorgiou and M. Gratzel, *Inorg. Chem.*, 1996, **35**, 1168.
44 J. Fuller, R. T. Carlin, C., D.H. and D. Hayworth, *J. Chem. Soc., Chem. Commun.*, 1994, 299.
45 H. Ohno and K. Ito, *Chem. Lett.*, 1998, 751.
46 M. Watanabe and T. Mizumura, *Solid State Ionics*, 1996, **86**, 353.
47 C. A. Angell, C.-L. Liu and E. Sanchez, *Nature*, 1993, **362**, 137.
48 L. A. Blanchard, D. Hancu, E. J. Beckman and J. F. Brennecke, *Nature*, 1999, **399**, 28.
49 E. J. Maginn and J. F. Brennecke, *J. Phys. Chem. B*, 2002, **106**, 7315.
50 W. Xu and C. A. Angell, *unpublished results, presented at ECS symposium, Honolulu*, **2008**.
51 M. J. Earle, J. M. S. S. Esperança, M. A. Gilea, J. N. Canongia Lopes, L. P. N. Rebelo, J. W. Magee, K. R. Seddon and A. W. J., *Nature*, 2006, **439**, 831.
52 D. R. MacFarlane, M. Forsyth, E. I. Izgorodina, A. P. Abbott and K. Fraser, *Phys. Chem. Chem. Phys.*, 2009, **11**, 4962.
53 E. I. Izgorodina, U. L. Bernard and P. M. Dean, *etal.*, *Cryst. Growth Des.*, 2009, **9**, 4834.
54 H. S. T. Tokuda, M. A. B. H. Susan, K. Hayamizu and M. Watanabe, *J. Phys. Chem. B*, 2006, **110**, 19593.
55 K. Ueno, H. Tokuda and M. Watanabe, *Phys. Chem. Chem. Phys.*, 2010, **12**, 1649.
56 J. M. S. S. Esperanca, J. N. Canongia Lopes, M. Tariq, L. M. N. B. F. Santos, J. W. Magee and L. P. N. Rebelo, *J. Chem. Eng. Data*, 2010, **55**, 3.
57 M. Yoshizawa, W. Xu and C. A. Angell, *J. Am. Chem. Soc.*, 2003, **125**, 15411.
58 R. Kanzaki, K. Uchida, X.-D. Song, Y. Umebayashi and S.-i. Ishiguro, *Anal. Sci.*, 2008, **24**, 1347.
59 R. Kanzaki, X.-D. Song, Y. Umebayashi and S.-i. Ishiguro, *Chem. Lett.*, 2011, **39**, 578.
60 H. Luo, G. A. Baker, J. S. Lee, R. M. Pagni and S. Dai, *J. Phys. Chem. Lett.*, 2009, **113**, 4181.
61 E. I. Cooper and C. A. Angell, *Solid State Ionics*, 1986, **18&19**, 570.
62 W. Xu and C. A. Angell, *Science*, 2003, **302**, 422.
63 Z.-F. Zhao, K. Ueno and C. A. Angell, *J. Phys. Chem. B*, 2011, DOI: 10.1021/jp206491z.
64 R. Hagiwara, T. Hirashige, T. Tsuda and I. Yasuhiko, *J. Electrochem. Soc.*, 2002, **149**, D1.
65 T. L. Greaves and C. J. Drummond, *Chem. Rev.*, 2008, **108**, 206.
66 J.-P. Belieres and C. A. Angell, *J. Phys. Chem. B*, 2007, **111**, 4926.
67 Y. Ansari; K. Ueno; C. A. Angell. to be published 2011.
68 N. Byrne and C. A. Angell, *Molecules*, 2010, **15**, 793.
69 W. Pfeil and P. L. Privalov, *Biophys. Chem.*, 1976, **4**, 23.
70 N. Byrne, J.-P. Belieres and C. A. Angell, *Aust. J. Chem.*, 2009, **62**, 328.
71 N. Byrne and C. A. Angell, *Chem. Commun.*, 2009, 1046.
72 J. A. Bautista-Martinez, L. Tang, J.-P. Belieres, C. A. Angell and C. Friesen, *J. Phys. Chem.*, 2009, **113**, 12586.
73 Y. Ansari, K. Ueno and C. A. Angell, *(unpublished work)*, 2011.
74 J.-Z. Suna, L. R. Jordana, M. Forsyth and D. R. MacFarlane, *Electrochim. Acta*, 2001, **46**, 1703.
75 M. A. B. H. Susan, A. Noda, S. Mitsushima and M. Watanabe, *Chem. Commun.*, 2003, 938.
76 S.-Y. Lee, T. Yasuda and M. Watanabe, *J. Power Sources*, 2010, **195**, 5909.
77 J.-P. Belieres, D. Gervasio and C. A. Angell, *Chem. Commun.*, 2006, 4799.

# Simulations of the structure and dynamics of nanoparticle-based ionic liquids†

**Bingbing Hong, Alexandros Chremos and Athanassios Z. Panagiotopoulos***

*Received 20th April 2011, Accepted 19th May 2011*
**DOI: 10.1039/c1fd00076d**

We use molecular dynamics simulations over microsecond time scales to study the structure and dynamics of coarse-grained models for nanoparticle-based ionic liquids. The systems of interest consist of particles with charged surface groups and linear or three-arm counterions, which also act as the solvent. A comparable uncharged model of nanoparticles with tethered chains is also studied. The pair correlation functions display a rich structure resulting from the packing of cores and chains, as well as electrostatic effects. Even though electrostatic interactions between oppositely charged ions at contact are much greater than the thermal energy, we find that chain dynamics at intermediate time scales are dominated by chain hopping between core particles. The uncharged core particles with tethered chains diffuse faster than the ionic core particles.

## 1. Introduction

Nanoscale ionic materials (NIMs) consist of inorganic core particles surface-functionalized with ionic groups and dispersed in a medium consisting of large organic counterions, which also act as the solvent.[1–6] With appropriate choice of the counterion groups they exist as liquids at room temperature.[7] They share with conventional ionic liquids the desirable properties of having a hybrid chemical character and negligible vapor pressures.[8] Their physicochemical characteristics can be tuned by varying the core nanoparticle composition and size, as well as the chemical character and molecular weight of the counterions.[9–11] A related class of materials, nanoscale organic hybrids[2,12–14] (NOHMs), share many chemical and structural characteristics with NIMs and can also exist as solvent-free liquids. Tethering of traditional ionic liquid constituent ions with nanoparticles has also been proposed.

The synthesis approach for NIMs has evolved over time. First-generation NIMs were made with cationic organosilanes grafted to the nanoparticle cores, with anionic organic compounds as counterions. Second-generation NIMs synthesis entails sulfonic-acid terminated organosilanes grafted to the nanoparticles, with primary or tertiary amine-functionalized oligomers as counterions. Typical second-generation NIMs with tertiary amine groups are shown schematically on the left side of Fig. 1. Each of the oligomer chains on the amines typically consists of a few dozen methylene or ethoxy groups. Finally, third-generation NIMs are made by taking advantage of inherent surface acidic groups (silanol) or grafted (fullerol) hydroxyl groups neutralized with amine-functionalized linear oligomers. A

*Department of Chemical and Biological Engineering and Institute for the Science and Technology of Materials, Princeton, NJ 08544, USA. E-mail: azp@princeton.edu*

† Electronic supplementary information (ESI) available: Movies showing the microscopic dynamics of chain, ion and core particle motions. See DOI: 10.1039/c1fd00076d

 

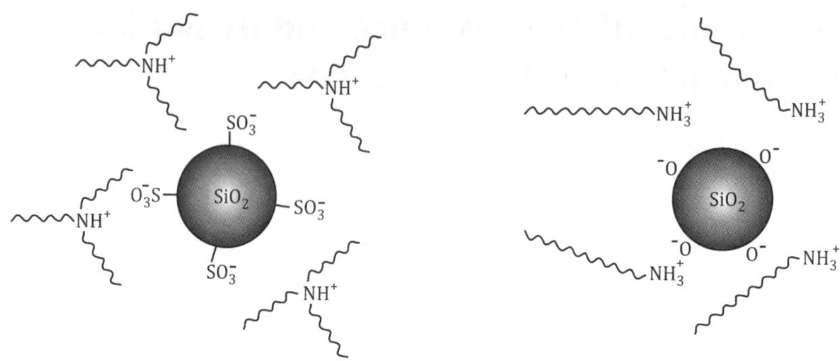


**Fig. 1** Left: second-generation NIMs with tertiary amine; right: third-generation NIMs with primary amine.

schematic representation of third-generation NIMs with primary amine counterions is shown on the right side of Fig. 1.

The structure and dynamics of NIMs have been investigated experimentally by techniques that include light scattering, dielectric spectroscopy, rheology, and NMR.[7,15,16] Water's high relative permittivity lowers effective interactions in ionic solutions, but water is not present in significant amounts in NIMs systems. Thus, oppositely charged ions are expected to interact strongly, with energies many times greater than the thermal energy. The structure and electrical conductivity of NIMs are clearly influenced by the mobility of the ionic groups. However, the reasons for their existence as liquids are not completely understood.[7] One of the objectives of the present work is to provide a microscopic picture of charged group dynamics in NIMs and possible explanations for their fluidity. Even though simulation studies of conventional ionic liquids (*e.g.* ref. 17–19) and polymer nanocomposites (*e.g.* ref. 12, 20–22) are available, simulations of NOHMs and NIMs have not been previously attempted, to the best of our knowledge. In a recent study,[23] NOHMs structural features were obtained by field-theoretic methods.

The structure of this paper is as follows. In section 2, we define the three model systems studied and quantify the parameters for the interactions taken into account in the models. Simulation methods and technical details are described in section 3. In section 4.1, structural characteristics of the model systems are analyzed through the behavior of the pair correlation functions. Section 4.2 deals with the dynamics of diffusion of nanoparticle cores and motion of chains on the surface of a core and from core to core. The paper closes with overall conclusions and possible directions for future work.

## 2. Models

Three different models were studied, to provide qualitative representations of NIMs with primary and tertiary amines, as well as NOHMs. The NOHMs model system is the simplest and forms the basis for the two NIMs models. It is shown schematically in Fig. 2; all computations were performed in three dimensions, but a two-dimensional representation of the model is shown for clarity.

The NOHMs model is comprised of spherical structureless nanoparticles with attached linear oligomer chains. Each chain consists of 15 beads connected *via* harmonic springs, but otherwise free of bond bending or torsional constraints. The overall length of the chains is within the range of experimentally relevant "canopy" oligomers used in NIMs synthesis.[10]

Interactions between oligomer beads are described by a cut-and-shifted Lennard-Jones potential with cutoff distance $r_c = 2.5\sigma$:

 

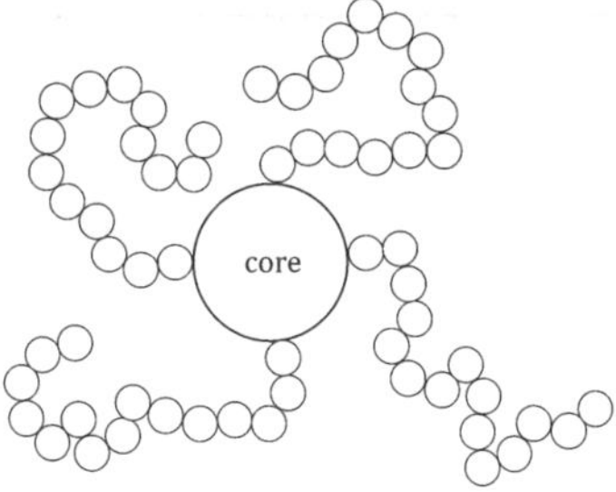


**Fig. 2** Schematic illustration of the (three dimensional) NOHMs model nanoparticle.

$$V(r) = \begin{cases} 4\varepsilon\left[\left(\frac{\sigma}{r}\right)^{12} - \left(\frac{\sigma}{r}\right)^{6}\right] - V_c & r \leq r_c \\ 0 & r > r_c \end{cases} \tag{1}$$

where $V_c$ is the value of the (unshifted) Lennard-Jones potential at $r = r_c$.

The size and energy parameters of the chain beads were obtained so as to represent approximately an ethoxy repeat unit ($-CH_2OCH_2-$), as follows. Since the ethoxy repeat unit does not exist as a free molecule, the group contribution method of Constantinou and Gani[24] was used to obtain an estimated critical temperature of $T_C = 409$ K and critical volume, $V_C = 0.124$ L mol$^{-1}$. These critical parameters are close to the experimentally measured values for dimethyl ether, $CH_3OCH_3$. The reduced critical parameters of the cut-and-shifted Lennard-Jones fluid with $r_C = 2.5\sigma$ have been obtained by Smit[25] as $k_BT_C/\varepsilon = 1.085$ and $\sigma^3/V_C = 0.317$, where $k_B$ is Boltzmann's constant. Thus, the energy parameter for the bead-bead interactions was obtained as $\varepsilon/k_B = 377$ K, and the size parameter was obtained as $\sigma = 0.40$ nm. The bead mass was set to $m_b = 44$ g mol$^{-1}$, appropriate for the ethoxy group.

The harmonic bond potential acting between adjacent beads is:

$$V_H(r) = k(r - \sigma)^2 \tag{2}$$

The spring constant was set to $k = 1000\varepsilon/\sigma^2$, which is sufficiently stiff that bond lengths remain within 10% of their equilibrium values.

The nanoparticle core diameter was set to $d = 5\sigma$ (= 2.0 nm), in the low range of values relevant for the experimental systems.[6,7] The mass of each core particle was set to $m_c = (d/\sigma)^3 m_b$, corresponding to a mass density for the particles of 2.2 g cm$^{-3}$, appropriate for $SiO_2$. For the core-core and core-bead interactions, we used the purely repulsive Weeks-Chandler-Andersen (WCA) potential,[26] modified to take into account the differences in particle size,[27]

$$V_{WCA}(r) = \begin{cases} 4\varepsilon\left[\left(\frac{\sigma}{r-\Delta_{ij}}\right)^{12} - \left(\frac{\sigma}{r-\Delta_{ij}}\right)^{6} + \frac{1}{4}\right] & r \leq r_m \\ 0 & r > r_m \end{cases} \tag{3}$$

where $r_m = 2^{1/6}\sigma + \Delta_{ij}$. The same values of parameters $\varepsilon$ and $\sigma$ were used for the core-core and core-bead interactions as for the bead-bead interactions. The shift distance was set to $\Delta_{cc} = d - \sigma$ for the core-core interactions, and $\Delta_{cb} = (d - \sigma)/2$ for the core-bead interactions. The potentials as a function of distance are shown in Fig. 3.

Each nanoparticle has 25 grafted oligomers, so that the grafting density is 2.0 chains nm$^{-2}$, within the range of experimentally achievable values.[2,9] The attachment points were fixed on the surface of the core particles for the production periods of the

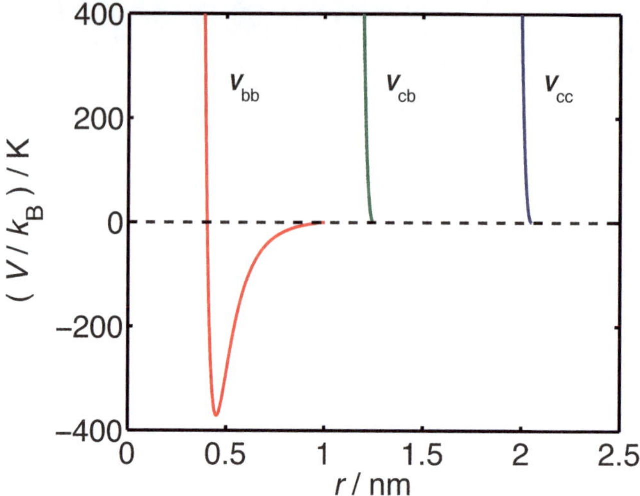


**Fig. 3** Potentials acting between beads ($V_{bb}$), between cores and beads ($V_{cb}$), and between cores ($V_{cc}$), as functions of distance.

simulations, so that the core particles move as rigid bodies together with the first beads of the chains. This was done to represent covalently grafted chemical groups. The attachment positions were obtained from initial simulations in which a random distribution of attachment points on the surface of the particles was allowed to evolve, subject to forces from a harmonic spring of the same constant as in eqn (2) and WCA repulsions of range $2.1\sigma$ (= 0.84 nm) between surface beads. This was done in order to avoid steric overlaps between chains near their attachment points.

The NIMs model with primary amines (termed "NIMs-L" for "linear" from here on) is identical to the model of NOHMs, except that the first bead on the surface of the cores is given a negative charge of $-e$ ($= -1.6 \times 10^{-19}$ C), and the bond between the first and second bead of the chain is severed so that the chains, now consisting of 14 beads, are no longer covalently attached to the nanoparticles. The surface sites are immobile, which is appropriate for "third-generation" NIMs systems. A charge of $+e$ is assigned to one of the two terminal beads of the chain to represent the ammonium group.

The NIMs model with tertiary amines is similar, except that the bond between second and third beads of the NOHMs model is now severed and the second bead given a negative charge. This results in a negative site with more freedom to move, appropriate for "second-generation" NIMs systems. The topology of the free chains is also changed to that of a three-armed star with four-bead chains attached to a central bead of charge of $+e$, to represent the tertiary ammonium group. This model is termed "NIMs-S" for "star" from this point on. The NIMs-L and NIMs-S models are shown schematically in Fig. 4.

Electrostatic interactions between charged groups $i$ and $j$ in the NIMs models are given by:

$$V_E(r) = \frac{1}{4\pi\varepsilon_0\kappa}\frac{q_i q_j}{r} \tag{4}$$

where $q$ is the charge of the corresponding group, $\varepsilon_0$ the dielectric permittivity of vacuum ($= 8.85 \times 10^{-12}$ F/m) and $\kappa$ is the relative permittivity of the medium, necessary because our coarse-grained models contain structureless core particles and do not take into account the polar character of the ethoxy groups. Polyethylene oxide has a relative permittivity that decreases with molecular weight, with $\kappa = 13.7$ for molecular weight of 600, from ref. 28. Silica and many non-conjugated polymers have relative permittivities near $\kappa = 4$,[29] which we adopt in this study. This value

 

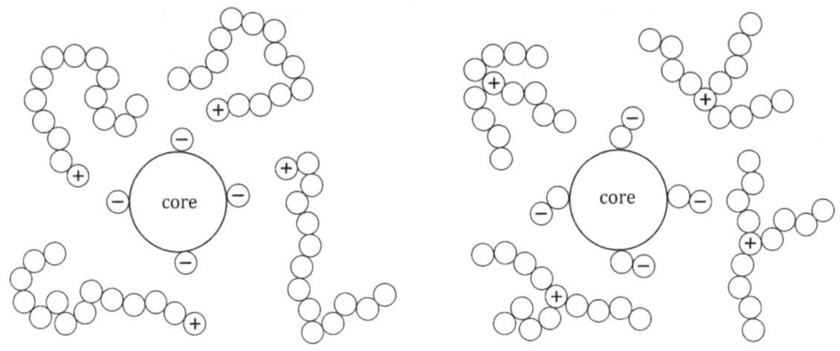


**Fig. 4** Schematic of the (three-dimensional) NIMs models: left: NIMs-L; right: NIMs-S.

of κ results in electrostatic energy of two oppositely charged beads at contact, $V_E(\sigma) = 35k_BT$ at the temperature of the calculations ($T = 300$ K).

## 3. Methods

The LAMMPS code,[30] available from http://lammps.sandia.gov, was used for the calculations. Electrostatic summations were performed with the particle-particle particle-mesh method with grid size of 72 × 72 × 72, resulting in each charge spanning 5 grid cells in each spatial dimension. The relative accuracy of the electrostatic energy calculation with these parameters was $10^{-4}$. After initial trials to determine the maximum allowable time step consistent with energy conservation and long-term stability of the simulations, the reduced time step was set to $\delta t^* = 0.008$. This corresponds to a real time step of $\delta t = \delta t^* \sigma\sqrt{m_b/\varepsilon} = 12$ fs. Energy was conserved to within 0.05% for the NIMs and within 0.08% for the NOHMs "production" simulation runs.

The base system size studied consisted of $N = 100$ nanoparticles, or 37 600 total interaction centers, of which 5000 were charged in the NIMs models. The box length was determined from a series of constant-temperature simulations at $T = 300$ K, so as to obtain near-zero pressure (corresponding to atmospheric conditions) for the NOHMs model system. This resulted in a box length of $L = 14.8$ nm, corresponding to an overall mass density of 1.1 g cm$^{-3}$. The volume fraction of core nanoparticles in the system, defined as the sum of volumes of the core particles divided by the total system volume, is 13%. The systems were initialized in the following way. The nanoparticles were first placed randomly in a cubic simulation box of size $L = 220$ nm so that no two particles were closer than 7 nm. The chains on each particle were then randomly grown such that there was no overlap between them and the particles. The simulation box was subsequently reduced at a rate of 0.001 nm per time step until the desired box length was reached. After the initial equilibration at constant temperature, the production simulations were performed in the *NVE* ensemble, to avoid any spurious effects of a thermostat on the dynamics. The kinetic temperature of the systems during the production runs averaged 306 K. A smaller system with $N = 50$ nanoparticles ($L = 11.7$nm) was also studied to investigate finite-size effects.

Simulations were performed on multiple Intel X5650 2.66 GHz 6-core processors connected by Infiniband switches. For the base system size of $N = 100$ nanoparticles with time step $\delta t^* = 0.008$, runs on 36 processor cores produced approximately 24 ns of simulation data per wall-clock day for NIMs. Runs for NOHMs were significantly faster, given the absence of electrostatic interactions; they produced 120 ns of simulation data per wall-clock day on 24 processor cores.

The mean-squared displacement and ion association correlation functions were determined from snapshots separated by 1000 time steps (12 ps). Separate origins corresponding to the initial 25% of a run were used, thus allowing correlation functions to be computed for times up to 75% of the total run time. We define association between positive ions and negative ions by a distance criterion with cutoff distance of 0.62 nm. This distance corresponds to the first minimum in the positive-negative pair correlation function. If multiple centers satisfy the distance criterion, the positive ion is considered associated with the center of lowest index. At any given time, we count how many of the ions that were originally within the association radius of a surface site or core particle still satisfy the distance criterion, irrespective of their trajectories at intermediate times.

## 4. Results and discussion

### 4.1. Structure

The pair correlation functions between core particles are shown in Fig. 5. For the NOHMs systems, the core-core correlation function, $g_{cc}$, has a broad first peak with fine structure, and a weaker second peak. The NIMs-L system has a sharp first peak at a distance of 2.4 nm, corresponding to a single bead separating two core particles. Since negative surface sites are rigidly attached to the core particles, this means that negative surface layers of the two core particles interpenetrate. This is energetically favored because of the presence of positive ions "bridging" nearby cores by interacting strongly with negatively charged surface sites. As will be discussed in Section 4.2, this structural feature has important implications on dynamics, allowing for chains to "hop" from core to core, and providing a path for conductivity. A strong "correlation hole" is present for the NIMs-L system around distances of 3 nm. The NIMs-S system has a structure for $g_{cc}$ similar to that of NOHMs, but with a slightly higher first peak with no fine structure, deeper trough near $r = 4.8$ nm, and slightly higher broad second peak. The negative ions are attached to a bead, rather than the surface of the core particles in the NIMs-S model, so they tend to be further away from their cores. Also, because of steric hindrances of the tertiary amines in the NIMs-S model, they are present in much lower amounts on the core surfaces relative to the linear chains of the NIMs-L model. The first peak of the NIMs-S $g_{cc}$ rises from zero at a larger distance than NOHMs, indicating that the cores are, on average, further apart from each other relative to the NOHMs or NIMs-L systems.

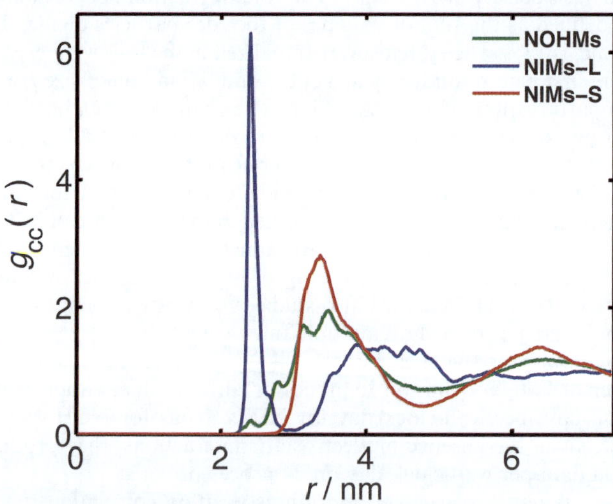


**Fig. 5** Pair correlation functions between core particles for the base system ($N = 100$). Green: NOHMs; blue: NIMs-L; red: NIMs-S.

 

Fine ripples are present in the NOHMs $g_{cc}$, with period equal to $\sigma$ (0.4 nm), suggesting subtle packing of beads around the cores. However, these ripples are absent in the NIMs-S system and much attenuated in the NIMs-L system. Their presence in NOHMs is a consequence of the tethered character of the chains. For NIMs-L, the positively charged beads at the end of the linear chains bridge two core particles, resulting in a strong first peak, as is also discussed in connection to the core-bead correlation functions. These strongly held chains have a partially tethered character and lead to a slight modulation of $g_{cc}$.

The pair correlation functions, $g_{cb}$, between core particles and the second beads of the chains are shown in Fig. 6. All three models display two strong main peaks, the first corresponding to the bead directly on the surface of the core particle (at $r = 1.2$ nm), and the second corresponding to one intervening bead (at $r = 1.6$ nm). Another common feature of these core-bead correlation functions is the existence of a "correlation hole" from 1.6 to 2 nm. However, the heights of the two first peaks are quite different for the three model systems. For NIMs-L, for which the second bead is positively charged and strongly attracted to the negative sites on the cores, the first peak is an order of magnitude higher than the second peak. The linear (primary) amines can penetrate near the core surface, allowing their charged ends to interact with the surface groups. For NOHMs, with an uncharged second bead, the first peak is about twice as high as the second one; both peaks are due to bonded interactions combined with packing effects. For NIMs-S, for which the second bead is negatively charged, the second peak is higher than the first, because the beads prefer to be in contact with counterions (tertiary amine groups) that cannot penetrate as easily as linear chains near the surface of the cores. Another reason for the higher second peak in the NIMs-S system is the mutual repulsion of negatively charged groups tethered to the cores.

Fig. 7 shows the pair correlation functions between core particles and the seventh bead of the chains. For NIMs-S, this is the central, positively charged bead, while for NOHMs and NIMs-L the seventh bead is uncharged. The first two peaks for NIMs-S, located at 1.3 nm and 1.5 nm, have almost the same positions as the first two peaks of core-second bead $g(r)$ shown in Fig. 6. This indicates that although they are subject to steric hindrance, the charged beads of the star chains are still able to get near the cores; however, the peak height is only around 3 for the tertiary amines, whereas the linear chains had a peak height of more than 20. The structure of $g_{cb}$ is

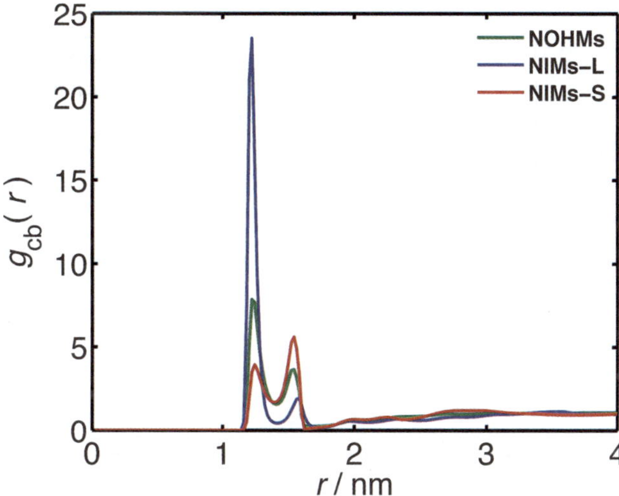


**Fig. 6** Pair correlation functions between core particles and bead 2 for the base system ($N$ = 100). Colors are the same as for Fig. 5; the range of $r$ distances has been reduced with respect to Fig. 5 for clarity.

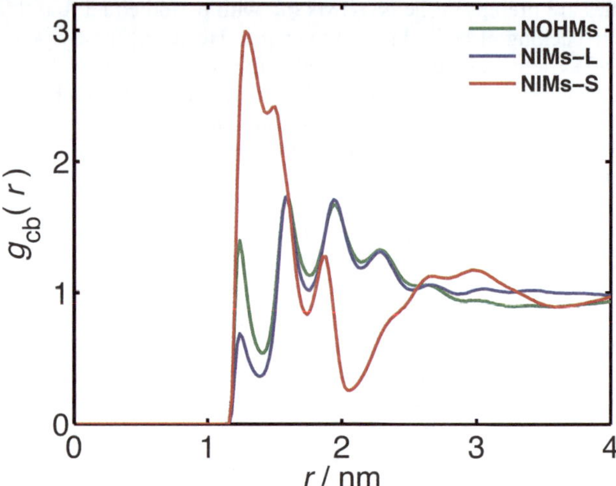


**Fig. 7** Pair correlation functions between core particles and bead 7 for the base system ($N = 100$). Colors and $r$ scale are the same as for Fig. 6.

quite similar for the NIMs-L and NOHMs systems, being determined primarily by packing effects of the chains. The only noticeable difference is a lower first peak (on the core surface) for NIMs-L relative to NOHMs, due to the crowding effect of the charged beads seen in Fig. 6.

The pair correlation functions between positive and negative ions, $g_{pn}$, are shown in Fig. 8. The functions are broadly similar, rising to a large value representing a contact ion pair at separation of 0.4 nm ($= \sigma$), but dropping quickly beyond this distance. The NIMs-S system has fewer counterions relative to the NIMS-L system at intermediate distances between 0.5 and 1 nm, because of steric repulsions of the tertiary amines. However, the increased mobility of the tethered negative charge for NIMs-S, over the surface-immobilized charge for NIMs-L, leads to a higher first peak for the $g_{pn}$, despite the increased steric hindrances from the chains.

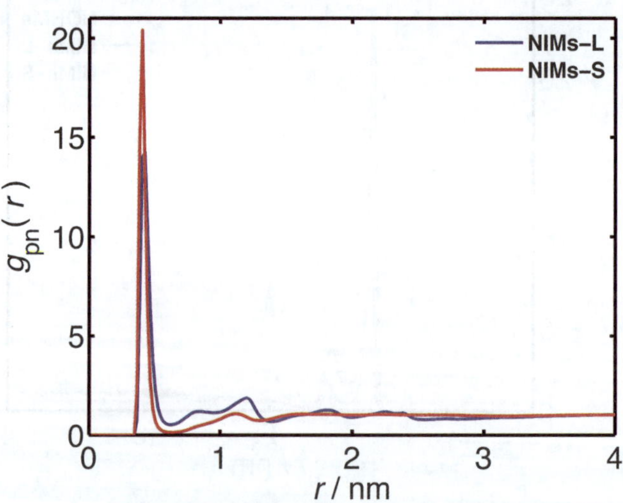


**Fig. 8** Pair correlation functions between positive and negative ions for the base system ($N = 100$). Colors and $r$ scale are the same as for Fig. 6.

 

The correlation functions for the small system ($N = 50$) are close to those of the base system, with some variations of up to 20% in peak heights (and up to 40% for the high first peak of $g_{cc}$ for NIMs-L), but nearly identical peak features and positions. Some of these variations are due to the influence of the initial configuration, rather than system size, as confirmed by duplicate runs from different initial configurations. However, most of the dependence of $g_{cc}$ is due to the fact that even the $N = 100$ system contains relatively few core particles. Larger systems, however, cannot currently be simulated for sufficiently long times to reach diffusive behavior.

### 4.2. Dynamics

A key dynamical feature that we have observed for our model NIMs systems is that, despite the strong electrostatic attractions between oppositely charged ions in the absence of a high-dielectric solvent, the chains have considerably mobility, initially on the surface of the core particles to which they are bound, and at later times between nearby cores. Movies showing the motion of chains and ions are provided as supplementary information to this article and illustrate this mobility quite clearly.† We quantify these chain motions by monitoring the fraction of chains, $f$, attached to their original surface sites or their original cores as a function of time (Fig. 9), calculated as described in Section 3. At time $t = 0$, the fraction of positive ions attached to negative sites is very close to 100%, reflecting the strong electrostatic interactions in the NIMs systems. The initial rapid drop-off of the fraction attached is likely a consequence of the assignment of any positive ion to a single negative surface site, even though the positive ions are frequently interacting strongly with multiple negative sites. A fraction of the cations shared between two such sites can get pulled from its original site within a short time. For later times, if we assume simple first-order kinetics for the detachment process, we anticipate that $\ln(f/f_0) = -t/\tau$, where $\tau$ is a characteristic time and $f_0$ the initial fraction attached. As seen in Fig. 9, in which the logarithm of $f$ is plotted *versus* linear time, this assumption becomes reasonably good at longer times for ion detachments. For core detachments, first-order kinetics are approximately followed at all times. Characteristic times $\tau$ obtained from the inverse slopes of the curves are on the order of hundreds of ns for core detachments and tens of ns for ion detachments. A more quantitative

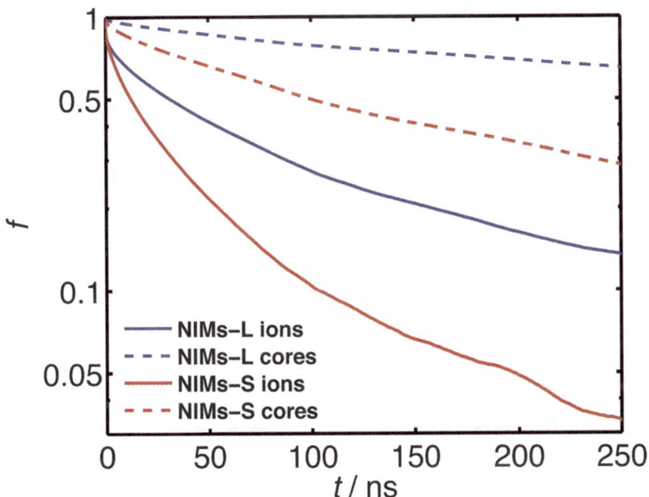


**Fig. 9** Detachments *versus* time for the base NIMs systems ($N = 100$). Continuous lines are for positive ions detaching from negative ions and dashed lines for positive ions detaching from cores. Colors are as for Fig. 5.

analysis is not warranted, given that first-order kinetics is only a rough approximation of the actual dynamics.

Detachment kinetics for chains in the NIMs-L system are slower than the NIMs-S system for both ions and cores. Most NIMs-L chains have their positively charged ions on the surface of the cores, with a smaller fraction one bead diameter from the cores (as evidenced from the height of the first and second peaks in Fig. 6). The NIMs-S chains do not have as many positive ions on the surface as NIMs-L chains, because of steric hindrances and the entropy loss of their arms when near the surface.[31] They can become detached from the original ions and original cores more readily than the NIMs-L chains, as reflected in the more rapid decay of the corresponding curves in Fig. 9. Another reason for the enhanced mobility is the tethering of negative sites in the NIMs-S system. This allows chains to move more readily from one core to another, without a need for the two cores to come very near each other.

The mean-squared displacements (MSD) of core particles and centers of mass of chains are plotted in Fig. 10. For reference, the line of unit slope (corresponding to diffusive behavior) is also shown in the figure. Statistical noise is more pronounced at longer times. NOHMs particles reach diffusive behavior after approximately 300 ns, when they have moved on average $\approx 1.5$ nm, a little less than a particle diameter away from their original positions. NIMs-L and NIMs-S core particles move more slowly than NOHMs particles, but reach near-diffusive behavior at comparable times. The estimated diffusion coefficients are $1.5 \pm 0.5 \times 10^{-8}$ cm$^2$ s$^{-1}$ for NOHMs, and $2 \pm 1 \times 10^{-9}$ cm$^2$ s$^{-1}$ for NIMs-L and NIMS-S cores.

We can obtain a rough approximation for the viscosity of these systems if we assume a hydrodynamic radius for the core particles along with their tethered or strongly held chains, and use the Stokes–Einstein equation to link diffusion coefficients of the core particles to the system viscosity. Using a hydrodynamic radius of $r_h = 1.3$ nm for both NIMs and NOHMs particles, we obtain a viscosity of 0.1 Pa s for our model NOHMs system and 1 Pa s for the model NIMs systems. We expect our coarse-grained models to have generally lower viscosities relative to their atomistic counterparts, because of the lack of torsional, bond bending and other orientational constraints. Experimentally measured NIMs viscosities[1,3] at room temperatures are typically between 10 and 100 Pa s.

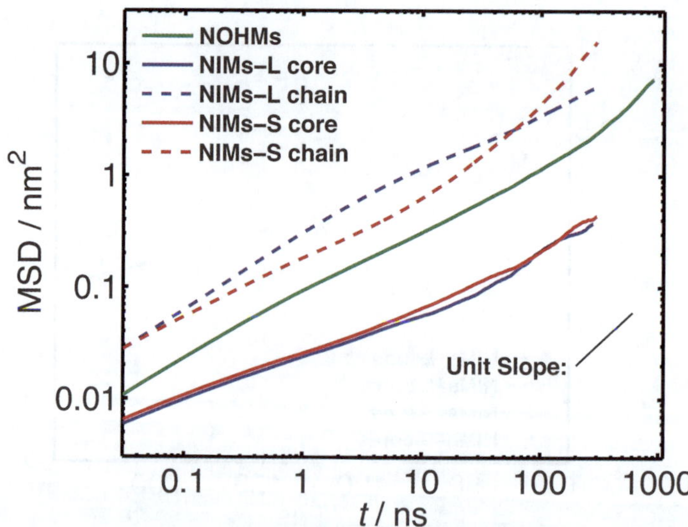


**Fig. 10** Mean-squared displacements *versus* time. Dashed lines are for the chain centers of mass and continuous lines are for the core particles. Colors are as for Fig. 5. The short, thin black line represents a line of unit slope.

 

In contrast to the behavior of the core particles, the motion of chains follows distinctly different behavior in the NIMs-S and NIMs-L systems. Chains travel faster than the core particles, but they do not display diffusive behavior until the cores themselves have become diffusive. This is a consequence of their motion primarily along the surfaces of the core particles until their detachment over time scales of hundreds of ns. The NIMs-S (star) chains move more slowly than NIMs-L chains at intermediate times between 0.1 and 50 ns. However, the NIMs-L linear chains are more strongly associated with the surface charges, and are slower in jumping from particle to particle, so their diffusion is eventually slowed down relative to their star counterparts. Diffusion coefficients of equal-sized uncharged linear and three-arm branched chains in solution[32] are within 10% of each other for chain lengths from 8 to 100. Thus, the main reason for the different diffusivities in our systems are electrostatic forces and differences in attachment intensities.

## 5. Conclusions

Two coarse-grained models for nanoscale ionic materials have been developed, consisting of core particles with attached charged sites and linear or three-arm branched chain counterions (NIMs-L and NIMs-S, respectively). The models are sufficiently simple so that their behavior can be simulated over sufficiently long (μs) time scales relevant for dynamics in these systems. A comparable model with tethered chains (NOHMs) has also been studied.

For the NIMs-L model of third-generation NIMs with primary amines, we find the amine groups preferentially on the surface of the cores, attracted to multiple negative sites on the same or different particles. As a result, core-core correlations for NIMs-L show a sharp first peak corresponding to two core particles with interpenetrating negative surface layers sharing contacts with a positive chain end. By contrast, the NIMs-S model of second-generation NIMs with secondary amines have significantly fewer positive groups near the core particle surface and do not show a contact peak in the core-core correlation function.

Despite the strong electrostatic forces that result in interaction energies at contact between positive groups on chains and core negative sites equal to $35k_BT$, chains are mobile. Over time scales of tens of ns, both linear and star chains detach from their original ions to nearby sites on the surface of a core particle. Over periods of hundreds of ns, chains move from one core particle to another, always keeping their positive sites in close contact with negative sites on the particles. Mobility of the NIMs-S tertiary chains is greater than of the NIMs-L linear chains. This is because tertiary chains have fewer of their positive ions on the core surfaces, because of steric hindrances and entropy loss, and also because tethering of the negative sites allows easier transfer from core to core.

Nanoparticles in both NIMs and NOHMs systems become diffusive after approximately 300 ns. Diffusivity of the cores is greater by a factor of ten in the NOHMs system relative to NIMs, also resulting in a lower estimated viscosity. Chains in NIMs-L and NIMs-S systems show distinct dynamical behavior. Linear chains move faster on the surface of the cores than the bulkier star chains at intermediate times up to 50 ns, but are more strongly associated with their original cores and thus slower at longer times.

Our findings on chain mobility are in agreement with a recent experimental study of canopy dynamics on NIMs,[15] that concluded that "the liquid-like behavior in NIMs is due to rapid exchange of the [canopy chains] between the ionically modified nanoparticles." Chain hopping between surface sites and from core particle to core particle provide a mechanism for cation-based superionic conductivity (that is, higher than anticipated from their fluidity), as hypothesized in ref. 1. These observations also provide the basis for understanding why replacement of small counterions with large organic ones turns NIMs from high-melting solids into liquids.[1]

Only a single set of model parameters were examined in this study. In the future, we plan to investigate the influence of core particle size and volume fraction, chain composition and length, and temperature, on the structure, fluidity, and conductivity of these systems. More quantitative comparisons to specific experimental systems should also be possible.

## Acknowledgements

This paper is based on work supported by Award KUS-C1-018-02 made by King Abdullah University of Science and Technology (KAUST) and by grants DE-SC-0002128 from the US Department of Energy, Office of Basic Energy Sciences and CBET-1033155 from NSF. Simulations were performed on the Della cluster of PICScIE, a facility supported by Princeton University. The authors would like to thank Prof. Fernando Escobedo for suggesting the NOHMs model, and Prof. Emmanuel Giannelis and Lynden Archer for many helpful discussions.

## References

1 A. B. Bourlinos, K. Raman, R. Herrera, Q. Zhang, L. A. Archer and E. P. Giannelis, *J. Am. Chem. Soc.*, 2004, **126**, 15358–15359.
2 A. B. Bourlinos, R. Herrera, N. Chalkias, D. D. Jiang, Q. Zhang, L. A. Archer and E. P. Giannelis, *Adv. Mater.*, 2005, **17**, 234–236.
3 A. B. Bourlinos, S. R. Chowdhury, R. Herrera, D. D. Jiang, Q. Zhang, L. A. Archer and E. P. Giannelis, *Adv. Funct. Mater.*, 2005, **15**, 1285–1290.
4 A. B. Bourlinos, A. Stassinopoulos, D. Anglos, R. Herrera, S. H. Anastasiadis, D. Petridis and E. P. Giannelis, *Small*, 2006, **2**, 513–516.
5 Q. Li, L. J. Dong, J. F. Fang and C. X. Xiong, *ACS Nano*, 2010, **4**, 5797–5806.
6 N. Fernandes, P. Dallas, R. Rodriguez, A. B. Bourlinos, V. Georgakilas and E. P. Giannelis, *Nanoscale*, 2010, **2**, 1653–6.
7 A. B. Bourlinos, E. P. Giannelis, Q. Zhang, L. A. Archer, G. Floudas and G. Fytas, *Eur. Phys. J. E*, 2006, **20**, 109–117.
8 J. F. Brennecke and E. J. Maginn, *AIChE J.*, 2001, **47**, 2384–2389.
9 P. Agarwal, H. B. Qi and L. A. Archer, *Nano Lett.*, 2010, **10**, 111–115.
10 R. Rodriguez, R. Herrera, L. A. Archer and E. P. Giannelis, *Adv. Mater.*, 2008, **20**, 4353–4358.
11 R. Rodriguez, R. Herrera, A. B. Bourlinos, R. P. Li, A. Amassian, L. A. Archer and E. P. Giannelis, *Appl. Organomet. Chem.*, 2010, **24**, 581–589.
12 Q. Zhang and L. A. Archer, *Langmuir*, 2002, **18**, 10435–10442.
13 J. L. Nugent, S. S. Moganty and L. A. Archer, *Adv. Mater.*, 2010, **22**, 3677–80.
14 D. Kim and L. A. Archer, *Langmuir*, 2011, **27**, 3083–3094.
15 M. L. Jespersen, P. A. Mirau, E. von Meerwall, R. A. Vaia, R. Rodriguez and E. P. Giannelis, *ACS Nano*, 2010, **4**, 3735–3742.
16 J. M. Oommen, M. M. Hussain, A. H. M. Emwas, P. Agarwal and L. A. Archer, *Electrochem. Solid-State Lett.*, 2010, **13**, K87–K88.
17 T. I. Morrow and E. J. Maginn, *J. Phys. Chem. B*, 2002, **106**, 12807–12813.
18 T. Y. Yan, C. J. Burnham, M. G. Del Popolo and G. A. Voth, *J. Phys. Chem. B*, 2004, **108**, 11877–11881.
19 E. J. Maginn, *J. Phys.: Condens. Matter*, 2009, **21**, 373101.
20 G. D. Smith, D. Bedrov, L. W. Li and O. Byutner, *J. Chem. Phys.*, 2002, **117**, 9478–9489.
21 F. W. Starr, J. F. Douglas and S. C. Glotzer, *J. Chem. Phys.*, 2003, **119**, 1777–1788.
22 V. Pryamtisyn, V. Ganesan, A. Z. Panagiotopoulos, H. J. Liu and S. K. Kumar, *J. Chem. Phys.*, 2009, **131**, 221102.
23 H. Y. Yu and D. L. Koch, *Langmuir*, 2010, **26**, 16801–16811.
24 L. Constantinou and R. Gani, *AIChE J.*, 1994, **40**, 1697–1710.
25 B. Smit, *J. Chem. Phys.*, 1992, **96**, 8639–8640.
26 J. D. Weeks, D. Chandler and H. C. Andersen, *J. Chem. Phys.*, 1971, **54**, 5237–47.
27 J. S. Smith, D. Bedrov and G. D. Smith, *Compos. Sci. Technol.*, 2003, **63**, 1599–1605.
28 A. M. Awwad, A. H. Al-Dujaili and H. E. Salman, *J. Chem. Eng. Data*, 2002, **47**, 421–424.
29 P. R. Gray, P. J. Hurst, S. H. Lewis and R. G. Meyer, *Analysis and Design of Analog Integrated Circuits*, p. 40, Wiley (New York), 5th edn, 2009.
30 S. Plimpton, *J. Comput. Phys.*, 1995, **117**, 1–19.
31 A. Chremos, P. J. Camp, E. Glynos and V. Koutsos, *Soft Matter*, 2010, **6**, 1483–1493.
32 C. L. Fu, W. Ouyang, Z. Y. Sun and L. An, *J. Chem. Phys.*, 2007, **127**, 044903.

 

# Using ethane and butane as probes to the molecular structure of 1-alkyl-3-methylimidazolium bis[(trifluoromethyl)sulfonyl] imide ionic liquids

**Margarida F. Costa Gomes,*[ab] Laure Pison,[ab] Alfonso S. Pensado[a] and Agilio A. H. Pádua[ab]**

*Received 19th April 2011, Accepted 1st June 2011*
**DOI: 10.1039/c1fd00074h**

In this work, we have studied the solubility and the thermodynamic properties of solvation, between 298 and 343 K and at pressures close to atmospheric, of ethane and *n*-butane in several ionic liquids based on the bis[(trifluoromethyl) sulfonyl]imide anion and on 1-alkyl-3-methylimidazolium cations, $[C_nC_1Im]$ $[NTf_2]$, with alkyl side-chains varying from two to ten carbon atoms. The solubility of butane is *circa* one order of magnitude larger than that of ethane with mole fractions as high as 0.15 in $[C_{10}C_1Im][NTf_2]$ at 300 K. The solubilities of both *n*-butane and ethane gases are higher for ionic liquids with longer alkyl chains. The behaviour encountered is explained by the preferential solvation of the gases in the non-polar domains of the solvents, the larger solubility of *n*-butane being attributed to the dispersive contributions to the interaction energy. The rise in solubility with increasing size of the alkyl-side chain is explained by a more favourable entropy of solvation in the ionic liquids with larger cations. These conclusions are corroborated by molecular dynamics simulation studies.

## 1 Introduction

Ionic liquids form a novel class of solvents with potential industrial applications in environmentally acceptable processes.[1] They can act as sophisticated solvation or reaction media as they can dissolve a wide range of polar or nonpolar, organic or inorganic molecules,[2] providing new paths to carry out chemical reactions or industrial separations.

Ionic liquids are composed of asymmetric and flexible ions interacting in a sophisticated manner that can exist in a huge variety of cation-anion combinations. In this context, it is important to understand, at a molecular level, the key factors determining the solubilisation of different classes of solutes. Molecular simulation studies have shown that ionic liquids exhibit medium range ordering with the existence of two persistent nanoscale domains in the liquid phase: one formed by aggregates of the non-polar side chains, and the other by a network composed by the charged head groups and dominated by electrostatic interactions. The segregation of polar and nonpolar domains changes the way in which solvation can be understood in these liquids. It was demonstrated by molecular simulation that, depending on the

[a]*CNRS, UMR6272 LTIM, BP 80026, 63171 Aubiere, France. E-mail: margarida.c.gomes@. univ-bpclermont.fr*
[b]*Laboratoire Thermodynamique et Interactions Moléculaires, Clermont Université, Université Blaise Pascal, BP 80026, 63171 Aubiere, France*

nature of the solute (polar, nonpolar or associative), it would be solvated preferentially in one of the two domains described.[3]

The aim of the present work is to use a family of simple gaseous solutes, check their behaviour in solution of ionic liquids with known molecular structures,[4] and determine how these molecules can serve as probes to the molecular structure and interactions in these media. For that, we have studied the solvation properties of ethane and butane in several 1-alkyl-3-methylimidazolium bis[(trifluoromethyl) sulfonyl]imide based ionic liquids with alkyl side chains of different lengths—$[C_nC_1Im][NTf_2]$. We expect that the experimentally observed variation of the solubility and of the thermodynamic properties of solvation can be interpreted by considering the nanoscale segregation in these room temperature ionic liquids, thus providing another indirect experimental evidence of this behaviour first discovered by molecular simulation studies. We have also investigated *in silico* the behaviour of the ethane and butane solutions in imidazolium-based ionic liquids in order to understand the experimental observations at a molecular level, namely by identifying the preferential solvation domains and how the variation of their solubility can be explained at a molecular level.

Several research teams have measured the solubility of ethane in different imidazolium based ionic liquids as a function of temperature, both at high pressures[5–7] and at pressures close to atmospheric.[8–11] Ethane is quite soluble in imidazolium based ionic liquids with mole fraction solubilities, at 300 K and 1 bar, of the order of $10^{-2}$. *n*-Butane is much less studied in ionic liquids with only two studies previously published concerning liquids with the 1-alkyl-3-methylimidazolium cation—one reporting the solubility of this gas in $[C_4C_1Im][NTf_2]$[12] and the other dealing with butane in ionic liquids with the $[C_2C_1Im]$ and $[C_4C_1Im]$ cations associated with several anions (tetrafluoroborate, hexafluorophosphate, dicyanamide, trifluoromethanesulfonate and $[NTf_2]^-$).[13] A third study concerns the measurement of the solubility of butane in quaternary ammonium based ionic liquids.[14] The solubility of butane is much larger than that of ethane in reaching, for imidazolium or quaternary ammonium based ionic liquids, mole fractions of 0.2 at 303 K and 1 bar. No systematic study of the solubility of these two gases as a function of the temperature and as the size of the alkyl side-chain in the cation has been found in the open access literature.

## 2 Experimental

### Materials

The samples of ionic liquids used were all synthesised at the QUILL Centre, Belfast, according to standard synthesis procedures described in the literature.[15] The minimum mole fraction purity is 0.99 as found by $^1H$, $^{13}C$ and $^{19}F$ NMR spectroscopy.

The ionic liquids were kept under vacuum for 15 h at 303 K before each measurement. The water contents of each degassed sample were determined, with a precision of ±5 ppm, using a coulometric Karl Fisher titrator (Mettler Toledo DL32). The water content of the degassed samples was found to lay between 50 ppm for $[C_4C_1im][NTf_2]$ and up to 200 ppm for $[C_{10}C_1im][NTf_2]$.

The gases used for this study were supplied by Linde Gas: *n*-butane 3.5, mole fraction purity of 0.9995 (with impurities other than $C_nH_m \leq 500$ ppm), and ethane 3.5, mole fraction purity of 0.9995 (with impurities other than $C_nH_m \leq 450$ppm). The gases were used as received from the manufacturer.

### Solubility measurements

The experimental method used for the gas solubility measurements is based on an isochoric saturation technique and has been described in previous publications.[11,16,17] In this technique, a known quantity of gaseous solute is put in contact with a precisely determined quantity of degassed solvent at a constant temperature inside an accurately known volume. When thermodynamic equilibrium is attained, the pressure above the liquid solution is constant and is directly related to the gas solubility in the liquid solvent.

The quantity of ionic liquid introduced in the equilibration cell is determined gravimetrically. This quantity is equal to the amount of solvent present in the liquid solution, $n_1^{liq}$, as the ionic liquid does not present a measurable vapour pressure in the temperature range covered. The amount of solute present in the liquid solution, $n_2^{liq}$ (subscripts 1 and 2 stand for solvent and solute, respectively), is calculated by the difference between two $pVT$ measurements: first when the gas is introduced in a calibrated bulb with volume $V_{GB}$ and second after thermodynamic equilibrium is reached:

$$n_2^{liq} = \frac{p_{ini} V_{GB}}{[Z_2(p_{ini}, T_{ini}) R T_{ini}]} - \frac{p_{eq}(V_{tot} - V_{liq})}{[Z_2(p_{eq}, T_{eq}) R T_{eq}]} \quad (1)$$

where $p_{ini}$ and $T_{ini}$ are the pressure and temperature in the first $pVT$ determination and $p_{eq}$ and $T_{eq}$ the pressure and temperature at the equilibrium. $V_{tot}$ is the total volume of the equilibration cell, $V_{liq}$ is the volume of the liquid solution and $Z_2$ is the compression factor for the pure gas. The solubility can then be expressed in mole fraction, or as Henry's law constant:

$$K_H = \lim_{x_2 \to 0} \frac{f_2(p, T, x_2)}{x_2} \cong \frac{\phi_2(p_{eq}, T_{eq}) p_{eq}}{x_2} \quad (2)$$

where $f_2$ is the fugacity of the solute and $\phi_2$ its fugacity coefficient.

We consider that the volume of the ionic liquid does not change when the gas is solubilised and so the volume of the liquid solution is equal to the molar volume of the pure ionic liquid. The molar volumes of the pure ionic liquids studied here were calculated from density measurements using a U-shaped vibrating-tube densimeter (Anton Paar, model DMA 512P) operating in a static mode, following the procedure described in previous publications.[18]

### Molecular simulations

The structure of the solutions of both $C_2H_{16}$ and $n$-$C_4H_{10}$ gases in the five ionic liquids considered was also investigated by molecular simulation. We used an all-atom force field based in the OPLS-AA framework, specifically developed to represent the 1-alkyl-3-methylimidazolium cations and the $[NTf_2]^-$ anion.[19,20] The OPLS-AA model[21,22] was used for the two gases.

Five simulation boxes containing eight $C_4H_{10}$ molecules and 275, 227, 229, 227 and 231 ion pairs of the liquids $[C_2C_1im][NTf_2]$, $[C_4C_1im][NTf_2]$, $[C_6C_1im][NTf_2]$, $[C_8C_1im][NTf_2]$ and $[C_{10}C_1im][NTf_2]$ were prepared, respectively. Additionally, we set three different simulation boxes containing eight $C_2H_6$ molecules and 281, 234 and 237 ion pairs of the liquids $[C_2C_1im][NTf_2]$, $[C_6C_1im][NTf_2]$, and $[C_{10}C_1im][NTf_2]$, respectively. We selected this reduced number of solute molecules in order to avoid solute-solute interactions while still obtaining a good sampling of configurations in condensed phase. The initial configurations were lattices with low density. Equilibrations starting from the low-density arrangement took 500 ps at constant $NpT$ and $T = 300$ K, with a timestep of 2 fs. Once the equilibrium density was attained, simulation runs of 500 ps were performed using the molecular dynamics method implemented in the DL_POLY package.[23] At the final densities of the systems, the length of the side of the simulation boxes was around 50 Å.

## 3 Results and discussion

The densities of the ionic liquid samples, necessary for the determination of the gas solubilities, were measured at atmospheric pressure, as a function of temperature, and are reported in Table 1.

For each system studied, multiple experimental gas solubility data points were obtained in the temperature interval between 303 K and 343 K in steps of approximately 10 K. The values of the second virial coefficients for the gases, necessary for the calculation of the compressibility factor, were taken from the compilation of Dymond and Smith.[24] The solubilities of ethane in $[C_nC_1Im][NTf_2]$ with $n = 4$, 8 and 10 and of *n*-butane in $[C_nC_1Im][NTf_2]$ with $n = 2$, 4, 6, 8 and 10 determined in this work are reported in Table 2 and represented in Fig. 1. The solubility results are given as Henry's law constants and as mole fractions of the solute. These last values are calculated from the experimental data of $K_H$ (at slightly different total pressures) assuming a partial pressure of the gaseous solute equal to 1 bar.

To correlate the solubility as a function of temperature, Henry's law constants obtained from the experimental data were adjuster to a power series in $1/T$:

$$\ln\left[K_H/10^5\text{Pa}\right] = \sum_{i=0}^{n} A_i (T/\text{K})^{-i} \tag{3}$$

The coefficients $A_i$ are listed in Table 3 together with the average absolute deviations, denoted AAD, obtained for each solute. These deviations can be regarded as an estimation of the precision of the experimental data, which is, in the present case, considered as 1%.

The solubility of butane in the imidazolium ionic liquids is significantly superior to that of other gases like carbon dioxide or ethane, often considered as very soluble in ionic liquid media. In Fig. 2 are compared the mole fraction solubilities of *n*-butane, ethane and carbon dioxide in several imidazolium based ionic liquids at 303 K and at a partial pressure of gas equal to one atmosphere.

The solubility of the three gases increases with the increasing size of the alkyl side chain on the imidazolium cation. For carbon dioxide, it has been shown that, although the solubility increases with the number of carbon atoms in the alkyl chain, the augmentation is less important for alkyl chains longer than six carbon atoms.[26] The solubility of carbon dioxide rises due to the more important van der Waals-type interactions between the gas and the ionic liquids, the solute–solvent electrostatic interactions being approximately constant.[27] The uptake of the gas does not increase linearly with the length of the alkyl chain in the cation because, for larger cations, the nonpolar domains probably become too large for a solute that is still preferentially solvated near the regions of the ionic liquid dominated by electrostatic interactions.[26]

In order to better visualise the variation of the solubility of ethane and butane as a function of the size of the alkyl side chain, in Fig. 3 they are represented as a function of the number of carbon atoms of the chain of the ionic liquid. The solubility of the gases in all the ionic liquids decrease with the temperature as can be observed in Fig. 2 and 3. In the case of ethane, the solubility at 323 K, expressed as mole fraction of gas at 0.1 MPa, changes from approximately $5 \times 10^{-3}$ for $[C_2C_1im][NTf_2]$ to

**Table 1** Density of the pure ionic liquids studied at atmospheric pressure and as a function of temperature from 303 to 343 K. The experimental values were fitted to functions of the form $\rho = a_0 - a_1 T$. $\sigma$ is the standard deviation of the fit

| Ionic liquid | $a_0$/kg m$^{-3}$ | $a_1$/kg m$^{-3}$ K$^{-1}$ | $\sigma$ |
|---|---|---|---|
| $[C_2C_1Im][NTf_2]$ | 1.8322 | $1.0460 \times 10^{-3}$ | 0.04 |
| $[C_4C_1Im][NTf_2]$ | 1.7418 | $1.0170 \times 10^{-3}$ | 0.03 |
| $[C_6C_1Im][NTf_2]$ | 1.6607 | $0.9627 \times 10^{-3}$ | 0.04 |
| $[C_8C_1Im][NTf_2]$ | 1.6051 | $0.9425 \times 10^{-3}$ | 0.03 |
| $[C_{10}C_1Im][NTf_2]$ | 1.5570 | $0.9292 \times 10^{-3}$ | 0.03 |

 

**Table 2** Experimental values for the solubility of $C_2H_6$ and $n$-$C_4H_{10}$ in $[C_nC_1Im][NTf_2]$ ionic liquids ($n = 2, 4, 6, 8$ and 10) expressed both as Henry's law constants, $K_H$, and as gas mole fraction, $x_2$, corrected for a partial pressure of solute of 0.1 MPa. $p$ is the experimental equilibrium pressure and the per cent deviation is relative to the correlations of the data reported in Table 2

| $T$/K | $p/10^2$ Pa | $K_H/10^5$ Pa | $x_2/10^{-3}$ | dev (%) | $T$/K | $p/10^2$ Pa | $K_H/10^5$ Pa | $x_2/10^{-3}$ | dev (%) |
|---|---|---|---|---|---|---|---|---|---|
| $C_2H_6$ + $[C_4C_1Im][NTf_2]$ | | | | | $C_2H_6$ + $[C_8C_1Im][NTf_2]$ | | | | |
| 302.42 | 634.9 | 104.3 | 9.516 | −0.8% | 303.40 | 766.8 | 59.42 | 16.71 | +0.8% |
| 303.40 | 757.8 | 102.8 | 9.659 | +2.1% | 303.40 | 782.7 | 60.04 | 16.54 | −0.3% |
| 303.40 | 800.0 | 105.4 | 9.425 | −0.4% | 303.41 | 513.4 | 59.50 | 16.69 | +0.6% |
| 312.87 | 658.8 | 120.9 | 8.217 | −0.6% | 303.41 | 717.1 | 60.04 | 16.54 | −0.3% |
| 313.33 | 828.5 | 123.0 | 8.079 | −1.6% | 303.44 | 717.3 | 60.32 | 16.46 | −0.7% |
| 313.34 | 785.3 | 120.2 | 8.270 | +0.7% | 313.37 | 795.2 | 69.24 | 14.35 | +0.0% |
| 322.82 | 681.4 | 139.1 | 7.149 | +0.2% | 313.40 | 744.6 | 69.95 | 14.21 | −0.9% |
| 323.31 | 812.5 | 139.9 | 7.106 | +0.3% | 313.41 | 534.1 | 68.99 | 14.40 | +0.4% |
| 323.34 | 856.9 | 142.7 | 6.967 | −1.6% | 323.40 | 771.9 | 81.45 | 12.21 | −0.6% |
| 333.22 | 705.2 | 162.5 | 6.121 | +0.4% | 323.41 | 554.7 | 80.08 | 12.42 | +1.1% |
| 333.51 | 839.6 | 158.9 | 6.261 | +3.1% | 333.30 | 869.8 | 95.19 | 10.45 | −0.1% |
| 343.17 | 728.2 | 196.0 | 5.078 | −3.0% | 333.36 | 799.0 | 95.24 | 10.45 | −0.1% |
| 343.47 | 866.8 | 188.2 | 5.289 | +1.5% | 343.33 | 826.1 | 112.53 | 8.845 | +0.0% |
| $C_2H_6$ + $[C_{10}C_1Im][NTf_2]$ | | | | | $C_4H_{10}$ + $[C_2C_1Im][NTf_2]$ | | | | |
| 303.41 | 725.29 | 52.84 | 18.79 | +0.3% | 283.59 | 642.8 | 15.09 | 63.90 | −0.4% |
| 303.43 | 746.26 | 52.67 | 18.96 | +1.2% | 293.29 | 693.6 | 20.16 | 48.04 | +0.4% |
| 303.44 | 519.44 | 52.95 | 18.75 | +0.2% | 303.41 | 742.2 | 26.94 | 36.07 | −0.5% |
| 303.44 | 726.02 | 53.81 | 18.45 | −1.4% | 303.45 | 741.9 | 26.80 | 36.27 | +0.2% |
| 313.37 | 752.74 | 61.22 | 16.23 | −0.8% | 313.30 | 813.4 | 33.67 | 28.96 | +2.2% |
| 313.44 | 539.88 | 60.72 | 16.36 | +0.1% | 323.21 | 823.8 | 43.69 | 22.37 | −1.2% |
| 313.48 | 775.52 | 60.77 | 16.45 | +0.7% | 333.16 | 860.9 | 54.41 | 18.01 | −2.2% |
| 322.31 | 777.60 | 70.69 | 14.06 | −1.9% | 343.08 | 936.2 | 63.33 | 15.50 | +1.6% |
| 323.46 | 560.36 | 69.96 | 14.21 | +0.8% | — | — | — | — | — |
| 333.21 | 832.70 | 81.16 | 12.35 | +2.3% | — | — | — | — | — |
| 333.34 | 807.45 | 82.99 | 11.99 | −0.5% | — | — | — | — | — |
| 343.42 | 835.14 | 98.30 | 10.12 | −0.7% | — | — | — | — | — |
| $C_4H_{10}$ + $[C_4C_1Im][NTf_2]$ | | | | | $C_4H_{10}$ + $[C_6C_1Im][NTf_2]$ | | | | |
| 303.49 | 687.7 | 16.44 | 60.01 | −0.1% | 307.40 | 312.1 | 10.32 | 94.34 | −0.5% |
| 303.71 | 687.7 | 16.44 | 60.84 | −0.1% | 312.37 | 327.7 | 11.78 | 82.72 | −0.0% |
| 313.76 | 740.1 | 21.27 | 47.02 | +0.5% | 312.37 | 543.5 | 11.67 | 83.53 | +0.9% |
| 322.90 | 784.5 | 26.63 | 37.56 | +0.4% | 323.33 | 359.7 | 15.49 | 63.12 | −0.8% |
| 323.30 | 796.5 | 27.17 | 36.81 | −0.7% | 332.26 | 628.2 | 18.46 | 53.22 | +0.4% |
| 332.88 | 830.1 | 33.79 | 29.60 | −0.4% | 343.00 | 665.8 | 22.32 | 43.98 | −0.1% |
| 333.22 | 784.5 | 33.79 | 29.59 | +0.3% | — | — | — | — | — |
| 342.97 | 872.9 | 42.30 | 23.64 | −1.1% | — | — | — | — | — |
| 343.16 | 879.9 | 41.46 | 24.12 | +1.3% | — | — | — | — | — |
| $C_4H_{10}$ + $[C_8C_1Im][NTf_2]$ | | | | | $C_4H_{10}$ + $[C_{10}C_1Im][NTf_2]$ | | | | |
| 303.45 | 548.9 | 8.265 | 117.6 | −0.1% | 303.49 | 575.3 | 6.350 | 153.1 | −1.5% |
| 303.46 | 550.8 | 8.309 | 117.0 | −0.6% | 303.55 | 618.5 | 6.233 | 155.9 | +0.5% |
| 303.48 | 548.4 | 8.235 | 118.0 | +0.4% | 303.68 | 619.5 | 6.257 | 155.4 | +0.5% |
| 313.33 | 597.9 | 10.74 | 90.82 | +0.2% | 313.39 | 637.8 | 8.330 | 117.0 | −1.1% |
| 313.35 | 597.4 | 10.69 | 91.17 | +0.6% | 313.88 | 688.8 | 8.265 | 118.0 | +1.0% |
| 313.35 | 604.6 | 10.81 | 90.21 | −0.4% | 314.01 | 689.2 | 8.274 | 117.9 | +1.2% |
| 323.25 | 643.1 | 13.75 | 71.09 | +0.1% | 323.00 | 745.2 | 10.41 | 93.87 | +1.1% |
| 323.28 | 653.6 | 13.77 | 71.01 | +0.0% | 323.33 | 695.5 | 10.76 | 90.82 | −1.4% |
| 333.21 | 758.9 | 17.28 | 56.69 | +0.2% | 333.14 | 802.7 | 13.24 | 74.01 | +1.0% |
| 333.22 | 685.3 | 17.45 | 56.15 | −0.7% | 333.21 | 748.1 | 13.66 | 71.71 | −1.9% |

**Table 2** (*Contd.*)

| $T$/K | $p/10^2$ Pa | $K_H/10^5$ Pa | $x_2/10^{-3}$ | dev (%) | $T$/K | $p/10^2$ Pa | $K_H/10^5$ Pa | $x_2/10^{-3}$ | dev (%) |
|---|---|---|---|---|---|---|---|---|---|
| 343.12 | 740.2 | 21.27 | 46.15 | +0.8% | 343.34 | 926.2 | 16.61 | 59.10 | +0.5% |
| 343.16 | 723.2 | 21.57 | 45.52 | −0.5% | — | — | — | — | — |

around $14 \times 10^{-3}$ for $[C_{10}C_1im][NTf_2]$. For *n*-butane, the mole fraction solubility at the same conditions varies from *circa* $2 \times 10^{-2}$ for $[C_2C_1im][NTf_2]$ to about $9 \times 10^{-2}$ for $[C_{10}C_1im][NTf_2]$. The relative variation of the solubility is similar for the two studied gases.

In the case of ethane and *n*-butane, it is expected that they will be solvated in the non-polar domains of the ionic liquids.[2] One way of assessing the molecular reasons that explain the difference of solubilities between the two gases and of the variation of the solubility with the size of the alkyl chain in the cation is to analyse the thermodynamic properties of solvation. The variation with temperature of the solubility of the gases studied in the ionic liquids, expressed in Henry's law constant, is directly related to the thermodynamic properties of solvation which, as explained above, in the case of gaseous solutes at low pressures, is practically identical to the thermodynamic properties of solution.[28,29] The Gibbs energy of solvation is given by:

$$\Delta_{solv}G^{\infty} = RT\ln(K_H/p^0) \tag{4}$$

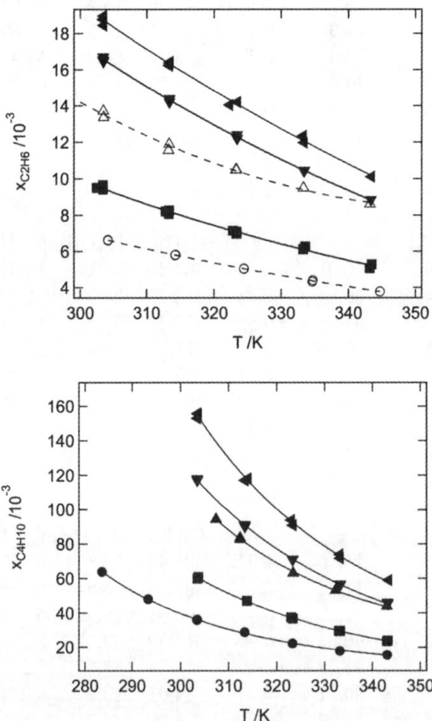


**Fig. 1** Solubility of ethane in $[C_nC_1Im][Ntf_2]$ ionic liquids (upper plot): ○, $n = 2$[10]; ■, $n = 4$; △, $n = 6$[11]; ▼, $n = 8$; ►, $n = 10$. Solubility of *n*-butane in $[C_nC_1Im][Ntf_2]$ ionic liquids (bottom plot): ●, $n = 2$; ■, $n = 4$; ▲, $n = 6$; ▼, $n = 8$; ►, $n = 10$.

 

**Table 3** Parameters of eqn (3) used to smooth the experimental results on $K_H$ from Table 1 along with the per cent average absolute deviation of the fit (AAD)

| Ionic liquid | $A_0$ | $A_1$ | $A_2$ | AAD (%) |
|---|---|---|---|---|
| $C_2H_6$ | | | | |
| $[C_4C_1Im][NTf_2]$ | +16.45 | $-5.867 \times 10^3$ | $+6.945 \times 10^5$ | 1.3 |
| $[C_8C_1Im][NTf_2]$ | +18.37 | $-7.371 \times 10^3$ | $+9.221 \times 10^5$ | 0.5 |
| $[C_{10}C_1Im][NTf_2]$ | +19.39 | $-8.176 \times 10^3$ | $+1.061 \times 10^6$ | 0.9 |
| $C_4H_{10}$ | | | | |
| $[C_2C_1Im][NTf_2]$ | +7.486 | $-1.150 \times 10^2$ | $-3.516 \times 10^5$ | 1.1 |
| $[C_4C_1Im][NTf_2]$ | +10.41 | $-2.124 \times 10^3$ | $-5.726 \times 10^4$ | 0.5 |
| $[C_6C_1Im][NTf_2]$ | −2.206 | $-1.676 \times 10^3$ | $-1.268 \times 10^6$ | 0.5 |
| $[C_8C_1Im][NTf_2]$ | +9.082 | $-1.676 \times 10^3$ | $-1.332 \times 10^5$ | 0.4 |
| $[C_{10}C_1Im][NTf_2]$ | +7.721 | $-9.036 \times 10^2$ | $-2.680 \times 10^5$ | 1.1 |

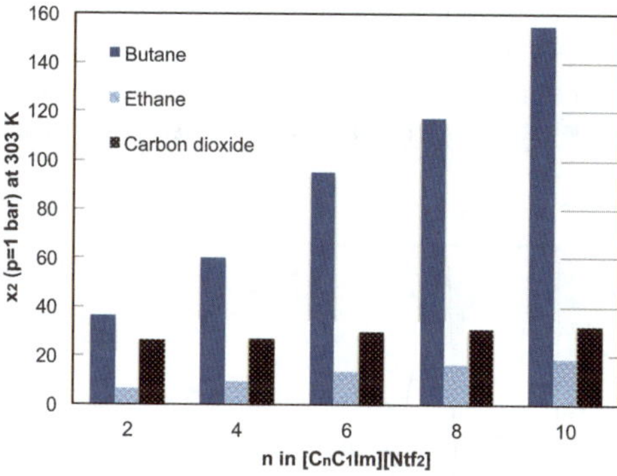


**Fig. 2** Mole fraction solubility of *n*-butane, ethane[10] and carbon dioxide[25] at 303 K and 0.1 MPa in $[C_nC_1Im][Ntf_2]$ ionic liquids with $n = 2$ to 10.

where $p^0$ is the standard-state pressure. The partial molar differences in enthalpy and entropy between the ideal gas and solution states can be obtained by calculating the corresponding partial derivatives of the Gibbs energy with respect to temperature

$$\Delta_{solv}H^{\infty} = -T^2\partial/\partial T(\Delta_{solv}G^{\infty}/T) = -RT^2\partial/\partial T\left[\ln\left(K_H/p^0\right)\right] \quad (5)$$

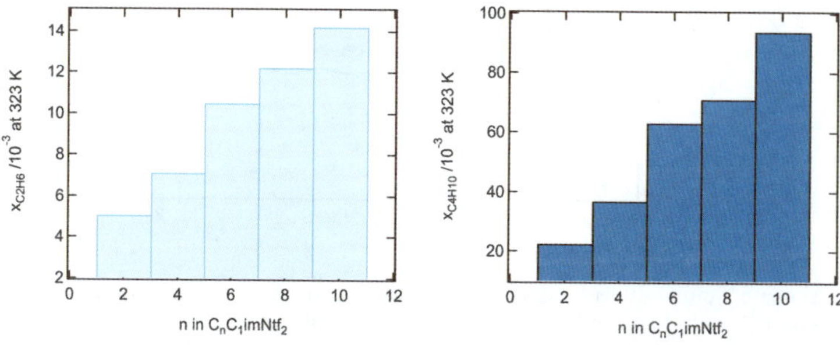


**Fig. 3** Solubility of ethane (left-hand plot) and of *n*-butane (right-hand plot) in $[C_nC_1Im][Ntf_2]$ ionic liquids at 323 K and 0.1 MPa.

**Table 4** Thermodynamic properties of solvation of ethane and *n*-butane in $[C_nC_1Im][NTf_2]$ ionic liquids in the temperature range studied

| Ionic liquid | $\Delta_{solv}H^\infty$ kJ mol$^{-1}$ | $\Delta_{solv}S^\infty$ J mol$^{-1}$ K$^{-1}$ | Ionic liquid | $\Delta_{solv}H^\infty$ kJ mol$^{-1}$ | $\Delta_{solv}S^\infty$ J mol$^{-1}$ K$^{-1}$ |
|---|---|---|---|---|---|
| | $C_2H_6$ | | | $C_4H_{10}$ | |
| $[C_2C_1Im][NTf_2]$ | $-12 \pm 1$ | $-80 \pm 3$ | $[C_2C_1Im][NTf_2]$ | $-20 \pm 2$ | $-92 \pm 4$ |
| $[C_4C_1Im][NTf_2]$ | $-13 \pm 1$ | $-81 \pm 3$ | $[C_4C_1Im][NTf_2]$ | $-20 \pm 2$ | $-90 \pm 4$ |
| $[C_6C_1Im][NTf_2]$ | $-10 \pm 1$ | $-69 \pm 2$ | $[C_6C_1Im][NTf_2]$ | $-20 \pm 2$ | $-83 \pm 3$ |
| $[C_8C_1Im][NTf_2]$ | $-13 \pm 1$ | $-76 \pm 3$ | $[C_8C_1Im][NTf_2]$ | $-21 \pm 2$ | $-86 \pm 4$ |
| $[C_{10}C_1Im][NTf_2]$ | $-12 \pm 1$ | $-74 \pm 2$ | $[C_{10}C_1Im][NTf_2]$ | $-21 \pm 2$ | $-85 \pm 4$ |

$$\Delta_{solv}S^\infty = (\Delta_{solv}H^\infty - \Delta_{solv}G^\infty)/T = -RT\partial/\partial T\left[\ln\left(K_H/p^0\right)\right] - R\ln\left(K_H/p^0\right) \quad (6)$$

The average values for the Gibbs energy, enthalpy and entropy of solvation in the temperature range studied for each gas in each ionic liquid are given in Table 4.

For the two gases, the enthalpy of solvation is negative, as expected from the variation of their solubility with temperature. In both cases, the process of dissolution of

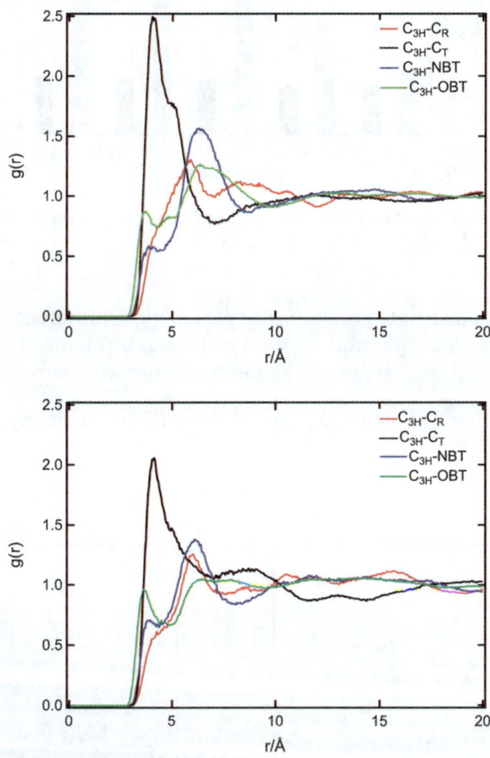


**Fig. 4** Site-site, solute–solvent radial distribution functions of ethane (upper plot) and *n*-butane (lower plot) in $[C_6C_1im][NTf_2]$: $C_{3H}$ are the terminal carbon atoms of the solute; $C_R$ is the carbon atom in position 2 on the imidazolium ring; $C_T$ is the terminal carbon atom of the alkyl side-chain on imidazolium cations; NBT stands for the nitrogen atom of the anion and OBT indicates the oxygen atoms of the anion.

 

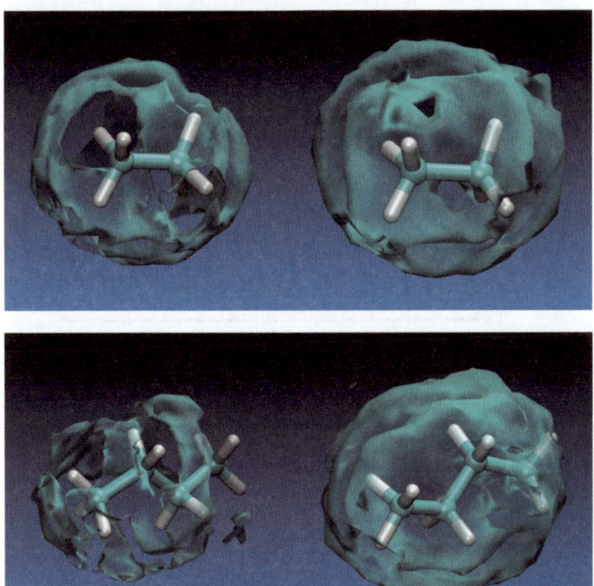

**Fig. 5** Spatial distribution functions of the terminal carbon atoms in the alkyl side chains of the imidazolium cations around ethane and *n*-butane solutes. The surface corresponds to an iso-probability density of 1.8 times the average density in the system. For ethane (upper plates) the ionic liquids are $[C_2C_1Im][NTf_2]$ (left-hand side) and $[C_6C_1Im][NTf_2]$ (right-hand side), and for *n*-butane (lower plates) $[C_2C_1Im][NTf_2]$ (left-hand side) and $[C_8C_1Im][NTf_2]$ (right-hand side).

the gas is exothermic. The enthalpy of solution is closely related to the cross gas–ionic liquid molecular interactions, the entropy of solvation giving indications about the structure of the solvent molecules surrounding the solute. In the case of *n*-butane, the enthalpy of solvation is much more negative than in the case of ethane, which can explain the difference of solubility between the two gases by the more favourable interactions between the heavier gaseous solute and the ionic liquids.

The variation of the solubility as a function of the size of the alkyl side chain can not be explained by more favourable interactions between the solute and the ionic liquid because the enthalpies of solvation do not significantly vary with the alkyl chain of the cation of the ionic liquid. The variation of the solubility can only be explained by a more favourable entropy of solvation for the dissolution of the gases in the ionic liquids based on the larger imidazolium cations.

One way of evaluating the molecular structure of the solutions is to calculate site-site solute–solvent radial distribution functions. As expected, the molecular simulations confirmed the solvation of both gases in the non-polar domains of the ionic liquids, close to the alkyl side-chains of the cations. This is rendered by the radial distribution functions in Fig. 4 that represent the probability of finding the terminal carbon atom of the solutes (the carbons of the $-CH_3$ groups) near different sites of the cation and of the anion of the ionic liquid $[C_6C_1im][NTf_2]$. For both ethane and *n*-butane, a larger probability of finding the solute near the terminal atom of the alkyl chain of the cation is observed.

A more detailed view is given by the spatial distribution functions of the alkyl side chain terminal carbon atoms of the imidazolium cations around ethane and *n*-butane. As can be seen in Fig. 5, in both solutes the first solvation shell is occupied by the alkyl chains of the cation, especially for the larger cations, typically with side-chains longer than six carbon atoms.

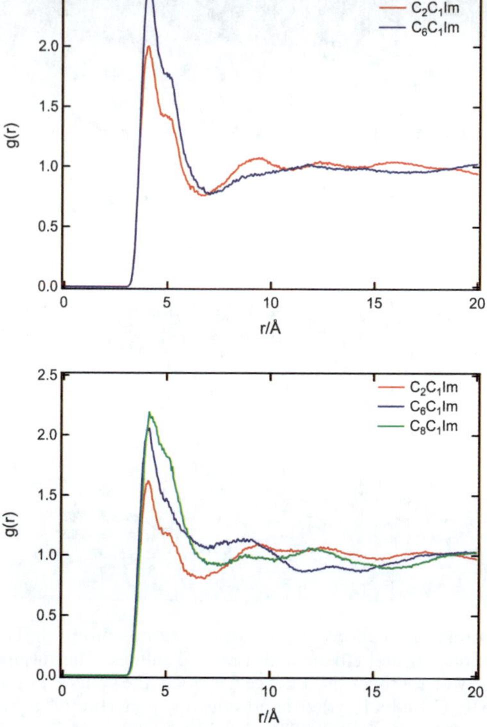


**Fig. 6** Solute–solvent radial distribution functions of ethane (upper plot) and *n*-butane (lower plot) in the ionic liquids $[C_2C_1im][NTf_2]$ and $[C_6C_1im][NTf_2]$, and $[C_2C_1im][NTf_2]$, $[C_6C_1im][NTf_2]$ and $[C_8C_1im][NTf_2]$. The peaks represent the probability of finding the terminal atom of the solute near the terminal atom of the alkyl-side chain on the imidazolium cation in the ionic liquids.

Molecular simulations also confirm the differences found experimentally between the two gases solvated in the different studied ionic liquids. In Fig. 6 are represented the radial distribution functions of the terminal atoms of the solutes and of the alkyl side-chains of the ionic liquids: $[C_2C_1im][NTf_2]$ and $[C_6C_1im][NTf_2]$ for ethane (upper plot in Fig. 6); and $[C_2C_1im][NTf_2]$, $[C_6C_1im][NTf_2]$ and $[C_8C_1im][NTf_2]$ for *n*-butane (lower plot in Fig. 6). The intensity of the peaks of the radial distribution functions increases when the number of carbon atoms in the alkyl chain length on the cation increases. These peaks also become broader, an indication of a less localized solute–solvent interaction, an observation compatible with the increase of the entropy for the solutions with the larger cations, found experimentally. This increase in entropy can be understood in terms of the greater mobility allowed in the larger regions of low charge density in the ionic liquids with longer alkyl side chains.

## 4 Conclusions

The molecular mechanisms of the solvation of ethane and *n*-butane in the family of the ionic liquids $[C_nC_1im][NTf_2]$ have been described. It has been proved by molecular simulation that, as expected, both gases interact preferentially with the low charge density regions of the ionic liquids. The larger solubility of butane found experimentally is explained by a more favorable solute–solvent interaction expressed by more negative enthalpies of solvation, calculated from the temperature derivative of the measured solubilities.

 

The solubility of both gases increases with the size of the alkyl chain of the imidazolium cation, a fact that can only be justified by more favorable entropies of solvation as calculated from the experimental solubility data and confirmed by the analysis of the site-site solute–solvent radial distribution functions.

Through this work, it has been shown that well-chosen gaseous solutes allow the study of the solvation in ionic liquid media and the identification of the molecular mechanisms responsible for the macroscopic properties measured in these solutions. Not only can the solubility be tuned by slightly modifying the structure of the ionic liquid cation, but also the molecular properties of the solutions can be adjusted allowing, for example, a greater mobility of the solutes within well determined regions of the nanostructured solvent. Herein is demonstrated, both experimentally and by molecular simulation, that differences in the entropy of solvation can explain the behavior of gaseous solutes in ionic liquid media.

## Acknowledgements

The authors thank Prof. C. Hardacre from Queen's University Belfast for providing the ionic liquid samples; Dr P. Husson from LTIM for assistance in the density measurements; and Dr J. Jacquemin from François Rabelais University in Tours for fruitful discussions about the present results.

## References

1 N. V. Plechkova and K. R. Seddon, *Chem. Soc. Rev.*, 2008, **37**, 123–150.
2 J. N. Canongia Lopes, M. F. Costa Gomes and A. A. H. Padua, *J. Phys. Chem. B*, 2006, **110**, 16816–16818.
3 A. A. H. Padua, M. F. Costa Gomes and J. N. Canongia Lopes, *Acc. Chem. Res.*, 2007, **40**, 1087–1096.
4 A. Triolo, O. Russina, H.-J. Bleif and E. Di Cola, *J. Phys. Chem. B*, 2007, **111**, 4641–4644.
5 L. J. Florusse, S. Raeissi and C. J. Peters, *J. Chem. Eng. Data*, 2008, **53**, 1283–1285.
6 M. Shiflett and A. Yokozeki, *J. Chem. Eng. Data*, 2006, **51**, 1931–1939.
7 J. L. Anderson, J. K. Dixon and J. F. Brennecke, *Acc. Chem. Res.*, 2007, **40**, 1208–1216.
8 J. Jacquemin, M. F. Costa Gomes, P. Husson and V. Majer, *J. Chem. Thermodyn.*, 2006, **38**, 490–502.
9 J. Jacquemin, P. Husson, V. Majer and M. F. Costa Gomes, *Fluid Phase Equilib.*, 2006, **240**, 87–95.
10 G. Hong, J. Jacquemin, M. Deetlefs, C. Hardacre, P. Husson and M. F. Costa Gomes, *Fluid Phase Equilib.*, 2007, **257**, 27–34.
11 M. F. Costa Gomes, *J. Chem. Eng. Data*, 2007, **52**, 472–475.
12 B.-C. Lee and S. L. Outcalt, *J. Chem. Eng. Data*, 2006, **51**, 892–897.
13 D. Camper, C. Becker, C. Koval and R. Noble, *Ind. Eng. Chem. Res.*, 2005, **44**, 1928–1933.
14 R. A. Condemarin and P. Scovazzo, *Chem. Eng. J.*, 2009, **147**, 51–57.
15 P. Bonhote, A. P. Dias, N. Papageorgiou, K. Kalyanasundaram and M. Gratzel, *Inorg. Chem.*, 1996, **35**, 1168–1178.
16 J. Jacquemin, M. F. Costa Gomes, P. Husson and V. Majer, *J. Chem. Thermodyn.*, 2006, **38**, 490–502.
17 J. Jacquemin, P. Husson, V. Majer and M. F. Costa Gomes, *Fluid Phase Equilib.*, 2006, **240**, 87–95.
18 J. Jacquemin, P. Husson, A. A. H. Padua and V. Majer, *Green Chem.*, 2006, **8**, 172–180.
19 J. N. Canongia Lopes, J. Deschamps and A. A. H. Padua, *J. Phys. Chem. B*, 2004, **108**, 2038–2047.
20 J. N. Canongia Lopes and A. A. H. Pádua, *J. Phys. Chem. B*, 2004, **108**, 16893–16898.
21 W. D. Cornell, P. Cieplak, C. I. Bayly, I. R. Gould, K. M. Merz, D. M. Fergusin, D. C. Spellmeyer, T. Fox, J. W. Caldwell and P. A. Kollman, *J. Am. Chem. Soc.*, 1995, **117**, 5179–5197.
22 W. L. Jorgensen, D. S. Maxwell and J. Tirado-Rives, *J. Am. Chem. Soc.*, 1996, **118**, 11225–11236.
23 W. Smith and T. R. Forester, *DL_POLY Molecular Simulation Package*, (2007).
24 J. H. Dymond and E. B. Smith, *The Virial Coefficients of Pure Gases and Mixtures*, Clarendon Press, Oxford, 1980.

 

25 J. Jacquemin, P. Husson, V. Majer and M. F. Costa Gomes, *J. Solution Chem.*, 2007, **36**, 967–979.
26 D. Almantariotis, T. Gefflaut, J. Y. Coxam, A. A. H. Padua and M. F. Costa Gomes, *J. Phys. Chem. B*, 2010, **114**, 3608–3617.
27 J. Deschamps, M. F. Costa Gomes and A. A. H. Pádua, *ChemPhysChem*, 2004, **5**, 1049–1052.
28 M. F. Costa Gomes and A. A. H. Pádua, *Pure Appl. Chem.*, 2005, **77**, 653–665.
29 M. F. Costa Gomes and A. A. H. Padua, in *Solubility, Developments and Applications*, Chapter 10, ed. T. Letcher, Royal Society of Chemistry, Cambridge, 2006.

 

# Critical behaviour and vapour-liquid coexistence of 1-alkyl-3-methylimidazolium bis-(trifluoromethylsulfonyl)amide ionic liquids *via* Monte Carlo simulations†

**Neeraj Rai and Edward J. Maginn***

*Received 5th May 2011, Accepted 1st June 2011*
**DOI: 10.1039/c1fd00090j**

Atomistic Monte Carlo simulations are used to compute vapour–liquid coexistence properties of a homologous series of [$C_n$mim][$NTf_2$] ionic liquids, with $n = 1, 2, 4, 6$. Estimates of the critical temperatures range from 1190 K to 1257 K, with longer cation alkyl chains serving to lower the critical temperature. Other quantities such as critical density, critical pressure, normal boiling point, and accentric factor are determined from the simulations. Vapour pressure curves and the temperature dependence of the enthalpy of vapourisation are computed and found to have a weak dependence on the length of the cation alkyl chain. The ions in the vapour phase are predominately in single ion pairs, although a significant number of ions are found in neutral clusters of larger sizes as temperature is increased. It is found that previous estimates of the critical point obtained from extrapolating experimental surface tension data agree reasonably well with the predictions obtained here, but group contribution methods and primitive models of ionic liquids do not capture many of the trends observed in the present study

## Introduction

Although ionic liquids (ILs) are similar to their alkali halide cousins in that they are characterised by strong Coulombic interactions among ions, their larger size, asymmetric shape and the presence of significant van der Waals attractive and repulsive forces play a major role in shaping their thermophysical properties. Consequently, the properties of ILs can be tuned over a wide range by changing the structure and chemical composition of the constituent ions. This tunability, along with the almost limitless number of ILs that can be formed by combining different cation and anion pairs, have made ILs the focus of intense research in areas such as $CO_2$ capture,[1] reaction media for synthesis,[2] electrolytes in high capacity batteries and fuel cells,[2] polymers,[3] plasticisers,[4] and even as a medium for liquid mirrors in space-based telescopes.[5,6] One of the properties of ILs that makes them especially attractive for these and other applications is their extremely low volatility. Not long ago, it was erroneously claimed that ILs were "non-volatile" and had "zero vapour pressure". These claims were overturned about six years ago by a fascinating series of studies.

*Department of Chemical and Biomolecular Engineering, University of Notre Dame, Notre Dame, IN, USA. E-mail: ed@nd.edu; Fax: +01 574 6318366; Tel: +01 574 6315687*

† Electronic supplementary information (ESI) available: The saturated densities, vapour pressure, and enthalpy of vapourisation along with uncertainties are tabulated in Supplementary Tables 1 to 4. See DOI: 10.1039/c1fd00090j

 

Rebelo and co-workers[7] showed that aprotic ILs could be thermally vapourised, and they extrapolated experimental surface tension and density data to estimate normal boiling points and critical temperatures. Soon after, Kabo and co-workers[8] reported the first indirect measurement of the vapour pressure of 1-*n*-butyl-3-methylimidazolium bis(trifluoromethylsulfonyl)amide ($[C_4mim][NTf_2]$). At about the same time, Earle *et al.*[9] provided conclusive proof that ILs not only have significant vapour pressures at high temperature, but that they could also be distilled without thermal degradation. This led to intense research in the field to measure vapour pressures and enthalpies of vapourisation,[10–15] investigate the separation of ILs *via* distillation,[16,17] and study the nature of the species present in the vapour phase.[18–23] Two reviews on the topic have appeared recently.[24,25]

Aside from the obvious practical utility of knowing the volatility of a given IL, understanding the vapour–liquid phase equilibrium of ILs is important from a fundamental standpoint. Quantities such as the enthalpy of vapourisation, vapour pressure, normal boiling point and critical point are essential properties used in developing intermolecular potential functions, formulating equations of state,[26] and corresponding states theories.[27,28] These theories provide a unifying framework to understand, correlate and predict the thermodynamic properties of fluids. They are used routinely for conventional fluids and would be extremely useful for ILs, given the chemical diversity and complexity of these liquids. Unfortunately, the low volatility of ILs makes experimental determination of vapour pressures and enthalpies of vapourisation challenging. The fact that decomposition temperatures of ILs tend to be much lower than their critical temperatures and normal boiling points makes experimental determination of these quantities virtually impossible.[7] Nevertheless, there have been direct and indirect measurements of vapour pressures and enthalpies of vapourisation reported.[8,10,13,14,18,29] Critical temperatures have been estimated for ILs by extrapolation of temperature-dependent surface tensions to the point where the surface tension vanishes.[7] Normal boiling points have been estimated either through use of an empirical relation that relates the normal boiling point to the critical temperature[7] or by extrapolating low temperature vapour pressures.[18] These extrapolations are performed over a wide temperature range and rely upon semi-empirical models developed for conventional fluids[30] that have not been validated for ILs, making accuracy a real concern. Note that even if the critical points and normal boiling points are beyond decomposition temperatures, "hypothetical" values are still extremely useful as reference states for developing equations of state and group contribution methods for ionic liquids.

Given the experimental difficulties mentioned above, a useful alternative for estimating these properties is to use molecular simulations, which have already been shown to provide an accurate treatment of the liquid state of ILs.[31] A number of groups have used molecular simulations to predict the enthalpy of vapourisation (or cohesive energy density) of ILs.[32–39] Results have generally agreed with experimental measurements. The standard procedure for doing this is to compute the average internal energy of the liquid state and a single ion pair, *which is assumed to be representative of the vapour phase*.[40] While this assumption appears to be valid at low temperatures, the aggregation state of ions in the vapour phase at higher temperatures and saturated vapour densities remains an open question. Certainly as the critical point is approached (and the ability to distinguish between the liquid and vapour phases vanishes) one expects to see clustering of ions in the vapour phase. Simulations by Ballone and co-workers[41] have suggested the presence of clusters in the vapour phase, which would affect the computed vapourisation enthalpy. More importantly, simulations of this type do not provide a direct means for estimating vapour pressures, boiling points or critical points.

Weiss and co-workers[42] computed the temperature dependence of the surface tension of 1-*n*-butyl-3-methylimidazolium hexafluorophosphate using interfacial molecular dynamics (MD) simulations, and then estimated the critical temperature using an extrapolation scheme similar to that employed by Rebelo and co-workers with experimental surface tension data. Very recently, Leroy and Weiss[43] performed interfacial MD simulations on an idealised IL model in which cations and anions were treated as charged Lennard-Jones (LJ) spheres with varying ratios of Coulombic-to-dispersion interactions. They computed coexisting densities, surface tensions and critical points for this model system and predicted that the critical temperature and critical density of ILs increase as the ratio of dispersive to Coulombic interactions increases.

The most direct way of simulating phase behaviour is through use of Gibbs ensemble Monte Carlo (GEMC)[44] or related methods[45] in which thermodynamic equilibrium between two phases (here, the vapour and liquid phases) is enforced through a series of Monte Carlo moves. This approach avoids the use of an explicit interface, which can interfere with the determination of bulk phase behaviour, especially for small systems with long range interactions. Unfortunately, these Monte Carlo methods are difficult to apply to ILs because they involve either insertion of molecules into a dense phase or exchange of molecules between the phases; such moves are notoriously difficult for large molecules with specific interactions,[46,47] and the presence of ions results in additional complications.[48] Orkoulas and Panagiotopoulos have developed Monte Carlo methods for treating ionic systems[49,48] and applied them to models of alkali halides. Application of these methods to realistic representations of ILs has so far proved challenging, thus necessitating the use of highly idealised models. Martin-Betancourt *et al.*[50] employed a primitive model of an IL in which the cation was modelled as a charged hard spherocylinder and the anion as a charged hard sphere. In contrast to the trend observed by Leroy and Weiss,[43] they found that the estimated critical temperature went down dramatically as the aspect ratio of the spherocylinder (representative of the length of the cation "alkyl tail") increased. Even for such a simplified model, the authors could not simulate vapour-liquid coexistence curves (VLCCs) beyond a hypothetical "alkyl chain length" of 3.5 units due to sampling difficulties.[50]

Recently, we reported results of a GEMC study for a homologous series of 1–alkyl–3–methylimidazolium tetrafluoroborate ([$C_n$mim][$BF_4$]) ILs, where we showed for the first time that one can compute vapour–liquid coexistence curves of ILs using a detailed atomistic model of the IL.[51] We found that for $BF_4$ based ionic liquids the vapour pressure increases and critical temperature and density decrease as the size of the cation increases. Moreover, the simulations show that the vapour state consists of a significant fraction of aggregated ion pairs at high temperature and pressure. As the critical behaviour of [$C_n$mim][$BF_4$] showed a clear but weak dependence on alkyl chain length of the cation, it is difficult to conclude if the same trends will be observed for larger anions with a significantly more diffuse charge distribution. To address this questions, we report an extensive study in which we compute VLCCs for 1–alkyl–3–methylimidazolium bis(trifluoromethylsulfonyl)-amide ILs, [$C_n$mim][$NTf_2$], where $n = 1, 2, 4, 6$ (see Fig. 1). We selected this system as [$NTf_2$] based ILs are very stable and one the most widely studied ionic liquids. Consequently, a large amount of low temperature experimental data are available.

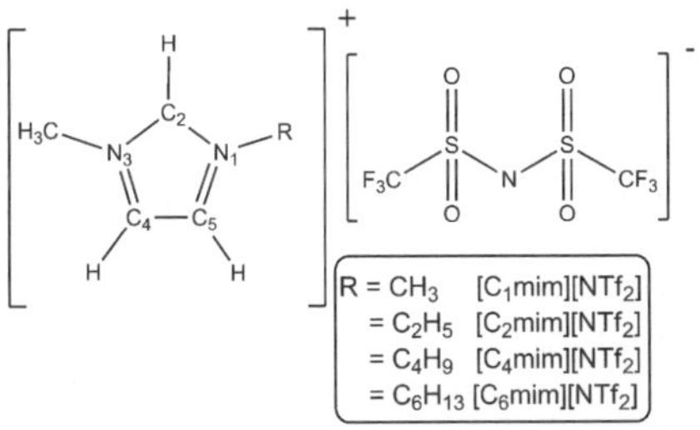

R = $CH_3$ [$C_1$mim][$NTf_2$]
= $C_2H_5$ [$C_2$mim][$NTf_2$]
= $C_4H_9$ [$C_4$mim][$NTf_2$]
= $C_6H_{13}$ [$C_6$mim][$NTf_2$]

**Fig. 1** Molecular structure of the ionic liquids under study.

Additionally, due to the presence of a large number of branch points and dihedral flexibility, [NTf$_2$] based ILs present significant sampling/computational difficulty compared to nearly "spherical" $BF_4$ or $PF_6$ based ionic liquids. In the following sections we examine how the larger, flexible [NTf$_2$] anion affects phase behaviour. We report estimated critical points, the temperature dependence of vapour pressures and enthalpies of vapourisation. We also examine the aggregation state of ions in the vapour phase and compare results with previous experimental and theoretical studies.

## Force field and simulation details

To model the ILs, a force field developed by Zhong *et al.*[52] was used. In this force field, alkyl groups ($CH_x$ and $CF_3$) are modelled as a single interaction site (united atom) while all other atoms are modelled explicitly. This force field reproduces both the temperature and alkyl chain length dependence of thermodynamic and transport properties reasonably well.[52] This was the main criterion for selecting this force field as we are interested in studying the temperature dependence and alkyl chain length dependence of phase behaviour of ionic liquids. In most force fields, interaction sites for a united atom coincide with the carbon atom. In the present force field, the centre of mass of the united atom was used as the interaction centre. This was essential in ensuring good transferability of united atom parameters, thus providing good structural and temperature dependence. For computational efficiency, the aromatic ring and bonds were treated as rigid, since the flexibility of these degrees of freedom has a negligible effect on VLCCs.[53]

A radial cutoff of 14 Å with analytic long range corrections[47] was used to compute the Lennard-Jones (LJ) potential. The Ewald summation method with tin foil boundary conditions was used to compute long range Coulombic interactions.[47,54,55] For the liquid phase, a cutoff of $r_{cut} = 14$ Å was used for the real space part of the Ewald summation while $r_{cut} = 0.45 \times boxlength$ was used for the vapour phase. As the density of ion pairs in the vapour phase is much less than the liquid phase, this choice of real space cutoff significantly reduces computational overhead. The Ewald sum convergence parameter, $\alpha$, was set to $3.2/r_{cut}$ while the reciprocal space cutoff, $k_{cut}$, was set to $2.0 \times \pi \times \alpha$.

We used the canonical version of Gibbs ensemble Monte Carlo[44,47,56] in this study. The system consisted of 200 to 400 ion pairs with larger numbers used for the higher temperatures. The total volume of the two phases was adjusted to have around 50 to 250 ion pairs in the vapour phase, with the larger numbers for the higher temperatures. Since we observed significant aggregation in the vapour phase at higher temperatures, we used a larger number of ions in the vapour phase to ensure proper sampling of larger aggregates. As we didn't know the critical point for these ILs *a priori*, we ran exploratory simulations between 600 and 1500 K to get a quick estimate of the critical points. Based on these runs, we carried out simulations at seven different temperatures spanning the range 900 to 1100 K. Decomposition temperatures of [C$_n$mim][NTf$_2$] ionic liquids, as measured by thermogravimetric analysis, are on the order of 700 K.[57,58] As the lowest temperature considered in the present study is higher than the decomposition temperature, the coexistence region presented here is not accessible *via* experiments. The computed boiling points and critical points are therefore hypothetical states of these ionic liquids.

Simulations were started by placing cations and anions randomly in the liquid box and the vapour box. The initial configurations were generated by inserting ions using a configurational-bias growth scheme, thus avoiding any overlaps in the initial configurations. The system was allowed to relax using only centre of mass translations, rotations about centre of mass, and configurational-bias growth moves[47,59,60] for $5 \times 10^3$ MC cycles (1 MC cycle consists of 2 × number of ion pairs moves). In the vapour phase, ion pair translation[48] and cluster moves[48] were also employed. The initial relaxation was followed by $10^4$ MC cycles where volume exchange moves were employed to get a reasonable density of the liquid phase. The final equilibration consisted of 5 × $10^5$ MC cycles where additional configurational-bias ion pair exchange moves were employed to equilibrate the chemical potential of the liquid and the vapour phases. During equilibration, maximum displacements for the translation, rotation, and volume moves were adjusted to achieve 50% acceptance rates. During production runs, one ion pair swap move was accepted approximately every 1000 MC cycles at $T = 900$ K for $[C_6mim][NTf_2]$ (the most difficult IL to swap) while at $T = 1100$ K the acceptance rate was one swap move accepted every 100 MC cycles. The configurational-bias growth and ion pair exchange moves used a dual cutoff scheme proposed by Vlugt *et al.*[61] In addition, bond angles were sampled by carrying out a pre-simulation of fragments at the simulation temperatures and populating a reservoir of fragment conformations used during configurational-bias moves.[62,63]

The production period consisted of three to five independent runs totalling 1.5–3.0 × $10^6$ cycles. The runs were divided into blocks of $10^5$ MC cycles to compute averages and standard deviations. The probabilities for centre of mass translations, rotations about an axis passing through the centre of mass of ions, configurational-bias regrowth,[47] volume moves, ion pair exchange moves, cluster moves,[48] (vapour phase) and ion pair translation moves[48] (vapour phase) were set to approximately 0.25, 0.25, 0.1, 0.004, 0.2, 0.1, and 0.1, respectively. The normal boiling point ($T_b$) and critical pressure ($P_c$) were determined by fitting the saturated vapour pressure data to the Clausius–Clapeyron equation, and $\rho_c$ and $T_c$ were computed by weighted linear fits of the five highest subcritical temperature data sets to the law of rectilinear diameters and the density scaling law, respectively.[47] The enthalpies of vapourisation ($\Delta H^{vap}$) at the boiling point were estimated by fitting $\Delta H^{vap}$ data to a third order polynomial using Matlab (version 7.7.0.0471(R2008b)). The uncertainties in critical pressures and enthalpies of vapourisation at the normal boiling point were estimated using Monte Carlo sampling where $\langle P^{sat} \rangle$ and $\langle \Delta H^{vap} \rangle$ were assumed to be normally distributed about their mean with variance obtained by block averaging.[64] All simulations were conducted using a locally developed Monte Carlo code called Cassandra.

## Results and discussions

Care must be taken in developing Monte Carlo procedures for simulating ILs. Standard Monte Carlo phase equilibrium sampling methods involve exchanges of a single molecule between the two coexisting phases. For salts, charge neutrality requires that ion pairs be exchanged between the vapour and liquid phases.[66] Moreover, selecting a cation and an anion for deletion in an unbiased random fashion is very inefficient, as most attempted moves are rejected due to a large energetic penalty associated with removing the ions from their neighbouring counterion environments.[48] The acceptance rates can be improved if one of the ions is selected at random while the counterion is selected in a biased manner based on the distance from the first ion (see ref. 48 for details). We implemented this procedure and carried out benchmark validation tests using a simple Lennard-Jones model for argon (Ar) as well as the restricted primitive model (RPM), which consists of charged hard spheres with opposite charges. Fig. 2A shows a comparison of the results obtained for the VLCCs of Ar obtained using the conventional one-molecule exchange and an unbiased two-molecule exchange move. The results are identical and agree with literature values,[65] indicating that the unbiased two-molecule exchange procedure is correct. The biased two-molecule exchange move was tested in the Ar system at 120 K and gave results identical to the unbiased two-molecule exchange method. We then tested our simulation procedures using the RPM, which requires that a cation and anion be exchanged simultaneously. The computed VLCC for the RPM is shown in Fig. 2B along with the VLCC data reported by Orkoulas and

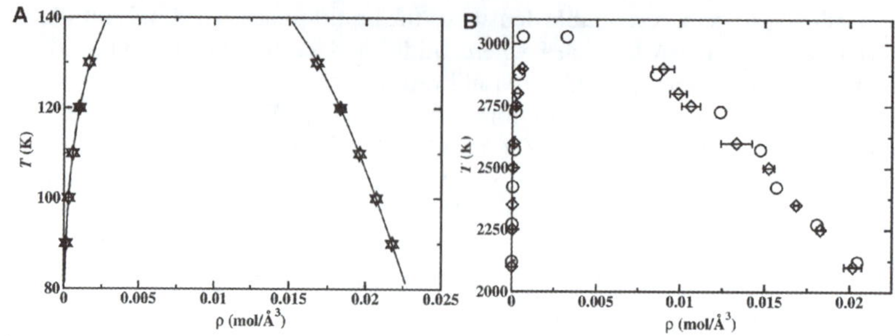


**Fig. 2** Validation of the GEMC procedures used in this work. (A): Vapour–liquid coexistence curve for argon. The up triangles and down triangles represent one and two molecule (unbiased) exchanges, respectively, between the liquid and vapour phases. The stars represent results when the second molecule is picked using a distance bias criterion similar to Ref.48. The lines are accepted reference data from NIST.[65] (B): Vapour–liquid coexistence curve for the RPM model. The circles represent data from the ref.48 while diamonds are our GEMC data.

Panagiotopoulos.[48] From Fig. 2B it is evident that our results are in good agreement with the literature values.

The VLCCs for the ILs are shown in Fig. 3 and normal boiling points and critical properties are provided in Table 1. The change in $T_c$ as a function of cation chain length $n$ in [$C_n$mim] is depicted in Fig. 4. The simulations predict that an increase in cation alkyl chain length results in decrease in the critical temperature ($T_c$) and critical density ($\rho_c$). The decrease in $T_c$ is approximately 14 K per methylene group ($CH_2$). The dependence of critical properties on cation alkyl chain length for [$C_n$mim][$NTf_2$] ILs is very similar to that observed in the case of [$C_n$mim][$BF_4$] ILs.[51] For the same cation, the critical temperature of [$NTf_2$] ILs is approximately 50 K lower than that of [$BF_4$] ILs, suggesting that intermolecular interactions present in [$BF_4$] ILs are somewhat stronger than that of the [$NTf_2$] ILs. This can be understood from the fact that the [$NTf_2$] ion is significantly larger than the [$BF_4$] ion, thus the charge on [$NTf_2$] is more delocalised. This results in weaker Coulombic interactions between the ions in [$NTf_2$]–based ILs. Although the dispersion part of the cohesive energy of [$NTf_2$]–based ILs is expected to be larger than that of [$BF_4$]–based IL, the Coulombic interactions appear to dominate. These results agree with the experimental findings that ILs having the [$NTf_2$] anion tend to be more volatile than those with smaller anions. The trends in $T_c$ and $\rho_c$ with the size of the cation conflict with the results obtained by Leroy and Weiss[43] using a simple spherical model of the ions. On the other hand, while the primitive model

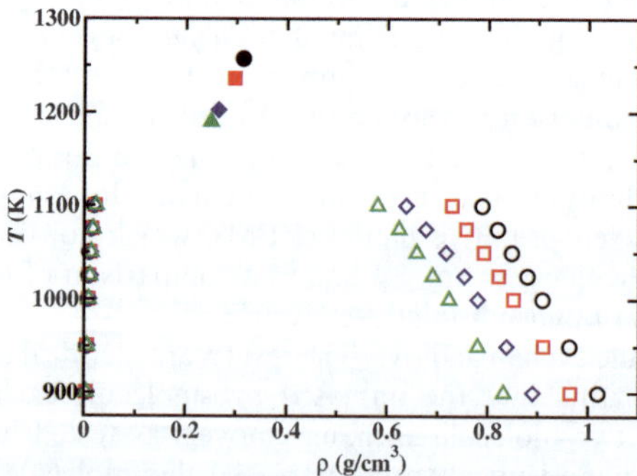


**Fig. 3** Vapour–liquid coexistence curves of [$C_n$mim][$NTf_2$] ionic liquids. The symbols refer to the following systems: black circles ($C_1$); red squares ($C_2$); blue diamonds ($C_4$); and green triangles ($C_6$). The filled symbols of the same shape and colour show corresponding critical points.

 

**Table 1** Computed critical temperature $T_c$, critical density $\rho_c$, critical pressure $P_c$, critical compressibility $Z_c$, normal boiling point $T_b$ and boiling point to critical temperature ratio $T_b/T_c$ for [$C_n$mim][$NTf_2$] ionic liquids. The subscripts are uncertainties in the last digit. Refer to the text for an explanation of the different methods

| $C_n$ | Method | $T_c$ (K) | $\rho_c$ (g cm$^{-3}$) | $P_c$ (bar) | $Z_c$ | $T_b$ (K) | $T_b/T_c$ |
|---|---|---|---|---|---|---|---|
| $C_1$ | GEMC | $1257_3$ | $0.313_4$ | $5.13_{13}$ | $0.058_1$ | $987_3$ | 0.79 |
| $C_2$ | GEMC | $1236_3$ | $0.297_4$ | $4.97_9$ | $0.064_1$ | $977_3$ | 0.79 |
| | Gug[a] | 1100 | | | | 660 | 0.6 |
| | Eöt[a] | 1209 | | | | 725 | 0.6 |
| | GC[67] | 1249 | 0.447 | 32.7 | 0.28 | 817 | 0.65 |
| | Expt[b] | | | | | 907 | |
| $C_4$ | GEMC | $1203_4$ | $0.265_6$ | $4.19_{15}$ | $0.066_2$ | $970_2$ | 0.81 |
| | Gug[a] | 1012 | | | | 607 | 0.6 |
| | Eöt[a] | 1077 | | | | 646 | 0.6 |
| | GC[67] | 1270 | 0.423 | 27.6 | 0.26 | 862 | 0.68 |
| | Expt[b] | | | | | 933 | |
| $C_6$ | GEMC | $1190_4$ | $0.251_5$ | $4.11_{10}$ | $0.074_2$ | $970_3$ | 0.82 |
| | Gug[a] | 932 | | | | 559 | 0.6 |
| | Eöt[a] | 967 | | | | 580 | 0.6 |
| | GC[67] | 1293 | 0.405 | 23.6 | 0.24 | 908 | 0.70 |
| | Expt[b] | | | | | 885 | |

[a] Guggenheim and Eötvos EOS estimates using low temperature experimental surface tension and density data, and $T_b/T_c = 0.6$ was assumed to compute normal boiling point.[7] [b] Estimated using low temperature experimental vapour pressure data.[10]

of Martin–Betancourt *et al.*[50] does predict a decrease in $T_c$ with cation alkyl chain length, the amount of the decrease (roughly 250 K per methylene increment) is much larger than that observed here. This suggests that such simple models are

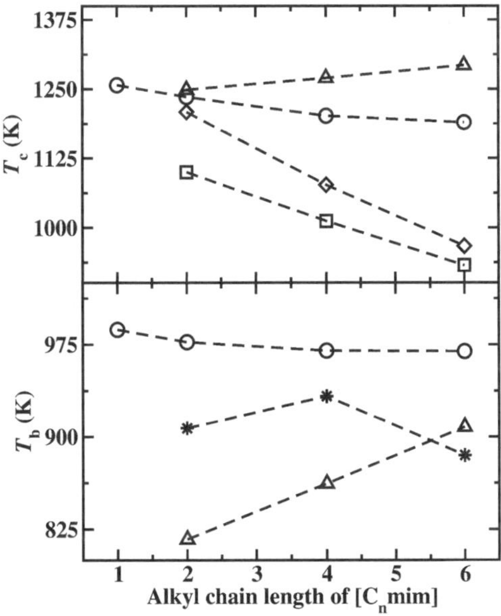


**Fig. 4** Critical temperatures ($T_c$) and normal boiling points ($T_b$) as a function of cation alkyl chain length. The symbols have the following meaning: circles (GEMC simulation results); squares (Guggenheim EOS[7]); diamonds (Eötvos EOS[7]); triangles (group contribution method[67]); and stars (experiment[10]). Dashed lines are guides to the eye.

unable to capture all of the subtle energetic and structural features of ILs that appear to be important in their phase equilibria.

Rebelo *et al.*[7] employed the Guggenheim (Gug) and Eötvos (Eöt) equations of state to estimate critical points of several ILs, including those studied here. The procedure involves an extrapolation of temperature-dependent surface tension and liquid density data. We find that both models yield critical temperatures that are in reasonable agreement with our simulation results, although the Eötvos model results are closest to the simulations. Importantly, both models also predict a decrease in the critical temperature with increasing cation alkyl chain length, although the value of 50 K per methylene increment is substantially higher than the predictions of the simulations (see Fig. 4). One should be cautious in attributing too much to this difference, as the experimental data used in the models were taken over a narrow range of temperatures close to ambient conditions, and the subsequent extrapolation results in relatively large uncertainties of approximately 100 K in the predicted critical temperatures.[7] Nevertheless, the fact that the models and the simulations show the same basic trends and predict roughly the same critical temperatures is intriguing. By way of comparison, a group contribution (GC) method developed by Valderrama and Rojas[67] to predict critical properties of ionic liquids gives a value of $T_c$ for [$C_2$mim] that is only 13 K higher than simulation predictions. However, the GC method predicts an *increase* in $T_c$ upon increasing cation alkyl chain length instead of a decrease (see Fig. 4), similar to what is observed with conventional liquids like alkanes. Given that this trend is inconsistent with vapour pressure trends, we suspect that the shortcoming of the GC method stems from the lack of experimental critical data to refine the methylene increment for $T_c$.

The saturated vapour pressures of [$C_n$mim][$NTf_2$] in a Clausius–Clapeyron representation are shown in Fig. 5. Antoine equation coefficients were fit to the data and are given in Table 2. We find that at a given temperature $P^{sat}$ increases from [$C_1$mim] to [$C_4$mim] while the vapour–pressure of [$C_6$mim] is nearly identical to [$C_4$mim]. The predicted normal boiling points are listed in Table 1 and the cation alkyl chain length dependence is plotted in Fig. 4. We find that $T_b$ decreases upon going from $C_1$ to $C_4$. The predicted boiling point for $C_6$ is almost identical to that of $C_4$. Zaitsau *et al.*[10] measured vapour pressures of [$C_n$mim][$NTf_2$] for temperatures up to approximately 500 K *via* the integral effusion Knudson method. For the temperature range considered in the study, volatility obeyed the following order: $C_8 < C_4 < C_6 < C_2$.[10] Additionally, the presented data suggest that there is a crossover in the volatility upon increasing the temperature. Thus the boiling points suggest $C_4$ to be least volatile followed by $C_2$ and $C_6$ (Fig. 4).[10] Except for the $C_4$ data point, the experimental results

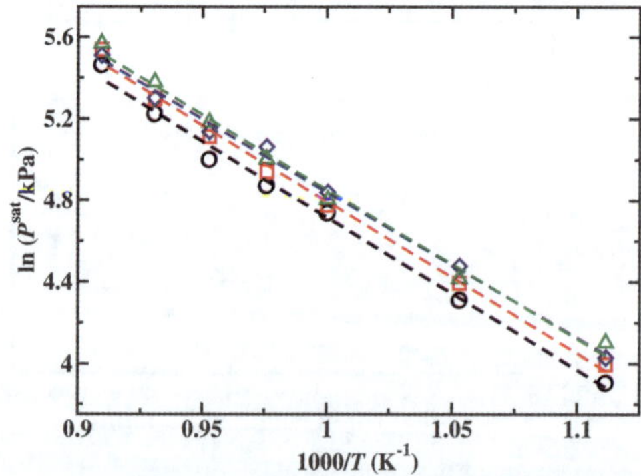


**Fig. 5** Clausius–Clapeyron representation of vapour pressure data of [$C_n$mim][$NTf_2$] ionic liquids. The symbols have the following meaning: black circles ($C_1$); red squares ($C_2$); blue diamonds ($C_4$); and green triangles ($C_6$). The dashed lines are linear fits to the simulation data.

 

**Table 2** Computed vapourisation properties of [$C_n$mim][$NTf_2$] ionic liquids. The vapour pressure data are fit to $\ln(P^{sat}/kPa) = A–B \times 10^3/T$. The subscripts are uncertainties in the last digit

| $C_n$ | $\Delta H^{vap}_{T_b}$ (kJ mol$^{-1}$) | $\Delta S^{vap}_{T_b}$ (J mol·K$^{-1}$) | $A$ | $B$ (K) | $P^{sat}{}_{298.15}$$^a$ ($10^{-3}$ Pa) | $\omega$ |
|---|---|---|---|---|---|---|
| $C_1$ | $69.9_3$ | $70.8_4$ | 12.1792 | 7.4660 | $3_3$ | $0.11_2$ |
| $C_2$ | $69.0_3$ | $70.6_4$ | 12.2297 | 7.4419 | $3_3$ | $0.12_1$ |
| $C_4$ | $66.5_4$ | $68.5_5$ | 11.9751 | 7.1405 | $6_6$ | $0.11_2$ |
| $C_6$ | $67.1_5$ | $69.2_6$ | 12.1683 | 7.3203 | $4_5$ | $0.14_1$ |

[a] Saturation pressure at $T = 298.15$ $K$ ($P^{sat}{}_{298.15}$) extrapolated based on high temperature simulation data and represent an upper bound (see text).

point to trends similar to our calculations. We note that experimental extrapolation is over a quite large range, thus trends in the boiling points are less reliable than the volatility trends observed for temperatures at which measurements were performed. Widegren *et al.*[16] measured relative volatility *via* distillation and found a volatility rank order of $C_1 \approx C_6 < C_2 < C_4$. Thus except for the anomalous behaviour of $C_6$ these trends are in agreement with our GEMC simulations.[16] In contrast to our predictions, the GC method predicts an increase in the boiling point with increasing cation alkyl chain length.[67] We used a linear fit (see Table 2 for regression coefficients) to extrapolate $P^{sat}$ at $T = 298.15$ K. We find that the average saturation pressure for these ILs at $T = 298.15$ K is approximately $4 \times 10^{-3}$ Pa (Table 2). Given the range of the extrapolation (600 K) this should be considered as a rough estimate. Additionally, given that at lower temperatures the slope of the Clausius–Clapeyron plot will be steeper due to an increase in the enthalpy of vapourisation, these values are an upper bound for saturation pressure at room temperature.

Given vapour pressures and critical temperatures, one can compute critical pressures ($P_c$) and critical compressibility factors ($Z_c$). The computed $P_c$ and $Z_c$ are presented in Table 1. We find that the critical pressure is a monotonically decreasing function of the number of carbon atoms in the cation alkyl chain. In the case of critical pressures, the GC method predicts a similar chain length dependence as our GEMC simulations,[67] but the absolute values are predicted to be nearly 5 to 6 times larger than our results and roughly similar to that observed for non-associating fluids.[67,68] $Z_c$ provides a good measure of the degree of non-ideality at the critical point. We find that $Z_c$ changes from 0.058 to 0.074 when going from $C_1$ to $C_6$. The $Z_c$ values predicted for these [$NTf_2$] ILs are larger than that of the corresponding [$BF_4$] ILs by approximately 50%[51] and 2 to 3 times larger than the RPM model.[69] This suggests that at the critical point, [$BF_4$] ILs and the RPM fluid are more non-ideal than [$NTf_2$] ILs. The GC method[67] predicts $Z_c$ to be approximately an order of magnitude larger than our results and again similar to that observed for traditional fluids.[67,70] Another quantity often used in the corresponding states analysis is the accentric factor ($\omega$), which roughly corresponds to deviations from "sphericity". Table 2 lists the computed $\omega$ values for [$C_n$mim][$NTf_2$] ILs. The values are relatively small compared to what might be expected from the size of the two ions. Interestingly, the GC approach predicts accentric factors for a range of ILs that vary widely from about 0.1 to values greater than 1.0.[71] For the [$C_n$mim][$NTf_2$] ILs, the GC accentric factors vary from 0.18 to 0.25 in going from [$C_2$mim][$NTf_2$] to [$C_6$mim][$NTf_2$].[72]

The ratio of the normal boiling point to the critical temperature, $T_b/T_c$, is of great practical utility as one can often estimate either the critical temperature or the boiling point conveniently, from which one can then predict the other unknown temperature. For most non-polar substances, it has been observed that $T_b/T_c \approx 0.6$,[30] but given the uncertainties in estimated values of $T_c$ and $T_b$, it is not clear what this ratio is for ILs. Some data even suggest that some ILs won't have a normal boiling point as the critical pressure will be lower than one atmosphere.[30] We believe GEMC simulations are well suited for computing this ratio as vapour pressures and critical temperatures can be computed for a given model in a consistent manner. For $[C_n mim][NTf_2]$ ILs we find this ratio to be about 0.8 (Table 1). This is identical to the $T_b/T_c$ ratio we observed for $[C_n mim][BF_4]$ ILs in our previous study,[51] and suggests that this may be a more suitable approximation to use for ILs than the value of the $T_b/T_c = 0.6$ and 0.7 used for non-polar and polar substances, respectively.[7,30]

The computed enthalpy of vapourisation at the normal boiling point ($\Delta H_{T_b}^{vap}$) and the entropy of vapourisation at the normal boiling point ($\Delta S_{T_b}^{vap}$) or "Trouton's constant" are given in Table 2. $\Delta H_{T_b}^{vap}$ is on the order of 67–70 kJ mol$^{-1}$, which is similar to the value computed for $[C_n mim][BF_4]$[51] and about twice the value of a hydrogen bonding fluid such as ethanol (38 kJ mol$^{-1}$).[68] Note that the energy required to completely dissociate two ions is estimated to be around 250–300 kJ mol$^{-1}$[40] suggesting that the vapour phase does not contain individual ions but rather exists as either ion pairs or larger aggregates. The fact that Trouton's constant is on the order of 70 J mol$^{-1}$ K also supports this conclusion; typical nonpolar compounds have larger Trouton's constants of approximately 85 J mol$^{-1}$ K. Polar compounds, due to the greater order they exhibit in the liquid phase, have even higher Trouton's constants on the order of 105 J mol$^{-1}$ K.[30] The fact that the Trouton's constants of these ILs are smaller than this suggests that some of the order in the liquid phase is maintained in the vapour phase through the formation of ion pairs and larger clusters. Other fluids which are known to associate in the vapour phase, such as acetic acid, also have smaller Trouton's constants.[73] Additional evidence for clustering in the vapour phase is provided below.

Fig. 6 shows the temperature dependence of the enthalpy of vapourisation ($\Delta H^{vap}$). In general, $\Delta H^{vap}$ decreases upon increasing the alkyl chain length of the cation, but the change in $\Delta H^{vap}$ corresponds to less than 1 kJ mol$^{-1}$ per methylene group. Consistent with saturated vapour pressures, at low temperatures $C_4$ and $C_6$ have very similar $\Delta H^{vap}$. The trends in $\Delta H^{vap}$ with chain length at these high temperatures is opposite to what is observed at low temperatures by different experimental techniques or MD simulations.[38] One possible explanation for this could be that at these elevated temperature the liquid expands considerably, such that intermolecular interactions are reduced just enough to lower the cohesive energy density of the liquid phase. It is known that the expansivity of ILs increases with increasing chain length, so this effect could be greater for the longer alkyl chains.

To further understand these trends in $\Delta H^{vap}$, we examined Coulombic ($\Delta U_{Coul}^{vap}$), LJ ($\Delta U_{LJ}^{vap}$), and $PV$ ($\Delta H_{PV}^{vap}$) contributions to the enthalpy of vapourisation. The temperature dependence of these terms is shown in Fig. 7, where one immediately notices

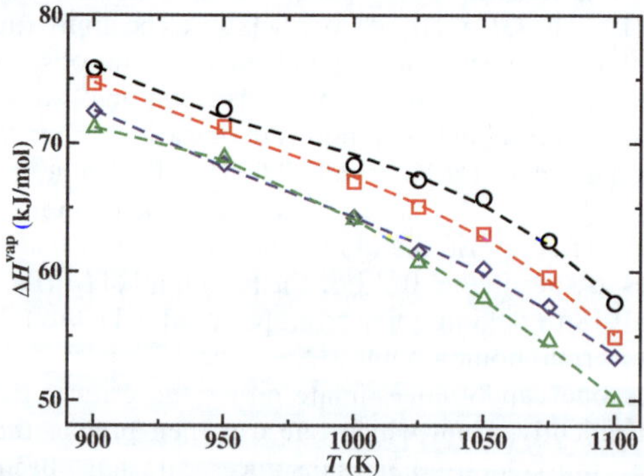


**Fig. 6** Enthalpy of vapourisation data for $[C_n mim][NTf_2]$ ionic liquids. The symbols have the following meaning: black circles ($C_1$); red squares ($C_2$); blue diamonds ($C_4$); and green triangles ($C_6$).The dashed lines of the corresponding colours are third order polynomial fits to the simulation data.

Coulombic and LJ contributions are about the same magnitude while $PV$ contributions, as expected, are significantly smaller. Additionally the Coulombic contributions decrease significantly with an increase in the number of carbon atoms in the cation alkyl chain. The decrease is approximately 8 kJ mol$^{-1}$ in going from [$C_1$mim] to [$C_6$mim], and is nearly identical to that observed in MD simulations by Köddermann *et al.* at $T = 298$ K.[38] It is interesting that low temperature MD simulations predict a slight increase in the LJ contribution in going from [$C_1$mim] to [$C_6$mim]; we find that LJ contributions show little variation as a function of chain length at elevated temperatures. The main reason for this difference is that we observe significant aggregation in the vapour phase at high temperature which serves to "cancel out" the effect of chain length observed between the liquid and vapour states. Because most of the $\Delta H^{vap}$ calculations so far have assumed that the vapour phase of ILs consists of only ion pairs, this effect has not been observed. Our results suggest that while the assumption that the vapour is comprised of single ion pairs is reasonable at room temperature, it is not appropriate at elevated temperatures. Note that at these conditions the $PV$ contributions are approximately 0.5R$T$ and decrease as the temperature increases; at these conditions, the ideal gas approximation clearly does not hold.

To understand non-ideality in the vapour phase, an aggregation analysis was carried out. A simple distance based criterion was used, such that two ions within a distance of $d_{ip} = 10$ Å were considered to be a pair. The value of $d_{ip}$ was found from a vapour-phase simulation of a single ion pair of [$C_6$mim][$NTf_2$] at 1100 K. During the course of simulation, the distance between the centre of mass of the cation, "modified" to only account for the atoms having a partial charge, and the centre of mass of [$NTf_2$] ions was recorded. It was found that 99.95% of the configurations had centres of mass separations smaller than 10.0 Å, so this was selected as

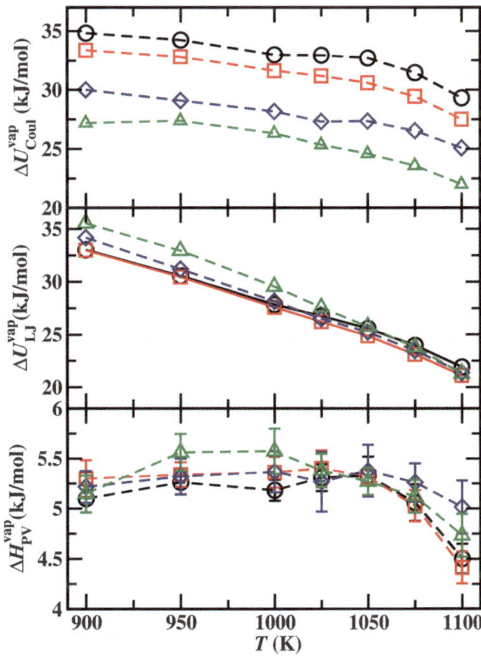


**Fig. 7** The Coulombic ($\Delta U^{vap}_{Coul}$), LJ ($\Delta U^{vap}_{LJ}$), and $PV$ ($\Delta H^{vap}_{PV}$) contributions to enthalpy of vapourisation for [$C_n$mim][$NTf_2$] ionic liquids. The symbols have the following meaning: black circles ($C_1$); red squares ($C_2$); blue diamonds ($C_4$); and green triangles ($C_6$). The dashed lines are guides to the eye.

the ion pair distance. The modified centre of mass of the cation was needed to eliminate spurious effects due to the floppy non-polar alkyl tail on the cation. The same distance criterion was used for all ILs and temperatures for consistency.

Fig. 8 shows the probability of observing a given aggregate of a particular size or aggregation number ($N_{aggr}$) for different ILs at $T = 1000$ K. It is evident that there is almost no variation in the cluster population among different ILs. The main feature of this plot is that there is very small population ($\approx 2\%$) of charged aggregates and that single ion pairs are the predominant (>60%) aggregation state of ions at $T = 1000$ K. As $N_{aggr}$ increases for the neutral clusters, there is nearly an exponential decay in their population. Although the largest aggregate observed during the course of the simulation at this temperature is 68, 62, 56, and 36 for $C_1$, $C_2$, $C_4$, and $C_6$, respectively, the population of aggregates with $N_{aggr} > 10$ is less than 2% irrespective of the cation.

To explore further the nature of ions in the vapour phase as a function of temperature, we considered only [$C_6$mim] as the trends are similar for all of the ILs considered in this study. Fig. 9 depicts the temperature dependence of the fraction of clusters ($f_{N_{aggr}}$) and fraction of ions ($f_{ion}$) in an aggregate of a given size. The $f_{N_{aggr}}$ is defined as the ratio of the number of clusters with given $N_{aggr}$ to the total clusters observed, while the $f_{ion}$ is the ratio of the number of ions in a given aggregate to the total number of ions. At all three temperatures examined (900 K, 1000 K and 1100 K) ion pairs and neutral aggregates are the predominant aggregation state. The fraction of isolated ions increases with increasing temperature, but even at the highest temperature considered in the present study this fraction is less than 0.5%. For the neutral clusters the population of larger aggregates ($N_{aggr} \geq 6$) increases with increasing temperature while the opposite trend is observed for smaller aggregates (see Fig. 9). When counting clusters, more than 60% of the clusters are ion pairs but when considering the number of ions in a given aggregate, only about 30, 45, and 50% of ions form ion pairs at 900, 1000, and 1100 K, respectively. Additionally, nearly 22, 32, and 48% of ions are part of clusters larger than $N_{aggr} \geq 6$ at $T = 900$, 1000, and 1100 K, respectively. This shows that upon increasing temperature there is significant increase in the number density of ions, such that larger aggregates are observed with higher frequency. This can be seen in Fig. 10, which shows snapshots of the vapour phase of [$C_6$mim][$NTf_2$] at $T = 900$, 1000, and 1100 K. Additionally, we observe an increase in the number of charged clusters with increasing temperature. Less than 0.5% of the clusters are non-neutral at $T = 900$ K, while the number of charged clusters increases to almost 9% at $T = 1100$ K. At temperatures close to $T_c$ the saturated vapour density increases such

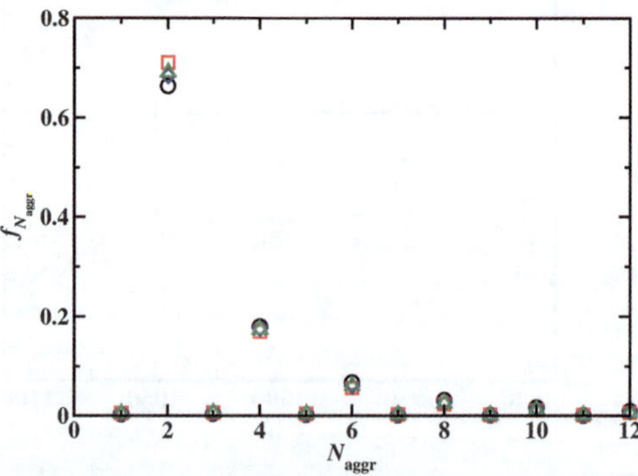


**Fig. 8** Fraction of clusters ($f_{N_{aggr}}$) with a given aggregation number ($N_{aggr}$) for ionic liquids at $T = 1000$ K. The symbols have the following meaning: black circles ($C_1$); red squares ($C_2$); blue diamonds ($C_4$); and green triangles ($C_6$).

 

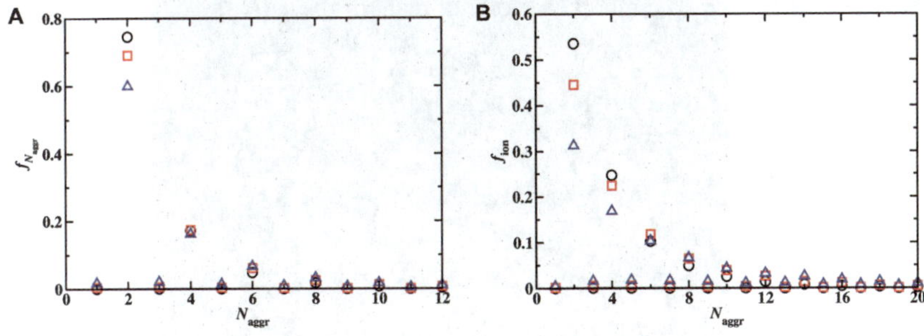


**Fig. 9** (A): Temperature dependence of fraction of clusters, $f_{N_{aggr}}$, with a given aggregation number; (B) fraction of ions, $f_{ion}$ in an aggregate with a given size for [$C_6$mim]. The black circles, red squares, and blue triangles represent data for $T = 900$, 1000, and 1100 K, respectively.

that the energetic penalty of forming these charged clusters is significantly smaller than that at low temperatures.

Recently, Akai *et al.*[74] were able to isolate ion pairs of [$C_2$mim][$NTf_2$] using a low-temperature rare-gas matrix-isolation technique. The authors found that the FTIR spectrum of the ion pairs was not only different from that of the liquid phase, but it also differed from the spectrum predicted for ion pairs using quantum chemical calculations.[22,74] To gain a better understanding of the local structure of vapour phase species, we computed three-dimensional spatial distribution plots (SDFs) of [$NTf_2$] ions around the imidazolium cation. Fig. 11 shows SDFs of [$NTf_2$] ions around [$C_1$mim] and [$C_6$mim] in the liquid and the vapour phase at $T = 1000$ K. The surfaces shown in these plots correspond to the 20% most likely positions of the centre of mass of [$NTf_2$] ions in the first solvation shell of a cation. The preferred position of [$NTf_2$] ions in the vapour phase is around the $C_2$ position (see Fig. 1) of the imidazolium cation, forming a horseshoe–like pattern above and below the ring. In the liquid phase, on the other hand, the [$NTf_2$] ions are localised around the $C_2$, $C_4$ and $C_5$ atoms (see Fig. 1) of the imidazolium ring. This suggests that differences in the local environment around the [$C_n$mim] ion in the liquid and the vapour phase can be responsible for the differences in the IR spectra. The results suggest that in the vapour phase, [$NTf_2$] ions are delocalised, exploring several different local environments around the cation. If a quantum chemical calculation is only performed on a single energy minimised structure, the effect this delocalisation has on the computed IR spectrum will not be captured. This suggests that an ensemble of ion pair configurations should be used to compute IR spectra in the vapour phase. Although delocalisation is perhaps exaggerated at the high temperatures considered here, this could be one possible explanation for the differences observed by Akai *et al.* Furthermore, these results reinforce the problem with simplified models of ILs; simple spherical interaction centres cannot capture the different association patterns observed between the liquid and vapour phases.

Increasing the alkyl chain length from $C_1$ to $C_6$ has almost no effect on the arrangement of [$NTf_2$] ions in the vapour phase but in the liquid phase there is a pronounced excluded volume for the alkyl chain of [$C_6$mim] which results in a depletion in the probability density of anions around the $C_5$ position. Similar depletion around the $C_5$ atom has been observed near ambient conditions in 1-*n*-butyl-3-methylimidazolium hexafluorophosphate.[75] This suggest that the microstructure of the liquid phase is different for cations with different alkyl chain lengths.

## Summary and conclusions

Atomistic Monte Carlo simulations employing a realistic force field have been used to compute the vapour–liquid coexistence properties of a homologous series of

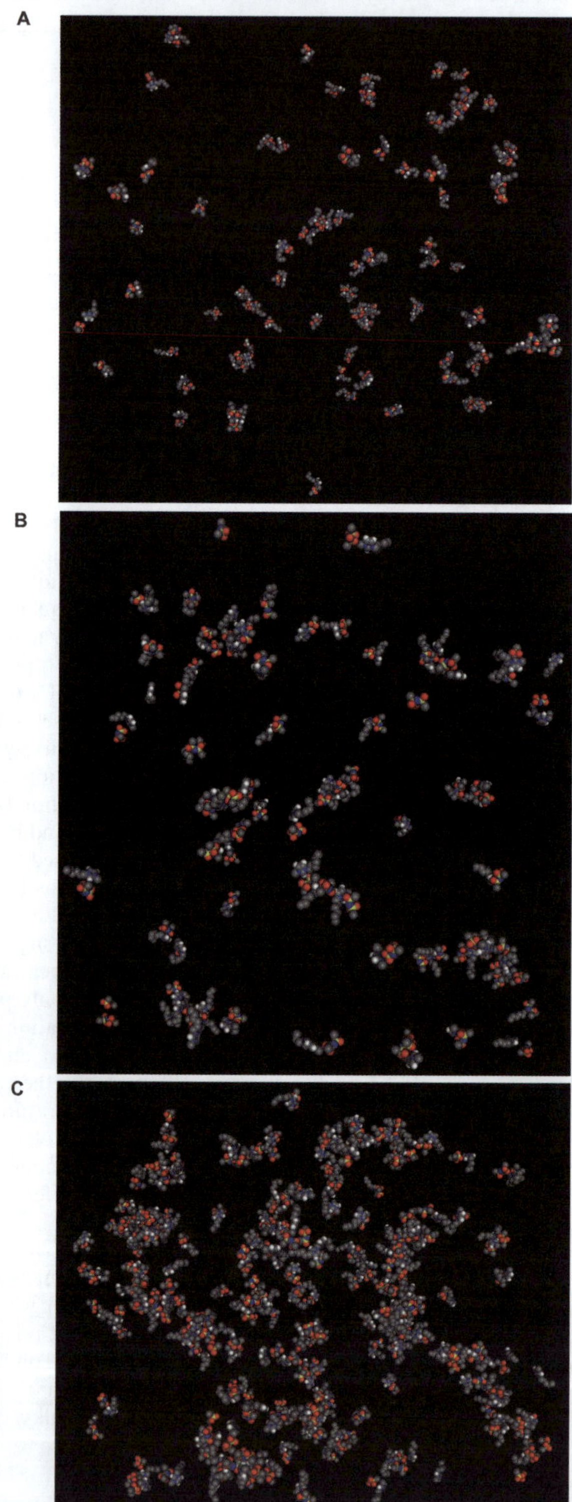


**Fig. 10** Snapshots of the vapour phase of [$C_6$mim][$NTf_2$]. (A) $T = 900$ K; (B) $T = 1000$ K; (C) $T = 1100$ K.

 

$[C_n mim][NTf_2]$ ionic liquids, with $n = 1, 2, 4, 6$. Critical temperatures, densities, pressures and compressibility factors are determined for each ionic liquid, in addition to normal boiling points, accentric factors, vapour pressure curves, and enthalpies of vapourisation. The simulations predict that the critical temperatures of these ionic liquids range from 1190 K to 1257 K, with the critical temperature decreasing as the alkyl chain length on the cation increases. These results generally agree with estimates obtained by extrapolation of experimental surface tension and density data *via* the Guggenheim and Eötvos models.[7] Normal boiling points and vapour pressures have a weak dependence on cation alkyl chain length, perhaps decreasing slightly as the cation size increases. These trends are explained in terms of a competition between Coulombic and van der Waals forces. The ratio of the normal boiling point to the critical temperature is around 0.8, which differs from the value of 0.6 typically found for non-polar liquids. An upper bound on the vapour pressures at 298 K are estimated to be on the order of $4 \times 10^{-3}$ Pa. Calculated accentric factors range from 0.11 for $[C_1mim][NTf_2]$ to 0.14 for $[C_6mim][NTf_2]$. Antoine equation coefficients are proposed to describe the high temperature vapour pressures of these liquids. The vapour phase consists mainly of neutral ion pairs, although the fraction of ions found in larger neutral clusters increases with increasing temperature. Consequently, Trouton's constant is about 70 J mol$^{-1}$ K, which is less than that of most conventional liquids but similar to that found with other associating fluids. Very few ions are found in charged clusters even at extremely elevated temperatures. The vapour phase becomes highly non-ideal as the critical temperature is approached.

The results of the present study show that atomistic Monte Carlo simulations can be used to compute vapour–liquid coexistence curves of ionic liquids with complex molecular structure. In addition to providing useful thermodynamic data and a microscopic understanding of the structure of the liquid and vapour phases, the

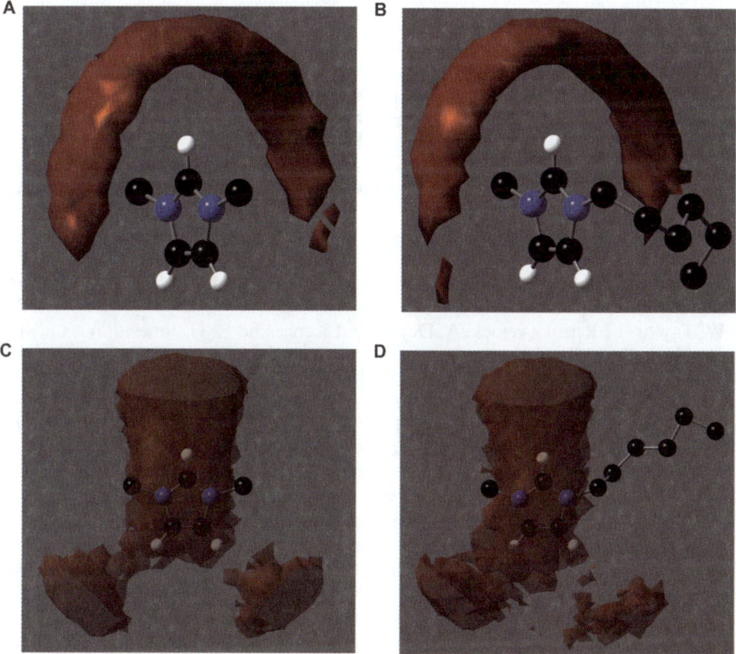


**Fig. 11** Spatial distribution functions for the liquid and vapour phase of $[C_1mim][NTf_2]$ and $[C_6mim][NTf_2]$ at $T = 1000$ K. The 20% most likely positions of $[NTf_2]$ ion around the cation are shown. (A): Vapour phase $[C_1mim][NTf_2]$; (B): vapour phase $[C_6mim][NTf_2]$; (C): liquid phase $[C_1mim][NTf_2]$; (D): liquid phase $[C_6mim][NTf_2]$.

estimated critical points, boiling points and related properties can be used to develop equation of state and corresponding states models for ionic liquids.


## Acknowledgements

The authors thank Prof. Zhiping Liu for providing force field parameters. Computational resources were provided by the Notre Dame Center for Research Computing. This material is based upon work supported by the Air Force Office of Scientific Research under AFOSR award number FA9550-10-1-0244.


## References


1 B. Gurkan, B. F. Goodrich, E. M. Mindrup, L. E. Ficke, M. Massel, S. Seo, T. P. Senftle, H. Wu, M. F. Glaser, J. K. Shah, E. J. Maginn, J. F. Brennecke and W. F. Schneider, *J. Phys. Chem. Lett.*, 2010, **1**, 3494–3499.
2 T. Torimoto, T. Tsuda, K. Okazaki and S. Kuwabata, *Adv. Mater.*, 2010, **22**, 1196–1221.
3 T. P. Lodge, *Science*, 2008, **321**, 50–51.
4 M. Scott, C. Brazel, M. Benton, J. Mays, J. Holbrey and R. Rogers, *Chem. Commun.*, 2002, 1370–1371.
5 R. D. Rogers, *Nature*, 2007, **447**, 917–918.
6 E. F. Borra, O. Seddiki, R. Angel, D. Eisenstein, P. Hickson, K. R. Seddon and S. P. Worder, *Nature*, 2007, **447**, 979–981.
7 L. P. N. Rebelo, J. N. C. Lopes, J. M. S. S. Esperanca and E. Filipe, *J. Phys. Chem. B Lett.*, 2005, **109**, 6040–6043.
8 Y. U. Paulechka, D. H. Zaitsau, G. J. Kabo and A. A. Strechan, *Thermochim. Acta*, 2005, **439**, 158–160.
9 M. J. Earle, J. M. S. S. Esperanca, M. A. Gilea, J. N. C. Lopes, L. P. N. Rebelo, J. W. Magee, K. R. Seddon and J. A. Widegren, *Nature*, 2006, **439**, 831–834.
10 D. H. Zaitsau, G. J. Kabo, A. A. Strechan, Y. U. Paulechka, A. Tschersich, S. P. Verevkin and A. Heintz, *J. Phys. Chem. A*, 2006, **110**, 7303–7306.
11 K. R. J. Lovelock, A. Deyko, J.-A. Corfield, P. N. Gooden, P. Licence and R. G. Jones, *ChemPhysChem*, 2009, **10**, 337–340.
12 A. Deyko, K. R. J. Lovelock, J.-A. Corfield, A. W. Taylor, P. N. Gooden, I. J. Villar-Garcia, P. Licence, R. G. Jones, V. G. Krasovskiy, E. A. Chernikova and L. M. Kustov, *Phys. Chem. Chem. Phys.*, 2009, **11**, 8544–8555.
13 K. R. J. Lovelock, A. Deyko, P. Licence and R. G. Jones, *Phys. Chem. Chem. Phys.*, 2010, **12**, 8893–8901.
14 K. Fumino, A. Wulf, S. P. Verevkin, A. Heintz and R. Ludwig, *ChemPhysChem*, 2010, **11**, 1623–1626.
15 S. Verevkin, V. Emel'yanenko, D. Zaitsau, A. Heintz, C. Muzny and M. Frenkel, *Phys. Chem. Chem. Phys.*, 2010, **12**, 14994–15000.
16 J. A. Widegren, Y.-M. Wang, W. A. Henderson and J. W. Magee, *J. Phys. Chem. B*, 2007, **111**, 8959–8964.
17 A. W. Taylor, K. R. J. Lovelock, A. Deyko, P. Licence and R. G. Jones, *Phys. Chem. Chem. Phys.*, 2010, **12**, 1772–1783.
18 J. P. Armstrong, C. Hurst, R. G. Jones, P. Licence, K. R. J. Lovelock, C. J. Satterley and I. J. Villar-Garcia, *Phys. Chem. Chem. Phys.*, 2007, **9**, 982–990.
19 D. Strasser, F. Goulay, M. S. Kelkar, E. J. Maginn and S. R. Leone, *J. Phys. Chem. A*, 2007, **111**, 3191–3195.
20 J. P. Leal, J. M. S. S. Esperana, M. E. Minas da Piedade, J. N. Canongia Lopes, L. P. N. Rebelo and K. R. Seddon, *J. Phys. Chem. A*, 2007, **111**, 6176–6182.
21 J. H. Gross, *J. Am. Soc. Mass Spectrom.*, 2008, **19**, 1347–1352.
22 N. Akai, D. Parazs, A. Kawai and K. Shibuya, *J. Phys. Chem. B*, 2009, **113**, 4756–4762.
23 C. Hogan Jr. and J. Fernandez de la Mora, *J. Am. Soc. Mass Spectrom.*, 2010, **21**, 1382–1386.
24 R. Ludwig and U. Kragl, *Angew. Chem., Int. Ed.*, 2007, **46**, 6582–6584.
25 J. M. S. S. Esperanca, J. N. C. Lopes, M. Tariq, L. M. N. B. F. Santos, J. W. Magee and L. P. N. Rebelo, *J. Chem. Eng. Data*, 2010, **55**, 3–12.
26 M. Shiflett and A. Yokozeki, *Ind. Eng. Chem. Res.*, 2005, **44**, 4453–4464.
27 K. S. Pitzer, *J. Chem. Phys.*, 1939, **7**, 583–590.
28 H. W. Xiang, *The Corresponding-States Principle and its Practice* (First Edition), Elsevier, Amsterdam, 2005.


 

29 S. D. Chambreau, A. Ghanshyam, L. V. To, C. Koh, D. Strasser, O. Kostko and S. R. Leone, *J. Phys. Chem. B*, 2010, **114**, 1361–1367.
30 V. C. Weiss, *J. Phys. Chem. B*, 2010, **114**, 9183–9194.
31 E. J. Maginn, *J. Phys.: Condens. Matter*, 2009, **21**, 373101.
32 J. Shah, J. Brennecke and E. Maginn, *Green Chem.*, 2002, **4**, 112–118.
33 T. Morrow and E. Maginn, *J. Phys. Chem. B*, 2002, **106**, 12807–12813.
34 Z. Liu, S. Huang and W. Wang, *J. Phys. Chem. B*, 2004, **108**, 12978–12989.
35 X. Liu, S. Zhang, G. Zhou, G. Wu, X. Yuan and X. Yao, *J. Phys. Chem. B*, 2006, **110**, 12062–12071.
36 C. Cadena and E. J. Maginn, *J. Phys. Chem. B*, 2006, **110**, 18026–18039.
37 L. M. N. B. F. Santos, J. N. C. Lopes, J. A. P. Coutinho, J. M. S. S. Esperana, L. R. Gomes, I. M. Marrucho and L. P. N. Rebelo, *J. Am. Chem. Soc.*, 2007, **129**, 284–285.
38 T. Koeddermann, D. Paschek and R. Ludwig, *ChemPhysChem*, 2008, **9**, 549–555.
39 M. Klähn, A. Seduraman and P. Wu, *J. Phys. Chem. B*, 2008, **112**, 13849–13861.
40 M. S. Kelkar and E. J. Maginn, *J. Phys. Chem. B*, 2007, **111**, 9424–9427.
41 P. Ballone, C. Pinilla, J. Kohanoff and M. G. Del Ppolo, *J. Phys. Chem. B*, 2007, **111**, 4938–4950.
42 V. C. Weiss, B. Heggen and F. Muller-Plathe, *J. Phys. Chem. C*, 2010, **114**, 3599–3608.
43 F. Leroy and V. C. Weiss, *J. Chem. Phys.*, 2011, **134**, 094703.
44 A. Z. Panagiotopoulos, *Mol. Phys.*, 1987, **61**, 813–826.
45 A. Panagiotopoulos, *J. Phys.: Condens. Matter*, 2000, **12**, R25–R52.
46 J. I. Siepmann, S. Karaboni and B. Smit, *Nature*, 1993, **365**, 330–332.
47 D. Frenkel and B. Smit, *Understanding Molecular Simulation From Algorithms to Applications*, Academic Press, New York, 2002.
48 G. Orkoulas and A. Z. Panagiotopoulos, *J. Chem. Phys.*, 1994, **101**, 1452–1459.
49 G. Orkoulas, M. E. Fisher and A. Z. Panagiotopoulos, *Phys. Rev. E: Stat. Phys., Plasmas, Fluids, Relat. Interdiscip. Top.*, 2001, **63**, 051507.
50 M. Martin-Betancourt, J. M. Romero-Enrique and L. F. Rull, *J. Phys. Chem. B*, 2009, **113**, 9046–9049.
51 N. Rai and E. J. Maginn, *J. Phys. Chem. Lett.*, 2011, **2**(12), 1439–1443.
52 X. Zhong, Z. Liu and D. Cao, *J. Phys. Chem. B*, 2011, **115**(33), 10027–10040.
53 N. Rai, D. Bhatt, J. I. Siepmann and L. E. Fried, *J. Chem. Phys.*, 2008, **129**, 194510.
54 P. Ewald, *Ann. Phys.*, 1921, **64**, 253–287.
55 M. P. Allen and D. J. Tildesley, *Computer Simulation of Liquids*, Clarendon Press: Oxford, 1987.
56 A. Z. Panagiotopoulos, N. Quirke, M. Stapleton and D. J. Tildesley, *Mol. Phys.*, 1988, **63**, 527–545.
57 P. Bonhte, A.-P. Dias, N. Papageorgiou, K. Kalyanasundaram and M. Grtzel, *Inorg. Chem.*, 1996, **35**, 1168–1178.
58 H. Tokuda, K. Hayamizu, K. Ishii, M. A. B. H. Susan and M. Watanabe, *J. Phys. Chem. B*, 2004, **108**, 16593–16600.
59 J. I. Siepmann and D. Frenkel, *Mol. Phys.*, 1992, **75**, 59–70.
60 J. J. de Pablo, M. Laso and U. W. Suter, *J. Chem. Phys.*, 1992, **96**, 2395–2403.
61 T. J. H. Vlugt, M. G. Martin, B. Smit, J. I. Siepmann and R. Krishna, *Mol. Phys.*, 1998, **94**, 727–733.
62 J. K. Shah and E. J. Maginn, submitted to *J. Chem. Phys.*
63 M. D. Macedonia and E. J. Maginn, *Mol. Phys.*, 1999, **96**, 1375–1390.
64 S. Jayaraman and E. J. Maginn, *J. Chem. Phys.*, 2007, **127**, 214504.
65 http://www.cstl.nist.gov/srs/LJ_PURE/sattmmc.htm.
66 J. P. Valleau and L. K. Cohen, *J. Chem. Phys.*, 1980, **72**, 5935–5941.
67 J. O. Valderamma and R. E. Rojas, *Ind. Eng. Chem. Res.*, 2009, **48**, 6890–6900.
68 E. W. Lemmon, M. O. McLinden & D. G. Friend, Thermophysical Properties of Fluid Systems, NIST Chemistry WebBook, NIST Standard Reference Database Number 69, ed. Linstrom, P. J. and Mallard, W. G., National Institute of Standards and Technology, Gaithersburg MD, 20899 (http://webbook.nist.gov), (retrieved April 25, 2011).
69 G. Orkoulas and A. Z. Panagiotopoulos, *J. Chem. Phys.*, 1999, **110**, 1581–1590.
70 J. M. Prausnitz, R. N. Lichtenthaler and E. G. de Azvedo, *Molecular Thermodynamics of Fluid Phase Equilibria*, Prentice Hall, Englewood Cliffs, 2nd ed., 1986.
71 J. O. Valderrama, W. W. Sanga and J. A. Lazzus, *Ind. Eng. Chem. Res.*, 2008, **47**, 1318–1330.
72 J. O. Valderrama and P. A. Robles, *Ind. Eng. Chem. Res.*, 2007, **46**, 1338–1344.
73 E. C. Bingham, *J. Am. Chem. Soc.*, 1906, **28**, 723–731.
74 N. Akai, A. Kawai and K. Shibuya, *Chem. Lett.*, 2008, **37**, 256–257.
75 W. Zhao, F. Leroy, B. Heggen, S. Zahn, B. Kirchner, S. Balasubramanian and F. Muller-Plathe, *J. Am. Chem. Soc.*, 2009, **131**, 15825–15833.

# Phase behaviour, transport properties, and interactions in Li-salt doped ionic liquids

**Jagath Pitawala,[a] Jae-Kwang Kim,[a] Per Jacobsson,[a] Victor Koch,[b] Fausto Croce[c] and Aleksandar Matic*[a]**

***Received 1st April 2011, Accepted 19th May 2011***
**DOI: 10.1039/c1fd00056j**

We report on the influence of lithium bis(trifluoromethanesulfonyl)imide (LiTFSI) doping on the glass transition temperature ($T_g$), the ionic conductivity, and Li-ion coordination of two dicationic ionic liquids (DILs) based on the TFSI anion. The results are compared to the behaviour of traditional mono-cationic ionic liquids. The cations of the DILs contain two imidazolium rings, connected by a decane hydrocarbon chain. Homogeneous mixtures of these ILs and LiTFSI can be obtained in a large concentration range. With increasing Li–salt concentration the ionic conductivity decreases whereas the glass transition temperature increases in both systems. However, the influence of the salt doping on the ionic conductivity and the glass transition temperature is low compared to typical mono–cationic ionic liquids, based on for example the pyrrolidinium cation and the TFSI anion. This behaviour is mirrored in the average coordination number of TFSI anions around Li-ions, determined by Raman spectroscopy. The coordination number is systematically lower in the DILs, suggesting a connection between the difference in the Li-ion environment and the behaviour of the glass transition and the ionic conductivity. A $T_g$-scaled Arrhenius plot of the ionic conductivity shows that the ionic conductivity for all LiTFSI concentrations has the same temperature dependence, *i.e.* the fragility of the liquid is the same. This implies that the conduction process is dominated by the viscous properties of the liquids over the entire concentration range. This provides further support for linking the local environment of the Li-ions to the glass transition and conduction process in the ionic liquid/salt mixtures.

## Introduction

The ability of ionic liquids (ILs) to dissolve lithium salts, for instance lithium bis(trifluoromethanesulfonyl)imide (LiTFSI), make them of high interest as new electrolyte materials for rechargeable lithium metal and lithium ion batteries.[1] The wide liquid range, low volatility and inflammability of ILs are the key advantages as compared to traditional organic solvents.[1] One important limitation for the use of IL/Li-salt mixtures as electrolytes is a large increase in viscosity with high Li-salt concentrations resulting in a low ionic conductivity. In addition a high viscosity of ILs with high Li-salt concentrations makes preparation of homogeneous materials difficult.

In order to further develop and optimize this new class of Li-ion conductors, an understanding of the ion transport mechanism and interactions of IL/Li-salt

[a]*Department of Applied Physics, Chalmers University of Technology, Göteborg, Sweden. E-mail: matic@chalmers.se; Fax: +46 31 772 2090; Tel: +4631772 5176*
[b]*Covalent Associates Inc., Corvallis, OR, 97330, USA*
[c]*Dipartimento di Scienze del Farmaco, Università "G.D' Annunzio", Chieti, Italy*

mixtures is important. We have previously shown how the addition of Li-salt influences the ionic conductivity, the phase behaviour, and Li-ion coordination of ionic liquids based on pyrrolidinium cations and the TFSI anion at low to intermediate Li-salt concentrations.[2,3] We have also shown the close connection between the conductivity and the glass transition temperature.[2,4] It is now of interest to extend the concentration range of Li-salt doping and to investigate the role of the cation structure on the properties of IL/Li-salt mixtures.

The ion transport properties and the glass transition temperatures are thoroughly studied for neat ILs.[4–8] There are also some studies that discuss the influence of the Li-salt doping on ILs related to their ionic conductivity and the phase behaviour,[2,9–15] however most of the ionic conductivity studies have been performed with rather low salt doping, $x \leq 0.3$. It is found that the decrease in ionic conductivity is accompanied by an increase in the glass transition temperature.[2,11] Molecular dynamics (MD) simulation suggest that the introduction of Li-salts into ILs causes aggregation of anions around the small Li-ion forming a more rigid structure and as a result viscosity increases and conductivity decreases.[16]

The complex ionic association of Li-ions and anions in ionic liquids has been experimentally investigated by Raman spectroscopy, in particularly for ILs with the TFSI anion.[17–19] The strong Raman band around 740 $cm^{-1}$ of the TFSI-anion, related to the expansion-contraction mode of the whole anion,[20,21] can be used to distinguish between free TFSI-anions and anions coordinated to Li-ions. In Li-salt/IL mixtures of imidazolium/TFSI ionic liquids a coordination number around 2, *i.e.* 2 TFSI anions coordinated to each Li-ion, is found for intermediate concentrations of LiTFSI ($x \approx 0.1$–$0.2$), only slightly decreasing for higher salt concentrations.[19] A coordination number of 2 could suggest the presence of $[Li(TFSI)_2]^-$ complexes,[17–19] but could also result from bigger clusters $[Li_{y/2}(TFSI)_y]^{(y/2)-}$, where the Li-ions crosslink between different TFSI anions. Other possible configurations, such as $[Li(TFSI)_3]^{2-}$ or $[Li(TFSI)_4]^{3-}$, have been suggested based on both experimental and simulation work.[22–24] We have recently shown that the coordination number of TFSI around Li in pyrrolidinium/TFSI ionic liquids is strongly dependent on salt concentration at low concentrations ($x < 0.08$, $x$LiTFSI + $(1 - x)$IL) and it is of interest to understand how this is reflected in the properties of the material.[3]

In this work, we present new data on the ionic conductivity, the glass transition temperature, and Li-ion coordination in two dicationic ILs, based on the TFSI anion and the imidazolium cation family,[4] over a wide range of LiTFSI concentrations. For these ILs homogeneous solutions can be obtained up to very high salt concentrations, even $x > 0.4$, $x$LiTFSI/$(1 - x)$IL. In addition, we compare these results to traditional mono-cationic ILs based on the TFSI anion and pyrrolidinium

**Table 1** Cation structures, acronyms, molecular weights ($M_w$), and glass transition temperatures ($T_g$) of the investigated ionic liquids

| Cation structure | Acronym | $M_w$ (g $mol^{-1}$) | $T_g$ (K) |
|---|---|---|---|
|  | $(MI)_2C_{10}(TFSI)_2$ | 864.77 | 212 |
|  | $(M2I)_2C_{10}(TFSI)_2$ | 892.82 | 225 |
| 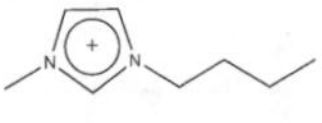 | BMImTFSI | 404.33 | 186 |
| 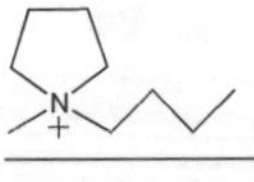 | $PyR_{14}TFSI$ | 422.41 | 187 |

 

$(PyR_{14}TFSI)$ and imidazolium (BMImTFSI) cations respectively. The cationic structures of the investigated dicationic and mono-cationic ILs are shown in Table 1. We show that in these dicationic ILs the ionic conductivity and the glass transition temperature is less affected by addition of Li-salt compared to the mono-cationic ILs and that this is reflected in a difference in the coordination of Li-ions to TFSI.

## Experimental details

### Materials

The two dicationic ILs, $(MI)_2C_{10}(TFSI)_2$: 1,10-bis(3-methylimidazolium)decane di-bis(trifluoromethanesulfonyl)imide and $(M2I)_2C_{10}(TFSI)_2$: 1,10-bis(2,3-dimethylimidazolium)decane di-bis(trifluoromethanesulfonyl)imide, were prepared by Covalent Associates, Inc., USA, with purities >99%. The synthesis procedure and spectroscopic properties of these DILs is described in detail in ref. 26.

The two mono-cationic ILs, BMImTFSI (1-butyl-3-methylimidazolium bis(trifluoromethanesulfonyl)imide) and $PyR_{14}TFSI$ (*N*-butyl-*N*-methylpyrrolidinium bis (trifluoromethanesulfonyl)imide) were purchased from Sigma Aldrich with purities >99%. LiTFSI (purity 99.95%) was also purchased from Sigma Aldrich.

The $x$LiTFSI/$(1 - x)$ IL mixtures were prepared by addition of the proper molar fraction ($x$) of vacuum dried LiTFSI to the ILs. The mixtures were stirred for few hours at 323–333 K until homogeneous and transparent solutions were obtained. Samples with six different concentrations, in the range $0.01 < x < 0.4$, were prepared. All sample preparation was performed in argon atmosphere (oxygen and water less than 1 ppm).

### Thermal analysis

Differential scanning calorimetry (DSC) measurements were performed on a TA Instrument Q1000. For each measurement, the sample ($\approx$10–15 mg) was placed in a hermetically sealed aluminium pan (in an Ar-filled glove box), and an empty pan was used as reference. The samples were first cooled from 298 to 153 K at a rate of 20 K $min^{-1}$, and the DSC traces were recorded during the subsequent heating scan up to 393 K at a rate of 10 K $min^{-1}$. During the measurements, a flow of helium gas was used in order to keep an inert and dry environment in the sample block. Universal Analysis 2000 software (TA instruments) was used to identify the thermal transitions of the $x$LiTFSI/$(1 - x)$IL mixtures. The glass transition temperatures were determined as the midpoint of the heat capacity change on the heating scan.

### Conductivity measurements

The temperature dependence of the ionic conductivity was measured by dielectric spectroscopy using a Novocontrol broadband dielectric spectrometer in the frequency range $10^{-1}$–$10^{7}$ Hz. The sample was placed between two stainless steel electrodes with a Teflon spacer (diameter-12.3 mm and thickness-0.97 mm) and loaded into a cryo-furnace. The cell was assembled in argon atmosphere. Data were collected in steps of 10 K in the temperature range 373 K to 253 K on cooling. A steady flow of nitrogen gas was supplied to the sample holder in order to maintain a dry atmosphere. The DC conductivity was extracted as the low frequency plateau in the frequency dependent conductivity plot.

### Raman spectroscopy

Raman spectra were recorded on a Bruker IFS66 Fourier Transform spectrometer, equipped with a FRA106 Raman module. The 1064 nm line of a Nd : YAG laser was used as excitation source. The laser power was set to 200 mW, the resolution was 2 cm$^{-1}$, and the spectra were obtained as the average of over 1000 scans. All the Raman spectra were recorded at room temperature.

## Results and discussion

Fig. 1 shows the ionic conductivity as a function of the inverse temperature for the different LiTFSI concentrations in the two dicationic ILs. The conductivity decreases with increasing LiTFSI concentration in agreement with previous studies.[2,10–12] However, the decrease is quite moderate compared to other systems, see further below. For the very high LiTFSI concentration $x = 0.4$, the room temperature ionic conductivity is of the order of $10^{-4}$ S cm$^{-1}$ in both dicationic ILs. The ionic conductivities of the $x$LiTFSI/(1 – $x$)$(MI)_2C_{10}(TFSI)_2$ system are slightly higher compared to the $x$LiTFSI/(1 – $x$)$(M2I)_2C_{10}(TFSI)_2$ system over the whole investigated temperature range.

The temperature dependence of the conductivity shows the typical non-Arrhenius behaviour of liquid electrolytes and is well described by the VFT (Vogel–Fulcher–Tamman) function:

$$\sigma = \sigma_0 e^{\frac{DT_0}{T-T_0}} \qquad (1)$$

where $\sigma_0$ is the conductivity at infinitely high temperatures, $D$ relates to the fragility of the material, and $T_0$ is related to an ideal glass transition temperature.[27,28] The VFT function is commonly used to describe the viscosity and relaxation time behaviour of glass forming liquids. The applicability of the function to the conductivity data illustrates the close relation between the conductivity and the viscosity of ILs and is discussed further below.

Fig. 2 shows the results from the DSC experiments during heating scans. For all samples a glass transition is observed and the glass transition temperatures, $T_g$, are summarized in Table 2. The $T_g$ values for the pure dicationic ILs are higher compared to mono-cationic ILs, compare for instance (BMImTFSI) with $T_g$ = 186 K to $(MI)_2C_{10}(TFSI)_2$ having $T_g$ = 212 K and see Table 1. With addition of LiTFSI, $T_g$ increases in both dicationic ILs, following the trend observed in Li-salt doped mono-cationic ILs.[2,11,12] The pure IL $(MI)_2C_{10}(TFSI)_2$ shows a slight tendency for crystallization following the glass transition on the heating scan (not shown). However, we observe that the crystallization is suppressed with increasing LiTFSI concentration and is not observed for $x > 0.01$. This is in agreement with what has previously been observed in other Li-salt doped ILs,[2,11,12] and shows that the glass forming ability increases with Li-salt doping. Arguably, the large

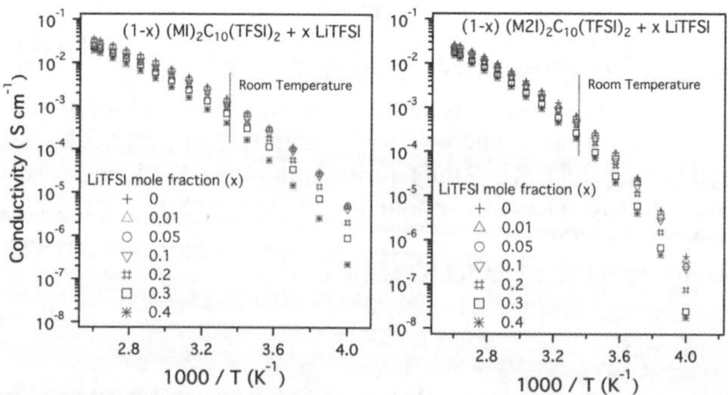


**Fig. 1** Ionic conductivity as a function of inverse temperature for $x$LiTFSI/(1 – $x$) $(MI)_2C_{10}(TFSI)_2$ (left) and $x$LiTFSI/(1 – $x$)$(M2I)_2C_{10}(TFSI)_2$ (right).

 

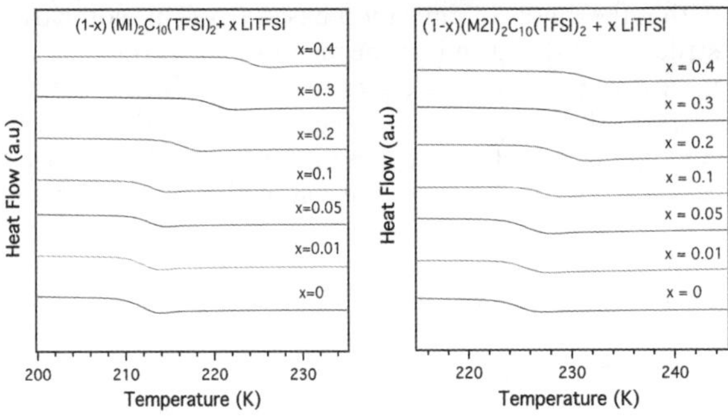


**Fig. 2** DSC results from the heating scan for the two dicationic ionic liquids for different salt concentration in the region around the glass transition.

**Table 2** Glass transition temperatures for the different LiTFSI concentrations

| $x$ | $(MI)_2C_{10}(TFSI)_2$ $T_g$ (K) | $(M2I)_2C_{10}(TFSI)_2$ $T_g$ (K) |
|---|---|---|
| 0.01 | 213 | 226 |
| 0.05 | 213 | 226 |
| 0.1 | 214 | 227 |
| 0.2 | 217 | 229 |
| 0.3 | 221 | 232 |
| 0.4 | 225 | 235 |

size difference between the Li-ion and the large organic cations in ILs prevents efficient packing, thus crystallization, in these LiTFSI/ILmixtures.

To compare our results with the behaviour of typical mono-cationic ILs, in Fig. 3, we plot the relative change in the room temperature ionic conductivity, $\sigma_{RT}(x)/\sigma_{RT}(0)$, and the glass transition temperature, $T_g(x)/T_g(0)$, as a function of LiTFSI concentration. In this comparison, we have included data for the mono-cationic ILs commonly used in studies of Li-electrolytes, BMImTFSI and $PyR_{14}TFSI$. All systems follow the same trend. However, the reduction of the room temperature ionic conductivity and the increase in the glass transition temperature at high salt concentrations is considerably lower for the imidazolium cation based ILs compared to the pyrrolidinium cation based IL. In addition, the effect is even less for the

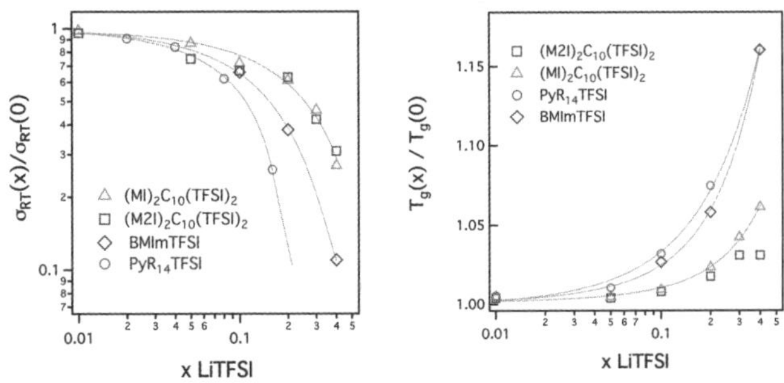


**Fig. 3** Relative change as a function of LiTFSI concentration in the room temperature ionic conductivity, $\sigma_{RT}(x)$, (left) and glass transition temperature, $T_g(x)$, (right) of the two dicationic ILs compared to the two mono-cationic ILs.

dicationic ILs within the imidazolium cation based ILs. This comparison points out the role of the structure of the cation on the macroscopic properties of Li-salt doped ILs. Both the cation type and the size are expected to influence the type of Li-ion : anion complexes formed as well as the average ionicity of the system.

To further investigate the relation between the ionic conductivity and the glass transition temperature, in Fig. 4, we plot the conductivity data with a scaled temperature axis, $T_g/T$.

We find that for the very large concentration range of Li-salt doping, up to $x = 0.4$, all data fall on a master curve in this scaled representation. This shows that the ionic conductivity for all LiTFSI concentrations has the same temperature dependence, *i.e.* the fragility of the system does not change with the addition of Li-salt even at high concentrations. This implies that the conduction process is dominated by the viscous properties of the liquids for all concentrations and for the whole investigated temperature range, all the way down to the glass transition temperature.

To compare the temperature dependence of the ionic conductivity in Fig. 5 we show the $T_g$-scaled Arrhenius plot of the conductivity for the two dicationic ILs and the two mono-cationic ILs at the same Li-salt concentration, $x = 0.2$. This plot reveals that there is a clear cation dependence of the shape of the temperature dependence of the ionic conductivity, *i.e.* that the fragility of the different ionic liquids is different with $PyR_{14}TFSI$ being the most fragile. It also reveals that the high temperature limit seems to be somewhat different. For small changes of the

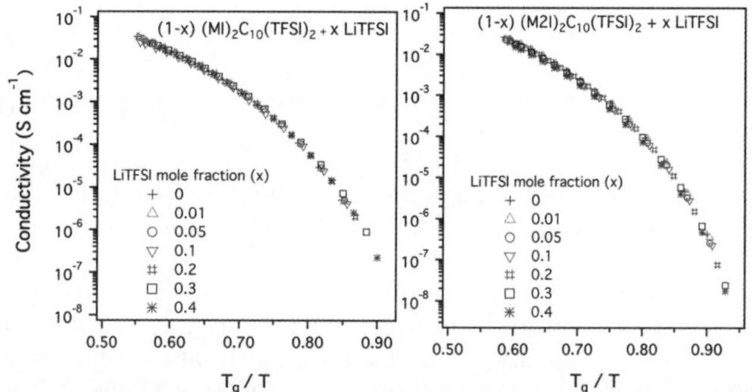


**Fig. 4** $T_g$-scaled Arrhenius plots of the ionic conductivities for different LiTFSI concentrations, $(MI)_2C_{10}(TFSI)_2$ (Left) and $(M2I)_2C_{10}(TFSI)_2$ (Right).

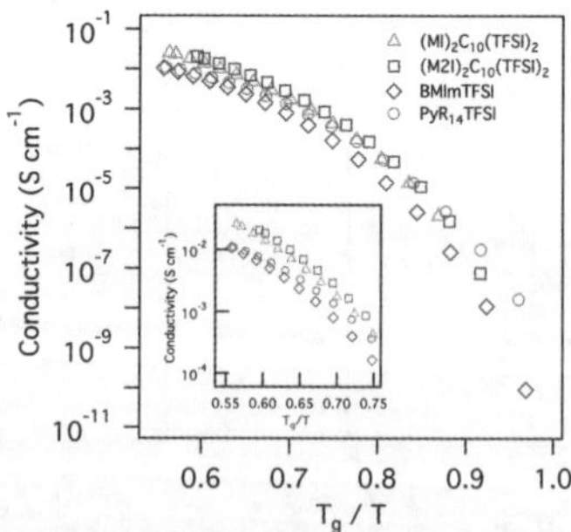


**Fig. 5** Comparison of $T_g$-scaled Arrhenius plots of ionic conductivities of 0.2 LiTFSI doped dicationic and mono cationic ILs (0.8IL + 0.2LiTFSI). Inset shows an enlargement of the conductivities in the high temperature regime.

 

cation structure, such as between the two dicationic ILs, the change is relatively small. This is in agreement with previous results found for changes in alkyl side-chain length for pyrrolidinium/TFSI ILs.[2]

The increase in the glass transition temperature with Li-salt concentration has been proposed to be related to the formation of Li-ion : anion complexes with reduced mobility.[24,29] It can also be understood in terms of the decrease in the average ionic volume and increase in ionicity of the system with increased Li-salt concentration.[8]

The different types of cluster will influence the physical properties of the Li-salt/IL mixtures in different ways. For instance comparing the $[Li(TFSI)_2]^-$ and the $[Li_{y/2}(TFSI)_y]^{(y/2)-}$ configurations, both having a coordination number 2, the latter can be expected to have a larger influence through the crosslinking of different TFSI anions. Such a crosslinking could be transferred into, for example, a higher glass transition temperature. Also configurations with high coordination numbers, such as $[Li(TFSI)_4]^{3-}$, crosslink different anions, but here one can expect that the life time of such a cluster is shorter since the interaction between Li and an individual TFSI is weaker then for lower coordination number. Thus, the local environment around the Li-ions relaxes quite fast and the influence on macroscopic properties is smaller.

In Fig. 6 we show the Raman spectra in the region of the expansion-contraction mode of the TFSI-anion, around 740 $cm^{-1}$, for the neat dicationic ionic liquid $(MI)_2C10(TFSI)_2$and doped with $x = 0.3$ LiTFSI (the spectra of $(M2I)_2C10$ $(TFSI)_2$ are very similar and are thus not shown). In the neat ionic liquid the 740 $cm^{-1}$ band should be described by two spectral components, related to the two conformers $C_1$ (cisoid, I) and $C_2$ (transoid, II) coexisting in the liquid.[21,25] With increasing salt concentration we observe the growth of a spectral component at the high frequency flank. This spectral component is a signature of the TFSI-anions coordinated to Li, as shown previously from *ab initio* calculations and verified by Raman experiments.[17–19] In Fig. 6 we also compare the 740$cm^{-1}$ Raman band from the two dicationic ILs to $PyR_{14}TFSI$ at the same salt concentration ($x =$ 0.3). Just by inspection a striking difference is observed with the coordinated component being much stronger in the case of $PyR_{14}TFSI$, showing that the fraction of TFSI ions coordinated to Li-ions is higher compared to the dicationic ILs.

To extract quantitative information from the Raman spectrum the 740 $cm^{-1}$ band has been fitted with three Lorentzian functions, corresponding to the $C_1$ and $C_2$

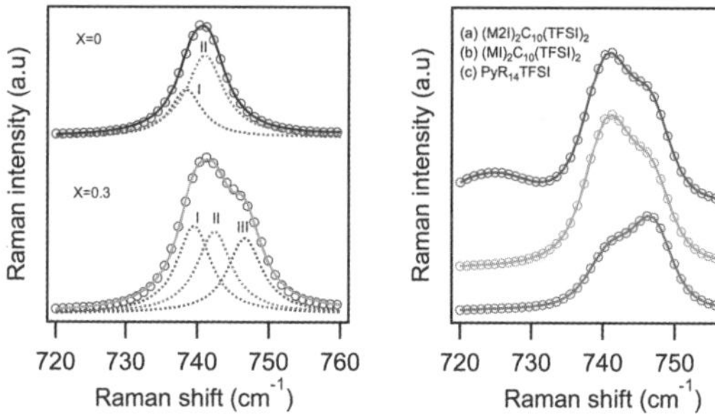


**Fig. 6** (Left) Raman spectrum in the region around the 740 $cm^{-1}$ mode of the TFSI-anion for neat $(MI)_2C_{10}(TFSI)_2$and doped with 0.3 LiTFSI. The spectra have been vertically offset for clarity. The experimental spectrum (open circles) is shown together with the best fit (solid line) and the individual Lorentzian components (dashed lines). Roman numbers indicate the components due to weakly coordinated TFSI anions (bands I and II, corresponding to the $C_1$ and $C_2$ conformations respectively), and $TFSI^-$ anions coordinated to Li-ions (band III). (Right) Comparison of the 740 $cm^{-1}$ mode of the TFSI-anion for the two dicationic ionic liquids and for $PyR_{14}TFSI$ at $x = 0.3$. The spectra have been vertically offset for clarity.

conformations of the non-coordinated TFSI-anions and to the Li-ion coordinated TFSI-anions respectively. From these fits one can extract the fraction of non-coordinated ($C_{nc}$) and coordinated ($C_c$) TFSI-anions from the areas of the three components, and also calculate the average coordination number, $N$, of TFSI around Li-ion from

$$N = \frac{C_c}{x} = \frac{A_{III}/A_{tot}}{x}. \tag{2}$$

In eqn (2) $A_{III}$ is the area of the high frequency component, corresponding to TFSI coordinated to Li-ions, and $x$ is the Li-ion concentration. The average coordination numbers as a function of Li-salt concentration for the two dicationic ILs are shown in Fig. 7 together with data previously obtained for $PyR_{14}TFSI$.[3] A strong concentration dependence is observed for all ionic liquids, in particular for low concentrations, where the average coordination number reaches quite high values, indicating that configurations of the type $[Li(TFSI)_N]^{(N-1)-}$ with $N \approx 4$–6 dominate. At intermediate concentrations $N$ is around 2, in agreement with previous studies on imidazolium/TFSI liquids,[17] pointing towards either $[Li(TFSI)_2]^-$ triplets or larger clusters $[Li_{y/2}(TFSI)_y]^{(y/2)-}$.

Qualitatively all the dicationic ILs and $PyR_{14}TFSI$ show the same trend for the concentration dependence of the coordination number. However, the coordination number for the dicationic ionic liquids is systematically lower than what is found for $PyR_{14}TFSI$, reflecting the smaller fraction of Li-coordinated TFSI-anions. One can speculate if this can be connected to the trends also observed in the concentration dependence of the conductivity and the glass transition temperature, where the dicationic ILs show a small decrease and increase respectively compared to the mono-cationic ILs, see Fig. 3. In the intermediate concentration regime this could be related to that in the case of dicationic ILs the formation of $[Li(TFSI)_2]^-$ triplets, with less crosslinking of the structure, are favoured, possibly due to the large size of the cation that also needs to be accommodated. Linking the local Li-ion environment to the glass transition and the conduction process is further supported by the master curve found in the $T_g$ scaling of the temperature dependence of the conductivity, Fig. 4. The scaling holds over the whole concentration range where we see large changes in the average Li-ion coordination. Thus, these changes are mirrored in the glass transition and the viscous properties of the liquids dominating the conduction process. The lack of $T_g$ scaling between dicationic and mono-cationic ionic liquids at the same salt concentration, Fig. 5, also points in this direction since the average coordination number is also different.

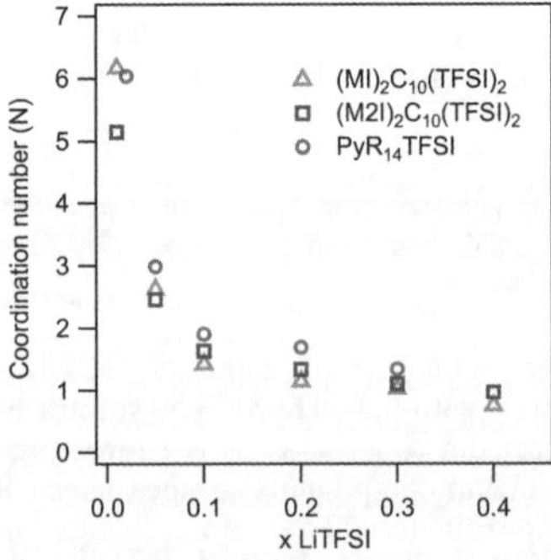


**Fig. 7** Calculated coordination numbers of TFSI-anion around Li-ions as a function of salt concentration. Data for PY14TFSI is taken from ref. 3.

## Conclusions

In this paper, we investigate the effect of Li-salt doping on the ionic conductivity, the glass transition temperature, and the Li-ion coordination of two dicationic ILs based on the imidazolium cation and the TFSI anion. In addition, we compare our results to two mono-cationic ILs. For the two dicationic ILs, homogeneous solutions can be obtained up to very high salt concentrations, even $x > 0.4$, $x$LiTFSI/$(1 - x)$IL. The ionic conductivity drop, and the corresponding increment in glass transition temperature, for high salt concentrations is quite moderate in the imidazolium cation based ILs compared to pyrrolidinium cation based IL. Within the imidazolium cation family, the dicationic ILs are less affected than the mono-cationic IL. We speculate that this can be related to different Li-ion coordination in the dicationic ILs compared to the mono-cationic ILs. The $T_g$-scaled Arrhenius plots of the ionic conductivity results show that the fragility of the systems does not change with the addition of LiTFSI, not even at high salt concentrations. The master curve found with $T_g$ scaling of the conductivity provides further support for linking the local Li-ion environment to the glass transition of the Li-salt/IL mixtures.

## Acknowledgements

The authors wish to acknowledge the funding support from the Swedish Research Council (VR), the Swedish Energy Agency (STEM), FORMAS, and the Chalmers Area of Advance Energy.

## Notes and references

1 M. Armand, F. Enders, D. R. McFarlane, H. Ohno and B. Scrosati, *Nat. Mater.*, 2009, **8**, 621.
2 A. Martinelli, A. Matic, P. Jacobsson and L. Börjesson, *J. Phys. Chem. B*, 2009, **113**, 11247.
3 J. Pitawala, A. Martinelli, A. Fernicola, B. Scrosati, P. Johansson, P. Jacobsson and A. Matic, *submitted for publication*, 2011.
4 J. Pitawala, A. Matic, A. Martinelli, P. Jacobsson, V. Koch and F. Croce, *J. Phys. Chem. B*, 2009, **113**, 10607.
5 H. Tokuda, S. Tsuzuki, Md. A. B. H. Susan, K. Hayamizu and M. Watanabe, *J. Phys. Chem. B*, 2006, **110**, 19593.
6 *Ionic Liquids: Physicochemical Properties*, ed. by Suojing Zhang, 2009.
7 H. Tokuda, K. Hayamizu, K. Ishii, Md. A. B. H. Susan and M. Watanabe, *J. Phys. Chem. B*, 2005, **109**, 6103.
8 W. Xu, E. I. Cooper and C. A. Angell, *J. Phys. Chem. B*, 2003, **107**, 6170.
9 D. R. MacFarlane, J. Huang and M. Forsyth, *Nature*, 1999, **402**, 792.
10 K. Ito, N. Nishina and H. Ohno, *Electrochim. Acta*, 2000, **45**, 1295.
11 K. Hayamizu, Y. Aihra, H. Nakagawa, T. Nukuda and W. S. Price, *J. Phys. Chem. B*, 2004, **108**, 19527.
12 E. Paillard, Q. Zhou, W. A. Handerson, G. B. Appetecchi, M. Montanino and S. Passerini, *J. Electrochem. Soc.*, 2009, **156**, A891.
13 A. Fernicola, F. C. Weise, S. G. Greenbaum, J. Kagimoto, B. Scrosati and A. Soleto, *J. Electrochem. Soc.*, 2009, **156**, A514.
14 J.-K. Kim, A. Matic, J.-H. Ahn and P. Jacobsson, *J. Power Sources*, 2010, **195**, 7639.
15 Q. Zhou, K. Fitzgerald, P. D. Boyle and W. A. Handerson, *Chem. Mater.*, 2010, **22**, 1203.
16 M. J. Monteiro, F. F. C. Bazito, J. A. Siqueira, M. C. C. Ribeiro and R. M. Torresi, *J. Phys. Chem. B*, 2008, **112**, 2102.
17 J. C. Lassègues, J. Grondin and D. Talaga, *Phys. Chem. Chem. Phys.*, 2006, **8**, 5629.
18 Y. Umebayashi, T. Mitsugi, S. Fukuda, T. Fugimori, K. Fugii, R. Kanzaki, M. Takeuchi and S. I. Ishiguro, *J. Phys. Chem. B*, 2007, **111**, 13028.
19 J. C. Lassègues, J. Grondin, C. Aupetit and P. Johansson, *J. Phys. Chem. A*, 2009, **113**, 305.
20 P. Johansson, S. P. Gejji, J. Tegenfeldt and J. Lindgren, *Electrochim. Acta*, 1998, **43**, 1375.
21 M. Herstedt, M. Smirnov, P. Johansson, M. Chami, J. Grondin, L. Servant and J. Lassegues, *J. Raman Spectrosc.*, 2005, **36**, 762.
22 O. Borodin, G. D. Smith and W. Henderson, *J. Phys. Chem. B*, 2006, **110**, 16879.
23 Y. Saito, T. Umecky, J. Niwa, T. Sakai and S. Maeda, *J. Phys. Chem. B*, 2007, **111**, 11794.

24 M. J. Monterio, F. F. C. Bazito, L. J. A. Siqueira, M. C. C. Ribeiro and R. M. Torresi, *J. Phys. Chem. B*, 2008, **112**, 2102.
25 A. Martinelli, A. Matic, P. Johansson, P. Jacobsson, L. Börjesson, A. Fernicola, S. Panero, B. Scrosati and H. Ohno, *J. Raman Spectrosc.*, 2011, **42**, 522.
26 A. B. McEwen, "Cyclic Delocalized Cations Connected by Spacer Groups", U.S. Patent No. 6, 531, 241, 2003.
27 H. Vogel, *Phys. –Z.*, 1921, **22**, 645.
28 G. Tamman and W. Hesse, *Z. Anorg. Allg. Chem.*, 1926, **156**, 245.
29 B. G. Nicolau, A. Sturlaugson, K. Fruchey, M. C. C. Ribeiro and M. D. Fayer, *J. Phys. Chem. B*, 2010, **114**, 8350.

 

# General discussion

**Professor MacFarlane** opened the discussion of the paper by Austen Angell: The properties of the ionic liquids (ILs) based on the guanidinium cation are fascinating. Is there the potential for a Grotthus type mechanism of proton conductivity in this case?

**Professor Angell** replied: It was exactly for this possibility that we commenced this study. The guanidinium cation has 6 exchangeable protons, and high symmetry suitable for spinning, so it seems to be a good candidate for passing protons along. The Walden plot, which I had no time to show, is supporting this, but not as strongly as we had hoped. GdmSCN yields the highest plot but it is still only just on the ideal line. A paper on this subject has been published elsewhere.[1]

1 Z. F. Zhao, K. Ueno and C. A. Angell, *J. Phys. Chem. B*, DOI: 10.1021/jp206491z.

**Professor Jung** asked: You have shown examples of protic ionic liquids where the conductivity and viscosity are described by the fractional Walden rule, instead of the normal Walden rule. This phenomenon is strikingly similar to what has been observed in many fragile glass forming liquids, called the fractional Stokes–Einstein relation or the decoupling phenomenon, that is, the breakdown of the inverse proportionality between the diffusion constant and viscosity. There, the underlying mechanism is believed to be the notion of dynamic heterogeneity. Do you think that the same mechanism is responsible for the fractional Walden rule observed in protic ionic liquids?

**Prof. Angell** answered: Generally, no. The fractional Walden rule observations that you are describing are mostly for molecular liquids which have no ionic or charge carrying component that can decouple from the viscous modes. That is what is necessary for the “superionic” behavior I was discussing.

**Professor Canongia Lopes** opened the discussion of the paper by Athanassios Panagiotopoulos: What is the range of silica core diameters used in the simulations? What is the minimum value for real systems?

**Professor Panagiotopoulos** responded: The model we studied has monodisperse particles of diameter equal to 2 nm. Typical experimental silica-based systems have particle diameters between 5 and 20 nm, values that would be impractical to simulate with the proposed model. However, a key parameter controlling structure and dynamics is the volume fraction of core particles, defined as the ratio of particle over total system volume. This ratio is 13% in our systems, comparable to that in real NIMs and NOHMs systems. Experimentally realizable systems with smaller core particles also exist, namely fullerene- and POSS-based systems.

**Professor Canongia Lopes** said: Some Keggin-based ionic liquids were developed in Belfast some years ago and have a striking morphological similarity to those shown during the presentation. They are composed of spherical anions typically with a −3 charge made of 12 tungsten-atoms interconnected by 36 oxygen atoms. The whole structure is maintained by an extra central phosphate ($PO_4$) group. The following figure shows an example of such compound where each Keggin “ball” is surrounded by three hexylmethylimidazolium cations.

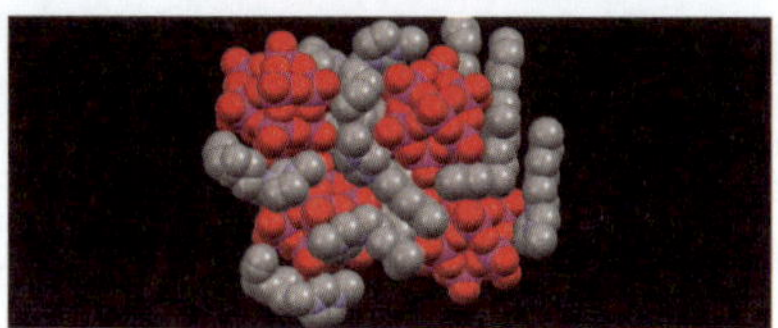

I have been involved in trying to model this system at an atomistic level and I was wondering if any kind of bridge to the systems you have shown in your presentation is possible (the size and number of charges of the spherical anion are both much smaller: around 1 nm across and with a −3 charge, respectively) or if a coarse-grained model like the one you have used would also yield realistic results.

**Professor Panagiotopoulos** replied: The Keggin-based ionic liquids clearly have similarities with the models studied in this paper. The small size (1 nm core) and charge mean that a coarse-grained model can be used to obtain dynamics and structure over long time and length scales. It should be feasible to tune the parameters of the coarse-grained model to smaller-scale atomistic simulations of a realistic model.

**Dr Turton** remarked: The hopping process described is an activated diffusional process. The two systems in Fig. 9 and 10 of the paper are very different: the primary amine appears far more strongly bound than the tertiary amine so its activation energy for hopping should be far higher. Why then do the diffusional dynamics of the chains (Fig. 10) appear to be very similar? I accept that this is not a realistic chemical system but my point was that Fig. 9 and 10 appear to be inconsistent (activation energy determines both detachment time and diffusional rate).

**Professor Panagiotopoulos** responded: The hopping process is responsible for long-time (>100 ns) dynamics in both cases studied, namely of linear and star chains. Because linear chains are more strongly attached to the cores, they diffuse more slowly at long times, as seen in Fig. 10 from the crossing over of the mean squared displacement (MSD) for linear chains below to the curve for star chains. The linear chains do not even reach diffusive behavior at the end of the simulation runs, contrary to the star chains. At shorter times (between 0.1 and 60 ns), the greater steric hindrances of the star chains are reflected in slower local motions of their centers of mass. Fig. 9, on the other hand, emphasizes longer times and reflects motion of the charged group of a chain from one core to a nearby different nanoparticle. This motion is faster for the more weakly held tertiary charged group. Thus, the apparent difference between Fig. 9 and 10 is due to the fact that the motion of the center of mass and of the charged group of a chain is different at short times.

**Professor Maroncelli** asked: I was surprised that you found only a factor of 10 difference in the viscosities you estimated for comparable NIMs and NOHMs systems. In similar comparisons we've made between idealized molecular ionic liquids and their neutral mixture counterparts we find much larger differences (greater than 100-fold).

**Professor Panagiotopoulos** replied: The viscosities in real NIMs and NOHMs systems depend strongly on core volume fraction, chemical character and chain length. A typical value for NIMs with tertiary amines reported in ref. 11 of the paper is 10–20 P at volume fractions of cores comparable to those of our simulations. Low volume-fraction NOHMs systems also have viscosities in the same range. Our simulations suggest that there is no large gap in viscosities between neutral and charged systems in our model systems.

 

**Professor Maginn** commented: Can you explain the physical difference between ion hopping and core displacement, and comment on the differences in timescale one sees in Fig. 9?

**Professor Panagiotopoulos** answered: The two sets of curves shown in Fig. 9 refer to two related phenomena. The "ion" (continuous) lines monitor detachment of a charged positive group from the negative surface site to which it was originally attached. This detachment is associated with hopping motion on the surface of a core particle. The "core" (dashed) curves refer to hopping that leads to detachment from the original nanoparticle to which a chain was attached. This second motion is responsible for long-range travel of chains, given that core motions in NIMs systems are rather slow.

**Professor Maginn** asked: $35kT$ is a very large energy for detachment, yet detachment appears to be facile. This must imply some sort of cooperative motion; is that correct? Is it possible to compute some sort of potential of mean force for this motion in order to better understand the actual barriers encountered?

**Professor Panagiotopoulos** replied: The hopping motion involves a positive ion that is "shared" between two negative surface groups, changing its "allegiance" from one neighbor to another. A potential-of-mean force can certainly be determined in the reaction coordinate of linear displacement along a line connecting two nearby surface groups. It will indeed be interesting to compute the barriers involved; they are clearly going to be lower than $35kT$, but high enough so that this is an activated process.

**Mr Frolov** asked: In which way should one change the structure of the nanoscale ionic materials (the length of the non-polar tails, particularly) to increase/decrease their melting point (crystallinity)?

**Professor Panagiotopoulos** responded: The only systems that have been systematically studied in this respect are uncharged (NOHMs) nanoparticles, but it is reasonable to assume that charged NIMs systems will behave comparably. For NOHMs systems, changing the grafting density and chain length leads to several thermally reversible structural transitions to isotropically/anisotropically aggregating systems, and glasses. Systems with shorter chains and high grafting density show a higher tendency towards crystallization.

**Professor Johansson** commented: If you would define a host–guest relationship for each particle and chain, respectively, then you would be able to see if there is indeed a plateau to be found within the data presented in Figure 10. I believe this would also make it clearer if there is true diffusive transport occurring—right now the MSD (Fig. 10) is of the order of 1 $nm^2$ which really can be just movement on the surface of each particle.

**Professor Panagiotopoulos** responded: We know from Fig. 9 that over 100s of ns, chains detach from their original cores in significant numbers, especially for the star chains. Unfortunately, the simulations do not extend long enough in time for the mean-square displacement to become diffusive for linear chains, as seen in Fig. 10. The cores, on the other hand, appear to move diffusively. This is clearly the case for the NOHM systems, which have MSDs of 8 $nm^2$ over the course of the simulations. NIM cores, on the other hand, diffuse more slowly (only by about 1 nm, about half their diameter over the course of the run). However, this seems to be sufficient for them to reach the diffusive regime. A plateau may be present at the longest times simulated for the linear NIMs chains, which (from Fig. 9) are still mostly attached to their original cores after 250 ns.

**Dr Mezger** commented: One key property of ILs is that they solely consist of ions, *i.e.* without solvent. However, the presented nanoparticle-based ionic liquids contain long and flexible aliphatic side chains that might act as a kind of effective non-polar solvent. How does this affect the analogy between conventional and nanoparticle-based ILs? Would an increase in chain length affect physical properties, such as viscosity, in a similar way to adding a solvent?

**Professor Panagiotopoulos** responded: These systems indeed can be considered to be "particles that carry their own solvent." However, their behavior is quite different from that of particles in a conventional solvent. In a separate study that is currently being written up for publication, we have investigated the impact of chain length on dynamics for NOHMs (uncharged) systems. Somewhat counter-intuitively, the dynamics can become faster for longer chains in the "solvent-free" systems, contrary to the behavior of particles in a medium composed of free polymer chains.

**Professor Lynden-Bell** opened the discussion of the paper by Margarida Costa Gomes: Do you have any explanation as to why the entropies of solvation that you showed during the discussion were surprisingly low in the [$C_6$mim] liquid compared to the trends in the other [$C_n$mim] liquids?

**Professor Costa Gomes** replied: The thermodynamic properties of solvation of ethane and n-butane in the [$C_n$mim][$NTf_2$] ionic liquids are represented in Fig. 1:

The larger solubility of n-butane, compared to that of ethane, is explained by the more negative enthalpy of solvation, meaning that the interactions between this gas and the ionic liquids are more favorable. The increase of the solubility of the two gases with the increase of the size of the alkyl side-chain of the cation of the ionic liquid can only be justified by an entropy of solvation that becomes less negative. This general trend observed in the average values of the entropy calculated from the experimental data is also confirmed by the analysis of the site–site solute–solvent radial distribution functions calculated using molecular dynamics simulations. The exceptionally low values for the entropy of solvation in the [$C_6$mim][$NTf_2$] can not easily be explained. An experimental imprecision or an impurity of the samples can not account for the values of $\Delta_{solv}S^{\infty}$ for both gases in [$C_6$mim][$NTf_2$] as the data for ethane were obtained in a previous study,[1] using a different ionic liquid sample. The authors cannot at present explain these surprising results.

1 M. F. Costa Gomes, *J. Chem. Eng. Data*, 2007, **52**, 472–475.

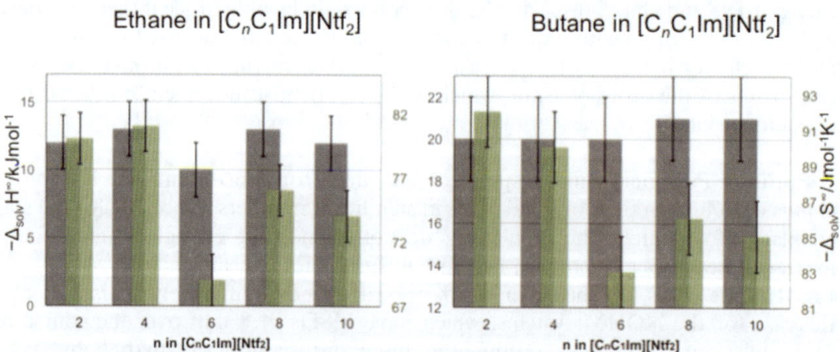


**Fig. 1** Thermodynamic properties of solvation for ethane in [$C_n$mim][$NTf_2$] ionic liquids (left-hand plot) and for n-butane in [$C_n$mim][$NTf_2$] ionic liquids (right hand plot) in the temperature range 303–343 K. The left hand side axis concerns the average enthalpies of solvation (bars in grey) and the right-hand side axis concerns the entropy of solvation (bars in green).

 

**Professor Panagiotopoulos** remarked: Can you comment on the feasibility of using Widom insertions, grand canonical, or Gibbs ensemble Monte Carlo simulations to predict solubilities for ethane or butane in the ionic liquids that you have studied experimentally?

**Professor Costa Gomes** replied: Our previous simulations[1] show that free volume ionic liquids are very small (probed by the probability of inserting hard spheres). Therefore Widom insertion only works for small, non-polar solutes. For gases like carbon dioxide and ethane in other ionic liquids based on imidazolium cations, we calculated solubility using thermodynamic integration and free-energy perturbation methods with reasonably good results.[2,3] Butane is larger and more complex in terms of conformations, so GEMC would likely be a good technique.

1 J. Deschamps, M. F. Costa Gomes and A. A. H. Pádua, *ChemPhysChem*, 2004, **5**, 1049–1052.
2 D. Almantariotis, T. Gefflaut, A. A. H. Pádua, J. Y. Coxam and M. F. C. Gomes, *J. Phys. Chem. B*, 2010, **114**, 3608–3617.
3 A. S. Pensado, A. A. H. Pádua and M. F. Costa Gomes, *J. Phys. Chem. B*, 2011, **115**, 3942–3948.

**Professor Castner** said: Could you speculate on the solubilities of ethane and 1-butane in 1-alkyl-3-methylimidazolium bis(trifluoromethylsulfonyl)amide ionic liquids when the alkyl chain is branched *vs.* straight chain? For example, could you compare the ethane and 1-butane solubilities in 1-octyl-3-methylimidazolium *vs.* 1-isooctyl-3-methylimidazolium ionic liquids?

**Professor Costa Gomes** answered: It is difficult to predict the solubility of n-butane in ionic liquids based in 1-alkyl-3-methylimidazolium cations with branched alkyl-side chains. One can speculate that the contribution of the solute–solvent interactions, expressed by the enthalpy of solvation, is similar to that of n-butane or of ethane in ionic liquids with linear alkyl-side chains. The difference of solubility of the gases in the two families of ionic liquids would then depend on the contribution of the entropy of solvation to the Gibbs energy. Different molecular mechanisms can be involved in the dissolution of ethane or n-butane in ionic liquids based in the imidazolium cation with branched alkyl side-chains. For example, the branched ionic liquid would have a lower free volume that can lead to a decrease of the mobility of the solute and so to a higher entropy of solvation. The solubility of ethane or n-butane would then be lower in 1-isooctyl-3-methylimidazolium than in 1-octyl-3-methylimidazolium based ionic liquids. It is also possible that the branching of the alkyl side-chain of the cation would have a destructuring effect of the non-polar domains of the solution, especially in the presence of linear non-polar solutes like butane. In that case, the contribution of the entropy of solvation to the Gibbs energy can be more favorable in 1-isooctyl-3-methylimidazolium than in 1-octyl-3-methylimidazolium based ionic liquids, leading to a larger solubility of n-butane in the ionic liquid with the branched alkyl side-chain. These analyses are mere speculations and prove the necessity of experimental studies in these systems. The authors would be glad to study the solubility of ethane and n-butane in 1-isooctyl-3-methylimidazolium bis(trifluoromethylsulfonyl)amide, provided a pure sample of the ionic liquid is available.

**Professor Maroncelli** commented: I was struck by the fact that the data in Fig. 3 of the paper vary so simply with alkyl chain length. One might have expected the solubilities to show some signature of the variation in the nanoscale structure predicted to occur as a function of chain length. We have been examining solvation in this same series of ionic liquids spectroscopically with larger probe solutes such as anthracene and we also fail to observe any clear signs of this structural evolution.

**Professor Costa Gomes** answered: It is shown in this work that both gaseous solutes—ethane and n-butane—are preferentially solvated in the low charge density regions of the ionic liquids and do not disrupt their characteristic structure. The dissolution of the gases surely leads to larger low-charge domains. The variation of the solubility with the size of the alkyl side chain of the cation follows the trend observed in the X-ray diffraction patterns of the same ionic liquids.[1] The structuring of the liquids is gradual and there seems to be no abrupt transition between the structure-dependent properties of the ionic liquids with shorter and longer alkyl side-chains.

1 O. Russina, A. Triolo, L. Gontrani, R. Caminiti, D. Xiao, L. G. Hines Jr, R. A. Bartsch, E. L. Quitevis, N. Plechkova and K. R. Seddon, *J. Phys.: Condens. Matter*, 2009, **21**, 424121.

**Professor Quitevis** said: In collaboration with Yang and Voth, we studied mixtures of $CS_2$ and $[C_5C_1im][NTf_2]$ as a function of the mole fraction of $CS_2$ using molecular dynamics (MD) simulations and optical Kerr effect (OKE) spectroscopy at room temperature.[1] The MD radial distribution functions (RDFs) of 5, 10, and 20 mol % mixtures indicate a greater probability of finding $CS_2$ near the terminal carbon of the pentyl side-chain of the cation than near the imidazolium ring or the anion. These results are consistent with the MD simulation results of Prof. Costa Gomes and coworkers for ethane and butane in $[C_nC_1im][NTf_2]$ ionic liquids ($n$ = 2, 4, 6, 8, 10) and provide further support for the solvation of nonpolar solutes in the nonpolar domains.[2,3] The reduced spectral densities of 10, 15, 20 and 25 mol% mixtures obtained from the OKE measurements can be explained in terms of an additivity model with components arising from the sub-picosecond dynamics of $CS_2$ and the ionic liquid. The $CS_2$ contribution to the reduced spectral densities of the mixtures is narrower and lower in frequency than that of neat $CS_2$. This result is similar to that of previous OKE studies of mixtures of $CS_2$ and n-alkane solvents where the spectral changes are due to the softening of the intermolecular potential seen by a $CS_2$ molecule upon dilution.[4–7] Just as the solubility behavior of ethane and butane as a function of the length of the alkyl side-chain can be rationalized in terms of preferential solvation of these gases in the nonpolar domains, the OKE results can be rationalized by the preferential solvation of $CS_2$ in the nonpolar domains. Interestingly, the MD simulations show that tail–tail and cation–anion RDFs are not strongly affected by the presence of $CS_2$ in the ionic liquid. This observation led us to approximate the ionic liquid contribution to the reduced spectral densities of the mixtures by the reduced spectral density of neat $[C_5C_1im][NTf_2]$. In the MD simulations of ethane and butane in $[C_nC_1im][NTf_2]$ ionic liquids, were the cation–anion RDFs strongly affected by the presence of these gases?

1 P. Yang, G. A. Voth, D. Xiao, L. G. Hines Jr, R. A. Bartsch and E. L. Quitevis, *J. Chem. Phys.*, 2011, **135**, 034502.
2 J. N. Canongia Lopes, M. F. Costa Gomes and A. A. H. Pádua, *J. Phys. Chem.*, 2006, **110**, 16816.
3 A. A. Pádua, M. F. Costa Gomes and J. N. Canongia Lopes, *Acc. Chem. Res.*, 2007, **40**, 1087.
4 C. Kalpouzos, D. McMorrow, W. T. Lotshaw and G. A. Kenney-Wallace, *Chem. Phys. Lett.*, 1988, **150**, 138.
5 D. McMorrow, N. Thantu, J. S. Melinger, S. K. Kim and W. T. Lotshaw, *J. Phys. Chem.*, 1996, **100**, 10389.
6 T. Steffen, N. A. C. M. Meinders and K. Duppen, *J. Phys. Chem. A*, 1998, **102**, 4213.
7 A. Scodinu and J. T. Fourkas, *J. Phys. Chem. B*, 2003, **107**, 44.

**Professor Costa Gomes** responded: In our simulations we only added a small number of solute molecules to the ionic liquid simulation box, namely 8 solute molecules in 500 ion pairs. We did not discern a significant effect due to the presence of the solutes at these concentrations on the solvent–solvent radial distribution functions.

**Dr Triolo** addressed Professor Costa Gomes, Professor Pádua and Professor Canongia Lopes: The alkyl chain conformational distribution in ILs is known to be characterised by a mixture of conformers.[1] This emerges, for example, from Raman spectroscopy measurements that indicate the coexistence of all-trans and trans–gauche conformations for butyl chains in a variety of $C_4$mimX salts. This is at odds with the morphology found in crystal phases for the same materials where only one of the accessible conformations is found (and hence the polymorphism in ILs).[1–3] Some indications exist that in the liquid state, the different conformers might segregate, thus leading to micro-environments where enhanced probability of finding A-T or T-G is found.[1] This issue was never, to my knowledge, explored with computational tools.

On the other hand if such a situation occurs, apolar compounds distributed in the neighborhood of the IL's alkyl chains might experience a conformationally-heterogeneous environment that would affect the equilibrium conformational distribution of the apolar compound (*e.g.* butane).

Can your simulation highlight any heterogeneity in the conformational distribution of the butane? Do these molecule experience such a conformational heterogeneous environment? Can you compare the conformational distribution of butane in ILs with that of neat (liquid?) butane?


1 S. Hayashi, R. Ozawa and H. Hamaguchi, *Chem. Lett.*, 2003, **32**, 498; R. Ozawa, S. Hayashi, S. Saha, A. Kobayashi and H. Hamaguchi, *Chem. Lett.*, 2003, **32**, 948; S. Saha, S. Hayashi, A. Kobayashi and H. Hamaguchi, *Chem. Lett.*, 2003, **32**, 740.
2 J. D. Holbrey, W. M. Reichert, M. Nieuwenhuyzen, S. Johnston, K. R. Seddon and R. D. Rogers, *Chem. Commun.*, 2003, 1636.
3 O. Russina, B. Fazio, C. Schmidt and A. Triolo, *Phys. Chem. Chem. Phys.*, 2011, **13**, 12067.


**Professor Canongia Lopes** answered: This is a very interesting question. We (myself, Agílio Pádua in France and Yasuhiro Umebayashi in Japan) have used MD to perform some conformational analyses of IL ions (cations like alkylmethylimidazolium and *N*,*N*-dialkylpyrrolidinium; anions like bistriflamide and its derivatives).[1–5] We have never combined such analyses with the fact that the same sort of MD simulations also show nano-segregated domains (we have always considered average conformer distributions over the entire fluid). I think it is possible to extract that kind of combined information from the existing MD data, especially for cases where the interconversion between conformers is not too easy.


1 J. N. A. Canongia Lopes, A. A. H. Pádua, *J. Phys. Chem. B*, 2006, **110**, 7485–7489.
2 J. N. Canongia Lopes, K. Shimizu, A. A. H. Pádua, Y. Umebayashi, S. Fukuda, K. Fujii and S.-i. Ishiguro, *J. Phys. Chem. B*, 2008, **112**, 9449–9455.
3 J. N. Canongia Lopes, K. Shimizu, A. A. H. Pádua, Y. Umebayashi, S. Fukuda, K. Fujii and S.-i. Ishiguro, *J. Phys. Chem. B*, 2008, **112**, 1465–1472.
4 Y. Umebayashi, T. Mitsugi, K. Fujii, S. Seki, K. Chiba, H. Yamamoto, J. N. Canongia Lopes, A. A. H. Pádua, M. Takeuchi, R. Kanzaki and S.-i. Ishiguro, *J. Phys. Chem. B*, 2009, **113**, 4338–4346.
5 Y. Umebayashi, H. Hamano, S. Tsuzuki, J. N. Canongia Lopes, A. A. H. Pádua, Y. Kameda, S. Kohara, T. Yamaguchi, K. Fujii and S.-i. Ishiguro, *J. Phys. Chem. B*, 2010, **114**, 11715–11724.


**Professor Pádua** answered: Indeed, simulation can be used to obtain distributions of conformers of butane dissolved in the different ionic liquids, but this has not been done yet (although, it is relatively straightforward). We will perform these calculations in further work on this subject.

**Professor Costa Gomes** replied: The distribution of conformers of n-butane in the ionic liquids with different lengths of alkyl side-chain were calculated from the molecular dynamics trajectories at 300 K. The systems contained 8 n-butane molecules in simulation boxes with 250 ion pairs. The populations of trans and gauche conformers of butane are shown in Fig. 2. Although the data are noisy, there appears to be a trend, with an increased probability of trans conformers in ILs with long alkyl chains, and an increased probability of gauche in ILs with short alkyl chains. At first sight this could be interpreted as a "less entropic" arrangement of n-butane in ILs with longer alkyl side chains, which is contrary to the experimental data. However, the analysis of just one internal degree of freedom of the solute is not sufficient to infer on the entropy change of the entire system, in which all the internal degrees of freedom of solute and solvent, plus intermolecular (residual) contributions, intervene. A detailed analysis of the conformational space of the solvent could not be performed in the scope of this discussion.

**Professor Canongia Lopes** addressed Professor Costa Gomes and Professor Maroncelli: I would like to comment on the issue raised by Mark Maroncelli concerning the apparent lack of correlation between some structural changes that can occur along certain homologous series of ionic liquids with increasing alkyl side chains and the thermodynamic (solubility) results obtained using molecular probes. Apart from possible cancellation effects, I think that one must also consider the fact that some of the molecules are as large or larger than the shorter alkyl side chains of the studied ILs (C2 or C4). This can create a "whale in a swimming pool" effect where the structure of the pure ionic liquids is somewhat modified when introducing the probes (this effect is probably not constant along the series). This does not mean that using probes is not a valid way to gather crucial structural information on these complex fluids (the same applies to the use of—even larger—solvatochromic probes); just that some analyses/correlations have to be addressed with due caution.

**Professor Maroncelli** responded: Your point is well taken. I do not doubt that the presence of a solute, even one of modest size like butane, will alter the structure of the neat ionic liquid to some extent. My comment was intended to point out that while I believe nanoscale domain formation to be a reality in many neat ionic liquids, I am struck by the rarity of clear evidence for the effects of this structure on solute-based observables.

**Professor Maginn** asked: Is it possible to understand the solubility trends you observe as a function of ionic liquid alkyl chain length on the basis of molar volume

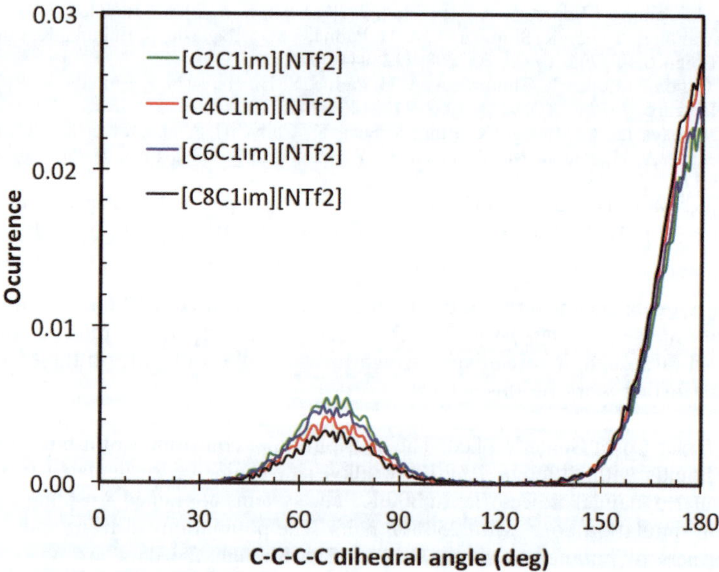


**Fig. 2** Distribution of dihedral angles of n-butane dissolved in $[C_nC_1Im][NTf_2]$ ionic liquids.

 

arguments? That is, the longer the alkyl chain on the ionic liquid, the more volume there is for a given gas molecule to explore in the liquid phase per mole of ionic liquid. If this is true, then one should be able to correlate mole fraction solubility with molar volumes. Have you looked into this?

**Professor Costa Gomes** answered: It is indeed possible to correlate the solubility of ethane or n-butane in 1-alkyl-3-methylimidazolium bis[(trifluoromethyl)sulfonyl] imide ionic liquids on the basis of molar volume of the ionic liquids or even on the basis of their molecular weight. Because it is known that non-polar gases like ethane or n-butane interact preferentially with the alkyl side-chains of the cations, it is correct to explain this fact using the presented arguments—for longer alkyl chains of the ionic liquids, more volume is available for a given (non-polar) gas molecule to explore per mole of ionic liquid.

This correlation of the solubility of gaseous solutes on the basis of the molar volume can fail when different families of ionic liquids are considered or when the gaseous solutes can simultaneously interact with different molecular sites of the ionic liquid. This is the case, for example, for carbon dioxide in 1-alkyl-3-methylimidazolium-based ionic liquids with both tetrafluoroborate[1] or bis[(trifluoromethyl) sulfonyl]imide[2] anions for which a correlation on the basis of the molar volume or on the molecular weight of the ionic liquid fails as the solubility of the gas in ionic liquids with alkyl side chains longer than 8 carbon atoms is too low.

1 Y. Chen, S. Zhang, X. Yuan, Y. M. Zhang, X. Zhang, W. Dai and R. Mori, *Thermochim. Acta*, 2006, **441**, 42.
2 D. Almantariotis, T. Gefflaut, A. A. H. Pádua, J.-Y. Coxam, M. F. Costa Gomes, *J. Phys. Chem. B*, 2010, **114**, 3608.

**Dr Wishart** asked: In addition to studying the effects of branched alkyl chains as suggested by Prof. Castner, or as an alternative to them, have you considered looking at whether branched alkane solutes, such as isobutane, show different behavior than linear ones? If there are differences, they could be the basis for energy-efficient separations systems.

**Professor Fox** commented: Your work is very interesting. The difference of solubilities for butane and ethane in the ionic liquid you attribute to the different contributions to the entropies of solvation of these small hydrocarbons. My first question is which butane was used, because n-butane is a linear hydrocarbon and iso-butane is a much more compact molecule which has a lower contribution to its entropy of solvation? My second question is whether the substantial differences in solubility between ethane and butane ionic liquid could be used as the basis for a separation process of lower hydrocarbons?

**Professor Costa Gomes** answered: Thank you very much for your questions and your comments. The present study concerns the solubility of n-butane in $[C_nC_1Im]$ $[NTf_2]$ ionic liquids. It would be interesting to study also the solubility of iso-butane and to calculate the thermodynamic properties of solvation for this solute. The enthalpies of solvation should be close to those calculated for n-butane, and so the differences in solubility between the two gases would be related to differences in the entropy of solvation. It is the authors' opinion that an experimental study is necessary to verify whether the contribution to the entropy of solvation is lower or higher in the case of iso-butane. Sufficiently large differences on gas solubility can be explored for the development of new, more efficient gas separation systems. The differences in solubility of the two gases measured in the present work—ethane and n-butane—could serve as the basis for a separation process for low molecular weight hydrocarbons. The ideal selectivity for ethane/butane in $[C_nC_1Im][Ntf_2]$ ionic liquids, calculated as the ratio of the Henry's law constants, increases with the size of the alkyl side-chain and decreases with temperature, being as high as 8.3 in $[C_{10}C_1Im][NTf_2]$ at 303 K. This is still a relatively low value but it suggests the possibility of developing specific ionic liquids as absorbents for gas separation processes.

**Professor Dr Roling** opened the discussion of the paper by Edward Maginn: I would suggest to use the term 'ion pair' in an ionic liquid only if the time scale for two ions being in close contact is considerably longer than the time scale of diffusive ion movements. Would you agree on this? Do such ion pairs exist in your simulated ILs?

**Professor Maginn** answered: The simulations we report here arc Monte Carlo simulations, and so do not have a time scale. However, we have previously conducted molecular dynamics simulations (which do have a time scale) and we see that in the liquid phase, ions are surrounded by several counterions at the same time in a first solvation shell. These distributions are consistent with the equilibrium distributions we see in the Monte Carlo calculations. The counterions are in turn surrounded by their own counterions, such that a charge ordered system results. Depending on the system and temperature, one or more counterions will leave the first solvation shell of an ion, but these counterions are quickly replaced by others. Of the systems we have examined, we have never seen a single ion pair all by itself in the liquid. So I would argue that, at least in the liquid phase, a better picture is one in which a given ion is surrounded by several counterions, each solvating the central ion. We do not observe an ion pair here, and an ion pair there. The picture is different in the vapor phase as I showed. In this case, it is possible for individual ion pairs to exist and the distances between adjacent ion pairs can be quite large. So if you are talking about the pure ionic liquid phase, I would say no, there are no formal ion pairs.

**Professor Ballone** asked: Recently, we (R. Cortes-Huerto, N. C. Forero-Martinez and myself) estimated the critical parameters of an IL force field model, using a method arguably simpler than the one implemented by N. Rai and E. J. Maginn. To this aim, we carried out molecular dynamics simulations for $[BMIm][NTf_2]$ and $[BMIm][PF_6]$ in the NPT ensemble, based on the force field developed by the group of A. A. H. Pádua and J. N. Canongia Lopes. We sampled the $T = 1200$ K isotherm, supposedly close to the critical one, covering a wide range of pressures (50 atm $< P <$ 1200 atm).

A fit of the average volume $V$ as a function of $P$ using the van der Waals equation of state gives a critical temperature of 1250 K for $[BMIm][NTf_2]$, and 1350 K for $[BMIm][PF_6]$. Admittedly, the method is over-simplified, and our aim was to obtain a quick estimate of critical parameters. Nevertheless, the results for $[BMIm][NTf_2]$ seem to be intriguingly close to those presented by E. J. Maginn, using a more sophisticated method, and, presumably, longer computations.

Our results are still unpublished. It would be interesting to investigate the dependence of the results on the force field, on the choice of the isotherm, and on the equation of state used to interpolate the simulation data.

**Professor Maginn** answered: I think this is a very nice approach to use, and would certainly be faster than the GEMC simulations we carried out. I look forward to reading more about the details when this work is published. It will be another important simulation tool to help us study these systems. It is encouraging that the critical points are close to those we computed.

**Professor Panagiotopoulos** remarked: Can you provide details on the cavity-bias and configurational-bias moves used for insertions and removals, in addition to the distance-biased selection of the oppositely charged ions?

 

**Professor Maginn** replied: As your question suggests, transfers of the large and bulky ions are difficult. In addition to the distance-biased cation/anion moves, we used a fragment-based configurational bias technique to "grow" the ions whenever swaps were made between a donating and receiving phase. The first fragment inserted into the receiving phase was placed using an energy-bias scheme where multiple trial insertions were made and a favorable location chosen with a Boltzmann weighting scheme. We also used configurational biasing techniques in thermal equilibration moves. In this case, a portion of the ion was randomly deleted and regrown in a biased sense.

**Professor Castner** commented: I was intrigued by your results on ion pairs in the gas phase. Although Monte Carlo simulations would not address this directly, have you learned anything about the mechanism for boiling?

**Professor Maginn** answered: As I mentioned, unfortunately the Monte Carlo simulations have no "time" and so mechanistic details are not directly observable. So we can't say from these simulations exactly what the mechanism is.

**Professor Castner** commented: Let me suggest a hypothetical mechanism that might be consistent with what we know from experimental measurements of clusters observed in electrospray ionization mass spectrometry (where singly charged clusters of N-cation, (N-1)-anion or N-anion, (N-1)-cation are typically observed for N up to 5 or more). Since the alkyl tails on either cations or anions are known to protrude from the liquid into the vacuum, does the ion with the alkyl tail leave the interface first, subsequently pulling the counterion with it?

**Professor Maginn** answered: Your hypothesis, that the cations with long tails emerge first and bring along their counterion, is certainly reasonable. I would note that at the interface of [$NTf_2$]-containing ILs, our calculations (and I believe experimental work as well) shows that the fluorine atoms also protrude into the gas region.[1] So one can't discount that the anion goes first either. Simulations could help shed light on this by the use of potential of mean force profiles in which a cation or an anion is "pulled" from the liquid. To my knowledge, this has not been done yet.

1 M. E. Perez-Blanco and E. J. Maginn, *J. Phys. Chem. B*, 2010, **114**(36), pp 11827–11837.

**Professor Canongia Lopes** commented: A previous study by Del Pópolo and co-workers[1] concerning the aggregation of 1-butyl-3-methylimidazolium triflate in the gas phase (also by simulation) showed that as temperature increases the size of the neutral aggregates decreases (above 400 K the dominant species is always the ion pair) and that above 600 K there is even the possibility of an increasing number of isolated ions. The big difference to the present work is, of course, that they worked at very low pressure/low density conditions (well below 0.1 Pa and at $10 \times 10^{-5}$ or $10 \times 10^{-7}$ mol m$^{-3}$), not over the saturation curve. This means that the designation "high-temperature aggregates" can be a bit misleading since the driving force for aggregation is not simply the temperature but the increasing pressure (and concomitant increase of the density of the gas phase) along the saturation curve.

1 P. Ballone, C. Pinilla, J. Kohanoff and M. G. Del Pópolo, *J. Phys. Chem. B*, 2007, **111**, 4938–4950.

**Professor Maginn** answered: I completely agree with your comment. The results we reported apply to systems in phase equilibrium, not to arbitrary high temperature states. Because the simulations we reported were in vapor-liquid phase equilibrium, as temperature increased the density of the vapor phase also increased. Under these conditions, we observed the formation of clusters. To be clear: the formation of these clusters is driven by increasing vapor density that occurs with increasing temperature, not by increasing temperature alone. I agree that if an ionic liquid is kept at extremely low density, ion pairs will be the dominant species and that if the temperature is increased while keeping the density very low, there will be a tendency for single ions to form.

**Professor Canongia Lopes** asked: At this point, I would like to focus on the issue of the location of the critical point of the studied ionic liquid. The first point to be considered is the relation between the critical compressibility factors ($Z_c$) of ionic liquids (at least this particular one) and molecular solvents. Since the $Z_c$ values are quite different, one can always question the validity of the approximations (based on RGT) generally used to determine the critical point from data on both sides of the saturation curves. A possible validation would be to consider two molecular fluids whose $Z_c$ values are different (or as different as possible) and check if their critical exponents behave as robust observables.

**Professor Maginn** answered: It is true that the $Z_c$ we computed for ionic liquids is quite a bit smaller than that for conventional ionic liquids. I would point out that it is not too dissimilar to the value obtained for the restricted primitive model some years ago, however, which suggests that for strongly ionic systems, $Z_c$ is small. There has been a lot of discussion on how one should estimate the critical point of ionic systems, and I agree that one should be very careful in this regard. Lacking clear guidance either way, we utilized a rather standard estimation procedure that is typically used for normal fluids. We did not perform a detailed study of finite size effects either. I agree that both of these things should be investigated more deeply, and the validation test suggested here is also a good idea. Additional work using different methods seems warranted in this case.

**Professor Johansson** said: From the more delocalized picture of anion around cation in the liquid phase than in the vapour phase (Fig. 11 of your paper), it seems that in the former the anion is almost everywhere but where the steric repulsion from the cation is present. This would, in fact, imply that within the liquid we could argue that there are no strong specific interactions controlling the local ion-ion features for TFSI-based imidazolium ILs. This of course changes a lot when going to the IL vapour.

From IR studies on the liquid phase and DFT calculations on various TFSI-based imidazolium ILs, we see just this—the liquid spectra is best described by independent ions.[1] Comment?

1 J. Grondin, J.-C. Lassègues, D. Cavagnat, T. Buffeteau, P. Johansson and R. Holomb, *J. Raman Spectrosc.*, 2011, **42**, 733–743.

**Professor Maginn** responded: If I understand your IR and Raman studies correctly, you observe in the neat liquid phase that anions are subjected to weak but anisotropic electrostatic fields, making them appear "independent". This is consistent with the results of our liquid phase simulations, where we observe that more than one anion surrounds a cation (coordination numbers can vary and depend on the definition, but we observe six or more counterions surrounding a given ion at any given time). We observe that there are preferential locations where anions like to sit about the cations, which would result in weak anisotropic electrostatic fields. For anions like TFSI, these anisotropies might be very weak due to thermal motion. We also observe regions where steric repulsion exists. This gives rise to the formation of depopulated areas, consistent with your comment. Instead of thinking that the anions are free to go anywhere, however, I think a more accurate picture is that the anions surround the cation in a sort of dynamic equilibrium (and conversely, anions are surrounded by a sea of cations in dynamic equilibrium). Some anions may have relatively strong, localized interactions with a cation for some time, while others experience a more homogeneous electrostatic attraction with cations surrounding them. These localized interactions tend to be averaged out in the liquid.

You also observe that in solvents with a low dielectric constant, ion pairs appear to form, and you see evidence of specific interactions between cations and anions. This is consistent with our gas phase (very low dielectric medium) simulations where we observed that the dominant configuration is where a single anion localizes around a single cation. Finally, you observe that in solvents of high dielectric constant, the anions appear to be free, indicating that the cations and anions are totally solvated by the solvent. While the simulation results we presented do not address this situation, in other simulations where we have simulated ionic liquids in water, we observe the formation of free ions and solvent separated ion pairs. So I believe the totality of these results are consistent with your experiments.

**Professor Panagiotopoulos** commented: When one talks about a system that "does not have a normal boiling point," this only implies that the critical pressure is lower than 1 atm.

**Prof. Angell** replied: That is a helpful clarification, but one must avoid saying that "the boiling point is above the critical point". That would be unphysical.

**Professor Quitevis** asked: The snapshots of $[C_6mim][NTf_2]$ at $T = 1000$ K and 1100 K show that large clusters are present in the vapor phase. Are the large clusters formed from the aggregation of smaller clusters in the vapor phase or are they formed directly by clusters leaving the liquid?

**Professor Maginn** answered: The simulations are only of the equilibrium distribution of clusters (we used Monte Carlo and so have no dynamic information). We do observe over the course of the simulation that small clusters grow and larger clusters break up, resulting in a distribution of cluster sizes at equilibrium. I would note that at this state point, the pressure is also fairly high (recall that the critical pressure for this IL is 4.1 bar) and so the vapor density is large. This is why clusters form. While these simulations do not give us mechanistic details for how the ions leave the liquid phase, one imagines that individual ion pairs leave the liquid and then recombine in the vapor phase until the equilibrium distribution of cluster sizes is formed. It would be interesting to carry out molecular dynamics simulations of the vaporization mechanism to see if this is the correct picture.

**Dr Hayamizu** opened the discussion of the paper by Aleksander Matic: The coordination number of two around a lithium ion by TFSA in ILs can be interpreted by the relations between lithium diffusion coefficient and viscosity in P13-TFSA and EMIm-TFSA ILs.[1,2] The concentration dependencies of Li diffusion coefficients in EMIm-$BF_4$ [3] and DEME-TFSA[4] were similar to those of the TFSA and cation diffusion coefficients and ionic conductivity. The steep increase of the coordination number with dilution is very doubtful. The exchange of TFSA between the coordination sphere and the bulk takes place in any salt concentration and temperature. In addition we measured the diffusion coefficients of Li, TFSA, $BF_4$ and organic solvents (PC and GBL) near infinitesimal concentration[5] and did not find the increase of the solvation number around a lithium ion with the dilution.

1 K. Hayamizu, S. Tsuzuki, S. Seki, K. Fujii, M. Suenaga and Y. Umebayashi, *J. Chem. Phys.*, 2010, **133**, 194505.
2 K. Hayamizu, S. Tsuzuki, S. Seki and Y. Umebayashi, *J. Chem. Phys.*, 2011, **135**, 84505.
3 P. Ballone, C. Pinilla, J. Kohanoff and M. G. Del Pópolo, *J. Phys. Chem. B*, 2004, **108**, 19527.

4 K. Hayamizu, Y. Aihara, H. Nakagawa, T. Nukuda and W. S. Price, *J. Phys. Chem. B*, 2008, **112**, 1189.
5 Y. Aihara, K. Sugimoto, W. S. Price and K. Hayamizu, *J. Chem. Phys.*, 2000, **113**, 1981.

**Professor Matic** responded: Most previous studies on the coordination dependence of Li-ions in ionic liquids are concerned with molar fractions >0.1 ($x$LiTFSI-($1 - x$)IL). What is seen in our study is that only when you go below this ($x < 0.1$) is there an increase in the coordination number. The situation in organic solvents might be completely different, since there the Li-ions also have a considerable coordination to other solvent molecules besides TFSI-anions. In ionic liquids, we expect that the Li-ion environment at low salt concentrations (having a high coordination number) is very dynamic. There is likely a rapid exchange of nearest neighbors and we do not expect that the Li-ion is diffusing together with a large number of TFSI ions, which should result in a low diffusion coefficient. Thus, we don't see any disagreements between our results and previous results from diffusion experiments.

**Professor Dr Roling** remarked: Is anything known about the lifetime of Li-TFSI complexes in your doped ILs or about the free energy barrier for lithium ions moving from one TFSI complex into a different one?

**Professor Matic** answered: One can get some insight into the lift time from NMR experiments. One indication that the configurations with coordination number around 2 are quite stable is that the diffusion of Li-coordinated TFSI follows the one of Li, as if they diffuse together. See also comment made by Patrik Johansson to this paper.

**Professor Johansson** said: It has been proven by Duluard *et al.*[1] that the $[Li(TFSI)_2]^-$ complexes are stable during relatively long times by PGSE-NMR. This is true for molar fractions of LITFSI 0.03–0.23.

1 S. Duluard, J. Grondin, J.-L. Bruneel, I. Pianet, A. Grélard, G. Campet, M.-H. Delville and J.-C. Lassègues, *J. Raman Spectrosc.*, 2008, **39**, 627.

**Dr Wishart** commented: Regarding the comparison of Raman spectra of the lithiated dicationic ILs and $PyR_{14}$TFSI in Fig. 6 and the associated calculation of lithium coordination number using eqn 2, it would be more appropriate to compare the lithiated dicationic ILs with lithiated BMIm–TFSI rather than with the pyrrolidinium salt. The pyrrolidinium cation is not a good H-bond donor compared to imidazolium, which would compete with $Li^+$ for interactions with the TFSI anion and lower the apparent coordination number compared to $PyR_{14}$TFSI. As it stands, there is no basis for a systematic comparison of the effects of di- versus monocations.

**Professor Matic** responded: We agree that it would have been valuable to also compare the behavior of the coordination number to the case of a mono-cationic IL based on a imidazolium cation, in particular BMIm. Unfortunately, such data were not available at that point. However, for the conductivity and the glass transition temperature such a comparison is made in the paper.

**Professor Quitevis** remarked: In your paper you showed that the temperature dependence of the ionic conductivity is the same for all LiTFSI concentrations based on the fact that all the data fall on a master curve when plotted with a $T_g$-scaled temperature axis. These results led you to conclude that the conduction process is dominated by the viscous properties of the IL-LiTFSI mixtures. Have you tried using a Walden plot of log ($\Lambda$) versus log ($\eta^{-1}$), where $\Lambda$ is the equivalent conductivity and $\eta$ is the viscosity, to analyze your mixture data? Such a plot should show a clear correlation between conduction and viscosity. Moreover, if the formation of Li-ion–anion complexes affects conduction but not viscosity, the lines for these mixtures should fall below the ideal Walden line.

**Professor Matic** replied: This is a very good suggestion. Unfortunately, we do not have viscosity data for these systems. But such experiments should be performed enabling the analysis you suggest.

**Professor Maroncelli** commented: In the absence of Li you decompose the 740 mode in Fig. 6 into two contributions for the cisoid and transoid conformations of TFSI. Why do you fit only a single band in the case of TFSI influenced by Li coordination? Also, are the spectral features in this spectral region in the presence of Li accurately predicted using electronic structure methods?

**Professor Matic** responded: Indeed one could think of using two or more contributions also for the TFSI influenced by Li coordination. In the case one has 2 TFSI coordinated to each Li there could be three configurations (*cis-cis*, *cis-trans*, and *trans-trans*). We have chosen to use just one band, mainly for the reason that there is no clear spectral signature telling us to use more bands. However, we have noted (not reported in the paper) that at high concentrations a better fit is obtained when using two bands also for the Li influenced TFSI (of course this fit also has more parameters). With this 4 band fit the coordination number decreases somewhat but not substantially. The influence of using more than one coordinated band should be more important at high concentrations. The spectral features have been predicted from *ab initio* calculations. These results are reported in, for example, ref. 13 and 17 in our paper.

**Professor Canongia Lopes** asked: Perhaps the comparisons between different mixtures of ionic liquid with lithium should take into account the monocationic or dicationic nature of the ionic liquid: instead of the molar fraction of the ionic liquid one could use equivalent fractions? In my opinion one mole of $(MI)_2C_{10}(TFSI)_2$ is more comparable to two moles of BMImTFSI than to just one mole of BMImTFSI. This would also solve the problem that solutions with the same lithium mole fraction have very different numbers of TFSI anions in the case of dicationic or monocationic IL-based solutions.

**Professor Matic** answered: It is true that having the di-cations (with 2+ charge) complicates things in the comparison. As suggested, comparing to two moles of a mono cationic IL is one method which might be more correct. The difference will be largest at small concentrations compared to the way we present the data.

**Professor Castner** asked: I am most interested in Fig. 7 from your article where the coordination numbers of the TFSI anions about lithium cations is plotted against the mole fraction of lithium. I'm confused about why very similar results are observed for both monocations and dications— shouldn't the stoichiometry differences alone lead to very different results? Could you please explain further?

**Professor Matic** replied: In the comparison of the coordination number between the two systems we have used eqn 2 to calculate $N$ and then plotted the result as a function of $x$. The overall trend, with the increasing coordination number, is correctly captured by this approach. However, we agree that it is not strictly correct. In eqn 2 one should have replaced the $x$ in the denominator by $x/(2-x)$, *i.e.* dividing by the concentration of Li vs. TFSI. However, to then compare the two systems one should also plot the result in the same manner, *i.e. vs.* $x$ for mono-cationic and *vs.* $x/(2-x)$ for the di-cationic case. Thus, the coordination number should be higher

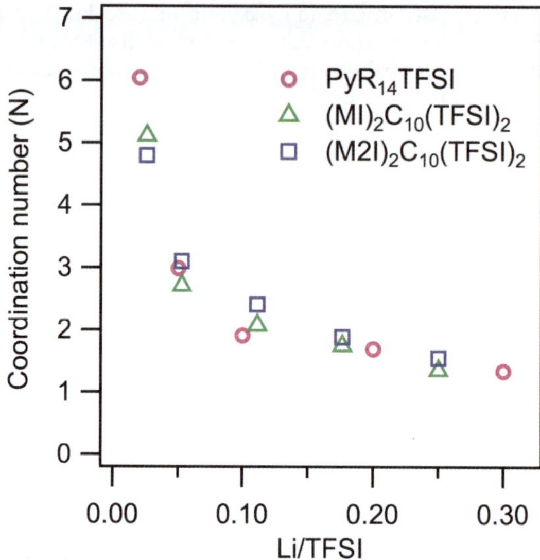


**Fig. 3** Average Li-ion coordination number as a function of Li/TFSI ratio for a mono-cationic and two di-cationic ionic liquids.

but also shifted to lower concentrations in Fig. 7 for the di-cationic case. These two factors make the comparison of the data look very similar, but we can agree that it is not quantitatively correct. In Fig. 3 below we report the data of Fig. 7 from the paper recalculated using $N = C_{\text{coord}}/x$ for mono-cationic and $N = C_{\text{coord}}/(2 - x)$ for di-cationic case and as a function of Li/TFSI ratio. The figure clearly shows that the overall trend is found for both systems but the slightly lower coordination number for the di-cationic systems is not really clear.

**Dr Mele** said: The comment concerns the model proposed for Li complexation in pyrrolidinium-based ionic liquids with anions like TFSI, BETI and $IM_{14}$. Raman data suggest the formation of clusters with two or more anions coordinating the Li cation. The data are consistent with what was found by my group *via* multinuclear NMR.[1] In particular, $^{1}H$–$^{7}Li$ NOE experiments showed selective contacts of Li ions with the H atoms of $PYR_{14}$ cation located around the $N^+$ cation, and no dipolar contacts with the butyl chain protons. The same selectivity was obtained by $^{1}H$–$^{19}F$ NOE correlation, thus demonstrating that the Li ion is strongly coordinated to the anion, in turn forming an ion pair with the pyrrolidinium cation. The coordination was recently demonstrated by X-ray crystallography.[2] Moreover, diffusion measurements as a function of $T$ showed that the activation energy for $Li^+$ diffusion is significantly larger than those for $PYR_{14}$ and TFSI components of the liquid. We have interpreted this difference as a clue for different diffusion mechanisms. In particular, the activation energy for $Li^+$ diffusion can be related to the "hopping" mechanism proposed by Borodin.[3]

1 F. Castiglione, E. Ragg, A. Mele, G. B. Appetecchi, M. Montanino and S. Passerini, *J. Phys. Chem. Lett.*, 2011, **2**, 153.
2 Q. Zhou, P. D. Boyle, L. Malpezzi, A. Mele, J.-H. Shin, S. Passerini and W. A. Henderson, *Chem. Mater.*, 2011, **23**, 4331.
3 O. Borodin, G. D. Smith and W. Henderson, *J. Phys. Chem. B*, 2006, **110**, 16879.

 

# New experimental evidence supporting the mesoscopic segregation model in room temperature ionic liquids

**Olga Russina**[*a] **and Alessandro Triolo**[*b]

***Received 19th April 2011, Accepted 13th June 2011***
**DOI: 10.1039/c1fd00073j**

The existence of a high degree of order over the mesoscopic spatial scale in room temperature ionic liquids is one of their most intriguing properties. Recently the possibility that such a feature, that is witnessed by the occurrence of peculiar low Q diffraction features, reflects nm-scale structural organization has been questioned on the basis of both experimental and computational studies. In this contribution we discuss these studies and present novel experimental evidence that confirm the existence of nm-scale spatial heterogeneities due to the segregation of apolar moieties dispersed in a polar network. The consequence of this scenario is that when the chain polarity gets closer to that of the charged head, the structural heterogeneities are no longer observed.

## 1 Introduction

Room temperature ionic liquids (RTILs) represent an interesting class of materials whose smart properties are being exploited for a wide range of applications. A detailed understanding of the correlation between bulk properties and their morphology and dynamics at a mesoscopic level is required to rationalize and improve the performances.[1–4]

RTILs are ionic compounds that are composed solely of ionic species and are, conventionally, liquid under 100 °C. Typically they are composed of asymmetric ions and/or bear chains that destabilize the crystalline phases.

RTILs are presently considered as more structurally complex than simple molten salts.[2,5] Some years ago Molecular Dynamics (MD) simulations highlighted the existence of a high degree of intermediate range order (IRO) in RTILs whose alkyl chain length is longer than a butyl group. Ribeiro *et al.* calculated the static structure factor, S(Q), for 1-alkyl-3-methylimidazolium chloride (alkyl = methyl, ethyl, butyl, octyl) salts and the longer chain salts were found to be characterized by the occurrence of a low Q diffraction peak, whose position and intensity depend on the alkyl chain length.[6]

Voth *et al.* used coarse grained MD simulations to detect the existence of spatial heterogeneities in ionic liquids with long enough alkyl chains[7] and Canongia-Lopes/Padua proposed a similar nanostructural organization in these materials using atomistic simulations.[8]

These studies can be considered as the first ones that highlighted the peculiar IRO taking place in common RTILs. Already at that stage it was proposed that the main driving force for the development of such an enhanced structuring is the self

[a]*Dipartimento di Chimica, Sapienza Università di Roma, Piazzale Aldo Moro 5, I-00185 Roma, Italy. E-mail: olga.russina@uniroma1.it*
[b]*Istituto di Struttura della Materia, Consiglio Nazionale delle Ricerche, Via del Fosso del Cavaliere 100, I-00133 Roma, Italy. E-mail: triolo@ism.cnr.it*

assembling of the side alkyl chains (provided they are long enough) due to the different nature of interaction between these apolar moieities and the polar portions (*i.e.* the cation's head and the anion).

Strong support for these findings has been provided on the basis of experimental data based on Raman spectroscopy,[9,10] OHD-RIKE spectroscopy,[11,12] viscosity and conductivity measurements,[13] dielectric and Kerr spectroscopies[14] and small angle x-ray[15–28] and neutron[29–35] scattering measurements.

Since the first simulation works, other groups have explored the morphology and dynamics of RTILs, providing further support to the mentioned structural scenario that considers a segregation of the alkyl chains into a three dimensional network that is built up by the charged portions of the system (see *e.g.* refs. 36–38). Currently the understanding is mostly based on the synergy between molecular dynamics simulations and diffraction experiments.

MD simulations are typically used to calculate static structure factors that highlight the appearance of a low Q peak in ILs with long enough alkyl chains. They are also used to calculate specific pair distribution functions, such as the CT-CT, where CT is the terminal alkyl carbon, that show an enhancement of the tail-tail structural correlation. Furthermore, using a successful color scheme, Canongia-Lopes and Padua provided color figures of selected snapshots where the charged and the alkyl portions of the system were colored in a different way, thus visually highlighting the existence of a mesoscale structural heterogeneity.[39] Recently it has been mentioned that the latter approach might be misleading, as it shows just a snapshot over a sequence of millions of simulation steps in a highly viscous and complex system and so this visualization might not be representative of the bulk behavior.

The use of X-ray or neutron scattering techniques provided a sound experimental confirmation for the IRO's existence in RTILs. These studies allowed accessing information on the static structure factor in a more systematic and reliable way than the corresponding MD simulations which were typically developed for a few selected samples from each research group and whose description of this specific mesoscopic structural feature is difficult to achieve with sufficient accuracy (presumably due to the large simulation boxes that are required to fully sample the system in order to describe such a long range structural feature). The experimental diffraction data indicate the existence of an amorphous halo at momentum transfer positions between 0.2 and 0.8 $\text{Å}^{-1}$.[15] Typically these peaks strongly depend on the alkyl chain length, as their position and amplitude change in a systematic way together with this parameter. In most of the studies the low Q peak is found only for chains longer than butyl, for shorter chains either it appears as a weak shoulder of the higher Q amorphous halo or it is fully submerged by the latter peak (if existing at all). Recent experiments of small angle neutron and x-ray scattering, however, succeeded in highlighting this diffraction feature also in RTILs with a very short chain, as short as ethyl, in the case of protic ionic liquids.[31,25,34] We also mention that recently Hardacre and co-workers described an interesting effect:[30] using the SANS technique on selectively deuterated samples, they highlighted that the low Q peak, which is very hard to detect in fully protiated $[C_4mim][PF_6]$, becomes very well defined in methyl- or butyl- deuterated samples. Such a finding can be helpful in rationalizing the apparent discrepancy between the behavior of protic ionic liquids (such as EAN and PAN) and the other RTILs (where the low Q peak is found only in salts with long alkyl chains). Presumably SANS experiments with properly selectively deuterated samples will help in detecting the low Q feature in salts bearing a chain as short as ethyl and also for the case of imidazolium-based ILs.

The structural model proposed on the basis of the existing SAXS data considers that the alkyl chains, because of the different nature of the interaction between them and the polar portions tend to segregate into domains whose size grows linearly with the alkyl chain length. These domains are embedded into the matrix composed by the long range interacting polar portions. A characteristic size for these structural heterogeneities has been proposed on the basis of the Bragg's law ($D = 2\pi/Q_{peak}$);

 

interestingly it has been found that $D$ depends on the chain length with a slope ($\partial D/\partial n$) equal to *ca.* 2.1 Å per $CH_2$ unit.[15,16,18] This is approximately twice the van der Waals diameter of a $CH_2$ unit and such a dependence prompts for a limited interdigitation between neighbor alkyl chains.

In 2010 the first systematic study[30] using the SANS technique on a (selectively deuterated) intermediate alkyl chain length RTIL, namely $[C_n\text{mim}][PF_6]$, with $n$ = 4,6,8, proposed that the previously reported linear trend of $D$ *vs.* $n$ is part of a more complex behavior when comparing the SANS (and the published SAXS) results together with other data for both shorter ($n = 1$) and longer ($10 \leq n \leq 20$) alkyl chains. Based on this observation, the authors proposed that the low Q peak does not fingerprint the existence of a specific mesoscopic order, but is just the consequence of the RTIL's geometric anisotropy, due to the presence of the alkyl chain.

Such a proposal has been recently shared by Margulis and coworkers in a simulation study where the authors explore in a detailed way the origin for such a structural feature.[40]

In this contribution, we will report new experimental results on the occurrence of the low Q peak in a series of new samples, and discuss these and previous data sets aiming to further extend the experimental basis for a more detailed rationalization of this phenomenology.

## The status of the understanding of mesoscopic order in RTILs: some recent proposals

At present, much of the understanding on the nature of the mesoscopic organization in RTILs is based on the synergic exploitation of both computational and scattering techniques.

Simulation studies indicate the existence of a segregated alkyl chain building up nm-scale domains that are kept separated from the charged matrix that is organized in a 3-dimensional network, the driving force for this segregation phenomenon being the polar–apolar dichotomy inherent in RTIL's chemical composition.

Recently Canongia-Lopes *et al.* further developed this model observing that, depending on the cation's geometrical nature different mesoscopic organization becomes possible: from the 'classical' globular-like that is found in imidazolium based RTILs, to the sandwich-like found in trialkylammonium salts,[41] to the filamentous-like that occurs in tetra-alkylphosphonium salts.[42,43]

While this point of view seems rather reasonable, recently both experimental and computational evidence has been reported against this interpretation, by proposing that the existence of the low Q peak is simply the consequence of the geometrical anisotropy of RTIL molecules.

In the following we will report on these points of view and present some issues that do not support them. In the following sections we will also provide new experimental data that will provide evidence for the understanding of the mesoscopic order phenomenon and in contrast with the mentioned proposal.

In their recent paper Hardacre *et al.*[30] report $D$ *vs.* $n$ for a series of imidazolium $PF_6$-based salts (Fig. 4 of their paper), together with data for other imidazolium based salts (with anions different from $PF_6$). In contrast with our original proposal that the characteristic size of the spatial heterogeneities follows a linear trend *vs.* $n$, with slope approximately independent of the anion and given as $\partial D/\partial n = 2.1$ Å per $CH_2$ unit, the authors report that $D = 3.7506 + 2.6717n - 0.04524n^2$, pointing towards a much more complex (non linear) trend. Focusing only on $PF_6$-based salts, together with data for $n = 4, 6, 8$, the authors considered data for the second shell (cation–cation and anion–anion) separation (for $n = 1$) and data for a long chain salt ($n = 16$) that was obtained by extrapolating to room temperature the approximately linear trend that was obtained for the low Q peak of the smectic phase (and not the isotropic one).[44] We mention here that, despite the fact that both the isotropic and the smectic phase are characterized by low Q peaks whose position shifts to higher Q values with increasing temperature (and this is at odds with the behavior found in crystalline phases), the latter extrapolation to room temperature procedure might not be a correct approach to correctly estimate the size of the structural heterogeneities in the isotropic phase. This is because, *a*) the smectic phase's *D* often has a non linear temperature trend (see *e.g.* ref. 45) and *b*) assuming the equivalence of the *D*s for the smectic and the isotropic phases at the smectic-isotropic transition, it is not yet known whether the temperature dependence of the two *D*s are the same. Furthermore the (negative) deviation from the linear trend that is observed for $n > 12$ might be the consequence of the higher flexibility of the longer alkyl chains (as compared to the more rigid shorter chains) that leads to more coiled chains. Accordingly we believe that the mentioned proposal for the alkyl chain length dependence of *D* should be better verified and tested.

As a direct consequence of the above mentioned proposal, Hardacre and coworkers further propose that the observed low Q diffraction feature in the isotropic phase 'does not demonstrate the presence of a nanostructure beyond an immediate correlation' between nearest neighbors. The authors then propose that the observed increasing characteristic size is just the effect of the introduction of a progressively larger amount of diluting $CH_2$ units. We mention here that the proposed role of the alkyl chains which would simply act as diluting agents for the strongly scattering polar units is at odds with the observed trend for the amplitude of the peak, which should decrease if the same scattering units (the polar portions, as shown by Margulis[40]) were more and more diluted by the increasing amount of a simply *diluting* agent. For a selected set of samples (the series of bis-triflamides for which we have a large number of members ($2 \leq n \leq 10$)),[18] we observe (for $6 \leq n \leq 10$, where the peak amplitude can be easily determined) that the low Q peak amplitude grows following a $n^2$ trend: this growing trend witnesses a scattering behavior typical of aggregates growing in size with increasing *n*.

The structural scenario proposed by Hardacre and coworkers was recently supported by a stimulating simulation work from Margulis and coworkers.[40] In this paper, the authors compare the crystalline structure of $[C_{10}mim][PF_6]$ with its liquid structure. They propose that the low Q peak that is found in the liquid state corresponds to the one found in the crystalline one.

This observation is consistent with findings from long alkyl chain salts such as $[C_nmim]X$ (with $n$ = 12–18 and X = $Tf_2N$, Br and Cl),[45] $[C_nmim][PdCl_4]$ ($n$ = 10–18),[46] $[C_nmim][PF_6]$ ($n$ = 14, 16),[44] where it is observed that a complex phase diagram exists for these materials, as, depending on temperature, several crystalline, one smectic and the isotropic phases exist (at increasing temperature). Each of these phases is characterized by low Q peaks: generally for crystal phases the low Q peak shifts to lower Q values with increasing *T*, while the smectic phase one (which is systematically at lower Q values than the crystalline one) shifts to higher Q values with increasing *T*. Although there is little evidence of this, it can be assumed that the low Q peak for the isotropic phase is centred approximately at the position of the smectic phase (although the latter is much narrower than the former). It is generally believed that the isotropic phase is very similar to the smectic phase, apart from the high periodicity occurring in the latter phase as a consequence of its quasi-lamellar structure imposed by the highly organized charged moieties. Similarly it is believed that the smectic phase has strong structural analogies with the crystalline phase, as they differ mostly on the fact that in the former phase the alkyl chains are molten. Moreover the structural analogy proposed by Margulis *et al.*, is consistent with the previous observations of a high degree of similarity between the structure of the liquid state and the crystalline one.[47,48]

Based on past experimental work, however, we believe that the structural organization in molten ILs is partly different from the one observed in crystalline samples (apart from the obvious lack of long range order in the liquid state).

 

Considering a given IL, such as $[C_{12}mim][PF_6]$, the low Q peak observed for the high temperature crystalline phase is centred at a position that is substantially different from the one where the low Q peak for the liquid state occurs. In Fig. 1, we show the alkyl chain dependence of characteristic size obtained from $D = 2\pi/Q$ for the low Q peak for both crystalline ($n$ = 12, 14, 16, 18)[49] and liquid ($n$ = 4, 6, 8)[16] salts at ambient temperature. In the same graph, the corresponding quantity is plotted for the linear alcohols from ethanol to decanol.[50] It can be appreciated that the trend related to the crystalline phase has a substantially different slope ($(\partial D/\partial n)_{crystal}$ = 1.1 Å per $CH_2$ unit) than the one related to the liquid phase ($(\partial D/\partial n)_{liquid}$ = 2.1 Å per $CH_2$ unit). In particular it is noteworthy that the crystalline phase trend is very similar to the one observed in linear alcohols ($(\partial D/\partial n)_{alcohol}$ = 1.2 Å per $CH_2$ unit). In the latter case, we rationalized the observed dependence proposing an almost complete interdigitation between neighbor alkyl chains.[16] This model would then also apply for crystalline $[C_nmim][PF_6]$ salts. On the other hand the already reported difference in slope found for liquid $[C_nmim][PF_6]$ salts prompts for assuming an almost non interdigitated interaction between the alkyl chains.[16] Further differences between the crystalline and the liquid state behavior concern the temperature dependence of the low Q peak position. Our data for a series of $[C_nmim][BF_4]$ salts,[15] as well as the recently published data from Aoun *et al.*[23] for $[C_nmim]Br$, indicate that upon increasing temperature the low Q peak shifts to higher Q values for the liquid state, while it shifts to lower Q values for the crystalline phase.

These observations prompt us to propose that despite the strong similitude between the morphologies in crystalline and liquid ILs, the two states are characterized by a different structural organization and a series of parameters, such as temperature or chain length, affect in different ways such a structure.

As mentioned by Margulis *et al.*[40] it is observed that, from the calculation of the different contribution to the static structure factor, the low Q peak is generated from the structural correlations involving both the cation heads and the anions rather than the carbon tails. This observation is consistent with the effect of isotopic substitution on the SANS patterns reported by Hardacre *et al.*[30]

By analogy with the diffraction pattern of the $[C_{10}mim][PF_6]$ crystalline phase that also shows peaks at 0.4 and 0.9 Å$^{-1}$, these peaks are interpreted, in the liquid state, as due to first neighbor structural correlations between the polar heads (that are considered as built up by the neighboring cation's head and anion) that are either in close contact (peak at 0.9 Å$^{-1}$) or at the closest contact when separated by two non-

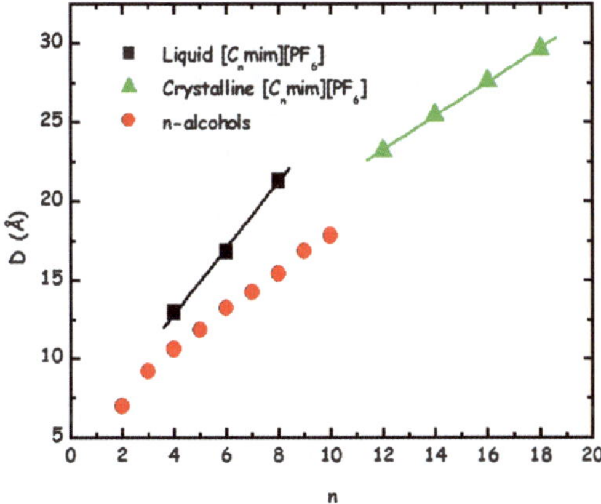


**Fig. 1** Alkyl chain dependance of the characteristic size, $D$, estimated with the Bragg's law from the low Q peak' position. The symbols refer to either liquid (black squares) or crystalline (green triangles) $[C_nmim][PF_6]$ salts and to linear alcohols (red circles). The lines are linear fits of the data.

interdigitating alkyl chains (peak at 0.4 Å$^{-1}$).[40] Such a scenario, if plausible in the crystalline state, seems quite unlikely for the liquid state, where it is hard to imagine that neighbor polar heads cannot get closer, by simply tilting the alkyl chains: such a scenario would then lead to a scattering amplitude also in the intermediate range between 0.4 and 0.9 Å$^{-1}$, where, instead there is a minimum in the experimental data. The conclusion from Margulis *et al.*, is that "*it is the intrinsic anisotropy in the geometry of the cation along the direction of the longer alkyl tail that prevents the close approach of other polar groups in that particular direction*". We will show in the following sections new experimental evidence supporting the view that the phenomenon is instead to be described in terms of segregation of non-polar moieties in an otherwise homogeneous matrix of charged parts.

## Experiments

[$C_6$mim], [$C_8$mim] and [$C_{10}$mim] chlorides were MERCK products with purity >98% and were used without further purification. The salts were kept under dynamic vacuum at 80 °C for two days before the measurements. Mixtures of [$C_6$mim]Cl and [$C_{10}$mim]Cl were prepared in a glove box with relative humidity < 1ppm, by weight.

1-ethyl-3-methylimidazolium octyl-sulfate, [$C_2$mim][$C_8OSO_3$] and 1-butyl-3-methylimidazolium octyl-sulfate, [$C_4$mim][$C_8OSO_3$], are products from Solvent Innovation, and were commercialised as ECOENG 218 and 418, respectively. The liquid salts were used as obtained (the chemicals were classified as purum (98%) or puriss. (99%)) without further purification. They were kept for several days at 40 °C under vacuum aiming to reduce the moisture content of the as-received samples.

1-hexyl,3-methylimidazolium bis(trifluoromethyl sulfonyl)imide, [$C_6$mim][$Tf_2N$], was purchased from IOLITEC; its declared purity is 99% and was used without further purification. Before measurements the sample was kept for several days under dynamic vacuum at 50 °C, to reduce the moisture content.

1-methoxyethoxymethyl-3-methyl-imidazolium bis(trifluoromethylsulfonyl) imide, [C1OC2OC1mim] [$Tf_2N$], was kindly donated by Prof. P. J. Dyson[51] (Ecole Polytechnique Federale de Lausanne) and was used after treatment at 50 °C for two days under vacuum.

The small-wide angle x-ray scattering (S-WAXS) experiment was conducted at the high brilliance beam line ID02, European Synchrotron Radiation Facility (ESRF), Grenoble, France,[52] using an instrumental setup which allows coverage of the momentum range Q between 1 and 20 nm$^{-1}$, with a wavelength $\lambda = 0.75$ Å (Energy = 16.5 keV). Measurements were collected at 25 °C, using a thermostated bath and the sample was kept inside a temperature controlled flow-through cell, with internal diameter of 1.9 mm. The corresponding empty cell contribution was subtracted. Calibration to absolute units (mm$^{-1}$) was obtained using a neat water sample in a 2 mm capillary.

## Results and discussion

Evidence of nano-scale segregation in 1-alkyl-3-methyl-imidazolium salts with chloride anions were the first to be reported in 2007.[15] In the inset of Fig. 2, we report some of the published SAXS data for the series of neat RTILs at ambient temperature ([$C_n$mim]Cl, $n = 6, 8, 10$), while in Fig. 2 data from a series of binary mixtures of [$C_6$mim]Cl and [$C_{10}$mim]Cl, over the whole concentration range are shown. The published data for the neat samples are characterized by the occurrence of a low Q peak whose amplitude and position depends on the alkyl chain length. The data for the C6/C10 mixtures are also characterized by a unique low Q peak whose positions depend on the mixture composition and are found between the two pure components' peaks (*i.e.* neat [$C_6$mim]Cl and [$C_{10}$mim]Cl).

 

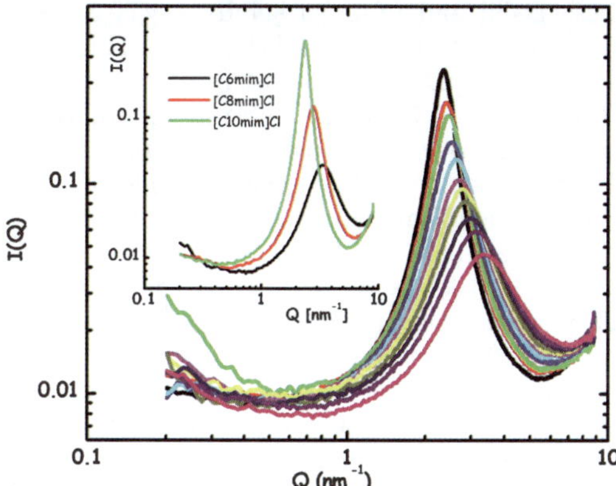


**Fig. 2** SWAXS data for binary mixtures of [$C_6$mim]Cl and [$C_{10}$mim]Cl, over the whole concentration range. In the inset the corresponding data for pure [$C_6$mim]Cl, [$C_8$mim]Cl and [$C_{10}$mim]Cl are presented.

This observation is noteworthy, as it implies that: *a*) the nano-scale segregation of the alkyl chains into non-polar domains occurs in binary mixtures of ILs as well and *b*) there is no differentiation between domains built up solely by the short (hexyl) and the long (decyl) chains (as the latter behavior would have led to two different low Q peaks centred at the positions of the neat components).

We estimated the size of the nano-segregated domains, *D*, by the Bragg law. In Fig. 3, D is plotted *versus* the molar concentration of [$C_6$mim]Cl in the mixture. In such a representation 50 % mol corresponds to an apparent number of $CH_2$ units in the side alkyl chain ($n = x_{C6}*6 + x_{C10}*10$) equal to the one found in neat [$C_8$mim]Cl. In this figure, we highlight the mentioned linear trend describing the behavior of D *versus* the number of carbon atoms in the side alkyl chain (for $n$ = 6, 8, 10, the black, red and green squares, respectively). The deviations observed in the case of the mixtures from this trend indicate that the mixture is not behaving ideally and a positive excess is observed (we stress however that in a study on mixtures of

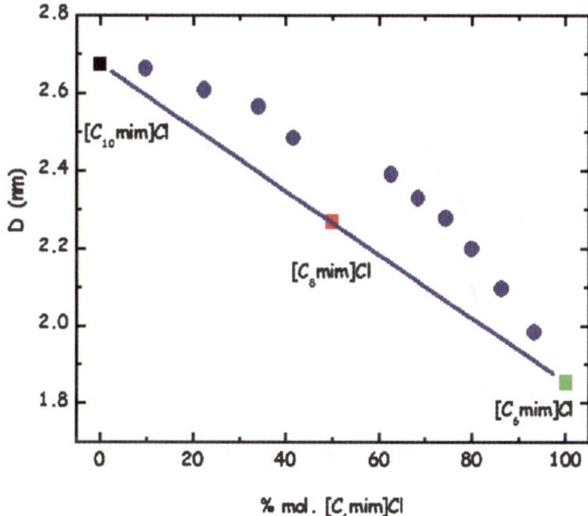


**Fig. 3** Dependence of the characteristic size *D* from the [$C_6$mim]Cl content in the binary mixture of [$C_6$mim]Cl and [$C_{10}$mim]Cl. Data are also presented for the two pure compounds ([$C_6$mim]Cl and [$C_{10}$mim]Cl) as well as for [$C_8$mim]Cl. The straight line describes the linear trend for *D vs. n* in the case of neat compounds ([$C_n$mim]Cl).

$[N_{1444}][Tf_2N]/[N_{1666}]$ $[Tf_2N]$ and $[N_{1666}][Tf_2N]/[N_{1888}][Tf_2N]$, Pott *et al.* did not observe significant deviations from the ideal behavior[24]). This experimental evidence supports recent simulation studies that focused on a related equimolar mixture, namely $[C_2mim][Tf_2N]/[C_6mim][Tf_2N]$, at ambient conditions.[53] In this computational study, Shimizu and coworkers observed that the ubiquitous polar/apolar nm-scale separation occurring in neat ILs, takes place also in such binary mixtures. They also find that short and long chains homogeneously mix forming the alkyl chains domains. Moreover their study indicates that the equimolar mixture has slightly different properties (such as interactions and morphology) than the effective system $[C_4mim][Tf_2N]$, in agreement with the differences that we observe in Fig. 3 between the experimental data and the ideal linear trend.

This study is a preliminary analysis of this kind of mixtures and now we plan to extend the study to mixtures where a larger difference exists between the alkyl chain lengths of the two components. However, we can already observe that a scenario that considers the low Q peak as the fingerprint of the shortest close neighbors between polar heads separated by extended alkyl chains,[54] would presumably imply the coexistence of two different characteristic sizes that correspond to neighbor either short or long chains (and hence two different low Q peaks centred at the positions of the two neat compounds), that is not the case.

Another pertinent computational study focused on $[C_4mim][C_8OSO_3]$, namely 1-octyl,3-methylimidazolium octyl-sulfate.[55] This system although a pure compound (*i.e.* not a binary mixture) is characterized by the coexistence of two ionic species that bear different alkyl chains (*i.e.* a butyl in the cation and an octyl in the anion). On the basis of their simulations, Dàvila *et al.* proposed that in this system nm-scale segregation of the alkyl chains occurs (our experimental data confirm this proposal, *vide infra*).[55] The clustering of the terminal methyl groups belonging either to the anion (identified as Cs) or to the cation (identified as Ct) was observed. Moreover the authors observed a clustering of the methyl groups belonging to both anion and cation (*i.e.* Cs-Ct clustering); this prompts for the formation of intimately mixed alkyl chains belonging to both cations and anions into apolar domains, in agreement with the previously reported findings for binary mixtures. In Fig. 4, we report a comparison between the experimental SAXS patterns from two alkyl-sulfate-based RTILs, namely ECOENG 218 and ECOENG 418, 1-ethyl-3-methylimidazolium octyl-sulfate and 1-butyl-3-methylimidazolium octyl-sulfate, respectively. Both salts bear an octyl chain linked to the sulfate group and differ for the alkyl side chain

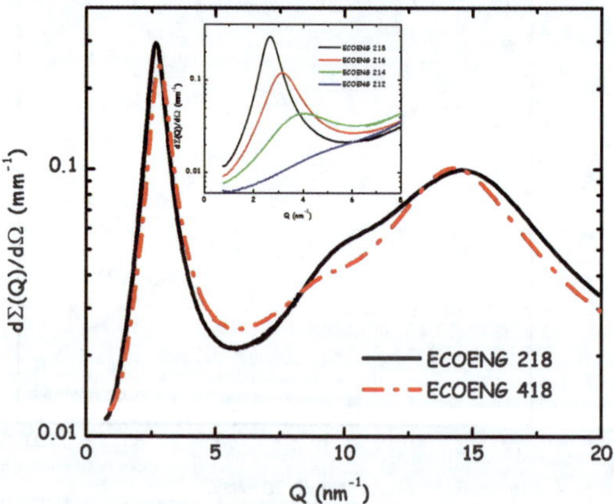


**Fig. 4** SWAXS data for 1-ethyl-3-methyl-imidazolium octyl-sulfate and 1-butyl-3-methyl-imidazolium octyl-sulfate (ECOENG 218 and ECOENG 418, respectively). In the inset the SWAXS data for 1-ethyl-3-methyl-imidazolium alkyl-sulfate, where alkyl = ethyl, butyl, hexyl and octyl (ECOENG 21*n*, with $n = 2, 4, 6, 8$, respectively) are presented.

 

in the cation that is either an ethyl or a butyl group, respectively. It is noteworthy that both salts show a low Q peak (at $Q < 5\ nm^{-1}$) and they are centred at approximately the same position (at $Q = 2.6\ nm^{-1}$). In the inset of Fig. 4 we also report the published data for the series ECOENG 21*n*, with $n = 2, 4, 6, 8$, corresponding to 1-ethyl,3-methyl-imidazolium alkyl-sulfate, where alkyl = ethyl, butyl, hexyl and octyl, respectively.[20] We already observed that also these salts are characterized by a well defined low Q feature, whose position linearly depends on the length of the (longer) *anion* alkyl chain. The comparison between ECOENG 218 and 418, whose low Q peak's position is only marginally affected by the *cation* alkyl chain length, indicates that only the *longer* chain is responsible for the size of the nano-segregated domains. Once more, confirming the results for the $[C_6mim]/[C_{10}mim]$ Cl mixtures, it appears that differently long alkyl chain tend to thoroughly mix when clustering and forming the non-polar domains.

Hardacre[30] proposed that the occurrence of the low Q peak does not demonstrate the existence of nanostructure beyond an immediate correlation between nearest neighbors. Such a point of view has been shared by Margulis and coworkers[40] (in a study on $[C_6mim]Cl$, $[C_8mim][PF_6]$ and $[C_{10}mim][PF_6]$) who stated that the low Q peak "exists because of the intrinsic anisotropy of the cation". The same authors agree in proposing the origin of the low Q peak as related to the existence of steric hinderance along given coordination directions (the ones involving the long alkyl tails) between neighbor cation molecules and the main conclusion of their works is that the low Q peaks are not due to long range morphologies. This interpretation has been further proposed in a study[56] by Castner and Margulis on $[N_{1444}][Tf_2N]$, although in the mentioned system only tiny indication of the existence of a low Q peak exists (as it was also shown by Pott *et al.*[24]) at the chosen instrumental conditions; coherently with this observation, the authors state that they find no evidence for any mesoscale ordering or interdigitated alkyl chains. In their paper the authors observe also that, given the relatively short alkyl chain length (butyl), the mentioned structural anisotropy is not found to play a major role, but is expected to do it when the chains are longer. In an almost simultaneous paper, Canongia Lopes and coworkers[41] simulated a wider set of related compounds (namely $[N_{1nnn}][Tf_2N]$, with $n = 4, 6, 8$) and, in agreement with Pott's experimental results, do find evidences for alkyl chains segregation in all their investigated systems as well as for strong structural correlation between the alkyl chains that was described as a sleeve-like wrapping of the alkyl chains around each chain.

In our opinion, while the mentioned anisotropy can play a minor role, on the other hand, we still believe that the occurrence of the low Q peak feature reflects the existence of a high degree of intermediate range order that is the direct consequence of the segregation of the alkyl chain, due to the strong difference in the nature of the interaction between polar and apolar portions in the IL. In order to test this proposal, we compared the SWAXS patterns from two relatively similar RTILs, namely $[C_6mim][Tf_2N]$[57] and $[C_1OC_2OC_1mim][Tf_2N]$, where the latter RTIL is 1-methoxyethoxymethyl-3-methylimidazolium bis triflamide (see Fig. 5). These RTILs share the same anion and polar head; however, they differ for the chemical nature of the side alkyl chain: the former salt bears an hexyl chain, while the latter IL bears a poly-ether chain; overall the two side chains are characterized by six groups ($CH_2$, O or $CH_3$). Moreover the O atoms bear the same number of electrons as the $CH_2$ group, so that the scattering from the polyether chain is similar to the one of the alkyl one. Above $0.8\ \text{Å}^{-1}$ (Fig. 5), the two diffraction patterns are rather similar, as they mostly reflect features that are related to the anion (that is the same for the two salts). However the two SWAXS patterns are dramatically different in the low Q portion, where the alkyl side chain salt is characterized by the occurrence of the well-known peak (centred at *ca.* $4\ nm^{-1}$), while the polyether side chain salt is essentially featureless in that Q range. This is an outstanding result and is, to our knowledge, the very first example of a long alkyl chain ($n > 4$) RTIL that does

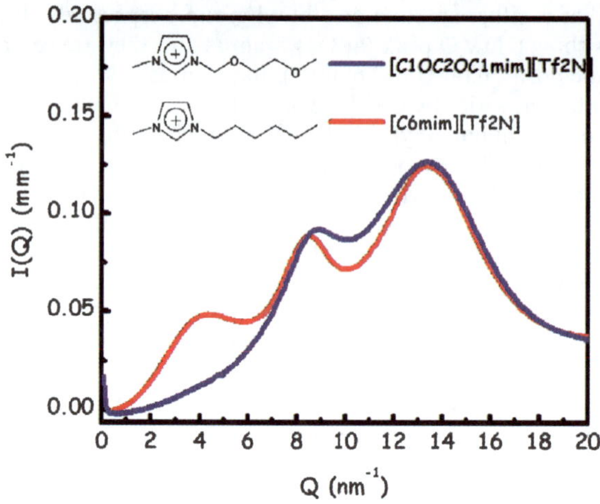


**Fig. 5** SWAXS data for 1-hexyl-3-methyl-imidazolium bis triflamide ($[C_6mim][Tf_2N]$) and 1-methoxyethoxymethyl-3-methylimidazolium bis triflamide ($[C_1OC_2OC_1mim][Tf_2N]$).

not show appreciable evidence of the occurrence of the low Q peak. Indirect indication of such a behavior has been recently reported by Shirota and coworkers on the basis of OHD-RIKES data on ammonium and phosphonium salts bearing either alkyl or poly-ether chains.[58] These authors observed that the ether chains lead to a reduction of the shear viscosity and this was rationalized considering the enhanced flexibility of the ether chains and proposing the possibility that no or limited segregation of the chain occurs when they are characterized by a polar nature comparable to the one of the polar head.

In Fig. 5 we report the first example of structural data supporting this hypothesis. Our data clearly show that the introduction of a polar chain, such as poly-ether in a IL, leads to the disappearance of the low Q peak that is instead found in the analogous IL bearing an apolar alkyl chain of the same length as the ether-like. We mention in passing that a manuscript is in preparation where additional experimental data will be reported further confirming these results.

This experimental evidence cannot, in our view, be rationalized in terms of the minimalist scenario proposed by Hardacre *et al.*[30] and supported by the work of Margulis *et al.*,[40] where the occurrence of specific structural correlations is just the consequence of steric hindrance. In the present experimental data set, the steric hindrance due to the ether chains is comparable to that of the alkyl chains, so one would expect to find comparable evidence of structural heterogeneities at the mesoscopic scale. The dramatic difference observed in the low Q portion of Fig. 5 prompts for a more active role played by the competition between long range, coulombic interactions and short range, dispersive ones in determining the overall morphology in this class of materials. Already in 2005, Voth mentioned that such a competition, together with the geometrical constraints imposed by the chemical bond between polar and apolar moieties, would lead to the formation of a "balanced liquid crystal-like structure".[7] Also in the works of Ribeiro[6] and Canongia-Lopes and Padua[8] the role of the different nature and spatial extent of the interactions involving either the polar parts or the aliphatic moieties was stressed as determining for the development of the mesoscopically heterogeneous morphology. In view of the present results we confirm the proposal that (apolar) alkyl chains segregation occurs in the bulk liquid state of these materials, thus building up domains that are kept separated from the (polar) charge network. When the difference in polarity between chains and charged parts decreases such a segregation is no longer effective and a more homogeneous morphology is found.

 

## Conclusion

In this contribution, we discuss the recent findings and proposals from several groups that are active in exploring the hot issue of the mesoscopic order in RTILs. Among the most original findings we discussed Hardacre's experimental work and Margulis' simulations. These authors provide experimental and computational evidence that prompt for a minimalistic role played by alkyl chains in determining characteristic correlation distances in the mesoscopic range. On the basis of new experimental results (among which we mention the comparison between salts bearing either an alkyl chain or a poly-ether one) we confirm the proposal that RTILs' IRO originates from the difference in the nature and spatial extent of the interactions of heads and tails. When a strong polar–apolar difference exists, this will eventually lead to the tail segregation; on the other hand when comparable polarity exists both in the heads and the tails, no evidence of such an IRO is found.

## Acknowledgements

We thank Prof. P. J. Dyson for kindly making available to us the sample of 1-methoxyethoxymethyl-3-methylimidazolium bis-triflamide. We acknowledge the European Synchrotron Radiation Facility for provision of synchrotron radiation facilities and would like to thank Dr E. Di Cola, M. F. Martinez and T. Narayanan for their kind and competent assistance in exploiting beamline ID02. A.T. acknowledges support from the FIRB "Futuro in Ricerca" research project (RBFR086BOQ_001, "Structure and dynamics of ionic liquids and their binary mixtures"). We also acknowledge the support of the University of Rome "Sapienza" (Progetto Ateneo 2010: Protic Ionic Liquids). Useful discussions with Prof. E. W. Castner and H. Shirota are acknowledged.

## References


1 E. W. Castner and J. F. Wishart, *J. Chem. Phys.*, 2010, **132**, 120901.
2 H. Weingaertner, *Angew. Chem., Int. Ed.*, 2008, **47**, 654–670.
3 R. D. Rogers and G. A. Voth, *Acc. Chem. Res.*, 2007, **40**, 1077–1078.
4 J. F. Wishart and E. W. Castner Jr, *J. Phys. Chem. B*, 2007, **111**, 4639–4640.
5 C. Chiappe, *Monatsh. Chem.*, 2007, **138**, 1035–1043.
6 S. M. Urahata and M. C. C. Ribeiro, *J. Chem. Phys.*, 2004, **120**, 1855–1863.
7 Y. T. Wang and G. A. Voth, *J. Am. Chem. Soc.*, 2005, **127**, 12192.
8 J. N. Canongia Lopes and A. A. H. Padua, *J. Phys. Chem. B*, 2006, **110**, 3330.
9 H.-O. Hamaguchi and R. Ozawa, in *ADVANCES IN CHEMICAL PHYSICS, VOL 131*, 2005, vol. 131, *pp. 85–104*.
10 K. Iwata, H. Okajima, S. Saha and H.-O. Hamaguchi, *Acc. Chem. Res.*, 2007, **40**, 1174–1181.
11 D. Xiao, J. R. Rajian, A. Cady, S. Li, R. A. Bartsch and E. L. Quitevis, *J. Phys. Chem. B*, 2007, **111**, 4669–4677.
12 D. Xiao, J. R. Rajian, S. Li, R. A. Bartsch and E. L. Quitevis, *J. Phys. Chem. B*, 2006, **110**, 16174–16178.
13 H. Tokuda, K. Hayamizu, K. Ishii, A. Bin, H. Susan and M. Watanabe, *J. Phys. Chem. B*, 2005, **109**, 6103–6110.
14 D. a. Turton, J. Hunger, A. Stoppa, G. Hefter, A. Thoman, M. Walther, R. Buchner and K. Wynne, *J. Am. Chem. Soc.*, 2009, **131**, 11140–6.
15 A. Triolo, O. Russina, H.-J. Bleif and E. Di Cola, *J. Phys. Chem. B*, 2007, **111**, 4641–4.
16 A. Triolo, O. Russina, B. Fazio, R. Triolo and E. D. Cola, *Chem. Phys. Lett.*, 2008, **457**, 362–365.
17 A. Triolo, O. Russina, B. Fazio, G. B. Appetecchi, M. Carewska and S. Passerini, *J. Chem. Phys.*, 2009, **130**, 164521.
18 O. Russina, A. Triolo, L. Gontrani, R. Caminiti, D. Xiao, L. G. Hines Jr, R. a. Bartsch, E. L. Quitevis, N. Pleckhova and K. R. Seddon, *J. Phys.: Condens. Matter*, 2009, **21**, 424121.
19 D. Xiao, L. G. Hines, S. Li, R. A. Bartsch, E. L. Quitevis, O. Russina and A. Triolo, *J. Phys. Chem. B*, 2009, **113**, 6426–33.

20 O. Russina, L. Gontrani, B. Fazio, D. Lombardo, A. Triolo and R. Caminiti, *Chem. Phys. Lett.*, 2010, **493**, 259–262.
21 E. Bodo, L. Gontrani, R. Caminiti, N. V. Plechkova, K. R. Seddon and A. Triolo, *J. Phys. Chem. B*, 2010, **114**, 16398–16407.
22 E. Bodo, L. Gontrani, a. Triolo and R. Caminiti, *J. Phys. Chem. Lett.*, 2010, **1**, 1095–1100.
23 B. Aoun, A. Goldbach, M. a. Gonzaćlez, S. Kohara, D. L. Price and M.-L. Saboungi, *J. Chem. Phys.*, 2011, **134**, 104509.
24 T. Pott and P. Méléard, *Phys. Chem. Chem. Phys.*, 2009, **11**, 5469–75.
25 T. L. Greaves, D. F. Kennedy, S. T. Mudie and C. J. Drummond, *J. Phys. Chem. B*, 2010, **114**, 10022–10031.
26 T. L. Greaves, D. F. Kennedy, A. Weerawardena, N. M. K. Tse, N. Kirby and C. J. Drummond, *J. Phys. Chem. B*, 2011, **115**, 2055–2066.
27 C. S. Santos, H. V. R. Annapureddy, N. S. Murthy, H. K. Kashyap, E. W. Castner and C. J. Margulis, *J. Chem. Phys.*, 2011, **134**, 064501.
28 C. S. Santos, N. S. Murthy, G. a. Baker and E. W. Castner, *J. Chem. Phys.*, 2011, **134**, 121101.
29 M. Macchiagodena, L. Gontrani, F. Ramondo, A. Triolo and R. Caminiti, *J. Chem. Phys.*, 2011, **134**, 114521.
30 C. Hardacre, J. D. Holbrey, C. L. Mullan, T. G. a. Youngs and D. T. Bowron, *J. Chem. Phys.*, 2010, **133**, 074510.
31 R. Atkin and G. G. Warr, *J. Phys. Chem. B*, 2008, **112**, 4164.
32 R. Hayes, S. Imberti, G. G. Warr and R. Atkin, *Phys. Chem. Chem. Phys.*, 2011, **13**, 3237–47.
33 K. Fujii, S. Seki, S. Fukuda, T. Takamuku, S. Kohara, Y. Kameda, Y. Umebayashi and S. ichi Ishiguro, *J. Mol. Liq.*, 2008, **143**, 64–69.
34 S.-i I. Y. Umebayashi, W.-L. Chung, T. Mitsugi, S. Fukuda, M. Takeui, K. Fujii, T. Takamuku and R. Kanzaki, *J. Comput. Chem., Jpn.*, 2008, **7**, 125–134.
35 T. Shimomura, K. Fujii and T. Takamuku, *Phys. Chem. Chem. Phys.*, 2010, **12**, 12316–24.
36 B. L. Bhargava, S. Balasubramanian and M. L. Klein, *Chem. Commun.*, 2008, 3339–3351.
37 B. L. Bhargava, R. Devane, M. L. Klein and S. Balasubramanian, *Soft Matter*, 2007, **3**, 1395–1400.
38 A. Seduraman, M. Klahn, and P. Wu, *CALPHAD-COMPUTER COUPLING OF PHASE DIAGRAMS AND THERMOCHEMISTRY*, 2009, 33, 605–613.
39 A. a H. Pádua, M. F. Costa Gomes and J. N. a. Canongia Lopes, *Acc. Chem. Res.*, 2007, **40**, 1087–96.
40 H. V. R. Annapureddy, H. K. Kashyap, P. M. De Biase and C. J. Margulis, *J. Phys. Chem. B*, 2010, **114**, 16838–46.
41 K. Shimizu, A. A. H. Paćdua and J. N. Canongia Lopes, *J. Phys. Chem. B*, 2010, **114**, 15635–15641.
42 L. Pison, J. N. Canongia Lopes, L. P. N. Rebelo, a. a. H. Padua and M. F. Costa Gomes, *J. Phys. Chem. B*, 2008, **112**, 12394–400.
43 L. Gontrani, O. Russina, F. L. Celso, R. Caminiti, G. Annat and A. Triolo, *J. Phys. Chem. B*, 2009, **113**, 9235–40.
44 J. De Roche, C. M. Gordon, C. T. Imrie, M. D. Ingram, A. R. Kennedy, F. Lo Celso and A. Triolo, *Chem. Mater.*, 2003, **15**, 3089–3097.
45 A. E. Bradley, C. Hardacre, J. D. Holbrey, S. Johnston, S. E. J. McMath and M. Nieuwenhuyzen, *Chem. Mater.*, 2002, **14**, 629–635.
46 C. Hardacre, J. D. Holbrey, P. B. McCormac, S. E. J. McMath, M. Nieuwenhuyzen and K. R. Seddon, *J. Mater. Chem.*, 2001, **11**, 346–350.
47 C. Hardacre, J. D. Holbrey, S. E. J. McMath, D. T. Bowron and A. K. Soper, *J. Chem. Phys.*, 2003, **118**, 273.
48 B. Aoun, A. Goldbach, S. Kohara, J.-F. Wax, M. A. Gonzalez and M.-L. Saboungi, *J. Phys. Chem. B*, 2010, **114**, 12623–12628.
49 C. M. Gordon, J. D. Holbrey, A. R. Kennedy and K. R. Seddon, *J. Mater. Chem.*, 1998, **8**, 2627–2636.
50 M. Tomsic, A. Jamnik, G. Fritz-Popovski, O. Glatter and L. Vlcek, *J. Phys. Chem. B*, 2007, **111**, 1738–51.
51 Z. Fei, W. H. Ang, D. Zhao, R. Scopelliti, E. E. Zvereva, S. A. Katsyuba and P. J. Dyson, *J. Phys. Chem. B*, 2007, **111**, 10095–10108.
52 T. Narayanan, O. Diat and P. Boesecke, *Nucl. Instrum. Methods Phys. Res., Sect. A*, 2001, **467**, 1005.
53 K. Shimizu, M. F. Costa Gomes, A. a H. Pádua, L. P. N. Rebelo and J. N. Canongia Lopes, *THEOCHEM*, 2010, **946**, 70–76.
54 C. Hardacre, J. D. Holbrey, C. L. Mullan, T. G. a. Youngs and D. T. Bowron, *J. Chem. Phys.*, 2010, **133**, 074510.


 

55 M. J. Dàvila, S. Aparicio, R. Alcalde, B. Garcìa and J. M. Leal, *Green Chem.*, 2007, **9**, 221.
56 C. S. Santos, H. V. R. Annapureddy, N. S. Murthy, H. K. Kashyap, E. W. Castner and C. J. Margulis, *J. Chem. Phys.*, 2011, **134**, 064501.
57 O. Russina, M. Beiner, C. Pappas, M. Russina, V. Arrighi, T. Unruh, C. L. Mullan, C. Hardacre and A. Triolo, *J. Phys. Chem. B*, 2009, **113**, 8469–8474.
58 H. Shirota, H. Fukazawa, T. Fujisawa and J. F. Wishart, *J. Phys. Chem. B*, 2010, **114**, 9400–12.

# Ionic liquids studied across different scales: A computational perspective

**Katharina Wendler,[a] Florian Dommert,[b] Yuan Yuan Zhao,[c] Robert Berger,[c] Christian Holm[b] and Luigi Delle Site*[a]**

*Received 30th March 2011, Accepted 4th July 2011*
**DOI: 10.1039/c1fd00051a**

For theoreticians, ionic liquids represent a major challenge. This is due to the fact that intermolecular interactions are particularly strong because of ionic liquids' ionicity. This, in turn, causes a subtle interplay between different scales which is encoded in the measured macro- and mesoscopic properties and also in the molecular electrostatic characteristics. Therefore, force fields have to describe the microscopic processes correctly in order to reproduce macroscopic properties accurately over a large range of state variables. Herein, imidazolium-based ionic liquids were studied at different scales, going from the detailed quantum electronic scale to the classical atomistic scale. It is indicated how the information gained at each level could be used for the other scales. In particular, the issue of deriving suitable partial charges for use in classical force fields is addressed. The Blöchl method was employed to generate partial charges reproducing the multipole distribution accurately for bulk systems. This led naturally to absolute ionic charges of less than $|1\ e|$, *i.e.* charge scaling. So, the monopole structure of the herein introduced force field mimics the quantum chemical behaviour observed in the liquid phase. This led to a substantial improvement in the description of dynamical properties of immediate experimental interest, such as electric conductivity. For further insight, the electric dipole moment of the ions was taken as physical indicator of their electronic structure. The electric dipole moment was found to fluctuate strongly and to depend on polarisation. Hence, our scale-combined study offers a gateway to rational design of models, based on the relevant underlying physics rather than on mere numerical parameterisation, and thereby to (possibly) more direct physical interpretation of experimental results.

## 1 Introduction

Ionic liquids (ILs) have sparked widespread interest due to their peculiar properties and the resulting possibility of manifold applications.[1–10] Given this popularity, there is an increasing call for further investigations. In this context, theoretical studies are particularly needed for the interpretation and prediction of experimental results.[11–17] A powerful theoretical tool of study, at molecular as well as statistical (thermodynamical) level, is computer simulation. Especially, classical molecular dynamics (MD) simulations have the potential to allow for a basic molecular understanding of the dynamical processes and yet describe the essential overall behaviour as

[a]*Max Planck Institute for Polymer Research, Ackermannweg 10, D-55128 Mainz, Germany. E-mail: dellsite@mpip-mainz.mpg.de; Fax: +49 6131 379 340; Tel: +49 6131 379 328*
[b]*Institut für Computerphysik, Universität Stuttgart, Pfaffenwaldring 27, D-70569 Stuttgart, Germany*
[c]*TU Darmstadt, Petersenstr. 22, D-64287 Darmstadt, Germany*

 

a liquid. Therefore, several MD studies elucidated the statics and dynamics of ILs.[18–44] They were restricted, however, to few species of ILs due to the lack of reliable force fields. As the quality of the force fields determines the accuracy of the MD studies, this is the most challenging aspect in the MD approach to ILs at the moment. Many force fields in current literature have problems in reproducing experimental data for both dynamic and static properties and often can describe only one aspect at the price of sacrificing the other.[17,25,45] This latter aspect can be interpreted as an indication that the basic physics of ILs involves several scales which are strongly connected and need to be explicitly taken into account in some way. From this perspective, a systematic multiscale study[46] analysing the microscopic as well as the mesoscopic scale is needed to identify the key principles to then consistently incorporate them in a model with predictive power.

In general, liquid systems are characterised by the interplay of different physical and chemical interactions which in ionic liquids have been identified with Coulomb[47–50] and van der Waals interactions[47,51–54] as well as hydrogen bonds.[23,38,46,53–58] Coulomb interactions are strong and long-ranged whereas van der Waals forces and hydrogen bonds are short-ranged and weaker. On the one side, the Coulomb interactions dominate in simple salts like sodium chloride in which van der Waals interactions are negligible in comparison; moreover, monoatomic ions do not possess any internal vibrational or rotational degrees of freedom. On the other side, in molecular solvents with neutral building blocks, like water for instance, van der Waals forces and hydrogen bonding are dominant. All three kinds of interactions play a crucial role in ILs and, indeed, this might be the reason for their unique properties and for the difficulties in developing force fields which can describe this subtle balance. The interplay of such interactions results in the structural properties of the liquid and in the electrostatic properties of the ions. This means that if a consistent description on all scales is to be reached, it is important to transfer the electrostatic and the structural properties consistently from one scale to another. This is the main intention of this work, *i.e.* to sketch a schematic protocol to study ILs at different scales. By using a bottom-up approach, the information gained at each scale shall be transferred to the next scale in a consistent way. The scales studied here are the quantum scale where electrons are considered explicitly and the classical scale in which each molecule carries its explicit chemical identity, but is regarded as a classical object. Quantum chemical calculations can treat only relatively small systems, but at an electronic level. Instead, rather large systems can be described by classical MD, but at the price of a less accurate, parameterised molecular description. Thus, it is clear why consistency between these two approaches is an ideal compromise between feasibility and accuracy. The quantum description in this work was accessed on two different levels: post-Hartree–Fock (post-HF) approaches and density functional theory (DFT) approaches. While the former can provide explicit, systematically improvable information on the electronic level, the latter are more approximate in practice, but have the potential to give a robust description of the electron density. Given the accuracy of post-HF calculations, these can be taken to gauge the quality of the DFT calculations. In particular, for a series of imidazolium-based ILs, the focus was on the calculation of two quantities that are particularly useful to modelling: atomic partial charges and molecular electric dipole moments. In fact, classically, the electrostatic properties can be expressed by the multipole moments. Thus, classical force field parameters should represent the electric monopole and dipole properties found in quantum chemical calculations which describe the electronic structure explicitly. The monopoles are given by the partial charges in a force field, while the electric dipole moment is a more complicated quantity. It is a physical indicator for charge displacement as well as local packing. So, the interplay of charge and shape of Coulomb forces, van der Waals forces, and hydrogen bonding is reflected in the electric dipole moment and its correlations.

In our previous work on 1,3-dimethyl-imidazolium chloride, [MMIM][Cl], we found that the electrostatic properties of isolated ions in the gas phase deviated from the ones of ions in small neutral gas phase clusters or in the liquid phase.[59,60] Following this indication, in section 3.1, the study of small clusters was extended to evaluate the transition from isolated ions to the liquid phase progressing also *via* charged clusters. To anticipate some of our results, the net absolute charges of the ions were found to be consistently smaller than |1 *e*| and show considerable and systematic dependence on the structure of the clusters and ion arrangements in the liquid, leading to fluctuations in the dynamical simulations. So, the electric dipole moment distribution of ions in the liquid phase was observed to be very broad for several different ILs.[60,61] The large dynamical fluctuations in the electrostatic molecular properties, observed to be connected to a certain locality[61,62] and polarisation (see section 3.2), seemed to be a general characteristic of ILs. The electric dipole moment distribution in liquid phase was shown to not depend heavily on temperature (within a range of 100 K), simulation times, or system sizes (see section 3.2). This leads to the interesting conclusion that the partial charge parameterisation appeared already to be possible using small liquid phase systems and short simulation times. Furthermore, some transferability across the range of temperature is suggested. Different force field models for ILs with a reduced net charge have already been published,[18,26,29,31,40,63] but the scaling factors were either set heuristically or determined from gas phase calculations. In contrast, our partial charges were derived by a Blöchl analysis from bulk-like DFT simulations, and hence termed Blöchl charges.[62] Recently, in the context of parameterising force fields for non-polar solvents, a charge scaling was justified on the basis of including electron polarizability with the help of an electronic dielectric constant, $\varepsilon_{el}$, within the MDEC theory of Leontyev and Stuchebrukhov.[64,65] In contrast to this MDEC method, the effective dipole of the molecule are not adjusted here and only the Lennard-Jones interactions remain for the final parameterisation of the force field. The new force field for [MMIM][Cl] is shown to describe several static and dynamic properties correctly. Additionally, Blöchl partial charges are reported for 1-ethyl-3-methyl-imidazolium chloride, [EMIM][Cl], and 1-ethyl-3-methyl-imidazolium dicyanoamide, [EMIM][DCA], and compared to the ones of [MMIM][Cl][46,62] (see section 3.2). The results indicate a certain transferability for the partial charges as well.

## 2 Computational details

### Post-Hartree–Fock calculations

Equilibrium structures of gas phase clusters were obtained by applying second-order Møller–Plesset perturbation theory (MP2) as implemented in the Turbomole program[66,67] within the resolution of the identity (RI) approximation. The basis set employed was an augmented correlation consistent polarised valence triple zeta set (aug-cc-pVTZ) for all atoms. All valence electrons were included in the electron correlation treatment. Harmonic vibrational frequencies were determined by using Turbomole. Coupled cluster single point energies on the CCSD(T)-F12a level (explicitly electron correlating coupled cluster with singles and doubles and non-iterative triples) were calculated with the program Molpro.[68] Partial charges were determined *via* the CHELPG[69] and RESP[70] protocols as well as with the help of the Bader analysis[71] using the programs Gaussian[72] and Bader.[73–75]

### Density functional theory

Static properties of isolated clusters in the gas phase were studied with DFT and compared to results obtained with post-HF methods. As the focus was on the electronic properties, MP2 and DFT studies were performed on exactly the same cluster structures that were obtained by energy optimization of the structures on the MP2 level (see above). Moreover, Car–Parrinello molecular dynamics[76] (CPMD) simulations were performed to study liquid phase properties of several ILs.

The DFT wavefunction optimisation was done for the eleven clusters by using as a reference the MP2 optimised structure. Technically, a general gradient-corrected DFT setup was used within the CPMD package[77] together with, in one case, the Perdew–Burke–Ernzerhof (PBE) functional[78] with norm-conserving Troullier–Martins type pseudopotentials[79] which were derived using the PBE functional.[78] In the other case, the gradient-corrected Becke-Lee–Yang–Parr (BLYP) functional,[80,81] corresponding pseudopotentials, and an empirical dispersion correction[82] were used for the DFT wavefunction optimisation of the MP2 optimised structures. This setup is denoted as BLYP-D. The electron density localisation *via* Wannier analysis[83–85] was carried out on the optimised wavefunctions as implemented in CPMD. The Blöchl analysis[86] was done by using an Gaussian plane wave (GPW) approach[87] with the PBE functional[78,88] and TZV2P basis sets and pseudopotentials in the Goedecker-Teter-Hutter format[89,90] as implemented in the CP2K package.[91] The charges from electrostatic potentials (CHELPG) analysis[92] were calculated by Gaussian[72] using a PBE functional and aug-cc-pVTZ basis sets for all atoms.

Several *ab initio* molecular dynamics (AIMD) simulations of the liquid phase were performed by using the CPMD approach.[76] In all simulations, the Kohn–Sham orbitals were expanded in a plane wave basis with a kinetic energy cutoff of 35 $E_h$ (70 Ry) and the fictitious electron mass was 400 $m_e$. Periodic boundary conditions, a timestep of 0.1 fs and standard Nose-Hoover thermostats[93,94] were applied. The simulations were started from a snapshot of a classical MD simulation and the equilibration time was chosen to be five ps. More details are summarised below:

(1) 8 ion pairs [MMIM][Cl],[61] 425 K, cubic box of (11.62 Å)$^3$, 84.66 ps, PBE functional and corresponding pseudopotentials,

(2) 30 ion pairs [MMIM][Cl],[60] 425 K, cubic box of (18.05 Å)$^3$, 20.00 ps, PBE functional and corresponding pseudopotentials,

(3) 8 ion pairs [EMIM][Cl], 400 K, cubic box of (12.15 Å)$^3$, 110.9 ps, PBE functional and corresponding pseudopotentials,

(4) 30 ion pairs [EMIM][DCA],[61] 400 K, cubic box of (20.37 Å)$^3$, 48.85 ps, BLYP functional[80,81] and corresponding pseudopotentials with empirical dispersion correction,[82] and

(5) 30 ion pairs [EMIM][DCA], 333 K, cubic box of (20.11 Å)$^3$, 21.18 ps, BLYP functional and corresponding pseudopotentials with empirical dispersion correction.[82]

In all five simulations, the electric dipole moments were calculated using the maximally localised Wannier analysis every 500th timestep. For the ionic dipole moment calculation (see Appendix), the geometric center of the imidazolium ring (see Fig. 1) is chosen as reference center for imidazolium cations. The geometric center of all atoms is selected for all other ions. For 100 snapshots per system, the Blöchl analysis was carried out by using the same functionals and TZV2P basis sets within CP2K.

### Classical molecular dynamics

Classical molecular dynamics simulations have been performed with the GROMACS simulation package.[95–97] Canonical *NpT* simulations with a time step of 1 fs were used applying a Nose-Hoover thermostat[93] in combination with a Parrinello-Rahman barostat.[98] Hydrogen bonds were kept fixed *via* the holonomic constraints that were solved using P–LINCS.[99] Electrostatic interactions were treated with the smooth Particle Mesh Ewald method.[100] They were optimised to an accuracy of $10^{-4}$ kJ mol$^{-1}$ nm$^{-1}$.[101] Short range interactions were cut off at 1.7 nm and a dispersion correction for the energy and pressure was applied.[102] For the optimisation and validation of the force field parameters, 239 [MMIM][Cl] ion pairs were used.

In this paper, the various atoms of the imidazolium cations are numbered as shown in Fig. 1.

 

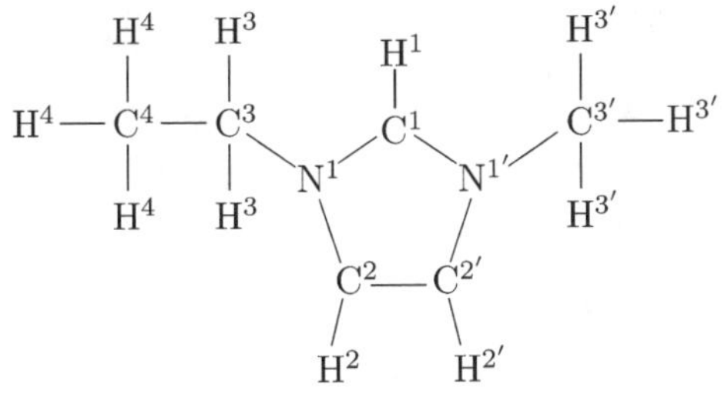

Fig. 1 Numbering of the atoms for $[EMIM]^+$.

## 3 Elucidating the relevant physics: electrostatic properties

When going from an isolated ion to clusters to the liquid phase, one increases the degree of complexity as the number of possible interactions between building blocks increases and cooperative effects may emerge. Indeed, the interactions present in clusters and in the liquid phase lead to a quantitative change of the electrostatic properties of the ions compared to isolated ions.[46,59,60,103] The study of clusters of increasing size is one possibility to elucidate how this change emerges from the ions' interactions. This may allow us to draw conclusions on the underlying physics. Furthermore, such small clusters can also be studied by accurate post-HF methods. This allows gauging if more approximate methods, such as DFT commonly employed in electronic structure based molecular dynamics calculations, are capable of getting the underlying physics quantitatively or at least qualitatively right. So, in the following section 3.1, small clusters, charged as well as uncharged, were examined by post-HF and DFT methods. The focus is specifically on the total electric dipole moments of the clusters as well as the partial charges of their ionic building blocks. Although partial charges and dipole moments of charged systems are not uniquely defined and depend on the protocol chosen for their computation, they are nevertheless important ingredients for judging differences between computational methods and can, as argued below, also be considered as reasonable indicators for probing the underlying physics. The results are discussed and compared to findings of liquid phase AIMD simulations in section 3.2.

### 3.1 Clusters in gas phase

In this work, several clusters (see Fig. 2) were studied at 0 K with MP2 and DFT methods:

- the three negatively charged clusters 1C2A of one cation and two anions $[MMIM][Cl]^-_2$ derived from systematic addition of an additional chloride to the previously found four structures of one ion pair (1C1A),[59,62]
- four energetically low-lying out of a total of 11 positively charged clusters 2C1A of two cations and one anion $[MMIM]_2[Cl]^+$ as obtained from combining various structural motifs of 1C1A and
- four selected neutral clusters 2A2C of two ion pairs $[MMIM]_2[Cl]_2$.

In earlier work on 1, 2, 4, and 8 ion pair clusters,[59,62] a huge variation in the ionic electric dipole moments was observed.[59] But the average electric dipole moments were in the range observed for the liquid phase.[60,103] Hence, a relatively small number of ions seemed sufficient to capture liquid-like charge distributions.

**Total energies and electric dipole moment of clusters.** The relative total energies and total dipole moments of the clusters found by the various methods agreed in the expected range. Thus, the overall electron distribution was found to be qualitatively similar by all methods. The relative energies obtained with MP2 were in good agreement with the CCSD(T)-F12a reference data (see Table 1). Further, if an empirical

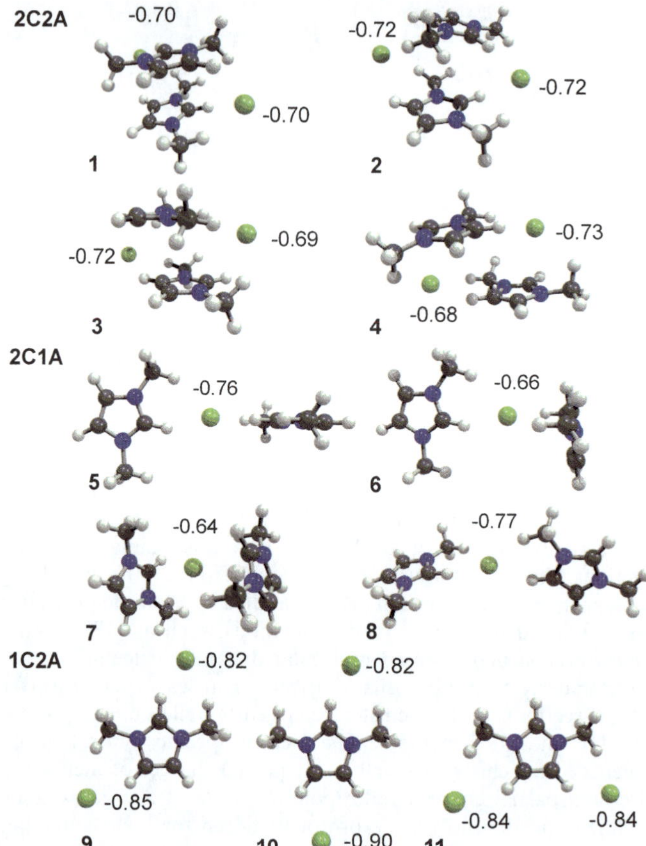


**Fig. 2** Cluster structures (C black, N blue, H white, Cl green) together with partial charges (in *e*) for chloride obtained *via* the CHELPG protocol with PBE/aug-cc-pVTZ (see Table 3).

dispersion correction was included in the DFT calculation, the energetic order of the clusters was the same as in the post-HF data. The worse performance of a pure DFT functional as PBE without dispersion correction pointed to the essential influence of van der Waals interactions in clusters already reported.[51,52] The total electric dipole moment of the clusters was calculated with respect to the geometric center of the clusters by MP2 and by the Wannier analysis as well as the Blöchl charges. The DFT based Wannier centers and Blöchl charges led to lower total electric dipole moments of the neutral clusters and of positively charged clusters compared to MP2. For the negatively charged clusters, the absolute deviations between DFT methods and MP2 were the highest. So, the electric dipole moments of negatively charged clusters were overestimated with a root mean square deviation of 1 D (11%) by the Wannier analysis and of 0.60 D (7%) by the Blöchl analysis. Altogether, the Wannier and Blöchl analysis performed similarly well as they were both based on the same DFT approach.

**Net charge of anions in clusters.** In a molecular picture, the absolute anion charge in the present study was always found to be less than |1 *e*| by the various methods used (see Table 2 and 3). For 1C1A, it was shown previously[62] that MP2/aug-cc-pVDZ results were in good agreement with CCSD/aug-cc-pVDZ data (see also Table 2). The basis set dependence was mild when going from MP2/aug-cc-pVDZ to MP2/aug-cc-pVTZ. This provided some justification for gauging the performance

 

**Table 1** Comparison of total energies (in kJ mol$^{-1}$) of clusters relative to the cluster with the lowest total energy. The total energies were calculated for MP2/aug-cc-pVTZ optimized structures by using CCSD(T)-F12a/aug-cc-pVDZ, MP2/aug-cc-pVTZ, PBE/plane wave basis set, and BLYP/plane wave basis set with an empirical dispersion correction.[82]

| Cluster | CCSD(T)-F12a | MP2 | PBE | BLYP-D |
|---|---|---|---|---|
| 2C2A | | | | |
| 1 | 0.00 | 0.00 | 0.00 | 0.00 |
| 2 | 7.76 | 11.1 | −0.13 | 11.9 |
| 3 | 7.71 | 7.91 | 5.71 | 6.57 |
| 4 | 0.86 | 3.12 | −5.12 | 0.10 |
| 2C1A | | | | |
| 5 | — | 0.00 | 0.00 | 0.00 |
| 6 | — | 5.32 | 16.7 | 9.02 |
| 7 | — | 12.6 | 34.3 | 19.2 |
| 8 | — | 15.5 | 17.7 | 19.9 |
| 1C2A | | | | |
| 9 | 0.00 | 0.00 | 0.00 | 0.00 |
| 10 | 20.2 | 21.3 | 26.5 | 24.5 |
| 11 | 36.5 | 31.4 | 29.6 | 29.1 |

of DFT methods by comparing them with the corresponding MP2/aug-cc-pVTZ data. Overall, the charge reduction was slightly overemphasized with (pure) density functionals such as PBE, as was noted for one ion pair previously[62] and was also

**Table 2** Comparison of partial charges (in *e*) of chloride ions obtained for one ion pair clusters 1C1A with four different setups of the Bader, RESP, CHELPG and Blöchl analysis: PBE/aug-cc-pVTZ (PBE/TZV2P for Blöchl), MP2/aug-cc-pVTZ, MP2/aug-cc-pVDZ, and CCSD/aug-cc-pVDZ. The values were taken from ref. 46 and 62 for comparison, except for the CCSD/aug-cc-pVDZ and MP2/aug-cc-pVDZ Bader charges and the CCSD/aug-cc-pVDZ RESP charges.

| Cluster | Method | Bader | RESP | CHELPG | Blöchl |
|---|---|---|---|---|---|
| 1C1A | | | | | |
| 12 | PBE/aug-cc-pVTZ | −0.80 | −0.70 | −0.70 | −0.68 |
| | MP2/aug-cc-pVTZ | −0.85 | −0.74 | −0.74 | — |
| | MP2/aug-cc-pVDZ | −0.85 | −0.75 | −0.75 | — |
| | CCSD/aug-cc-pVDZ | −0.85 | −0.75 | −0.75 | — |
| 13 | PBE/aug-cc-pVTZ | −0.72 | −0.65 | −0.64 | −0.63 |
| | MP2/aug-cc-pVTZ | −0.81 | −0.72 | −0.71 | — |
| | MP2/aug-cc-pVDZ | −0.80 | −0.73[a] | −0.72[a] | — |
| | CCSD/aug-cc-pVDZ | −0.81 | −0.73 | −0.72 | — |
| 14 | PBE/aug-cc-pVTZ | −0.86 | −0.80 | −0.80 | −0.80 |
| | MP2/aug-cc-pVTZ | −0.91 | −0.85 | −0.85 | — |
| | MP2/aug-cc-pVDZ | −0.91 | −0.86[a] | −0.85 | — |
| | CCSD/aug-cc-pVDZ | −0.91 | −0.86 | −0.86 | — |
| 15 | PBE/aug-cc-pVTZ | −0.81 | −0.74 | −0.73 | −0.73 |
| | MP2/aug-cc-pVTZ | −0.86 | −0.79 | −0.78 | — |
| | MP2/aug-cc-pVDZ | −0.86 | −0.79 | −0.78 | — |
| | CCSD/aug-cc-pVDZ | −0.86 | −0.80 | −0.79 | — |

[a] Corrected value compared to ref. 46.

**Table 3** Comparison of partial charges (in *e*) of the chloride ions found by Bader, RESP, CHELPG and Blöchl analysis on the PBE level with the aug-cc-pVTZ basis set (Bader, RESP, CHELPG) and TZV2P (Blöchl). Values in italic correspond to MP2/aug-cc-pVTZ benchmark results.

| | Bader | | RESP | | CHELPG | | Blöchl | |
|---|---|---|---|---|---|---|---|---|
| Cluster | Cl1 | Cl2 | Cl1 | Cl2 | Cl1 | Cl2 | Cl1 | Cl2 |
| 2C2A | | | | | | | | |
| 1 | −0.80 | −0.80 | −0.71 | −0.71 | −0.70 | −0.70 | −0.63 | −0.63 |
| 2 | −0.79 | −0.79 | −0.75 | −0.75 | −0.72 | −0.72 | −0.66 | −0.66 |
| 3 | −0.79 | −0.81 | −0.71 | −0.74 | −0.69 | −0.72 | −0.62 | −0.66 |
| 4 | −0.81 | −0.80 | −0.74 | −0.69 | −0.73 | −0.68 | −0.66 | −0.58 |
| 2C1A | | | | | | | | |
| 5 | −0.81 | — | −0.82 | — | −0.76 | — | −0.78 | — |
| 6 | −0.79 | — | −0.68 | — | −0.66 | — | −0.55 | — |
| 7 | −0.78 | — | −0.66 | — | −0.64 | — | −0.56 | — |
| 8 | −0.81 | — | −0.85 | — | −0.77 | — | −0.79 | — |
| 1C2A | | | | | | | | |
| 9 | −0.87 | −0.89 | −0.82 | −0.85 | −0.82 | −0.85 | −0.75 | −0.78 |
| | *−0.90* | *−0.92* | *−0.85* | *−0.88* | *−0.85* | *−0.87* | — | — |
| 10 | −0.87 | −0.92 | −0.82 | −0.89 | −0.82 | −0.90 | −0.75 | −0.86 |
| | *−0.90* | *−0.94* | *−0.85* | *−0.91* | *−0.85* | *−0.92* | — | — |
| 11 | −0.87 | −0.87 | −0.84 | −0.84 | −0.84 | −0.84 | −0.78 | −0.78 |
| | *−0.90* | *−0.90* | *−0.87* | *−0.87* | *−0.86* | *−0.86* | — | — |

confirmed herein for the charged 1C2A clusters (see Table 3). While RESP, CHELPG and Blöchl provided almost identical partial charges on chloride in one ion pairs 1C1A, they deviated for the charged clusters and the two ion pair structures. The Blöchl method yielded typically a significantly smaller absolute value of the chloride partial charge than the other methods. The Bader analysis, instead, gave rise to typically larger absolute values of the chloride partial charge in the various clusters and showed a less pronounced dependence on structural changes. The other methods indicated reduced absolute values of the anion's charge specifically for structures with (partial) cation–anion π-complex character (*e.g.* 6, 7, 13) as compared to those structures involving hydrogen bond like arrangements (*e.g.* 5 and 8).

For all methods, the chloride partial charges were influenced by the interionic arrangements. For instance, chlorides involved in a strong hydrogen bond to the most acidic proton $H^1$ (see Fig. 1 and 2) had a lower partial charge than chlorides forming a hydrogen bond with less acidic hydrogens $H^2$ and $H^{2'}$. The lowest partial charges were found if the chloride was above the imidazolium ring plane (see Fig. 2). The same dependences were observed for the chloride net charges in one ion pairs[46,62] (see Table 2) which underlined the quite general behaviour. Despite the significant surface effects in gas phase clusters, these findings foreshadow already a broad distribution of chloride charges to be expected in simulations of the liquid phase. This was, indeed, observed for Blöchl partial charges in liquid phase studies reported recently.[62] Furthermore, in the liquid phase, it was found that chlorides forming a strong hydrogen bond had a higher mean electric dipole moment (see Fig. 3). These fluctuations of the chloride partial charges for varying structures point already to limitations to be met when assigning a common (scaled) charge to the individual ions (see below) and indicate room for future refinements.

 

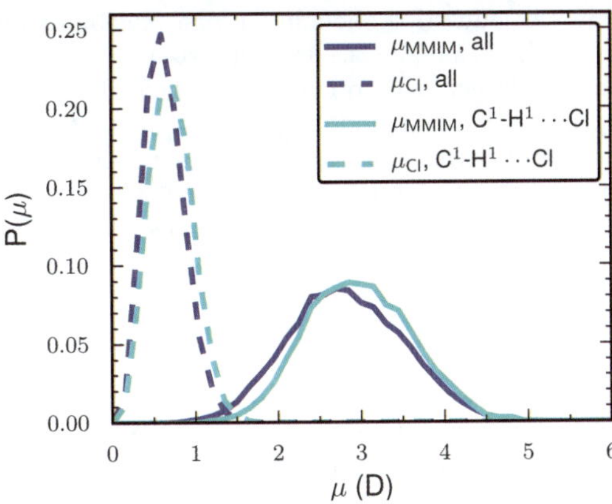


**Fig. 3** Molecular electric dipole moment distributions of all ions (blue) and hydrogen-bonded ions (cyan) in 30 ion pairs [MMIM][Cl] based on the Wannier analysis. The chosen hydrogen bond cutoffs were a distance $r_{\mathrm{H^1 \cdots Cl}}$ of 3.5 Å and an angle $\alpha_{\mathrm{C^1-H^1 \cdots Cl}}$ of 30°.

However, the chloride net charges seemed to be influenced only locally as two ion pairs were found to have very similar electric monopoles as ions in the liquid phase. So, the mean chloride net charge observed by the Blöchl analysis decreased from one ion pair clusters ($-0.71\ e$[46,62]) to the positively charged clusters 5 to 8 ($-0.67\ e$, see Table 3) to the two ion pair clusters 1 to 4 ($-0.64\ e$, see Table 3). Thus, the mean chloride net charge of the two ion pair clusters was about the same as in liquid phase ($-0.63\ e \pm 0.08\ e$[62]).

Overall, quantum chemical calculations consistently yielded reduced chloride point charges what suggested to employ reduced partial charges also in classical MD simulations. For these, reduced ion net charges were already used[8,18,26,29,31,63] and reported to accelerate the dynamics. But these studies were based on the partial charges gained from the isolated ion which were scaled down subsequently to a somewhat arbitrarily chosen net charge. Conversely here, we found reduced net charges by all used quantum chemical methods covering DFT, MP2 and highly accurate CCSD(T), implying that non-integer charges do have a physical origin and are a fundamental characteristics of IL. Indeed, experimental evidence for the charge scaling has been reported.[104,105]

**Net charge of cations in clusters.** As for the anions, the Blöchl analysis gave reduced cation charges. The cations in neutral cluster had net charges of $0.62\ e$ to $0.66\ e$, cations in positively charged clusters $0.73\ e$ to $0.92\ e$, and cations in negatively charged clusters showed the lowest net charges with charges of $0.53\ e$ to $0.61\ e$. Thus, cations in two ion pairs clusters and negatively charged clusters had a mean net charge in the range of the one found in the liquid phase ($0.63\ e \pm 0.15\ e$[61]). Hence, the presence of two counterions seemed sufficient to reproduce liquid-like net charges as was discussed for chloride above. If an ion was surrounded by only one counterion and one coion, as cations in 2C1A and anions in 1C2A, the described electrostatic properties differed most from the ones of the liquid phase. In the liquid phase, ions are surrounded by mostly counterions in the closest neighborhood. Indeed, clusters in which ions were surrounded by more counterions than coions seemed to approximate the liquid structure better and were associated with electrostatic properties more similar to the ones observed in liquid phase.

### 3.2 Liquid phase

While computationally typically prohibitive for post-HF methods, system sizes being representative of the liquid phase can often be accessed with DFT. In principle, this is not only a question of computational feasibility, but also of efficient analysis. In the liquid phase, a highly accurate electronic structure may not be necessary for understanding larger scale properties. In fact, the huge amount of data produced (*e.g.* by a post-HF method) might even cloud the essential statistical and thermodynamic properties. Thus, a DFT type approach might be considered appropriate to get quantum mechanical aspects described and to extract the relevant large scale liquid behaviour. In any case, in the sections above, it was shown that the functionals employed gave results in reasonable to fair agreement with MP2. This is at least a basic indication that the (correct) underlying electronic structure is sufficiently well captured by DFT.

**Electric dipole moment distribution of different Ils.** In general, the electric dipole moment distributions were found to be very broad for several ILs including a protic one.[60,61,103] In particular, they were broader than the electric dipole moment distribution of a non ionic liquid like water.[106] So for ILs, there were more pronounced fluctuations in the electronic structure which seemed to be an intrinsic property of ILs.[61] Fig. 4 gives the electric dipole moment distributions of three ILs, [MMIM][Cl], [EMIM][Cl], and [EMIM][DCA]. In [MMIM][Cl] and [EMIM][Cl], the cations $[MMIM]^+$ and $[EMIM]^+$ had essentially the same electric dipole moment distribution (see Fig. 4(a)). Also, the chloride electric dipole moment distribution did not depend on the counterion. Thus, the influence of the side chain on the general electronic structure seemed to be small if changing from $[MMIM]^+$ to $[EMIM]^+$. If, instead of the cation, the anion was varied, the effect on the counterion's electric dipole moment distribution was higher. In [EMIM][DCA], the $[EMIM]^+$ electric dipole moment distribution was less broad than in [EMIM][Cl] while the mean electric dipole moment varied only by 0.04 D. So, the anion seemed to have a larger influence on the properties of imidazolium-based ILs than the cation as it has been recently proposed.[61]

**Impact of ionic arrangements.** These large fluctuations were found to be the product of very local, but diverse molecular packing.[61] Similar to the findings in clusters (see Section 3.1), the mean electric dipole moments of cations involved in a hydrogen bond at the most acidic proton were higher than the mean electric dipole moments of all cations (see Fig. 3). The same occurs also for the anion electric dipole moments. Thus, this hydrogen bond was connected with an additional polarisation

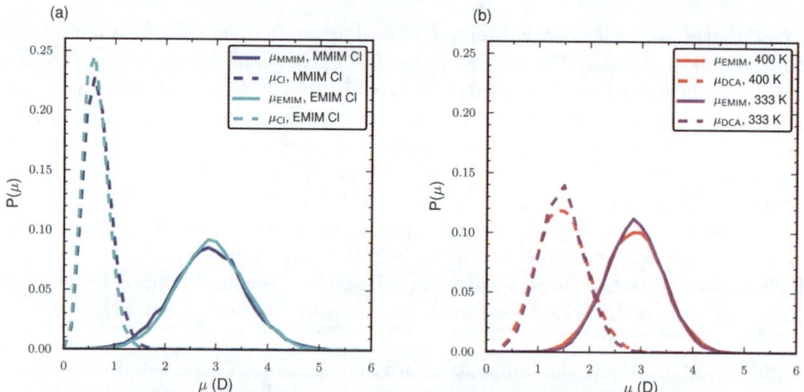


**Fig. 4** Molecular electric dipole moment distributions in imidazolium ILs: (a) [MMIM][Cl] (blue) and [EMIM][Cl] (cyan) and (b) [EMIM][DCA] at two temperatures 400 K (red) and 333 K (violet). Electric dipole moments of imidazolium ions are calculated based on the Wannier analysis with respect to the center of ring (COR), all other with respect to the geometric center (COG) (see also Appendix).

 

as expected and also reported for non-imidazolium based ionic liquids.[107] While the mean electric dipole moments of hydrogen-bonded ions were increased, the electric dipole moment distribution stayed as broad as the distribution of all ions. This implied that while these hydrogen bonds provided a preferential tendency to increase the ion's electric dipole, the environment is still dominant in determining the broadness of the electric dipole distribution. In general, other ions within only a short ranged vicinity around the ions influence the electrostatic properties. We found that solely ions closer than 8 Å[61] influenced the electrostatic properties of other ions. This local behaviour seems to be consistent with the hypothesis of an ion rattling in a long-living ion cage.[20–22,44,57,108,109]

**Practical relevance of the results found.** These results are relevant in the multiscale modelling because the electric dipole moment distribution was shown to be largely insensitive to system size as well as shorter simulation times. Thus, comparatively small simulations suffice typically to extract information for modelling at coarser scale. This is important because it implies not only that the computational resources needed are not prohibitive, but also that one can study many more IL species due to the modest request of computational resources. In fact regarding the system size, it was shown that eight ion pair simulations gave essentially the same electric dipole moment distributions and Blöchl charges as simulations with at least 30 ion pairs.[61] Furthermore, it was found that a simulation time of only 3 ps sampled an electric dipole moment distribution similar to the one obtained from much longer runs as shown for [EMIM][DCA] in Fig. 5. Analogously, short simulation times provided a sufficiently exhaustive representation of the electrostatic properties for any of the studied ILs. Moreover, the results above do not depend on changes in temperature within a range of 333 K to 400 K. Thus, on one hand, CPMD calculations can be performed at higher temperature in order to explore faster a larger portion of the phase space. On the other hand, force fields might be transferable with respect to the thermodynamic state point. The electric dipole moment distributions of liquid phase [EMIM][DCA] at 400 K and 333 K were nearly identical (see Fig. 4(b)). The one at 333 K was slightly narrower which seemed to be due to less excited vibrations associated with smaller mean bond lengths. The electronic structure appeared to be rather insensitive to the temperature change.

To further ensure the significance of the AIMD based results, two tests were carried out based on classical MD trajectories which sampled much longer time and length scales. Classical MD simulations can probe phenomena which are

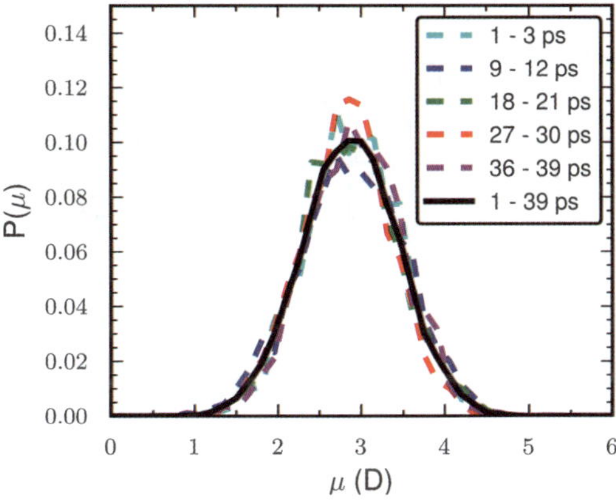


**Fig. 5** Molecular electric dipole moment distributions in 30 ion pairs [EMIM][DCA] at 400 K averaged over the whole simulation (black thick) and over 3 ps intervals (coloured thin) based on the Wannier analysis.

inaccessible with AIMD, such as density fluctuations. First, ten uncorrelated configurations of 64 [EMIM][DCA] ion pairs were chosen in 1 ns time steps from the last 10 ns of a classical MD trajectory of 40 ns. The dipole moment distribution found after wavefunction optimisation on these configurations agreed with the one found in the AIMD simulations. Second, 240 [EMIM][DCA] ion pairs were simulated classically for 40 ns after equilibration and the dipole moment distribution based on the point charges was calculated and found to be identical to the one given in Fig. 6(a). Thus, both tests gave the same dipole moment distributions as the ones based on AIMD trajectories. This implied that the time and size of the AIMD studies were sufficient to sample all decisive configurations for the molecular electric properties.

In conclusion, it seems appropriate to carry out AIMD simulations that are restricted to a small number of ion pairs and a short time in order to obtain partial charges. Moreover, these simulations can be done at higher temperature and, thus, sample faster the phase space without a significant change in the observed electrostatic properties, at least for the systems studied herein.

**Partial charge assignment.** The partial charges obtained by the Blöchl analysis differed from the CHELPG partial charges used by the CLaP force field (see Fig. 7). In general, the net cation charge found by CAB was not 1 *e* as in CHELPG-CLaP, but reduced as in the clusters. This can be understood to reflect the electronic polarizability of the molecules within the MDEC theory of Leontyev and Stuchebrukhov[64] that has been successfully applied to water.[64,65] While the scaling factor for most of the existing force fields with a reduced ionic net-charge[18,22,26,29,31,40,63] was derived empirically or on small cluster calculations, our approach yielded internally consistent partial charges $q^{eff}$ reproducing the electric far field, and also an *ab initio* value for the corresponding electronic dielectric constant $\varepsilon_{el}$ that described the polarization effects arising in the liquid phase implicitly. Since $\varepsilon_{el}$ is related to the refractive index $n$ by $\varepsilon_{el} = n^2$ and $n$ is of the order of 1.4 for typical ILs,[110] a charge scaling of $\frac{q}{\sqrt{\varepsilon_{el}}} \approx 0.7q$ can be expected. This is strong experimental and theoretical evidence that the reduced charges are not an artefact of the method for the calculation of the partial charges. Further experimental evidence is found by NMR and X-ray spectroscopy.[105]

For our investigated ILs, reduced net charges of 0.63 $e \pm 0.15$ $e$ for [MMIM][Cl],[62] 0.61 $e \pm 0.16$ $e$ for [EMIM][Cl], and 0.67 $e \pm 0.21$ $e$ for [EMIM][DCA][61] were obtained. In this context, reduced ion charges were considered to be a general characteristic of ILs[61] and consistent within the framework of MDEC theory[64,65] and experiments.[105]

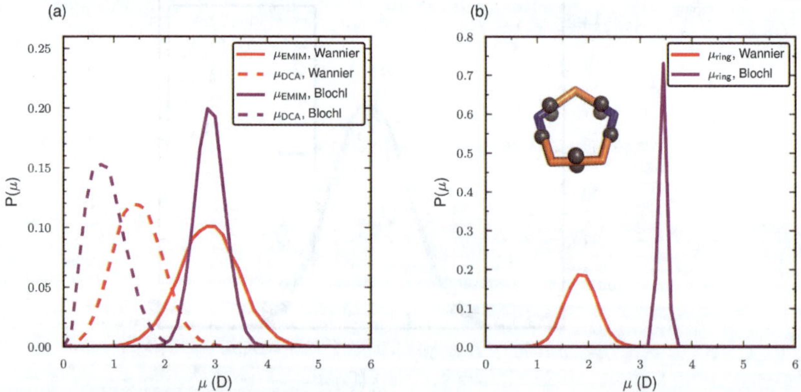


**Fig. 6** Electric dipole moment distributions obtained with rescaled Blöchl partial charges compared to the Wannier center based electric dipole moment distributions of 30 ion pairs [EMIM][DCA] (400 K): (a) for the ions and (b) for the imidazolium ring fragment of $[EMIM]^+$.

 

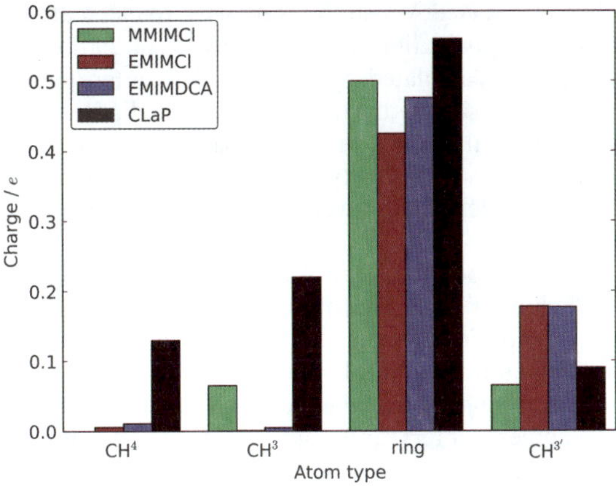


**Fig. 7** Net charges of atom groups in $e$ based on Blöchl partial charges (see Fig. 9).

Apart from the net ionic charge, the Blöchl partial charge distribution differed in general from the one of CLaP force field charges. In the CLaP force field, the charges of the side chains and the nitrogen are taken to be symmetric. For the Blöchl charges instead, there was no constraint. Results show that, for the $[EMIM]^+$ cation (see Fig. 1), the partial charges of $C^3$ and $C^{3'}$ as well as $N^1$ and $N^{1'}$ were different. The net charge of the imidazolium ring was very similar to the Blöchl partial charges and the CHELPG-CLaP charges, but the distribution was different. The Blöchl analysis gave a less negative partial charge on the $C^1$ than the one used in the CLaP force field, but more negative charges on the $C^2$ and $C^{2'}$. The different charge distributions on the ring might have implications on the liquid structure found in a classical simulation.

The obtained partial charges of $[EMIM]^+$ were very similar in [EMIM][Cl] and [EMIM][DCA] (see Fig. 9). Thus, there seemed to be a certain transferability of the parameters being important in the force field development (see section 4.2). Furthermore, the partial charges of the ring atoms $C^1$, $C^2$, $C^{2'}$ (see Fig. 1) and their hydrogens were found to be alike in $[MMIM]^+$ and $[EMIM]^+$. But the partial charges of the side chain atoms and of the nitrogen atoms $N^1$ and $N^{1'}$ being bonded to the side chains differed between $[MMIM]^+$ and $[EMIM]^+$. For more insight into the possibility of having transferable information, other imidazolium-based ILs need to be studied. This is subject of work in progress.

**Partial charge based electric dipole moments.** The broad electric dipole moment distribution was found to be of electronic origin as the electric dipole moment distribution based on Blöchl partial charges was much narrower. Below, these findings are discussed. For the simplest case of an anion as the monoatomic chloride, a classical point charge representation has always a zero electric dipole moment while the quantum-based approaches found a non-zero electric dipole moment for chloride in clusters as well as in snapshots of the liquid phase (see previous sections). Thus, it may be important to include polarisation in the classical force field to reproduce the right electrostatic properties for the chloride anion. More relevant is the case of the larger ions. For these, a broad electric dipole moment distribution might be caused mostly by internal vibrational degrees of freedom and, thus, by atomic motion only. If this is the case, electric dipole moments based on point charges should reproduce reasonably well the Wannier center based electric dipole moment distribution of the quantum chemical calculations. If there is no possibility to reproduce the electric dipole moment distribution by a set of partial charges, it means that the quantum scale plays a key role in determining molecular electrostatic properties.

Rescaled Blöchl charges (see Appendix) were used to calculate the electric dipole moments of the cation configurations from the AIMD simulation for which the electric dipole moments were calculated by the Wannier center analysis, *i.e.* the same cation configurations were sampled. The cation electric dipole moment distribution based on the rescaled Blöchl charges was less broad than the one based on Wannier centers, but had the same mean value (see Fig. 6(a)). For anions, the electric dipole moments based on Wannier centers were on average higher than the one of the Blöchl charges.

In particular, for highly polarisable molecular parts like the anions or the imidazolium ring, Blöchl charges based electric dipole moments failed to reproduce the large spread of the distribution. The most polarisable part of the imidazolium cation is the aromatic imidazolium ring. If only this part of the cation is considered and polarisation is important, the difference between the electric dipole moment distributions based on Wannier centers and that based on partial charges should be particularly large. In fact, the imidazolium ring does not have many vibrational degrees of freedom which could cause broad electric dipole moment distributions in partial charge systems. As shown in Fig. 6(b), the electric dipole moment distribution of the imidazolium ring was very narrow if the partial charge description was used. In the case of electric dipole moments based on Wannier centers, the imidazolium ring was characterised by a considerable fluctuation of the corresponding electric dipole moment. This means that the electronic polarisation played a key role and could not be reproduced by partial charges. It might be possible to assign partial charges in a way that the electric dipole moment distribution of the whole cation is as broad as the one based on Wannier centers. But the width of the electric dipole moment distribution of the imidazolium ring itself can not be reproduced by partial charges. This implies that polarisable models would be needed to model this effect at the classical level.[21,25,32–34,37,41,111] In general, the ion dynamics was faster if polarisable force fields were employed and local packing differed from non-polarisable models as firstly shown by Yan *et al.*[21] for [EMIM][$NO_3$]. However, the electrostatic properties were found to be rather local.[41,61] So, the explicit inclusion of polarisation in the force field and the detailed reproduction of the local quantum behaviour might not be necessary for large scale properties. Bulk and long time properties, as density fluctuations or diffusion behaviour, might require only an "average" effective model which makes use of the very essential information gained at the quantum level. In the next section we will show that, by a combination of MDEC theory[64] and an appropriate optimization procedure of the short-range interactions, a non-polarizable model is also capable of reproducing the long-time scale dynamical behaviour given by a polarisable model.

## 4 Including charge scaling: Force field development

Although *ab initio* methods allow us to study molecular properties based on electronic structure calculations, the system size and time scale is quite limited. Especially for the prediction of dynamical variables as the diffusion constant or conductivity, simulations of systems of hundreds of ion pairs and tens or even hundreds of nanoseconds are required to calculate the desired property within a reasonable level of accuracy. However, an accurate classical force field is indispensable to perform such simulations. Though the force field of Canongia Lopes *et al.*[19,112,113] (CLaP) covers a wide range of ILs and describes static properties consistently, it fails in a reasonable description of the dynamic ones.[45] For instance, as shown in Table 4, the conductivity predicted by CLaP for [MMIM][Cl] was an order of magnitude too low whereas the density was given correctly (see Table 4 and Fig. 8). The CLaP force field partial charges were parameterised on isolated ions by using CHELP. As discussed, the electrostatic properties of isolated ions differed from ions with adjacent ions. Thus, the gained knowledge of liquid phase electrostatic properties was applied to the CLaP force field using partial charge values that most accurately reproduce the electrostatic potential.

### 4.1 Force field based on CAB

Firstly, the partial charges of the CLaP force field were replaced by the CAB ones (see Fig. 7 and Ref. 115). All other parameters were left unchanged. As can be seen in Fig. 8 and Table 4, this force field, denoted as BLFF, led to much lower density and an increased conductivity. The conductivity was above the experimental value. Unfortunately, little experimental data is available for [MMIM][Cl],[114] except for the conductivity which was compared here. As reported earlier,[46] the liquid structure found by BLFF resembled the one obtained by AIMD simulations more than the one based on the CLaP force field. The radial distribution functions had lower maxima hinting at a higher disorder. The reduced net charges seemed to cause a weaker ionic interaction and, hence, a lower density and a less pronounced liquid structure. Also, if the ions interacted less strongly, they could move more freely and the conductivity was increased. If the Coulomb interactions are weaker, the van der Waals interactions become more decisive. The critical influence of van der Waals interactions in ILs was reported in several studies.[47,51–53]

Thus, secondly, a parameterisation of the Lennard-Jones parameters was conducted to recover the intermolecular energies and to adapt the short-range interactions. An iterative procedure[115] was applied tuning the short-range parameters to a desired level of accuracy. The experimental measured mass density at three temperatures, needed anyway to setup the AIMD simulation, and radial distribution functions obtained by the AIMD simulations were employed as targets. Thus, the mass densities at three temperatures were the only experimental input used throughout this multiscale study. The gained optimised force field is denoted as BTFF. Obviously with this, the mass density was recovered over a wide range of temperature as shown in Fig. 8. However, the BTFF gave not only static properties correctly, but also dynamic properties were predicted reasonably (see Table 4). The electric conductivity obtained deviated less than 10% from the experimental value. Hence, the BTFF force field based on CAB partial charges and optimised Lennard-Jones parameters was an improvement with regard to the CLaP force field. Therefore, a dynamic property, the electric conductivity, was described correctly without a polarisable force field. Thus, for [MMIM][Cl], it seemed sufficient to include mean electrostatic properties to simulate bulk properties. The exact procedure of how the BTFF parameterisation was obtained together with all necessary parameters can be found in a forthcoming publication.[115] Unfortunately, no experimental value is available for the diffusion constants in [MMIM][Cl]. However, comparing the results given by BTFF to similar charge-scaling approaches of Youngs and Hardacre[31] and Liu *et al.*[40] revealed an interesting difference of the ratio $\gamma = D^+/D^-$. Although all the diffusion coefficients derived with different parameterisations

**Table 4** Density $\rho$, self diffusion constants $D$, and the conductivity $\sigma$ obtained by the simulations with the parameters of CLaP, BLFF, and the final tuning stage BTFF at $T = 425$ K. The conductivity was calculated with either the Nernst-Einstein equation (NE) or the Einstein-Helfand equation (EH). The estimated errors are given in parentheses.

| | | $D/10^{-5}$ cm$^2$ s$^{-1}$ | | $\sigma$/S m$^{-1}$ | |
|---|---|---|---|---|---|
| | $\rho$/kg m$^{-3}$ | $D^+$ | $D^-$ | NE | EH |
| CLaP | 1114.1(4.5) | 0.02(0.00) | 0.01(0.00) | 0.5(0.0) | 0.5(0.0) |
| BLFF | 949.7(6.1) | 1.39(0.10) | 1.46(0.10) | 21.5(0.0) | 14.8(0.9) |
| BTFF | 1123.3(4.7) | 0.48(0.03) | 0.62(0.04) | 10.0(0.5) | 9.7(0.6) |
| exp.[114] | 1123.4 | n.a. | | 10.7 | |

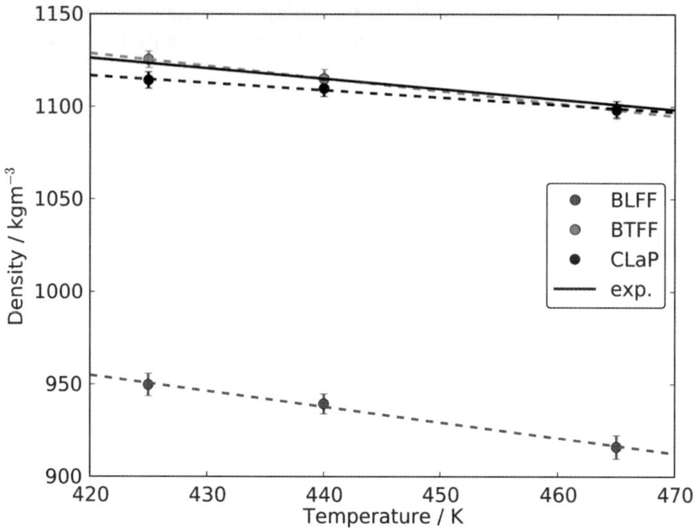


**Fig. 8** Density of mass for different parameterisations.

were of the same order of magnitude, BTFF predicted a value of $\gamma < 1$ while the parameterisation of Liu *et al.*[40] yielded $\gamma > 1$. Youngs and Hardacre[31] performed simulations with different scaling factors and observed a "cross-over" of $\gamma > 1$ to $\gamma < 1$ if they decreased the ionic net-charge. The value $\gamma_{BLFF} = 0.95$ for the untuned BLFF force field corresponded extremely well if results of Youngs and Hardacre[31] were interpolated to derive a number corresponding to the charge-scaling factor of BLFF and BTFF. The tuned force field BTFF even gave a ratio of $\gamma = 0.77$ which corresponded to a scaling-factor of 0.5 in the parameterisation scheme of Youngs and Hardacre.[31] However, the absolute values of $D^+$ and $D^-$ corresponded to a scaling factor between 0.6 and 0.7 as suggested by Youngs and Hardacre.[31] Since polarisable FF models allow a more detailed study of dynamical properties, we extended our comparison to the results of Yan *et al.*[21] and Borodin.[111] The first IL studied with a polarisable force field was [EMIM][$NO_3$][21] and an extensive study by Borodin[111] broadened the range of studied ILs. With regard to the diffusion, our results are comparable to the polarisable models of [EMIM][$NO_3$][21] and [EMIM][$C(CN)_3$][111] with $\gamma^{21} = 0.96$ and $\gamma^{111} = 0.90$, respectively. Like in the case of increased charge scaling, Yan *et al.*[21] observed the "cross-over" of $\gamma$ by switching from the non-polarisable to the polarisable model. Though it is known that different ILs behave differently,[111] this is a nice indicator that our implicit description of the polarizability with the Blöchl charges allows an improved treatment of the IL dynamics in the diffusive regime by a non-polarisable force field.

### 4.2 Further development of CAB based force fields

While the multiscale approach finally allowed us to adapt an accurate force field for [MMIM][Cl], how well it performs for other ILs still has to be studied. Although the approach was based only on one experimental data point, it was computationally demanding to obtain the BTFF force field parameters. Due to the large number of ILs, a stable and efficient technique is desired to allow studies on their properties. Thus, the transferability of the force field parameters is crucial. For instance, the parameters for a certain ion should be independent of the counterion. Furthermore, for ions belonging to the same ion type, *e.g.* all imidazolium-based ions, atoms at corresponding positions should have the same parameters. That means the

 

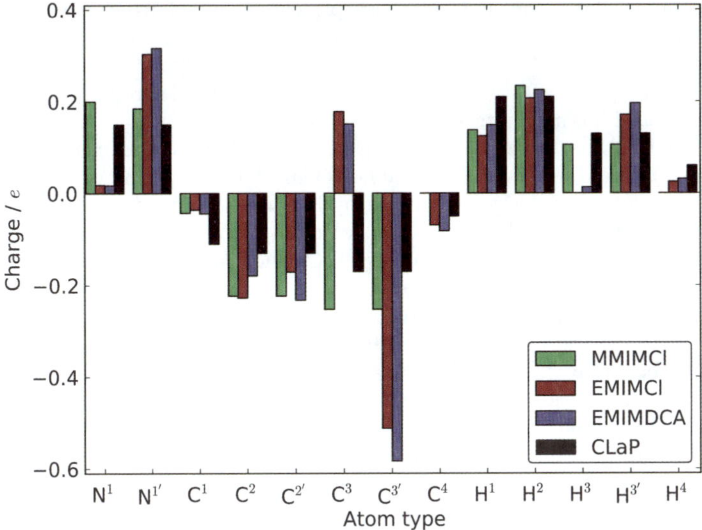


**Fig. 9** Mean Böchl partial charges of atom types in *e* obtained from liquid phase snapshots. The CAB analysis was done by using the same functional as in the AIMD simulations of 30 ion pairs [MMIM][Cl], eight ion pairs [EMIM][Cl], and 30 ion pairs [EMIM][DCA], *i.e.* PBE/TZV2P for [MMIM][Cl] and [EMIM][Cl] and BLYP/TZV2P for [EMIM][DCA].

parameters of the $C^1$ should be unrelated to the side chain length. So, the number of atom types used in the force field can be as low as possible.

Therefore, the CAB method would be a convenient approach, if it gave similar partial charges to alike atoms in different imidazolium-based cations. The strong fluctuation of the Blöchl charges observed[62] indicated a certain flexibility in the choice of the actual force field charges. This may allow us to keep the same parameters for corresponding atoms in different ions. To establish a transferable force field within a certain class of ILs, polar and apolar domains of the side chains have to be identified at first. As shown in Fig. 7, the net charge of the side chains was quite similar for $[EMIM]^+$ in both $[EMIM]^+$ containing liquids. Also, the charge of the ring deviated less than 0.1 *e* and corresponding atoms were always charged in the same order of magnitude (see Fig. 9). In the case of [EMIM][DCA] and [EMIM][Cl], the long ethyl side chain was an apolar domain, in contrast to the methyl group that was a polar one. In $[MMIM]^+$, both methyl groups carried a charge about half as large as the methyl group in $[EMIM]^+$. Thus, there seemed to be a certain limited transferability of the partial charges. The study of other ILs, for instance [BMIM]-based, will be necessary. Apart from the electric monopoles, the electric dipole moments may have to be represented also in a reasonable manner. Indeed, while the mean electric dipole moment derived by Wannier centers and rescaled CAB charges agreed for $[EMIM]^+$, disagreement for the anions was observed (see Fig. 6). In lieu of constructing a polarisable model, the partial charge distribution could be changed such that the average electric dipole moment found by Wannier centers is recovered.

## 5 Conclusion

A multiscale approach, based on a sequential bottom-up idea from post-HF to DFT to classical MD, was applied to the study and modelling of [MMIM][Cl]. The results were compared to the findings for [EMIM][Cl] and [EMIM][DCA]. This study focused on the electrostatic properties at each scale, *i.e.* electric monopoles, dipoles and their molecular origin, and some basic principles characteristic of ILs were identified. Results confirmed the strong screening of Coulomb interactions.[18,20,26,31,41,51,61–63,104,116,117] The range of locality corresponded to the molecular size. This was connected to very high fluctuations in the electronic properties and overall reduced ion net charges.

For small clusters, the Blöchl charges gave ion net charges significantly reduced to less than |1 *e*| as did well-established methods like CHELPG, RESP or Bader in post-HF calculations. The charge assignment by Blöchl, CAB, is a reliable method that was used to assign partial charges to bulk (liquid-like) systems for which a similar reduction of the net ionic charge was seen. Considering that the environment has a strong effect on the values of partial charges, it is advisable to parameterise partial charges on liquid phase configurations.

The broad electric dipole moment distribution of ions in the liquid phase was shown to be rather insensitive to temperature changes, system size, simulation time or exchange of ions. Hence, these high fluctuations in the electronic structure of ions seemed to be a general characteristic of ILs. Furthermore, the fluctuations were shown to be connected to polarisation as the electric dipole moment distributions based on partial charges were not as broad as the ones based on Wannier centers. In general, electronic polarisation can not be reproduced by partial charges and polarisable models[21,32–34,37,111] might be necessary at the classical level. However, since the electrostatic properties were so local, an "average" effective model might be sufficient to simulate macroscopic bulk properties. Indeed, the satisfactory performance of the introduced classical force field strengthened this argument.

The findings, *i.e.* the reduced partial charges, were incorporated into a [MMIM][Cl] force field which gave reliable predictions for static and dynamic properties. This implies that CAB based force fields may have a large potential. A crucial test in this direction will be the transferability of the force field parameters which turned out to be sufficient in first tests. In conclusion, the multiscale approach appeared to open a valuable gateway to rational force field design.

## Appendix: Deriving local properties from the electron density

In general, the electron density is distributed continuously over the whole system and a sophisticated analysis has to be conducted to assign local electronic properties to an atom or a molecule. Such an analysis can be done in various ways, leading to different results. Here, two methods were used to localise the electron density around atoms. They are introduced shortly in the following.

### Maximally localised Wannier functions

The Wannier functions[118–120] are an alternative representation of (plane wave) basis sets and can be obtained from these by unitary transformations. Although the Wannier functions are non-unique in principle, a protocol to obtain well defined functions has been introduced.[121] The resulting functions are called "maximally localised" Wannier functions and are those used in this work. Within this localisation algorithm, the second moment of the Wannier functions around their centers is minimised. The charge center of a maximally localised Wannier function corresponds to the classical location of an electron pair.[122] Hence, it can provide an insight into the chemical bond structure[122] as well as shed light on the local polarisation. For instance, molecular electric dipole moments can be calculated based on the nuclei charges and the Wannier centers (see *e.g.* ref. 106 for water). The Wannier centers are assigned to the closest nuclei and its molecule and each molecule has an integer net charge. The electric dipole moment is then calculated as:

$$\mu = \sum_{i=1}^{N} q_i \cdot \vec{r}_i \qquad (1)$$

 

in which $N$ is the number of nuclei and Wannier centers considered. $q_i$ is the charge and $\vec{r}_i$ the position of the $i$th nucleus or Wannier center. Here, the charge corresponding to one Wannier center is $-2\ e$ as each Wannier function describes two electrons. In general, the electric dipole moment of an ion is not defined uniquely, but depends on the center of reference.[123] In this work, the geometric center of the imidazolium ring (see Fig. 1) is chosen as reference center for imidazolium cations. The geometric center of all atoms is selected for all other ions. Still, the dependence on an arbitrary reference center implies that the absolute value of ionic electric dipoles does not have a physical meaning. However, it was found that the results presented here did not depend qualitatively on the choice of the reference point.

### Charge assignment by Blöchl (CAB)

In the charge assignment by Blöchl (CAB),[86] it is assumed that the interaction between separated charge densities can be expressed entirely through their multipole moments. It can be shown that the electrostatic interactions can be treated efficiently in the reciprocal representation for systems with periodic boundary conditions. Thus, the fitting in the CAB method is done in reciprocal space in contrast to common electrostatic potential fitting procedures like RESP or CHELPG in which the fitting is carried out in real space. Fitting in real space demands excluding regions close the atoms to avoid the singularity of the potential. The CAB approach being performed in the reciprocal space does not suffer from this problem. The CAB method fits charge densities by projecting it on atom-centered Gaussians and, so, describes the effective charge of an ion. Hence, the net charge of ions might deviate from the integer value. If the electric dipole moment shall be calculated by using partial charges of CAB, the charge of the ion might be rescaled to the integer value. Otherwise, there is a prefactor in the electric dipole formula if compared to the Wannier electric dipole moment which describes ions with integer charges. This can be done by dividing each Blöchl partial charge by the net charge of the ion. Here, the so-obtained partial charge set gives an ion net charge of $\pm 1\ e$ and is denoted as rescaled Blöchl charges.

The CAB method was applied for [MMIM][Cl].[46,62] The net ion charges obtained were significantly lower than $|1\ e|$.[62] In the liquid phase of [MMIM][Cl], the ion net charge was $\pm(0.63\ e \pm 0.08\ e)$. This means that, indeed, ions did not experience the full charge of an adjacent ion, but an effective (screened) charge.


## Acknowledgements

The authors thank the DFG priority program "Ionic liquids" for funding, the RZG Garching of the Max Planck Society, the HLRS in Stuttgart, as well as the Center for Scientific Computing (CSC) Frankfurt for computational resources, and Denis Andrienko for critical reading.


## References


1 R. D. Rogers and K. R. Seddon, *Science*, 2003, **302**, 792–793.
2 *Ionic Liquids in Synthesis*, ed. P. Wasserscheid and T. Welton, Wiley-VCH, 2007.
3 N. V. Plechkova and K. R. Seddon, *Chem. Soc. Rev.*, 2008, **37**, 123–150.
4 M. Armand, F. Endres, D. R. MacFarlane, H. Ohno and B. Scrosati, *Nat. Mater.*, 2009, **8**, 621–629.
5 A. Pinkert, K. N. Marsh, S. Pang and M. P. Staiger, *Chem. Rev.*, 2009, **109**, 6712–6728.
6 E. W. Castner and J. F. Wishart, *J. Chem. Phys.*, 2010, **132**, 120901–120901.
7 R. Giernoth, *Angew. Chem., Int. Ed.*, 2010, **49**, 5608–2839.
8 H. Liu, Y. Liu and J. Li, *Phys. Chem. Chem. Phys.*, 2010, **12**, 1685–1697.
9 H. Olivier-Bourbigou, L. Magna and D. Morvan, *Appl. Catal., A*, 2010, **373**, 1–56.
10 S. Werner, M. Haumann and P. Wasserscheid, *Annu. Rev. Chem. Biomol. Eng.*, 2010, **1**, 203–230.
11 P. A. Hunt, *Mol. Simul.*, 2006, **32**, 1–1.


 

12 E. J. Maginn, *Acc. Chem. Res.*, 2007, **40**, 1200–1207.
13 A. A. H. Pádua, M. F. Costa Gomes and J. N. A. Canongia Lopes, *Acc. Chem. Res.*, 2007, **40**, 1087–1096.
14 Y. Wang, H. Pan, H. Li and C. Wang, *J. Phys. Chem. B*, 2007, **111**, 10461–10467.
15 Y. Wang, W. Jiang, T. Yan and G. A. Voth, *Acc. Chem. Res.*, 2007, **40**, 1193–1199.
16 B. L. Bhargava, S. Balasubramanian and M. L. Klein, *Chem. Commun.*, 2008, 3339–3351.
17 B. Kirchner, *Top. Curr. Chem.*, 2009, **290**, 213–262.
18 T. I. Morrow and E. J. Maginn, *J. Phys. Chem. B*, 2002, **106**, 12807–12813.
19 J. N. Canongia Lopes, J. Deschamps and A. A. H. Padua, *J. Phys. Chem. B*, 2004, **108**, 2038–2047.
20 M. G. Del Popolo and G. A. Voth, *J. Phys. Chem. B*, 2004, **108**, 1744–1752.
21 T. Yan, C. J. Burnham, M. G. Del Popolo and G. A. Voth, *J. Phys. Chem. B*, 2004, **108**, 11877–11881.
22 B. L. Bhargava and S. Balasubramanian, *J. Chem. Phys.*, 2005, **123**, 144505.
23 B. Bhargava and S. Balasubramanian, *Chem. Phys. Lett.*, 2006, **417**, 486–491.
24 T. Yan, S. Li, W. Jiang, X. Gao, B. Xiang and G. A. Voth, *J. Phys. Chem. B*, 2006, **110**, 1800–1806.
25 A. Bagno, F. D'Amico and G. Saielli, *J. Mol. Liq.*, 2007, **131–132**, 17–23.
26 A. L. Bhargava and S. Balasubramanian, *J. Chem. Phys.*, 2007, **127**, 114510.
27 T. Köddermann, D. Paschek and R. Ludwig, *ChemPhysChem*, 2007, **8**, 2464–2470.
28 R. M. Lynden-Bell, M. G. Del Popolo, T. G. A. Youngs, J. Kohanoff, C. G. Hanke, J. B. Harper and C. C. Pinilla, *Acc. Chem. Res.*, 2007, **40**, 1138–1145.
29 W. Zhao, H. Eslami, W. L. Cavalcanti and F. Müller-Plathe, *Z. Phys. Chem.*, 2007, **221**, 1647–1662.
30 C. Schröder and O. Steinhauser, *J. Chem. Phys.*, 2008, **128**, 224503–224507.
31 T. G. A. Youngs and C. Hardacre, *ChemPhysChem*, 2008, **9**, 1548–1558.
32 O. Borodin, *J. Phys. Chem. B*, 2009, **113**, 12353–12357.
33 T. M. Chang and L. X. Dang, *J. Phys. Chem. A*, 2009, **113**, 2127–2135.
34 P. Lopes, B. Roux and A. MacKerell, *Theor. Chem. Acc.*, 2009, **124**, 11–28.
35 R. M. Lynden-Bell and T. G. A. Youngs, *J. Phys.: Condens. Matter*, 2009, **21**, 424120.
36 B. Aoun, M. A. Gonzalez, J. Ollivier, M. Russina, Z. Izaola, D. L. Price and M.-L. Saboungi, *J. Phys. Chem. Lett.*, 2010, **1**, 2503–2507.
37 D. Bedrov, O. Borodin, Z. Li and G. D. Smith, *J. Phys. Chem. B*, 2010, **114**, 4984–4997.
38 M. Costa Gomes, J. N. Canongia Lopes and A. Padua, *Top. Curr. Chem.*, 2009, **290**, 161–183.
39 D. Kerlé, R. Ludwig, A. Geiger and D. Paschek, *J. Phys. Chem. B*, 2009, **113**, 12727–12735.
40 Z. Liu, T. Chen, A. T. Bell and B. Smit, *J. Phys. Chem. B*, 2010, **114**, 10692–10692.
41 R. M. Lynden-Bell, *Phys. Chem. Chem. Phys.*, 2010, **12**, 1733–1740.
42 B. S. Mallik and J. I. Siepmann, *J. Phys. Chem. B*, 2010, **114**, 12577–12584.
43 S. S. Sarangi, W. Zhao, F. Müller-Plathe and S. Balasubramanian, *ChemPhysChem*, 2010, **11**, 2001–2010.
44 M. Kohagen, M. Brehm, J. Thar, W. Zhao, F. Müller-Plathe and B. Kirchner, *J. Phys. Chem. B*, 2011, **115**, 693–702.
45 F. Dommert, J. Schmidt, B. Qiao, Y. Zhao, C. Krekeler, L. Delle Site, R. Berger and C. Holm, *J. Chem. Phys.*, 2008, **129**, 224501.
46 F. Dommert, J. Schmidt, C. Krekeler, Y. Y. Zhao, R. Berger, L. Delle Site and C. Holm, *J. Mol. Liq.*, 2010, **152**, 2–8.
47 P. Ballone, C. Pinilla, J. Kohanoff and M. G. Del Popolo, *J. Phys. Chem. B*, 2007, **111**, 4938–4950.
48 H. V. Spohr and G. N. Patey, *J. Chem. Phys.*, 2009, **130**, 104506.
49 H. V. Spohr and G. N. Patey, *J. Chem. Phys.*, 2010, **132**, 154504.
50 H. V. Spohr and G. N. Patey, *J. Chem. Phys.*, 2010, **132**, 234510.
51 S. Zahn, G. Bruns, J. Thar and B. Kirchner, *Phys. Chem. Chem. Phys.*, 2008, **10**, 6921–6924.
52 S. Zahn, F. Uhlig, J. Thar, C. Spickermann and B. Kirchner, *Angew. Chem., Int. Ed.*, 2008, **47**, 3639–3641.
53 S. B. C. Lehmann, M. Roatsch, M. Schöppke and B. Kirchner, *Phys. Chem. Chem. Phys.*, 2010, **12**, 7473–7486.
54 S. Zahn, J. Thar and B. Kirchner, *J. Chem. Phys.*, 2010, **132**, 124506–124506.
55 K. Fumino, A. Wulf and R. Ludwig, *Angew. Chem., Int. Ed.*, 2008, **47**, 3830–3834.
56 B. Qiao, C. Krekeler, R. Berger, L. Delle Site and C. Holm, *J. Phys. Chem. B*, 2008, **112**, 1743–1751.
57 J. Thar, M. Brehm, A. P. Seitsonen and B. Kirchner, *J. Phys. Chem. B*, 2009, **113**, 15129–15132.

 

58 A. Stark, *Top. Curr. Chem.*, 2009, **290**, 41–81.
59 C. Krekeler, J. Schmidt, Y. Y. Zhao, B. F. Qiao, R. Berger, C. Holm and L. Delle Site, *J. Chem. Phys.*, 2008, **129**, 174503.
60 C. Krekeler, F. Dommert, J. Schmidt, Y. Y. Zhao, C. Holm, R. Berger and L. Delle Site, *Phys. Chem. Chem. Phys.*, 2010, **12**, 1817–1821.
61 K. Wendler, S. Zahn, F. Dommert, R. Berger, C. Holm, B. Kirchner and L. Delle Site, J. Chem. Theory Comput., 2011, DOI: 10.1021/ct200375v.
62 J. Schmidt, C. Krekeler, F. Dommert, Y. Zhao, R. Berger, L. Delle Site and C. Holm, *J. Phys. Chem. B*, 2010, **114**, 6150–6155.
63 N. Sieffert and G. Wipff, *J. Phys. Chem. B*, 2006, **110**, 13076.
64 I. V. Leontyev and A. A. Stuchebrukhov, *J. Chem. Phys.*, 2009, **130**, 085102.
65 I. V. Leontyev and A. A. Stuchebrukhov, *J. Chem. Theory Comput.*, 2010, **6**, 1498–3161.
66 R. Ahlrichs, M. Bär, M. Häser, H. Horn and C. Kölmel, *Chem. Phys. Lett.*, 1989, **162**, 165–169.
67 TURBOMOLE V6.3 2011, a development of University of Karlsruhe and Forschungszentrum Karlsruhe GmbH, 1989–2007, TURBOMOLE GmbH, since 2007; available from http://www.turbomole.com.
68 H.-J. Werner, P. J. Knowles, F. R. Manby, M. Schütz , *et al.*, MOLPRO, version 2010.1, a package of ab initio programs, http://www.molpro.net, 2010.
69 C. I. Bayly, P. Cieplak, W. D. Cornell and P. A. Kollman, *J. Phys. Chem.*, 1993, **97**, 10269–10280.
70 C. M. Breneman and K. B. Wiberg, *J. Comput. Chem.*, 1990, **11**, 361–373.
71 R. Bader, *Atoms in Molecules: A Quantum Theory*, Oxford University Press, 1990.
72 M. J. Frisch, G. W. Trucks, H. B. Schlegel, G. E. Scuseria, M. A. Robb, J. R. Cheeseman, J. A. Montgomery, Jr., T. Vreven, K. N. Kudin, J. C. Burant, J. M. Millam, S. S. Iyengar, J. Tomasi, V. Barone, B. Mennucci, M. Cossi, G. Scalmani, N. Rega, G. A. Petersson, H. Nakatsuji, M. Hada, M. Ehara, K. Toyota, R. Fukuda, J. Hasegawa, M. Ishida, T. Nakajima, Y. Honda, O. Kitao, H. Nakai, M. Klene, X. Li, J. E. Knox, H. P. Hratchian, J. B. Cross, V. Bakken, C. Adamo, J. Jaramillo, R. Gomperts, R. E. Stratmann, O. Yazyev, A. J. Austin, R. Cammi, C. Pomelli, J. Ochterski, P. Y. Ayala, K. Morokuma, G. A. Voth, P. Salvador, J. J. Dannenberg, V. G. Zakrzewski, S. Dapprich, A. D. Daniels, M. C. Strain, O. Farkas, D. K. Malick, A. D. Rabuck, K. Raghavachari, J. B. Foresman, J. V. Ortiz, Q. Cui, A. G. Baboul, S. Clifford, J. Cioslowski, B. B. Stefanov, G. Liu, A. Liashenko, P. Piskorz, I. Komaromi, R. L. Martin, D. J. Fox, T. Keith, M. A. Al-Laham, C. Y. Peng, A. Nanayakkara, M. Challacombe, P. M. W. Gill, B. G. Johnson, W. Chen, M. W. Wong, C. Gonzalez and J. A. Pople, *GAUSSIAN 03*, Gaussian, Inc., Wallingford, CT, 2004.
73 G. Henkelman, A. Arnaldsson and H. Jónsson, *Comput. Mater. Sci.*, 2006, **36**, 354–360.
74 E. Sanville, S. D. Kenny, R. Smith and H.G., *J. Comput. Chem.*, 2007, **28**, 899–908.
75 W. Tang, E. Sanville and G. Henkelman, *J. Phys.: Condens. Matter*, 2009, **21**, 084204.
76 R. Car and M. Parrinello, *Phys. Rev. Lett.*, 1985, **55**, 2471.
77 CPMD, Copyright IBM Corp 1990–2008. Copyright MPI für Festkörperforschung Stuttgart 1997–2001., http://www.cpmd.org/.
78 J. P. Perdew, K. Burke and M. Ernzerhof, *Phys. Rev. Lett.*, 1996, **77**, 3865–3865.
79 N. Troullier and J. L. Martins, *Phys. Rev. B: Condens. Matter*, 1991, **43**, 1993–1993.
80 A. D. Becke, *Phys. Rev. A: At., Mol., Opt. Phys.*, 1988, **38**, 3098–3098.
81 C. Lee, W. Yang and R. G. Parr, *Phys. Rev. B*, 1988, **37**, 785–785.
82 S. Grimme, *J. Comput. Chem.*, 2006, **27**, 1787–1799.
83 G. Berghold, C. J. Mundy, A. H. Romero, J. Hutter and M. Parrinello, *Phys. Rev. B: Condens. Matter*, 2000, **61**, 10040–10040.
84 P. L. Silvestrelli, M. Bernasconi and M. Parrinello, *Chem. Phys. Lett.*, 1997, **277**, 478–482.
85 P. L. Silvestrelli, N. Marzari, D. Vanderbilt and M. Parrinello, *Solid State Commun.*, 1998, **107**, 7–11.
86 P. E. Blöchl, *J. Chem. Phys.*, 1995, **103**, 7422–7422.
87 J. VandeVondele, M. Krack, F. Mohamed, M. Parrinello, T. Chassaing and J. Hutter, *Comput. Phys. Commun.*, 2005, **167**, 103–128.
88 Y. Zhang and W. Yang, *Phys. Rev. Lett.*, 1998, **80**, 890–890.
89 S. Goedecker, M. Teter and J. Hutter, *Phys. Rev. B: Condens. Matter*, 1996, **54**, 1703–1703.
90 C. Hartwigsen, S. Goedecker and J. Hutter, *Phys. Rev. B: Condens. Matter*, 1998, **58**, 3641–3641.
91 J. Hutter, Computer code CP2K, 2000–2008, http://cp2k.berlios.de.
92 L. E. Chirlian and M. M. Francl, *J. Comput. Chem.*, 1987, **8**, 894–905.
93 S. Nosé, *J. Chem. Phys.*, 1984, **81**, 511–511.

 

94 W. G. Hoover, *Phys. Rev. A: At., Mol., Opt. Phys.*, 1985, **31**, 1695–1695.
95 H. J. C. Berendsen, D. van der Spoel and R. van Drunen, *Comput. Phys. Commun.*, 1995, **91**, 43–56.
96 E. Lindahl, B. Hess and D. van der Spoel, *J. Mol. Mod.*, 2001, **7**, 306–317.
97 B. Hess, C. Kutzner, D. van der Spoel and E. Lindahl, *J. Chem. Theory Comput.*, 2008, **4**, 435–447.
98 M. Parrinello and A. Rahman, *J. Appl. Phys.*, 1981, **52**, 7182–7190.
99 B. Hess, *J. Chem. Theory Comput.*, 2008, **4**, 116–122.
100 U. Essmann, L. Perera, M. L. Berkowitz, T. Darden, H. Lee and L. Pedersen, *J. Chem. Phys.*, 1995, **103**, 8577.
101 H. Wang, F. Dommert and C. Holm, *J. Chem. Phys.*, 2010, **133**, 034117.
102 M. P. Allen and D. J. Tildesley, *Computer Simulation of Liquids*, Clarendon Press, Oxford, 1st edn, 1987.
103 C. E. R. Prado, M. G. Del Popolo, T. G. A. Youngs, J. Kohanoff and R. M. Lynden-Bell, *Mol. Phys.*, 2006, **104**, 2477–2483.
104 J. P. Hallett, C. L. Liotta, G. Ranieri and T. Welton, *J. Org. Chem.*, 2009, **74**, 1864–1868.
105 T. Cremer, C. Kolbeck, K. R. J. Lovelock, N. Paape, R. Wölfel, P. S. Schulz, P. Wasserscheid, H. Weber, J. Thar, B. Kirchner, F. Maier and H.-P. Steinrück, *Chem.–Eur. J.*, 2010, **16**, 9018–9033.
106 P. L. Silvestrelli and M. Parrinello, *Phys. Rev. Lett.*, 1999, **82**, 3308–3308.
107 S. Zahn, K. Wendler, L. Delle Site and B. Kirchner, *Phys. Chem. Chem. Phys.*, 2011, **13**, 15083–15093.
108 M. Buhl, A. Chaumont, R. Schurhammer and G. Wipff, *J. Phys. Chem. B*, 2005, **109**, 18591–18599.
109 D. A. Turton, J. Hunger, A. Stoppa, G. Hefter, A. Thoman, M. Walther, R. Buchner and K. Wynne, *J. Am. Chem. Soc.*, 2009, **131**, 11140–11146.
110 M. Tariq, P. Forte, M. C. Gomes, J. Canongia Lopes and L. Rebelo, *J. Chem. Thermodyn.*, 2009, **41**, 790–798.
111 O. Borodin, *J. Phys. Chem. B*, 2009, **113**, 11463–11478.
112 J. N. Canongia Lopes and A. A. H. Padua, *J. Phys. Chem. B*, 2004, **108**, 16893–16898.
113 J. N. Canongia Lopes and A. A. H. Padua, *J. Phys. Chem. B*, 2006, **110**, 19586–19592.
114 A. A. Fannin, D. A. Floreani, L. A. King, J. S. Landers, B. J. Piersma, D. J. Stech, R. L. Vaughn, J. S. Wilkes and W.J.L., *J. Phys. Chem.*, 1984, **88**, 2614–2621.
115 F. Dommert, K. Wendler, L. Delle Site, R. Berger and C. Holm, to be submitted, 2011.
116 S. Kossmann, J. Thar, B. Kirchner, P. Hunt and T. Welton, *J. Chem. Phys.*, 2006, **124**, 174506.
117 S. Zahn and B. Kirchner, *J. Phys. Chem. A*, 2008, **112**, 8430–8435.
118 G. H. Wannier, *Phys. Rev.*, 1937, **52**, 191–197.
119 W. Kohn, *Phys. Rev.*, 1959, **115**, 809–821.
120 J. des Cloizeaux, *Phys. Rev.*, 1963, **129**, 554–566.
121 N. Marzari and D. Vanderbilt, *Phys. Rev. B: Condens. Matter*, 1997, **56**, 12847–12847.
122 N. Marzari, I. Souza and D. Vanderbilt, *Psi-K Scient. Highlight of the Month*, 2003, **57**, 129–168.
123 M. N. Kobrak and H. Li, *Phys. Chem. Chem. Phys.*, 2010, **12**, 1922–1932.

 

# Temperature-dependent structure of ionic liquids: X-ray scattering and simulations

Hemant K. Kashyap,[a] Cherry S. Santos,[b] Harsha V. R. Annapureddy,[a] N. Sanjeeva Murthy,[c] Claudio J. Margulis*[a] and Edward W. Castner, Jr*[b]

*Received 5th April 2011, Accepted 15th June 2011*
**DOI: 10.1039/c1fd00059d**

In this article we determine the temperature-dependent structure of the tetradecyltrihexylphosphonium bis(trifluoromethylsulfonyl)amide ionic liquid using a combination of X-ray scattering and molecular dynamics simulations. As in many other room-temperature ionic liquids three characteristic intermolecular peaks can be detected in the structure function $S(q)$. A prepeak or first sharp diffraction peak is observed at about $q = 0.42$ Å$^{-1}$. Long range anion-anion correlations are the most important contributors to this peak. In all systems we have studied to date, this prepeak is a signature of solvation asymmetry. The peak in $S(q)$ near $q = 0.75$ Å$^{-1}$ is the signature of ionic alternation and arises from the charge ordered separation of ions of the same charge. The most intense diffraction peak near $q = 1.37$ Å$^{-1}$ arises from short-range separation between ions of opposite charge combined with a significant contribution from cationic carbon-carbon interactions, indicating that cationic hydrophobic tails have significant contacts.

## 1. Introduction

Ionic liquids (ILs) are increasingly becoming the electrolyte of choice for a wide range of energy applications.[1] A number of recent special issues[2–4] and reviews have been devoted to the physical chemistry of ionic liquids.[5–9] Phosphonium based ILs have been shown to have interesting transport and solvation properties; supercritical $CO_2$ as well as other solutes appear to have improved solubilities in them.[10–15]

Structures of high-temperature molten salts with monatomic and small, symmetric ions are reviewed by Tosi *et al.*[16] Room-temperature ionic liquids (RTILs) generally have glass transition temperatures and melting points that are much lower than molten salts from the alkali halide family. One reason for this is that RTIL ions are generally larger, more asymmetric, and flexible than the corresponding alkali halide ions.

A number of reports have been made on the structural properties of RTILs based on molecular simulations and theoretical results[17–25] as well as experiments using X-ray and neutron scattering.[26–39] In common with the high-temperature molten salts,[16] the RTILs show a strong signature of charge-ordering.[22,40]

[a]*Department of Chemistry, University of Iowa, Iowa City, IA, 52242-1294. E-mail: claudio-margulis@uiowa.edu*
[b]*Department of Chemistry and Chemical Biology, Rutgers, The State University of New Jersey, Piscataway, NJ, 08854-8066. E-mail: ed.castner@rutgers.edu*
[c]*New Jersey Center for Biomaterials, Rutgers, The State University of New Jersey, Piscataway, NJ, 08854-8066*

In addition to the charge-ordering that is found to be in common between molten salts and RTILs, both types of liquids can display a first sharp diffraction peak (FSDP) in the structure function for low values of the scattering vector $q$.[16,17,20,25,30,38,39,41,42] (The FSDP is also referred to as a 'prepeak'). This body of work shows that the FSDP in both molten salts and RTILs can be caused by the subtle interplay of different contributions to the structure function.[25,30,38,39,42]

In this article we report on the structural properties of liquid tetradecyltrihexylphosphonium bis(trifluoromethylsulfonyl)amide ($P_{14,666}^+/NTf_2^-$) using results from X-ray scattering and molecular dynamics simulations in the temperature range between 150–400 K. The structure of $P_{14,666}^+/NTf_2^-$ is given in Fig. 1, showing the four alkyl tails arrayed in a tetrahedral geometry. The tetradecyltrihexylphosphonium cation ($P_{14,666}^+$) is frequently used in ILs because this bulky yet flexible cation leads to overall lower viscosities while containing a larger fraction of hydrocarbon than most other IL cations. We note that tetraalkylphosphonium ILs have lower viscosities than the homologous tetraalkylammonium ILs.[43] A detailed structural study of $P_{14,666}^+/Cl^-$ has been reported by Gontrani *et al.*[44] Molecular dynamics simulations have been reported for $P_{14,666}^+/NTf_2^-$ by Liu *et al.*[45] showing three-dimensional isodensity surfaces around the cation as well as radial distribution functions.

## 2. Experimental methods

The $P_{14,666}^+/NTf_2^-$ ionic liquid was a gift from our collaborators Xiang (Lillian) Li and Prof. Mark Maroncelli at Penn State University. The sample was dried on a Schlenk vacuum line at 323 K for 48 h prior to the experiment and then back-filled with argon. The water content was determined to be 18 ppm using Karl Fischer coulometric titration. X-ray experiments were performed at the Advanced Photon Source (APS) at Argonne National Laboratory (USA). The WAXS data used to calculate the structure functions were collected at APS beam line 11-ID-C.

### 2.1 Wide-angle X-ray scattering and the liquid structure function

The WAXS experimental methods have been previously described in detail.[40,42] X-ray scattering experiments were performed in transmission geometry with the sample sealed in an X-ray quartz capillary. The 2.0 mm (outer diameter) X-ray quartz capillaries were obtained from Hampton Research (HR6-150) and filled with $P_{14,666}^+/NTf_2^-$ to a sample height of approximately 2 cm in an argon glovebox and sealed with capillary wax from Hampton Research (HR4-328). A Cryostream Plus open-flow cooling system (Oxford Cryosystems) was used for sample temperature control between 150 and 400 K. The 115 keV X-ray beam ($\lambda = 0.1078$ Å) from a Si(311) Laue monochromator was collimated to a beam size of about 0.2 × 0.2 mm.

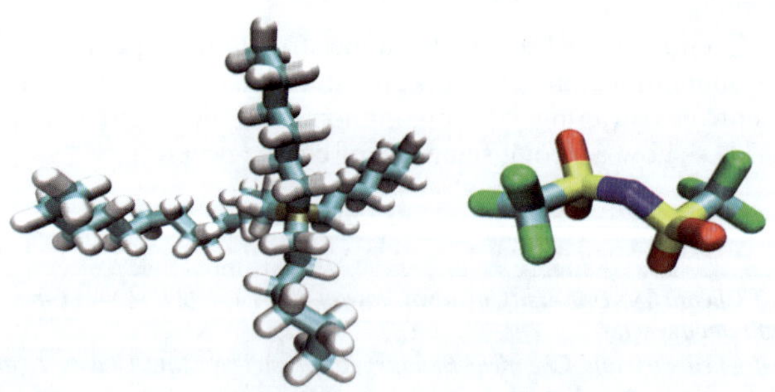

**Fig. 1** Chemical structures of tetradecyltrihexylphosphonium ($P_{14,666}^+$, left) and bis(trifluoromethylsulfonyl)amide ($NTf_2^-$, right).

 

Diffraction patterns were collected using a Perkin Elmer amorphous Silicon 1621 AN3 detector with a sample-to-detector distance of 661 mm and a $q$ range from 0.2 to 20 $\text{Å}^{-1}$. X-ray scattering data were obtained with a total of 10 min exposure time (5 s/frame, 24 frames/file and 5 files/data set). All raw data were integrated and converted to intensity *versus* scattering wave vector $q$ using the Fit2D software from Hammersley *et al.*[46,47] The scattering vector magnitude is defined as $q = |\vec{q}| = \frac{4\pi}{\lambda}\sin(\theta)$, where $2\theta$ is the scattering angle and $\lambda$ is the X-ray wavelength. The structure function is calculated from the observed total scattering intensity $I_{coh}(q)$ using eqn (1).

$$S(q) = \frac{I_{coh}(q) - \sum_i x_i f_i^2(q)}{\left[\sum_i x_i f_i(q)\right]^2} \tag{1}$$

where for atomic species $i$, $x_i$ and $f_i$ are the atomic fraction and X-ray form factor, respectively. The PDFgetX2 program from Qui and Billinge[48] was used to obtain the total scattering structure function $S(q)$ after corrections for sample absorption, Compton scattering and multiple scattering, as described previously.[40,42]

## 3. Simulation and theoretical methods

All molecular dynamics (MD) simulations were performed using the GROMACS[49,50] simulation package with 216 pairs of $P_{14,666}^+$ cations and $NTf_2^-$ anions in a cubic box. OPLS-AA force field parameters[51,52] along with modifications introduced by Lopes and Padua[53,54] were used to model cations and anions. Several short simulations of about 100 ps were run in which partial charges and temperature were varied in order to equilibrate the system at target temperatures and pressures. Systems were coupled to a Nose-Hoover[55–57] thermostat and a Parinnello-Rahman[58] barostat. The pressure was set to 1 bar for all the temperatures studied. For higher target temperatures, production trajectories starting from a pre-equilibrated configuration were run at constant temperature and pressure in the isothermal-isobaric (NPT) ensemble for at least 5 ns, otherwise trajectories were at least 7 ns in duration. The final 1 ns of the production runs was saved in each case at 100 fs intervals for the calculation of average liquid properties. Equations of motion were solved using the leap-frog algorithm with a time step of 1 fs as coded in GROMACS.[49,50] Proper periodic boundary conditions and minimum image convention were applied. Cutoffs for the Lennard-Jones and real space part of the electrostatic interactions were set to 15 Å. Electrostatic interactions were calculated using the Particle Mesh Ewald (PME)[59,60] summation technique with interpolation order of 6 and Fourier grid spacing of 0.8 Å.

The total structure function $S(q)$ of the liquid was calculated using eqn (2).

$$S(q) = \frac{\rho_o \sum_i^n \sum_j^n x_i x_j f_i(q) f_j(q) \int_0^\infty 4\pi r^2 (g_{ij}(r) - 1) \frac{\sin qr}{qr} dr}{\left[\sum_i^n x_i f_i(q)\right]^2} \tag{2}$$

In eqn (2), $g_{ij}(r)$ is the radial pair distribution function for atoms of type $i$ and $j$; $x_i$ is the fraction of atom of type $i$; $f_i(q)$ is the X-ray atomic form factor for the $i^{th}$ type atom; and $\rho_o$ is the number density. The atomic form factors are taken from the International Tables for Crystallography.[61]

Different partitionings of the above structure function may be advantageous in order to better understand the origin of the features in $S(q)$.[25,40] In this study, we have chosen to partition the structure function either into atomic pair contributions such that $S(q) = \sum_i^n \sum_j^n S_{ij}(q)$, where $S_{ij}(q)$ is given by

$$\frac{S_{ij}(q) = \rho_o x_i\, x_j\, f_i(q) f_j(q) \int_0^\infty 4\pi r^2 (g_{ij}(r) - 1) \frac{\sin qr}{qr} dr}{\left[\sum_i x_i f_i(q)\right]^2} \tag{3}$$

or $S(q)$ can also be partitioned into cationic and anionic subcomponents and their cross correlations. In this case, $S(q) = S^{c-c}(q) + S^{a-a}(q) + S^{c-a}(q) + S^{a-c}(q)$, where the different ion-type subcomponents of $S(q)$ are given by[40]

$$S^{\alpha-\alpha}(q) = \frac{\rho_o \sum_i \sum_j x_i\, x_j f_i(q) f_j(q)}{\left[\sum_i x_i f_i(q)\right]^2} \times \int_0^\infty 4\pi r^2 \frac{\sin qr}{qr} dr \left(g_{ij}^{\alpha-\alpha}(r) - \lim_{r\to\infty} g_{ij}^{\alpha-\alpha}(r)\right) \tag{4}$$

where $\alpha$ represents either the cation or the anion and

$$S^{\alpha-\beta}(q) + S^{\beta-\alpha}(q) = \frac{\rho_o \sum_i \sum_j x_i\, x_j f_i(q) f_j(q)}{\left[\sum_i x_i f_i(q)\right]^2} \times \left\{ \int_0^\infty 4\pi r^2 \frac{\sin qr}{qr} dr \left(g_{ij}^{\alpha-\beta}(r) - \lim_{r\to\infty} g_{ij}^{\alpha-\beta}(r)\right) + \int_0^\infty 4\pi r^2 \frac{\sin qr}{qr} dr \left(g_{ij}^{\beta-\alpha}(r) - \lim_{r\to\infty} g_{ij}^{\beta-\alpha}(r)\right) \right\}. \tag{5}$$

In eqn (5), $g_{ij}^{\alpha-\beta}(r)$ is the partial pair distribution function of the $i^{th}$ type of atom in $\alpha$ and $j^{th}$ type of atom in $\beta$.

## 4. Results and discussion

Simulated and experimental densities (in parenthesis) for $P_{14,666}^+/NTf_2^-$ at 225, 295, 353 and 400 K are 1.123(1.121), 1.084(1.068), 1.041(1.026) and 1.003(0.993) g cm$^{-3}$ respectively. The agreement between computation and experiment is quite good with a maximum deviation of about 2%.[10,62,63]

Fig. 2 shows a comparison of the temperature dependent structure functions obtained from experiments and simulations. Similar to other $NTf_2^-$ based liquids showing a pre-peak, liquid $P_{14,666}^+/NTf_2^-$ shows three characteristic peaks at values of $q$ below 2 Å$^{-1}$. These peaks, which are predominantly intermolecular in nature, appear in this case at about 0.42, 0.75 and 1.37 Å$^{-1}$. As expected, the three intermolecular peaks shift towards lower $q$ values with increasing temperature due to the accompanying decrease in density.

With some variations across ionic liquids, these three peaks are quite characteristic of their structural landscape. The peak at higher values of $q$ is always indicative of close contact neighbors. In the case of RTILs, because of their ionic nature, it is often associated with the adjacency of species of different charge. It will become apparent from our analysis below that because of the length of the cationic alkyl tails in $P_{14,666}^+/NTf_2^-$, there is a significant component of carbon-carbon (or non-polar)

 

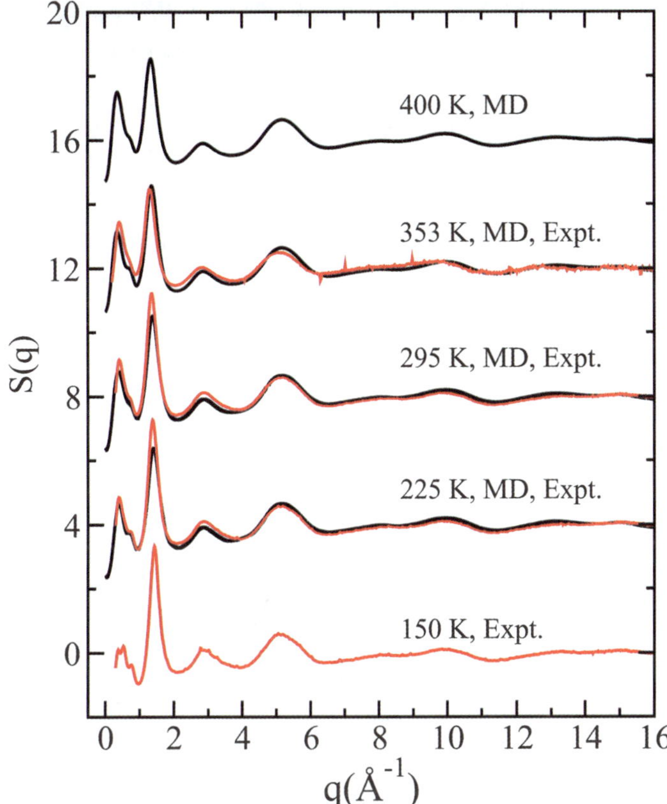


**Fig. 2** The X-ray structure functions $S(q)$ for $P_{14,666}{}^{+}/NTf_2{}^{-}$ at five different temperatures. MD simulation and experimental results are plotted with black and red solid lines, respectively. The $S(q)$ functions have offsets added for clarity: 225 K (+ 4), 295 K (+ 8), 353 K (+ 12) and 400 K (+ 16).

adjacency contributing to this peak. The intermolecular peak at intermediate values of $q$, in our case a shoulder at about 0.75 Å$^{-1}$, is a signature of charge alternation. This peak is associated with the length scale for cation-cation and anion-anion interactions. Independent of the IL studied, this peak always appears as the result of cancellation of large positive same-charge contributions and large negative opposite-charge contributions. This simply means that at locations where one expects to find ions of the same charge there is a systematic absence of ions having the opposite charge. Clearly this should be a feature common to most ionic liquids. Since RTILs are often asymmetric, it is common to find at least two characteristic length scales of separation between anions or cationic polar parts. The longer of these two asymmetry length scales is often associated with a prepeak.

In order to identify the atomic pair correlation functions responsible for these three peaks, in Fig. 3 we show all partial atomic structure functions calculated using eqn (3). These partial structure functions are split into three sub-graphs for visual clarity. From Fig. 3 it is evident that atomic pairs O–F, O–S and F–S are the most significant contributors to the prepeak at $q = 0.42$ Å$^{-1}$. All of these atoms belong only to the anions. The fact that the anions are the main contributors to the prepeak can be further verified by partitioning $S(q)$ into ion types instead of atom types as in eqn (4) and (5). Fig. 4 shows that the anion-anion partial structure function dominates the prepeak region.

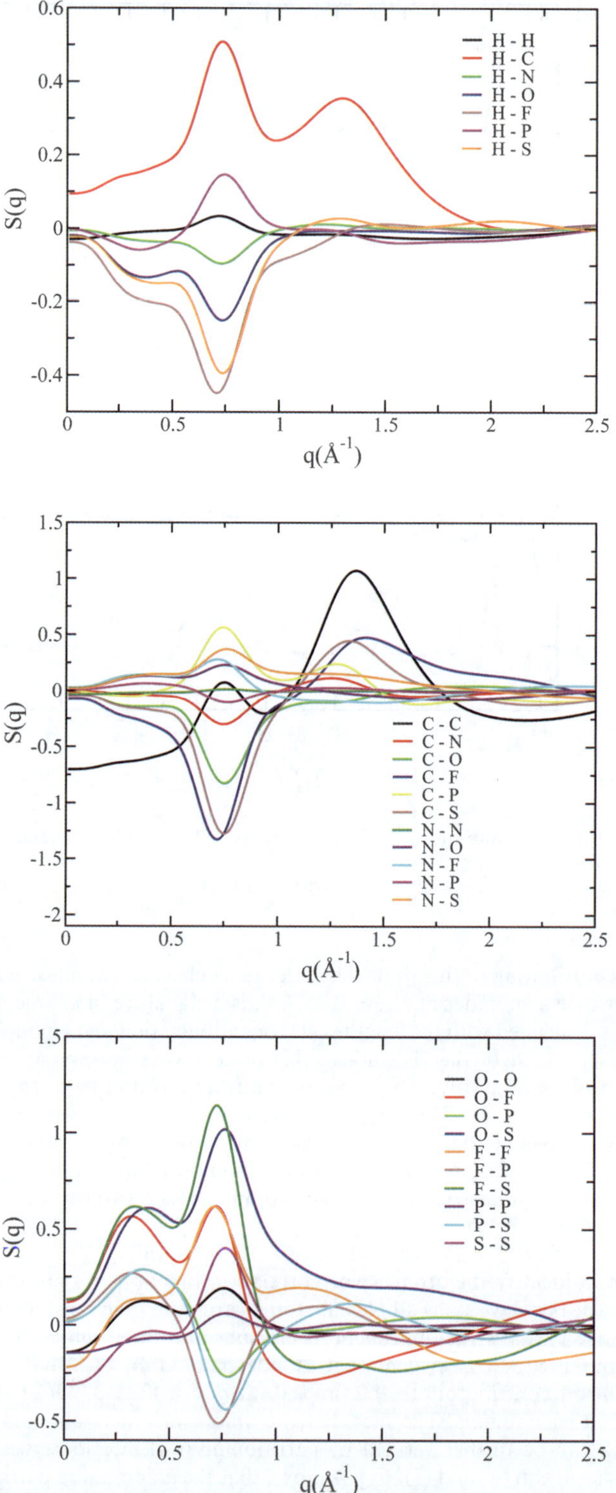


**Fig. 3** Atomic structure functions $S_{ij}(q)$ for $P_{14,666}^+/NTf_2^-$, from MD simulations at 400 K.

 

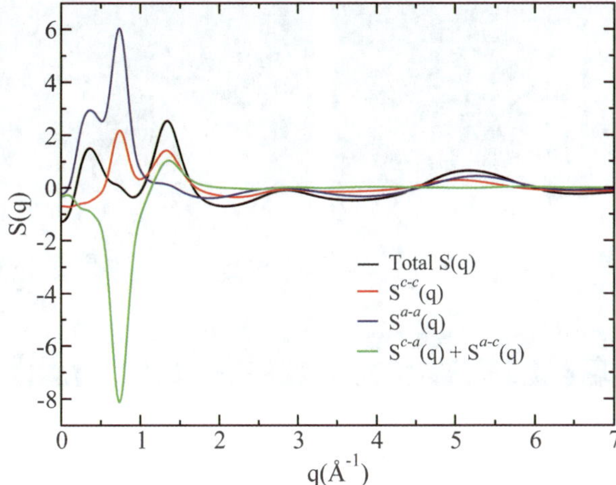


**Fig. 4** Total and partial structure functions $S(q)$ for $P_{14,666}{}^+/NTf_2{}^-$ at 400 K from MD simulations.

Fig. 4 unequivocally demonstrates that the shoulder peak in the total $S(q)$ at about 0.75 Å$^{-1}$ arises from positive cation-cation and anion-anion partial structure factors canceling large negative cation-anion contributions. This proves that the shoulder peak near $q = 0.75$ Å$^{-1}$, which corresponds to a real space length of about 8.5 Å, is due to typical cation-cation and anion-anion distances. The out of phase behavior of cation-anion cross terms with respect to cation-cation and anion-anion self terms is a signature of the charge ordering in ionic liquids (or molten salts).[25,40]

The most intense peak in $S(q)$ below 2 Å$^{-1}$ appears around $q = 1.37$ Å$^{-1}$ (see Fig. 2). From Fig. 4 we see that this peak arises mainly from cation-cation and cation-anion correlations but not from anion-anion correlations. This peak is due to ions in close contact. In this particular liquid these close contact interactions are between ions of opposite charge and between cationic hydrophobic regions. Fig. 5 confirms that it is the carbon tails that are the main contributors to the cation-cation intermolecular adjacency correlation.

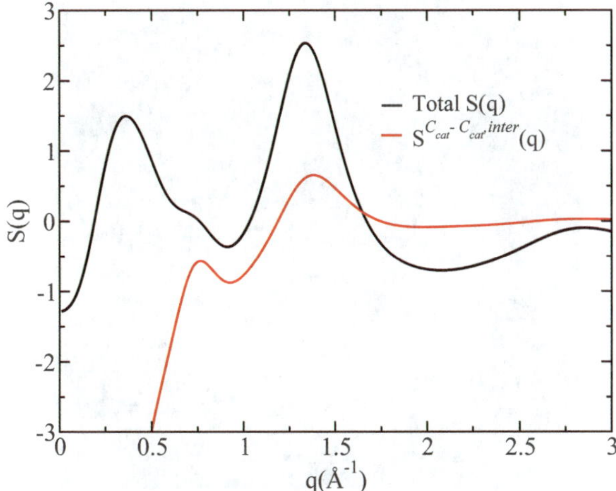


**Fig. 5** Total $S(q)$ compared with the partial intermolecular structure function between cationic carbon atoms.

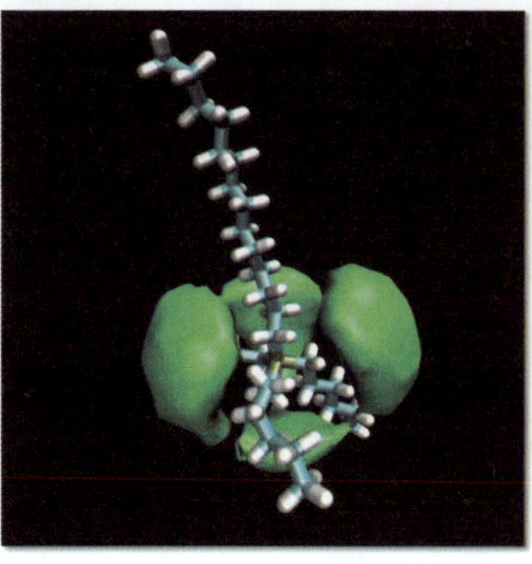

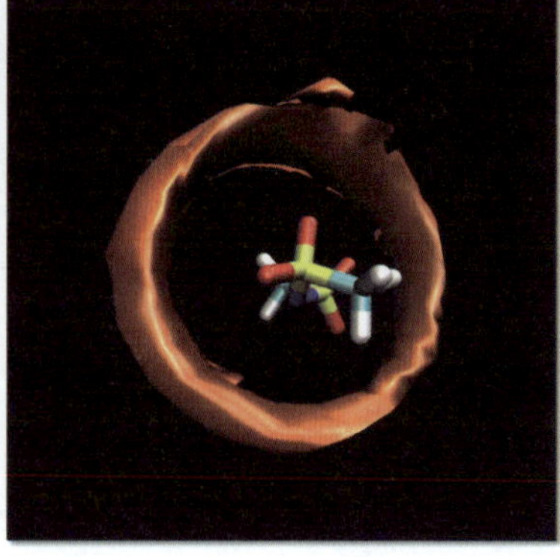

**Fig. 6** Left: Average isodensity for all anion atoms surrounding a central cation; this shows the anion packing in the spaces between the tetrahedral alkyl substituents on $P_{14,666}^{+}$. Right: Average isodensity for all cation atoms distributed around a central anion. Results are averaged over 25 liquid snapshots from the MD trajectory at 400 K.

To illustrate the most probable arrangement of ions in the liquid, in Fig. 6 we show the isodensity surfaces for anion atoms surrounding a cation and for cation atoms about an anion. It is clear in Fig. 6(left) that the distribution of anions around a cation is asymmetric. Anions surround a cation close to the P atom, adjacent to the shorter alkyl tails, but are absent along the direction of the longer tail. The distribution of cations around an anion shown in Fig. 6(right) has the form of a ring about the center of the anion.

Fig. 7 shows cationic and anionic isodensities around an anion. This figure clearly exemplifies the charge alternation that gives rise to the peaks at 0.75 and 1.37 $\mathrm{\AA}^{-1}$. Fig. 6(left) hints at the type of asymmetry that may be the origin of the prepeak in this liquid. While assigning the exact origin of the prepeak to a particular liquid feature is difficult, we note that values of $2\pi/q$ match a distance corresponding to the second anionic solvation shell of the anions. By this we mean the second layer of anions around an anion (see Fig. 8). This figure also shows that the N–N distribution is almost identical to P–P distribution.

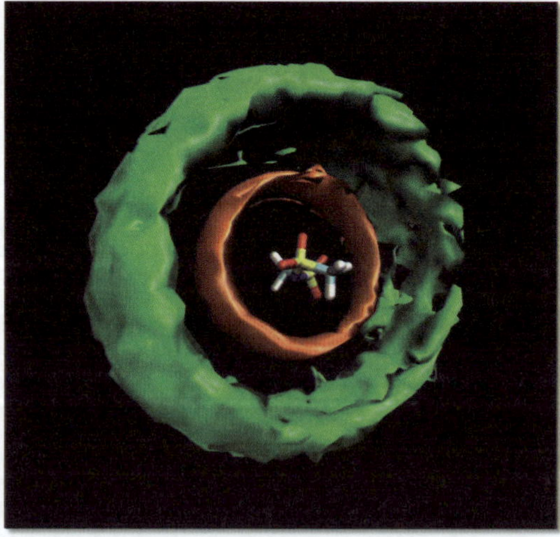

**Fig. 7** Averaged isodensity surface for all cation atoms (orange) surrounding a central anion and for all anion atoms (green) surrounding an anion from the MD simulation at $T = 400$ K.

 

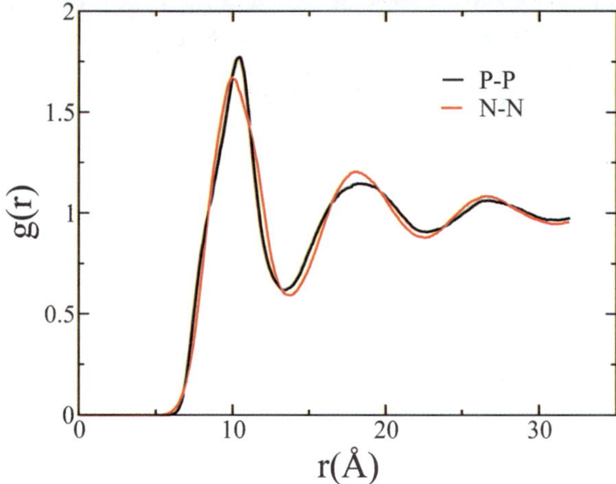


**Fig. 8** Cation-cation (P–P atoms) and anion-anion (N–N atoms) partial radial distribution functions.

The $P_{14,666}^+/NTf_2^-$ ionic liquid shows significant complexity and hydrophobic aggregation. This can be appreciated by examining the significant contribution of the C–C terms to the total $S(q)$ in Fig. 5 to the peak around 1.37 Å$^{-1}$. The anionic contribution to the prepeak is likely to correlate with the distribution of hydrophobic tails. While tails do not contribute significantly to the prepeak, they are the spacers between electron-rich anions that solvate cationic polar parts and thus determine the spacing between these groups.

## 5. Conclusions

X-ray scattering experiments and molecular dynamics simulations show that $NTf_2^-$ based ionic liquids often display three characteristic peaks (two if there is no FSDP) in the intermolecular region of the X-ray structure function for values of $q$ in the range from 0.3 to 1.5 Å$^{-1}$. These peaks are signatures of the structures present in this class of RTILs. For $P_{14,666}^+/NTf_2^-$, the peak at 1.37 Å$^{-1}$ results from adjacency of oppositely charged ions as well as hydrophobic interactions. While close contact of oppositely charged species is common to most ionic liquids, the hydrophobic interactions appear because of the large alkyl substituents on the cations.

The shoulder at about 0.75 Å$^{-1}$ appears as a result of cancellation of large structural correlations having different signs. The positive contribution is due to ions of the same charge while the negative contribution is due to ions of opposing charge. This feature appears in every ionic liquid we have studied to date and is in fact the signature of charge alternation in RTILs.

A prepeak in the X-ray structure function is often observed in ionic liquids with longer alkyl tails. In the system studied here the prepeak mainly arises from the second anion-anion layer; there are cations separating each anionic layer and therefore their asymmetry and patterns of hydrophobic adjacency determine the position of the prepeak.

## Acknowledgements

CSS, NSM and EWC would like to recognize the help extended by Dr Chris Benmore on the data analysis and for assisting in the data collection at ANL-APS beamline 11-ID-C. This work was supported by the U. S. Department of Energy, Office of Basic Energy Sciences, Division of Chemical Sciences, Geosciences, and Biosciences under SISGR Grant No. DE-FG02-09ER16118 awarded to EWC and CJM. Use of the Advanced Photon Source at Argonne National Laboratory was supported by the U. S. Department of Energy, Office of Science, Office of Basic Energy Sciences, under Contract No. DE-AC02-06CH11357.

## References

1 J. F. Wishart, *Energy Environ. Sci.*, 2009, **2**, 956–961.
2 R. D. Rogers and G. A. Voth, *Acc. Chem. Res.*, 2007, **40**, 1077–1078.
3 J. F. Wishart and E. W. Castner Jr., *J. Phys. Chem. B*, 2007, **111**, 4639–4640.
4 F. Endres, *Phys. Chem. Chem. Phys.*, 2010, **12**, 1648.
5 Z. Hu and C. J. Margulis, *Acc. Chem. Res.*, 2007, **40**, 1097–1105.
6 E. W. Castner, Jr., J. F. Wishart and H. Shirota, *Acc. Chem. Res.*, 2007, **40**, 1217–1227.
7 K. R. J. Lovelock, I. J. Villar-Garcia, F. Maier, H.-P. Steinrück and P. Licence, *Chem. Rev.*, 2010, **110**, 5158–5190.
8 E. W. Castner Jr. and J. F. Wishart, *J. Chem. Phys.*, 2010, **132**, 120901.
9 E. W. Castner, Jr., C. J. Margulis, M. Maroncelli and J. F. Wishart, *Annu. Rev. Phys. Chem.*, 2011, **62**, 85–105.
10 K. J. Fraser, E. I. Izgorodina, M. Forsyth, J. L. Scott and D. R. MacFarlane, *Chem. Commun.*, 2007, 3817–3819.
11 D. R. MacFarlane, M. Forsyth, E. I. Izgorodina, A. P. Abbott, G. Annat and K. Fraser, *Phys. Chem. Chem. Phys.*, 2009, **11**, 4962–4967.
12 H. Jin, G. Baker, S. Arzhantsev, J. Dong and M. Maroncelli, *J. Phys. Chem. B*, 2007, **117**, 7291–7302.
13 N. Ito, S. Arzhantsev, M. Heitz and M. Maroncelli, *J. Phys. Chem. B*, 2004, **108**, 5771–5777.
14 J. Huang and T. Ruether, *Aust. J. Chem.*, 2009, **62**, 298–308.
15 P. J. Carvalho, V. H. Alvarez, I. M. Marrucho, M. Aznar and J. A. P. Coutinho, *J. Supercrit. Fluids*, 2010, **52**, 258–265.
16 M. P. Tosi, D. L. Price and M.-L. Saboungi, *Annu. Rev. Phys. Chem.*, 1993, **44**, 173–211.
17 S. M. Urahata and M. C. Ribeiro, *J. Chem. Phys.*, 2004, **120**, 1855–1863.
18 M. G. Del Popolo and G. A. Voth, *J. Phys. Chem. B*, 2004, **108**, 1744–1752.
19 Y. Shim, M. Choi and H. Kim, *J. Chem. Phys.*, 2005, **122**, 44510.
20 J. N. A. Canongia Lopes and A. A. H. Pádua, *J. Phys. Chem. B*, 2006, **110**, 3330–3335.
21 C. Schröder, T. Rudas and O. Steinhauser, *J. Chem. Phys.*, 2006, **125**, 244506.
22 C. Schröder, C. Wakai, H. Weingärtner and O. Steinhauser, *J. Chem. Phys.*, 2007, **126**, 084511.
23 B. L. Bhargava, R. Devane, M. L. Klein and S. Balasubramanian, *Soft Matter*, 2007, **3**, 1395–1400.
24 C. Schröder and O. Steinhauser, *J. Chem. Phys.*, 2008, **128**, 224503.
25 H. V. R. Annapureddy, H. K. Kashyap, P. M. De Biase and C. J. Margulis, *J. Phys. Chem. B*, 2010, **114**, 16838–16846.
26 A. E. Bradley, C. Hardacre, J. D. Holbrey, S. Johnston, S. E. J. McMath and M. Nieuwenhuyzen, *Chem. Mater.*, 2002, **14**, 629–635.
27 C. Hardacre, J. D. Holbrey, S. E. J. McMath, D. T. Bowron and A. K. Soper, *J. Chem. Phys.*, 2003, **118**, 273–278.
28 A. Triolo, A. Mandanici, O. Russina, V. Rodriguez-Mora, M. Cutroni, C. Hardacre, M. Nieuwenhuyzen, H.-J. Bleif, L. Keller and M. A. Ramos, *J. Phys. Chem. B*, 2006, **110**, 21357–21364.
29 C. Hardacre, J. D. Holbrey, M. Nieuwenhuyzen and T. G. A. Youngs, *Acc. Chem. Res.*, 2007, **40**, 1146–1155.
30 A. Triolo, O. Russina, H.-J. Bleif and E. Di Cola, *J. Phys. Chem. B*, 2007, **111**, 4641–4644.
31 R. Atkin and G. G. Warr, *J. Phys. Chem. B*, 2008, **112**, 4164–4166.
32 C. Hardacre, J. D. Holbrey, C. L. Mullan, M. Nieuwenhuyzen, T. G. A. Youngs and D. T. Bowron, *J. Phys. Chem. B*, 2008, **112**, 8049–8056.
33 S. Fukuda, M. Takeuchi, K. Fujii, R. Kanzaki, T. Takamuku, K. Chiba, H. Yamamoto, Y. Umebayashi and S.-i. Ishiguro, *J. Mol. Liq.*, 2008, **143**, 2–7.
34 K. Fujii, Y. Soejima, Y. Kyoshoin, S. Fukuda, R. Kanzaki, Y. Umebayashi, T. Yamaguchi, S.-i. Ishiguro and T. Takamuku, *J. Phys. Chem. B*, 2008, **112**, 4329–4336.
35 K. Fujii, S. Seki, S. Fukuda, T. Takamuku, S. Kohara, Y. Kameda, Y. Umebayashi and S.-i. Ishiguro, *J. Mol. Liq.*, 2008, **143**, 64–69.
36 R. Kanzaki, T. Mitsugi, S. Fukuda, K. Fujii, M. Takeuchi, Y. Soejima, T. Takamuku, T. Yamaguchi, Y. Umebayashi and S. Ishiguro, *J. Mol. Liq.*, 2009, **147**, 77–82.
37 K. Fujii, T. Mitsugi, T. Takamuku, T. Yanaguchi, Y. Umebayashi and S. Ishiguro, *Chem. Lett.*, 2009, **38**, 340–341.

 

38 O. Russina, A. Triolo, L. Gontrani, R. Caminiti, D. Xiao, L. G. Hines Jr., R. A. Bartsch, E. L. Quitevis, N. Pleckhova and K. R. Seddon, *J. Phys.: Condens. Matter*, 2009, **21**, 424121.
39 C. Hardacre, J. D. Holbrey, C. L. Mullan, T. G. A. Youngs and D. T. Bowron, *J. Chem. Phys.*, 2010, **133**, 74510–74517.
40 C. S. Santos, H. V. R. Annapureddy, N. S. Murthy, H. K. Kashyap, E. W. Castner, Jr and C. J. Margulis, *J. Chem. Phys.*, 2011, **134**, 064501.
41 A. Triolo, O. Russina, B. Fazio, G. B. Appetecchi, M. Carewska and S. Passerini, *J. Chem. Phys.*, 2009, **130**, 164521.
42 C. S. Santos, N. S. Murthy, G. A. Baker and E. W. Castner, Jr, *J. Chem. Phys.*, 2011, **134**, 121101.
43 H. Shirota, H. Fukazawa, T. Fujisawa and J. F. Wishart, *J. Phys. Chem. B*, 2010, **114**, 9400–9412.
44 L. Gontrani, O. Russina, F. L. Celso, R. Caminiti, G. Annat and A. Triolo, *J. Phys. Chem. B*, 2009, **113**, 9235–9240.
45 X. Liu, G. Zhou, S. Zhang and G. Yu, *Mol. Simul.*, 2010, **36**, 79–86.
46 A. Hammersley, S. Svensson, M. Hanfland, A. Fitch and D. Hausermann, *Int. J. High Pressure Res.*, 1996, **14**, 235–248.
47 A. P. Hammersley, *FIT2D V10.3 Reference Manual V4.0*, European Synchrotron Radiation Facility, 1998.
48 X. Qui, J. W. Thompson and S. J. L. Billinge, *J. Appl. Crystallogr.*, 2004, **37**, 678.
49 B. Hess, C. Kutzner, D. van der Spoel and E. Lindahl, *J. Chem. Theory Comput.*, 2008, **4**, 435–447.
50 D. van der Spoel, E. Lindahl, B. Hess, G. Groenhof, A. E. Mark and H. J. C. Berendsen, *J. Comput. Chem.*, 2005, **26**, 1701–1718.
51 W. L. Jorgensen, D. S. Maxwell and J. Tirado-Rives, *J. Am. Chem. Soc.*, 1996, **118**, 11225–11236.
52 M. L. P. Price, D. Ostrovsky and W. L. Jorgensen, *J. Comput. Chem.*, 2001, **22**, 1340–1352.
53 J. N. C. Lopes and A. A. H. Pádua, *J. Phys. Chem. B*, 2004, **108**, 16893–16898.
54 J. N. C. Lopes and A. A. H. Pádua, *J. Phys. Chem. B*, 2006, **110**, 19586–19592.
55 S. Nosé, *J. Chem. Phys.*, 1984, **81**, 511–519.
56 S. Nosé, *Mol. Phys.*, 1984, **52**, 255–268.
57 W. G. Hoover, *Phys. Rev. A: At., Mol., Opt. Phys.*, 1985, **31**, 1695.
58 M. Parrinello and A. Rahman, *J. Appl. Phys.*, 1981, **52**, 7182–7190.
59 T. Darden, D. York and L. Pedersen, *J. Chem. Phys.*, 1993, **98**, 10089–10092.
60 U. Essmann, L. Perera, M. L. Berkowitz, T. Darden, H. Lee and L. G. Pedersen, *J. Chem. Phys.*, 1995, **103**, 8577–8593.
61 *International Tables for Crystallography*, ed. E. Prince, International Union of Crystallography, 2006, vol. C.
62 J. Jacquemin, R. Ge, P. Nancarrow, D. W. Rooney, M. F. Costa Gomes, A. A. H. Pádua and C. Hardacre, *J. Chem. Eng. Data*, 2008, **53**, 716–726.
63 M. Tariq, P. Forte, M. C. Gomes, J. C. Lopes and L. Rebelo, *J. Chem. Thermodyn.*, 2009, **41**, 790–798.

# Structure and dynamics in protic ionic liquids: A combined optical Kerr-effect and dielectric relaxation spectroscopy study

**David A. Turton,**[*a] **Thomas Sonnleitner,**[b] **Alex Ortner,**[c] **Markus Walther,**[c] **Glenn Hefter,**[d] **Kenneth R. Seddon,**[e] **Simona Stana,**[e] **Natalia V. Plechkova,**[e] **Richard Buchner**[b] **and Klaas Wynne**[a]

*Received 1st April 2011, Accepted 1st June 2011*
**DOI: 10.1039/c1fd00054c**

The structure and dynamics of ionic liquids (ILs) are unusual due to the strong interactions between the ions and counter ions. These microscopic properties determine the bulk transport properties critical to applications of ILs such as advanced fuel cells. The terahertz dynamics and slower relaxations of simple alkylammonium nitrate protic ionic liquids (PILs) are here studied using femtosecond optical Kerr-effect spectroscopy, dielectric relaxation spectroscopy, and terahertz time-domain spectroscopy. The observed dynamics give insight into more general liquid behaviour while comparison with glass-forming liquids reveals an underlying power-law decay and relaxation rates suggest supramolecular structure and nanoscale segregation.

## 1 Introduction

Protic ionic liquids (PILs) have particular importance due to their ongoing development as electrolytes for advanced fuel cells,[1–5] in addition to a wide range of potential applications including as acidic catalysts. An understanding of structural properties and dynamics are crucial to the full realisation of such applications and there is increasing interest in the presence of nanostructure.[6,7] Femtosecond optical Kerr-effect (OKE) and dielectric relaxation spectroscopies are both highly sensitive probes of low-frequency diffusional and librational motions.[8–10] By comparing OKE spectra with dielectric spectra (that cover a range extended to 10 THz, by combining dielectric relaxation spectroscopy (DRS) with time-domain terahertz and FTIR spectroscopies) we recently identified the mesoscopic structure in 1,3-dialkylimidazolium-based room-temperature ionic liquids (RTILs)[11] that had been predicted by molecular simulations.[12,13] This structure appeared even for short-chain RTILs and its presence is critical to understanding their macroscopic properties—viscosity, thermal conductivity, and solute diffusion. Ethylammonium nitrate (EAN) is the most studied, prototypical, PIL.[14–18]

[a]*School of Chemistry and WestCHEM, University of Glasgow, Glasgow, G12 8QQ, UK. E-mail: david.turton@glasgow.ac.uk; klaas.wynne@glasgow.ac.uk*
[b]*Inst. Phys. and Theor. Chemistry, Uni. Regensburg, 93040 Regensburg, Germany. E-mail: Thomas.Sonnleitner@chemie.uni-regensburg.de; Richard.Buchner@chemie.uni-regensburg.de*
[c]*Department of Molecular and Optical Physics, Albert-Ludwigs-Universität Freiburg, 79104 Freiburg, Germany. E-mail: alex.ortner@email.de; walther@physik.uni-freiburg.de*
[d]*Chemistry Department, Murdoch University, Murdoch, W.A. 6150, Australia. E-mail: g.hefter@murdoch.edu.au*
[e]*QUILL, Queen's University, Belfast, BT9 5AG, UK. E-mail: quill@qub.ac.uk*

For EAN and propylammonium nitrate (PAN) the presence of nanoscale segregation is controversial. It has been suggested recently that the ability of each ion to form multiple hydrogen bonds promotes a heterogeneous structure,[16] similar to that of water,[17] and that the hydrogen bond network has on average one H-bond donor and acceptor position free.[19] Small angle neutron scattering (SANS) and large angle X-ray scattering (LAXS) measurements on EAN (and PAN) have been interpreted as evidence of a disordered, locally smectic or sponge-like structure arising partially from electrostatic and hydrogen bonding attractions between the amine nitrogen and the nitrate anion rather than through the ordering of imidazolium ring cations assumed in the imidazolium-based ILs.[20,21] Further evidence of structure through SANS was shown to be consistent with calculations that imply a nanometre-scale heterogeneity.[22] However, computational and X-ray scattering studies have been interpreted to show that low-$q$ scattering peaks do not necessarily imply mesoscale order.[23]

The objective of this study is to interpret the transport dynamics of these liquids and identify, if present, structural heterogeneities. Limited measurements of the dielectric spectra have been made previously;[15,24] this work presents the first continuous spectrum for the entire intermolecular region and is the first OKE study. Wideband temperature-dependent OKE and dielectric spectra for EAN are presented, with additional OKE measurements for PAN.

OKE spectroscopy measures the derivative over time of the two-point time-correlation function of the anisotropic part of the many-body polarisability tensor $\mathbf{\Pi}$ in the time domain, $S_{\mathrm{OKE}}(t) \propto (\mathrm{d}/\mathrm{d}t)\langle \mathbf{\Pi}_{xy}(t)\mathbf{\Pi}_{xy}(0)\rangle$.[25] A Fourier-transform deconvolution yields the spectrum and this is equivalent to the Bose–Einstein corrected depolarised Raman spectrum.[26] Dielectric spectroscopy similarly measures the derivative over time of the two-point time-correlation function of the dipole moment vector, $S_{\mathrm{DRS}}(t) \propto (\mathrm{d}/\mathrm{d}t)\langle \boldsymbol{\mu}(t)\boldsymbol{\mu}(0)\rangle$. The two techniques are therefore, to a degree, complementary, but measure the same dynamics resulting in spectra that differ fundamentally only in the amplitudes.[27] One exception to this rule is that rotational motions generally appear faster in OKE than in DRS due to the higher symmetry of the polarisability tensor[28–30] (as discussed below).

## 2 Experimental

The high viscosities of these strongly interacting liquids result in slow relaxation timescales. Hence, even at room temperature, wide time spans are required to capture the entire intermolecular dynamics. Two OKE set-ups were therefore employed; the faster dynamics were measured using a 7 nJ, 20 fs pulse laser oscillator as described previously,[31] while—to increase the signal-to-noise of the weaker relaxation measured at longer times—a similar set-up employing a regeneratively-amplified laser (Coherent Legend USX) with a 1 μJ pulse stretched to a duration of *ca.* 1 ps was used. The two measured time-domain signals could be overlapped by greater than an order of magnitude in time allowing accurate concatenation.[32] The liquids were contained in a fused-quartz cuvette with a path length of 2 mm. The temperature was controlled by a cryostat (Oxford Instruments, Optistat DN, ±0.01 K) at sub-ambient temperature or, at higher temperatures, by a lab-built copper cell holder thermostatically controlled to ±0.1 K.

Broadband dielectric spectra were measured by a frequency-domain reflectometer (Agilent E8364B VNA combined with Agilent 85070E-020 (0.2–20GHz) and 85070E-050 (0.5–50 GHz) probes)[33] and two waveguide interferometers (IFMs) at 27 to 89 GHz.[34] Raw VNA data were corrected for calibration errors with a Padé approximation. For selected samples, two further IFMs at 8.5 to 17.5 GHz[34] were used to crosscheck the reliability of the VNA results and showed excellent agreement, within the precision of the instruments. The temperature was controlled to within 0.05 K. A transmission/reflection terahertz time-domain spectrometer (THz-TDS) recorded data from 0.3 to 3 THz,[35] with temperature control using a lab-built cell holder with a precision of ±0.5 K. The FIR data were recorded from 0.9 to 12 THz on a Bruker Vertex 70 FTIR spectrometer with the liquids held between a pair of polymethylpentene (TPX) windows with a path length of 20 μm in a temperature controlled mount (Harrick TFC-S25 & ATC-024). An average of about 4000 scans was taken for each sample and a background spectrum taken from the windows alone was subtracted. Complex permittivity spectra were then derived by Kramers-Kronig transformation,[36] and the data were resampled to a logarithmic frequency scale to match the lower frequency data. These data were then summed to give a nearly continuous spectrum to 10 THz.

Temperature-dependent viscosities were measured for EAN on an Anton-Paar AMVn rolling ball micro-viscometer over the temperature range 5 °C to 100 °C (±0.05 K).

The EAN sample was prepared in Regensburg by the reaction of equimolar amounts of ethylamine with nitric acid. Water was removed by rotary evaporation followed by lyophilisation. The crude EAN was recrystallized thrice from acetonitrile and the resulting colourless liquid dried at <10 nbar and 50 °C for one week and then stored under nitrogen. The water content determined by coulometric Karl Fischer titration of the final product was 192 ppm. The PAN sample was provided by QUILL at Queen's University Belfast and was prepared by a dropwise addition of concentrated nitric acid to a cold solution of propylamine. The product was dried by lyophilisation for 24 h at 0.38 mbar to give a pale yellow liquid. The water content was <1 wt%. For both samples, purity was confirmed by $^1$H NMR spectroscopy.

## 3 Results and analysis

The OKE and dielectric data for EAN at 25 °C are compared directly in Fig. 1. There are both remarkable differences and similarities in the observed intermolecular dynamics. In both cases, the higher frequencies are dominated by a peak that is familiar from studies of molecular liquids, where it is generally associated with librations (hindered rotational motions). In EAN, surprisingly, the librational band appears to be 2–3 times higher in frequency for the dielectric measurement. At low frequency a strong structural or α-relaxation mode is observed that appears to have a Cole-Davidson (CD) lineshape $S_{CD}(\omega) = (1 + i\omega\tau)^{-\alpha}$.[10,24] With respect to

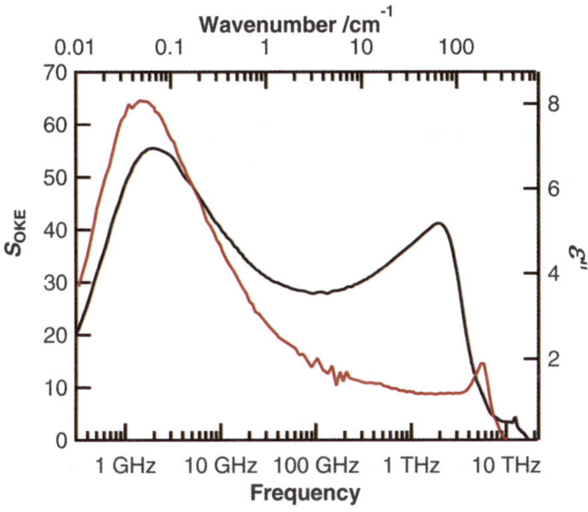


**Fig. 1** OKE (black) and dielectric loss, ε'' (red) data for EAN at 25 °C (the relative amplitude scaling is arbitrary).

a Debye function, the CD lineshape is asymmetrically broadened by an additional high frequency contribution that, in the time domain, results in an initial power law decay $S \propto t^{1-\alpha}$.[37] The CD function is often used as a model of heterogeneous relaxation in deeply supercooled liquids.[38] The measured α-relaxation is generally associated with molecular reorientations (*via* the dipole moment or polarisability tensor). Due to the higher symmetry of the polarisability tensor, the (2nd rank) OKE signal relaxes three times faster than the (1st rank) DRS signal in simple cases of diffusional relaxation.[9,28] However, here the frequencies are quite similar.

For EAN, OKE and DRS data were recorded from 5 °C to 65 °C in steps of 10 °C with the terahertz and FIR data measured at 20 °C steps (Fig. 2).

Arguably, the least well understood region of the spectrum is where the librational modes give way to relaxations. The mode-coupling theory predicts low frequency behaviour above the glass transition $T_g$ but does not yet provide a practical fitting model. In the absence of a more meaningful physical model, such data are typically analysed through phenomenological lineshapes applied in order to parameterise the behaviour allowing comparison to other systems and ideally to give insight into the physical origin of the dynamics. However, the intermolecular dynamics (measured below a few terahertz) often cannot be decomposed into simple functions. Furthermore, data measured in the time domain often are analysed using a different set of models (exponential decay, stretched exponential, power law, *etc.*) to those employed in the frequency domain (Debye, Havriliak-Negami, Gaussian, *etc.*). Several of these functions have no symbolic Fourier transform which leads to confusion as apparently different phenomena are observed. The reconciliation of these two views is overdue.[38]

As both techniques measure the decay in anisotropy following an impulse, there is an inertial contribution to the signal that occurs through the librational motions. Therefore, the librational process and the relaxation should be somehow combined. This is to some extent achieved when the relaxation is modified by an inertial rise function,[29,39] in which the inertial frequency is determined by the librational frequencies. However, the spectrum of EAN is typical of viscous liquids in that the librations are well resolved from the α relaxation. The broad region that "connects" the two is poorly understood. In studies of glass-forming liquids, this region has in the time domain been approximated as a series of power laws.[40] Alternatively, in the frequency domain the sum of a number of Debye, damped harmonic (Brownian), or similar lineshapes are typically employed.[11,41] Here, there is no actual evidence of individual modes (with the exception of a poorly resolved shoulder in the OKE spectrum close to 1 THz). A solution to this problem appeared with the PAN sample

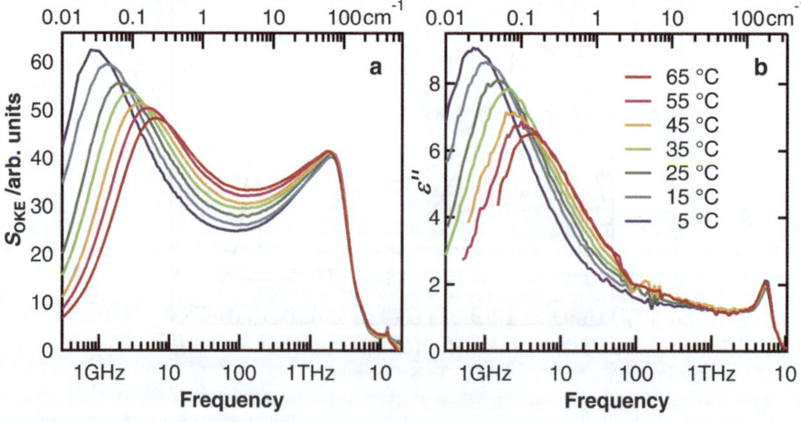


**Fig. 2** Spectra for EAN: **(a)** OKE and **(b)** dielectric loss measured from 5–65 °C.

 

that proved to be a relatively good glass former. OKE data were measured down to 140 K allowing the relaxations to be frozen out and the line-shape of the non-diffusional response to be revealed (Fig. 3(a)).

In the time domain, it can be seen that as the temperature falls, the response approaches a line (on logarithmic scales). This is characteristic of a power-law decay, $S \propto t^{-p}$. At 200 K, $p$ is measured at 0.82. At lower temperatures, but above the glass transition, the sample typically crystallised after several minutes preventing reliable measurements of the relaxation, but it appears that $p$ tends to unity, *i.e.* to a logarithmic decay.[40] In the frequency domain, this appears as a constant loss or white noise response.[42] A convenient method of generating such a response is given by the Cole-Cole function $S_{CC}(\omega) = 1/(1 + (i\omega\tau)^{\beta})$. The Cole-Cole function is a Debye function broadened symmetrically by the exponent $\beta$ for $0 < \beta < 1$. We have shown previously how this function can be made consistent with spectroscopic dynamics by terminating it in the $\alpha$ relaxation at low frequency, and by applying an inertial rise modification at high frequency.[29] In the limit of $\beta \rightarrow 0$, a logarithmic decay results that for the modified function extends appropriately from the librations to the $\alpha$ relaxation (Fig. 4 and 5). This fit was then used as the basis for fitting the temperature dependent data for both the OKE and dielectric measurements.

The fit for the two data sets at 25 °C is shown in Fig. 5. A concise fit was given by a Cole-Davidson function for the $\alpha$ relaxation with the librational modes described by a Gaussian function, the width of which might imply the heterogeneity of the cage structures. The weak mode at ~1 THz is fitted by a damped-harmonic oscillator. The logarithmic decay (Cole-Cole function) links the librations to the $\alpha$ relaxation, and is defined by a single amplitude parameter combined with the inertial rise rate and $\alpha$ relaxation time.

The $\alpha$-relaxation time constants are also shown on an Arrhenius plot in Fig. 6. The viscosity, which extends to 100 °C, is non-Arrhenius and is fitted by the Vogel-Fulcher-Tammann function.[43,44] The two measures of the $\alpha$ relaxation do appear to be Arrhenius over the measured temperature range, but have activation energies that differ by ~10% (21.7 kJ mol$^{-1}$ for DRS and 23.9 kJ mol$^{-1}$ for OKE).

## 4 Discussion

Both the difference of librational frequencies and, to some extent, the anomalous relaxation rates can be explained simply. The nitrate ion is planar with delocalised

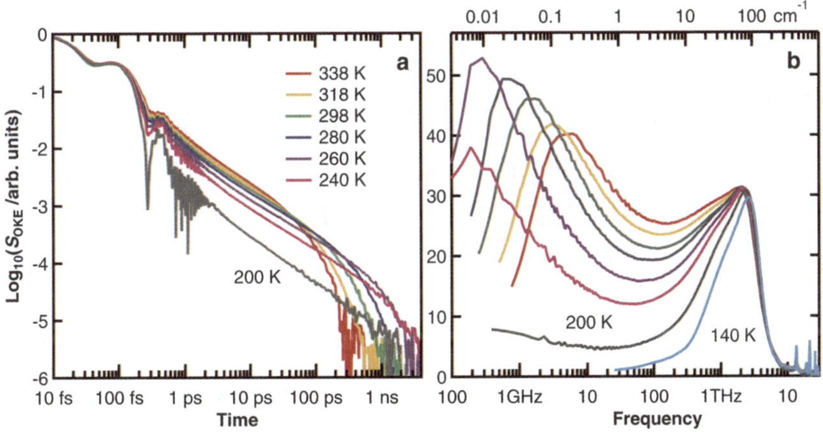


**Fig. 3** **(a)** Time-domain OKE signal for PAN for seven temperatures between 200 K and 338 K. The fast and slow dynamics signals are concatenated at *ca.* 5 ps. **(b)** The imaginary part of the spectrum is yielded by a Fourier transform deconvolution of (a).

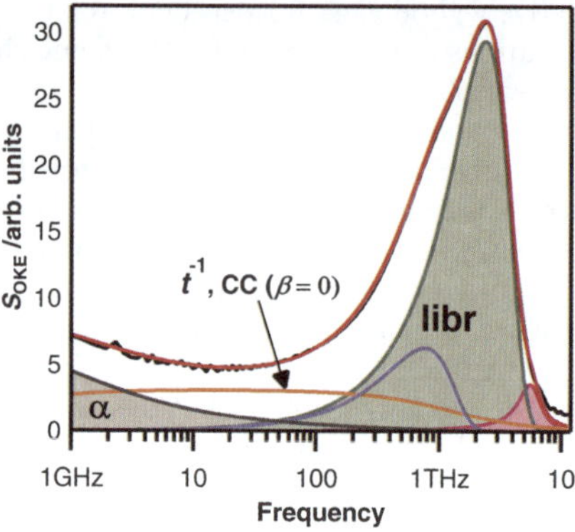


**Fig. 4** Fit (red) of the OKE spectrum (black) for PAN at 200 K. CC is the Cole-Cole function with $\beta = 0$ as described in the text.

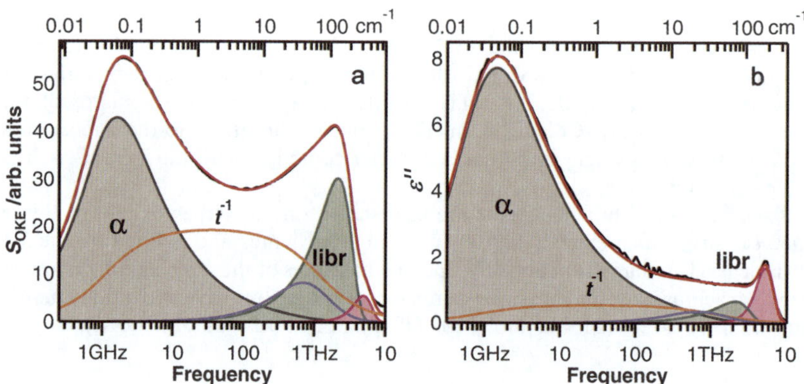


**Fig. 5** Fit to **(a)** the OKE data and **(b)** the dielectric loss for EAN at 25 C.

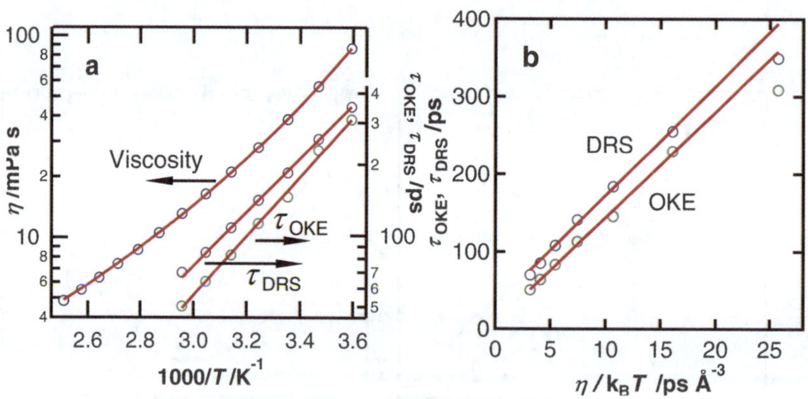


**Fig. 6** **(a)** Arrhenius plot of the alpha relaxation time constant for the fits to both OKE and DRS data for EAN compared to viscosity. **(b)** Stokes–Einstein–Debye relationship.

π-bonding and an overall negative charge. This results in a highly polarisable ion that has a high degree of anisotropy ($\alpha_{\parallel} \approx 5.6$ Å$^3$, $\alpha_{\perp} \approx 3.1$ Å$^3$ [45]), but no intrinsic dipole moment. Conversely, the alkylammonium ion does have a permanent dipole moment, and also has a tightly-bound depleted charge system with no aromatic character and is therefore only weakly polarisable. This system therefore has almost perfect complementarity: the dielectric spectrum is sensitive to motions of the cation while OKE is sensitive to the anion. It has been suggested, with reference to DFT calculations,[19] that the two strong bands, at ~2 and ~6 THz, are directly equivalent to the hydrogen bond bend and stretch of water.[28] However, since the dielectric and OKE signals are both primarily sensitive to rotations, the dominant contribution in this region is likely to be librational, in which case the shoulder observed just below 1 THz would be a likely candidate for the hydrogen bond stretch.

Both OKE and dielectric measurements for EAN reveal a well defined α-relaxation mode but there is no evidence of the sub-α mode indicative of mesoscopic structure that is present in the more complex planar imidazolium-based ILs.[11] Furthermore, the temperature-dependent data do not reveal a clear change from structured to unstructured as might be expected to appear as a break in the Arrhenius dependence or an anomalous change in the terahertz dynamics.

However, in contrast to single-molecule relaxation, which follows Arrhenius behaviour, the viscosity becomes "super-Arrhenius" at low temperature, even over the limited range of 5–65 °C. This behavior is referred to as "fragile". Fragility is not unusual in ILs[44] but this decoupling of viscosity and diffusivity implies a Stokes–Einstein–Debye violation. The super-Arrhenius behaviour is often interpreted as due to the presence of clusters that result in increasing length scales as a liquid is supercooled. However, then it would be usual for the single-molecule dynamics to be similarly affected. The Stokes–Einstein–Debye (SED) expression $\tau_{rot} = V_{eff}\eta/k_BT$, relates the rotational relaxation time constant $\tau_{rot}$ to viscosity $\eta$ through the effective molecular volume $V_{eff}$. Although derived for spherical particles diffusing in a homogeneous fluid, the expression has proven applicable to a much wider range of systems. In Fig. 6(b), $\tau_{rot}$ is plotted against $\eta/k_BT$, to yield $V_{eff}$. Values of *ca.* 15 Å$^3$ are found for both OKE and DRS, but for the latter the additional factor of 3 gives 5 Å$^3$. Corrections can be applied for the molecular aspect ratio and slip or stick conditions[36] but this simple estimate suggests that $V_{eff}$ for the ethylammonium ion is underestimated. The active motion for the nitrate ion (in OKE) is out-of-plane rotation and for the alkylammonium ion (in DRS) end-over-end rotation which does not suggest that $V_{eff}$ for the ammonium ion should be particularly small.

The most obvious explanation for these disparities is that the liquid is not a simple one. Rather it is a strongly-interacting cooperatively-relaxing mixture. In particular, it is reasonable to expect that the multiple hydrogen bonds result in relaxation through a molecular jump mechanism as described for water.[46] The relaxation time is therefore determined, in each case, by the rate of hydrogen bond breaking rather than the diffusional process intrinsic to the SED relation.

The divergence of viscosity from single-molecule diffusivity at low temperatures, visible in Fig. 6(a) and 6(b), is consistent with increasing inhomogeneity due to cluster formation. If so, these clusters do not introduce additional low-frequency modes in either spectrum, which suggests aggregation of the alkyl groups on the cation consistent with the previous experimental observations and simulations of weakly-defined structure. Such aggregates would have zero net dipole moment (if approximately symmetrical), would not result in an appreciable OKE signal due to the very weak polarisability of the cation, and would not significantly influence the relaxation of the nitrate ion, which hydrogen bonds to the $-NH_3$ site. The degree of aggregation must be low according to the small deviation from Arrhenius behaviour and since the DRS signal is not severely attenuated at low temperature.

## Acknowledgements

We are grateful to the EPSRC and DFG for funding.

## References


1 W. Xu and C. A. Angell, *Science*, 2003, **302**, 422–425.
2 M. Yoshizawa, W. Xu and C. A. Angell, *J. Am. Chem. Soc.*, 2003, **125**, 15411–15419.
3 J. Le Bideau, L. Viau and A. Vioux, *Chem. Soc. Rev.*, 2011, **40**, 907–925.
4 H. Nakamoto and M. Watanabe, *Chem. Commun.*, 2007, 2539–2541.
5 S. Y. Lee, A. Ogawa, M. Kanno, H. Nakamoto, T. Yasuda and M. Watanabe, *J. Am. Chem. Soc.*, 2010, **132**, 9764–9773.
6 T. L. Greaves, D. F. Kennedy, S. T. Mudie and C. J. Drummond, *J. Phys. Chem. B*, 2010, **114**, 10022–10031.
7 T. L. Greaves, D. F. Kennedy, A. Weerawardena, N. M. K. Tse, N. Kirby and C. J. Drummond, *J. Phys. Chem. B*, 2011, **115**, 2055–2066.
8 G. Giraud, C. M. Gordon, I. R. Dunkin and K. Wynne, *J. Chem. Phys.*, 2003, **119**, 464–477.
9 D. A. Turton, J. Hunger, G. Hefter, R. Buchner and K. Wynne, *J. Chem. Phys.*, 2008, **128**, 161102.
10 J. Hunger, A. Stoppa, S. Schrödle, G. Hefter and R. Buchner, *ChemPhysChem*, 2009, **10**, 723–733.
11 D. A. Turton, J. Hunger, A. Stoppa, G. Hefter, A. Thoman, M. Walther, R. Buchner and K. Wynne, *J. Am. Chem. Soc.*, 2009, **131**, 11140–11146.
12 Y. T. Wang and G. A. Voth, *J. Am. Chem. Soc.*, 2005, **127**, 12192–12193.
13 J. Lopes and A. A. H. Padua, *J. Phys. Chem. B*, 2006, **110**, 3330–3335.
14 J. N. C. Lopes and L. P. N. Rebelo, *Phys. Chem. Chem. Phys.*, 2010, **12**, 1948–1952.
15 M. Kruger, E. Brundermann, S. Funkner, H. Weingartner and M. Havenith, *J. Chem. Phys.*, 2010, **132**, 101101.
16 S. Ishiguro, Y. Umebayashi, R. Kanzaki and K. Fujii, *Pure Appl. Chem.*, 2010, **82**, 1927–1941.
17 K. Fumino, A. Wulf and R. Ludwig, *Angew. Chem., Int. Ed.*, 2009, **48**, 3184–3186.
18 T. L. Greaves and C. J. Drummond, *Chem. Soc. Rev.*, 2008, **37**, 1709–1726.
19 S. Zahn, J. Thar and B. Kirchner, *J. Chem. Phys.*, 2010, **132**, 124506.
20 R. Atkin and G. G. Warr, *J. Phys. Chem. B*, 2008, **112**, 4164–4166.
21 Y. Umebayashi, W.-L. Chung, T. Mitsugi, S. Fukuda, M. Takeuchi, K. Fujii, T. Takamuku, R. Kanzaki and S.-i. Ishiguro, *J. Comput. Chem., Jpn.*, 2008, **7**, 125–134.
22 R. Hayes, S. Imberti, G. G. Warr and R. Atkin, *Phys. Chem. Chem. Phys.*, 2011, **13**, 3237–3247.
23 C. S. Santos, H. V. R. Annapureddy, N. S. Murthy, H. K. Kashyap, E. W. Castner and C. J. Margulis, *J. Chem. Phys.*, 2011, **134**, 064501.
24 H. Weingartner, A. Knocks, W. Schrader and U. Kaatze, *J. Phys. Chem. A*, 2001, **105**, 8646–8650.
25 C. J. Fecko, J. D. Eaves and A. Tokmakoff, *J. Chem. Phys.*, 2002, **117**, 1139–1154.
26 N. A. Smith and S. R. Meech, *Int. Rev. Phys. Chem.*, 2002, **21**, 75–100.
27 G. Giraud and K. Wynne, *J. Chem. Phys.*, 2003, **119**, 11753–11764.
28 T. Fukasawa, T. Sato, J. Watanabe, Y. Hama, W. Kunz and R. Buchner, *Phys. Rev. Lett.*, 2005, **95**, 197802.
29 D. A. Turton and K. Wynne, *J. Chem. Phys.*, 2008, **128**, 154516.
30 C. J. F. Böttcher, *Theory of electric polarization*, Elsevier, Amsterdam, 1973.
31 D. A. Turton, D. F. Martin and K. Wynne, *Phys. Chem. Chem. Phys.*, 2010, **12**, 4191–4200.
32 D. A. Turton, C. Branca, C. Corsaro, M. Candelaresi, K. R. Seddon, F. Mallamace and K. Wynne, *Faraday Discuss.*, 2011, in press.
33 A. Placzek, G. Hefter, H. M. A. Rahman and R. Buchner, *J. Phys. Chem. B*, 2011, **115**, 2234–2242.
34 J. Barthel, K. Bachhuber, R. Buchner, H. Hetzenauer and M. Kleebauer, *Ber. Bunsen-Ges. Phys. Chem.*, 1991, **95**, 853–859.
35 P. Uhd Jepsen, B. M. Fischer, A. Thoman, H. Helm, J. Y. Suh, R. Lopez and R. F. Haglund, *Phys. Rev. B: Condens. Matter Mater. Phys.*, 2006, **74**, 205103.
36 J. Hunger, A. Stoppa, A. Thoman, M. Walther and R. Buchner, *Chem. Phys. Lett.*, 2009, **471**, 85–91.
37 P. Lunkenheimer, U. Schneider, R. Brand and A. Loidl, *Contemp. Phys.*, 2000, **41**, 15–36.
38 A. Brodin and E. A. Rössler, *J. Chem. Phys.*, 2006, **125**, 114502.
39 C. Kalpouzos, W. T. Lotshaw, D. McMorrow and G. A. Kenneywallace, *J. Phys. Chem.*, 1987, **91**, 2028–2030.
40 H. Cang, V. N. Novikov and M. D. Fayer, *J. Chem. Phys.*, 2003, **118**, 2800–2807.
41 D. Xiao, J. R. Rajian, A. Cady, S. F. Li, R. A. Bartsch and E. L. Quitevis, *J. Phys. Chem. B*, 2007, **111**, 4669–4677.
42 J. Wiedersich, T. Blochowicz, S. Benkhof, A. Kudlik, N. V. Surovtsev, C. Tschirwitz, V. N. Novikov and E. Rössler, *J. Phys.: Condens. Matter*, 1999, **11**, A147–A156.


 

43 R. Richert and C. A. Angell, *J. Chem. Phys.*, 1998, **108**, 9016–9026.
44 M. Anouti, M. Caillon-Caravanier, C. Le Floch and D. Lemordant, *J. Phys. Chem. B*, 2008, **112**, 9412–9416.
45 P. Salvador, J. E. Curtis, D. J. Tobias and P. Jungwirth, *Phys. Chem. Chem. Phys.*, 2003, **5**, 3752–3757.
46 D. Laage and J. T. Hynes, *Science*, 2006, **311**, 832–835.

# 2D or not 2D: Structural and charge ordering at the solid-liquid interface of the 1-(2-hydroxyethyl)-3-methylimidazolium tetrafluoroborate ionic liquid†

**Karina Shimizu,[a] Alfonso Pensado,[bc] Patrice Malfreyt,[bc] Agílio A. H. Pádua*[bc] and José N. Canongia Lopes*[ad]**

***Received 15th March 2011, Accepted 13th April 2011***
**DOI: 10.1039/c1fd00043h**

Molecular dynamics simulations of a 5 nm-thick layer of the ionic liquid 1-(2-hydroxyethyl)-3-methylimidazolium tetrafluoroborate, $[(OH)C_2C_1im][BF_4]$, over silica, alumina and boro-silicate glass substrates have been performed. The structure of the ionic liquid at the solid-liquid interface has been interpreted taking into account the corresponding normal density profiles, lateral interfacial structure, orientational ordering and planar density contours. Comparisons with experimental data suggest that the adsorption and stratification process of ionic liquids over solid substrates can be correctly modeled using a realistic rendition of a non-uniform amorphous substrate such as a glass material.

## Introduction

Ionic liquids and their mixtures with conventional molecular solvents can be regarded as very versatile electrolyte solutions that—in the cases where the ionic liquid is completely miscible with the selected molecular solvent—can span the entire concentration range, from very diluted electrolyte solutions to the most concentrated ones, *i.e.* the pure ionic liquid. The use of ionic liquids in electrochemistry entails the study of the electrolyte solutions at the surface of the electrodes. On the other hand, the study of solid/liquid interfaces involving ionic liquids is also relevant for other applications where ionic liquids can be used as thin films deposited on solid surfaces, such as in tribology, heterogeneous catalysis, or microelectromechanical and microelectronic devices. The need for a correct structural characterization of these solid/liquid interfaces at a molecular level is still not matched by an adequate amount of data in this research front, where, nevertheless, new experimental and simulation results have been appearing apace in recent years.

Some of these studies include sum-frequency generation spectroscopy,[1] X-ray reflectivity,[2] atomic force microscopy[3–9] and also molecular simulations.[10,11] In general, these studies confirmed the adsorption of an ionic liquid layer at the solid substrate surface and, in some cases, indicated the presence of lamellar-like structures parallel to the surfaces.

[a]*Centro de Química Estrutural, Instituto Superior Técnico, 1049 001 Lisboa, Portugal*
[b]*Clermont Université, Université Blaise Pascal, Laboratoire Thermodynamique et Interactions Moléculaires, BP 10448, F-63000 Clermont-Ferrand, France*
[c]*CNRS, FRE3099, LTSP, F-63177 Aubiere, France. E-mail: agilio.padua@univ-bpclermont.fr*
[d]*Instituto de Tecnologia Química e Biológica, UNL, AV. República Ap. 127, 2780 901 Oeiras, Portugal. E-mail: jnlopes@ist.utl.pt*

† Electronic supplementary information (ESI) available. See DOI: 10.1039/c1fd00043h

In the present study we have focused on the modeling of a 5 nm-thick layer of a neat ionic liquid—1-(2-hydroxyethyl)-3-methylimidazolium tetrafluoroborate, $[(OH)C_2C_1im][BF_4]$, *cf.* Fig. 1—over different substrates (silica, alumina and boro-silicate glasses).

Unlike previous simulation studies, where the solid substrate was modeled by a regular lattice corresponding to one of the crystalline planes of the solid[4,10] or by non-atomistic charged sheets,[11] the solid substrates modeled in the present work consist of 2.5 nm-thick slabs corresponding to amorphous (glass) materials. The choice of a functionalized ionic liquid with a short alkyl side chain was based on the experimental fact that in spite of its high surface tension value at room temperature (65 mN $m^{-1}$), $[(OH)C_2C_1im][BF_4]$ is also able to interact strongly with a glass substrate (corresponding to contact angles below 50° over BK7 pyrex glass).[12]

## Experimental

### Ionic liquid simulation

The molecular force field used to model $[(OH)C_2C_1im][BF_4]$ is based on the OPLS-AA framework,[13] with additional parameters specifically tailored for ionic liquid ions.[14,15] The function has the general form given in eqn (1) with the traditional decomposition of the potential energy into covalent bonds, valence angles, torsion dihedral angles, and atom-atom repulsive, dispersive, and electrostatic contributions. The repulsive and dispersive terms are described by the Lennard-Jones 12-6 potential. Following the spirit of OPLS-AA, intramolecular terms related to covalent bonds and angles are taken from the AMBER force field,[16] and efforts are concentrated on carefully describing conformational and intermolecular terms. Details concerning the development and implementation of the force-field can be found elsewhere.[14]

All MD simulations were performed using the DL_POLY code.[17] The simulation runs started with 400 ion pairs (*ca.* 10 000 atoms) initially placed in low-density cubic boxes with boundary conditions in all directions and equilibrated under constant *NpT* conditions for 500 ps at 300 K using the Nosé-Hoover thermostat and isotropic barostat with time constants of 0.5 and 2 ps, respectively. The equilibrium density was attained after about 50 ps, corresponding to a cubic box with a side of approximately 5 nm. Further simulation runs of 100 ps each were used to produce

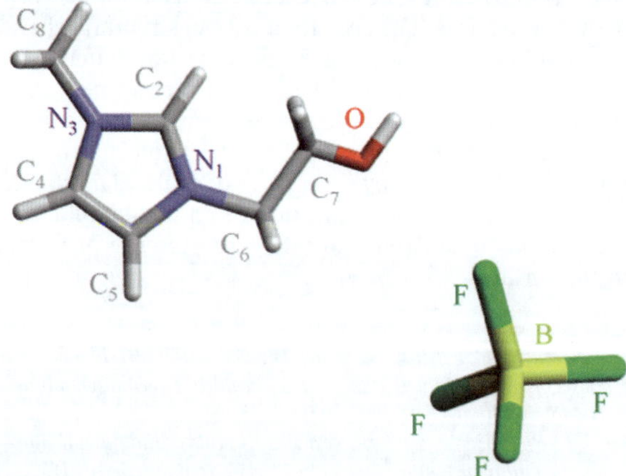


**Fig. 1** Adopted nomenclature for the sites of the ionic liquid 1-(2-ethanol)-3-methylimidazolium tetrafluoroborate, $[(OH)C_2C_1im][BF_4]$.

consecutive equilibrated systems at 300 K, in order to check the attainment of true equilibrium conditions. Long-range electrostatic interactions were handled using the Ewald summation method considering six reciprocal-space vectors, and repulsive-dispersive interactions were explicitly calculated below cutoff distances of 1.6 nm. Finally, 4000 configurations were stored from production runs of 400 ps. This bulk ionic liquid was then "sliced" in order to be placed over the glass slab (*cf.* 'Glass simulation' and 'Glass-liquid interface' sections).

$$u_{\alpha,\beta} = \sum_{ij}^{\text{bonds}} \frac{k_{ij}}{2}\left(r_{ij-}r_{0,ij}\right)^2 + \sum_{ijk}^{\text{angles}} \frac{\theta_{ij}}{2}\left(\theta_{ijk-}\theta_{0,ijk}\right)^2$$

$$\sum_{ijkl}^{\text{dihedrals}} \sum_{m=1}^{4} \frac{V_{m,ijkl}}{2}\left[1-(-1)^m\cos\left(m\phi_{ijkl}\right)\right]+$$

$$\sum_{ij}^{\text{nonbonded}} \left\{4\varepsilon_{ij}\left[\left(\frac{\sigma_{ij}}{r_{ij}}\right)^{12} - \left(\frac{\sigma_{ij}}{r_{ij}}\right)^{6}\right] + \frac{1}{4\pi\varepsilon_0}\frac{q_i q_j}{r_{ij}}\right\} \qquad (1)$$

## Glass simulation

For the glass phase, a different model was adopted. The molecular force field used in the simulations was obtained from a parameterization included in the DL_POLY MD code package. The interactions between ions were described by a two-body Buckingham potential (eqn (2)) complemented by a three-body truncated Vessal potential (eqn (3)). The latter includes the description of the bond-angle potentials for Si–O–Si and O–Si–O in the case of silica and BK7 glasses, O–Al–O for alumina glasses, and Si–O–H, B–O–H and Al–O–H in the case of all the above mentioned glasses with surface hydroxy groups (wet glasses). All parameters are given as Supplementary Information.†

$$U\left(r_{ij}\right) = A\exp\left(\frac{r_{ij}}{\rho}\right) - \frac{C}{r_{ij}^6} \qquad (2)$$

$$U\left(\theta_{jik}\right) = k\left[\theta_{jik}^a\left(\theta_{jik}-\theta_0\right)^2\left(\theta_{jik}+\theta_0-2\pi\right)^2 - \frac{a}{2}\pi^{a-}\right.$$

$$\left.\left(\theta_{jik}-\theta_0\right)^2\left(\pi-\theta_0\right)^3\right]\exp\left[\left(r_{ij}^8+r_{ik}^8\right)/\rho^8\right] \qquad (3)$$

The glass simulations were carried out using systems of 4200, 4320 and 4720 particles for silica ($SiO_2$), BK7 (a borosilicate glass containing small amounts of potassium and sodium oxides, with approximate formula $(SiO_2)_{66}(B2O_3)_{15}(K_2O)_3(Na_2O)_8$), and alumina ($Al_2O_3$) glasses, respectively. Each glass was first pre-equilibrated under constant *NVT* conditions at 4000 K, using a Nosé-Hoover thermostat with time constant of 0.5 ps, and a density set to the corresponding room-temperature experimental value,[18–20] using cubic simulation boxes with sides of around 2.5 nm. These high temperature/high energy simulation runs lasted for just 10 ps with a 1 fs timestep and were immediately followed by a temperature quench from 4000 K to 300 K at a rate of 10 K $ps^{-1}$. Long-range electrostatic interactions were treated using the Ewald summation method beyond a cutoff distance of 1 nm. The equilibration of the glass material continued by allowing the simulation to evolve through several temperature annealing cycles between 1000 K and room temperature.

The glass surface was formed by joining four (identical) (2.5 × 2.5 × 2.5 nm) simulation boxes in a square formation, inserting a large void space in the $z$ axis direction (15 nm), and allowing the resulting simulation box (5 × 5 × 17.5 nm) to equilibrate under additional temperature annealing cycles. The glass slabs inside this box had about (5 × 5 × 2.5 nm) dimensions.

To model the hydration layer of real glasses the equilibrated "dry" glass slabs were allowed to "react" with water molecules in order to form silanol groups at the glass surface. Water molecules (SPC model[21]) were inserted in the empty space of the simulation box over the glass slab and an initial simulation run under $NVT$ conditions at 1000 K was allowed to evolve for 100 ps. This was immediately followed by a temperature quench to 300 K at a rate of 12 K $ps^{-1}$. The distribution of the water molecules "condensed" at the glass surface was then analyzed and the formation of silanol groups *via* glass/water hydroxilation was allowed to occur if the distance between the different types of atom of the glass and the hydrogen or oxygen atoms of the water molecules (HW or OW, respectively) met the pre-established criteria specified in Table 1.

After the removal of all non-bonded water molecules, the obtained "wet" glass surfaces were covered with silanol groups with surface densities very close to the experimental values.[22,23] Further simulation runs of 200 ps were used to produce equilibrated systems at 300 K. Fig. 2 summarizes the steps that led to the formation of the glass slabs used in the solid-liquid interface studies (*cf.* next section).

## Solid-liquid interface

The glass-ionic liquid interface was modeled by merging the equilibrated bulk ionic liquid and the glass slab (*cf.* previous sections) into a single simulation box. Since all further simulations were to be conducted at low temperatures (always well below 1000 K) the dimensions of the box containing the glass were fixed, and the slightly smaller ionic liquid box was "imported" into the empty space over the glass slab (*cf.* Fig. 3). The system was then allowed to relax under constant $NVT$ conditions for 50 ps at 350 K, with cutoff distances set to 1.8 nm in order to eliminate the gaps caused by the fitting of the smaller ionic liquid simulation box to the larger glass slab box and to allow the adsorption of the liquid to the solid substrate. Consecutive 200 ps production runs on the equilibrated systems at different temperatures yielded the configurations (1000 per run) needed to perform the structural and energy analyses. In some of the simulations the glass atoms were set to fixed positions in order to optimize the time performance of the simulation runs, with no noticeable difference in the results.

Unlike liquid-vacuum interface studies, where a lot of effort is generally dedicated to the estimation of the surface tension at the interface and to the calculation of charge and density profiles in the direction normal to the surface ($z$-axis), the solid-liquid interface of the present glass-ionic liquid system is characterized by the heterogeneity of the modeled substrate and how it affects the structure of the ionic liquid layers placed above it. This means that average charge and density profiles normal to the surface had to be complemented by structural information at each surface layer. In order to achieve this type of analyses, the electric field at the glass surface and at surfaces placed along $z$-axis (0.5 nm apart) were calculated using point-grids with 0.02 nm mesh.

It must also be noted that the present work also includes a "free" ionic liquid surface (vacuum liquid interface). Some of the structural features analyzed near the solid substrate will be compared with results obtained at this other interface.

**Table 1** Bonding parameters

| | Si–OW | Al–OW | O–HW |
|---|---|---|---|
| rX–Y (nm) | 0.160 | 0.165 | 0.100 |

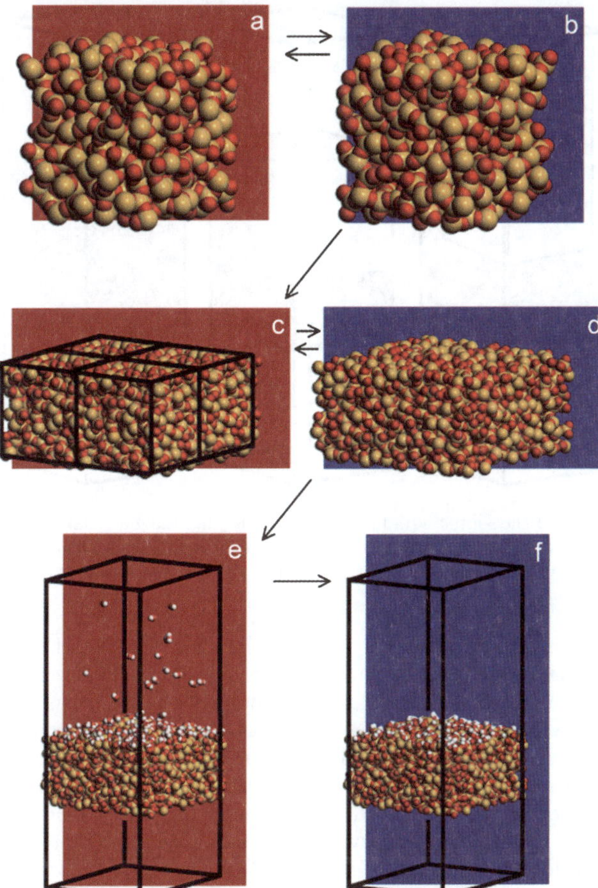


**Fig. 2** Modeling the amorphous glass slabs: (a) to (b) Temperature annealing cycles (red = high temperature, blue = low temperature) in $2.5 \times 2.5 \times 2.5$ nm simulation boxes (bulk glass); (b) to (c) creation of a $5 \times 5 \times 2.5$ nm slab with a large void volume in the $z$ direction; (c) to (d) temperature annealing of the glass slab; (d) to (e) and (f) insertion of water molecules at high temperature, followed by their condensation at the glass surface and *ad hoc* reaction to form silanol groups.

## Results and discussion

Unlike the vacuum-liquid distribution where the ionic liquid surface adjoins a completely isotropic medium (a vacuum), the solid substrate at the glass-liquid interface has been modeled to reflect the structure and heterogeneities at the atomic level of three different types of amorphous glass. This is a different scenario from previous simulation work at the solid-ionic liquid interface where the solid substrate has been modeled as a regular lattice of atoms or simply as a charged homogeneous (non-atomistic) boundary.

### Normal density profiles

The discussion on IL surface structure often involves two different but interrelated aspects, surface composition and molecular orientation at the surface. Surface composition involves the identification of the molecules present in the near-surface region and whether there is enhancement of certain parts of ions and depletion of other parts of ions with respect to the bulk composition.

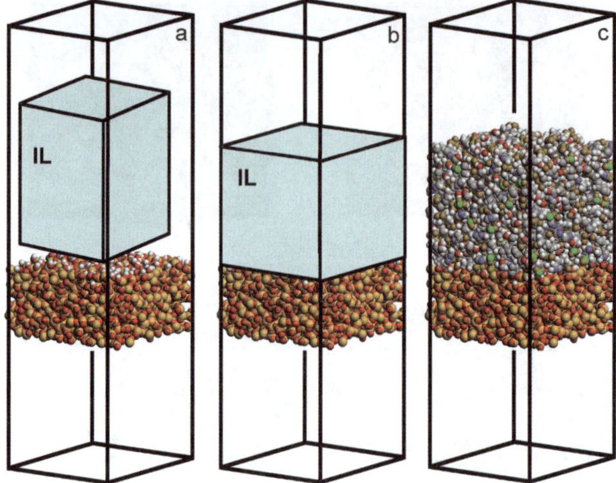


**Fig. 3** Building the glass-ionic liquid interface: (a) merging the glass slab simulation box (*cf.* Fig. 2f) with a slightly smaller (exaggerated in the scheme) simulation box containing the pure ionic liquid; (b) and (c) adsorption and equilibration of the ionic liquid on the glass surface.

In order to analyze the structure at the solid-liquid interface, the corresponding ion center-of-mass number density and charge density profiles were calculated in the direction normal to the glass slab. Examples for the BK7 glass/[(OH)$C_2C_1$im][$BF_4$] system are presented in Fig. 4 and 5. Analogous conclusions can be drawn from similar representations obtained for the other two types of glass.

The comparison of Fig. 4 with similar density profiles obtained for neat ionic liquid slabs with two free surfaces (in the absence of the glass substrate)[24] shows that the average ion number density is not greatly affected by the presence of the glass slab: the ion density slightly increases from 6.9 ± 0.2 ions $nm^{-3}$ to 7.2 ±

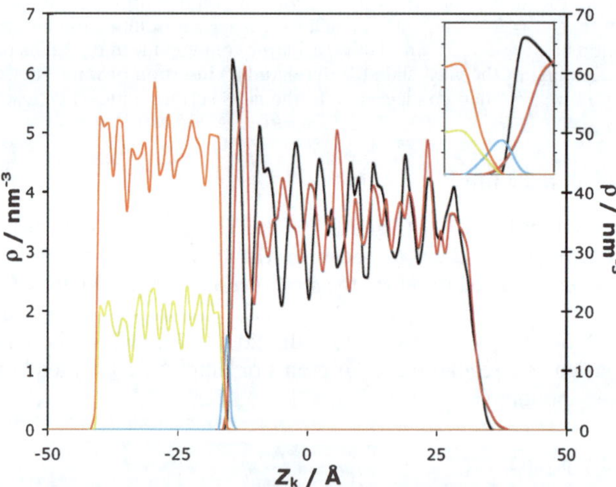


**Fig. 4** Number density profiles of [(OH)$C_2C_1$im]$^+$ (red line), [$BF_4$]$^-$ (black), BK7 Si atoms (yellow), BK7 O atoms (orange) and, silanol OH groups (light blue) in the BK7 glass/[(OH)$C_2C_1$im][$BF_4$] system. The inset shows a zoom of the interface region ($-1.8 < z_k$ /nm $< -1.3$). The expanded scale on the right corresponds to the numerical densities of the glass atoms (orange and yellow lines) and silanol groups (light blue).

 

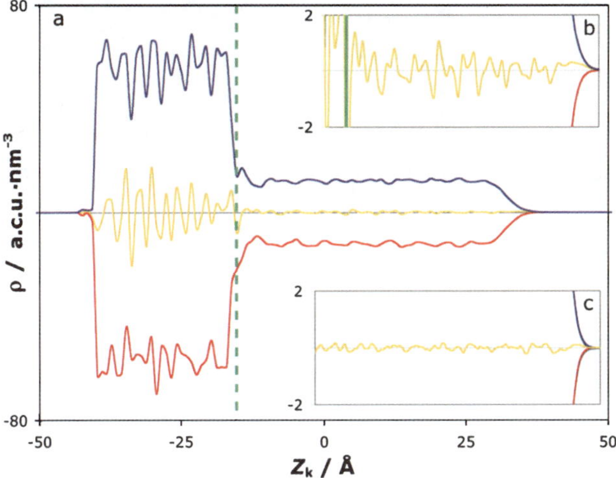


**Fig. 5** (a) Electric charge density profiles in the BK7 glass/[(OH)$C_2C_1$im][$BF_4$] system. The top inset (b) shows a zoom of the charge density profile at the ionic liquid layer from $z_k = -20$ Å (glass) to $z_k = 40$ Å (vacuum); the bottom inset (c) shows a similar charge density profile in a layer of ionic liquid bounded only by vacuum. Blue lines: positive charge density; red lines: negative charge density; yellow lines: total charge density; dashed/solid green line: glass-ionic liquid boundary.

0.9 ions nm$^{-3}$ when the 5 nm-thick layer of ionic liquid adsorbs at the glass surface. This small increase in density can be attributed on one hand to the compression of the ionic liquid near the glass due to the strong interactions between the ions and the glass and on the other hand to the 2D structural ordering of the ionic liquid caused by the glass surface (see discussion below). However, the most conspicuous feature of the data shown in Fig. 4 are the larger numerical density fluctuations of the ionic liquid ions when compared with the corresponding fluctuations present in density profiles obtained without the glass layer.[24] This is also translated in quantitative terms by the quite different numerical density standard deviations (0.9 ions nm$^{-3}$ for the glass/ionic liquid system against 0.2 ions nm$^{-3}$ for the vacuum/ionic liquid system).

The large density fluctuations shown in Fig. 4 are in fact a first indication of the layering of the ionic liquid at the glass surface. Those fluctuations are not only large but, in the case of the small quasi-spherical anion [$BF_4$]$^-$, also exhibit a periodical behavior near the glass surface (the density profile periodicity of the [(OH)$C_2C_1$im]$^+$ cation is not so clear-cut due to the inherent asymmetry of the ion). Interestingly, such periodicity can be correlated with the results reported by Atkin and Warr[9] that have produced solvation force profiles for different ionic liquids confined between the $Si_3N_4$ tips of an atomic-force microscope (AFM) and mica, silica, and graphite surfaces.

Fig. 6 shows the overlap of the anion density profile obtained in this work for the [(OH)$C_2C_1$im][$BF_4$]/BK7 glass system with the solvation force profile obtained from ref. 9 for the [$C_2C_1$im][$CH_3COO$]/mica glass system. The comparison between these two systems is warranted in this case due to the similarity between the numerical densities of the two ionic liquids (both around 7 ions nm$^{-3}$).

The figure allows three different types of conclusion: i) the density fluctuations mimic the layer separation distances given by the breaks in the force profile of the AFM experiments, *i.e.* the glass surface is inducing the structural rearrangement of the ionic liquid into layers parallel to its surface that can be "punctured" by the AFM tip as it approaches the glass surface; ii) the thickness of the layers depends on the dimensions of the ionic pair (cation + anion). The correlation between the two

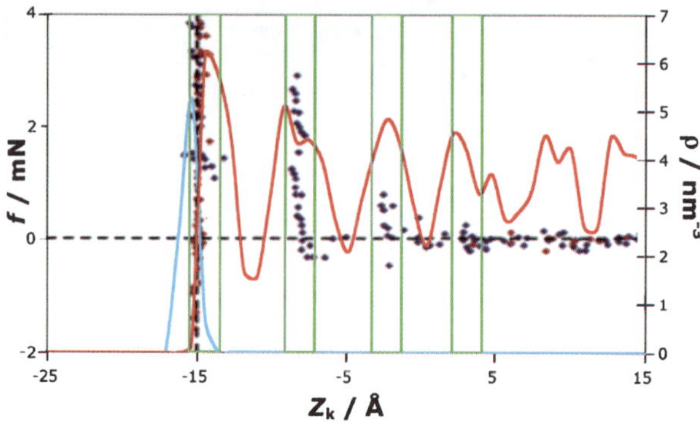


**Fig. 6** Number density profiles of the $[BF_4]^-$ anions of the ionic liquid (red, right axis) and the silanol groups at the glass surface (light blue, right axis) overlaid on top of the force profile of an AFM tip approaching and retracting from a mica surface covered with the $[C_2C_1im]$ $[CH_3COO]$ ionic liquid (blue and red data points from ref. 9, left axis). The green vertical lines were drawn around the separation distances between layers given in the same reference. The silanol number density was divided by a factor of three to fit the scale on the right.

different systems displayed in Fig. 6 ($[(OH)C_2C_1im][BF_4]$ *versus* $[C_2C_1im]$ $[CH_3COO]$) is quite good due to the similarity between the size of the ions and the corresponding densities. Other systems with higher density (like the previously reported ethylammonium or propylammonium nitrate systems[9]) exhibit smaller inter-layer distances and would not correlate with the present data; iii) The degree of structuration/layering depends on the nature of both the ionic liquid and the surface. The AFM studies[9] concluded that smooth, high-charge mica surfaces and smaller, less flexible ions produce layered structures further away from the surface and deeper into the ionic liquid bulk. In the present case the density fluctuations are much clearer in the case of the smaller, quasi-spherical anions and are noticeable at relatively large distances from the glass. However, the apparently high activity of the BK7, silica an alumina glasses studied in this work, can be just an effect introduced by the model used to characterize each surface. On the other hand the functionalized $[(OH)C_2C_1im][BF_4]$ has been selected in this study because it has been shown experimentally[9] that it partially wets different types of glass and, therefore, it was expected to interact strongly with those surfaces.

## Lateral interface structure

Another way to check the structural differences between the layers of ionic liquid near a glass surface and the bulk of an isotropic ionic liquid phase is to use zone-resolved tangential pair distribution functions[25,26] (TRDF). These can be calculated using the relation

$$g_{ij}(r) = \frac{\sum_{i,j} \delta\left(r - r_{ij}\right)}{2\pi r dr\, \rho_{region} \Delta z}; \quad z_{ij} < \Delta z \tag{4}$$

where $\rho_{region}$ is the average number density in each region which normalizes the corresponding TRDF to unity at infinite distance, and $r_{ij_\circ} = (x_{ij}^2 + y_{ij}^2)^{1/2}$ is a two dimensional distance parallel to the surface plane. $\Delta z = 5$ Å is chosen in order to achieve significant statistical averages.

Fig. 7 shows different TRDFs obtained for 0.5 nm-thick layers of ionic liquid at different distances from the glass surface. For comparison purposes we have also included the corresponding TRDFs for an isotropic ionic liquid sample. Most TRDFs shown in Fig. 7 exhibit clear differences between the tangential structure

 

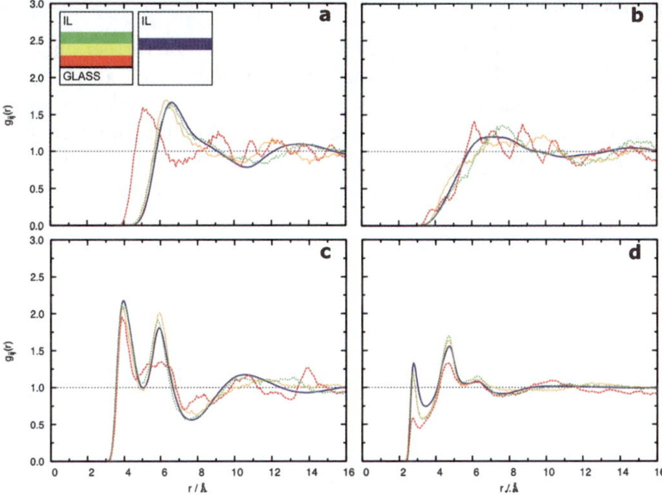


**Fig. 7** Tangential radial distribution functions (TRDF) of several representative pairs of atoms for different ionic liquid layers: (a) B-B anion-anion TRDFs$^-$; (b) $C_2$-$C_2$ cation-cation TRDFs; (c) B-$C_2$ anion-cation TRDFs; (d) H-B hydroxyl-anion TRDFs. The continuous blue line represents TRDFs obtained in a 0.5 nm-thick layer of a bulk ionic liquid; the dashed red, yellow and green lines represent TRDFs in 0.5 nm-thick layers of ionic liquid positioned at 0.25, 0.75 and 1.25 nm average distances from the glass surface (see inset).

of the ionic liquid 0.5 nm-layer closest to the glass surface, the two next layers and a 0.5 nm-layer taken from the ionic liquid bulk. The differences are more noticeable in the distribution functions between anions.

The modifications in the structure of the ionic liquid layer closest to the glass surface can be summarized as follows: i) the glass-ion interactions partially break the tridimensional nature of the polar network of the ionic liquid. This can be seen in Fig. 7c, where the closest cation($C_2$)-anion(B) interatomic peak at around 0.4 nm is preserved but that at 0.6 nm is disrupted. This last peak corresponds to $C_2$-B distances typical of cases when the cation and anion are not interacting directly *via* the $C_2$ atom but through the $C_4$ or $C_5$ atoms. In other words, the branched polar network of imidazolium-based ionic liquids (where each cation can interact simultaneously with different anions at the $C_2$, $C_4$ or $C_5$ positions) is disturbed, with the substitution of some anion-cation interactions by glass-ion interactions; ii) in response to this perturbation caused by the new strong interactions with the glass surface the ions reorient/reposition themselves at the superficial layer. This is much more noticeable in the case of the smaller, quasi-spherical anions that are found at typical interionic distances almost 0.1 nm smaller than in the bulk ionic liquid (Fig. 7a). The positional/orientational modifications of the cations closer to the glass surface are harder to interpret (*cf.* the jagged shape of the corresponding TRDF shown in Fig. 7b) and will be analyzed in the next paragraph; finally, iii) the specific interactions between the hydroxyl group of the cation and the fluorine atoms of the anion are also perturbed by the proximity of the glass (*cf.* the depression of the first peak in the corresponding TRDF of Fig. 7d). This is a consequence not only of the strong interactions between the anions and the glass but also of those between the hydroxyl groups of the cation and the silanol groups of the glass (hydrogen bond-type interactions). This latter type of interaction probably justifies the fact that ionic liquids based on cations with hydroxyl-substituted alkyl side-chains exhibit improved glass-adhesion properties relative to ionic liquids with normal alkyl side-chains.

## Orientational ordering

The reorientation of the cations at the glass surface can be further analyzed taking into account the orientation of the imidazolium rings and their alkyl chains relative to the surface of the glass. Fig. 8 shows the orientational ordering parameters—defined by eqn (5) as the average of the second Legendre polynomial in the direction normal to the glass-ionic liquid interface—at distances corresponding to the three 0.5 nm-thick ionic liquid layers closest to the glass surface.

$$\langle P_2(\theta)\rangle = \left\langle \frac{1}{2}(3\cos^2\theta - 1)\right\rangle \tag{5}$$

In eqn (5), $\theta$ is taken as the angle between a specific direction vector in the molecule-fixed frame and the surface normal $z$. The Legendre polynomial functions enable us to investigate the range and extent of orientation preferences at the interface. $P_2(\theta)$ ranges from 1 to $-0.5$. A value of 1 implies that the two considered vectors are parallels, whereas a value of $-0.5$ indicates that they are perpendicular. In the interfacial region, the longer lateral chain of the imidazolium shows a tendency to lie in the direction normal to the interface. The ***NN*** vector (the ***NN*** vector points from the nitrogen carrying the ethanol group to the nitrogen carrying the methyl group) tends to be perpendicular to the surface normal whereas the normal to the imidazolium ring shows a tendency to lie parallel to the surface normal. The packing of the imidazolium rings that are parallel to the interface causes the enhancement in the local density in the region near to the interface. In the bulk region, all of the vectors become randomized.

The most obvious observation is that the cations adopt distinct orientation patterns only at the layer closest to the glass surface. Within this 0.5 nm-thick layer the imidazolium rings tend to lie perpendicular to the surface, with the two nitrogen atoms at the same distance from it; an orientation that favors interactions *via* the aromatic hydrogen atoms of the cation. Interestingly, the short chains of the $[(OH)C_2C_1im]^+$ cations in the vicinity of the glass surface tend to lie parallel to the surface—they are "anchored" by the $C_6$ carbon atom attached to the

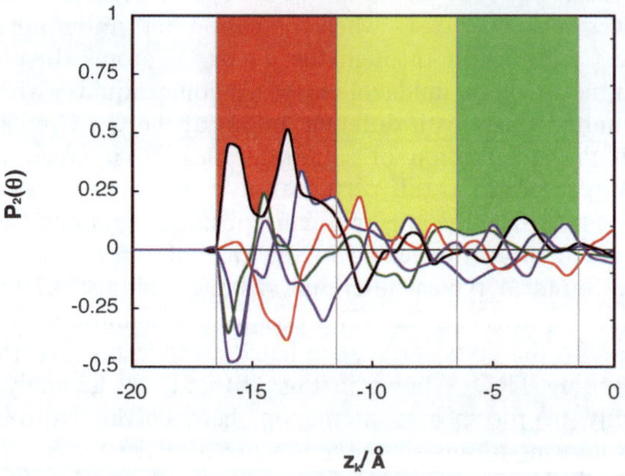


**Fig. 8** Orientational ordering parameter, $P_2(\theta)$, defined by the angle, $\theta$, between the direction vectors and the normal to the glass surface: red for the ***NN*** vector on $[(OH)C_2C_1im]^+$, purple for the vector normal to the imidazolium ring on $[(OH)C_2C_1im]^+$, black for the vector normal to the ***NN*** vector that points in direction of $C_2$ on $[(OH)C_2C_1im]^+$, blue for the $C_6O$ vector on $[(OH)C_2C_1im]^+$ and green for vector normal to the direction $C_6O$ on $[(OH)C_2C_1im]^+$. The three background colours correspond to the three 0.5 nm-thick layers of ionic liquid closest to the glass (*cf.* Fig. 7 inset).

imidazolium ring and the oxygen atom of the hydroxyl group (whose atoms can interact strongly with the silanol groups of the glass).

The overall picture that emerges from the data already discussed is that the glass surface induces a layering pattern in the ionic liquid (as attested by the density and charge profiles presented in Fig. 4 to 6 and the experimental data reported by Atkin and Warr[9]), but that only the first layer closer to the glass surface exhibits strikingly different local structuration. This somewhat surprising state of affairs can be clarified if we take into account the fact that glass is an amorphous material and its surface should not be treated as a homogeneous and regular boundary. In other words, by restricting our analyses to density profiles and tangential distribution functions we have dealt only with quantities averaged-out over the planes parallel to the surface. Now it is time to check the charge and density fluctuations within those planes.

## Planar density contours

Fig. 9 shows the unevenness of the glass surface (in terms of the electrical field produced at $Z_k = 0$) and how that uneven distribution of charge density modifies the structure of the adsorbed ionic liquid: the orange/red areas of the glass with more negative charge density (where the creation of a free surface produced a slight shift towards more oxygen atoms near the surface) correspond to the cyan/blue areas on the ionic liquid side with a higher concentration of cations; conversely the cyan/blue areas of the glass are matched by orange/red areas on the ionic liquid side. In other words, the surface charge of the glass (that overall tends to be negative but has negative and positive localized patches) is used as a template by the ionic liquid to distribute/orient its ions at the interface.

Fig. 9 also shows the position of the ions of the ionic liquid contained in the first 0.5 nm-thick layer of ionic liquid closest to the glass surface. This representation

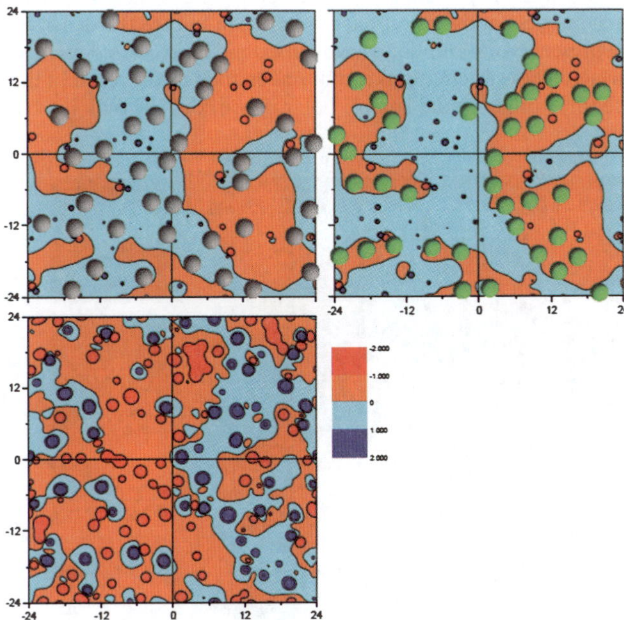



**Fig. 9** Top view of the glass-ionic liquid interface plane showing the electric fields generated by the glass atoms (contour graph at the bottom) and the ionic liquid ions (contour graphs at the top). All $C_2$ (grey spheres) and B atoms (green spheres) of the ionic liquid contained in the 0.5 nm-thick layer closest to the glass surface are also presented in the left and right top graphs, respectively. Distance units in Å, electric field in $1.44 \times 10^{11}$ J $C^{-1}$ $m^{-1}$.

explains the shifts observed in the corresponding TRDFs shown in Fig. 7: the ions (especially the anions) are packed closer together due to the strong interactions with the glass atoms. It is also quite obvious that the cations are mostly confined to the areas of the glass with a negative charge density whereas the anions are segregated to those areas with positive charge densities; the glass surface effectively breaks the three-dimensional polar network and its inherent local electroneutrality criteria that are found in the bulk ionic liquid. It must be stressed that the AFM tip used to determine the force profiles and infer the existence of ionic liquid layering above glass and other surfaces[9] cannot distinguish the different patches shown in Fig. 9 (the very end of the tip has a radius of about 20 nm): like the density and charge profiles discussed above it can only measure averaged-out quantities at a given distance (plane) from the surface. This means that layering occurs but the layers are not necessarily homogeneous; it also explains the fact reported by Atkin that the thickness of each layer is determined purely by the physical size of the IL ion pair—homogeneous sub-layers formed exclusively by cations or anions (and whose dimensions would be determined by the size of each individual ion) have never been observed.

Finally, the structural differences between the first and subsequent layers of ionic liquid over different types of glass are presented in Fig. 10.

The previously stated overall picture can now be analyzed from a molecular point of view. Flat and unevenly charged glass surfaces induce the layering of the ionic liquid adsorbed on them. This stratification process penetrates a few nm into the ionic liquid phase (*cf.* the density and charge density fluctuations presented in Fig. 4 and 5) but, due to the bulk, asymmetry, flexibility and charge delocalization of the ionic liquid ions (mainly the cations), only the ionic liquid layer in direct contact with the glass shows important reorganization its ions.

This was a concept previously explored by other authors[9,27,28] including the possibility of a first electrostatically bound layer of ionic liquid on substrates with high-charge densities. In fact two types of influence over two different distance scales seem to be at work in the present case: Pinilla *et al.*[28] have shown by MD simulation that even structureless and neutral flat surfaces can induce the density fluctuations found in the present work and that these can extend more than 3 nm away from the inducing surface. These fluctuations can be understood as the result of the accommodation of the three-dimensional flexible polar network of the bulk ionic liquid to the

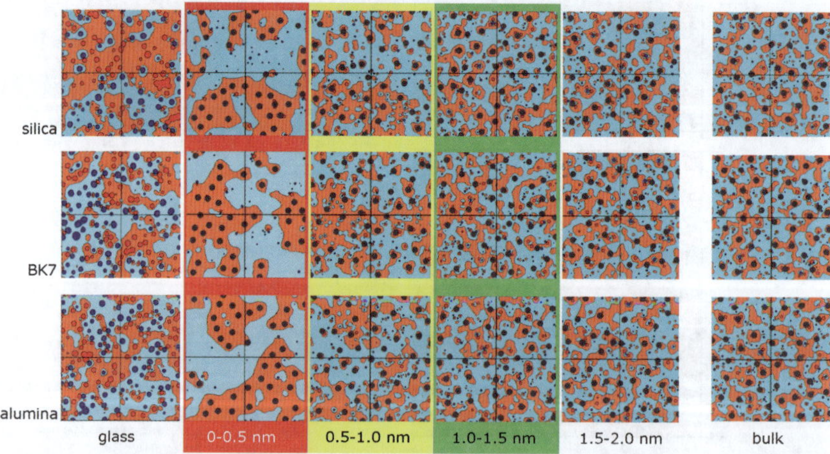


**Fig. 10** Top view of different planes showing the electric fields generated by the glass atoms (three contour graphs on the left) and the ionic liquid ions (all other contour graphs). All boron atoms of the ionic liquid contained in 0.5 nm-thick layers at different distances to the glass surface are represented as gray spheres. The side of each square is 4.8 nm, the charge density colors/units are the same as in Fig. 9.

 

two-dimensional nature of the boundary imposed by the flat surface; on the other hand when the surface is charged (unevenly charged as in the case of an amorphous glass) a second type of reorganization takes place at the layer in direct contact with the charged surface: Fitchett and Conboy[27] used sum-frequency vibrational spectroscopy to infer the (re)orientation of the imidazolium rings of 1-alkyl-3-methylimidazolium-based ionic liquids at the silica interface. Interestingly, in their case the ring appears to be tilted relative to the surface whereas the long alkyl side chains (hexyl, octyl and decyl) lie normal to the surface. This contrasts with the present study where the ring and the small substituted ethyl group lie normal and parallel to the surface, respectively. However this difference can be easily understood if one takes into account the interactive nature of the hydroxyl group attached to the end of the ethyl chain, that not only keeps the chain more or less parallel to the surface but also forces the ring to adopt a more normal orientation relative to the glass surface.

## Conclusions

The main conclusion of the present work is that the adsorption and stratification process of ionic liquids over solid substrates that has been reported experimentally can be correctly modeled using a realistic rendition of a non-uniform, amorphous substrate such as a glass material. An ionic liquid that interacts strongly with a given amorphous substrate will orient its ions in order to mirror the (uneven) local electrical field felt by it at the different regions of the surface. This is the driving force for stratification: the strong adsorption of the ionic liquid ions on a flat surface that is not necessarily uniform at a molecular level. The use of the solid substrate as a charge template by the ionic liquid ions reflects the versatility of ionic liquids as adsorbates of solvation media—even in a bulk phase, the ions of ionic liquids are known to interact selectively with different molecular solutes, adopting different spatial arrangements around them and promoting unique fluid phase demixing phenomena.[29,30]

The obtained normal density profiles (both in terms of number density and charge, Fig. 4 and 5) also show that above this first adsorbed layer the ionic liquid tends to retain some memory of the two-dimensional arrangement of its ions imposed by the flat substrate. The existence of 2-3 superimposed layers inferred by the fluctuations in the number density of the anions reflects quite well the experimental force profile results obtained by AFM.[9] The normal density profiles also show that the number and charge density fluctuations are much larger in the presence of the (highly interactive) solid surface than in the case of ionic liquid layers bounded only by a vacuum. Finally it was also shown that in spite of those larger fluctuations the in-plane profiles tend quite rapidly to a bulk distribution.

## Acknowledgements

Work supported by projects PTDC/QUI-QUI/101794/2008 (FCT/Portugal) and PICS 3090 (CNRS/France and GRICES/Portugal). K.S. acknowledges the support of grant SFRH/BPD/38339/2007 (FCT/Portugal) and the MSDG bursary for participating in the MSGD 2010 Summer Research Meeting.

## References

1 J. B. Rollins, B. D. Fitchett and J. C. Conboy, Structure and Orientation of the Imidazolium Cation at the Room-Temperature Ionic Liquid/$SiO_2$ Interface Measured by Sum-Frequency Vibrational Spectroscopy, *J. Phys. Chem. B*, 2007, **111**, 4990–4999.
2 M. Mezger, S. Schramm, H. Schröder, H. Reichert, M. Deutsch, E. J. De Souza, J. S. Okasinski, B. M. Ocko, V. Honkimäki and H. Dosch, Layering of $[BMIM]^+$-Based Ionic Liquids at a Charged Sapphire Interface, *J. Chem. Phys.*, 2009, **131**, 094701.

3 S. Bovio, A. Podestà, C. Lenardi and P. Milani, Evidence of Extended Solidlike Layering in [Bmim][NTf2] Ionic Liquid Thin Films at Room-Temperature, *J. Phys. Chem. B*, 2009, **113**, 6600–6603.

4 S. Bovio, A. Podestà, P. Milani, P. Ballone and M. G. Del Pópolo, Nanometric Ionic-liquid Films on Silica: a Joint Experimental and Computational Study, *J. Phys.: Condens. Matter*, 2009, **21**, 424118.

5 R. Hayes, G. G. Warr and R. Atkin, At the Interface: Solvation and Designing Ionic Liquids, *Phys. Chem. Chem. Phys.*, 2010, **12**, 1709–1723.

6 Y. Liu, Y. Zhang, G. Wu and J. Hu, Coexistence of Liquid and Solid Phases of Bmim-PF6 Ionic Liquid on Mica Surfaces at Room Temperature, *J. Am. Chem. Soc.*, 2006, **128**, 7456–7457.

7 Y. Yokota, T. Harada and K. Fukui, Direct Observation of Layered Structures at Ionic Liquid/Solid Interfaces by Using Frequency-Modulation Atomic Force Microscopy, *Chem. Commun.*, 2010, **46**, 8627–8629.

8 R. Hayes, S. Z. El Abedin and R. Atkin, Pronounced Structure in Confined Aprotic Room-Temperature Ionic Liquids, *J. Phys. Chem. B*, 2009, **113**, 7049–7052.

9 R. Atkin and G. G. Warr, Structure in Confined Room-Temperature Ionic Liquids, *J. Phys. Chem. C*, 2007, **111**, 5162–5168.

10 S. Wang, S. Li, Z. Cao and T. Yan, Molecular Dynamic Simulations of Ionic Liquids at Graphite Surface, *J. Phys. Chem. C*, 2010, **114**, 990–995.

11 M. Sha, G. Wu, Q. Dou, Z. Tang and H. Fang, Double-Layer Formation of [Bmim][$PF_6$] Ionic Liquid Triggered by Surface Negative Charge, *Langmuir*, 2010, **26**, 12667–12672.

12 J. Restolho, L. José, J. L. Mata and B. Saramago, On the Interfacial Behavior of Ionic Liquids: Surface Tensions and Contact Angles, *J. Colloid Interface Sci.*, 2009, **340**, 82–86.

13 W. L. Jorgensen, D. S. Maxwell and J. Tirado-Rives, Development and testing of the opls all-atom force field on conformational energies and properties of organic liquids, *J. Am. Chem. Soc.*, 1996, **118**, 11225–11236.

14 J. N. Canongia Lopes, L. Deschamps and A. A. H. Pádua, Modeling Ionic Liquids Using a Systematic All-Atom Force Field, *J. Phys. Chem. B*, 2004, **108**, 2038–2047.

15 J. N. Canongia Lopes and A. A. H. Pádua, Molecular force field for ionic liquids composed of triflate or bistriflylimide anions, *J. Phys. Chem. B*, 2004, **108**, 16893–16898.

16 W. D. Cornell, P. Cieplak, C. I. Bayly, I. R. Gould, K. M. Merz, D. M. Fergusin, D. C. Spellmeyer, T. Fox, J. W. Caldwell and P. A. Kollman, A second generation force field for the simulation of proteins, nucleic acids, and organic molecules, *J. Am. Chem. Soc.*, 1995, **117**, 5179–5197.

17 W. Smith and T. R. Forester, *The DL-POLY package of molecular simulation routines, Version 2.13*, The Council for the Central Laboratory of Research Councils, Daresbury Laboratory, Warrington, UK, 1999.

18 J. Du and A. N. Cormack, The medium range structure of sodium silicate glasses: a molecular dynamics simulation, *J. Non-Cryst. Solids*, 2004, **349**, 66–79.

19 S. Izman and V. C. Venkatesh, Gelling of chipsduring vertical surface diamond grinding of BK7 glass, *J. Mater. Process. Technol.*, 2007, **185**, 178–183.

20 V. V. Hoang and S. K. Oh, Simulation of pressure-induced phase transition in liquid and amorphous $Al_2O_3$, *Phys. Rev. B: Condens. Matter Mater. Phys.*, 2005, **72**, 054209.

21 M. Praprotnik, D. Janezic and J. Mavri, Temperature Dependence of Water Vibrational Spectrum: A Molecular Dynamics Simulation Study, *J. Phys. Chem. A*, 2004, **108**, 11056–11062.

22 B. P. Feuston and S. H. Garofalini, Water-induced Relaxation of the Vitreous Silica Surface, *J. Appl. Phys.*, 1990, **68**, 4830–4836.

23 K. C. Hass, W. F. Schneider, A. Curioni and W. Andreoni, The Chemistry of Water on Alumina Surfaces: Reaction Dynamics from First Principles, *Science*, 1998, **282**, 265–268.

24 A. S. Pensado, M. F. Costa Gomes, J. N. Canongia Lopes, P. Malfreyt and A. A. H. Pádua, Effect of Alkyl Chain Length and Hydroxyl Group Functionalization on the Surface Properties of Imidazolium Ionic Liquids, *Phys. Chem. Chem. Phys.*, 2011, **13**, 13518–13526.

25 T. Yan, S. Li, W. Jiang, X. Gao, B. Xiang and G. A. Voth, Structure of the Liquid – Vacuum Interface of Room-Temperature Ionic Liquids: A Molecular Dynamics Study, *J. Phys. Chem. B*, 2006, **110**, 1800–1806.

26 A. Aguado, M. Wilson and P. A. Madden, Molecular dynamics simulations of the liquid–vapor interface of a molten salt. I. Influence of the interaction potential, *J. Chem. Phys.*, 2001, **115**, 8603–8611.

27 B. D. Fitchett and J. C. Conboy, Structure of the room-temperature ionic liquid/$SiO_2$ interface 1studied by sum-frequency vibrational spectroscopy, *J. Phys. Chem. B*, 2004, **108**, 20255–20262.


 

28 C. Pinilla, M. G. Del Pópolo, R. M. Lynden-Bell and J. Kohanoff, Structure and dynamics of a confined ionic liquid. topics of relevance to dye-sensitized solar cells, *J. Phys. Chem. B*, 2005, **109**, 17922–17927.
29 K. Shimizu, M. F. C. Gomes, A. A. H. Padua, L. P. N. Rebelo and J. N. Canongia Lopes, On the Role of the Dipole and Quadrupole Moments of Aromatic Compounds in the Solvation by Ionic Liquids, *J. Phys. Chem. B*, 2009, **113**, 9894–9900.
30 F. J. Deive, A. Rodriguez, A. B. Pereiro, K. Shimizu, P. A. S. Forte, C. C. Romão, J. N. Canongia Lopes, J. M. S. S. Esperanca and L. P. N. Rebelo, Phase Equilibria of Haloalkanes Dissolved in Ethylsulfate- or Ethylsulfonate-Based Ionic Liquids, *J. Phys. Chem. B*, 2010, **114**, 7329–7337.

# From molten salts to room temperature ionic liquids: Simulation studies on chloroaluminate systems

**Mathieu Salanne,*[a] Leonardo J. A. Siqueira,[b] Ari P. Seitsonen,[c] Paul A. Madden[d] and Barbara Kirchner[e]**

*Received 1st April 2011, Accepted 3rd June 2011*
**DOI: 10.1039/c1fd00053e**

An interaction potential including chloride anion polarization effects, constructed from first-principles calculations, is used to examine the structure and transport properties of a series of chloroaluminate melts. A particular emphasis was given to the study of the equimolar mixture of aluminium chloride with 1-ethyl-3-methylimidazolium chloride, which forms a room temperature ionic liquid $EMI^+$–$AlCl_4^-$. The structure yielded by the classical simulations performed within the framework of the polarizable ion model is compared to the results obtained from entirely electronic structure-based simulations: An excellent agreement between the two flavors of molecular dynamics is observed. When changing the organic cation $EMI^+$ by an inorganic cation with a smaller ionic radius ($Li^+$, $Na^+$, $K^+$), the chloroaluminate speciation becomes more complex, with the formation of $Al_2Cl_7^-$ in small amounts. The calculated transport properties (diffusion coefficients, electrical conductivity and viscosity) of $EMI^+$–$AlCl_4^-$ are in good agreement with experimental data.

## 1 Introduction

When monovalent chloride salts are dissolved in liquid $AlCl_3$, ionic liquids with low melting points are formed. When the added cation is inorganic, the system becomes a molten salt, like $NaAlCl_4$ with $T_m = 152$ °C.[1] When it is organic, the system is likely to be liquid below 100 °C, thus potentially becoming a room temperature ionic liquid (RTIL); for example $T_m$ ($EMIAlCl_4$) = 7 °C[2] (where $EMI^+$ stands for 1-ethyl-3-methylimidazolium cation).

A number of investigations using molecular dynamics simulations have been performed for ionic liquids directed at either structure or dynamical properties.[3] Given the molecular nature of cations or anions forming ionic liquids, earlier molecular dynamics simulations were performed with pairwise additive potentials based on non-polarizable force fields such as OPLS or AMBER.[4–11] These simulations have been able to reproduce the structural properties of ionic liquids, for instance showing that a somewhat unusual hydrogen bond can take place in imidazolium-based ionic between

[a]*UPMC Univ Paris 06, CNRS, ESPCI, UMR 7195, PECSA, F-75005 Paris, France. E-mail: mathieu.salanne@upmc.fr; Fax: +33 1 4427 3228; Tel: +33 1 4427 3265*
[b]*Laboratório de Materiais Híbridos, Universidade Federal de São Paulo, CEP 04024-002 São Paulo, SP, Brazil*
[c]*Physikalisch-Chemisches Institut, University of Zurich, Winterthurerstrasse 190, CH-8057 Zurich, Switzerland*
[d]*Department of Materials, University of Oxford, Parks Road, Oxford, OX1 3PH, UK*
[e]*Wilhelm-Ostwald Institute of Physical and Theoretical Chemistry, University of Leipzig, Linnéstr. 2, D-04103 Leipzig, Germany*

the imidazolium ring proton and the anion species.[12] This observation is in agreement with IR and NMR spectroscopy studies.[13–15] At intermediate spatial range (0.5 to 0.8 $Å^{-1}$), a chain-length dependent pre-peak appears either in the static structure factor calculated from molecular dynamics trajectories[16,17] or obtained by small wide angle X-ray scattering (SWAXS) and/or small angle neutron scattering (SANS).[18–20] This feature has been assigned to structural segregation or inhomogeneity, *i.e.* the imidazolium rings and the alkyl chain regions of cations respectively cluster to form polar and apolar domains in ionic liquids where the cations have long alkyl chains.[21] Despite their capability to reproduce the structure, the non-polarizable force fields often begin to fail when predicting dynamical properties, such as diffusion coefficient, electrical conductivity, and viscosity.[22] In general the dynamical properties calculated with non-polarizable models are slower than is observed in experiments: The diffusion coefficient and electrical conductivities are lower while the viscosity is greater than the measured values.[8,23–26] Note that pairwise additive potentials are often parameterised to *implicitly* include some aspects of the polarization effects. For example, the simulated fluid may become less viscous if the magnitudes of the partial charges are reduced,[27,28] but sometimes at the price of a poorer representation of other properties. These trends are particularly well-documented in the case of the inorganic systems.[29]

Disregarding the polarization method used, molecular dynamics simulations of molten salts in which polarization effects are included have shown that the liquids become more mobile.[30,31] Moreover, molecular dynamics simulations of ionic liquids including polarization have shown better agreement of dynamical properties with experiment, that is, the simulated liquids are either as mobile as the experimental sample or more mobile than the liquids simulated with non-polarizable model.[32–36] Among the studies using a polarizable model for ionic liquids, we would cite the recent and comprehensive work by Yan *et al.*,[35,36] where the authors carefully investigated the role-played by polarization in the ionic liquid $EMI^+$–$NO_3^-$. The authors found that when polarization effects are included, the short-range interactions are enhanced due to the additional charge–dipole and dipole–dipole interactions, while long-range electrostatic interactions become more screened. Accordingly, the imidazolium ring proton–$NO_3^-$ hydrogen bond strength is increased upon switching the polarization on. Another consequence of the stronger polar short-range interactions achieved with the polarizable model for the structure is the enhancement of the tendency towards structural inhomogeneity, which acts by pulling the apolar chains away from the polar center and reinforces the assembly of ethyl chain into the apolar domains. The attenuation of long-range electrostatic interaction is responsible for the increase of diffusion, conductivity and structural relaxation and the reduction of viscosity.

In this work, we study a series of chloroaluminate liquids. A particular emphasis is given to the room temperature ionic liquid consisting of an equimolar mixture of EMICl with $AlCl_3$. We employ two flavors of molecular dynamics simulations, which differ by the method involved for the calculation of the interactions: In the first, they are obtained from a full density functional theory (DFT) calculation, and in the second an analytical force field (including polarization effects) is used. Both methods have already been used successfully for the study of pure $AlCl_3$.[37,38] In a first step we detail how the parameters of the interaction potential for EMICl–$AlCl_3$ have been obtained. The structure and dynamics of this melt are then analyzed and compared to a series of molten salts consisting of equimolar mixtures of $AlCl_3$ with LiCl, NaCl and KCl.

## 2 Interaction potentials

### 2.1 Functional form

In recent work on the simulation of inorganic ionic systems, interaction potentials that were used commonly included many-body effects such as polarization.[39–43] The most generally retained analytical expression[44] is

 

$$V = \sum_{i<j}\left(\frac{q^i q^j}{r^{ij}} + B^{ij}\mathrm{e}^{-a^{ij}r^{ij}} - f_6^{ij}\left(r^{ij}\right)\frac{C_6^{ij}}{\left(r^{ij}\right)^6} - f_8^{ij}\left(r^{ij}\right)\frac{C_8^{ij}}{\left(r^{ij}\right)^8}\right) + V_{\mathrm{pol}}, \quad (1)$$

where the three first terms are pairwise additive. The first term corresponds to the electrostatic interactions, and formal integer charges are used. The two following terms are expressed by a Born-Huggins-Mayer type potential; they consist of an exponentially decaying term accounting for the overlap repulsion of the electronic clouds at short distances, and a term which represents the dispersion effects. The $C_6^{ij}$ and $C_8^{ij}$ are the dipole–dipole and dipole–quadrupole dispersion coefficients; $f_n$ are Tang-Toennies dispersion damping functions[45] describing the short-range penetration correction to the asymptotic multipole expansion of dispersion. These functions take the form

$$f_n^{ij}\left(r^{ij}\right) = 1 - \mathrm{e}^{-b_n^{ij}r^{ij}}\sum_{k=0}^{n}\frac{\left(b_n^{ij}r^{ij}\right)^k}{k!} \quad (2)$$

where the $b_n^{ij}$ parameter sets the range of the damping effect. The polarization term reflects the distortion of the electronic density in response to the electric fields due to all the other ions. It is written

$$V_{\mathrm{pol}} = \sum_{i,j}\left[\left(q^i\mu_\alpha^j g^{ij}\left(r^{ij}\right) - q^j\mu_\alpha^i g^{ji}\left(r^{ij}\right)\right)\mathbb{T}_\alpha^{ij} - \mu_\alpha^i\mu_\beta^j\mathbb{T}_{\alpha\beta}^{ij}\right] + \sum_i\left(\frac{1}{2\alpha^i}\left|\boldsymbol{\mu}^i\right|^2\right) \quad (3)$$

where $\mathbb{T}_\alpha$ and $\mathbb{T}_{\alpha\beta}$ are the charge–dipole and dipole–dipole interaction tensors and $\alpha^i$ is the polarizability of ion $i$. Again, Tang-Toennies functions are included to account for the short-range damping effects:

$$g^{ij}\left(r^{ij}\right) = 1 - c^{ij}\mathrm{e}^{-b^{ij}r^{ij}}\sum_{k=0}^{4}\frac{\left(b^{ij}r^{ij}\right)^k}{k!} \quad (4)$$

These functions differ from the previously defined $f_n^{ij}$ due to the presence of an additional parameter $c^{ij}$, which measures the strength of the ion response to the short-range effects.

The set of induced dipoles $\{\boldsymbol{\mu}^i\}_{i\,\in[1,N]}$ is treated as $3N$ additional degrees of freedom of the systems. The dipoles are determined at each time step by minimization of the total polarization energy and they depend on the positions of all the atoms at the corresponding time; therefore the polarization part of the potential is a many-body term.

In the case of room-temperature ionic liquids, molecular species are involved, and additional intramolecular interactions accounting for covalent bonding therefore have to be included. Moreover, the electrostatic interactions now involve *partial* charges for each atom and the Lennard-Jones type is more commonly used than the Born-Huggins-Mayer one for the remaining pairwise additive interactions.[4–9] The repulsion and dispersion terms are then cast together in the following way:

$$V_{\mathrm{LJ}} = \sum_{i<j}4\times\varepsilon^{ij}\left[\left(\frac{\sigma^{ij}}{r^{ij}}\right)^{12} - \left(\frac{\sigma^{ij}}{r^{ij}}\right)^6\right] \quad (5)$$

## 2.2 Parameterization

As soon as a functional form has been chosen for the interaction potential, all the atomic and atom pair parameters have to be carefully determined. In all cases, the overall charges on the $Cl^-$, $Al^{3+}$, and $EMI^+$ are fixed as −1, +3, and +1 respectively.

Our approach consists in parameterizing the remaining terms on the basis of first-principles electronic structure calculations.[44] This is achieved by a "force-matching" method, which is generalized by using the first-principles calculated induced dipoles in addition to the forces in the fitting procedure. Amongst molten chlorides, this method has already been applied successfully to the case of LiCl, NaCl and KCl.[46] Here we extend this set of parameters in order to include the $Al^{3+}$ and 1-ethyl-3-methylimidazolium cations.

In the present work, we have chosen to describe the $EMI^+$ cation as a non-polarizable species for two reasons. Firstly, polarization effects are dominated by the chloride anions (in systems like $EMI–BF_4$ involving less polarizable anions the situation is opposite, with the polarization of the cation playing the main role[47]). Secondly, from a practical point of view, the computer time associated with the minimization procedure of the polarization term increases substantially with the number of polarizable species. For all the intramolecular/intermolecular interactions involving the $EMI^+$ cation only, the parameters were taken from the well-tested force field of Pádua and co-workers.[7] The topology of the molecule is shown on Fig. 1, and the partial charges and Lennard-Jones individual parameters are given in Table 1. The pair parameters involved in eqn (5) are obtained from the usual Lorentz-Berthelot mixing rules:

$$\varepsilon^{ij} = \sqrt{\varepsilon^i \times \varepsilon^j} \tag{6}$$

$$\frac{\sigma^{ij} = \sigma^i + \sigma^j}{2} \tag{7}$$

We have then applied the force-matching method to fit all the other interactions, which involve the chloride anion on the one hand and either the aluminium or the $EMI^+$ cation atoms on the other hand. In order to ensure that the force-matching really refines the cation-anion terms, we have held fixed the parameters for the Cl–Cl interactions at the values previously determined for the pure alkali halides.[46] To this end, we extracted several configurations from some trajectories obtained from Car–Parrinello molecular dynamics simulations of the liquids EMICl and $EMICl–AlCl_3$. Each of these configurations were then used again as an input to a plane wave density functional theory (DFT) calculation (the details of these calculations are given in the next section). We first performed an energy minimization in order to find the ground-state electronic structure. The forces on each chloride and aluminium (if present) ions were extracted at this stage. Then we transformed the Kohn–Sham orbitals into a set of maximally localized Wannier functions,[48] from

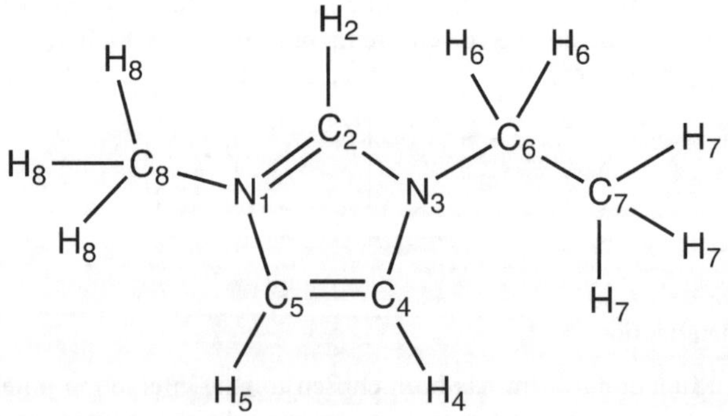

**Fig. 1** Topology of the 1-ethyl-3-methylimidazolium molecular ion.

 

**Table 1** Interaction potential parameters (partial charges, Lennard-Jones parameters) used for the $EMI^+$ cation.[7] Note that the $\varepsilon^i$ and $\sigma^i$ parameters were used for the cation-cation van der Waals interactions only. $e$ is the elementary charge

| Atom $i$ | $\varepsilon^i$/kJ mol$^{-1}$ | $\sigma^i$/Å | $q^i/e$ |
|---|---|---|---|
| $N_1$ | 0.71128 | 3.25 | 0.15 |
| $C_2$ | 0.29288 | 3.55 | −0.11 |
| $N_3$ | 0.71128 | 3.25 | 0.15 |
| $C_4$ | 0.29288 | 3.55 | −0.13 |
| $C_5$ | 0.29288 | 3.55 | −0.13 |
| $C_6$ | 0.27614 | 3.50 | −0.17 |
| $C_7$ | 0.27614 | 3.50 | −0.05 |
| $C_8$ | 0.27614 | 3.50 | −0.17 |
| $H_2$ | 0.12552 | 2.42 | 0.21 |
| $H_4$ | 0.12552 | 2.42 | 0.21 |
| $H_5$ | 0.12552 | 2.42 | 0.21 |
| $H_6$ | 0.12552 | 2.50 | 0.13 |
| $H_7$ | 0.12552 | 2.50 | 0.06 |
| $H_8$ | 0.12552 | 2.50 | 0.13 |

which we could determine the induced dipole moment on the $Cl^-$ anions. The parameters in the polarizable potentials are then optimized by matching the dipole and forces from the potential to the DFT calculated ones.[44,49] Two distinct functions are minimized. The first one corresponds to the polarization part:

$$\chi_D^2 = \frac{1}{N_{Cl^-}} \sum_{i=1}^{N_{Cl^-}} \frac{|\boldsymbol{\mu}_{DFT}^i - \boldsymbol{\mu}_{PIM}^i|^2}{|\boldsymbol{\mu}_{DFT}^i|^2} \tag{8}$$

where the subscript PIM stands for "Polarizable Ion Model". Since the chloride anion is the only polarizable species, for a given atom type $\alpha$ the two short-range damping parameters $b^{Cl\alpha}$ and $c^{Cl\alpha}$ were optimized with this function. The second function,

$$\chi_F^2 = \frac{1}{N_{Cl^-}} \sum_{i=1}^{N_{Cl^-}} \frac{|\mathbf{F}_{DFT}^i - \mathbf{F}_{PIM}^i|^2}{|\mathbf{F}_{DFT}^i|^2} + \frac{1}{N_{Al^{3+}}} \sum_{i=1}^{N_{Al^{3+}}} \frac{|\mathbf{F}_{DFT}^i - \mathbf{F}_{PIM}^i|^2}{|\mathbf{F}_{DFT}^i|^2}, \tag{9}$$

is used to parameterize the repulsion term of the interaction potential, *i.e.* for determining the various $B^{ij}$ and $a^{ij}$ values in eqn (1). A well known limitation of the PBE functional that we used in our DFT calculation is that it does not take into account properly the dispersion interactions,[50] so that the $C_6^{ij}$ and $C_8^{ij}$ cannot be fitted by this procedure. Instead, we have used our previously determined $Cl^-$–$Cl^-$ dispersion coefficients[46] for calculating the various $C_n^{ij}$ ($n = 6,8$) by using the Slater-Kirkwood relationship.

The quality of the representation of the forces is illustrated in Fig. 2 where we compare the three components of the forces acting on the aluminium (32 first atoms) and chloride ions calculated with our potential with the DFT calculated data, for one of the EMICl–$AlCl_3$ configurations. The agreement is seen to be uniformly good. For both RTIL systems, we obtained values of 0.13 for $\chi_F^2$, which is similar to our previous work on molten LiCl, NaCl and KCl. Concerning the dipoles, the values of $\chi_D^2$ are respectively 0.39 and 0.03 on average for the EMICl and EMICl–$AlCl_3$ configurations. The poorer agreement in the first case is due to the fact that the electric field felt by the $Cl^-$ anions, which is only due to the $EMI^+$ cation, is only partially reproduced by the partial charges of the atoms. When aluminium cations are added to the melt, this electric field becomes dominated by the trivalent

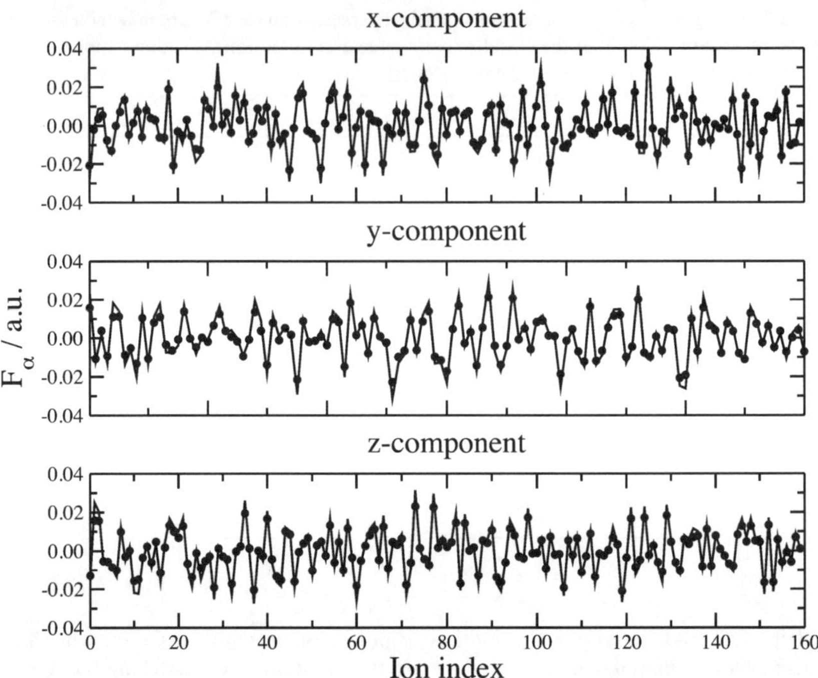


**Fig. 2** Results of force matching for one EMICl–AlCl$_3$ configuration. The three components of the force acting on each aluminium and chloride ions are compared (dots: DFT, line: PIM).

**Table 2** Parameters for the pairwise additive terms of the interaction potential. $b_6^{ij} = b_8^{ij} = 3.21$ Å$^{-1}$ for all pairs

| Atom pair *i*–*j* | $B^{ij}$ ×10$^{-6}$/J mol$^{-1}$ | $a^{ij}$/Å$^{-1}$ | $C_6^{ij}$ ×10$^{-6}$/J mol$^{-1}$ Å$^6$ | $C_8^{ij}$ ×10$^{-6}$/J mol$^{-1}$ Å$^8$ |
|---|---|---|---|---|
| Cl–Cl | 722.0 | 3.396 | 8.07 | 4.52 |
| Cl–Al | 95.8 | 2.955 | 0.0 | 0.0 |
| Cl–Li | 154.5 | 3.685 | 0.0 | 0.0 |
| Cl–Na | 177.2 | 3.262 | 2.16 | 2.70 |
| Cl–K | 547.5 | 3.458 | 1.52 | 1.91 |
| Cl–$C_i$ | 331.8 | 3.376 | 2.39 | 1.34 |
| Cl–$N_i$ | 260.4 | 4.401 | 1.44 | 0.81 |
| Cl–$H_i$ | 10.8 | 2.901 | 0.41 | 0.23 |

charges they carry, and this effect is much more easily captured by our model. The observation suggests that further refinement of the charges of the EMI$^+$ model would be a good target to further improve the predictive power of the simulations.

The parameters which were determined from the fitting procedure are summarized in Tables 2 and 3. No distinction has been made between the various $H_i$, $C_i$ and $N_i$ atom types because it did not lead to any improvement of the fit quality.

## 3 Simulation details

### 3.1 Electronic structure-based molecular dynamics

For the *ab initio*-molecular dynamics simulations we used density functional theory with the generalised gradient approximation of Perdew, Burke and Ernzerhof,

 

**Table 3** Parameters for the polarization part of the interaction potential. The polarizability of the $Cl^-$, $Na^+$ and $K^+$ ions were respectively set to 2.96, 0.13 and 0.70 Å$^3$

| Atom pair $i$–$j$ | $b^{ij}$/Å$^{-1}$ | $c^{ij}$ | $c^{ji}$ |
|---|---|---|---|
| Cl–Al | 3.802 | 2.953 | — |
| Cl–Li | 3.517 | 2.079 | — |
| Cl–Na | 3.326 | 3.000 | 0.697 |
| Cl–K | 3.084 | 3.000 | 0.917 |
| Cl–$H_i$ | 2.901 | 0.301 | — |

PBE[51] as the exchange–correlation term in the Kohn–Sham equations, and we replaced the action of the core electrons on the valence orbitals with norm-conserving pseudo potentials of the Troullier–Martins type.[52,53] We expanded the wave functions in plane waves up to the cut-off energy of 70 Ry. We sampled the Brillouin zone at the Γ point, employing periodic boundary conditions.

We performed the simulations in the NVT ensemble, employing a Nosé-Hoover thermostat at a target temperature of 300 K and a characteristic frequency of 595 cm$^{-1}$, a stretching mode of the $AlCl_3$ molecules. We propagated the velocity Verlet equations of motion with a time step of 5 a.t.u. = 0.121 fs, and the fictitious mass in the Car–Parrinello molecular dynamics[54] for the electrons was chosen as 700 a.u.; we note that the fictitious dynamics increases the temperature seen by the ions somewhat.[55] A cubic simulation cell with a edge length of 22.577 Å was used, containing both 32 anionic and cationic molecules, equalling the experimental density of 1.293 g cm$^{-3}$. The length of the trajectory was 20 ps. The CPMD simulation code was used.[56]

### 3.2 Polarizable ion model-based molecular dynamics

For the inorganic molten salts (LiCl–$AlCl_3$, NaCl–$AlCl_3$ and KCl–$AlCl_3$) the simulations were performed in the NVT ensemble, employing a Nosé-Hoover thermostat at a target temperature of 573 K, with a time step of 0.5 fs, for a total simulation time of 3 ns. For EMICl–$AlCl_3$, several simulations were performed at the temperatures of 298, 433, 473, 523 and 573 K, with a time step of 1 fs. Note that the higher temperatures employed may be above the thermal decomposition temperature of the ionic liquid (no data could be found in the literature); these simulations were performed in order to make a fair comparison of the structure and dynamics with the inorganic molten salts. The total simulation time was of 1.5 ns for each temperature except 298 K, for which it was of 3 ns. The FIST module of CP2K simulation package was used.[57] In all cases, cubic simulation cells were employed. The total number of atoms and the cell edge length, which were chosen according to the experimental

**Table 4** Number of atoms and cell edge length ($L$) employed in the simulations ($M^+$ = $Li^+$, $Na^+$, $K^+$ or $EMI^+$)

| System | $T$/K | $N_{M^+}$ | $N_{Al^{3+}}$ | $N_{Cl^-}$ | $L$/Å |
|---|---|---|---|---|---|
| LiCl–$AlCl_3$ | 573 | 120 | 120 | 480 | 28.426 |
| NaCl–$AlCl_3$ | 573 | 120 | 120 | 480 | 28.790 |
| KCl–$AlCl_3$ | 573 | 120 | 120 | 480 | 29.728 |
| EMICl–$AlCl_3$ | 298 | 200 | 200 | 800 | 41.576 |
| | 433 | 200 | 200 | 800 | 42.807 |
| | 473 | 200 | 200 | 800 | 43.200 |
| | 523 | 200 | 200 | 800 | 43.713 |
| | 573 | 200 | 200 | 800 | 44.251 |

densities (an extrapolation of the room temperature data was made for EMICl–$AlCl_3$ when necessary),[2,58] are provided in Table 4. The cut-off distance for the real space of the Ewald sum and the short range potential was set to 13 Å for EMICl–$AlCl_3$ and to half the size of the simulation cell otherwise.

## 4 Results and discussion

### 4.1 Chloroaluminate speciation

Unlike previous classical MD simulations of chloroaluminate-based ionic liquids, in which the $AlCl_4^-$ ion was treated as a molecular species,[4,59] an advantage of the two simulation methods involved in the present study is that they both treat the chloride and aluminium ions as independent species. This is done intrinsically in electronic structure based simulations on one hand, and by construction of the polarizable ion model (PIM) on the other. This capability has been used, for example, in pure $AlCl_3$ liquid, in which structure is characterized by the formation of $AlCl_4^-$ tetrahedra sharing an edge to form $Al_2Cl_6$ dimers or larger clusters. MD simulation studies either employing similar interaction potentials as in the present study[60] or based on electronic structure calculations[37,61] were able to confirm quantitatively this picture, in agreement with X-ray diffraction experiments.[62] The formation of large chloroaluminate species was also investigated in simulations of a system consisting of one EMICl pair dissolved in a $AlCl_3$ solvent.[63]

In equimolar mixtures of $AlCl_3$ with monovalent chloride $M$Cl salts, the main "chemical reaction" involving chloroaluminate species which is expected to occur[64–66] is:

$$AlCl_4^- + AlCl_4^- \rightarrow Al_2Cl_7^- + Cl^- \tag{10}$$

Here in the case of the equimolar EMICl–$AlCl_3$ the two simulation methods agree on the formation of isolated $AlCl_4^-$ anions only, in agreement with experimental results.[65] Small differences are observed in the first neighbour distances, which are shorter in the simulation employing the polarizable model: The first peaks of the Al–Cl radial distribution functions respectively appear at 2.18 Å and 2.22 Å for PIM-MD and CPMD, while the corresponding Cl–Cl functions display some first peak maxima at 3.53 Å and 3.61 Å. These differences may be attributed to the choice of keeping the $Cl^-$–$Cl^-$ interaction potential parameters similar to the values which had been previously determined for pure LiCl, NaCl and KCl molten salts.[46] Another source of difference could be the presence of attractive dispersion effects in the classical MD only.

When $M^+$ is an inorganic cation ($Li^+$, $Na^+$, $K^+$), we observe a weakening of the Al–Cl links. This may be quantified from the corresponding radial distribution functions (RDF), for which we observe a decrease of the intensity of the first peak, which switches from a value of 47 in EMICl–$AlCl_3$ to one of 18 in LiCl–$AlCl_3$ (not shown). Some $Cl^-$ ions begin to jump from one aluminium coordination sphere to another. The signature of these exchanges can also be tracked in the RDFs which are given on Fig. 3: The minimum occurring after the first maximum no longer corresponds to a value of zero as in the case in EMICl–$AlCl_3$. The value of this minimum also increases when the ionic radius of the $M^+$ cation decreases in the series. The position of the second peak of the RDF is slightly shifted toward larger distances when passing from $Li^+$ to $Na^+$ and then $K^+$, but this distance increases by as much as ~3 Å when passing from $K^+$ to $EMI^+$ (note that the peak also becomes much wider, with a pronounced shoulder). The absence of exchange of $Cl^-$ between $AlCl_4^-$ coordination tetrahedra in EMICl–$AlCl_3$ can therefore easily be explained by the longer distance that has to be crossed during a jump.

Going back to the chemical reaction defined in eqn (10), we observe some signature of the formation of $Al_2Cl_7^-$ ions in our systems by examining the Al–Al RDFs

 

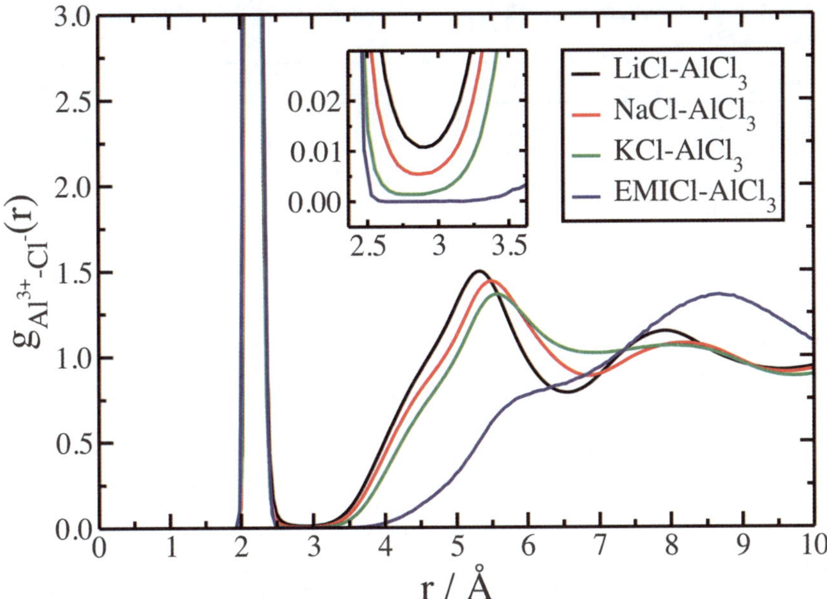


**Fig. 3** $Al^{3+}$–$Cl^{-}$ radial distribution functions for the *M*Cl–$AlCl_3$ mixtures at 573 K. Inset: Enlargement of the first minimum zone.

(shown on Fig. 4). A first peak with a small intensity is obtained in all the systems except EMICl–$AlCl_3$. The corresponding distance (ranging from 3.5 Å in LiCl–$AlCl_3$ to 3.7 Å in KCl–$AlCl_3$) is smaller than the length of two Al–Cl bonds, indicating pairs of $Al^{3+}$ ions linked by common $Cl^{-}$ anions.[60] To quantify the

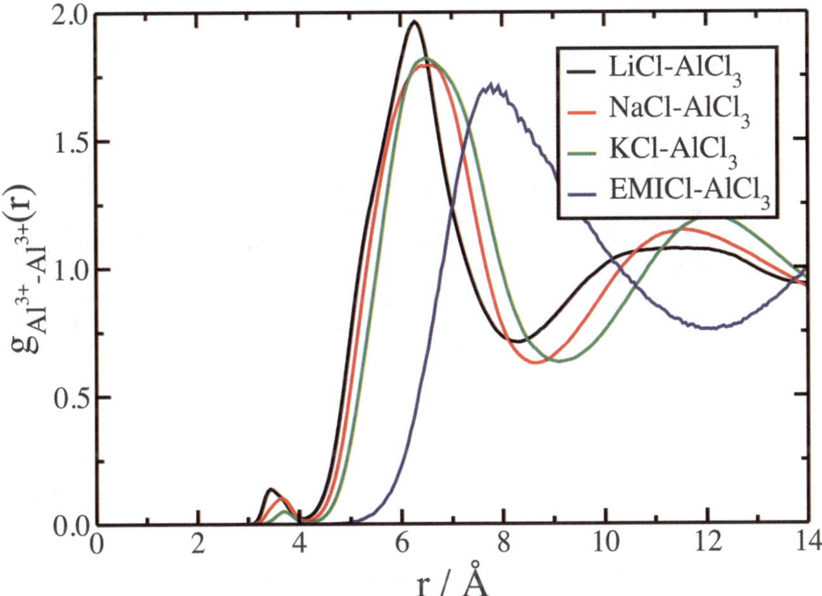


**Fig. 4** $Al^{3+}$–$Al^{3+}$ radial distribution functions for the *M*Cl–$AlCl_3$ mixtures at 573 K.

**Table 5** Proportions (given in percentage of aluminium ions) of the various chloroaluminate species at 573 K

| System | $[AlCl_4^-]$ | $[Al_2Cl_7^-]$ | $[Al_3Cl^-{}_{10}]$ |
|---|---|---|---|
| LiCl–$AlCl_3$ | 94.1 | 3.5 | 0.7 |
| NaCl–$AlCl_3$ | 95.7 | 3.4 | 0.2 |
| KCl–$AlCl_3$ | 98.2 | 1.6 | 0.0 |
| EMICl–$AlCl_3$ | 100.0 | 0.0 | 0.0 |

proportion of each chloroaluminate species present in the melt, we have followed the same procedure as in our previous work on LiF–$BeF_2$ mixtures, where we could provide a quantitative picture of the speciation in the melt for a wide range of composition.[67,68] Here we have chosen to define an Al–Cl–Al bond when two Al–Cl and the corresponding Al–Al distances are shorter than the corresponding RDFs minima. By extending the analysis of the linkages between ions, we can calculate the proportion of any different polynuclear ionic species present in the melt. The main species that we observed apart from the isolated $AlCl_4^-$ anions were the $Al_2Cl_7^-$ and $Al_3Cl^-{}_{10}$ species; their proportions are provided in Table 5 for a temperature of 573 K. Note that these numbers have to be taken cautiously: Due to the finite size of the simulation cells and to the relatively low number of chloride anions jumping or bridging events, some refinement may be necessary. We also observed some small proportions of transient chemical species such as $AlCl_5^{2-}$ or $Al_2Cl_8^{2-}$ in negligible amounts.

### 4.2 Effect of the $M^+$ cation nature on the intermolecular structure

From the analysis of speciation in the chloroaluminate species we can try to describe the equimolar mixtures of $M$Cl with $AlCl_3$ as generic $MX$ molten salts (*e.g.* alkali halides), where $M^+$ is the monovalent cation and $X^-$ is the $AlCl_4^-$ anion. In simple $MX$ molten salts, the structure around a given ion consists of alternating layers of opposite charges. As a result, the RDFs are characterized by strong oscillations, with the like-like RDFs maxima (minima) which are located at the same positions as the cation-anion RDF minimum (maximum). In the present work, we can compare the $M^+$–$Al^{3+}$ RDF to the $Al^{3+}$–$Al^{3+}$ one; the $Al^{3+}$ was chosen because it gives the $AlCl_4^-$ ion center of mass; for the $EMI^+$ cation the $M^+$–$Al^{3+}$ was calculated with respect to the imidazolium ion center of mass position. These functions are shown on Fig. 4 and 5. On the latter we also report the position of the $Al^{3+}$–$Al^{3+}$ RDF first maximum (omitting the small peak due to the formation of $Al_2Cl_7^-$) with an arrow on the abscissa in order to facilitate the comparison with the position of the first minimum in the cation-anion term. We observe that these arrows are not located at the same position as the corresponding minima, but the difference remains small. This means that the structure deviates slightly from that of a simple $MX$ molten salt. This conclusion still holds for EMICl–$AlCl_3$ at lower temperatures: The main features of the $M^+$–$Al^{3+}$ RDF remain unchanged when passing from 573 K to 298 K.

From this figure we can observe somewhat different behaviour depending on the size of the $M^+$ cation. In the case of the LiCl–$AlCl_3$ melt, the $M^+$–$Al^{3+}$ RDF minimum is located at a shorter distance than the corresponding $Al^{3+}$–$Al^{3+}$ RDF first maximum; this is due to the small size of the $Li^{3+}$ cation, which allows it to "penetrate" slightly into the $AlCl_4^-$ anion van der Waals sphere. The difference becomes smaller for the NaCl–$AlCl_3$ and KCl–$AlCl_3$ melts for which the extrema of the two functions almost coincide, and the situation is reversed for EMICl–$AlCl_3$ where closer packing of the anions is allowed. The $M^+$–$Al^{3+}$ RDF first peak is also broader in the case of EMICl–$AlCl_3$ compared to the alkali cations containing systems, but

 

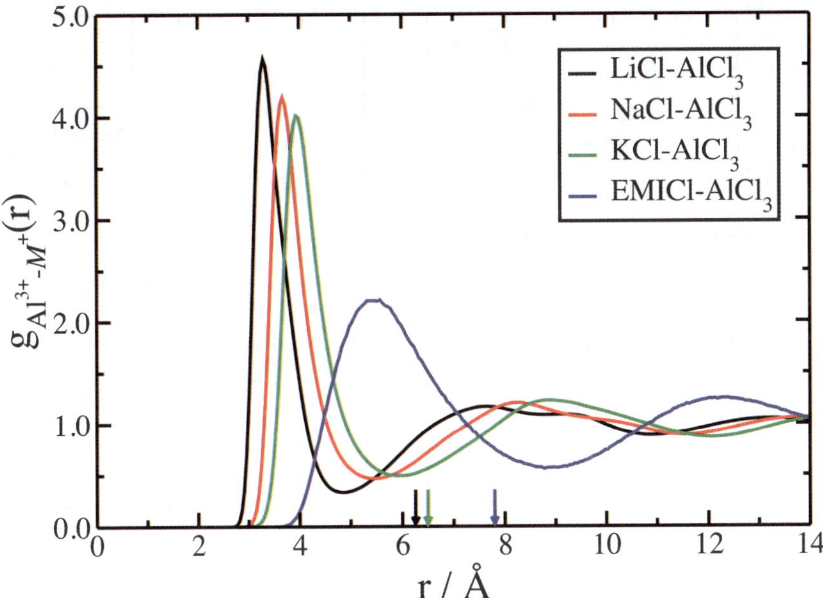


**Fig. 5** $AlCl_4^-$–$M^+$ radial distribution functions at 573 K. In the case where $M^+$ is the $EMI^+$ cation, the position of the center of mass of the molecule was used to calculate the function. Small arrows indicate the position of the corresponding $Al^{3+}$–$Al^{3+}$ RDF first maximum (omitting the first peak due to the formation of $Al_2Cl_7^-$ ions); the $Na^+$ one is not visible because it occurs at the same position as the $K^+$ one.

the difference is not dramatic. The two latter observations suggest that in the present system the $EMI^+$ cation behaves mostly like a monoatomic cation with a very large ionic radius. In the next section we will investigate its coordination structure on a more local scale, in order to understand if the charge delocalization as well as the presence of different atom types induce a particular ordering in the first solvation shell.

### 4.3 EMICl–$AlCl_3$ equimolar mixture

In order to characterize further the intermolecular structure of the EMICl–$AlCl_3$ system, the radial distribution functions of the chloride anions and the atoms from the $EMI^+$ cations have been calculated. They are plotted on Fig. 6 (hydrogen atoms) and 7 (carbon atoms). The CPMD and PIM results obtained at room temperature are compared; an excellent agreement is observed for all the atoms. The small differences observed are more likely due to the poorer sampling in the CPMD simulation rather than to a deficiency of the classical interaction potential. In a previous attempt to obtain classical force field parameters through a force-matching procedure similar to ours on the ionic liquid dimethylimidazolium chloride (DMICl), Youngs *et al.* have been less successful in reproducing the structure obtained from CPMD.[69] Although they were able to reproduce the peak positions better than with previous classical MD potentials, the intensities of the first peaks where largely overestimated (especially in the case of the ring protons). The overstructuring obtained by these authors could partly be cancelled by multiplying all the $\varepsilon^i$ parameters by a factor of 2, at the price of a worse reproduction of the initial set of DFT forces (therefore leading to higher $\chi_F^2$ values). The better agreement observed here can be attributed to two factors. The main one is the use of a more complicated functional form for the analytical interaction potential, or in other

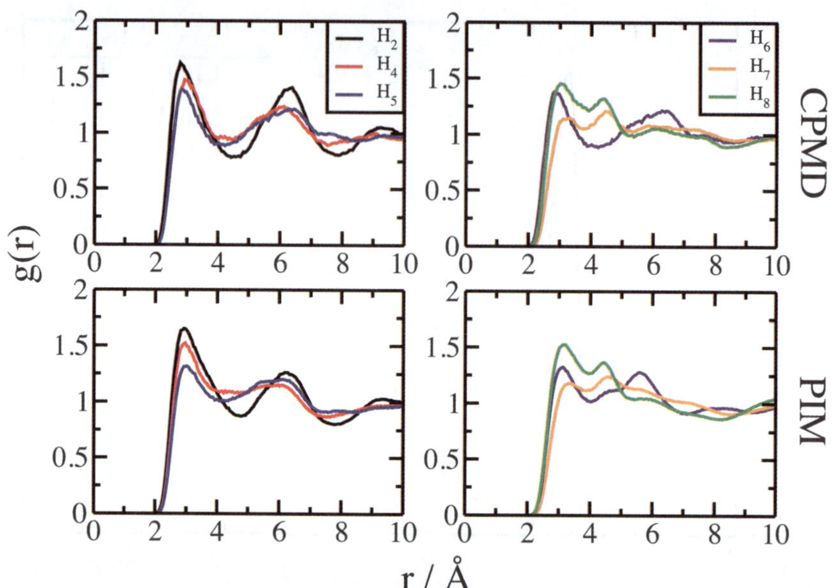


**Fig. 6** Radial distribution functions of the chloride anions and the protons from the $EMI^+$ cations. Top: CPMD results; bottom: PIM results. Left: Protons from the imidazolium ring. Right: Protons from the alkyl chains.

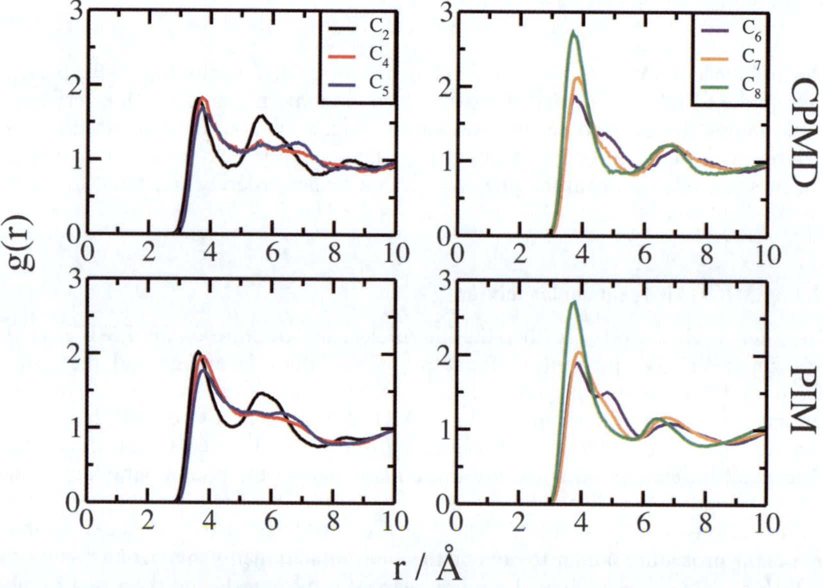


**Fig. 7** Radial distribution functions of the chloride anions and the carbon atoms from the $EMI^+$ cations. Top: CPMD results; bottom: PIM results. Left: Carbon atoms from the imidazolium ring. Right: Carbon atoms from the alkyl chains.

words to the inclusion of anion polarization effects. The second is the fact that we focused on the chloride (and aluminium) ion forces in our force-fitting process; in Youngs *et al.* study all the parameters, including the intramolecular ones, were fitted during the procedure. The intramolecular contribution may have somewhat "masked" the intermolecular ones, thus leading to a smaller precision for the latter. It is worth noticing that the better agreement observed here was not obtained by fitting more parameters. Due to the choice of using the set of partial charges of Padua and co-workers for the $EMI^+$ ion atoms, only twelve parameters involving the $Cl^-$ and $EMI^+$ species had to be fitted in the parameterization process: The eight Cl–$C_i$, $N_i$, $H_i$, Al repulsion parameters ($B^{ij}$ and $a^{ij}$) and the four parameters $b^{ij}$ and $c^{ij}$ involved in the damping of the chloride ion-induced dipoles by the hydrogen and aluminium atoms.

A striking feature of the radial distribution functions involving the chloride ions and the $EMI^+$ atoms is that the most intense peaks are observed for the carbon atoms. The peak is more pronounced for the $C_7$ and $C_8$, while its intensity is almost the same for the other carbons. This result contrasts sharply with some other RTILs, like the above-mentioned DMICl, for example, for which the anions tend to stay in the vicinity of the imidazolium ring protons.[69,70] On the contrary, it resembles the case of the 1-*n*-butyl-3-methylimidazolium hexafluorophosphate, for which the hydrogen bonds were found to be weaker than expected, as indicated by their short lifetimes.[71] In the present system the situation is due to the huge electrostatic interaction between $Cl^-$ anions and the (Lewis) acidic $Al^{3+}$ cation. To illustrate this the electrostatic potential mapped onto the isosurface of the electron density of the two individual ions is shown in Fig. 8. It provides insight into the charge distribution of each ion. For the $AlCl_4^-$ we recognize in the red color (low electrostatic potential) the consequence of the negative charge that is associated with this ion. The negative charge is distributed all over the chlorine atoms, with a higher negative charge on the inner side (the sphere-like surface around the chloride is coloured in dark red on the interior and in lighter red at the exterior of the $AlCl_4^-$ ion). Towards the center, close to the aluminium atom, we find a decreased negative charge shown as the blue color in the left panel of Fig. 6. The opposite is the case for the $EMI^+$. Here the positive charge leads to the blue color (high electrostatic potential) around this ion. Upon closer inspection we find that the ring protons show the same blue color as most of the molecule. A slight decrease of the charges can be found in the methyl group protons and a stronger decrease (green color) can be found at the terminal ethyl protons. This is in accordance with the observation from the radial pair distribution functions and with chemical intuition that these protons are less acidic than the other protons of the cation.

In the PIM simulations, the chloride anions therefore have their induced dipoles oriented along the Al–Cl axis, which hinders the formation of hydrogen bonds. If we look in further detail at the chloride-proton radial distribution functions, we see that the $H_2$–$Cl^-$ one shows the most pronounced peak, associated with the shortest separation. It is therefore clear that this proton is the most acidic coordination site. The

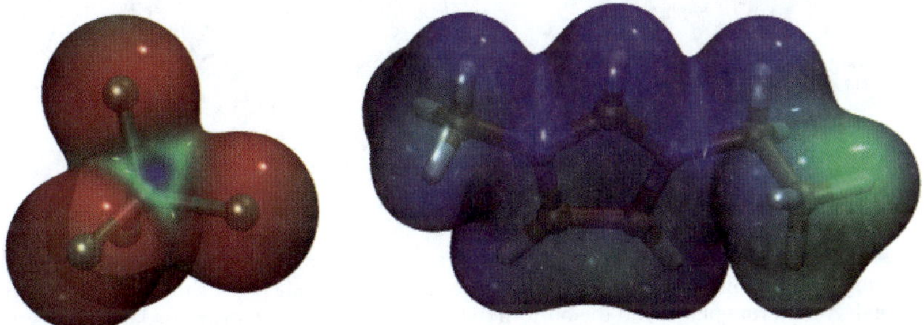

**Fig. 8** The electrostatic potential mapped onto the electron density with surface values of 0.067 $e^-$ $Å^{-3}$. The colour scale ranges from −0.1 (red) to +0.1 (blue) atomic units. Left: $AlCl_4^-$, right: $EMI^+$.

functions involving the other two protons of the imidazolium ring ($H_4$ and $H_5$) also display some peaks at slightly larger distances. However, considering the protons from the ethyl and the methyl group, small pronounced peaks can also be observed. The methyl group hydrogen ($H_8$) function first maximum has the same height as the $H_4$ one; a similar observation is made for the terminal hydrogen atoms of the ethyl group ($H_6$) compared to the ring $H_5$ protons.

The arrangement of the chloride anions around the imidazolium cations therefore seems to correspond more to an involved network including the alkyl chains as well as the ring of the imidazolium cation, rather than to individual ion pairs, confirming the results obtained for many imidazolium-based ionic liquids.[71,72] In order to confirm this picture, the three-dimensional density map was calculated for the chloride anions located in the first solvation shell of the $EMI^+$. Views of these maps perpendicular and parallel to the imidazolium ring are provided on Fig. 9. Red, green and blue regions correspond respectively to high, intermediate and low probability of finding an anion around a given cation. By high probability we mean higher than 80% of the maximum density of occurrences in these figures, and low probability means in between 40% and 60% of the maximum density of occurrences (probability lower than 40% is not shown). The important features of the radial distribution functions, *i.e.* a preference for some regions close to the $C_7$ and $C_8$ carbons and an absence of important correlations with the protons, are recovered. In general, localization effects are much less important than in the DMICl case, and the formation of hydrogen bond is not the main influence governing the structure of EMICl–$AlCl_3$. It is worth mentioning that in a study involving imidazolium cations with a series of halide anions of different sizes, the analysis of density maps showed that the larger the anion the lower the probability of the existence of hydrogen bond between the anion and the cation protons.[6] The present results agree with this view when considering that the relevant anion in our system is the full $AlCl_4^-$ entity.

The structural analysis of the equimolar EMICl–$AlCl_3$ mixture shows that it can almost be viewed as a classical *MX* molten salt, where $M^+$ = $EMI^+$ and $X^-$ = $AlCl_4^-$, with a structure characterized by a competition between close packing and charge ordering effects occurring between the two species. Small deviations from the *MX* system are observed due to the existence of weak spatial correlations arising from the formation of hydrogen bonds between the ring protons and the chloride anions.

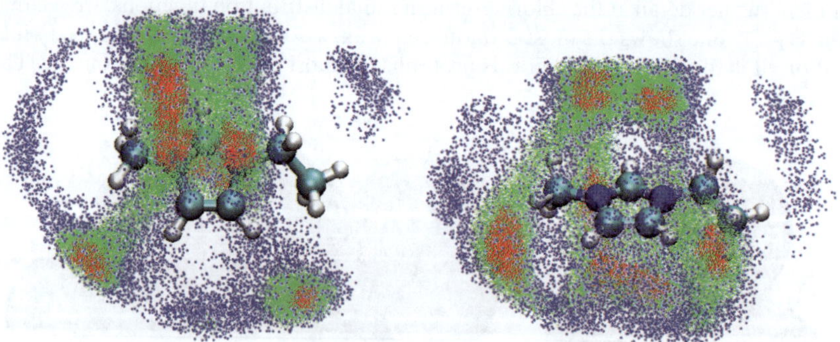

**Fig. 9** Probability density maps of location of nearest-neighbor anions around the imidazolium ring in equimolar EMICl–$AlCl_3$ mixture at 298 K. The left panel shows the view above the imidazolium ring plane, and the bottom panel shows the view on the ring plane. Red, green and blue regions correspond respectively to high, intermediate and low probability of finding an anion around a given cation. By high probability we mean higher than 80% of the maximum density of occurrences, and low probability means in between 40% and 60% of the maximum density of occurrences (probability lower than 40% is not shown).

 

## 4.4 Transport

In the recent simulation work on the room temperature ionic liquids, it was shown that polarization effects are important for reproducing accurately the transport properties.[32–34,36,47] Here from the Einstein relation

$$D_\alpha = \lim_{t\longrightarrow\infty} \frac{1}{6t}\left\langle \left|\delta\mathbf{r}_i(t)\right|^2 \right\rangle \tag{11}$$

where $\delta\mathbf{r}_i(t)$ is the displacement of a typical ions of species $\alpha$ in time $t$, we obtained a diffusion coefficient of $1.04 \times 10^{-10}$ m$^2$ s$^{-1}$ for the cation at 298 K. This compares very well with the experimental value measured by pulse-field gradient NMR spectroscopy, which is $0.95 \times 10^{-10}$ m$^2$ s$^{-1}$ at 293 K.[73] For the anion we obtained a diffusion coefficient of $4.45 \times 10^{-10}$ m$^2$ s$^{-1}$ at 298 K.

From the mean squared displacement of the overall charge density it is also possible to calculate the electrical conductivity of a melt according to the following formula:

$$\sigma = \frac{e^2}{k_BTV} \lim_{t\longrightarrow\infty} \frac{1}{6t}\left\langle \left| \sum_i q_i \delta\mathbf{r}_i(t)\right|^2 \right\rangle, \tag{12}$$

where $e$ is the elementary charge and $q_i$ the magnitude of the (partial or formal) charge on atom $i$. Notice that this expression involves the correlations between the displacements of different ions, which is essential in this case as the motion of the $Al^{3+}$ and $Cl^-$ ions within a given $AlCl_4^-$ is strongly correlated. The viscosity is calculated by integration of the correlation function of the stress tensor:

$$\eta = \frac{1}{k_BTV}\int_0^\infty \langle \sigma_{\alpha\beta}(0)\sigma_{\alpha\beta}(\tau)\rangle \mathrm{d}\tau, \tag{13}$$

where $\sigma_{\alpha\beta}$ is one of the components of the stress tensor. The two latter quantities, which are *collective* transport coefficients (compared to the diffusion coefficients, which are *individual*), require much longer simulation times to get converged values. Such quantities may therefore only be calculated with the analytic force-field model; they are well out of reach of full *ab initio* simulations. In the case of EMICl–$AlCl_3$, we could extract their values for all the temperatures except at 298 K. They are shown in Fig. 10 in the form of an Arrhenius plot. On the same figure we also show the experimental values;[2] although those were obtained at lower temperatures, we observe a good match between the two sets of data, which confirms that our polarizable ion model yields the correct dynamics of the system.

At the temperature of 573 K, which is above the melting point of the inorganic chloroaluminates, it is also possible to compare the diffusion coefficients of the

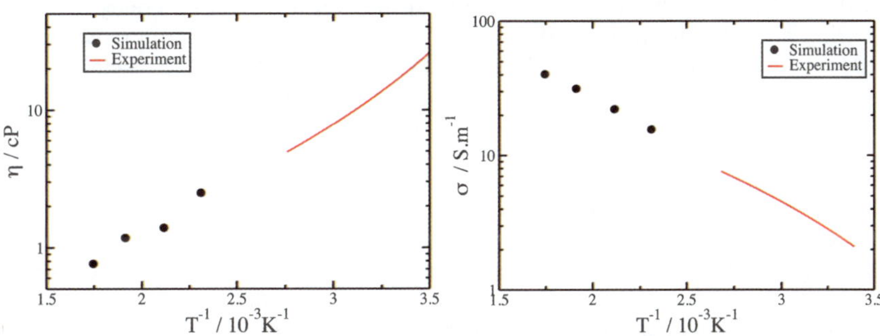


**Fig. 10** Viscosities (left) and electrical conductivities (right) of EMICl–$AlCl_3$ in the form of an Arrhenius plot. The simulated values, although obtained in a different temperature regime, compare well with the experimental ones.[2]

**Table 6** Diffusion coefficients of the equimolar $M$Cl–$AlCl_3$ systems at 573 K ($M^+ = Li^+$, $Na^+$, $K^+$ or $EMI^+$). All values are in $10^{-10}$ $m^2$ $s^{-1}$

| System | $D_{M^+}$ | $D_{Al^{3+}}$ | $D_{Cl^-}$ |
|---|---|---|---|
| LiCl–$AlCl_3$ | 16.4 | 6.19 | 6.28 |
| NaCl–$AlCl_3$ | 20.6 | 6.63 | 6.65 |
| KCl–$AlCl_3$ | 24.3 | 8.62 | 8.62 |
| EMICl–$AlCl_3$ | 35.5 | 23.2 | 23.2 |

various ions in all the systems. They are reported in Table 6. For all the species, we observe an increase of the diffusion coefficient with the $M^+$ cation size. Note that the number density also changes when switching from one system to another, which is likely to be the main cause for this evolution. Still it is possible to come to the conclusion that in all systems the chloroaluminate anion diffuses more slowly than the corresponding $M^+$ cation. We can also conclude that the organic cation-based ionic liquid, EMICl–$AlCl_3$, fits into the same pattern of behaviour as the inorganic cation-based compounds with regard to the variation of transport properties with cation size.

## 5 Conclusion

We have shown in this study that the polarizable ion model developed for simple molten salts[46] can successfully be transferred to chloroaluminate ionic liquids. The interaction potential was parameterized purely from first-principles through a force and dipole-matching procedure, and a common set of parameters could be obtained for the mixtures of $AlCl_3$ with several inorganic salts (LiCl, NaCl and KCl) as well as with the organic 1-ethyl-3-methylimidazolium chloride. An advantage of our model is that it allows us to quantify the chloroaluminate speciation due to the choice of treating $Al^{3+}$ and $Cl^-$ as independent ionic species.

A particular emphasis was given to the study of the equimolar mixture EMICl–$AlCl_3$, for which the structure yielded by the classical simulations performed within the framework of the polarizable ion model is compared to the results obtained from entirely electronic structure-based simulations: An excellent agreement between the two flavors of molecular dynamics is obtained, which enables us to conclude that the structure of EMICl–$AlCl_3$ can almost be viewed as the one of a classical $MX$ molten salt, where $M^+ = EMI^+$ and $X^- = AlCl_4^-$. The structure is characterized by a competition between close packing and charge ordering effects occurring between the two species. Small deviations from the $MX$ system are observed due to the existence of weak spatial correlations arising from the formation of hydrogen bonds between the ring protons and the chloride anions. When decreasing the cation ionic radius ($EMI^+ \rightarrow K^+ \rightarrow Na^+ \rightarrow Li^+$) small proportions of $Al_2Cl_7^-$ due to the reaction

$$AlCl_4^- + AlCl_4^- \rightarrow Al_2Cl_7^- + Cl^- \quad (14)$$

start to be observed ($Al_3Cl_{10}^-$ is also observed in equimolar LiCl–$AlCl_3$). An interesting extension of this work would be to study the same series of systems under non-stoichiometric conditions. The influence of the $M^+$ cation chemical nature in the Lewis acidity of the melt can also be quantified using a recently developed method.[74] In future work we will try to perform such calculations in order to relate it to the chloroaluminate speciation determined in the present study.

The only many-body effect included in the interaction potential for EMICl–$AlCl_3$ is the anion polarization, which enabled us to perform long enough runs to get the dynamic properties of the system, which are out of reach of first-principles simulations. The calculated transport properties were compared to the experimental data, showing again a good agreement for both the individual (diffusion coefficients) and collective (electrical conductivities and viscosities) dynamics of the system.

## Acknowledgements

MS is grateful to the FAPESP and the Federal University of São Paulo for a travel grant which enabled this collaboration. LJAS also thanks FAPESP, proc. 08/08670-7.

## References

1 H. A. Oye, *Metall. Mater. Trans. B*, 2000, **31B**, 641–650.
2 J. A. A. Fannin, D. A. Floreani, L. A. King, J. S. Landers, B. J. Piersma, D. J. Stech, R. L. Vaughn, J. S. Wilkes and J. L. Williams, *J. Phys. Chem.*, 1984, **88**, 2614–2621.
3 E. J. Maginn, *Acc. Chem. Res.*, 2007, **40**, 1200–1207.
4 J. de Andrade, E. S. Böes and H. Stassen, *J. Phys. Chem. B*, 2002, **106**, 3546–3548.
5 C. J. Margulis, H. A. Stern and B. J. Berne, *J. Phys. Chem. B*, 2002, **106**, 12017.
6 S. M. Urahata and M. C. C. Ribeiro, *J. Chem. Phys.*, 2004, **120**, 1855–1863.
7 J. N. Canongia Lopes, J. Deschamps and A. A. H. Pádua, *J. Phys. Chem. B*, 2004, **108**, 2038.
8 L. J. A. Siqueira and M. C. C. Ribeiro, *J. Phys. Chem. B*, 2007, **111**, 11776.
9 M. Salanne, C. Simon and P. Turq, *J. Phys. Chem. B*, 2006, **110**, 3504–3510.
10 S. L. Price, C. G. Hanke and R. M. Lynden-Bell, *Mol. Phys.*, 2001, **99**, 801–809.
11 T. I. Morrow and E. J. Maginn, *J. Phys. Chem. B*, 2002, **106**, 12161–12164.
12 B. L. Bhargava and S. Balasubramanian, *J. Chem. Phys.*, 2007, **127**, 114510.
13 A. Wulf, K. Fumino, D. Michalik and R. Ludwig, *ChemPhysChem*, 2007, **8**, 2265.
14 K. Fumino, A. Wulf and R. Ludwig, *Angew. Chem., Int. Ed.*, 2008, **47**, 3830.
15 K. Fumino, A. Wulf and R. Ludwig, *Angew. Chem., Int. Ed.*, 2008, **47**, 8731.
16 S. M. Urahata and M. C. C. Ribeiro, *J. Chem. Phys.*, 2006, **124**, 074513.
17 H. V. R. Annapuredy, H. K. Kashyap, P. De Biase and C. J. Margulis, *J. Phys. Chem. B*, 2010, **114**, 16838–16846.
18 A. Triolo, O. Russina, B. Fazio, R. Triolo and E. Di Cola, *Chem. Phys. Lett.*, 2008, **457**, 362.
19 C. Hardacre, J. D. Holbrey, C. L. Mulland, T. G. A. Youngs and D. T. Bowron, *J. Chem. Phys.*, 2010, **133**, 074510.
20 O. Russina, A. Triolo, L. Gontrani, R. Caminiti, D. Xiao, L. G. Hines, R. A. Bartsch, E. L. Quitevis, N. Plechkova and K. R. Seddon, *J. Phys.: Condens. Matter*, 2009, **21**, 424121.
21 B. L. Bhargava, R. Devane, M. L. Klein and S. Balasubramanian, *Soft Matter*, 2007, **3**, 1395–1400.
22 R. M. Lynden-Bell and T. G. A. Youngs, *J. Phys.: Condens. Matter*, 2009, **21**, 424120.
23 C. Cadena, Q. Zhao, R. Q. Snurr and E. J. Maginn, *J. Phys. Chem. B*, 2006, **110**, 2821.
24 S. Tsuzuki, W. Shinoba, H. Saito, M. Mikami, H. Tokuda and M. Watanabe, *J. Phys. Chem. B*, 2009, **113**, 10641.
25 T. Koddermann, D. Paschek and R. Ludwig, *ChemPhysChem*, 2007, **8**, 2464.
26 N.-T. Van-Oanh, C. Houriez and B. Rousseau, *Phys. Chem. Chem. Phys.*, 2010, **12**, 930–936.
27 F. Müller-Plathe and W. F. van Gunsteren, *J. Chem. Phys.*, 1995, **103**, 4745–4756.
28 B. L. Bhargava and S. Balasubramanian, *J. Chem. Phys.*, 2005, **123**, 144505.
29 D. Marrocchelli, M. Salanne, P. A. Madden, C. Simon and P. Turq, *J. Mol. Phys.*, 2009, **107**, 443.
30 P. A. Madden and M. Wilson, *J. Phys.: Condens. Matter*, 2000, **12**, A95.
31 L. J. A. Siqueira, S. Urahata and M. C. C. Ribeiro, *J. Chem. Phys.*, 2003, **119**, 8002.
32 T. Yan, C. J. Burnham, M. G. Del Pópolo and G. A. Voth, *J. Phys. Chem. B*, 2004, **108**, 11877–11881.
33 C. Schroder and O. Steinhauser, *J. Chem. Phys.*, 2010, **133**, 154511.
34 D. Bedrov, O. Borodin, Z. Li and G. Smith, *J. Phys. Chem. B*, 2010, **114**, 4984–4997.
35 T. Yan, Y. Wang and C. Knox, *J. Phys. Chem. B*, 2010, **114**, 6905.
36 T. Yan, Y. Wang and C. Knox, *J. Phys. Chem. B*, 2010, **114**, 6886.
37 B. Kirchner, A. Seitsonen and J. Hutter, *J. Phys. Chem. B*, 2006, **110**, 11475–11480.
38 F. Hutchinson, M. Wilson and P. A. Madden, *Mol. Phys.*, 2001, **99**, 811–824.
39 P. Tangney and S. Scandolo, *J. Chem. Phys.*, 2002, **117**, 8898–8904.

40 C. Merlet, P. A. Madden and M. Salanne, *Phys. Chem. Chem. Phys.*, 2010, **12**, 14109–14114.
41 M. Salanne, D. Marrocchelli, C. Merlet, N. Ohtori and P. A. Madden, *J. Phys.: Condens. Matter*, 2011, **23**, 102101.
42 V. Bitrian, O. Alcaraz and J. Trullas, *J. Chem. Phys.*, 2011, **134**, 044501.
43 L. Giacomazzi, P. Carrez, S. Scandolo and P. Cordier, *Phys. Rev. B: Condens. Matter Mater. Phys.*, 2011, **83**, 014110.
44 R. J. Heaton, R. Brookes, P. A. Madden, M. Salanne, C. Simon and P. Turq, *J. Phys. Chem. B*, 2006, **110**, 11454–11460.
45 K. Tang and J. Toennies, *J. Chem. Phys.*, 1984, **80**, 3726–3741.
46 N. Ohtori, M. Salanne and P. A. Madden, *J. Chem. Phys.*, 2009, **130**, 104507.
47 O. Borodin, *J. Phys. Chem. B*, 2009, **113**, 11463–11478.
48 N. Marzari and D. Vanderbilt, *Phys. Rev. B: Condens. Matter*, 1997, **56**, 12847–12865.
49 A. Aguado, L. Bernasconi, S. Jahn and P. A. Madden, *Faraday Discuss.*, 2003, **124**, 171–184.
50 S. Zahn and B. Kirchner, *J. Phys. Chem. A*, 2008, **112**, 8430–8435.
51 J. P. Perdew, K. Burke and M. Ernzerhof, *Phys. Rev. Lett.*, 1996, **77**, 3865–3868.
52 N. Troullier and J. L. Martins, *Phys. Rev. B: Condens. Matter*, 1991, **43**, 001993.
53 N. Troullier and J. L. Martins, *Phys. Rev. B: Condens.Matter*, 1991, **43**, 008861.
54 R. Car and M. Parrinello, *Phys. Rev. Lett.*, 1985, **55**, 2471–2474.
55 P. Tangney, *J. Chem. Phys.*, 2006, **124**, 044111.
56 The CPMD consortium, *CPMD Version 3.13.2*, http://www.cpmd.org.
57 CP2K developers group, http://cp2k.berlios.de.
58 G. Janz, *J. Phys. Chem. Ref. Data*, 1988, **17**, 1–309.
59 J. de Andrade, E. S. Böes and H. Stassen, *J. Phys. Chem. B*, 2002, **106**, 13344–13351.
60 F. Hutchinson, M. K. Walters, A. J. Rowley and P. A. Madden, *J. Chem. Phys.*, 1999, **110**, 5821–5830.
61 A. L. L. East and J. Hafner, *J. Phys. Chem. B*, 2007, **111**, 5316–5321.
62 R. L. Harris, R. E. Wood and H. L. Ritter, *J. Am. Chem. Soc.*, 1951, **73**, 3151–3155.
63 B. Kirchner and A. Seitsonen, *Inorg. Chem.*, 2007, **46**, 2751–2754.
64 J. A. A. Fannin, L. A. King, J. A. Levisky and J. S. Wilkes, *J. Phys. Chem.*, 1984, **88**, 2609–2614.
65 M. Lipsztajn and R. A. Osteryoung, *J. Electrochem. Soc.*, 1985, **132**, 1126–1130.
66 C. L. Hussey, T. B. Scheffler, J. S. Wilkes and J. A. A. Fannin, *J. Electrochem. Soc.*, 1986, **133**, 1389–1391.
67 M. Salanne, C. Simon, P. Turq, R. J. Heaton and P. A. Madden, *J. Phys. Chem. B*, 2006, **110**, 11461–11467.
68 M. Salanne, C. Simon, P. Turq and P. A. Madden, *J. Phys. Chem. B*, 2007, **111**, 4678–4684.
69 T. G. A. Youngs, M. G. Del Pópolo and J. Kohanoff, *J. Phys. Chem. B*, 2006, **110**, 5697–5707.
70 M. G. Del Pópolo, R. M. Lynden-Bell and J. Kohanoff, *J. Phys. Chem. B*, 2005, **109**, 5895–5902.
71 W. Zhao, F. Leroy, B. Heggen, S. Zahn, B. Kirchner, S. Balasubramanian and F. Muller-Plathe, *J. Am. Chem. Soc.*, 2009, **131**, 15825–15833.
72 M. Kohagen, M. Brehm, J. Thar, W. Zhao, F. Müller-Plathe and B. Kirchner, *J. Phys. Chem. B*, 2011, **115**, 693–702.
73 C. K. Larive, M. Lin, B. J. Piersma and W. R. Carper, *J. Phys. Chem.*, 1995, **99**, 12409–12412.
74 M. Salanne, C. Simon and P. A. Madden, *Phys. Chem. Chem. Phys.*, 2011, **13**, 6305–6308.

 

# General discussion

**Professor Castner** opened the discussion of the paper by Alessandro Triolo: Your results for the several series of ionic liquids having the 1-alkyl-3-methylimidazolium cations paired with different anions, such as you have shown in Fig. 1 in your paper, have been most influential in influencing our thinking about the nature of aggregation of the alkyl tails in hydrophobic domains. For one case, the effective domain size from a Bragg analysis of the "prepeak" or first sharp diffraction peak in the X-ray structure function increases with a slope of 2.1 Å for each methylene group. However, the slope was much less when you considered a series of ionic liquids having the piperidinium cation. Can you explain the substantial differences between these two ionic liquids, one having the aromatic dialkylimidazolium cation with the other having the dialkylpiperidinium cation?

**Dr Triolo** replied: We think that the mentioned difference between aromatic (represented by the imidazolium family) and non-aromatic (*e.g.* piperidinium,[1] but also pyrrolydinium[2] and tetraalkylammonium[3] salts) is a significant one: The latter class of salts is characterised by a slope of $D$ (size of structural heterogeneity) *vs.* $n$ systematically lower than 2, while the former class is generally above that value. Presumably this is related to the cation head conformation (imidazolium is flat, piperidinium and pyrrolidinium are not) and/or positive charge localization. These features presumably affect the way that the alkyl chains organise in the nano-segregated domains, including their interdigitation.

1 A. Triolo, O. Russina, B. Fazio, G. B. Appetecchi, M. Carewska and S. Passerini, *J. Chem. Phys.*, 2009, **130**, 164521.
2 C. S. Santos, N. S. Murthy, G. A. Baker and E. W. Castner, Jr., *J. Chem. Phys.*, 2011, **134**, 121101.
3 T. Pott and P. Méléard, *Phys. Chem. Chem. Phys.*, 2009, **11**, 5469.

**Professor Matic** asked: In the mixture between $C_6$mimCl and $C_{10}$mimCl you find length scales larger than what is expected from just an average extrapolation between the two extremes. Can you comment on how you expect the morphology to change as a function of concentration?

**Dr Triolo** replied: We collected several experimental observations on the effect of mixing ions with different alkyl chain length: not only the mentioned $C_6$mimCl–$C_{10}$mimCl, but also, for example, the neat ionic liquids $C_2$mim–$C_8SO_4$ and $C_4$mim–$C_8SO_4$ (1-ethyl-3-methylimidazolium octylsulfate and 1-butyl-3-methylimidazolium octylsulfate, respectively) (see the paper). These experiments indicate that when mixing such ions, one does not observe different low $Q$ peaks corresponding to the longer and the shorter chain salts segregating into differentiated domains; rather one detects a unique diffraction feature centred at a position intermediate between the pure components ones. Accordingly chains with different lengths originating from different ions tend to thoroughly mix, without evidences of micro-separation (however we plan to further explore this issue for the case of larger differences between the chain lengths, *e.g.* C2–C10).

Moreover, we find that the role played by the longer chain is predominant in determining the overall size of the structural heterogeneity than the one played by the shorter chain and, interestingly, we observe a positive deviation of the heterogeneities' size with respect to the linear trend expected for the C6–C8–C10 trend (see figure in the paper).

**Professor Hardacre** asked: From our paper (ref. 30), we show that for short alkyl chain ionic liquids (*e.g.* butyl chain lengths), the contribution to the low $Q$ feature from the alkyl chain–alkyl chain contacts is negligible. The contribution increases with chain length increasing as you would expect. Our data was not consistent with a micellar type of structure. We do not dispute that there are areas of apolar character and areas of polar character. However, we believe that the apolar regimes are very fluid and thus space filling. The structure is likely to be defined by the polar regimes, especially at the short chain lengths, rather than the van der Waals interactions. In fact, although the cohesive energy becomes dominated by van der Waals contributions at longer chain length, this does not necessarily mean that the alkyl parts are structure forming. As the chain length increases there are just more contacts and, therefore, increased van der Waals energy. The polar headgroup interactions are still a critical factor in determining the structure.

**Dr Triolo** answered: I would like to take this opportunity to acknowledge that my choice[1] of presenting a "micelle-like" cartoon to describe the morphology in bulk ILs was a very unfortunate one! This led, in my view, to a number of misunderstandings with other groups. My only goal on that occasion was to highlight the existence—in space and time—of domains where alkyl chains tend to segregate. From my point of view there is a substantial difference between segregation and aggregation. The latter process involves a sort of active role played by the interaction forces. In the case of ILs, I share the opinion that the major driving force is of electrostatic nature and this determines the long ranged network. On the other hand, when longer and longer apolar chains are covalently attached to the polar heads, they tend to, so to say, 'passively' segregate wherever they can minimise their interaction with charges and maximise dispersive interactions. This mechanism is to be considered as highly dynamic (the bulk material is liquid!) and the domains are ill-shaped.

I would add that, while of course the chain length plays a role in the stabilization of these heterogeneities (for example asymmetric cations show better defined domains and higher viscosity than symmetric cations), it is clearer and clearer that even something as short as ethyl chains can lead to segregation (*e.g.* in ethylammonium nitrate). Accordingly, from my point of view, a low $Q$ peak is a fingerprint of an excess order that might well be determined just by the exclusion mechanism that I mentioned above and leads to a structural organization that differs from random distribution: this is my approach when mentioning just 'mesoscopic heterogeneities' characterising the structure in ILs.

I am aware of the difficulty in describing the diffraction data in terms of a classical 'micelle-like' model. Honestly I do not possess a conclusive reply to this problem, yet. Of course the use of selectively deuterated salts is great in highlighting any such kind of structural organization. It seems to me that your recent SANS data are not fully reliable at low $Q$ (I mean below the peak), so one cannot fully rely on them for such an analysis. I am aware of other data sets that might be of help, but for sure some further effort should be paid in this direction.

On the other hand, if we agree on the difference in the nature of interaction between polar and apolar portions in ILs, exactly in the terms that you mention, I would find it difficult to imagine a structural model that does not involve the structural heterogeneities over a spatial scale of the order of $2\pi/Q_{peak}$. Presumably they will be ill-shaped, they will live not that long and mutate their shape, but I find it hard to imagine the IL's bulk morphology without them.

1 A. Triolo, O. Russina, B. Fazio, R. Triolo, E. Di Cola, *Chem. Phys. Lett.*, 2008, **457**, 362.

**Professor Hardacre** remarked: The data on the ether functionalised ionic liquids is very interesting. Is the disappearance of the low $Q$ feature due to the disruption of the anion–cation interaction *via* hydrogen bonding of the ether chain with the imidazolium ring?

**Dr Triolo** answered: I try to see the glass as half full, rather than half empty. Yes, you are right: there are various reports where it emerges how the ether functionalization induces morphological changes that are related to interactions of the oxygen atoms with the imidazolium moiety. On the other hand, I have difficulty in imagining (and even more in synthesising them) chain moieties that are polar and would not interact with the imidazolium. But I believe that this is exactly the point that I am trying to make: alkyl chains that do not interact electrostatically (H-bonding can be described in these terms, after all) with the imidazolium will eventually segregate from the matrix built up by the charged moieties. However, on the other hand, whatever chain moiety that can electrostatically interact will eventually structurally correlate with the imidazolium or the anion: this will eliminate the driving force towards segregation and the formation of structural heterogeneities.

**Professor Lynden-Bell** addressed Dr Triolo and Professor Hardacre: You have presented an interesting comparison of ionic liquids up to [$C_{10}$mim] and I wondered whether anyone has looked at ionic liquids with longer chain lengths—in particular near the liquid crystal (LC) phase transition.

**Professor Hardacre** answered: Ionic liquid crystalline materials with chains lengths of C12 and above have been studied by SAXS and show low $Q$ features in the liquid which are slightly shifted with respect to both the crystal and the liquid crystalline phases.

**Dr Triolo** answered: The comparison between liquid state and liquid-crystalline state morphology is very pertinent in order to better understand if and how peculiar an IL's structure is.

For example Hardacre's[1,2] and, to a smaller extent, our[3] groups have reported on LC phases in long chain ILs. Hardacre noticed that after melting the LC phase, a residual low $Q$ peak could be observed in the isotropic phase. This peak has the same origin as the ones that we observe for shorter chains in the liquid state.

It is noteworthy however that the spacing associated with the LC-phase low $Q$ peak is characterised by a dependence from the alkyl chain length that is different from the one observed for the isotropic-phase peak. We believe that this observation might indicate a substantially different organization of the two phases. To our knowledge this issue has not been further explored to rationalise these observations.

1 A. E. Bradley, C. Hardacre, J. D. Holbrey, S. Johnston, S. E. J. McMath and M. Nieuwenhuyzen, *Chem. Mater.*, 2002, **14**, 629–635.
2 A. Downard, M. J. Earle, C. Hardacre, S. E. J. McMath, M. Nieuwenhuyzen, S. J. Teat, *Chem. Mater.*, 2004, **16**, 43–48.
3 J. De Roche, C. M. Gordon, C. T. Imrie, M. D. Ingram, A. R. Kennedy, F. Lo Celso and A. Triolo, *Chem. Mater.*, 2003, **15**, 3089–3097.

**Professor Jung** said: It is very intriguing to see that there exists a high degree of order over the mesoscopic spatial scale in RTILs. However, the mesoscopic segregation may not be static. Instead, there should be dynamical change of the segregation patterns if you watch them over a long time, which leads to the notion of dynamical heterogeneity. In fact, we have investigated that issue in our recent paper based on a coarse-grained model of RTILs.[1] I wonder if you can investigate the dynamic heterogeneity in RTILs using your experimental technique.

1 D. Jeong, M. Y. Choi, H. J. Kim, Y. Jung, *Phys. Chem. Chem. Phys.*, 2010, **12**(8), 2001

**Dr Triolo** answered: We believe that such structural heterogeneities should be characterised by a dynamical fingerprint, such as a slowing down of diffusion or similar effects. So far, we could not highlight these effects.

In your recent study, if I understand correctly, your focus was on a salt bearing an ethyl chain that, to our knowledge, leads to only minor segregation effects; presumably, then, the effects of segregation would be minor also on dynamics.

**Professor Quitevis** asked: Although the prepeak in small wide angle X-ray scattering data is commonly invoked as being evidence for mesoscopic segregation in room temperature ionic liquids, indirect evidence for this segregation can also be found in the physical properties of ionic liquids. For example, Watanabe and coworkers showed for $[C_nC_1im][NTf_2]$ ionic liquids ($n$ = 1, 2, 4, 6, and 8) that the viscosity decreases in going from $n = 1$ to 2, but then increases in going from $n = 2$ to 8.[1] Molecular dynamics (MD) simulations indicate that this behavior can be rationalized in terms of the clustering of the alkyl side chains as indicated by the site–site radial distribution functions (RDFs) between the terminal carbon atoms on the alkyl side chains.[2,3] To further understand the role of alkyl tail segregation in determining the physical properties of ionic liquids, we recently compared the densities and viscosities of $[C_{N-1}C_1im][NTf_2]$ to that of $[(C_{N/2})_2im][NTf_2]$ ($N$ = 4, 6, 8, and 10).[4] These ionic liquids only differ in the symmetry of the alkyl substitution on the imidazolium ring. Whereas for a given $N$, the density of $C_{N-1}C_1$ is nearly the same as that of $C_{N/2}C_{N/2}$, the viscosity of $C_{N-1}C_1$ is greater than that of $C_{N/2}C_{N/2}$. Following a suggestion by Dr Canongia Lopes,[5] we compared the tail–tail RDFs of symmetric and asymmetric cation ionic liquids obtained from MD calculations of Raju and Balasubramanian.[6] Such a comparison (Fig. 1) shows a stronger first peak in the tail–tail RDF of $[C_5C_1im][NTf_2]$ than in the tail–tail RDF of $[(C_3)_2im][NTf_2]$, indicating a greater probability of clustering of the alkyl side chains in $[C_5C_1im][NTf_2]$ than in $[(C_3)_2im][NTf_2]$. That the viscosity of the asymmetric $C_{N-1}C_1$ ionic liquid is greater than that of the symmetric $C_{N/2}C_{N/2}$ ionic liquid with the same $N$ therefore provides further support for nano-segregation in ionic liquids.


1 H. Tokuda, K. Hayamizu, K. Ishii, M. A. B. Susan and M. Watanabe, *J. Phys. Chem. B*, 2005, **109**, 6103.
2 J. N. A. Canongia Lopes and A. A. H. Pádua, *J. Phys. Chem. B*, 2006, **110**, 3330.
3 Y. Wang and G. A. Voth, *J. Phys. Chem. B*, 2006, **110**, 18601.


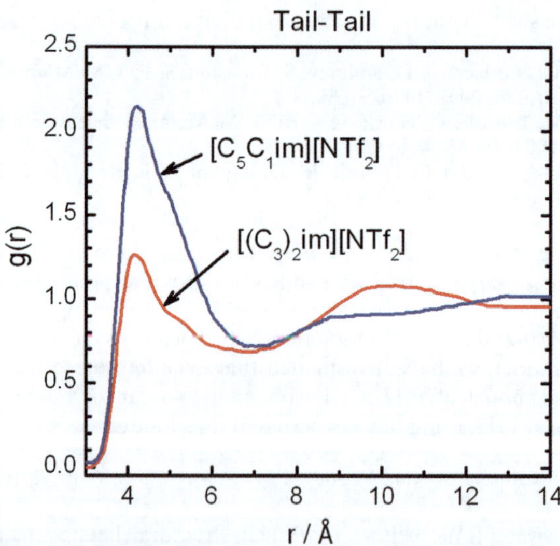


**Fig. 1** Comparison of radial distribution functions between terminal carbon atoms on the long alkyl side chains from MD simulations of Raju and Balasubramanian[6] for asymmetric $[C_5C_1im][NTf_2]$ and symmetric $[(C_3)_2im][NTf_2]$.

4 W. Zheng, A. Mohammed, L. G. Hines Jr, D. Xiao, O. J. Martinez, R. A. Bartsch, S. L. Simon, O. Russina, A. Triolo and E. L. Quitevis, *J. Phys. Chem. B*, 2011, **115**, 6572.
5 J. N. A. Canongia Lopes, private conversations.
6 R. Raju and S. Balasubramanian, *J. Phys. Chem. B*, 2010, **114**, 6455.

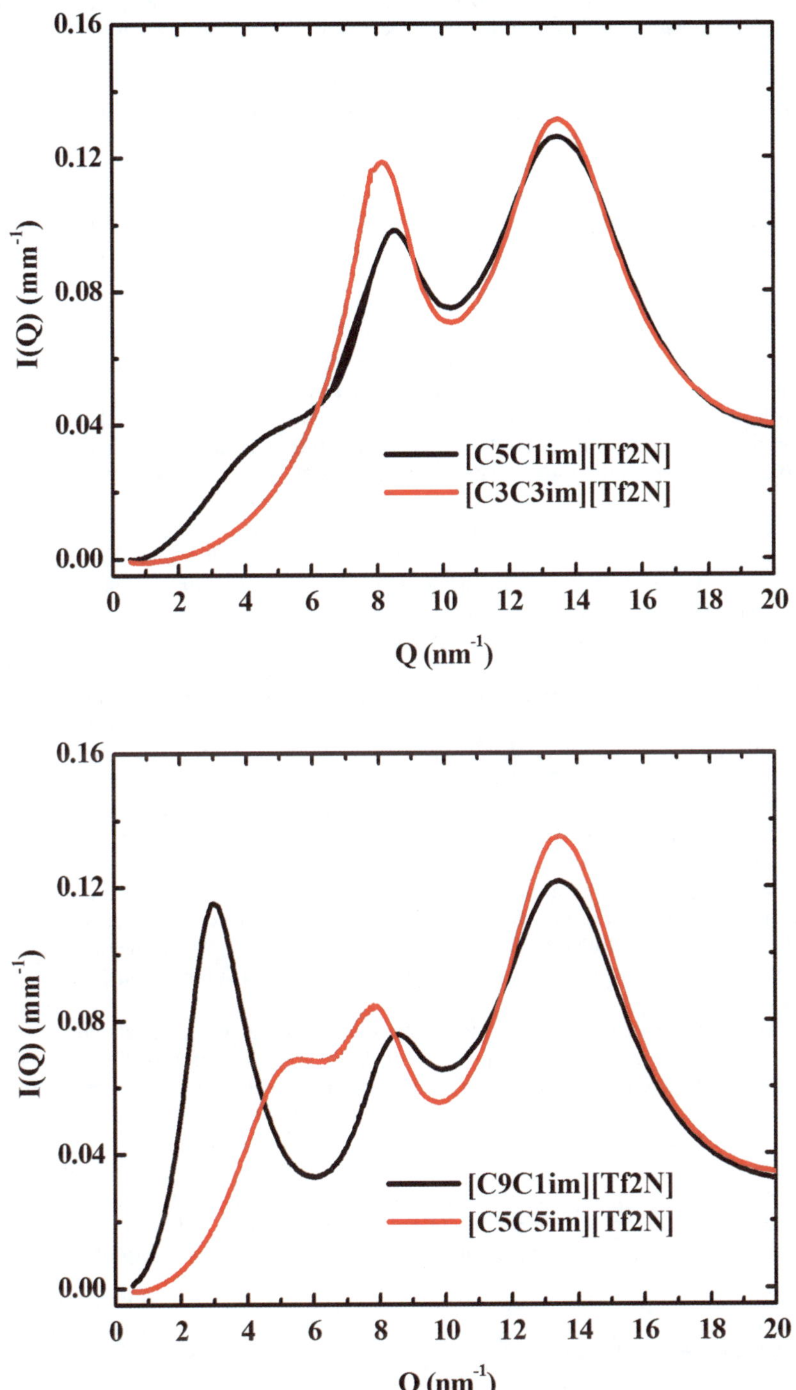

**Dr Triolo** replied: I fully agree with this comment.

The figures below contain some of the data that were presented in the mentioned *Journal of Physical Chemistry B* paper in 2011, where it emerges that asymmetric cation-based ILs are characterised by low $Q$ peaks of higher amplitude than symmetric cation-based ones. In my view, as discussed in a previous comment, longer chains are better in succeeding to creating stable micro-segregated domains that might eventually hinder diffusive process and thus affect viscosity properties.

**Professor Maroncelli** addressed Professor Hardacre and Dr Triolo: I am puzzled about Prof. Hardacre's comment on timescales of domain reorganization and the relationship between what is reported by X-ray experiments and simulations. Isn't it appropriate to think of X-ray experiments as taking instantaneous snapshots of liquid configurations which are averaged over long time periods? Apart from questions of complete sampling, isn't this precisely the same sort of averaging that is done in structure determinations by computer simulation?

**Professor Hardacre** answered: My comment concerning timescales was concerned with the longevity of specific apolar domains. The diffraction data does produce a snapshot of the liquid configurations but the broadness of such diffraction features will depend on how long lived these domains are. This has significant implications if the liquids are to be considered as nanostructurtured materials which can direct chemical reactions, for example. In this case the structure directing properties have to be long lived enough to direct the reactants.

**Dr Triolo** responded: As we are dealing with elastic X-ray/neutron scattering experiment, we are probing instantaneous snapshots of the sample at time = 0, *i.e.* we do not probe the consequent decay of the structural correlation that we are observing. Of course we average both spatially and temporally over large enough volumes/times so that we obtain an average of the morphology. It would interesting to probe how this time-zero structural correlation relaxes in time. Ways to do this exist and are so called in/quasi-elastic neutron/X-ray scattering techniques. At present, probing the time decay of the slow relaxations in ILs with QEXS is not feasible, as the technique does not reach high enough temporal resolution. On the other hand QENS (*i.e.* quasi-elastic neutron scattering) can probe relaxation times as long as hundreds of ns, so in principle this technique can access the relaxation decay of the mentioned structural correlations.

We collected this kind of data on deuterated samples, but so far obtained contradictory results. Our preliminary interpretation is that the relaxation of the structural heterogeneities occurs following a temperature law that differs from the commonly encountered Vogel–Fulcher–Tamman trend and is Arrhenius-like. The time scale for these processes is also of the order of several ns at ambient conditions.

**Professor Wynne** said: Your experiments show a pre-peak that is related to mesoscopic structure in the ionic liquids studied. In your presentation[1] and previous work[2–4] you have shown that the magnitude of this peak depends on temperature, alkyl chain length, and polarity of the alkyl tail. Can you *quantify* from your results what fraction of the liquid is in a structured state *vs.* a completely randomised state? Could this then be related quantitatively to other measurements, such as the shear viscosity of the liquids as a function of temperature, chain length, *etc.*? Could it be related to the strength of the fractional Stokes–Einstein–Debye effect observed[5] in di-alkylimidazolium RTILs or the degree of inhomogeneous reaction[6] dynamics?

1 O. Russina and A. Triolo, *Faraday Discuss.*, 2012, **154**, DOI: 10.1039/C1FD00073J.
2 O. Russina, L. Gontrani, B. Fazio, D. Lombardo, A. Triolo and R. Caminiti, *Chem. Phys. Lett.*, 2010, **493**, 259–262.

 

3 O. Russina, A. Triolo, L. Gontrani, R. Caminiti, D. Xiao, L. G. Hines, R. A. Bartsch, E. L. Quitevis, N. Pleckhova and K. R. Seddon, *J. Phys.: Condens. Matter*, 2009, **21**, 424121.
4 A. Triolo, O. Russina, B. Fazio, R. Triolo and E. Di Cola, *Chem. Phys. Lett.*, 2008, **457**, 362–365.
5 D. A. Turton, J. Hunger, A. Stoppa, G. Hefter, A. Thoman, M. Walther, R. Buchner and K. Wynne, *J. Am. Chem. Soc.*, 2009, **131**, 11140–11146.
6 K. Sahu, S. J. Kern and M. A. Berg, *J. Phys. Chem. A*, 2011, **115**, 7984–7993.

**Dr Triolo** answered: At the present stage, we did not succeed in analytically modelling the mentioned scattering features. We cannot then quantify the ratio between the amounts of ordered and randomised structure, to use your terminology. We can, however, observe that the longer the alkyl chain the stronger the amplitude of the peak, following approximately a $n^2$ law. This would imply that longer chains tend to impose a higher degree of order to the morphology than shorter chains. At this stage, this is, however, still qualitative.

**Dr Mezger** asked: For an interpretation of $D$ *vs.* $n$ curves (*e.g.* Fig. 1) data for different compounds was taken at a fixed temperature. Is the slope different if $D$ values were compared at temperatures with a constant degree of side chain ordering or a fixed fraction of gauche conformers in the aliphatic chains? While this would be complicated experimentally, comparing data at temperatures slightly above the respective melting points might provide a first estimate of the differences between these two approaches. Do you think that they contribute to the deviation from the linear trend when comparing ILs with short and long alkyl chains?

**Dr Triolo** responded: In principle I agree with the observation. As far as I know there are *no* data concerning the low $Q$ peak position for molten long chain salts. I think Hardacre's data[1] refer to the liquid-crystal phase spacing. Moreover, unless these data were extrapolated to ambient $T$, they underestimate (and hence a plausible origin for the negative deviation from the linear trend) the spacing, due to its temperature dependence. As a matter of fact it is known that the low $Q$ peak of the smectic phase (as well as of the isotropic phase) shifts to *higher* $Q$ values when increasing $T$. I would like to stress that the data in Fig. 1 of the paper refer to either liquid (black symbols) or crystalline (green symbols) phases, so their difference, in my view, is more to be related to a different structural organization in liquid and crystalline states rather than to the effect that is mentioned in the question.

1 C. Hardacre, J. D. Holbrey, C. L. Mullan, T. G. A. Youngs, D. T. Bowron, *J. Chem. Phys.*, 2010, **133**, 074510.

**Dr Ribeiro** addressed Dr Triolo and Professor Castner: We calculated $S(k)$ by MD simulations of ionic liquids based on the $[NTf_2]$ anion, and $N$-alkyl-$N$-ethyl-$N$,$N$-dimethylammonium cations, $[C_4C_2C_1C_1N]$, $[C_7C_2C_1C_1N]$, and $[C_{10}C_2C_1C_1N]$, and analogous systems in which the long carbon chain is modified by including oxygen atoms, $[C_3O_1C_2C_1C_1N]$, $[C_5O_2C_2C_1C_1N]$, $[C_7O_3C_2C_1C_1N]$, $[C_9O_4C_2C_1C_1N]$, and $[C_{11}O_5C_2C_1C_1N]$. Some comments relevant to the papers of A. Triolo[1] and E. W. Castner[2] follow below.

We consider that the more appropriate tool to unravel charge ordering in ionic liquids is the charge – charge structure factor:

$$S_{\text{charge}}(k) = \left\langle \sum_{i=1}^{N} \sum_{\alpha} \sum_{j=1}^{N} \sum_{\beta} q_{i\alpha} q_{i\beta} e^{i\mathbf{k}\cdot(\mathbf{r}_{i\alpha}-\mathbf{r}_{j\beta})} \right\rangle,$$

where $q_{i\alpha}$ is the partial charge of atom $\alpha$ in ion $i$. It is clear from the red curves in Fig. 2 below that the peak at k $\sim$ 0.8 Å$^{-1}$ observed in $S(k)$ of ionic liquids is due to charge–charge correlations. We agree that only the occurrence of a pre-peak at low $k$ in $S(k)$ does not necessarily imply polar/non-polar structural heterogeneity. The signature of polar/non-polar segregation is the simultaneous findings that atoms of the long alkyl

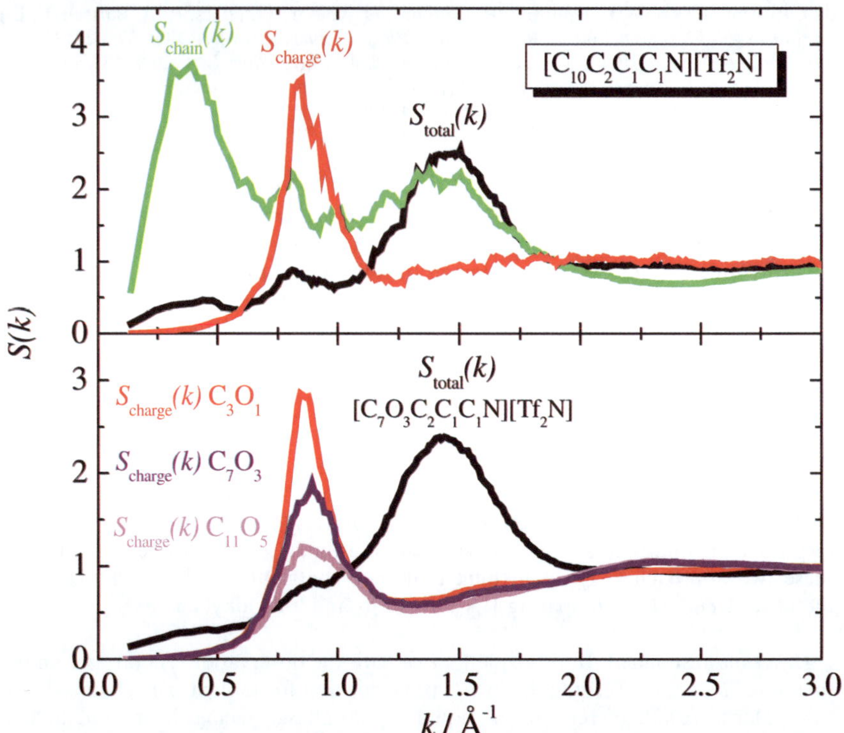


**Fig. 2** Black curves: Total static structure factor of the ammonium ionic liquid $[C_{10}C_2C_1C_1]$ $[NTf_2]$ (top panel), and the ether derivative $[C_7O_3C_2C_1C_1][NTf_2]$ (bottom panel). The green line in the top panel is the partial $S_{chain}(k)$, which considers only the carbon atoms of the decyl chain of $[C_{10}C_2C_1C_1][NTf_2]$. The red line in the top panel is the charge–charge $S_{charge}(k)$ of $[C_{10}C_2C_1C_1][NTf_2]$. In the bottom panel, the intensity of the peak of $S_{charge}(k)$ decreases in the sequence $[C_3O_1C_2C_1C_1N][NTf_2]$, $[C_7O_3C_2C_1C_1N][NTf_2]$, and $[C_{11}O_5C_2C_1C_1N][NTf_2]$.

chain contribute to the small and the high wave-vector ranges of the total $S(k)$, but less to the range of the peak in $S_{charge}(k)$ (green line in top panel of the figure). The total $S(k)$ of $[C_{10}C_2C_1C_1N][NTf_2]$ and $[C_7O_3C_2C_1C_1N][NTf_2]$ are similar (black curves in both panels). We found the same intensity of the peak in $S_{charge}(k)$ of the tetraalkylammonium ionic liquids with different chain lengths (not shown). However, in the ether derivatives, the intensity of the peak in $S_{charge}(k)$ decreases with increasing chain length (bottom panel of the figure). Therefore, the ether function is contributing to charge sites distributed throughout the liquid, and a more polar environment in the former non-polar domain of the tetraalkylammonium counterpart.


1 O. Russina and A. Triolo, *Faraday Discuss.*, 2012, **54**, DOI: 10.1039/C1FD00073J.
2 H. K. Kashyap, C. S. Santos, H. V. R. Annapureddy, N. S. Murthy, C. J. Margulis and E. W. Castner, Jr, *Faraday Discuss.*, 2012, DOI: 10.1039/C1FD00059D.


**Professor Castner** answered: I thank Prof. Ribeiro for his comments on elucidating the difference between the polar *vs.* non-polar domain segregation using the charge–charge structure function. It will be most illuminating to compare both the total and partial structure functions to the charge–charge structure functions for a number of classes of ionic liquids.

**Dr Triolo** responded: Of course it is very instructive and useful to have access to the functions $S_{charge}(k)$ and $S_{chain}(k)$ that nicely complement the total (and experimentally accessible) $S(k)$.

 

The $S_{chain}$ peaks alternating with the $S_{charge}$ one seem to be a valid tool to verify polar–apolar segregation. My main concern at the present stage is that, from our experimental observations, oxygen containing chains lead to ILs that do *not* show the low $Q$ peak. The finding of Ribeiro is also interesting in that it shows a chain–chain correlation at low $Q$, where recent findings from Hardacre indicate that no such correlations exist in the case of alkyl chains. I think that further experiments and simulations are due in order to better clarify the issue.

**Professor Panagiotopoulos** opened the discussion of the paper by Luigi Delle Site: Can you comment on the importance of including polarizability in classical force fields for ionic liquids?

**Dr Delle Site** replied: Polarisability appears to be important in the description of ionic liquids. As has been found before, polarisable force fields were able to speed up the dynamics considerably, hinting already towards the importance of polarization effects (see also other comments to our paper). The question is, does one need to include polarization *via* more expensive polarisable force field simulations, or can one get away with adding some sort of average polarisation effect to the force fields? Our approach of using Blöchl charges lead naturally to a representation of ions with reduced charge and to some non-standard partial charge assignments that are clearly the effects of polarization effects of the electronic clouds. Moreover, we found that the electrostatic properties of the ions are rather local, which lead us to propose that in the bulk case it might be sufficient to include polarization effects implicitly in this way. However, we believe that at interfaces or during phase changes that are associated with changes in polarization, these effective descriptions are of limited use. For water, the role of polarization is still a matter of on-going research and the conclusions reached in that research might be applicable also to ILs.

**Professor Canongia Lopes** said: The present charge-scaling scheme is a way to account for the lack of polarizability in many of the models used for ionic liquids. I am wondering if such a scheme would be robust enough to be used in a systematic and transferable way along homologous series of ionic liquids and between different anion–cation combinations. Also, I would like your comment on the fact that parts of the ions (their alkyl side chains) are not polarizable. Does this mean that the scaling scheme should be applied to only the charged part of the ion or to the whole?

**Dr Delle Site** responded: We want to point out that the set of partial charges is not a result of a globally applied charge scaling (see ref. 18, 26, 29, 31, 40, 63 in the paper), but given by a fit of the electron density of an *ab initio* MD (see ref. 62). For this reason, the method should allow, in principle, to capture the electrostatic properties of any ionic liquid in the liquid-like phase. If parts of an ion do not experience an overall polarization in the liquid, they will carry similar charges in the gas and liquid phase. This is already observed for the ethyl terminal group for which the set of our and your charges are within the standard error (see Fig. 9 in the paper). The aspect of transferability is currently under investigation. BMIM-based ILs with different anions should provide the basis to understand the influence of the side chain length on the polarisability and charges. Comparable to your force field, it seems that the atom groups separated more than two atoms from the ring do not carry a net charge. Hence, the approach should be robust enough to be applicable to at least a homologous series of an IL with a certain cation.

However, the obtained partial charges of the side chain atoms were associated with significant standard deviations. Also, the fragment dipole moment of the ethyl side chain was characterized by a large spread. Thus, the alkyl side chain seems to have a certain polarisability even though no overall polarisation was observed compared to the gas phase.

**Professor Dr Roling** asked: Do you use the reduced charges for calculating the conductivity *via* the Nernst–Einstein equation? This would imply that the ions carry only the reduced charge while diffusing in the liquid. In solid electrolytes, there is experimental evidence (*e.g.* from optical basicity measurements) that the mobile cations (*e.g.* Li, Na) carry only reduced charges while being at their equilibrium positions, but it is clear from electrochemical experiments that the ions carry their full charge when diffusing through the solid.

**Dr Delle Site** answered: Normally, we compute the conductivity *via* the Green–Kubo approach (*i.e.* the current autocorrelation functions) or use the equivalent integrated version, the Einstein–Helfand approach. In both cases, one has to use the effective charges in order to capture the linear response to an internally applied electric field. Both ways go beyond the Nernst–Einstein approximation by including all cross-correlations of the ions and, thus, should be preferred to the Nernst–Einstein approach.

However, if one likes to test the Nernst–Einstein approximation, one would employ the diffusion constant as measured in the simulations (again *via* Green–Kubo or MSD), and use the effective reduced charges since they determine the local interactions. The difference between these two values would give us the degree of ionic correlations. If our force field gives a good representation of the internal dynamics, the experiments should agree with the Green–Kubo values.

Experimental measurements that probe the ionic interactions in solution should also yield a reduced charge due to the various interactions, such as polarization effects and electron–phonon, electron–polaron or electron–electron interactions. Of course, the electron charge is still quantized. Hence, in electrochemical measurements, only complete electrons will hop from the electrode to the solution or *vice versa*. Thus, in the interpretation of such measurements, the ions are mostly assumed to carry integer charges. However, the processes occurring at an electrode are more complicated, since the environment around the electrode is different and the interactions between ions are likely to change as well.

**Professor Lynden-Bell** commented: You have shown that scaling the charges gives good agreement with dynamical properties such as self diffusion constants. However, including polarisability also reduces values of viscosity, diffusion constants, *etc.* My question is whether including polarisability would destroy the good agreement with experiment or whether one would then reparametrise the charges. It would be interesting to see whether including polarisability would restore the ion charges to unity. The argument that the charge scaling is proportional to $\varepsilon_{el}^{-1/2}$ suggests that this should be so.

**Dr Delle Site** replied: The reduced net charge is a result of polarisation and possible charge transfer. The Bader analysis, a topological electron density partitioning scheme, also gives reduced ion net charges in gas phase clusters (see Table 2 and Table 3 in the paper). Thus, the ion net charges might be reduced due to some charge transfer in addition to polarisation effects. This reflects that interactions between ions are not purely ionic, but contain some limited covalent contributions. Such a charge transfer could not be described by polarisation models, but would lead to reduced ion net charges. The Blöchl method captures both effects, charge transfer and polarisation, simultaneously, since it constructs a point charge distribution (*i.e.* the partial charges) on the atomic centers that captures all multipole moments of the *ab initio* reference electron density. However, this is done only on the average since those partial charges depend on the specific liquid configuration and usually were averaged over 100 snapshots. The strong fluctuations observed in our partial charge distribution suggest that one needs to go beyond the dielectric continuum explanation, where ions having integer net charges in a polarisable medium with $\varepsilon_{el} > 1$ experience electrostatic interactions reduced by $1/\varepsilon_{el}$.

 

Polarisable force fields, such as the commonly used Drude oscillator model, add additional degrees of freedom to the system capturing the influence of polarisation. If reduced partial charges are used, the induction effect on adjacent ions should be smaller than the one of fully charged ions. Thus, we would expect that the observed polarisation correction is smaller and that the average charges stay at the reduced ones. Indeed, Schröder *et al.* (private communication) found that including polarization by Drude oscillators corresponds to a partial charge distribution similar to our model. Probably, the description of dynamics would improve if polarisation is included in force fields with reduced ion net charges.

**Professor Maroncelli** addressed Dr Delle Site, Professor Canongia Lopes and Dr Schröder: The similar effects produced by adopting reduced charge representations *versus* explicitly including polarizability in simulations of ionic systems has been noted by several people, but I question how far this similarity extends. It is easy to envision situations, for example crystal lattices, where symmetry dictates no net polarization but where charge reduction will have a large effect on energetics.

**Dr Schröder** responded: Reduced charges cannot reproduce the effect of polarizability in all situations.

First, due to charge neutrality, the charge scaling factor of the cations and anions has to be equal. This may pose problems in cases where the polarizability of the cations and anions differs significantly. Closely related to this problem is the question of the transferability of force field parameters. Here, a uniform scaling factor has to be applied to all cations and anions under consideration.

Second, polarizable effects may affect high frequency regimes not accessible by classical molecular dynamics simulations with reduced charges.

Third, in contrast to polarizability, scaled partial charges reduce the Coulomb interaction independently of the distance.

**Dr Delle Site** answered: It might be helpful to distinguish between an apparent charge reduction caused by polarization and (chemical or topological) charge transfer. The first is a way to describe that an ion experiences electrostatic interactions reduced by $1/\varepsilon_{el}$ in a polarisable medium. The second one is a real electron density transfer. The Bader analysis captures that by partitioning the electron density spatially. The other three methods used here to assign partial charges—CHELPG, RESP, Blöchl—fit the electrostatic potentials and, hence, can in principle include both effects. (CHELPG and RESP may suffer from linear dependence if they are applied to much larger clusters, such that partial charges of inner atoms can become ill-defined). Indeed, the net charges found by the Bader analysis are reduced, but closer to the integer charge than for the other methods. One might speculate that the Bader analysis gives an estimate of the real physical charge transfer. While polarisation can be expressed by reduced charges, not all charge reduction is due to polarisation of the medium. Indeed in crystals, it is often observed that the electron density between atoms does not go down to zero. Thus, there seems to be some charge transfer which might explain the charge reduction affecting the energetics. However, in ionic liquids, the ions are also subject of polarisation in the bulk phase. This can be depicted from the broad dipole moment distributions.

**Professor Maginn** commented: I appreciate that your approach is designed to get at an approximate representation of the partial charges in the condensed phase without resorting to the use of polarizable force fields. Am I correct to think that the distribution of the charges is as important as the overall magnitude? Also, for atoms that are formally equivalent, does your method automatically ensure that these have the same partial charge, or do you have to do some sort of manual averaging procedure? Also I have a comment. We have performed melting point calculations on ionic liquids, and we find that the results are much improved if we follow a procedure similar to what you discussed. That is, if we derive partial charges from *ab initio* simulations of the crystal instead of the gas phase, we get better results. So I am very pleased to see how you have formalized this procedure.

**Dr Delle Site** answered: Indeed, both the absolute magnitude and the distribution of partial charges are important. Reduced net charges are associated with an increase in the dynamic properties in general. However, the distribution of partial charges (see *e.g.* ref. 44 and 45 of the paper) influences not only static properties, such as radial distribution functions, but also dynamic properties, such as the lifetime of hydrogen bonds.

The partial charges we observe are associated with a significant standard deviation. Partial charges of chemical equivalent atoms were the same within these uncertainties. However, to reduce the number of parameters in the force field, the partial charges of chemical equivalent atoms were chosen such that they were exactly identical.

**Dr Triolo** remarked: We are all aware of the structural complexity of the liquid state in ILs: this leads to non-negligible structural correlations, whatever their origin, far beyond the accessible spatial range to *ab initio* methods. Moreover, the life time of these heterogeneities is also very long, at least several ns. I agree that, by using AIMD, one aims to extract some core detail of the structural correlation that can be exported to the larger spatial scales.

However, the peculiar organization in ILs over mesoscopic spatial scales is not explicitly considered in AIMD and accordingly the investigated cluster might describe a situation that is not representative of the bulk state, not only because of the reduced size, but especially because specific anisotropies/heterogeneities/asymmetries are not explictly included in the cluster. Apart from the nm-scale heterogeneities that we have discussed in other contributions and are related to microphase separation effects (polar–apolar or sterically driven anisotrpies), I am also thinking of specific effects such as conformational heterogeneities (*e.g.* in the $Tf_2N$ anion or in the alkyl chain). Indications exist that such conformers might micro-segregate, so that one might find specific space–time domains where/when conformation clustering occurs, thus creating a very organised micro-environemnt. Do you think this is a real possibility? Can AIMD simulation tackle such effects?

**Dr Salanne** responded: When adopting a coarse-grained strategy, the idea (and certainly the most difficult part!) is to extract the relevant information from the small scale (here, *ab initio* calculations) to the larger one (classical calculations). Of course structural heterogeneities on nanometer scales cannot be studied (yet) from *ab initio* calculations, but it is not a problem because these heterogeneities do not impact on the electronic structure. Therefore, the information which is needed in order to build a force field remains correct, and the final model should be able to reproduce the good structure on the long-range scale (I would even say that it is much more difficult to reproduce the short-range structure). As an example, the force field of Canongia Lopes and Pádua, which was built based on *ab initio* calculations performed on single molecules in the gas phase, was able to account for the polar–apolar anisotropy that you mentioned.[1]

1 J. N. A. Canongia Lopes and A. A. H. Pádua, *J. Phys. Chem. B*, 2006, **110**, 3330.

**Dr Delle Site** answered: We agree fully with Dr Salanne. At the AIMD level, we study here only electrostatic properties which are connected to a very high locality (see also ref. 61 of the paper) in ILs. However, AIMD simulations might be set up in a way to sample certain micro-environments whose short time stability and dynamics might be studied. Still, the time and length scales accessible to AIMD in the moment are not sufficient to elucidate if such micro-environments are present in the liquid at equilibrium. Only classical or even coarse-grained simulations may address these issues.

**Dr Salanne** commented: How well does the structure yielded by the model compare to *ab initio* molecular dynamics results?

**Dr Delle Site** responded: In general, the agreement is rather good. But as described in the paper, the Lennard-Jones parameters need to be adjusted to obtain the correct density and while doing this we also took some cation-anion RDFs as target quantities in the fitting procedure. Hence, the agreement is partly forced. In forthcoming publications, we will provide more details on these points. However, an article has already been published (see ref. 46) that introduces our multiscaling approach and offers first insight into the effect of reduced charges on the local structure. For some RDFs, simulations with reduced charges gave improved agreement between MD and AIMD, while, for other RDFs, an adaptation of the short range parameters seemed inevitable to achieve consistent structures in MD and AIMD.

**Professor Johansson** commented: As we know that all the MP2, DFT and CCSD(T) calculations are deterministic, *i.e.* always following a path lowering the energy, it is a must to really scan the potential energy surface properly from the starting configurations. It is not clear to me how many starting configurations were used to arrive at the clusters in Fig. 2 in your paper, nor how these starting configurations were generated.
It is also mentioned that 4 clusters were selected (2A2C)—what was the criteria used?

**Dr Delle Site** replied: As outlined in ref. 46, the primary objective of our quantum chemical studies on isolated clusters in the gas is not to identify necessarily all local minima on the various potential energy hypersurfaces (PES), let alone to provide a thorough sampling of the entire PES. The main purpose is instead to provide high-level benchmark data (such as structural parameters, relative energies, harmonic vibrational frequencies, charge distributions) against which the more approximate methods such as density functional theory (DFT) can be compared. The latter methods, provided they agree sufficiently well with high-level post-Hartree–Fock methods, can then subsequently be employed in bulk simulations to gauge the quality of force field methods. Initial structures for structural energy optimisations of the clusters were in general obtained from the four known equilibrium structures of one ion pair, to which additional ions were subsequently added by adhering to the structural motifs found for the anion and cation in one ion pair (chloride above the ring; chloride near methyl and $C^1$–$H^1$; chloride near methyl and either $C^2$–$H^2$ or $C^{2'}$–$H^{2'}$; chloride near $C^2$–$H^2$ and $C^{2'}$–$H^{2'}$; for the numbering, see Fig. 1 in the paper, which shows emim instead of Mmim, however). Where possible, point group symmetry higher than $C_1$ was exploited to accelerate the energy optimisation of the structures. Selection of initial structures for two ion pairs was additionally guided by arrangements found within an MD simulation which were subsequently also energy optimised with DFT and used (where applicable after symmetrisation) as input data in post-Hartree–Fock structure optimisations. We considered in the paper explicitly four clusters composed of two ion pairs, four energetically low-lying clusters out of eleven clusters composed of two cations and one anion that we have optimised, as well as three clusters of one cation and two anions. For the latter, we have not localised structures with one anion above (or even one anion above and one below) the ring, although such an arrangement might give rise to minimum structures, at least for some given combination of method and basis set. These structures might then be expected to correspond to quite shallow minima on the corresponding PES. Although, it would certainly be desirable to obtain even more cluster benchmark data with highly accurate *ab initio* methods, such as explicitly correlated coupled cluster methods (here CCSD(T)-F12), the steep increase in computational cost with increasing cluster size limits significantly the range of applicability of such high-level methods. As is quite common in quantum chemistry, benchmarking is thus done sequentially. In our case, we use CCSD(T)-F12 or CCSD on one ion pair to assess the quality of second order Møller-Plesset perturbation theory (MP2) in describing relative energies or partial charges, respectively. MP2 in turn is then used to calibrate DFT approaches on slightly larger clusters (here two ion pairs) in the gas phase so that DFT can finally be employed to compare to force field results on much larger clusters or for simulations with periodic boundary conditions.

**Mr Frolov** said: Non-integer partial charges on ionic liquid ions obtained by the quantum chemical calculations of ion pairs suggest an importance of the explicit incorporation of polarizability in the classical force fields of ionic liquids. Which polarizable classical model (out of the most popular: polarizable dipoles, Drude particles or fluctuating charge models) is best suited to model bulk ionic liquids? Is there any model that is able at the same time to reasonably reproduce densities, diffusion coefficients as well as dielectric properties of ionic liquids?

**Dr Salanne** responded: Concerning the second part of the question, in the case of molten salts there is now important literature in which it was showed that polarizable models are able to predict quantitatively all the physical properties of the systems (we have just finished a review on this topic which has been published in *Molecular Physics*[1]). In the case of room temperature ionic liquids, the literature is less extensive but the recent results of Bedrov *et al.*[2] confirm this trend. Now, as for the method used for including the polarization, the polarizable point method is particularly interesting because it is very easy to include the short-range damping effects which are known to be of primary importance. It is also possible to go beyond the inclusion of the dipole polarization effect: inclusion of quadrupole polarization, ion deformation, *etc.* It has therefore been the most studied approach for molecular polarization.

1 M. Salanne and P. A. Madden, *Mol. Phys.*, 2011, **109**, 2299–2315.
2 D. Bedrov, O. Borodin, Z. Li and G. D. Smith, *J. Phys. Chem. B*, 2010, **114**, 4984.

**Dr Delle Site** answered: Up to now, we have not used polarisable force fields and, thus, do not have any experiences on the performance of the different polarisation models. However, there are reports in the literature (see *e.g.* ref. 21, 24, 111 within our paper). Also, Schröder *et al.* succeeded in reproducing dynamics and dielectric spectra with the Drude oscillator model (see *e.g.* ref. 1 below).

1 C. Schröder and O. Steinhauser, *J. Chem. Phys.*, 2010, **132**, 244109.

**Dr Schröder** noted: The computational effort of simulation with a polarizable force field is, of course, higher compared to a simple molecular dynamics simulation with reduced charges. Based on the model of polarization, the effective computational afford is more than doubled.

For example, in the Drude polarization model each non-hydrogen atom is accompanied by a Drude particle.

In other words, the number of particles is almost doubled. Furthermore, the time step of the polarizable simulation is often reduced compared to simulations with reduced charges. This smaller time step was necessary for Drude oscillators with high force constants. Additionally, polarizable simulations frequently use Thole screening and an extended 1-4 interaction scheme prolonging the computation of the simulation.

All in all, our polarizable simulations took us 3–5 times longer than the corresponding non-polarized simulations.

**Dr Delle Site** responded: Indeed, the main disadvantage of polarisable force fields is the increase in computational cost. Thus, for any simulation study, it has to be decided if polarisation is really needed for an accurate description of the properties of interest. In same cases, such as bulk or long-time properties, it might be sufficient to include polarisation in an implicit way by applying reduced charges. Thus, larger or longer simulations are possible, providing better sampling. However, if the polarization of an ion can change as, for example, at an interface or close to a solute, it might be unavoidable to treat polarization explicitly. In general, further studies on the various polarization descriptions, implicit or explicit, are necessary to learn more about their limits in applicability and accuracy.

**Professor Ballone** said: The definition of atomic charges based on the analysis of charge distributions obtained from *ab initio* approaches represents an improvement with respect to a purely empirical (global) fitting of parameters. However, the partitioning of *ab initio* charges is still plagued by conceptual ambiguities and by computational uncertainties. I think that the parametrisation of atomistic and coarse-grained models could greatly benefit from concepts and analytical tools developed long ago by the solid state community to describe/predict dynamical properties of crystals and, based on perturbation theory, response functions and sum rules. These concepts and tools, exemplified by the Born effective charges entering the definition of the dynamical matrix, could provide an *a priori* separation of short and long range interactions, and could also help in the identification of genuine polarisation effects. The separation of these qualitatively different contributions to potential energy is a prerequisite for the transferability of potential parameters from one system to a similar one.

Transferability will continue to be only approximate, but it could be significantly better than for the models we are using right now.

**Dr Delle Site** responded: Indeed, the assignment of partial charges from DFT electron density is associated with conceptual ambiguities. However, for gas phase clusters, we find similar net charges with different methods. Most importantly, we find that the electrostatic properties of the ions are different in the bulk phase than in other phases. In the bulk phase, many different configurations are found and the entropy is large. Thus, it is recommended to parametrize partial charges based on the bulk phase. Still, a certain arbitrariness in the partial charge assignment remains. Surely, there might be other, possibly more convenient ways, to parametrize them. We would very much welcome to see how other concepts such as those of the solid state community can be applied. Unfortunately, we are not aware of any work in which these tools have been applied for ILs. Furthermore, in the liquid phase, the effects of various local packing structures and spatial density fluctuations are important and can hardly be described by an ordered, space "fixed" system such as a solid. So, it is doubtful if solid state concepts will be applicable.

**Professor Pádua** said: The cohesive energy of ionic liquids contains significant contributions both from Coulomb and van der Waals interactions, as shown by the decomposition of the enthalpy of vaporization into these terms. In fact, in imidazolium ionic liquids with alkyl side chains of intermediate length, the van der Waals terms (represented by Lennard-Jones potentials) dominate coulombic ones.

In your work you used DFT methods in your quantum simulations, and most functionals give a poor representation of dispersion forces. How much can you trust these methods to be the foundation of an accurate force field?

**Dr Delle Site** responded: Indeed, conventional pure and hybrid density functionals do not account for dispersion interactions which can be decisive for ILs. That is why we evaluated the relative energetic order and the geometry of gas phase clusters. The energetic order found in post-HF reference calculations was reproduced by DFT if an empirical dispersion correction was employed (see Table 1 in the paper). Furthermore, in our NVT simulations, the mass density was chosen to be the experimentally measured one. So, the average electron density and interionic distances should be consistent. In this work, we focus on static properties, especially electrostatic properties which are not expected to be significantly influenced by van der Waals interactions.

**Dr Lovelock** said: Bearing in mind the importance of charge transfer and polarisability, I have highlighted work that provides experimental evidence that charge transfer occurs between ions in ionic liquids.[1–3] The paper from Erlangen/Leipzig[1] focuses upon imidazolium-based ionic liquids and involves X-ray photoelectron spectroscopy (XPS), NMR spectroscopy and theory, and the Nottingham papers[2,3] focus upon pyrrolidinium- and amino acid-based ionic liquids using XPS. All three papers demonstrate that charge transfer occurs to different extents in ionic liquids, depending upon the nature of the anion. For both imidazolium- and pyrrolidinium-based ionic liquids the more basic/co-ordinating the anion, the more charge is transferred from the anion to the cation. For example, the nitrogen atoms on the cation for $[Tf_2N]^-$-containing ionic liquids are more positively charged than the nitrogen atoms on the cation for halide-containing ionic liquids. The charge on the cation can in addition be tuned by using mixtures of anions.[2]

1 T. Cremer, C. Kolbeck, K. R. J. Lovelock, N. Paape, R. Wölfel, P. S. Schulz, P. Wasserscheid, H. Weber, J. Thar, B. Kirchner, F. Maier and H. P. Steinrück, *Chem.–Eur. J.*, 2010, **16**, 9018–9033.
2 S. Men, K. R. J. Lovelock and P. Licence, *Phys. Chem. Chem. Phys.*, 2011, **13**, 15244–15255.
3 B. Birru Hurisso, K. R. J. Lovelock and P. Licence, *Phys. Chem. Chem. Phys.*, 2011, **13**, 17737–17748.

**Dr Delle Site** replied: Indeed, it is very helpful that charge transfer is also seen in experiments and not only in our Bader partial charge analysis. However, the reduced partial charges found by us are most likely the result of both charge transfer and polarisation. Thus, it is delicate to separate both processes. For the description in a classical force field, an effective reduced charge might be sufficient, disregarding its physical origin.

**Professor Wynne** opened the discussion of the paper by Edward W. Castner: In Fig. 6 of your paper, there are no anions visible around the alkyl tail. Is this an indication that the alkyl tails aggregate as discussed in the presentation[1] by Dr Triolo?

1 O. Russina and A. Triolo, *Faraday Discuss.*, 2012, **154**, DOI: 10.1039/C1FD00073J.

**Professor Castner** replied: Prof. Wynne raises an interesting point. While it is certainly clear in our simulation results that the hydrocarbon chains do aggregate, as is consistent with the previous simulations from Lopes, Pádua and colleagues, it is also clear that there is no preferential aggregation of the longer tetradecyl chains with each other, but rather, a combination of tetradecyl chains interacting with both the shorter hexyl and longer tetradecyl tails on the phosphonium cations.

**Dr Triolo** addressed Professor Castner, Professor Costa Gomes, Professor Pádua and Professor Canongia Lopes: During the meeting, you seemed to propose a sort of privileged separation between shorter hexyl and longer tetradecyl chains in separate nano-clusters. Can you confirm this observation? In our work on P66614Cl[1] (as well as in the one from CG/P/CL[2]), this issue was not explored, but I would expect that if such a behaviour would be real, one would expect to find the corresponding bi-modal characteristic size distribution in the diffraction pattern and this is not the case either for $Tf_2N$ or for Cl salts.

1 L. Gontrani, O. Russina, F. Lo Celso, R. Caminiti, G. Annat and A. Triolo, *J. Phys. Chem. B*, 2009, **113**, 9235.

 

2 L. Pison, J. N. Canongia Lopes, L. P. N. Rebelo, A. A. H. Pádua and M. F. Costa Gomes, *J. Phys. Chem. B*, 2008, **112**, 12394

**Professor Canongia Lopes** answered: I have performed some simulations of equimolar mixtures of $C_2$mim and $C_6$mim (with $NTf_2$ as counter-ion) and compared the results with those of bulk [$C_4$mim][$NTf_2$].[1] In terms of volumetric properties the mixture is almost ideal (an almost null VE) and its density is also very close to that of [$C_4$mim][$NTf_2$]. However slight differences were noticeable in terms of the RDFs between sites belonging to the alkyl side chains. It seems that not only the alkyl side chains like to segregate from the polar domains, but C2 slightly prefers to stay with C2 and C6 with C6. However, such preference is very subtle (there are no C2 "islands" and C6 "islands"—just a distribution that is not completely random inside the non-polar domains). Perhaps the differences are larger with alkyl side chains that exhibit larger differences in terms of length.

1 K. Shimizu, M. Tariq, L. P. N. Rebelo, J. N. Canongia Lopes, *J. Mol. Liq.*, 2010, **153**, 52–56.

**Professor Castner** replied: As we have shown in Fig. 8 of our manuscript, the first peak in the pair distribution functions for the central phosphorus–phosphorus (cation to cation center distances) matches that for the distances between anion nitrogen centers, with both values being 10 Ångstroms. This distance lies in between the lengths of hexyl and tetradecyl chains, indicating that there is likely to be a statistical distribution of both tetradecyl and hexyl chains in the non-polar regions between the ions. Thus, I would not say that this confirms a separation between hexyl *vs.* tetradecyl chains into separate domains.

**Professor Canongia Lopes** asked: Our own MD work on phosphonium-based ILs (see ref. 1 and 2 and also the image below) has shown that the structure of these systems can be understood as a string-like (sometimes branched) polar network composed of the anions and the positive core of the phosphonium cations embedded in a vast continuous non-polar domain composed by the long (and longer) alkyl side chains of the cations. Is it possible to estimate from your data the coordination number of the ions? It would also be interesting to check (from the MD data) if the characteristic distances inferred from the structure factors are compatible with the length of the longer alkyl side chain (stretched C14) or with a somewhat lower value corresponding to a kinked C14 chain (as inferred from your data).

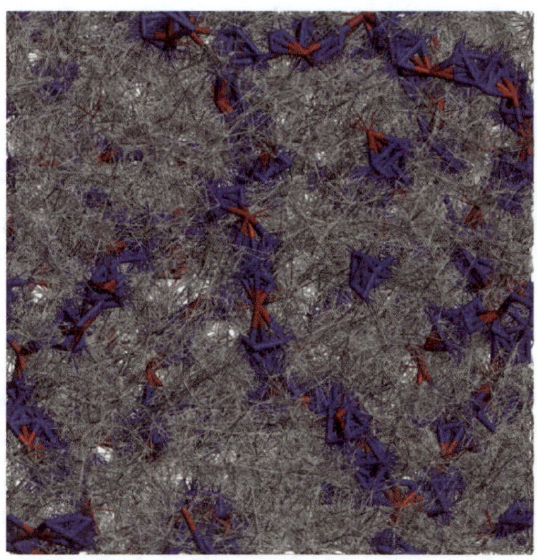

1 M. Blesic, J. N. Canongia Lopes, M. F. Costa Gomes and L. P. N. Rebelo, *Phys. Chem. Chem. Phys.*, 2010, **12**, 9685–9692.
2 K. Shimizu, M. F. Costa Gomes, A. A. H. Pádua, L. P. N. Rebelo and J. N. Canongia Lopes, *THEOCHEM*, 2010, **946**, 70–76.

**Professor Castner** responded: Prof. Lopes reminds us of the elegant string-like network that appears in their snapshots of the tetradecyltrihexylphosphonium cation paired with a number of different anions. Of course, the total number of hydrophobic alkyl chain atoms is substantially larger than the atoms in the polar groups. As is commonly found for other liquids with tetrahedral centers, the coordination number is slightly less than 4. The coordination number is 3.5 both for cations surrounding an anion and for anions surrounding a cation, considering the centers of mass for the ions. Inspection of the tetradecyl chains found in the box shows that none were in the *trans* conformation, but typically had 2 or 3 *gauche* kinks. The peaks for the cation–cation and anion–anion pair distributions shown in Fig. 8 of our manuscript are both at 10 Ångstroms, which is substantially shorter than an all-*trans* tetradecyl chain, but much longer than all-*trans* hexyl chains.

**Professor Hardacre** asked: In Fig. 2 in the paper, the low $Q$ feature in the experimental data at low temperature is much smaller than at the higher temperatures. Is this due to crystallisation/solidification of the ionic liquid?

**Professor Castner** replied: As Prof. Hardacre points out, the lowest observed temperature of 150 K in our X-ray scattering data is well below both the melting and glass transition temperatures. The reduced intensity at values of $Q \sim 0.4$ Å$^{-1}$ could arise from solidification. However, since we can always observe crystallization for this and many other ionic liquids when the samples are warmed from well below $T_g$, we believe that the 150 K data shown is from a vitrified rather than crystallized sample.

**Dr Harris** asked: This is very nice work. The pattern in the partial structure factors at low $q$, which nearly cancel in your figure, but are the sum of terms of opposite sign, appears to be very similar to that observed in velocity correlation functions derived from ion self-diffusion coefficients and the molar conductivity, $\Delta = c(2f_{+-} - f_{++} - f_{--})/(D_+ + D_-)$ There the cation–anion term is not quite cancelled by the sum of the anion–anion and cation–cation terms, giving the Nernst–Einstein deviation parameter, Δ.

In this context, given the possibility of ion-pairing in those ILs that lie in the "subionic" region of the Walden plot, I wonder if your methods could be extended to the "poor ionic liquids".

**Professor Castner** replied: As Prof. Harris' comment implies, there is no reason why either our X-ray scattering or molecular dynamics methods for obtaining the liquid structure functions and/or pair distributions functions cannot be applied to any and all ionic liquids, including "poor" ionic liquids (*i.e.* those prone to ion-pairing). I find it particularly intriguing to learn that the dynamical velocity correlation functions show a similar near-cancellation in the Nernst–Einstein deviation parameter to what we have observed for the peak in our structure functions at about $q = 0.75$ Å$^{-1}$.

**Professor Canongia Lopes** asked: My comments refer to the possibility of engineering a particularly weak aprotic ionic liquid—an ionic liquid that, in spite of its nature, would tend to favor the formation of ion pairs instead of a polar network. This perhaps could be achieved by promoting extreme steric hindrance on both ions so that only one interaction site was available on each of them. Tetra-alkyl phosphonium cations already represent a move in the right direction (extensive alkyl side chains already promote the existence of string-like polar networks with low interionic coordination numbers (2–3)) but this could be further extended by using three branched alkyl side chains and leaving just one "open" position like a methyl group. The anion could be a tetra-alkyl borate with the same built-in type of hindrance (another possibility would be to use extra fluorinated alkyl side chains in hexa-coordinated phosphorus atoms like in the case of tris(perfluoroalkyl)trifluorophosphate (FAP) ionic liquids).

**Professor Castner** answered: Prof. Lopes makes an excellent point that synthetic control of steric hindrance, such as by using one, two or three branched side chains on a tetrahedral or octahedral central atom may lead us towards design of "poor" ionic liquids.

**Dr Wishart** commented: Apropos to the discussion about trihexyl(tetradecyl) phosphonium bis(triflyl)amide being a poor ionic liquid, its Walden product at room temperature is 0.185 P S $cm^2$ $mol^{-1}$ (our unpublished results). Since we have been speaking of how specific interactions could lead to discrete ion pairing, if you take the same cation and make the bromide salt, that Walden product is 0.082 P S $cm^2$ $mol^{-1}$ at 40 ℃ (our results). A systematic study of ion pairing in ILs containing this cation and the tetrabutylphosphonium cation was published by Fraser, MacFarlane and coworkers.[1,2] In the bromide case, it's believed that the anion is actually tucked into the second shell of the phosphonium center (between the alkyl arms) and is hydrogen bonded to the relatively acidic alpha $CH_2$ groups.[1] So there are some examples.

1 K. J. Fraser, E. I. Izgorodina, M. Forsyth, J. L. Scott and D. R. MacFarlane, *Chem. Commun.*, 2007, 3817–3819.
2 K. J. Fraser and D. R. MacFarlane, *Aust. J. Chem.*, 2009, **62**, 309–321.

**Professor Quitevis** opened the discussion of the paper by David A. Turton: Given the difficulty in achieving the dynamic range of the time-domain OKE measurements needed to obtain the OKE spectra of EAN and PAN over a broad frequency range, these studies are very impressive. Your OKE spectra are similar to susceptibility spectra of supercooled molecular liquids above the glass transition obtained in previous depolarized light scattering experiments.[1] Just as in the case of the light scattering data, the OKE data are characterized by a weakly temperature-dependent high-frequency peak near 2 THz and a strongly temperature-dependent low-frequency peak (the α-peak) that shifts to lower frequency with decreasing temperature.[1] Although you are able to fit these peaks, as well as the region between these two peaks by phenomenological functions, have you considered using mode coupling theory (MCT) to analyze your OKE spectra? The OKE spectra are strikingly similar to susceptibility spectra predicted by schematic MCT models, with the minimum between the high- and low-frequency peaks being the β-relaxation region.[2] Fayer and coworkers[3] showed that the time-domain OKE data of supercooled ionic liquids above the glass transition can be explained using MCT. With the high quality of your OKE spectra you should also be able to obtain a consistent explanation of the dynamics of these ionic liquids near the glass transition using MCT.

1 H. Z. Cummins, G. Li, W. M. Du, J. Hernandez and N. J. Tao, *Transport Theor. Stat. Phys.*, 1995, **24**, 981.
2 W. Gotze and L. Sjogren, *Rep. Prog. Phys.*, 1992, **55**, 241.
3 J. Li, I. Wang, K. Fruchey and M. D. Fayer, *J. Phys. Chem. B*, 2006, **110**, 10384.

**Dr Turton** responded: Mode-coupling theory describes the (temperature-dependent) time correlation of density fluctuations in supercooled liquids. Due to the similarity of form of the susceptibility spectra measured in OKE experiments, it has been proposed that these can also be interpreted in the same framework. There is evidence that intermolecular spectra, including the librational modes (at *ca.* 2 THz), should not be treated as a simple superposition of independent modes. MCT can, in principle, provide a complete model of this behaviour, but, without making broad approximations, doesn't yet provide a tractable model with which OKE spectra can be fit.

MCT does make predictions that can be interpreted as simple lineshape functions including the power law decay[1] and Cole-Davidson function employed here, and also the stretched exponential (KWW) function. Comparisons of the temperature dependence of such functions with experimental measurements have favourably supported MCT. However such analyses remain fundamentally phenomenological fits and are therefore of interest but do not yet provide particular insight into the physical, *i.e.* structural, basis of the motions.

A "proper" mode-coupling analysis requires for input an accurate static structure factor, which as yet cannot be experimentally determined, and is still under development even for the simplest case of density fluctuations in a Lennard-Jones fluid.[2] The difficulties of applying it to even the very simplest ionic liquid are illustrated by Yamaguchi and Koda.[3]

1 A. P. Sokolov, A. Kisliuk, V. N. Novikov, and K. Ngai, *Phys. Rev. B: Condens. Matter Mater. Phys.*, 2001, **63**, 172204.
2 B. Schmid and W. Schirmacher, *J. Phys.: Condens. Matter*, 2011, **23**, 254211.
3 T. Yamaguchi and S. Koda, *J. Chem. Phys.*, 2010, **132**, 114502.

**Dr Triolo** addressed Dr Buchner, Dr Turton, Professor Wynne and Dr Ribeiro: The experimental evidences that you provided in your *Journal of the American Chemical Society* paper[1] are very interesting and can be directly related to the existence of mesoscopic dynamics that might reflect mesoscopic order. Such an enhanced degree of order can reasonably be considered the one that can be observed with diffraction experiments. I am fine with this. In the present report, however, you report on the same kind of measurements on a different kind of ILs and, despite the experimental evidence that these ILs are showing enhanced order of the same nature of the imidazolium ones, you do not find any indication of sub-alpha process. I am puzzled by this result. It seems that your point of view is that the specific joint techniques that your team is using are not adequate to probe the mentioned sub-alpha process. This is very disappointing! Can you confirm this explanation and speculate on possible experimental alternatives to them? Also would there be some different process that is present in imidazolium salts and *not* in protic ones that might more directly explain the mentioned discrepancy?

Can you envisage an alternative protic IL that, thanks to its different chemical composition, might lead to the emergence of the sub-alpha peak so to confirm that whenever there is mesoscopic order, the sub-alpha process is present?

1 D. A. Turton, J. Hunger, A. Stoppa, G. Hefter, A. Thoman, M. Walther, R. Buchner and K. Wynne, *J. Am. Chem. Soc.*, 2009, **131**, 11140–11146.

**Dr Buchner** responded: We have some rather indirect indications for some kind of ordering also in ethylammonium nitrate (EAN): for instance, the temperature dependence of relaxation time and shape parameter of the alpha mode. However, this evidence is far less conclusive than the sub-alpha peak observed for the OKE spectra of imidazolium ILs. Clearly, there is no indication for a slow collective mode of EAN that is associated with large-amplitude fluctuations of polarizability (OKE) or macroscopic dipole moment (DRS). This may be due to the specific nature of EAN. It would certainly be premature to conclude that all protic ILs lack a sub-alpha mode.

 

**Dr Turton** replied: For the imidazolium ion-based liquids it can be assumed that both OKE and dielectric spectroscopy are primarily sensitive to motions of the imidazolium moiety. Therefore, direct comparison of the measured relaxation rates can be made (with some caveats).[1] In contrast, for the protic liquids EAN and PAN, the OKE and dielectric spectra are apparently dominated by motions of the anion and cation, respectively, and because the OKE signal is only weakly sensitive to the cation, association of the cation would be unlikely to result in the enhanced very low frequency (sub-alpha) relaxation observed for imidazolium ILs. Therefore, although mesoscopic structure has been observed in EAN, and the anomalous temperature dependences presented here may be an indication of that, we cannot apply the same methodology. The sub-alpha mode cannot be taken as a universal indicator of mesoscopic structure as such structure would impact on the observable differently for different liquids, *i.e.* depending on the symmetry of the aggregates and how this influenced the total polarisibility and dipole. No doubt, there are protic ionic liquids for which the cation motions would dominate both spectra, but these, more complex liquids, would not necessarily be relevant to the structure of *e.g.* EAN.

1 D. A. Turton, J. Hunger, A. Stoppa, G. Hefter, A. Thoman, M. Walther, R. Buchner and K. Wynne, *J. Am. Chem. Soc.*, 2009, **131**, 11140.

**Professor Castner** said: Molecular dynamics simulations have revealed that both vibrational and diffusive types of translational motions make very significant total contributions to the observed solvent reorganization dynamics in ionic liquids, in contrast with strongly dipolar neutral solvents, where the solvation energy relaxation arises predominantly from orientational fluctuations. Given that the femtosecond Kerr experiment can be set up to measure not only the anisotropic part of the Raman signal, but also the isotropic part, what do you think the isotropic Raman signal could tell you about the intermolecular dynamics in ionic liquids?

**Dr Turton** answered: The OKE signal is sensitive to the total of the many-body polarisation fluctuations as reported through changes to the anisotropic part of the polarisability tensor.[1] Hence fundamentally only rotational motions are measured. Isotropic contributions (including translational motions) do contribute through interactions (that result in transient rotational coupling). In special cases these can be measured,[2] but normally make only a weak contribution to the spectrum that cannot be resolved from the stronger contributions.

In other experimental configurations, (SMOKE *etc.*[3,4]) the isotropic spectrum, and hence translational dynamics, can potentially be measured directly.

Unfortunately, even using OKE, measurements of the diffusive modes are challenging as a dynamic range of several orders of magnitude (in both time and amplitude) are essential if meaningful parameters are to be measured. The more complicated setups that are able to measure the isotropic part yield much poorer sensitivities. This is typically sufficient for measuring higher frequency modes, such as symmetric, intra- and intermolecular, stretches, but not for the relaxational modes essential to this study. Furthermore, at the lowest frequencies, motions are expected to be concerted and simple decompositions by symmetry are expected to break down. In fact, simulations show that the intermolecular translational and rotational dynamics have very similar spectral contributions for the diffusional motions. The higher-frequency modes are of course sensitive to local structure but only weakly sensitive to mesoscopic structure, but local structure is probably better probed by more direct, *e.g.*, scattering, techniques.

1 C. J. Fecko, J. D. Eaves, and A. Tokmakoff, *J. Chem. Phys.*, 2002, **117**, 1139.
2 D. A. Turton and K. Wynne, *J. Chem. Phys.*, 2009, **131**, 201101.
3 C. J. Fecko, J. D. Eaves and A. Tokmakoff, *J. Chem. Phys.*, 2002, **117**, 1139.
4 G. D. Goodno, G. Dadusc and R. J. D. Miller, *J. Opt. Soc. Am. B*, 1998, **15**, 1791.

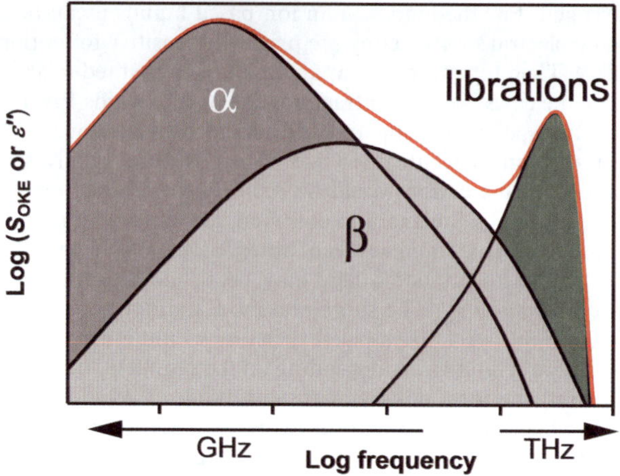


**Fig. 3** Characteristic OKE or dielectric spectrum for a viscous liquid.

**Professor Jung** said: What is the physical reason that the low frequency mode shows a strong temperature dependence in the spectra while the high frequency mode does not?

**Dr Turton** replied: The spectra of molecular liquids are typically continuous throughout the intermolecular region. At the lowest frequencies appears the α-relaxation with a second pronounced peak generally at around 1–3 THz. Whilst for many liquids the intervening region is featureless (see *e.g.* ref. 1 and 2 below), for glass-forming liquids the terahertz peak becomes well resolved and additional higher frequency (β) relaxations may appear (see Fig. 3 below).

The terahertz peak is often incorrectly termed the Boson peak, but in fact is generally assigned to librational motions, *i.e.* the rocking motions defined by the surrounding solvent cage. A Gaussian lineshape is often measured and this may be associated with an inhomogeneous distribution of the cage structures. With increasing temperature the librational mode typically broadens and shifts slightly to lower frequency as expected for an inhomogeneously broadened vibrational lineshape.

In contrast, the α-relaxation represents the complete disappearance of structural correlation and therefore implies rotation (or translation) over at least a molecular scale. This is therefore a diffusional, and hence, activated process that in the simplest cases follows the Arrhenius equation. In consequence the α-relaxation appears strongly temperature dependent whilst the librational mode is not.

In fact, for these complex liquids super-Arrhenius behaviour is often seen meaning that as the liquid is supercooled the viscosity increases more steeply than predicted by the Arrhenius expression. This is frequently associated with the presence of cooperatively rearranging regions of increasing size.


1 D. A. Turton and K. Wynne, *J. Chem. Phys.*, 2008, **128**, 154516.
2 D. A. Turton, D. F. Martin and K. Wynne, *Phys. Chem. Chem. Phys.*, 2010, **12**, 4191.


**Professor Angell** asked Dr Turton: In Figure 4 and also Figure 5, the decomposition into components gives you a quite large band with peak around 1 THz, or 30 $cm^{-1}$. Would it be reasonable to ascribe that to a boson peak (excess over Debye intensity in the vibrational density of states) and express interest in the temperature dependence of its intensity. Some of us think it should grow stronger with increase in temperature. The direct question would be, does this feature of the spectrum increase or decrease with increasing temperature?

 

**Dr Turton** replied: In many studies over the last decade that refer to the "Boson peak", the term has been used to describe the equivalent of the entire band that appears at around 1 THz here. At high temperature, the band is often subsumed by the relaxation modes but for strong liquids, which can be readily supercooled into the glassy state, a well-resolved peak is seen (see Fig. 3). The origin of the band is scattering due to the sub-wavelength inhomogeneities characteristic of disordered media and which are the counterparts of the crystalline phonons. Here we describe this region as librational. Both descriptions generally refer to hindered rotations, but the term Boson peak arises from a theoretical treatment of the subject and therefore for it to be correct to describe this as the Boson peak requires that the peak intensity has a temperature dependence that is determined by Bose–Einstein statistics as the question implies.

For these spectra the amplitude of the band is approximately independent of temperature and to the best estimate for the present OKE data the relative amplitudes (of the individual contributions) appear to be constant. Whether it is possible to make more detailed measurements closer to the glass transition is something that we will investigate.

**Professor Maroncelli** addressed Dr Turton and Dr Buchner: Although I do not doubt your coarse association of various frequency regions to different types of molecular motions, the "modes" you employ for simultaneous analysis of OKE and DR data are only phenomenological fitting functions. Are you aware of any simulation work using instantaneous normal mode (INM) analysis which seeks a more molecular description of the underlying mode structure? I am particularly curious about the spatial extent of INMs as a function of frequency.

**Dr Turton** responded: INM analyses have been carried out for simple liquids, *e.g.* benzene[1] and water[2] but ILs are very complex and INM calculations generate broad distributions that, due to the difficulties in obtaining accurate frequencies with meaningful amplitudes, are difficult to relate directly to spectra. Christian Schröder is currently collaborating with Richard Buchner in calculating the origin of these modes; perhaps he could describe this work.

1 S. Ryu and R. M. Stratt, *J. Phys. Chem. B*, 2004, **108**, 6782.
2 K. H. Tsai, T.-M. Wu, *Chem. Phys. Lett.*, 2006, **417**, 389

**Dr Buchner** replied: We are fully aware that the fitting models used by us are only phenomenological. Assignment of the resolved modes to physical processes must therefore be taken with a grain of salt. Life is far more complicated than these models, as can be seen on the poster by Christian Schröder and Othmar Steinhauser, which compares the simulated dielectric spectrum of [emim][TfO] with our data. These simulations clearly reveal that rotational and translational contributions strongly overlap. A clear cut decomposition of the spectrum into individual modes is virtually impossible. However, as long as theory and/or computer simulations do not achieve quantitative agreement with experiment (and this within reasonable time compared to experiment) we are forced to use our phenomenological approach. To my knowledge no attempt to extract wave-vector dependent information from simulations has been undertaken so far.

**Dr Schröder** replied: Responding to Dr Turton, in our approach to compute the dielectric spectrum, intra- and intermolecular vibrations are included in a different way. Intramolecular vibrations affect the collective rovibrational dipole moment. The relaxation of the autocorrelation function of this collective dipole moment is usually multi-exponential with time constants from subpicoseconds to nanoseconds. The corresponding peaks may overlap in the imaginary part of the spectrum of the generalized dielectric constant.

In particular, the shape of this spectrum at high frequencies is influenced by intramolecular vibrations, *e.g.* internal rotations. The intermolecular vibrations are caused by the motion of the center of mass of the molecules. These contributions are calculated by the current autocorrelation function and the mean-squared displacement of the collective translational dipole moment as demonstrated in ref. 1 below. The corresponding peaks in the imaginary part of the generalized dielectric spectrum span over a large range of frequencies. Consequently, rovibrational and translational contributions to the dielectric spectrum overlap.

In the case of [emim][OTf],[2] the overall peak structure is a superposition of both types of motions. Thereby, a direct assignment of a peak to a single type of motion is impossible. For more details on the computation of the generalized dielectric spectrum, the reader is referred to ref. 3 below.


1 C. Schröder, *J. Chem. Phys.*, 2011, **135**, 024502.
2 C. Schröder, T. Sonnleitner, R. Buchner and O. Steinhauser, *Phys. Chem. Chem. Phys.*, 2011, **13**, 12240.
3 C. Schröder and O. Steinhauser, *J. Chem. Phys.*, 2010, **132**, 244109.


**Professor Quitevis** asked: It is not surprising that the effective volume obtained from the Stokes–Einstein–Debye (SED) equation does not agree with the volumes of the ions in these ionic liquids, given that the equation was originally developed for macroscopic particles rotating in a continuum fluid. The interesting question is why rotational relaxation times obtained from OKE and DRS measurements correlate at all with $\eta/T$ in these complex fluids in a manner described by the SED equation.

**Dr Turton** answered: That is, of course, an interesting question, and it is also a long-standing and still open question.[1,2] Although it is perhaps inappropriate to invoke it for such a complex system, the SED (along with the Stokes–Einstein) relationship has become a widely used method of characterising molecular diffusion, for example, as SED, fractional-SED, and breakdown-of-SED types of behaviour. We still cannot say with certainty why one particular system should agree with SED over a limited temperature range, but we can, within limitations, infer from this fact a change in degree of the heterogeneity of the system.[3,4]


1 T. Köddermann, R. Ludwig and D. Paschek, *ChemPhysChem*, 2008, **9**, 1851.
2 Z. Li, *Phys. Rev. E: Stat., Nonlinear, Soft Matter Phys.*, 2009, **80**, 061204.
3 D. A. Turton, J. Hunger, A. Stoppa, G. Hefter, A. Thoman, M. Walther, R. Buchner and K. Wynne, *J. Am. Chem. Soc.*, 2009, **131**, 11140.
4 D. A. Turton, J. Hunger, A. Stoppa, A. Thoman, M. Candelaresi, G. Hefter, M. Walther, R. Buchner, K. Wynne, *J. Mol. Liq.*, 2011, **159**, 2.


**Professor Maroncelli** said: Your observation of anomalously rapid rotation of ions is consistent with the behavior observed in a several simulation studies, where it was found that reorientation often occurs *via* non-diffusive large-amplitude angular jumps. Many simulations also show predominantly rotational diffusion of the sort expected from simple hydrodynamic models. Similarly, experimental studies have reported both "normal" and "anomalous" rotations. The conditions giving rise to these different behaviors remain to be clarified.

**Dr Driver** asked: It is common to invoke a diffusional jump mechanism to explain observed decreases of the ratio $\tau_1/\tau_2$ from the limit of the Debye isotropic/spherical free diffusional model. In that case it is assumed that a large jump angle induces a decay of the orientational time correlation functions of all rank $L$ such that time constants associated with both $L = 1$ and $L = 2$ (*i.e.* $\tau_1$ and $\tau_2$) become similar due to the waiting time for the jump to occur. We would not really expect the isotropic Debye limit to hold for non-spherical $[C_2H_5NH_3]^+$ and $[C_3H_7NH_3]^+$

 

(both belonging to the $C_s$ symmetry point group) and planar $[NO_3]^-$ (with $D_{3h}$ symmetry) ions because by virtue of their asymmetric geometries, they must give rise to anisotropic reorientational motion.[1] In other words, while we expect $\tau_1/\tau_2 < 3$ to signal a breakdown of the Debye free rotational diffusion model, we realise this only applies to spherical/isotropic rotors. What value of $\tau_1/\tau_2$ would we similarly expect to signal a breakdown of anisotropic free rotational diffusion for the case of the asymmetric rotor?

1 J. W. T. Huntress, *J. Phys. Chem.*, 1968, **48**, 3524–3533.

**Dr Turton** answered: The factor that relates the rotational diffusion times arises from the different symmetry of the observable. In this case, the lower symmetry of the dipole (vector) results in an anisotropy decay that is slower than that of the polarizability (3rd order tensor). For spherical particles and for symmetric rotors where both the dipole and the major polarizability axis are co-aligned on the principal symmetry axis, a factor of three results. Unfortunately, for lower symmetry systems no general rules apply and without detailed calculations only very broad approximations can be made.[1,2] The diffusional jump model complicates the picture further with again a complex dependence of the relative relaxation times on the properties. For small jump angles a reduction in the $\tau_2/\tau_1$ ratio is predicted but for jump angles greater than 90° the OKE relaxation time can become even slower than that of the dielectric measurement.[2]

1 D. A. Turton, J. Hunger, A. Stoppa, A. Thoman, M. Candelaresi, G. Hefter, M. Walther, R. Buchner, K. Wynne, *J. Mol. Liq.*, 2011, **159**, 2.
2 D. A. Turton, J. Hunger, A. Stoppa, G. Hefter, A. Thoman, M. Walther, R. Buchner and K. Wynne, *J. Am. Chem. Soc.*, 2009, **131**, 11140.

**Professor Lynden-Bell** remarked: Are not dielectric and OKE spectroscopy "blunt instruments" for determining the origin of molecular motions?

**Dr Turton** answered: Although these techniques are limited, in that they measure a signal averaged over the whole liquid, they excel in providing excellent sensitivity and yielding information that more sophisticated techniques cannot. Here, by contrasting the two different measurements, we can shed light on the origins of the measured modes. Through (even phenomenological) data analysis, useful measures of the microscopic dynamics can certainly be made. The highly detailed low noise spectra are a very good test for both simulations and analysis. Here we describe a low parameter-number fitting function that closely reproduces the measured dynamics over a large range of frequency and temperature for both spectra. In the past, similar methodologies have often revealed the underlying physical mechanisms as well as stimulating theoretical progress.

**Professor Lynden-Bell** remarked: Could you make clearer the nature of the power law decay employed in the analysis?

**Dr Turton** replied: The power law decay $t^{-n}$ $(0 < n < 1)$ has been observed in light scattering and OKE studies of glass-forming liquids where it is used to describe the intermediate region that appears when the timescale of relaxation is well separated from the librations.[1,2] It is also a prediction of the mode coupling theory that describes the dynamics of supercooled liquids.[3,4] In the limit $n = 1$ a constant loss (white noise) response is reproduced (which is presumably a fundamental process). Applying such a response has only been done previously in a somewhat piecemeal manner. Here we show how it can be incorporated into a complete dynamical model that describes the whole intermolecular dynamics and which applies well to both the OKE and dielectric responses. In his contribution,[5] Luigi Delle Site describes the need for a theoretical model of dynamics, which is connected throughout the microscopic to mesoscopic range. In a necessarily much more modest approach, this is what we are trying to achieve from the experimental perspective.


1 A. P. Sokolov, A. Kisliuk, V. N. Novikov and K. Ngai, *Phys. Rev. B: Condens. Matter Mater. Phys.*, 2001, **63**, 172204.
2 H. Cang, V. N. Novikov and M. D. Fayer, *J. Chem. Phys.*, 2003, **118**, 2800.
3 J. Li, I. Wang, K. Fruchey and M. D. Fayer, *J. Phys. Chem. A*, 2006, **110**, 10384.
4 W. Gotze and L. Sjogren, *Rep. Prog. Phys.*, 1992, **55**, 241.
5 K. Wendler, F. Dommert, Y. Y. Zhao, R. Berger, C. Holm and L. Delle Site, *Faraday Discuss.*, 2012, **54**, 10.1039/C1FD00051A.


**Professor Jung** opened the discussion of the paper by Jose Nuno Canongia Lopes: In this work, you have only discussed the spatial heterogeneity of segregation of ionic liquid ions, arranged in a layer-by-layer structure, which is caused by the inhomogeneous electric fields generated by glass atoms as shown in Fig. 10 of the paper. However, if one watches the spatial inhomogeneity over a long time, one may see that ions move around along the $z$-direction, which will result in dynamical behavior of the sample. Have you ever studied this effect?

**Professor Canongia Lopes** replied: We have not explicitly studied the dynamics of the system along the $z$-axis (nor along the other two axes). However, we have monitored the movement of selected ions to check for possible simulation problems associated with finite size effects or ergodicity. Basically the ions that are directly adsorbed at the glass surface are only released from the surface during temperature annealing cycles. The counter-ions of those "first-layer" ions are already more mobile and the ions in subsequent layers have a mobility already resembling that of the bulk phase. It must be stressed that the ions used in this work do not have a long alkyl side chain, which means that further dynamical complexities derived from possible stratifications of non-polar domains are not present.

**Professor Jung** commented: Can you give us any estimate of the timescale with which spatially heterogeneous patterns of ions change along the $z$-direction?

**Professor Canongia Lopes** responded: At this stage it is not possible to establish a timescale. However, the dynamics vary from those typical of a bulk ionic liquid at the layers removed more than 2 nm from the glass surface to completely "frozen" and "charge-templated" ions at the surface of the glass.

**Professor Panagiotopoulos** asked: You indicated that the (bulk) equilibrium density is achieved after 50 ps and your simulations were 400 ps long. What is the characteristic time for attaining equilibrium for the orientations and 2D (near-wall) pair correlations? What is the average distance travelled by a typical ion in your system over a time period of 400 ps?

**Professor Canongia Lopes** answered: The 50 and 400 ps simulation times correspond to runs used to obtain the bulk ionic liquid (Ionic liquid section of the Experimental part of the paper). In fact the 50ps mentioned in the text correspond to the time elapsed (from the simulation box containing the initial configuration of the ionic liquid) until the bulk density levels off. Before the final 400 ps simulation (to obtain the bulk IL data), several intermediate 100 ps simulations were performed to check if the properties of the bulk IL were constant. Thus, the total bulk IL was equilibrated for 1.25 ns.

The IL that was sliced and placed at the solid–liquid interface was further equilibrated using consecutive 200 ps runs. At least 20 such runs were performed with some intermediate temperature annealing cycles. As stated in the Solid-liquid interface section of the Experimental part, the system evolved to similar configurations after such equilibration/annealing cycles. A total equilibration/annealing time of around 4 ns was used to make sure no ergodicity problems were present.

There is a large difference between the mobility of the ions that are directly adsorbed at the surface and those that are further away in the bulk. Those in the bulk can travel a few molecular diameters between each 200 ps interval (and their diffusion increases during an temperature annealing cycle). Almost all of the ions directly adsorbed at the layer only move in the directions parallel to the surface (and most of the time they just adopt different orientations). Only when performing annealing cycles we have observed important ionic exchange at the surface. It must be noted that the "first layer" of ionic liquid contains the directly adsorbed ions (both cations and anions, depending on the local surface charge of the glass) and their counterions just above them. The latter are able to move much more frequently and perform exchanges with ions in the layers above them.

**Professor MacFarlane** commented: Are you able to artificially manipulate the charge density on the surface? This would simulate an electrochemical experiment on an oxide semiconductor surface such as $TiO_2$ or ITO.

**Professor Canongia Lopes** responded: It is rather easy to manipulate the charges at the surface. However the most challenging aspect of this very interesting issue is to establish a model that on one hand is able to reproduce the surface electrostatic field (through the adequate charge distribution and intensity) and at the same time is realistic from a physical perspective. In the case of semiconductor surfaces (or even metallic surfaces) one should also pay close attention to the polarization of the surface.

**Mr Drueschler** asked: Fig. 9 and Fig. 10 of your paper show that the uneven distribution of charge density across the glass surface has an obvious templating effect on the ionic liquid which is in closest contact with it. However, having a look only at the innermost layer, can areas be found in which anions are exclusively next to anions or cations next to cations, respectively, depending on the local charge of the surface the ions are in direct contact with? Or in other words, is it also possible to find cations in contact with a positively charged part of the surface or anions in contact with a negatively charged part of it, which would be comparable with the classical model of specific adsorption? This would mean that the distribution of the ions in this innermost layer also depends on non-coulombic interactions with the surface and that the distribution can be influenced by modifying the chemical structure of the cation and/or the anion, respectively. For instance, if you use cations which are able to form hydrogen bonds, like $[(OH)C_2C_1im]^+$ used for your present study, this may also lead to a higher possibility of finding cations in contact with the more negatively charged parts of the surface since more oxygen atoms can be found there. Hence, do you expect similar ion distributions within the innermost layer compared with the results of the present simulation if you use the same cation with a $CH_3$ group instead of a OH group? Or do you think that this could lead to a higher possibility of finding anions in contact with a negatively charged part of the glass surface?

**Professor Canongia Lopes** replied: The answer to the first question (can areas be found in which exclusively anions are next to anions or cations next to cations, respectively, depending on the local charge of the surface the ions are in direct contact with?) is yes, as documented in Fig. 7a, 9 and 10, where the anions adsorbed at the surface exhibit anion–anion distances considerably shorter than those in the bulk: the anions adsorbed at a positive patch of the glass can lie closer together without breaking local charge neutrality (the positive glass below and a cation just above ensure that such criteria is fulfilled). The second part of the comment refers to the importance of other non-coulombic interactions in terms of the definition of the adsorbed layer. I think the two types of interaction can and do co-exist. In fact although the glass acts as a powerful charge template and defines the location of anions and cations I think that other factors (OH groups in the present case, long alkyl side chains in other cases) will influence the orientation of the ions at the surface. Such a notion is also visible in Fig. 7b and d, where the orientation of the cations is conditioned by the factors mentioned above.

**Professor Hardacre** addressed Professor Canongia Lopes and Professor Jones: Would it be worth performing the simulations on single crystals of silica? In these cases if the ionic liquid templates or is commensurate with the surface then it may be expected that the surface influenced structure would be felt further into the liquid as compared with an amorphous surface.

**Professor Canongia Lopes** answered: I think this is a pertinent issue and that the present system could be compared with simulations performed in a crystalline silica surface. I agree that the degree of "layering" is expected to increase not only because of the presence of a 2D solid surface (already present in the amorphous glass case) but also due to the possible ordered stratification of ions promoted by the exposed regular crystalline lattice of the substrate. It would also be interesting to see if different exposed surfaces would lead to different results.

**Professor Jones** responded: Assuming the question is asking what may happen if we adsorbed gases (*e.g.* water) onto just a few monolayers of ionic liquid which are themselves adsorbed on a single crystal of silica, we would suggest that the behaviour is likely to be significantly different to water adsorption/absorption using the surface of the bulk ionic liquid. Silica (and other surfaces) is likely to strongly adsorb the ionic liquid and both orientate and order the ions at the monolayer interface. The monolayer may well be better regarded as crystalline or glassy than liquid. Multilayers (*i.e.* a few layers) of ionic liquid above this monolayer would probably contain ions which are partially orientated, and possibly layered and ordered to some degree by the layers below them, such that an incoming water molecule would encounter a surface significantly different in structure to that of the surface of the bulk liquid. This will affect both the adsorption energy and the kinetics for transport across the surface and into the multilayer ionic liquid film. For ionic liquid layers consisting of just a few monolayers, there could conceivably be several absorption planes at different distance from the surface, due to the formation of layers of ions interspersed with layers of hydrocarbon chains. Clearly for sufficiently thick films the ionic liquid would attain the structure of the bulk liquid, sandwiched between specific, possibly multilayer structures for the solid/liquid and liquid/gas interfaces.

**Professor Endres** commented: What 2D structures would you expect if just 1 monolayer was deposited on the surface? Any temperature dependence?

**Professor Canongia Lopes** answered: I think the glass surface would continue to act as a charge template for the ions. We have not attempted such simulations due to the possibility of strong ergodicity problems (after adsorption at a given location it would be very hard for the ions to diffuse and equilibrate into more stable configurations). Nevertheless it would be interesting to attempt such kind of simulation.

**Professor Dr Roling** opened the discussion of the paper by Mathieu Salanne: Is it really realistic that the aluminium ions carry a charge of +3 in the chloroaluminate melts? Isn't it more reasonable to assume that the Al–Cl bonds exhibit a certain degree of covalency reducing the positive charge of Al?

**Dr Salanne** responded: The question of covalency in ionic melts has been extensively discussed in a paper by Madden and Wilson.[1] It was shown that many effects, which are often attributed to 'covalency', may in fact be very easily explained by treating carefully the intermolecular interactions (by the inclusion of induction effects and by accounting for the changes of size and shape of ions in response to the changes in their environment). From a more practical point of view, in the present work the use of partial charges on the aluminium and the chloride ions together with the inclusion of an harmonic bond term would hinder the possibility of observing some exchange of the chloride anions between two $AlCl_4^-$ units and the study of the chloroaluminate speciation. Finally, we would also like to point out that the DFT forces that act on both the aluminium and chloride atoms are very well reproduced by our model.

1 P. A. Madden and M. Wilson, *Chem. Soc. Rev.*, 1996, **25**, 339.

**Mr Frolov** said: What is the reason for verifying the structure of the melts (particle–particle radial distribution functions (RDFs)) predicted by your newly parameterized interaction potential by the RDFs calculated from *ab initio* molecular dynamics (MD) simulation trajectories? *Ab initio* MD contains many approximations and, thus, in principle, should be verified as well. Are there preliminary studies which show that the particular setup of the *ab initio* MD described in the presentation is accurate for description of the structure of melts? Are there experimental data to verify the simulations?

**Dr Salanne** answered: First we would like to emphasize that it is the DFT functional which is used in the *ab initio* MD which contains some approximations. This having been said, there have been two preliminary studies which showed that the particular setup used here was accurate for the description of the structure of melts.[1,2] The PBE functional has also been well tested in the case of many room-temperature ionic liquids[3–5] and interaction potentials based on this functional have also been shown to perform very well for reproducing the experimental properties of molten salts.[6–8] The main problem concerns the dispersion interactions, but we try to treat them carefully in our approach. The transferability of *ab initio* MD relies on the fact that it is not fitted to experimental data.

For these reasons, it is now well-established that *ab initio* MD is a powerful tool for studying the structure of liquids, and that reaching its accuracy from classical MD is a very difficult task. In our study the advantage of looking at the partial RDFs is that we can directly compare all the local structural properties. When the neutron (or X-ray) diffraction data is used for the comparison, all the partial structure factors are included, and only global information, which includes a strong intramolecular contribution, is sampled: In the case of RTILs, it is impossible to extract all the atomic structure factors. This does not mean that the comparison should not be done, but we believe that reproducing the *ab initio* RDFs is a more difficult test (in the case of inorganic compounds with fewer atomic types, the situation is different: there is a much lower number of contributions to the diffraction pattern and the comparison is very useful; see ref. 9 below where we show that a polarizable ion model of $GeO_2$ is able to reproduce the neutron diffraction data over a wide range of pressures).

1 B. Kirchner, A. P. Seitsonen and J. Hutter, *J. Phys. Chem. B*, 2006, **110**, 11475.
2 B. Kirchner and A. P. Seitsonen, *Inorg. Chem.*, 2007, **46**, 2751.
3 C. Krekeler, F. Dommert, J. Schmidt, Y. Y. Zhao, C. Holm, R. Berger and L. Delle Site, *Phys. Chem. Chem. Phys.*, 2010, **12**, 1817.
4 S. Zahn and B. Kirchner, *J. Phys. Chem. A*, 2008, **112**, 8430.
5 E. I. Izgorodina, U. L. Bernard and D. R. MacFarlane, *J. Phys. Chem. A*, 2009, **113**, 7064.
6 B. Rotenberg, M. Salanne, C. Simon and R. Vuilleumier, *Phys. Rev. Lett.*, 2010, **104**, 138301.
7 N. Ohtori, M. Salanne and P. A. Madden, *J. Chem. Phys.*, 2009, **130**, 104507.
8 R. J. Heaton, R. Brookes, P. A. Madden, M. Salanne, C. Simon and P. Turq, *J. Phys. Chem. B*, 2006, **110**, 11454.

9 D. Marrocchelli, M. Salanne and P. A. Madden, *J. Phys.: Condens. Matter*, 2010, **22**, 152102.

**Professor Panagiotopoulos** commented: In your paper, you compare experimental diffusivities and viscosities to simulation results, but there was no equivalent comparison for the densities. Can you comment on how well your model reproduces experimentally measured densities?

**Dr Salanne** answered: The difference between the experimental and calculated densities is of around 5%. But it should be noted that this problem can easily be corrected: In fact the dispersion interactions play a crucial role in the evaluation of the density. This was shown for example in the case of liquid alumina by Jahn *et al.*[1] or of liquid water by Schmidt *et al.*[2] In DFT calculations, as soon as standard functionals such as the PBE or BLYP ones are used, this interaction is not taken into account well, which results in an overestimation of the density. When we perform our force-fitting procedure, we therefore take care of treating this term independently. In the present work, we have simply determined all the $C_6$ and $C_8$ coefficients following the Slater-Kirkwood relationship and by using the Cl–Cl coefficients taken from our previous work[3] as a starting point. We are now trying to develop a method for calculating systematically these coefficients from DFT calculations (see ref. 4 below) which should allow for a better estimation of the density. Note that from our experience in ionic liquids, if the system is simulated at the correct density, the other properties (structure, diffusion coefficients) are very slightly affected (unlike the very special case of liquid water).

1 S. Jahn, P. A. Madden and M. Wilson, *Phys. Rev. B: Condens. Matter Mater. Phys.*, 2006, **74**, 024112.
2 J. Schmidt, J. VandeVondele, I.-F. W. Kuo, D. Sebastiani, J. I. Siepmann, J. Hutter and C. J. Mundy, *J. Phys. Chem. B*, 2009, **113**, 11959.
3 N. Ohtori, M. Salanne and P. A. Madden, *J. Chem. Phys.*, 2009, **130**, 104507.
4 B. Rotenberg, M. Salanne, C. Simon and R. Vuilleumier, *Phys. Rev. Lett.*, 2010, **104**, 138301.

**Mr Frolov** asked: Are there experimental data (presumably NMR measurements) for how long these species ($Al_2Cl_7^-$) exist in the melts? Can you compare your results with these data?

**Dr Salanne** answered: Up to now there has not been any experimental estimation on the lifetime of these species. From NMR experiments, the only information that can be extracted is that this lifetime is much shorter than a millisecond. Note, however, that our result concerning the speciation of chloroaluminate anions in the equimolar mixture of aluminium trichloride with 1-ethyl-3-methylimidazolium chloride, *i.e.* the absence of $Al_2Cl_7^-$ anions, is in agreement with the experimental work from Hussey *et al.*[1]

1 C. L. Hussey, T. B. Scheffler, J. S. Wilkes and A. A. Fannin, Jr., *J. Electrochem. Soc.*, 1986, **133**, 1389.

**Mr Frolov** said: Could it be possible to model the effects of the partial decomposition of the imidazolium cations in ionic liquids at elevated temperatures with the currently available *ab initio* molecular dynamics methods, or do these processes require a better level of theory and/or much longer simulation times that are currently affordable by supercomputers?

**Dr Salanne** replied: This question is discussed in the paper by Ballone and Cortes-Huerto in the same *Faraday Discussion* volume.[1] One of us (B. Kirchner) is currently carrying out some *ab initio* molecular dynamics simulations of decomposition reactions and the preliminary results show that it is possible to have correct results with the standard levels of theory. Of course the simulation time remains an important issue.

 

1 P. Ballone and R. Cortes-Huerto, *Faraday Discuss.*, 2012, **154**, DOI: 10.1039/C1FD00064K.

**Professor Ballone** said: The modelling of organic/inorganic ionic mixtures discussed in the contribution of Mathieu Salanne and collaborators largely relies on density functional computations using the PBE approximation for exchange and correlation.

I would like to emphasise that the DFT-PBE description of inorganic ionic systems such as aluminium tri-halides is quite unsatisfactory, and, in the case of inorganic/organic ionic mixtures, the quality of DFT-PBE is untested (as far as I know).

A few years ago, I carried out DFT-PBE computations for single molecules and dimers of aluminium tri-halides. The results (unpublished) are briefly discussed in a paper by Akdeniz and Tosi.[1] For molecules and dimers, geometrical and dynamical properties are known from gas phase spectroscopy. The comparison of computational and experimental data turned out to be relatively poor. More recently, DFT-PBE computations have been carried out for crystals ($T = 0$ K) of organic (room temperature) ionic "liquids".[2] Also in this case, a quantitative comparison of computational and experimental data is possible, and, again, the comparison is not very favourable. These results suggest that the DFT-PBE computations, which underlie the potential energy modelling, need to be validated by some strict and quantitative comparison with experimental data.

1 Z. Akdeniz and M. P. Tosi, *Z. Naturforsch., A: Phys. Sci.*, 1999, **54**, 180.
2 M. G. Del Pópolo, C. Pinilla and P. Ballone, *J. Chem. Phys.*, 2007, **126**, 144705.

**Dr Salanne** responded: It is important to keep in mind that our parameterization procedure relies on condensed-phase calculations on these species, and not on the gas phase. It is well known that negative ions are problematic in the gas phase.[1] We have now made many tests which have validated the use of the PBE functional for pure inorganic compounds, such as solid and molten chlorides, fluorides and oxides. In the case of $AlCl_3$, two different *ab initio* molecular dynamics based studies using either the BLYP[2] or the PBE[3] functionals have showed that this method provides the correct structure for the liquid when compared to experimental data (note that this was also the case in another simulation study involving classical molecular dynamics with a polarizable force field[4]). Concerning room temperature ionic liquids, besides the dispersion problem which was already discussed (see the answer to the earlier question from Professor Panagiotopoulos), several functionals have been tested in two studies.[5,6] In both cases, the authors concluded that the PBE yields the best results amongst GGA functionals. Of course, it is expected that using either hybrid functionals or dispersion corrected GGA ones will improve the results. Based on these results, we do not see why the PBE functional would not give a correct description of the inorganic/organic systems of interest here.

1 C. Domene, P. W. Fowler, P. Jemmer and P. Madden, *Chem. Phys. Lett.*, 1999, **299**, 51.
2 A. L. L. East and J. Hafner, *J. Phys. Chem. B*, 2007, **111**, 5316.
3 B. Kirchner, A. P. Seitsonen and J. Hutter, *J. Phys. Chem. B*, 2006, **110**, 11475.
4 F. Hutchinson, M. K. Walters, A. J. Rowley and P. A. Madden, *J. Chem. Phys.*, 1999, **110**, 5821.
5 S. Zahn and B. Kirchner, *J. Phys. Chem. A*, 2008, **112**, 8430.
6 E. I. Izgorodina, U. L. Bernard and D. R. MacFarlane, *J. Phys. Chem. A*, 2009, **113**, 7064.

**Professor Pádua** remarked: In approximate terms how important is the contribution due to polarization in your simulations?

**Dr Salanne** responded: The short answer is that if we don't include polarization effects, it is not possible to keep the $AlCl_4^-$ ion (or other chloroaluminate species) stable unless partial charges and harmonic bond terms are introduces, which goes back to the problems evoked in the answer of Prof. Roling's question.

To go further into details, we observed that the induced dipoles carried by the chloride anions are directed along the Al–Cl 'bond', which is at the origin of the weakening of the interaction between the chloride anion and the imidazolium protons (compared to pure EMICl, for example). This probably has strong effects on the thermodynamic and dynamic properties of the liquids, but they cannot really be quantified.

 

# The interface ionic liquid(s)/electrode(s): *In situ* STM and AFM measurements

**Frank Endres,[*ab] Natalia Borisenko,[ab] Sherif Zein El Abedin,[abc] Robert Hayes[d] and Rob Atkin[d]**

*Received 30th March 2011, Accepted 13th April 2011*
**DOI: 10.1039/c1fd00050k**

The structure of the interfacial layer(s) between the extremely pure air- and water-stable ionic liquid 1-butyl-1-methylpyrrolidinium tris(pentafluoroethyl) trifluorophosphate and Au(111) has been investigated using *in situ* scanning tunneling microscopy (STM) at electrode potentials more positive than the open circuit potential. The *in situ* STM measurements show that layers/islands form with increasing electrode potential. According to recently published atomic force microscopy (AFM) data the anion is adsorbed even at low anodic overvoltages and adsorption becomes slightly stronger with increasing electrode potential. Furthermore, the number of interfacial layers increases with increasing electrode potential. The present discussion paper shows that these layers are not uniform and have a structure on the nanoscale, supporting earlier results that the interface electrode/ionic liquid is highly complex. It is also shown that the addition of solutes changes this structure considerably. AFM results reveal that in the pure liquid, interfacial layers lead to a repulsive force but the addition of 10 wt% of LiCl leads to an attractive force close to the surface. These preliminary results show that solutes strongly alter the interfacial structure of the ionic liquid/ electrode interface.

## Introduction

Ionic liquids (ILs) have attracted considerable research interest in surface science and in physical chemistry in recent years. Apart from wide electrochemical and thermal windows they have high ionic conductivities, acceptable viscosities and extremely low vapor pressures. The latter properties make them suitable for ultra-high vacuum studies, and their wide electrochemical windows of up to $\pm 3$ V *vs.* NHE (including thermodynamic and kinetic contributions) are attractive for electrochemistry. The prospects and challenges of ILs in the field of electrochemical energy science were discussed in.[1] The interface of diluted electrolyte solutions/electrodes is usually described by the Helmholtz model, the Gouy-Chapman model, and the Stern model, the latter being more or less a combination of the first two. The measurable double layer capacitances of aqueous electrolytes/electrodes are around 100 μF $cm^{-2}$, and usually have a minimum at the potential of zero charge. For a description of ionic liquids/electrodes one has to bear in mind that these models were developed for dilute electrolytes, thus application to ILs which have a high density of positive

[a]*Institute of Particle Technology, Clausthal University of Technology, Arnold-Sommerfeld-Strasse 6, 38678 Clausthal-Zellerfeld, Germany. E-mail: frank.endres@tu-clausthal.de*
[b]*EFZN Goslar, Am Stollen 19, 38640 Goslar, Germany*
[c]*Electrochemistry and Corrosion Laboratory, National Research Centre, Dokki, Cairo, Egypt*
[d]*Centre for Organic Electronics, Chemistry Building, The University of Newcastle, Callaghan, New South Wales, 2308, Australia. E-mail: Rob.Atkin@newcastle.edu.au*

and negative charges is not appropriate. An IL represents a dense system of cations and anions with no solvent, so that the individual interactions between neighbouring ions cannot be neglected.[2–4] Thus, for ILs the situation is much more complicated. Several groups studied the double layer behaviour of ILs. Lockett *et al.* demonstrated hysteresis effects in the differential double layer capacitance, as well as maxima and minima in the capacitance curve, referred to as the *camel shape* capacitance curve.[5] Similar results were presented in reference 6 and Fedorov and Kornyshev suggested a theoretical model, which explains the results:[7] at an open circuit potential both the charged groups and neutral groups (chains) on the ions interact with the electrode surface. With increasing cathodic or anodic electrode potentials a reorientation occurs with the neutral groups repelled from the surface and replaced by the ion charged groups. Consequently an increase in double layer capacitance results until saturation is reached at high electrode potentials. This model was further developed and quite a recent theoretical study shows that ion multilayers can be described.[8] This is consistent with previous AFM[9] and X-ray[10] reflectivity measurements that revealed ILs are strongly adsorbed on solid surfaces and that several layers are present adjacent to the surface. In a joint paper we (Atkin and Endres *et al.*) showed that 1-butyl-1-methylpyrrolidinium bis(trifluoromethylsulfonyl) amide and 1-ethyl-3-methylimidazolium bis(trifluoromethylsulfonyl)amide behave quite differently on Au(111). Both liquids are adsorbed at the open circuit potential but the first one approximately 4 times more strongly.[11] It should be mentioned that the adsorption of ILs on a solid surface is not at all surprising as water and organic solvents can also be adsorbed.[12–14] However, the force to rupture IL layers is quite surprising as it is one order of magnitude higher than for aqueous/organic solvents. Such a strong adsorption must influence the image quality in an *in situ* STM experiment.[15] In a recently published joint paper the structure and dynamics of the interfacial layers between the air- and water-stable IL 1-butyl-1-methylpyrrolidinium tris (pentafluoroethyl)trifluorophosphate ($[Py_{1,4}]$FAP) and Au(111) were investigated using STM, cyclic voltammetry (CV), electrochemical impedance spectroscopy (EIS), and AFM measurements. *In situ* STM measurements reveal that the Au (111) surface undergoes a reconstruction, and at −1.2 V (*vs.* Pt quasi-reference) the $(22 \times \sqrt{3})$ "herringbone" superstructure is probed. AFM force–distance profiles showed that the IL becomes more structured at higher cathodic potentials and both the number of detectable layers and the push-through forces increase. EIS measurements showed a capacitive process at the IL/Au(111) interface which is considerably slower than electrochemical double layer formation and seems to be related to the herringbone reconstruction of Au(111).[16] It should be mentioned that the interface (electro-)chemistry of ILs is altered by solutes and even low amounts of impurities can have a considerable effect.[17] In the present paper we focus on *in situ* STM experiments in the anodic regime of $[Py_{1,4}]$FAP/Au(111). It will be shown that structures form on the nanoscale for electrode potentials of up to +2 V *vs.* the open circuit potential (ocp). These structures disappear when the electrode potential is returned to the ocp, and the original surface is recovered. Thus, the observed processes on the surface are reversible and not due to impurities which in this liquid are all below 10 ppm. Preliminary AFM results show that the addition of 10 wt% of LiCl strongly alters the near surface structure and in contrast to repulsive forces measured for the pure IL-interfaces (see references 18–20) the AFM tip experiences attractive forces on approach to the gold surface.

## Experimental

1-Butyl-1-methylpyrrolidinium tris(pentafluoroethyl)trifluorophosphate ($[Py_{1,4}]$ FAP) was purchased from Merck in the highest available quality. The liquid was custom-made and the purity protocol delivered by Merck showed that all detectable impurities were below 10 ppm. HF and oxide levels were below the detection limit of 10 ppm. It is important to mention that impurities in ILs, which are often present in commercial ILs (even in apparently ultrapure quality), can strongly alter the surface processes easily leading to misinterpretations.[17] There can be numerous impurities including decomposition products of anions and/or cations, side products from synthesis, $Li^+$, $Na^+$, $K^+$, and halides in the 1000 ppm range even in apparently ultrapure ILs. Even worse, purification procedures may introduce $SiO_2$ or $Al_2O_3$ particles.[21] Prior to use we analyze all ILs using CV, XPS and *in situ* STM to ensure the purity. If there are metallic impurities they are usually detected by STM on Au(111) in the cathodic regime and by underpotential deposition processes visible in the cyclic voltammogram. We should mention that a definite exclusion of impurities is quite challenging and needs the combination of several techniques together with a good amount of experience. Therefore, for fundamental studies we require custom-made ILs of the highest possible purity to avoid misinterpretations. Additionally our liquids are dried under vacuum ($10^{-3}$ mbar) at 100 °C to $H_2O$ contents of well below 1 ppm and stored in a closed bottle in an Ar-filled glove box with $H_2O$ and $O_2$ contents of below 2 ppm (OMNI-LAB from Vacuum-Atmospheres). *In situ* STM and AFM measurements were carried out using samples of the same IL which were sent in sealed ampoules to the Newcastle group. The substrates for AFM and STM experiments and the working electrode (WE) were Au(111) (a 300 nm thick film on mica) purchased from Agilent. Directly before use, the substrates were carefully heated in a $H_2$-flame to minimize possible surface contaminations. CV measurements were carried out in the glove box using a Parstat 2263 potentiostat/galvanostat (Princeton Applied Research) controlled by a PowerCV software. The electrochemical cell was made of polytetrafluoroethylene (Teflon) and clamped over a Teflon-covered Viton O-ring onto the substrate, thus yielding a geometric surface area of the WE of 0.3 $cm^2$. Pt wires (Alfa Aesar, 99.99%) of 0.5 mm diameter were applied as quasi-reference (RE) and counter (CE) electrodes, respectively. From our experience, Pt has a sufficiently stable electrode potential under *in situ* STM/AFM conditions. Directly before use the Pt wires were cleaned for 15 min in an ultrasonic bath in acetone followed by heating in a $H_2$-flame to red glow for a few minutes to minimize surface contaminations.

LiCl was sourced from Sigma-Aldrich (99%) and dried in a vacuum oven at 100 °C for 24 h prior to use. AFM force measurements were acquired continuously using a Digital Instruments NanoScope IIIa Multimode AFM in contact mode in an incubator at 21 °C. The scan rate and scan size were kept between 0.1 and 0.5 Hz and 10 and 50 nm, respectively. The $Si_3N_4$ cantilever was carefully rinsed in Milli-Q $H_2O$, dried and irradiated with ultraviolet light for 40 min prior to use. The IL was held in an AFM fluid cell, sealed using a silicone O-ring. Both of these were cleaned by sonication for 30 min, rinsed copiously in distilled ethanol and Milli-Q $H_2O$, and then dried using filtered $N_2$. A modified AFM cell setup was used to acquire force curves as a function of potential. A thin cylindrical strip of Cu metal and 0.25 mm Pt wire were used as the CE and RE, respectively. The CE and RE were cleaned firstly in diluted HCl acid solution and then washed with distilled ethanol and Milli-Q $H_2O$ and dried using filtered $N_2$. The CE was mounted with the O-ring in the groove of the fluid cell to establish an equipotential WE surface. The RE was located directly above the centre of the WE surface by securing the Pt wire through the outlet valve of the fluid cell. The electrodes were connected to an EG & G Princeton Applied Research Model 362 Scanning Potentiostat.

The features of the AFM force curves at a given surface potential did not alter over a 48 h period. Typical start distances for force scans were 30–50 nm from the Au(111) surface. The maximum applied force in contact was between 30 and 500 nN.

STM experiments were performed at 23 °C using in-house-built STM heads and scanners under inert gas conditions ($H_2O$ and $O_2$ < 2 ppm) with a Molecular Imaging PicoScan 2500 STM controller in feedback mode. Assembling of the STM head and filling of the electrochemical cell were performed in an Ar-filled glove box solely reserved for assembling of STM heads. The STM head was placed inside an Ar-filled vacuum-tight stainless steel vessel, to ensure inert gas atmosphere during the STM experiments, transferred to the air-conditioned laboratory ($T = 23 \pm 1$ °C) and placed onto a vibration damped table from IDE (Germany). STM tips were made by electrochemical etching of Pt-Ir wires (90/10, 0.25 mm diameter) with a 4 mol $L^{-1}$ NaCN solution and subsequently electrophoretically coated with an electropaint (BASF ZQ 84-3225 0201). During the STM experiments the potential of the WE was controlled by the PicoStat from Molecular Imaging/Agilent.

## Results and discussion

Before presenting new *in situ* STM results we will first review the cyclic voltammograms and our previous STM/AFM data on the interface structure of Au(111) in $[Py_{1,4}]$FAP in the cathodic regime to provide context. For details we refer the reader to references 16 and 22. Fig. 1 shows the cyclic voltammogram of Au(111) in $[Py_{1,4}]$FAP at a scan rate of 10 mV $s^{-1}$ between −3 and +1 V *vs.* the Pt quasi reference electrode. This voltammogram remains stable for subsequent scans and in[16] we showed that the cathodic peaks C1–C4 and their anodic counterparts are not due to Faradaic reactions. The AFM data presented in reference 16 show that there is a cation-rich layer with a thickness of 0.35 nm at the ocp. At −1 V the cation-rich layer is 0.25 nm thick and in both cases several layers of ionic liquid are on top of this layer. At −2 V the cation is so strongly adsorbed that it cannot be displaced by the AFM tip, and only an anion layer is detected. STM at the ocp shows a more or less normal terraced Au(111) surface. With decreasing electrode potential the herringbone superstructure is probed; this structure disappears at more negative electrode potentials. *In situ* STM did not give a clear picture of the adsorbed IL layers but did reveal restructuring of the gold surface. Fig. 2 shows the cyclic voltammogram of Au(111) in $[Py_{1,4}]$FAP at a scan rate of 10 mV $s^{-1}$ between −3 and +2 V *vs.* the Pt quasi reference electrode. The current response between −3 and +1 V is almost identical to the CV presented in Fig. 1. At electrode potentials of above 1 V the current remains almost constant and starts to rise steeply with a shoulder at ~ +1.5 V. In the backscan 2–3 cathodic waves are observed between ~ +0.8 and −0.2 V. The steeply rising anodic current up to +2 V is difficult to allocate to a reaction. In the STM experiments described in more detail below we lost the contrast at +2 V but there were no definite hints for gold oxidation. Therefore we might conclude that this anodic reaction is mainly due to an oxidation of the IL. Only at higher electrode potentials of up to +3 V on this scale we observed the oxidation of gold. The cyclic voltammogram alone implies a featureless surface electrochemistry in the anodic regime whereas in the cathodic regime definite processes

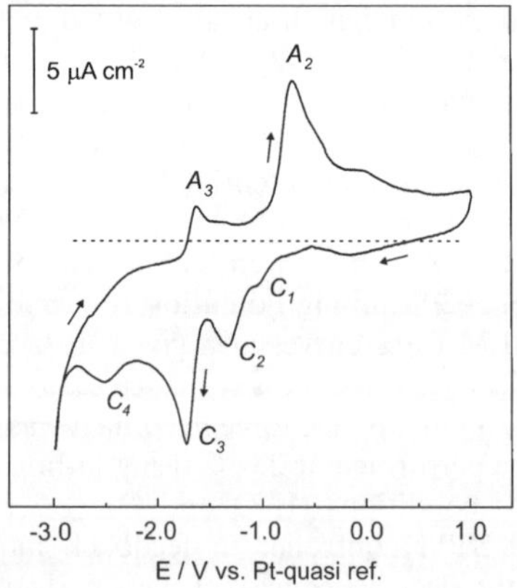


**Fig. 1** Cyclic voltammogram of Au(111)/$[Py_{1,4}]$FAP at a scan rate of 10 mV $s^{-1}$ between −3 V and +1 V (see also reference 16).

 

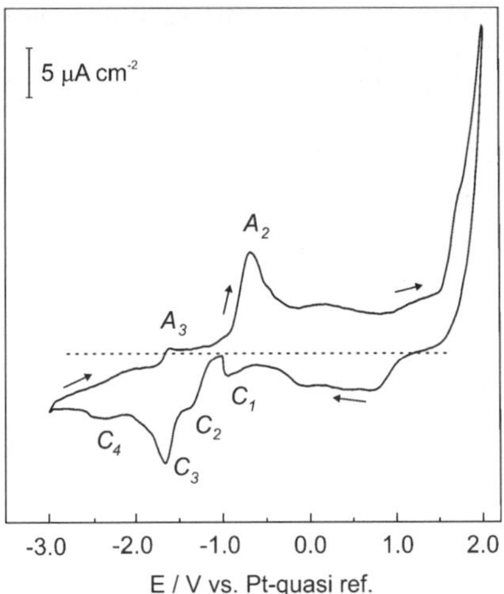


**Fig. 2** Cyclic voltammogram of Au(111)/[$Py_{1,4}$]FAP at a scan rate of 10 mV $s^{-1}$ between −3 V and +2 V.

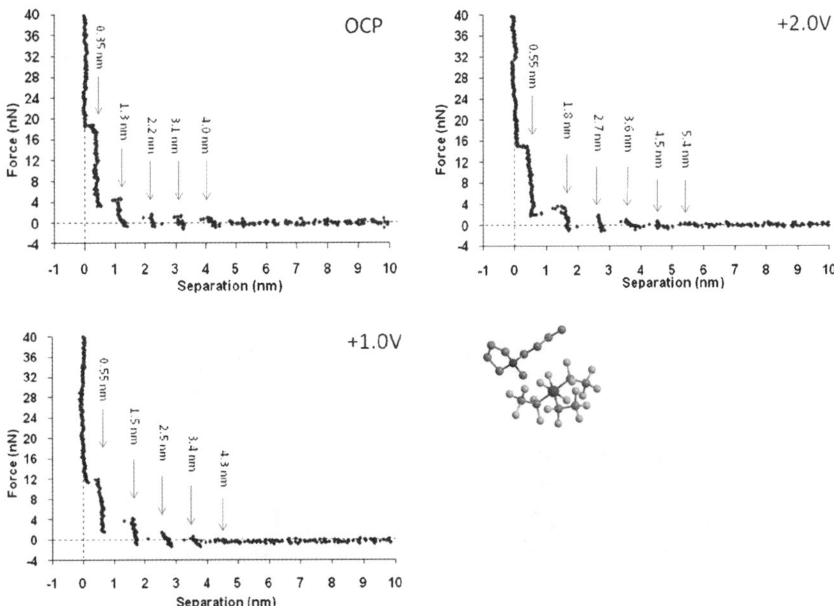


**Fig. 3** AFM force/distance profiles of Au(111)/[$Py_{1,4}$]FAP in the anodic regime, from reference 22. At the ocp cation adsorption prevails whereas with more anodic electrode potentials a preferential anion adsorption sets in.

are observed. Fig. 3 shows AFM approach curves at open circuit potential, +1 V and +2 V *vs.* the Pt quasi reference electrode.[22] As there is a negligible Faradaic process, the composition of the electrolyte is not altered, thus the reference electrode is stable. At the ocp there are 5 layers. Beginning 4 nm from the surface there are 4 steps each 0.9 nm, consistent with the size of the ion pair. The last layer, however, is only 0.35 nm thick indicative of the layer in contact with the surface being enriched in the cation. According to Fig. 3 the interface structure changes when more positive electrode potentials are applied. At + 1 V five layers are detected and the layer closest to the surface is 0.55 nm thick, indicative of anion adsorption. At +2 V the situation is similar, but the adsorbed anion layer appears slightly compressible. The AFM data alone would lead to the conclusion that these layers are more or less uniform, as the AFM tip probes reproducibly the same layer structure. The situation is slightly more complicated as our recent *in situ* STM results will show in this manuscript. In the past we mainly focused on the cathodic regime. On the one hand this is a consequence of our activities in the field of metal and semiconductor deposition, but on the other hand because the quality of the STM images in the anodic regime is usually lower. In the present study liquids of the highest available quality are used, allowing confidence that the observed surface modifications are due to the ionic liquid and not due to impurities. Fig. 4a shows a typical Au(111) surface in contact with $[Py_{1,4}]FAP$ with the typical terraces and 250 pm high steps at the open circuit potential. Zooming in does not give a better resolution and the images get rather blurred, which is a hint at the interaction of the STM tip with the surface. The AFM data show that there are at least 5 well-formed layers adjacent to the surface, and the layer closest to the surface is enriched in the cation and requires a push-through force of 20 nN, which is quite a high binding force. Nevertheless, electrons tunnel between the STM tip and surface, through this interfacial layer; the electronic structure in this layer will also determine exactly what the STM tip probes. As the typical distance between tip and sample in an STM experiment is 0.2–1 nm, the STM tip must move laterally through these adsorbed layers, meaning that ions are pushed away laterally. This — in our opinion — leads to the "blurred" image quality we often observe with these liquids. As these interfacial layers vary with liquid and electrode potential a simplification and a prediction for other liquids is not possible. There might be liquids that are much less strongly adsorbed allowing higher quality images to be obtained. We also expect that even low water concentrations will have an influence on these layers and thus alter the STM image quality. Fig. 4b shows the same site at +0.2 V (ocp + 0.4 V). Apart from a slight thermal drift the surface looks quite similar but is slightly rougher. Zooming in (Fig. 4c) shows

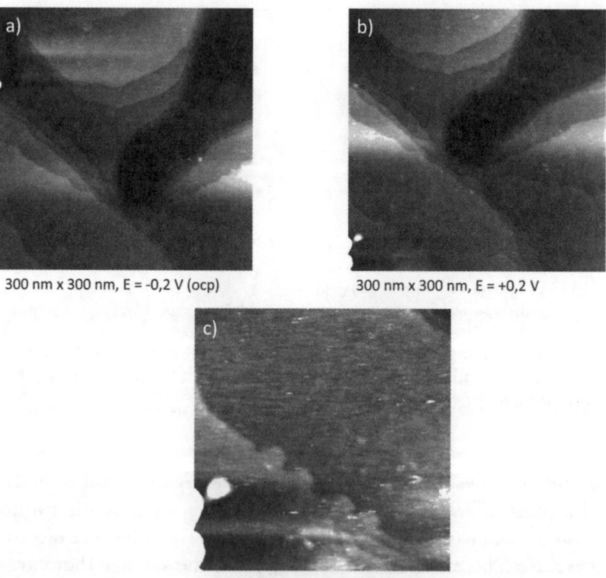


**Fig. 4** *In situ* STM images of Au(111)/$[Py_{1,4}]FAP$ at ocp and at slightly more positive electrode potentials (ocp + 0.4 V). There is a layer on the surface and islands evolve.

 

clearly a layer on the surface and the step between 2 terraces (arrow) looks "buried". In the case of aqueous solutions such image quality would strongly hints for organic contaminations which makes an atomic resolution experiment very difficult. Fig. 5a shows the same site with 300 nm × 300 nm resolution at +0.5 V (ocp + 0.7 V). In comparison to Fig. 4a there are now more islands visible and a 120 nm × 120 nm image reveals a layer on the surface with some structure. The measured height differences between these islands is between 0.2 and 0.5 nm, but for reasons discussed above, the error in these values is quite high. The surface structure does not vary appreciably between +0.5 and +1.2 V but Fig. 6a shows a layer with islands on top at 1.2 V. Because of the poor image contrast the steps are less clearly visible than at −0.2 V (Fig. 4a). The islands on the surface are sharper on the 120 nm × 120 nm scale, *cf.* Fig. 6b, but the steps between the terraces remain "buried". The surface is not uniform, but 0.3–0.6 nm high islands can be distinguished as indicated by the arrow. The AFM data in Fig. 3 show that at this electrode potential a slightly compressible anion-rich layer is present. The force to rupture this layer is 12 nN which is also fairly high.

Given that there is a cation-rich layer adsorbed at the ocp and an anion layer at +1 and +2 V there must be a transition between these states. We believe that Figs. 4–6 show this transition. At the first glance it is surprising that the AFM gives more or less discrete signals whereas the STM shows a structured surface. However, these results are not inconsistent as the maximum height differences observed (*e.g.* in Fig. 6b) are not more than 0.5 nm which is significantly less than the dimension of the ion pair. It is possible that the non-vertical force signals in Fig. 3 are due to the non-uniformity of the surface observed by the STM, in accordance with previous results.[18] Figs. 7a and 7b show the Au(111) surface at +1.6 and +1.7 V. Compared to +1.2 V the surface is more uniform and smoother, and the islands have almost completely disappeared. Steps are less clearly visible than at ocp which is a strong indication of an adsorbed layer, but can be seen more clearly when the image size is reduced to 120 nm × 120 nm, *cf.* Fig. 7c, where the measurable height contrast is less than 0.1 nm. This situation does not change when the electrode potential approaches +2 V where, however, the contrast is lost which hints at an electrochemical reaction. According to the AFM data there is a slightly compressible adsorbed layer of the anion at this potential on top of which 4 ionic liquid layers lie. As the maximum height differences observed with the STM were ≈ 0.5 nm, and as the STM tip is positioned between 0.5 and 1 nm above the surface (independently determined by current/distance tunneling spectroscopy), we believe that the STM probes this anion-rich layer. It should be noted that the IL layer on top of this anion-rich layer still requires a force of 4 nN to rupture. Together these layers are 1.5 nm thick which is a considerable distance for the STM tip to penetrate.

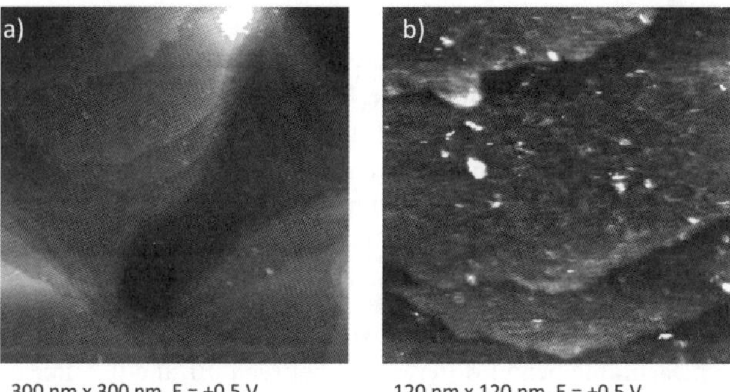


**Fig. 5** *In situ* STM images of Au(111)/[$Py_{1,4}$]FAP at ocp + 0.7 V. The evolving islands become more visible.

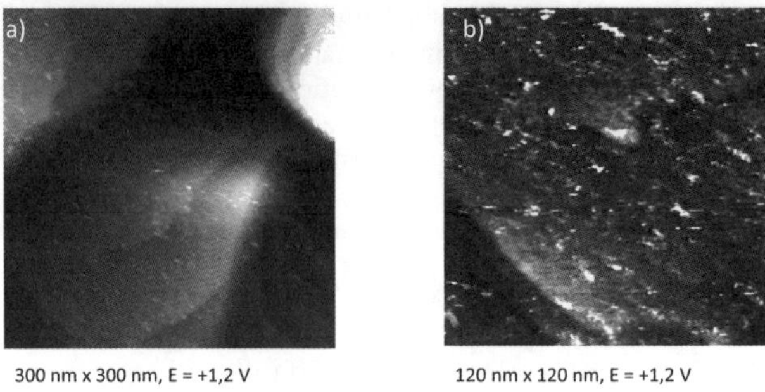


**Fig. 6** *In situ* STM images of Au(111)/[$Py_{1,4}$]FAP at ocp + 1.4 V. The number of islands increases and the probed layer is not uniform.

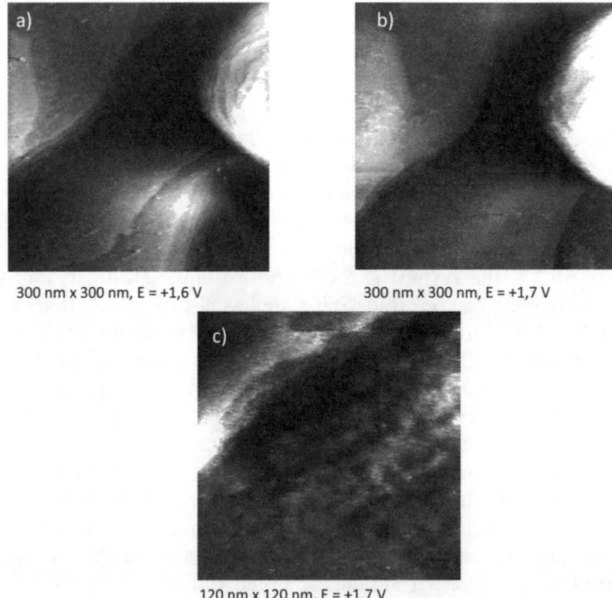


**Fig. 7** *In situ* STM images of Au(111)/[$Py_{1,4}$]FAP at ocp + 1.8 V (a) and at ocp + 1.9 V. The terrace steps look "buried" and the layer shows a slight contrast with height differences of ≈ 0.1 nm.

Adsorption of cations and anions to the STM tip is also expected, which also reduce the image quality and force resolution. Nevertheless we think that the STM probes the real surface, *i.e.* the anion and the IL layers.

Electrochemical reactions at this potential will require that the interfacial IL structure is disrupted to allow the reactant to make contact with the electrode surface. In the following we reduced the electrode potential from +1.8 V to +1.5 V (Fig. 8a) and subsequently from +1.5 V to +1.2 V (Fig. 8b) then we left the electrode potential at +1.2 V for 20 min on the 300 nm × 300 nm scale (Figs. 8c–d). There is still a layer on the surface but the steps become more visible, with some islands noted on the gold surface. The measurable height difference is between 0.1 and 0.2 nm, thus, an allocation to cation or anion is hardly possible. Figs. 9a and 9b show on the 200 nm × 200 nm scale that the surface very slowly changes. The gold surface is difficult to probe in high quality and according to the AFM data adsorption leads to the

 

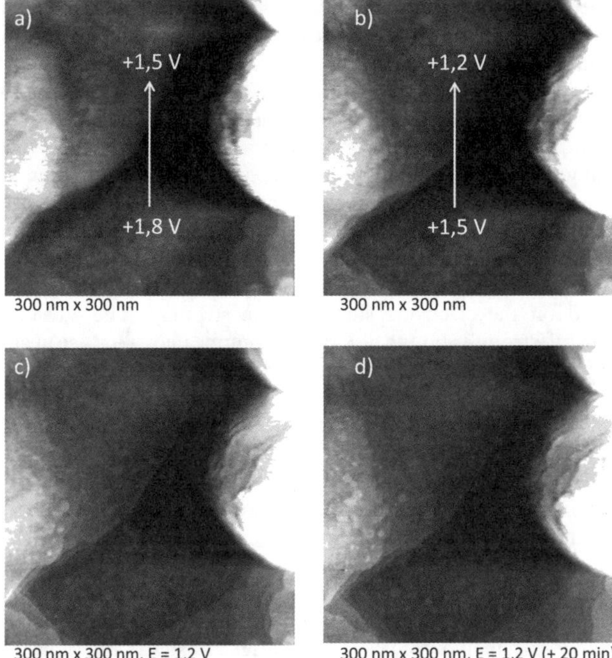


**Fig. 8** *In situ* STM images of Au(111)/[$Py_{1,4}$]FAP: The electrode potential has been set to less positive values and islands with a height difference of ≈ 0.2 nm appear on the surface. This is indicative of a reorientation and hints at cations being involved.

interfacial layer being enriched in the anion. We cannot allocate the islands to cation/anion or an IL layer, but it can be concluded that the surface is non-uniform. There is no contradiction with the AFM data, as the height variation of 0.1–0.2 nm is beyond the resolution of the AFM experiment. It is therefore possible that the slight compressibility measured in the AFM curve is due to surface inhomogenity. Fig. 10 shows the surface of Au(111) at the former ocp, and the individual terraces are again clearly visible on the 300 nm × 300 nm scale. There are some horizontal stripes in the STM image which disappear with time. We do not believe that these

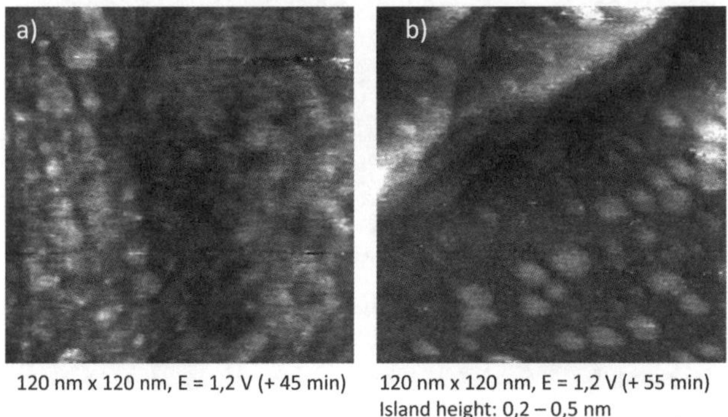


**Fig. 9** *In situ* STM images of Au(111)/[$Py_{1,4}$]FAP. The surface very slowly changes (within hours) and is not probed in high quality. The adsorbed layer might disturb the tunneling process.

stripes are due to a bad tip, rather we think that the very slow reorientation of the surface leads to these blurred images. The images presented here were reproduced in several experiments and the time elapsed between the images is ≈ 10 h.

These STM images show that the interface Au(111)/[$Py_{1,4}$]FAP is complex and more or less featureless at electrode potentials positive from the ocp. The high quality of the liquid, with all impurities present at below 10 ppm, and the recovering of the original gold surface after an anodic step up to +2 V make it unlikely that the STM experiment shows anything else other than the true potential dependent interaction of the IL with the gold surface. We must clearly say that the image quality shown in this study is far away from what is known in aqueous solutions or from other studies of our own in the cathodic regime, which is a consequence of the IL being adsorbed on the electrode surface quite strongly. The STM tip interacts with this layer of strongly bound ions: this complicated interface results in STM images of lower quality than might be expected in aqueous solutions. In our opinion, this will be the challenge with *in situ* STM experiments using ILs; detailed analysis of different liquids is necessary to shed more light on the potential and liquid dependent interface electrode(s)/ionic liquid(s).

In the last section of this paper we would like to present our first results on the influence of solutes on the interface structure Au(111)/[$Py_{1,4}$]FAP. In reference 23 we reported that the electrodeposition of tantalum in 1-butyl-1-methylpyrrolidinium bis(trifluoromethylsulfonyl)amide ([$Py_{1,4}$]TFSA) from $TaF_5$ is easier if LiF is added to the IL and an influence of LiF on the interface could not be excluded as an explanation. Stimulated by this experiment we added — for a first approach — 10 wt% of LiCl to the same extremely pure [$Py_{1,4}$]FAP. Fig. 11a shows the typical AFM force distance curve of [$Py_{1,4}$]FAP/Au(111) at the open circuit potential. The results are identical to the ones published before. However, if 10 wt% of LiCl is added to the liquid the situation is drastically changed at the open circuit potential, Fig. 11b. We still see 5 ion pair steps, each ~0.9 nm wide, but there is no clear repulsive

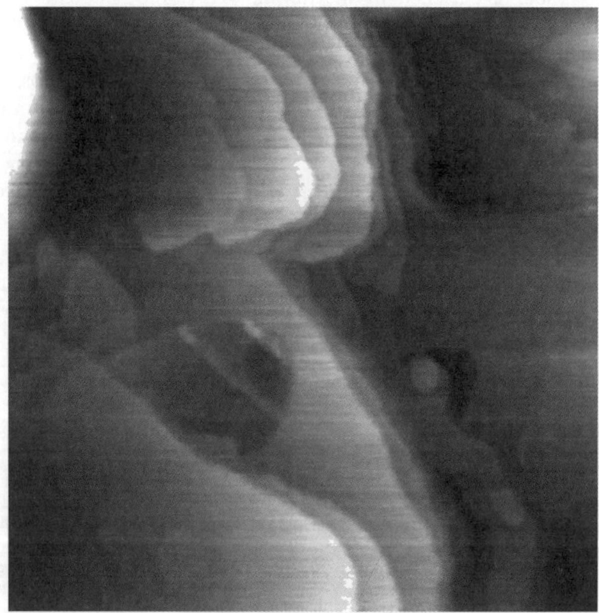


**Fig. 10** *In situ* STM image of Au(111)/[$Py_{1,4}$]FAP set back to the former ocp. The original gold surface is recovered.

 

step at the surface before the AFM tip touches the gold surface. Rather the force has become attractive, and a jump to contact occurs starting $\approx$ 2 nm from the surface, reminiscent of data obtained previously for the pure ethylammonium nitrate–graphite system[18] or when water was added.[24,25] The forces are attractive and there is jump into contact through the final layer. Fig. 11c is an overlay of both curves showing the considerably different behaviour. Although 10% of LiCl seems to be quite high at the first glance, such a concentration is close to that which would be feasible *e.g.* for an IL based battery electrolyte, and it has to be expected that solutes (and impurities) strongly alter the interface structure. A detailed picture of the interface processes in the presence of solutes will require experiments with different concentrations of solutes at different electrode potentials. However, these preliminary experiments suggest that the added inorganic ions are preferentially located immediately adjacent to the electrode surface, as the structure of the layer farther from the surface (*i.e.* beyond 2 nm) appears largely unaffected, with the number of steps and rupture force the same. The inorganic ions appear to weaken the near surface IL nanostructure, possibly by disrupting solvophobic interactions between cation alkyl chains,[26] such that the van der Waals attractions between the tip and surface dominate. This produces a jump to contact. An alternative explanation is that the level of structure remains constant, but that the Hamaker constant increases in the vicinity of the surface. In either case, the fact that the steps in the force curve are present through the jump to contact means that substantial near surface structure remains.

An *in situ* STM study of the Li underpotential deposition in $[Py_{1,4}]$TFSA containing 0.5 mol $L^{-1}$ LiTFSA showed, compared to the pure IL, a considerably different surface behavior on Au(111), and hints at a Solid Electrolyte Interface where the image quality in the STM experiment was altered.[27] The results presented here show without doubt that solutes can strongly alter the ionic liquid(s)/electrode(s)

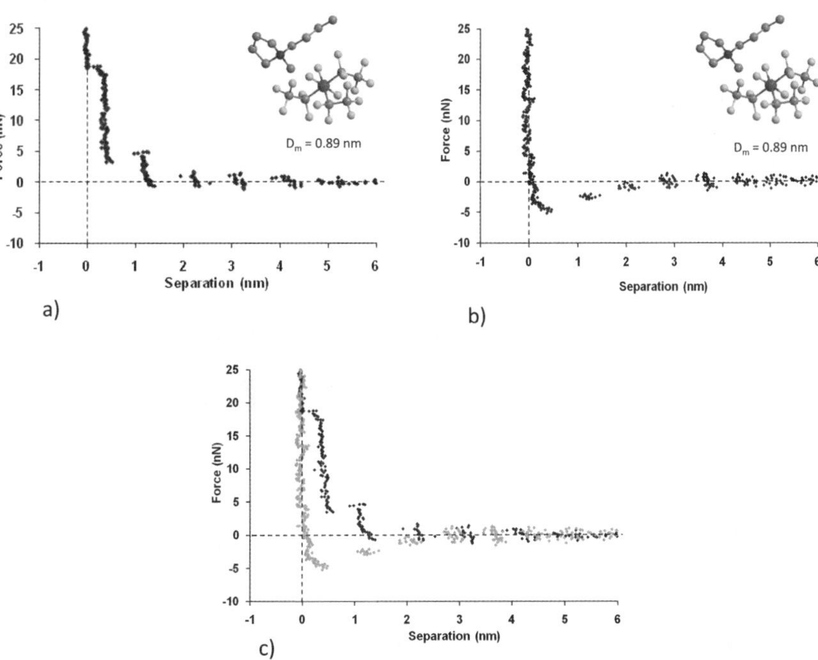


**Fig. 11** AFM force/distance profiles of Au(111)/$[Py_{1,4}]$FAP (a) and of Au(111)/$[Py_{1,4}]$FAP with 10 wt% of LiCl (b). The addition of LiCl changes the force profile from repulsive to attractive, thus the interface structure is strongly altered. c) shows both curves overlayed.

interface, and it has to be expected that these effects vary with liquid, solute, solute concentration, electrode, and temperature. In order to get a fundamental picture of electrochemical processes in ionic liquids a combination of electrochemical and *in situ* STM/AFM experiments will be required.

## Conclusions

In this paper we have presented our recent results on the interface Au(111)/[$Py_{1,4}$]FAP, probed with the *in situ* STM at electrode potentials more positive than open circuit potential. In contrast to the cathodic regime, where we were able to probe the "famous" herringbone superstructure of Au(111) in a limited potential regime with the STM, the image quality is much lower in the anodic regime, where AFM data suggest surface adsorption of the FAP anion. Nevertheless, it is clearly visible that a layer forms on top of the gold surface with maximum height differences of $\approx$ 0.6 nm. The combination of STM and AFM data reveals that the structure of this layer is not uniform and might be a mixture of cations and anions around +1 V. At +1.8 V a smooth surface is probed by the STM, and the steps between different terraces appear buried under an anion-rich layer. When the electrode potential is set back to the ocp the original gold surface is recovered, thus the adsorption/desorption of the ions seems to be reversible. The observed processes all occur very slowly over several hours.

We conclude that an *in situ* STM investigation of this liquid on Au(111) is not trivial and will require careful adjustment of tunneling parameters. Preliminary results on the addition of solutes (10 wt% of LiCl) to [$Py_{1,4}$]FAP reveal that the near surface structure at the open circuit potential is considerably altered. Instead of repulsive forces, attractive forces are measured, which must influence the STM experiment and more generally, it appears that dissolved solutes will substantially influence IL interface electrochemistry. At a minimum we can conclude that the situation for ionic liquids is much more complicated than the situation for aqueous solutions.


## Acknowledgements

This work was financially supported by the Deutsche Forschungsgemeinschaft (DFG) within the Priority Program SPP 1191 - Ionic Liquids and by an Australian Research Council Discovery Project (DP0986194). R.H. thanks the University of Newcastle for a PhD stipend.


## References


1 M. Armand, F. Endres, D. R. MacFarlane, H. Ohno and B. Scrosati, *Nat. Mater.*, 2009, **8**, 621.
2 M. V. Fedorov and A. A. Kornyshev, *J. Phys. Chem. B*, 2008, **112**, 11868.
3 M. V. Fedorov and A. A. Kornyshev, *Electrochim. Acta*, 2008, **53**, 6835.
4 S. A. Kislenko, I. S. Samoylov and R. H. Amirov, *Phys. Chem. Chem. Phys.*, 2009, **11**, 5584.
5 V. Lockett, R. Sedev, J. Ralston, M. Horne and T. Rodopoulos, *J. Phys. Chem. C*, 2008, **112**, 7486.
6 T. R. Gore, T. Bond, W. Zhang, R. W. J. Scott and I. J. Burgess, *Electrochem. Commun.*, 2010, **12**, 1340.
7 M. V. Fedorov, N. Georgi and A. A. Kornyshev, *Electrochem. Commun.*, 2010, **12**, 296.
8 M. Z. Bazant, B. D. Storey and A. A. Kornyshev, *Phys. Rev. Lett.*, 2011, **106**, article number 046102.
9 R. Atkin and G. G. Warr, *J. Phys. Chem. C*, 2007, **111**, 5162.
10 M. Mezger, H. Schröder, H. Reichert, S. Schramm, J. S. Okasinski, S. Schröder, V. Honkimäki, M. Deutsch, B. M. Ocko, J. Ralston, M. Rohwerder, M. Stratmann and H. Dosch, *Science*, 2008, **322**, 424.
11 R. Atkin, S. Zein El Abedin, R. Hayes, L. H. S. Gasparotto, N. Borisenko and F. Endres, *J. Phys. Chem. C*, 2009, **113**, 13266.


 

12 S. J. O'Shea, M. E. Welland and J. B. Pethica, *Chem. Phys. Lett.*, 1994, **223**, 336.
13 R. Lim and S. J. O'Shea, *Langmuir*, 2004, **20**, 4916.
14 S. Jeffery, P. M. Hoffmann, J. B. Pethica, C. Ramanujan, H. O. Ozer and A. Oral, *Phys. Rev. B: Condens. Matter Mater. Phys.*, 2004, **70**, 054114.
15 F. Endres, O. Höfft, N. Borisenko, L. H. S. Gasparotto, A. Prowald, R. Al-Salman, T. Carstens, R. Atkin, A. Bund and S. Zein El Abedin, *Phys. Chem. Chem. Phys.*, 2010, **12**, 1724.
16 R. Atkin, N. Borisenko, M. Drüschler, S. Zein El Abedin, F. Endres, R. Hayes, B. Huber and B. Roling, *Phys. Chem. Chem. Phys.*, 2011, **13**, 6849, DOI: 10.1039/c0cp02846k.
17 F. Endres, S. Zein El Abedin and N. Borisenko, *Z. Phys. Chem.*, 2006, **220**, 1377.
18 R. Hayes, G. G. Warr and R. Atkin, *Phys. Chem. Chem. Phys.*, 2010, **12**, 1709.
19 D. Wakeham, R. Hayes, G. G. Warr and R. Atkin, *J. Phys. Chem. B*, 2009, **113**, 5961.
20 R. Hayes, S. Zein El Abedin and R. Atkin, *J. Phys. Chem. B*, 2009, **113**, 7049.
21 B. R. Clare, P. M. Baylay, A. S. Best, M. Forsyth and D. R. MacFarlane, *Chem. Commun.*, 2008, 2689.
22 R. Hayes, N. Borissenko, M. K. Tam, P. C. Howlett, F. Endres and R. Atkin, J. Phys. Chem. C , DOI: 10.1021/jp200544b.
23 S. Zein El Abedin, H. K. Farag, E. M. Moustafa, U. Welz-Biermann and F. Endres, *Phys. Chem. Chem. Phys.*, 2005, **7**, 2333.
24 J. Smith, O. Werzer, G. B. Webber, G. G. Warr and R. Atkin, *J. Phys. Chem. Lett.*, 2010, 1.
25 R. G. Horn, D. F. Evans and B. W. Ninham, *J. Phys. Chem.*, 1998, **92**, 3531.
26 R. Hayes, S. Imberti, G. G. Warr and R. Atkin, *Phys. Chem. Chem. Phys.*, 2011, **13**, 3237.
27 L. H. S. Gasparotto, N. Borisenko, N. Bocchi, S. Zein El Abedin and F. Endres, *Phys. Chem. Chem. Phys.*, 2009, **11**, 11140.

# Molecular-scale insights into the mechanisms of ionic liquids interactions with carbon nanotubes†

**Andrey I. Frolov, Kathleen Kirchner, Tom Kirchner and Maxim V. Fedorov***

*Received 22nd April 2011, Accepted 5th July 2011*
**DOI: 10.1039/c1fd00080b**

By means of fully atomistic molecular simulations we study basic mechanisms of carbon nanotube interactions with several different room temperature ionic liquids (RTILs) in their mixtures with acetonitrile. To understand the effects of the cation molecular geometry on the properties of the interface structure in the RTIL systems, we investigate a set of three RTILs with the same TFSI (bis (trifluoromethylsulfonyl)imide) anion but with different cations, namely, EMIm (1-ethyl-3-methylimidazolium), BMIm (1-butyl-3-methylimidazolium) and OMIm (1-octyl-3-methylimidazolium) ions. The cations have identical charged methylimidazolium 'heads' but different nonpolar alkyl 'tails' where the length of the tail increases from EMIm to OMIm. The analysis of the simulation data results in the following conclusions:

• There is an enrichment of all molecular components of ionic liquids under study at the CNT surface with formation of several distinct layers even at the non-charged CNT surface.
• Mixing RTIL with acetonitrile decreases ion-counterion correlations in the electric double layer.
• Increase of the length of the non-polar 'tail' of cations increases the propensity of imidazolium-based cations to lay parallel to the CNT surface.
• At the CNT cathode TFSI anions and molecular cations are preferentially oriented parallel to the surface.
• At the CNT anode the TFSI anions are oriented parallel to the surface, however the preferred orientations of cations depend on the length of non-polar tail: EMIm cations are oriented perpendicular to the surface, BMIm cations can be in both parallel as well as perpendicular orientations, OMIm cations are oriented parallel to the surface.

As a result, by applying an electric potential on the CNT electrode and/or varying the structure of molecular ions it is possible to change molecular ion orientations at the surface and, consequently, the structure of the electrical double layer at the CNT-RTIL interface.

## 1 Introduction

Interest in carbon nanotubes (CNTs) dispersed in room temperature ionic liquids (RTILs) is rapidly growing.[1–4] One of the main reasons is the extraordinary electrochemical and bioelectrochemical properties of RTILs[5–9] as well as of carbon nanotube composite materials.[1,2,10] Such, carbon nanotubes (nanotube forests) with

*Max Planck Institute for Mathematics in the Sciences, D 04103 Leipzig, Germany. E-mail: fedorov@mis.mpg.de; Tel: +49 341 9959 804*

† Electronic supplementary information (ESI) available. See DOI: 10.1039/c1fd00080b

ionic liquids have promising supercapacitor applications.[2,3,11] We also note that in many emerging electrochemical applications, RTILs are actively used in mixtures with organic solvents (*e.g.* acetonitrile) to avoid problems related with the high viscosity of neat RTILs.[2,12,13]

It has been shown in several experimental[7,14–23] and theoretical[24–36] studies that the molecular structure of RTIL ions strongly influences the RTIL interface properties at different charged and uncharged interfaces. For instance, Lockett *et al.*[15] showed that asymmetry in the size and shape of molecular ions results in unequal distribution of molecular cations and anions in direction normal to the RTIL-vacuum interface. These results suggest that the molecular structure of RTIL ions should make significant effects at the CNT-RTIL interface. However, still, there is a lack of molecular level information on the mechanisms of the RTIL interactions with nanocarbon electrodes. Even less is known about the molecular effects of organic solvent on the interface properties of RTIL-solvent mixtures at the nanocarbon surfaces. The basic mechanisms of the electrical double layer formation in such systems (particularly, in RTIL-acetonitrile mixtures) are also not sufficiently explored.

Due to the recent progress in molecular-level experimental techniques very important information has been obtained on the molecular orientations of RTIL molecules and their layering structure at neutral and charged interfaces.[14–19,21–23,37–39]

Endres *et al.*[19] quoted an "undoubted" formation of at least three solvation layers of EMIm-TFSI on metal electrodes detected by atomic force microscopy (AFM) (note, that they name EMIm-TFSI as 1-ethyl-3-methylimidazolium bis[(trifluoromethylsulfonyl]amide, [EMIm]TFSA)). Recently, Hayes *et al.*[21] investigated the influence of the electric potential on the interface solvation layers in [EMIm]FAP (1-ethyl-3-methylimidazolium tris(pentafluoroethyl)trifluorophosphate) and [Py]FAP (1-butyl-1-methylpyrrolidinium tris(pentafluoroethyl)trifluorophosphate) by AFM. Applying an electric potential, they found that the innermost layer changes its structure and becomes more strongly bound to the surface. At the cathode, for "the first time an interfacial (innermost) anionic layer at a solid interface has been detected by AFM". We will discuss these results in Section 3.5 in more detail.

However, still many details of the interface structure of RTILs and their mixtures are not easily accessible by direct experiments. Molecular simulations can provide complimentary information to the experimental data that should help to obtain a detailed picture of the interface phenomena in RTIL systems (see more discussion on this subject in ref. 2). Therefore, we decided to use fully atomistic simulation methods to study basic mechanisms of the RTIL interactions with the CNT surface.

It has been shown that RTILs based on combination of imidazolium-based cations with hydrophobic anions (*e.g.* $BF_4$ or TFSI (bis(trifluoromethylsulfonyl)imide) are moisture stable and have very promising electrochemical applications.[8] Among those, the TFSI-based RTILs have good stability (such that, in ref. 8 it has been discussed that the $BF_4$ anion is not stable against carbon electrodes) and have a large electrochemical window.[40,41] Ion conductivity of EMIm-TFSI is comparable to the best of organic electrolyte solutions,[1] and this liquid is stable (no decomposition) up to 300–400 °C. The TFSI-based ionic liquids are practically not miscible with water but they are well miscible with several organic solvents, *e.g.*, acetonitrile.[42]

In this report we study basic mechanisms of CNT interactions with several different TFSI-based RTILs in their mixtures with acetonitrile (AN). To understand the effects of the cation molecular geometry on the properties of the interface structure in the RTIL systems, we investigate a set of three RTILs with the same anion (TFSI) but with different cations, namely, EMIm (1-ethyl-3-methylimidazolium), BMIm (1-butyl-3-methylimidazolium) and OMIm (1-octyl-3-methylimidazolium). The cations have identical charged methylimidazolium 'heads' but different nonpolar alkyl 'tails' where the length of the tail increases from EMIm to OMIm.

In the present study we would like to focus on a set of the following questions:

• What is the interfacial structure of RTIL-AN mixture at the *neutral* CNT surface?

 

• How does the interfacial structure change at the *positively* charged CNT surface?

• How does the interfacial structure change at the *negatively* charged CNT surface?

• Does the length of the cation alkyl tail affect the interfacial RTIL-AN structure and preferred orientation of the RTIL ions at the CNT surface?

• What is the role of acetonitrile solvent in these interfacial effects?

We note, that in this paper we would like to focus on the interface effects at the *outer* surface of CNT. In this study we use a CNT with (6,6) chirality which has a narrow pore. Taking into account the molecular volume of the investigated ions (see Table 2) we assume that the probability of finding an ion inside the CNT pore is low and we can neglect these effects in this particular case. Investigation of molecular mechanisms of RTIL interactions with the *internal* surface of subnanometer CNT pores is the subject of our ongoing research and will be considered elsewhere.

The paper is organized as follows. In the Simulation methods section we describe our simulation's methods. The Results and discussion section contain five parts, (i) Discussion of the simulation results for the neutral CNT surface; (ii) Structural reorganizations in the interfacial layer at the charged CNT surface; (iii) Effects of acetonitrile on the interfacial structures; (iv) Effects of the length of the cation alkyl chain on the structure of the double layer; (v) Correlations with the experimental data. In the Conclusions section we briefly summarize our results.

## 2 Simulation methods

To reveal basic molecular mechanisms of ion interactions with carbon nanotube (CNT) surfaces we performed Molecular Dynamics (MD) simulations of CNT with (6,6) chirality dissolved in several mixtures of RTILs with AN, namely, EMIm-TFSI, BMIm-TFSI and OMIm-TFSI RTILs (see Fig. 1). The RTIL molar concentrations in all RTIL-AN mixtures was 2M. To understand the role of AN solvent on the studied interfacial effects we performed an additional simulation of

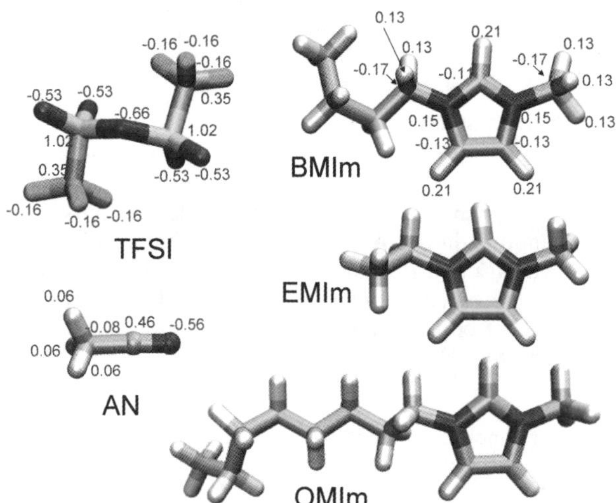


**Fig. 1** Sketch representation of the molecular species considered in the study. Cyan color – carbon atoms, white color – hydrogen atoms, blue color – nitrogen atoms, green color – fluorine atoms, yellow color – sulfur atoms, red color – oxygen atoms. Numbers at the atoms show the corresponding atomic partial charges in [e] units: the numbers highlighted in the blue color show positive partial charges and the numbers highlighted in the red color show negative partial charges (only some partial charges are represented).

the CNT dissolved in neat EMIm-TFSI. In our simulations we employed the fully atomistic OPLS-AA force field,[43,44] with partial charges and dihedral angles potential parameters developed by Lopes *et al.*[45,46] We used the Particle-Mesh Ewald method to evaluate the electrostatic interactions.[47] We used the Gromacs 4.5 software.[48]

We used the TubeGen program to generate coordinates of the CNT atoms.[49] A segment of CNT with length of 3.94 nm was placed in a rectangular box and was oriented along the $z$ axis. Then, the RTIL molecular ions and AN molecules were randomly placed inside the simulation boxes with the help of Packmol program.[50] We provide the resulting number of molecular species in the simulation boxes in Table 1 (the numbers are shown for the zero CNT surface charge). All the systems were investigated at different CNT surface charge densities: $\sigma = -0.5$, 0.0, +0.5 [e nm$^{-2}$]. Similar to ref. 31 in the simulations of the charged CNT surface the non-zero charge on the carbon nanotube was neutralized by addition of an extra number of cations (negatively charged surface) or anions (positively charged surface): in our case we used 5 extra ions for systems with $\sigma = -0.5$, +0.5 [e nm$^{-2}$].

The generated initial molecular configurations were optimized by the energy minimization algorithm implemented in Gromacs.[48] Then the systems were equilibrated in the NPT-ensemble at $T = 343.15$ K and $P = 1$ bar until the density becomes constant (it took at least 0.3 ns of the simulation time). After the NPT simulations we fixed the geometry of the simulation boxes and heated the systems up to 1000–1500 K and then annealed the final configurations over 2 ns simulation time with a gradual decrease of temperature from 1000–1500 K to 343.15 K. Starting from the resulting configurations we then performed production 30 ns simulations for each system in the NVT ensemble at 343.15 K to collect statistics. During the production runs we stored the atomic coordinates of the systems every 0.3 ps each for further analysis.

To facilitate the analysis of the excluded volume effects on the formation of the interfacial structures we calculated the volumes of cavities of molecular ions and AN molecules dissolved in AN with the help of Gaussian03 software.[51] The geometries of the species were optimized at the B3LYP/6-31g(d,p) level of theory. In the quantum mechanics calculations we used the Self Consistent Isodensity Polarizable Continuum Model (SCI-PCM)[52] to model the acetonitrile solvent.

See the Supporting Information for further details.†

## 3 Results and discussion

### 3.1 Neutral CNT

Firstly, we analyzed density distributions of the molecular ions around the non-charged carbon nanotube in 2M AN EMIm-TFSI solution. We present the radial density profiles (RDPs) of EMIm and TFSI molecular ions on Fig. 2. All RDPs were calculated for the centers of mass of the molecules. The pronounced peaks on the RDPs show that both EMIm and TFSI ions form distinct solvation shells around the CNT (see Fig. 2).

**Table 1** Numbers of the molecular species in the simulation boxes of the studied systems. The numbers are shown for the simulations with the neutral CNT surface.

| | $N$ (RTIL ion pairs) | $N$ (AN) |
|---|---|---|
| 2M AN EMIm-TFSI | 200 | 600 |
| Neat EMIm-TFSI | 340 | — |
| 2M AN BMIm-TFSI | 200 | 480 |
| 2M AN OMIm-TFSI | 200 | 250 |

 

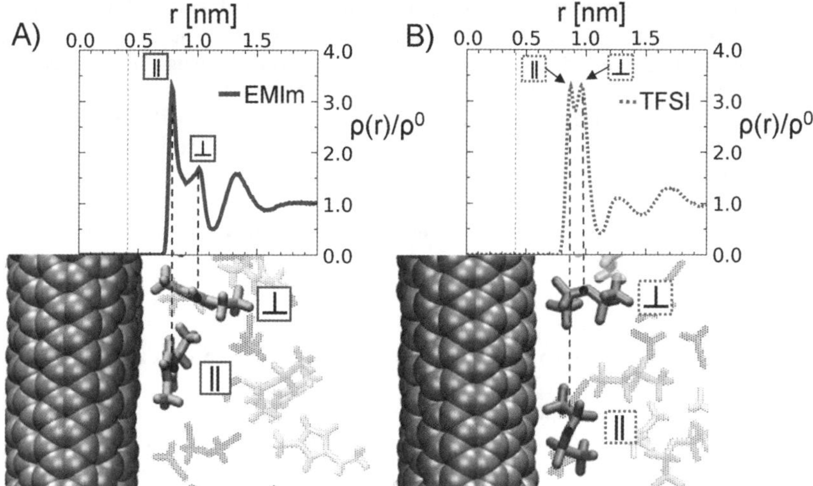


**Fig. 2** Radial density profiles of EMIm and TFSI molecular ion centers of mass around CNT in 2M AN EMIm-TFSI solution at the neutral CNT surface. The asymmetry of ion geometry (shape, size *etc.*, see Fig. 1) results in a complicated molecular structure of the electrolyte at the CNT surface. Molecular snapshots represent orientations of the molecules consistent with the peaks on the corresponding RDPs. For simplicity, all atoms of EMIm ions are represented by the cyan color, TFSI ions are represented by the magenta color and AN molecules are represented by the black color.

EMIm cations tend to lay parallel to the surface: there is a strong peak on the EMIm RDP around CNT at $r = 0.75$ nm (marked as $\|$). We consider the vector connecting the $\alpha$-carbons at the imidazolium ring as a "molecular vector" of the imidazolium-based cations and the angle between this "molecular vector" and the CNT surface determines orientation of molecular cations at the surface (see also Fig. S2 of the Supporting Information†). The second peak (at $r = 1.0$ nm) corresponds to a perpendicular to the surface orientation of the EMIm cations ($\perp$). However, the density peak for cations with perpendicular orientation is less pronounced compared to the $\|$ orientation. Overall, there is an enrichment of EMIm cations in the region from $r = 0.75$ to 1.0 nm from the CNT axis (the relative density here is about 1.5 the bulk density). This region corresponds to all intermediate orientations in between two characteristic ones: $\|$ and $\perp$. The hollow at about 1.1 nm indicates the boundary of the first EMIm solvation shell around the CNT. We note, that the EMIm cations also form at least one more solvation shell around the CNT as indicated by the distinct peak at $r = 1.3$ nm.

The first TFSI solvation shell around CNT is narrower compared to the EMIm solvation shell (Fig. 2B). TFSI ions may be thought of as dumbbell particles where each bead consists of a $CF_3$-$SO_2$ group. We consider the line connecting the centers of sulphur atoms as the "molecular vector" of the TFSI anions. TFSI ions (dumbbells) orient in 2 different ways around CNT: parallel ($\|$) and perpendicular to CNT surface ($\perp$). This is indicated by the 'fine structure' of the first large peak on the TFSI RDP that has two small peaks (see Fig. 2B). However, we note that due to the internal flexibility of TFSI anions, the "parallel" to the surface orientation of TFSI dumbbells includes a variety of slightly non-parallel to the surface orientations of the chosen "molecular vector" of TFSI (sulphur-sulphur vector) (see, for instance, the molecular picture of the $\|$ orientation on Fig. 2B: the $CF_3$–$SO_2$ groups are twisted one with respect another). In this study for simplicity we consider all these orientations as "parallel" to the surface orientations of TFSI ions (see Parts 4 and 5 of the Supporting Information for more details†).

The parallel (||) and perpendicular (⊥) orientations of TFSI are almost equally probable (the heights of the small peaks are similar, Fig. 2B). The second solvation shell is not as pronounced as for the EMIm distribution (the height of the peak at $r$ = 1.3 nm is around 1.1). However, there is an evident third solvation shell of TFSI around CNT (peak at $r$ = 1.7 nm), presumably due to the attraction of TFSI molecular ions to the second solvation shell of EMIm around CNT (at $r$ = 1.3 nm, see Fig. 2A).

### 3.2 Charging the CNT: changes in the interfacial structures in response to the external field

In this section we describe the changes in the double layer in 2M AN EMIm-TFSI electrolyte upon charging the CNT surface. The RDPs for different surface charge densities on CNT ($\sigma$ = −0.5, 0.0, +0.5 [e/nm$^{-2}$]) are presented on Fig. 3, first row.

**$\sigma$ = −0.5 [e nm$^{-2}$].** As expected, negatively charged CNT strongly attracts EMIm cations: there is a high peak on the corresponding RDP (see Fig. 3). At the negatively charged surface the EMIm cations orient parallel to the surface (|| EMIm orientation). The perpendicular to the surface orientation (⊥ EMIm) becomes unfavorable. The second shell of EMIm ions around CNT (at $r$ = 1.35 nm) does not change much compared to the system with neutral CNT.

The distributions of TFSI anions at the negatively charged CNT surface change considerably compared to the neutral CNT. The CNT-TFSI RDP shows only one peak at $r$ = 1.0 nm compared to the two peaks at $r$ = 0.9 and 1.0 nm in the case of neutral CNT. This peak corresponds to a layer of TFSI anions with preferred parallel orientation to the first dense layer of EMIm cations (see Supporting Information for more details†).

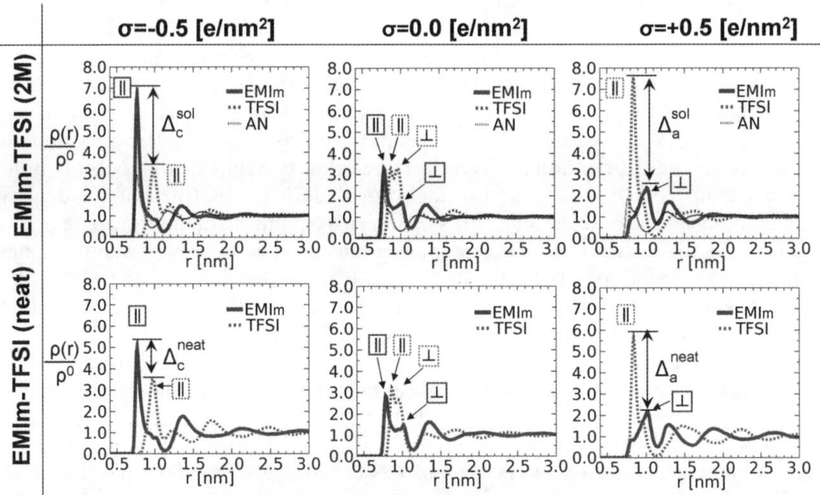


**Fig. 3** First row: radial density profiles of EMIm, TFSI and AN around CNT in 2M AN EMIm-TFSI solution for three surface charge densities on CNT: $\sigma$ = −0.5, 0.0, +0.5 [e nm$^{-2}$]. There is an asymmetry of molecular ions response to the potential created by charged CNT. The potential causes structural changes in the interfacial region. Second row: radial density profiles of EMIm, TFSI around CNT in neat EMIm-TFSI solution for the three surface charge densities on CNT. We note, that the bulk concentration of neat EMIm-TFSI is about 1.7 times higher than the concentration of EMIm-TFSI in the bulk 2M AN EMIm-TFSI solution. Due to this, to compare both rows in terms of absolute values, the graphs in the second row have to be multiplied by a factor 1.7. (See Figure S7 of the supporting information†).

 

We note that acetonitrile is not substituted by the cations adsorbed to the negatively charge surface. There is even a little enhancement of AN concentration compared to the system with neutral CNT (see Fig. 4). This can be explained by the strong interactions of AN (having strong dipole moment) with the charged surface. Overall, there is an enrichment of the density of the RTIL-AN solvent at the negatively charged surface.

The polarization of the electrolyte spreads up to 2–2.5 nm from the CNT axis (about 2–3 solvation shells of CNT). The range of polarization remains almost the same for all the three surface charge densities.

**$\sigma$ = +0.5 [e nm$^{-2}$].** The positively charged CNT surface attracts TFSI anions, which prefer to lay parallel to the surface (|| TFSI orientation). In turn, the EMIm ions become preferentially oriented perpendicular to the surface ($\perp$ EMIm orientation). For cations, orientation parallel to the surface (|| EMIm orientation) becomes unfavorable (the height of the first peak on the RDP decreased more than 3.5 times).

We note, that the number of AN molecules at the anode surface drops down compared to the system with neutral CNT (see Fig. 4). This is presumably due to the excluded volume effects because the anode interfacial region becomes occupied by the bulky TFSI anions. Indeed, the TFSI anions are about 1.4 times larger than EMIm cations and about 2.9 times larger than AN molecules (see Table 2). Thus, accommodation of large number of TFSI anions at the surface results in transfer of considerable amount of AN molecules from the CNT interface to the bulk solution.

### 3.3 Effects of acetonitrile on the electric double layer

In this section we compare the structures of the double layer in 2M AN EMIm-TFSI electrolyte and neat EMIm-TFSI ionic liquid to see whether the acetonitrile solvent makes an effect on the interfacial structures at the neutral and charged CNT surfaces. The RDPs of the molecular species for neat EMIm-TFSI at different surface charge densities on the CNT surface ($\sigma$ = −0.5, 0.0, +0.5 [e nm$^{-2}$]) are presented in Fig. 3, second row.

In neat EMIm-TFSI ionic liquid the polarization of the electrolyte spreads deep into the bulk of the neat RTIL for more than 2.5 nm. There are at least two to three solvation layers at the electrodes.

**$\sigma$ = 0.0 [e nm$^{-2}$].** The structures of the interfacial regions at the neutral CNT for the neat RTIL and its mixture with AN are almost identical: addition of AN has minor effect on the RDPs of molecular ions (see Fig. 3, second column). This

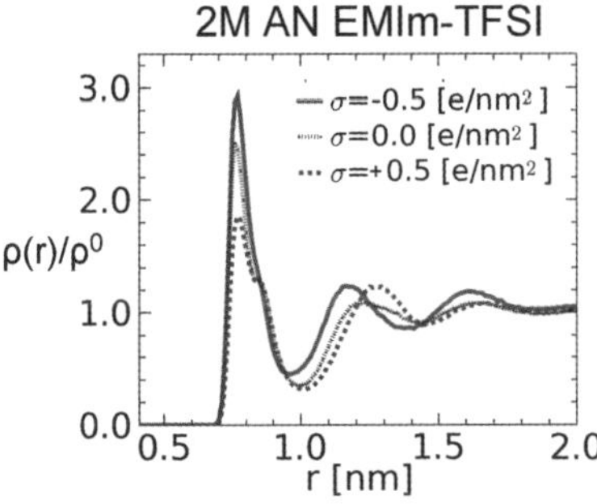


**Fig. 4** Radial density profiles of AN around CNT in 2M AN EMIm-TFSI electrolyte at different surface charge densities on the CNT.

**Table 2** Volumes of solute cavities in acetonitrile.

| | $V$/Å$^3$ |
|---|---|
| AN | 77.12 |
| TFSI | 224.39 |
| EMIm | 157.66 |
| BMIm | 203.03 |
| OMIm | 323.18 |

indicates that the interface structure is determined mainly by the CNT-ion interactions and ion-ion interactions.

**$\sigma = -0.5$ [e nm$^{-2}$].** At the negatively charged CNT the number of parallel oriented EMIm ions (|| EMIm orientation) and TFSI ions (|| TFSI orientation) increased, whereas only few ions oriented perpendicular to the surface are left (see the Supporting Information for more details†). The peak which corresponds to the || TFSI orientation in the first layer is pushed away from the surface by 0.1 nm. Also, in contrast to the solution in AN, the amplitude of the second and subsequent layers is increased, indicating that the surface charge enhances the layering structure of the interfacial ionic liquid. Addition of AN decreases the relative number of TFSI ions near the dense layer of EMIm cations attached to the surface (see the change in the peak height differences ($\Delta_c^{sol}$ and $\Delta_c^{neat}$) on Fig. 3, first column). However, we note, that there are no changes in the peak positions in the neat RTIL and 2M RTIL-AN RDPs.

**$\sigma = +0.5$ [e nm$^{-2}$].** The positively charged CNT surface has similar effect on the neat EMIm-TFSI as on its mixture with AN. Parallel orientation for EMIm ions (|| EMIm orientation) becomes unfavorable. But unlike in the AN solution, again the second and subsequent solvation layer peaks ($r > 1.2$ nm) increase in their amplitude, indicating the strengthening of the layering pattern. AN has a similar effect on the double layer as at the negatively charged CNT. AN increases the difference between the relative number of the TFSI anions at the surface and the number of the EMIm cations attracted to the first layer of the TFSI anions (see the change in the peak height differences ($\Delta_a^{sol}$ and $\Delta_a^{neat}$) on Fig. 3, first column). Similar to the negatively charged surface, AN does not change the structural organization of the double layer (the peak positions remain the same in the neat RTIL as well as in the RTIL-AN mixture).

That observation suggests that AN affects the interionic correlations making them less strong (presumably due to the high polarity of AN), but does not lead to the structural reorganization of the double layer.

### 3.4 Effects of the length of the cation alkyl chain on the structure of the electrical double layer

To understand the role of the cation molecular geometry on the interfacial structure and on the molecular orientation at the CNT surface we analyzed the simulation results for AN mixtures with RTILs varying the length of the cation alkyl chains. Fig. 5 shows the RDPs of molecular species at the charged and neutral CNT surfaces for EMIm-TFSI, BMIm-TFSI and OMIm-TFSI mixtures with AN. We also analyzed the preferred orientation of the molecules in the interfacial layers that reveal themselves in the RDP peaks (see more details in the Supporting Information†). The preferred orientations assigned to the peaks are shown by the || and $\perp$ symbols on Fig. 5.

**$\sigma = -0.5$ [e nm$^{-2}$].** The length of the cation alkyl chain does not make a significant effect on the orientation of the RTIL molecules at the CNT cathode. The RDPs and

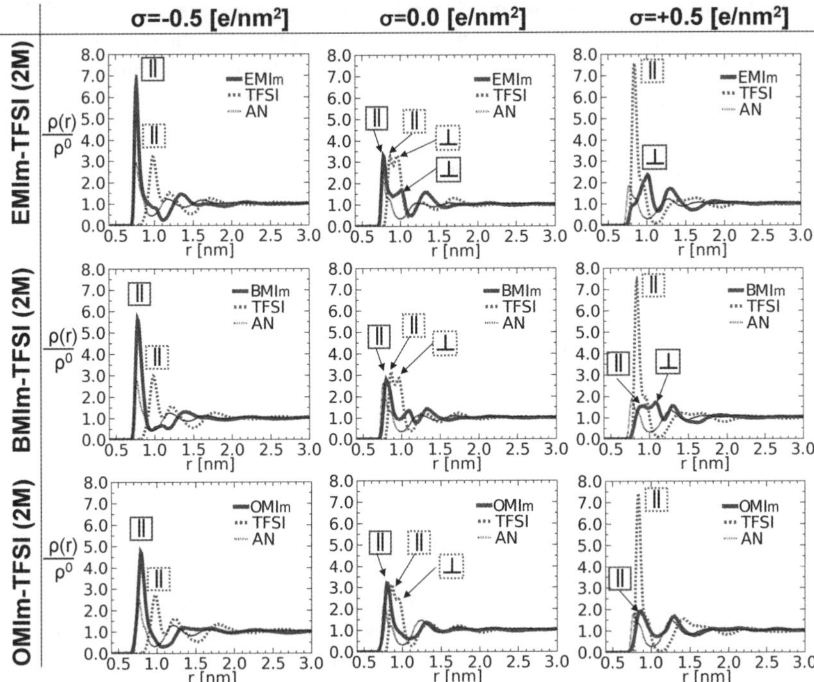


**Fig. 5** First row: radial density profiles of EMIm, TFSI and AN around CNT in 2M AN EMIm-TFSI solution for the three surface charge densities on CNT: $\sigma = -0.5, 0.0, +0.5$ [e nm$^{-2}$]. Second row: radial density profiles of BMIm, TFSI and AN around CNT in 2M AN BMIm-TFSI solution for the three surface charge densities on CNT: $\sigma = -0.5, 0.0, +0.5$ [e nm$^{-2}$] Third row: radial density profiles of OMIm, TFSI and AN around CNT in 2M AN OMIm-TFSI solution for the three surface charge densities on CNT: $\sigma = -0.5, 0.0, +0.5$ [e nm$^{-2}$]. The length of the cation alkyl chain does not make significant effects on the orientation of the RTIL molecular ions at the cathode (negatively charge electrode). However, it affects the structure of the interfacial layer at the neutral interface: the tendency for cations to lay parallel to the CNT surface increases with increase of the length of the 'tail'. In addition, the length of the 'tail' affects the orientation of the cations in the double layer at the anode (positively charge electrode). For the EMIm system with the shortest 'tail', the EMIm cations form a second layer after the first dense layer of TFSI anions and they are preferentially oriented perpendicular to the surface. However, with an increase of the 'tail' length the concentration of the cations in the first layer increases and they tend to be oriented parallel to the surface.

orientations of RTIL ions do not change *qualitatively* for different cations. However, the height of the first peak on the CNT-cation RDP decreases with lengthening of the alkyl chain. We attribute this to the corresponding increase of the excluded volume of the cations (see Table 2) (the larger the molecular volume the more difficult it is to 'pack' the ions to the dense first layer).

**$\sigma = 0.0$ [e nm$^{-2}$].** The length of the cation alkyl chain considerably affects the structure of the interfacial layer at the neutral interface: the tendency for cations to lay parallel to the CNT surface increases with an increase of the length of the 'tail'.

**$\sigma = +0.5$ [e nm$^{-2}$].** TFSI anions are strongly bound to the positively charged surface for all the studied electrolytes. The preferred orientation of TFSI anions for all electrolytes remains parallel to the surface due to the strong electrostatic attraction between these anions and the anode surface. However, similar to the neutral surface, the preferred orientation of cations at the anode surface depends on the length of the cation non-polar 'tail': the longer the 'tail' the more the

RTIL cations prefer to be oriented parallel to the surface. For instance, EMIm cations (the shortest 'tails') are preferentially oriented perpendicular to the CNT surface, BMIm cations have both perpendicular and parallel orientations, OMIm cations (the longest 'tail') tend to be oriented mostly parallel to the surface.

### 3.5 Correlations with the experimental data

Atkin *et al.*[17] published an atomic force microscopy (AFM) study of the gold interface solvated in EMIm-TFSI (note, that they name this as 1-ethyl-3-methylimidazolium bis[(trifluoromethylsulfonyl]amide, [EMIm]TFSA). They reported several steps in the force curve,[17] which coincide well with the observed layered structure in MD simulations of neat EMIm-TFSI at the carbon nanotube surface in this study (see Fig. 3). There is no pronounced third solvation layer of CNT in our study, as it is indicated by the AFM at the gold surface. We attribute the "weaker" layering pattern in our simulations to the differences in temperatures: temperature in the simulations was about 70 °C, while "all force curves were acquired continuously at room temperature (22 °C)" for the AFM measurements.[17]

There are some experimental studies on the orientation of IL molecules on the liquid–vacuum interface, see for example refs. 15, 53 and 54. For the liquid–solid interface these data are difficult to obtain. Nakajima *et al.*[53] investigated the liquid–vacuum interface of different 1-alkyl-3-methylimidazolium–TFSI ionic liquids using high-resolution Rutherford backscattering spectroscopy. They showed that, due to their solvophobic nature, long alkyl chains of cations point away from the bulk liquid to vacuum and therefore stimulate the imidazolium ring to stay perpendicular to the surface. In our simulations on the RTIL–carbon interface we observe an opposite effect: increase of alkyl chains increases the tendency for the imidazolium ring to lay parallel on the surface. We attribute the differences between RTIL–vacuum and RTIL–carbon nanotube interfaces to the strong Van der Waals attraction between the non-polar alkyl chains and carbon nanotube surface. Contrary to the RTIL–vacuum interface, at the RTIL–CNT interface the alkyl chains of imidazolium-based cations tend to lay parallel on the CNT surface and force the imidazolium rings also to lay flat on the carbon nanotube surface.

Hayes *et al.*[21] investigated the structure of different ionic liquids (1-ethyl-3-methylimidazolium tris(pentafluoroethyl)trifluorophosphate and 1-butyl-1-methylpyrrolidinium (Py) tris(pentafluoroethyl)trifluorophosphate at the charged Au(111) electrode. They showed that IL layering is more pronounced at charged Au(111) surface compared to the neutral surface. They showed that increase of the potential leads to flattening of the tightly bound cation layer, indicating possible reorientation of cations (EMIm and Py) to lay flat on the surface.[21] We observe similar effects: increase of the potential at the CNT cathode significantly increases the tendency of EMIm cation to lay flat on the surface (see Fig. 3 and Figure S6 of the supporting information†).

In this work, we show that by applying an electric potential on the CNT electrode and/or varying the chemical structure of RTIL molecular ions it is possible to change ion orientations and thus the structure of the CNT-RTIL interface shell. These simulation results support the experimental observations of Hayes *et al.* discussed above.

## 4 Conclusions

The analysis of the simulation data results in the following conclusions:

1. There is an enrichment of all molecular components of ionic liquids under study at the CNT surface with formation of several distinct layers even at the non-charged CNT surface.

2. Mixing RTIL with acetonitrile decreases ion-counterion correlations in the electric double layer.

 

3. Increase of the length of the non-polar cation 'tail' increases propensity of imidazolium-based cations to lay parallel to the CNT surface.

4. At the CNT cathode TFSI anions and molecular cations are preferentially oriented parallel to the surface.

5. At the CNT anode the TFSI anions are oriented parallel to the surface, however the preferred orientations of cations depend on the length of non-polar tail: EMIm cations are oriented perpendicular to the surface, BMIm are in both parallel and perpendicular orientations, OMIm are oriented parallel to the surface.

6. By applying electric potential on the CNT electrode or/and varying the chemical structure of RTIL molecular ions it is possible to change the interfacial orientation of RTIL ions and, consequently, the structure of the CNT-RTIL interface shell.

## Acknowledgements

The authors thank Alexei A. Kornyshev, Ruth M. Lynden-Bell, Aleksey G. Rozhin, Nikolaj Georgi, Vera Lockett, Sherif Z. El Abedin, Oliver Höfft and Natalia Borisenko for useful discussions. The work is influenced by inspiring discussions with Frank Endres and Yury Gogotsi. The authors are very thankful to Rob Atkin for providing the numerical data of the AFM measurements. The authors are grateful to Margarida C. Gomes for the information on the ionic liquid solubility in acetonitrile and Ekaterina L. Ratkova for the help in QM calculations. We acknowledge the supercomputing support from the John von Neumann-Institut für Computing (NIC), Juelich Supercomputing Centre (JSC), Forschungszentrum Juelich GmbH, Germany. Project ID: HLZ16. We acknowledge Deutsche Forschungsgemeinschaft (DFG) - German Research Foundation, Research Grant FE 1156/2-1. Some part of the work has been performed under the HPC-EUROPA2 project (project number: 228398) with the support of the European Commission - Capacities Area - Research Infrastructures. Part of the computations were performed with the facilities of HECToR, the UK's national high-performance computing service, which is provided by UoE HPCx Ltd at the University of Edinburgh, Cray Inc and NAG Ltd, and funded by the Office of Science and Technology through EPSRC's High End Computing Programme. We used VMD software for visualization of simulation results.[55] AIF is very thankful to the Royal Society of Chemistry for the travel funds and the financial support of the International Max Planck Research School Mathematics in the Sciences (supported by the Klaus Tschira Stiftung, Germany).

## References

1 M. Armand, F. Endres, D. R. MacFarlane, H. Ohno and B. Scrosati, *Nat. Mater.*, 2009, **8**, 621–629.
2 P. Simon and Y. Gogotsi, *Nat. Mater.*, 2008, **7**, 845–854.
3 G. P. Pandey, S. A. Hashmi and Y. Kumar, *J. Electrochem. Soc.*, 2010, **157**, A105–A114.
4 N. P. Tarasova, Y. V. Smetannikov and A. A. Zanin, *Russ. Chem. Rev.*, 2010, **79**, 463–477.
5 D. Silvester and R. Compton, *Z. Phys. Chem.*, 2006, **220**, 1247–1274.
6 A. A. Kornyshev, *J. Phys. Chem. B*, 2007, **111**, 5545–5557.
7 V. Lockett, R. Sedev, J. Ralston, M. Horne and T. Rodopoulos, *J. Phys. Chem. C*, 2008, **112**, 7486–7495.
8 *Electrodeposition from ionic liquids*, ed. F. Endres, D. MacFarlane and A. Abbott, Wiley-VCH, 2008, p. 410.
9 S. Z. El Abedin, M. Polleth, S. A. Meiss, J. Janek and F. Endres, *Green Chem.*, 2007, **9**, 549–553.
10 A. Izadi-Najafabadi, S. Yasuda, K. Kobashi, T. Yamada, D. N. Futaba, H. Hatori, M. Yumura, S. Iijima and K. Hata, *Adv. Mater.*, 2010, **22**, E235–E241.
11 P. Simon and Y. Gogotsi, *Philos. Trans. R. Soc. London, Ser. A*, 2010, **368**, 3457–3467.
12 H. Nakagawa, Y. Fujino, S. Kozono, Y. Katayama, T. Nukuda, H. Sakaebe, H. Matsumoto and K. Tatsumi, *J. Power Sources*, 2007, **174**, 1021–1026.
13 R. Lin, P. Huang, J. Segalini, C. Largeot, P. L. Taberna, J. Chmiola, Y. Gogotsi and P. Simon, *Electrochim. Acta*, 2009, **54**, 7025–7032.
14 R. Atkin and G. G. Warr, *J. Phys. Chem. C*, 2007, **111**, 5162–5168.

15 V. Lockett, R. Sedev, S. Harmer, J. Ralston, M. Horne and T. Rodopoulos, *Phys. Chem. Chem. Phys.*, 2010, **12**, 13816–13827.
16 V. Lockett, R. Sedev, C. Bassell and J. Ralston, *Phys. Chem. Chem. Phys.*, 2008, **10**, 1330–1335.
17 R. Atkin, S. Z. El Abedin, R. Hayes, L. H. S. Gasparotto, N. Borisenko and F. Endres, *J. Phys. Chem. C*, 2009, **113**, 13266–13272.
18 R. Hayes, S. Z. El Abedin and R. Atkin, *J. Phys. Chem. B*, 2009, **113**, 7049–7052.
19 F. Endres, O. Hofft, N. Borisenko, L. H. Gasparotto, A. Prowald, R. Al-Salman, T. Carstens, R. Atkin, A. Bund and S. Z. El Abedin, *Phys. Chem. Chem. Phys.*, 2010, **12**, 1724–1732.
20 M. Druschler, B. Huber and B. Roling, *J. Phys. Chem. C*, 2011, **115**, 6802–6808.
21 R. Hayes, N. Borisenko, M. K. Tam, P. C. Howlett, F. Endres and R. Atkin, *J. Phys. Chem. C*, 2011, **115**, 6855–6863.
22 M. Mezger, H. Schroder, H. Reichert, S. Schramm, J. S. Okasinski, S. Schoder, V. Honkimaki, M. Deutsch, B. M. Ocko, J. Ralston, M. Rohwerder, M. Stratmann and H. Dosch, *Science*, 2008, **322**, 424–428.
23 M. Mezger, S. Schramm, H. Schroder, H. Reichert, M. Deutsch, E. J. De Souza, J. S. Okasinski, B. M. Ocko, V. Honkimaki and H. Dosch, *J. Chem. Phys.*, 2009, **131**, 094701.
24 R. M. Lynden-Bell, J. Kohanoff and M. G. Del Popolo, *Faraday Discuss.*, 2005, **129**, 57–67.
25 C. Pinilla, M. Del Pópolo, J. Kohanoff and R. Lynden-Bell, *J. Phys. Chem. B*, 2007, **111**, 4877–4884.
26 R. M. Lynden-Bell, M. G. Del Popolo, T. G. A. Youngs, J. Kohanoff, C. G. Hanke, J. B. Harper and C. C. Pinilla, *Acc. Chem. Res.*, 2007, **40**, 1138–1145.
27 L. Yang, B. H. Fishbine, A. Migliori and L. R. Pratt, *J. Am. Chem. Soc.*, 2009, **131**, 12373–12376.
28 M. V. Fedorov and A. A. Kornyshev, *J. Phys. Chem. B*, 2008, **112**, 11868–11872.
29 M. V. Fedorov, N. Georgi and A. A. Kornyshev, *Electrochem. Commun.*, 2010, **12**, 296–299.
30 N. Georgi, A. A. Kornyshev and M. V. Fedorov, *J. Electroanal. Chem.*, 2010, **649**, 261–267.
31 Y. Shim and H. J. Kim, *ACS Nano*, 2010, **4**, 2345–2355.
32 J. Vatamanu, O. Borodin and G. D. Smith, *J. Am. Chem. Soc.*, 2010, **132**, 14825–14833.
33 J. Vatamanu, O. Borodin and G. D. Smith, *J. Phys. Chem. B*, 2011, **115**, 3073–3084.
34 Q. Dou, M. L. Sha, H. Y. Fu and G. Z. Wu, *J. Phys.: Condens. Matter*, 2011, **23**, 175001.
35 M. H. Ghatee and F. Moosavi, *J. Phys. Chem. C*, 2011, **115**, 5626–5636.
36 S. S. Sarangi, S. G. Raju and S. Balasubramanian, *Phys. Chem. Chem. Phys.*, 2011, **13**, 2714–2722.
37 S. Perkin, T. Albrecht and J. Klein, *Phys. Chem. Chem. Phys.*, 2010, **12**, 1243–1247.
38 R. Atkin, N. Borisenko, M. Druschler, S. Z. El Abedin, F. Endres, R. Hayes, B. Huber and B. Roling, *Phys. Chem. Chem. Phys.*, 2011, **13**, 6849–6857.
39 R. Hayes, G. G. Warr and R. Atkin, *Phys. Chem. Chem. Phys.*, 2010, **12**, 1709–1723.
40 F. Endres, *ChemPhysChem*, 2002, **3**, 144–154.
41 C. G. Liu, Z. N. Yu, D. Neff, A. Zhamu and B. Z. Jang, *Nano Lett.*, 2010, **10**, 4863–4868.
42 D. H. Lin, N. Liu, K. Yang, L. Z. Zhu, Y. Xu and B. S. Xing, *Carbon*, 2009, **47**, 2875–2882.
43 W. L. Jorgensen, D. S. Maxwell and J. TiradoRives, *J. Am. Chem. Soc.*, 1996, **118**, 11225–11236.
44 M. L. P. Price, D. Ostrovsky and W. L. Jorgensen, *J. Comput. Chem.*, 2001, **22**, 1340–1352.
45 J. N. Canongia Lopes and A. A. H. Pádua, *J. Phys. Chem. B*, 2004, **108**, 16893–16898.
46 J. N. Canongia Lopes, J. Deschamps and A. A. H. Pádua, *J. Phys. Chem. B*, 2004, **108**, 2038–2047.
47 U. Essmann, L. Perera, M. L. Berkowitz, T. Darden, H. Lee and L. G. Pedersen, *J. Chem. Phys.*, 1995, **103**, 8577–8593.
48 B. Hess, C. Kutzner, D. van der Spoel and E. Lindahl, *J. Chem. Theory Comput.*, 2008, **4**, 435–447.
49 J. T. Frey and D. J. Doren, *TubeGen 3.4*, web-interface: http://turin.nss.udel.edu/research/tubegenonline.html, 2011.
50 L. Martínez, R. Andrade, E. G. Birgin and J. M. Martínez, *J. Comput. Chem.*, 2009, **30**, 2157–2164.
51 M. J. Frisch, G. W. Trucks, H. B. Schlegel, G. E. Scuseria, M. A. Robb, J. R. Cheeseman, J. A. Montgomery, Jr., T. Vreven, K. N. Kudin, J. C. Burant, J. M. Millam, S. S. Iyengar, J. Tomasi, V. Barone, B. Mennucci, M. Cossi, G. Scalmani, N. Rega, G. A. Petersson, H. Nakatsuji, M. Hada, M. Ehara, K. Toyota, R. Fukuda, J. Hasegawa, M. Ishida, T. Nakajima, Y. Honda, O. Kitao, H. Nakai, M. Klene, X. Li, J. E. Knox, H. P. Hratchian, J. B. Cross, V. Bakken, C. Adamo, J. Jaramillo, R. Gomperts, R. E. Stratmann, O. Yazyev, A. J. Austin, R. Cammi, C. Pomelli, J. Ochterski, P. Y. Ayala, K. Morokuma, G. A. Voth, P. Salvador, J. J. Dannenberg,

 

V. G. Zakrzewski, S. Dapprich, A. D. Daniels, M. C. Strain, O. Farkas, D. K. Malick, A. D. Rabuck, K. Raghavachari, J. B. Foresman, J. V. Ortiz, Q. Cui, A. G. Baboul, S. Clifford, J. Cioslowski, B. B. Stefanov, G. Liu, A. Liashenko, P. Piskorz, I. Komaromi, R. L. Martin, D. J. Fox, T. Keith, M. A. Al-Laham, C. Y. Peng, A. Nanayakkara, M. Challacombe, P. M. W. Gill, B. G. Johnson, W. Chen, M. W. Wong, C. Gonzalez and J. A. Pople, *GAUSSIAN 03 (Revision E.01)*, Gaussian, Inc., Wallingford, CT, 2004.

52 J. B. Foresman, T. A. Keith, K. B. Wiberg, J. Snoonian and M. J. Frisch, *J. Phys. Chem.*, 1996, **100**, 16098–16104.

53 K. Nakajima, A. Ohno, H. Hashimoto, M. Suzuki and K. Kimura, *J. Chem. Phys.*, 2010, **133**, 044702.

54 H. Hashimoto, A. Ohno, K. Nakajima, M. Suzuki, H. Tsuji and K. Kimura, *Surf. Sci.*, 2010, **604**, 464–469.

55 W. Humphrey, A. Dalke and K. Schulten, *J. Mol. Graphics*, 1996, **14**, 33–38.

# Graphene-based supercapacitors in the parallel-plate electrode configuration: Ionic liquids *versus* organic electrolytes

**Youngseon Shim,[a] Hyung J. Kim*[bc] and YounJoon Jung*[a]**

***Received 2nd May 2011, Accepted 11th May 2011***
**DOI: 10.1039/c1fd00086a**

Supercapacitors with two single-sheet graphene electrodes in the parallel plate geometry are studied *via* molecular dynamics (MD) computer simulations. Pure 1-ethyl-3-methylimidazolium tetrafluoroborate ($EMI^+BF_4^-$) and a 1.1 M solution of $EMI^+BF_4^-$ in acetonitrile are considered as prototypes of room-temperature ionic liquids (RTILs) and organic electrolytes. Electrolyte structure, charge density and associated electric potential are investigated by varying the charges and separation of the two electrodes. Multiple charge layers formed in the electrolytes in the vicinity of the electrodes are found to screen the electrode surface charge almost completely. As a result, the supercapacitors show nearly an ideal electric double layer behavior, *i.e.*, the electric potential exhibits essentially a plateau behavior in the entire electrolyte region except for sharp changes in screening zones very close to the electrodes. Due to its small size and large charge separation, $BF_4^-$ is considerably more efficient in shielding electrode charges than $EMI^+$. In the case of the acetonitrile solution, acetonitrile also plays an important role by aligning its dipoles near the electrodes; however, the overall screening mainly arises from ions. Because of the disparity of shielding efficiency between cations and anions, the capacitance of the positively-charged anode is significantly larger than that of the negatively-charged cathode. Therefore, the total cell capacitance in the parallel plate configuration is primarily governed by the cathode. Ion conductivity obtained *via* the Green-Kubo (GK) method is found to be largely independent of the electrode surface charge. Interestingly, $EMI^+BF_4^-$ shows higher GK ion conductivity than the 1.1 M acetonitrile solution between two parallel plate electrodes.

## 1 Introduction

Electric double layer capacitors (EDLCs), also known as supercapacitors, have received extensive attention due to their intrinsically fast charging and discharging process that allows high power density, compared to batteries.[1–5] High electric field generated in supercapacitor double layers also enables better energy storage than conventional capacitors. Various carbon-based materials, such as activated carbons, carbon nanofibers and carbon nanotubes, have been the object of intensive scrutiny as electrode materials.[4] In particular, graphitic materials made of atomic carbon sheets[6] show great promise thanks to their excellent properties, including large surface area, superior stiffness, high electrical conductivity, and chemical and

[a]*Department of Chemistry, Seoul National University, Seoul, 151-747, Korea. E-mail: yjjung@snu.ac.kr*
[b]*Department of Chemistry, Carnegie Mellon University, Pittsburgh, PA, 15213, USA. E-mail: hjkim@cmu.edu*
[c]*School of Computational Sciences, Korea Institute for Advanced Study, Seoul, 130-722, Korea*

 

thermal inertness.[7–9] For example, the surface area of graphene accessible *via* ions is substantially larger than that of activated carbon, so that the former can yield considerably higher energy density than the latter. Furthermore, compared to complicated pore structures of activated carbon, flat surface structures of graphene electrodes can facilitate ion transport significantly, thereby increasing the power density of EDLCs.

Aqueous solutions such as sulfuric acid and potassium hydroxide and organic electrolytes based on acetonitrile or propylene carbonate are commonly used for EDLCs. However, low decomposition voltage of the former and low ion conductivity of the latter impose a significant limitation on, respectively, the energy and power densities of EDLCs. As such, there has been considerable interest in finding alternatives to these conventional electrolytes. Room-temperature ionic liquids are very promising candidates, which have many desirable properties, including a wide electrochemical window, high ion density, good thermal stability and non-volatility. For instance, the large electrochemical window of RTILs allows high cell voltage that can lead to a substantial increase in energy storage capacity of EDLCs, compared to conventional electrolytes.[10,11] Also their high ion density can compensate to a large degree for low diffusivity of individual ions arising from high viscosity.

Recently there has been significant progress in microscopic understanding of RTILs as electrolytes in supercapacitors *via* computer simulations.[12–20] For example, MD simulations correctly predict the specific capacitance of microporous electrode systems that varies significantly with the pore size,[13,16] consonant with measurements.[21,22] A detailed analysis[16] shows that the relative size of ions with respect to pores plays an important role in the determination of specific capacitance, including its anomalous behavior. For supercapacitors in the parallel plate electrode configuration, the capacitance exhibits a nonlinear dependence on cell voltage although its detailed functional form is somewhat controversial.[15,20] Our own effort has focused on the comparative analysis of RTIL and organic electrolytes, in particular, their performance in energy and power densities in supercapacitors.[23] We investigated the specific capacitance and ion conductivity of half-cell supercapacitors composed of a single-sheet graphene electrode immersed in an electrolyte by employing pure $EMI^+BF_4^-$ and a 1.1 M solution of $EMI^+BF_4^-$ in acetonitrile as a prototypical RTIL and an organic electrolyte, respectively. It was found that the former performs considerably better in energy density than the latter; the single-electrode capacitance in the former was higher than in the latter by 55–60%. Strong and efficient screening of electrode charges by ions close to the electrode surface is mainly responsible for high capacitance in pure RTILs. We also found that despite the high viscosity of $EMI^+BF_4^-$, its ion conductivity is indeed comparable to that of the acetonitrile solution, confirming the important role played by high ion density.

In this article, we extend our prior study on single-electrode supercapacitors[23] to analyze RTILs and organic electrolytes confined between two parallel, single-sheet graphene electrodes. One of the key differences from the previous single-electrode work is that only one side of the electrode is exposed to electrolytes in the present case. We vary both the surface charge density of, and separation between, the two electrodes and examine the specific capacitance. We also investigate ionic conductivity to gain insight into power density.

The outline of this paper is as follows: In section 2, we give a brief description of the models and methods employed in the present study. MD results for electrolyte structure and charge distributions and resulting electric potential are discussed in section 3. The capacitance and ion conductivity of the supercapacitors are also analyzed there. Section 4 concludes.

## 2 Simulation methods

Our supercapacitor system is composed of an electrolyte confined between two graphene electrodes in the parallel plate configuration (Fig. 1). Outside of the electrodes

 

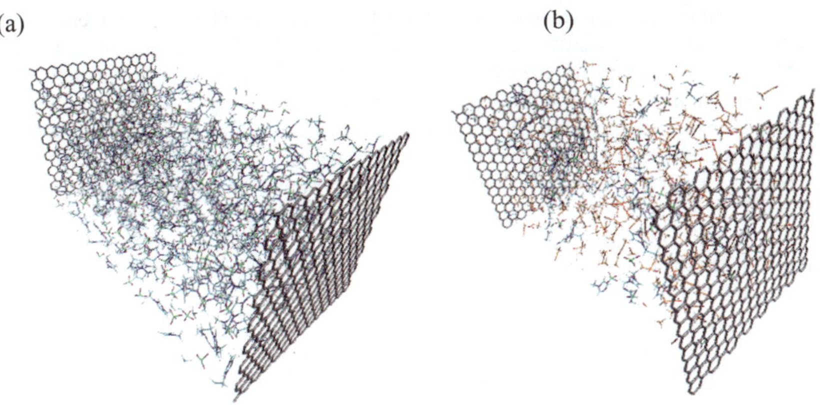

**Fig. 1** Model supercapacitor system with (a) pure $EMI^+BF_4^-$ and (b) their acetonitrile solution employed as an electrolyte. In both cases, the electrolyte is confined between two parallel single-sheet graphene electrodes, separated by 6.4 nm and electrified with surface charge density $\sigma_S = \pm 0.86e\ nm^{-2}$.

is a vacuum. Two different electrolytes, pure $EMI^+BF_4^-$ and a 1.1 M solution of $EMI^+BF_4^-$ with a mol fraction of 0.089 in acetonitrile, are considered. Hereafter, these two systems will be referred to as the RTIL and organic electrolyte supercapacitors, respectively. In both cases, each electrode consists of a rigid and flat graphene sheet with dimensions $3.432 \times 3.398\ nm^2$. For an efficient description of supercapacitor configurations and their MD results, we employ a cartesian coordinate system, where the two electrodes are placed parallel to each other with their graphene sheets in $xy$ -planes and their centers on the $z$ -axis. For convenience, we assume that the positively- and negatively-charged electrodes are located, respectively, at $z = -z_0$ and $z = +z_0$ ($z_0 > 0$). Two different electrode separations, $d(=2z_0) = 6.4$ nm and 2.0 nm, are examined (see Table 1). For $d = 6.4$ nm, the RTIL electrolyte comprises 256 pairs of $EMI^+$ and $BF_4^-$, while the organic electrolyte is composed of 50 pairs of $EMI^+$ and $BF_4^-$, and 512 $CH_3CN$ molecules. The corresponding compositions for the $d = 2.0$ nm case are 80 pairs of $EMI^+$ and $BF_4^-$ for the RTIL supercapacitor and 16 pairs of ions and 164 acetonitrile molecules for the organic electrolyte supercapacitor. For later use, we introduce a distance measured from each electrode $\Delta z = |z \pm z_0|$.

**Table 1** Results of electric potential drop and specific capacitance[a,b]

| Solvent | $\sigma_S$ | $d$ | $\Delta\Delta\Phi^{(+)}$ | $\Delta\Delta\Phi^{(-)}$ | $\Delta\Delta\Phi$ | $c^{(+)}$ | $c^{(-)}$ | $c^{tot}$ |
|---|---|---|---|---|---|---|---|---|
| $EMI^+BF_4^-$ | ±0.86 | 6.4 | 1.3 | −4.8 | 6.1 | 10 | 2.9 | 2.3 |
| | ±0.43 | 6.4 | 0.65 | −1.8 | 2.4 | 11 | 3.9 | 2.9 |
| | ±0.86 | 2.0 | 1.8 | −5.4 | 7.1 | 7.9 | 2.6 | 1.9 |
| | ±0.43 | 2.0 | 0.65 | −2.1 | 2.7 | 11 | 3.4 | 2.6 |
| $CH_3CN/EMI^+BF_4^-$ | ±0.86 | 6.4 | 1.5 | −6.1 | 7.6 | 9.2 | 2.3 | 1.8 |
| | ±0.43 | 6.4 | 0.70 | −2.8 | 3.5 | 9.8 | 2.5 | 2.0 |
| | ±0.86 | 2.0 | 1.5 | −6.0 | 7.5 | 9.2 | 2.3 | 1.8 |
| | ±0.43 | 2.0 | 0.60 | −2.7 | 3.3 | 12 | 2.6 | 2.1 |
| $CH_3CN$ | ±0.86 | 6.4 | | | 35 | | | 0.39 |
| | ±0.86 | 2.0 | | | 12 | | | 1.2 |

[a] $\Delta\Delta\Phi^S = \Delta\Phi^S(\text{charged}) - \Delta\Phi^S(\text{discharged})$. [b] Units: electric potential (V), capacitance ($\mu F/cm^2$), $\sigma_S(e/nm^2)$ and $d$ (nm).

The cathode and anode are represented by uniformly charged graphenes with partial negative and positive charges assigned to all of their carbon atoms, respectively. We considered three different anode/cathode surface charge densities, *viz.*, $\sigma_S = \pm 0.86\ e\ \text{nm}^{-2}$, $\pm 0.43\ e\ \text{nm}^{-2}$ and 0, where $e$ is the elementary charge. The corresponding total electrode surface charges are $\pm 10e$, $\pm 5e$, and 0. Hereafter, we will denote different supercapacitor configurations by their electrode charge and separation. For example, $\sigma_S = 0.86e\ \text{nm}^{-2}$, $0.43\ e\ \text{nm}^{-2}$ and 0 with $d = 6.4$ nm will be referred to as C(charged)6.4, H(half-charged)6.4 and N(neutral)6.4, respectively.

All other details of model descriptions employed here are the same as in our prior study.[23] Specifically, we used $\varepsilon = 43.2$ K and $\sigma = 0.34$ nm as Lennard-Jones (LJ) parameters for graphene carbon atoms.[24] For RTIL, the flexible all-atom description of ref. 16 and 25, based on the $EMI^+$ model in ref. 26 and 27 and $BF_4^-$ in ref. 28 and 29, were used. For $CH_3CN$, we employed the fully flexible six-site description of ref. 30. Electronic polarizability for both graphene and the electrolytes was ignored in our present study.[31,32]

We performed molecular dynamics simulation studies in the canonical ensemble at 350 K using the DL_POLY program.[33] The simulation cell was composed of the supercapacitor system described above, placed in an orthorhombic box of $3.432 \times 3.398 \times 30.0\ \text{nm}^3$. The long-range electrostatic interactions were computed *via* the Ewald method with account of shape-dependent correction[34] to properly describe two-dimensional systems. The trajectories were integrated *via* the Verlet leapfrog algorithm using a time step of 1 fs. Simulations were carried out with a 5 ns equilibration, followed by a 5 ns trajectory from which ensemble averages were computed. During the simulations, the graphene was held rigid with a carbon–carbon bond length $l_{CC} = 0.1415$ nm.[35] For comparison, we also studied conventional parallel plate capacitors consisting of two graphene electrodes and pure acetonitrile.

## 3 Results and discussion

### Structure

We begin by investigating electrolyte structures and their variations with the electrode surface charge density $\sigma_S$ and their separation $d$. As mentioned above, we use a cartesian coordinate system, where each graphene electrode is located in the $xy$ -plane ($-x_0 < x < x_0$ and $-y_0 < y < y_0$) with its center on the $z$ -axis. We introduce the ion densities averaged over $x$ and $y$

$$\bar{n}_\alpha(z) = A_0^{-1} \int_{-x_0}^{x_0} \int_{-y_0}^{y_0} dx'dy' n_\alpha(x', y', z);\ A_0 = 4x_0 y_0, \tag{1}$$

where $A_0$ is the surface area of the graphene electrode with $x_0 = 1.716$ nm and $y_0 = 1.699$ nm, $n_\alpha(x,y,z)$ is the local number density of solvent species $\alpha$ ($\alpha = EMI^+$, $BF_4^-$, or $CH_3CN$) at $(x,y,z)$, and the center-of-mass of ions and molecules is used to represent the position of $\alpha$.

We first consider the $\bar{n}_\alpha(z)$ results for the RTIL supercapacitor in Fig. 2. In the fully discharged cases in (a) and (c), ion number densities are symmetric under $z \to -z$. We note that ions develop a pronounced structure in the vicinity of the electrode surface, $\Delta z \lesssim 0.4$ nm. This will be referred to as the first solvation shell (or layer) hereafter. As $\Delta z$ increases beyond this first solvation layer, both cation and anion densities show significant fluctuations, regardless of the electrode separation $d$. This is in good agreement with previous simulation results that layered structures of RTIL ions are present over a considerable distance from the confining surface.[12,15,20,23,36] The oscillatory character of $\bar{n}_\alpha(z)$ varies with $\sigma_S$ and $d$. In the case of N6.4 (discharged electrodes with $d = 6.4$ nm), cation and anion structures in general oscillate in phase, *viz.*, their peak positions along $z$ generally coincide, while the layered structures tend to alternate between cations and anions for C6.4

 

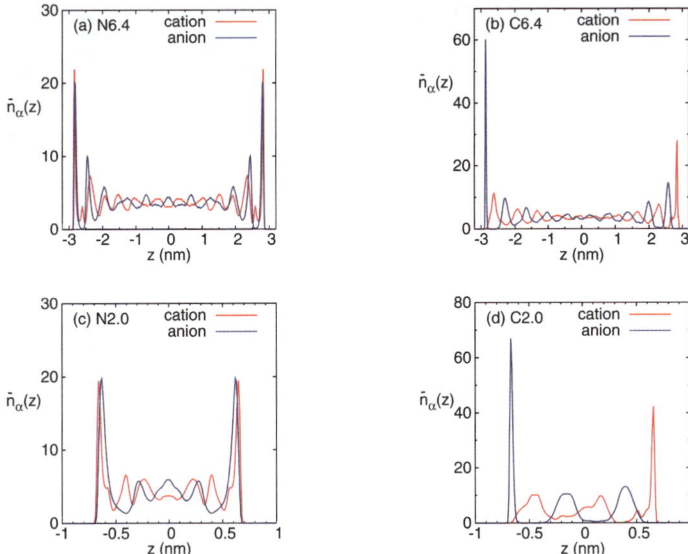


**Fig. 2** Average number density $\bar{n}_\alpha(z)$ (eqn (1)) of $EMI^+$ and $BF_4^-$ ions in the RTIL supercapacitor at 350 K in the (a) N6.4, (b) C6.4, (c) N2.0, and (d) C2.0 configurations. (a) and (c) represent the fully discharged case with electrically neutral electrodes, while (b) and (d) are for the oppositely-charged electrode configurations with $\sigma_S = \pm 0.86e$ nm$^{-2}$. In each panel, two vertical lines of the frame, which define the $y$ range of the plot, represent the positions of the two electrodes.

(charged electrodes with $d = 6.4$ nm). As expected, the structures of counter-ions and co-ions near the charged electrode becomes, respectively, enhanced and reduced, compared to those near the neutral electrode. Thus while there are no $BF_4^-$ ions present as co-ions within 0.5 nm of the negatively-charged cathode of C6.4, about 26 $BF_4^-$ ions participate as counter-ions in the formation of the first solvation layer the positively-charged anode at $\Delta z \lesssim 0.4$ nm. We notice in (c) and (d) that the general features of $\bar{n}_\alpha(z)$ remain unchanged for the smaller electrode separation $d = 2.0$ nm.

We turn to the organic electrolyte supercapacitor case in Fig. 3. Due to the low ionic concentration of 1.1 M, the neutral electrode is covered almost exclusively by acetonitrile molecules. This is revealed by prominent first solvation layer peaks of acetonitrile at $\Delta z \approx 0.34$ nm, whereas discernible ion structures begin to appear only around $\Delta z \approx 0.8$ nm. In the case of charged supercapacitors, electrostatic interactions between the electrode surface charge and ions lead to distinct first solvation layer structures of counter-ions at $\Delta z = 0.4$–$0.5$ nm. In the case of C6.4, for instance, four $EMI^+$ and six $BF_4^-$ ions form these structures near the cathode and anode, respectively. Despite these counter-ion peaks, however, acetonitrile is still the major component near the electrodes. Compared to the RTIL supercapacitors, the alternating aspect of the cation and anion distributions in the presence of charged electrodes becomes weakened. We further notice that their populations shift towards the oppositely-charged electrodes, resulting in charge separation, especially in C2.0. This is ascribed to the combined effect of low ion density and high surface charge density.

## Electric potential and capacitance

We proceed to the average electric potential of the supercapacitors, in particular, their variations along $z$ and dependence on $d$. For convenience, we decompose the

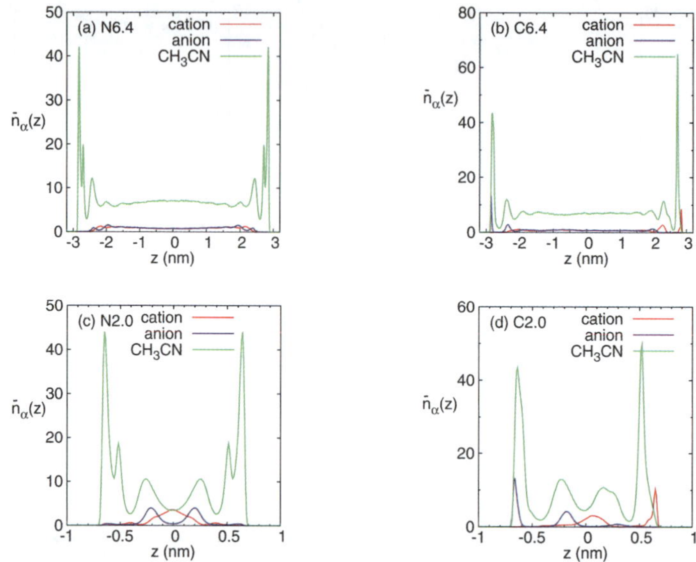


**Fig. 3** Average number density $\bar{n}_\alpha(z)$ of $EMI^+$, $BF_4^-$ and acetonitrile in the organic electrolyte supercapacitor at 350 K in the (a) N6.4, (b) C6.4, (c) N2.0, and (d) C2.0 configurations. As in Fig. 2, two vertical lines of each frame, which define the *y* range of the plot, represent the positions of the two electrodes.

total electric potential $\Phi$ into two components, $\Phi_\sigma$ and $\Phi_{IL}$, arising from the graphene surface charge density and RTIL ions, respectively. In the case of the organic electrolyte supercapacitor, there is an additional contribution $\Phi_{CH_3CN}$ from acetonitrile. We calculated $\Phi_{IL}(z)$ by integrating the Poisson equation

$$\Phi_{IL}(z) = -4\pi \sum_\alpha \int_{-z_0}^{z} (z - z')\bar{\rho}_\alpha(z')dz'; \quad \bar{\rho}_\alpha(z) = A_0^{-1}\int_{-x_0}^{x_0}\int_{-y_0}^{y_0} dx'dy'\rho_\alpha(x',y',z), \quad (2)$$

where $\rho_\alpha(x,y,z)$ is the local charge density arising from the atomic charge distribution of ionic species $\alpha$, $\bar{\rho}_\alpha(z)$is the average charge density at $z$ obtained by averaging $\rho_\alpha(x, y,z)$ over $x$ and $y$, and $\sum_\alpha$ denotes sum over ionic species.

The results for $\bar{\rho}_\alpha(z)$ between two charged electrodes are presented in Fig. 4. We notice in (a) and (c) that total charge density of the pure RTIL is characterized by multiple charge layers along $z$ near the electrode surface. Counter-ions near the charged surface, especially $BF_4^-$, develop a strong oscillatory behavior for $\Delta z \lesssim 1$ nm. It is noteworthy that oscillations of the anion charge density $\bar{\rho}\_(z)$ are accompanied by sign changes even though the overall anion charge is $-e$. This is attributed to the extended charge distribution of $BF_4^-$, in particular, its charge separation character arising from F with partial negative charge and P with partial positive charge.[16,23] Since the degree of charge separation of $EMI^+$ is weaker than $BF_4^-$, the charge density variations of the former along $z$ near the negatively-charged cathode are not as pronounced as those of the latter close to the anode. As in our previous study,[23] charge density oscillations nearly disappear for $\Delta z \gtrsim 1$ nm even though structural order tends to persist well beyond 1 nm from the electrode surface (Fig. 2). This is also ascribed to the extended nature of the ion charge distributions.

The charge densities in the organic electrolyte in Fig. 4 (b) and (d) also exhibit rapid oscillations with sign changes near the cathode and anode. It should be noticed that acetonitrile plays an important role in governing the electrolyte charge density near the electrodes. Because of low ionic concentration, the electrode surfaces are mainly wetted by acetonitrile molecules. Their dipoles align near the electrified

 

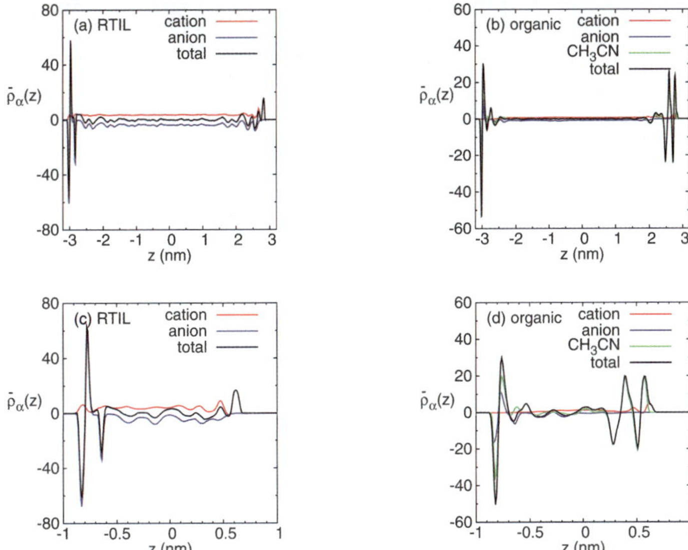


**Fig. 4** Average local charge density $\bar{\rho}_{\alpha}(z)$ (units: $e$ nm$^{-3}$) in the RTIL and organic electrolyte supercapacitors at 350 K. The electrode configuration/electrolyte is (a) C6.4/RTIL, (b) C6.4/acetonitrile solution, (c) C2.0/RTIL and (d) C2.0/acetonitrile solution. In each panel, two vertical lines of the frame, which define the $y$ range of the plot, represent the positions of the two electrodes.

surfaces, thereby resulting in oscillatory charge density. As in the RTIL supercapacitor case, the contribution of $BF_4^-$ as counter-ions to the electrolyte charge density oscillations is much greater than that of $EMI^+$.

The profile of the total electric potential $\Phi(z)$ for both RTILs and organic electrolyte supercapacitors is displayed in Fig. 5. For comparison, we have also analyzed conventional capacitors with pure acetonitrile employed as a dielectric material. Among the most notable features is the difference between the supercapacitors and conventional capacitors in $\Phi(z)$. The average electric potential in the former remains essentially constant in the entire eletrolyte region except for the close neighborhood of the electrodes. Thus both the RTILs and organic electrolyte supercapacitors behave like an ideal electric double layer capacitor regardless of the anode–cathode separation considered here. By contrast, $\Phi(z)$ for conventional capacitors varies nearly linearly with $z$ between the cathode and anode.

Another noteworthy aspect is that there is a significant disparity in the $\Phi(z)$ behavior near the anode and cathode of the supercapacitors. Specifically, the $\Phi(z)$ change near the cathode is much bigger than that near the anode for both the RTIL and organic electrolytes. If we define the potential drop $\Delta\Phi^{S}$ associated with the anode and cathode as

$$\Delta\Phi^{(+)} = \Phi(-z_0) - \Phi(0); \; \Delta\Phi^{(-)} = \Phi(+z_0) - \Phi(0), \tag{3}$$

we obtain $\Delta\Phi^{(+)}/|\Delta\Phi^{(-)}| = 1.5/4.7$ V and 1.6/6.0 V for the RTIL and organic electrolyte supercapacitors in the C6.4 configuration, respectively (*cf.* Table 1). We note that the relative $\Delta\Phi^{S}$ difference between the cathode and anode obtained here, *viz.*, $\Delta\Phi^{(+)}/|\Delta\Phi^{(-)}| \approx 4$, is somewhat larger than other RTIL systems at lower voltage.[15,20,37] As discussed below, the cathode–anode asymmetry in $\Delta\Phi^{S}$ is attributed to the difference in screening efficiency between the cations and anions. Interestingly, this asymmetry is not that sensitive to the electrode separation $d$. As a result, the overall potential drop of the supercapacitor cell, $\Delta\Phi \equiv \Delta\Phi(-z_0) - \Delta\Phi(z_0)(= \Delta\Phi^{(+)} - \Delta\Phi^{(-)})$, does not vary much with $d$. In the RTIL electrolyte

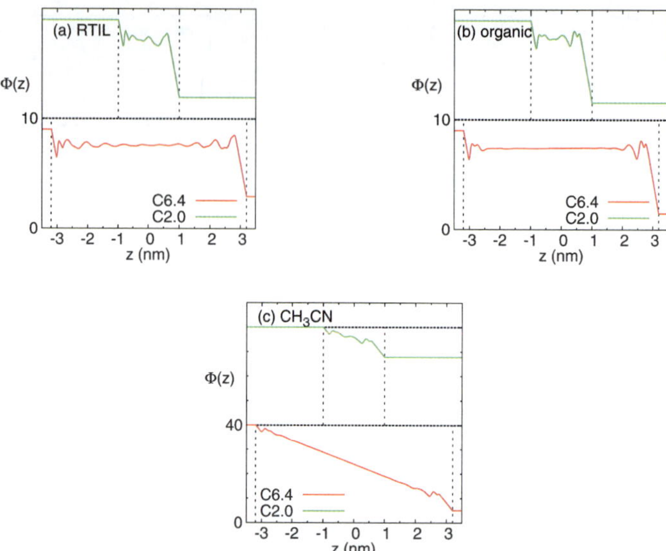


**Fig. 5** Profile of total electric potential $\Phi(z)$ (in units of V) in (a) RTILs and (b) organic electrolyte supercapacitors. MD results for a conventional capacitor with pure acetonitrile used as a dielectric material are displayed in (c). For clear comparison, the results for $d = 2$ nm are shifted upwards with respect to the $d = 6.4$ nm case. The positions of the anode and cathode are denoted as dashed vertical lines.

case, we find $\Delta\Phi = 6.1$ V and 7.1 V for C6.4 and C2.0, respectively. For the organic electrolyte, $\Delta\Phi \approx 7.5$ V irrespective of the $d$ values we considered. $\Delta\Phi$ of the conventional supercapacitor in Fig. 5 (c), on the other hand, increases linearly with $d$ and as a result, their capacitance decreases inversely proportional to $d$.

To understand the average electric potential better, its two components, $\Phi_{\rm IL}(z)$ and $\Phi_\sigma(z)$, are compared in Fig. 6. For organic electrolyte supercapacitors, $\Phi_{\rm CH_3CN}$is also presented there. For clarity, the results for $-\Phi_\sigma(z)$ (rather than $\Phi_\sigma(z)$) are exhibited. In the RTIL supercapacitor case in (a) and (c), $\Phi_{\rm IL}(z)$ tracks $-\Phi_\sigma(z)$ very closely except in the immediate vicinity of the electrodes, *i.e.*, $\Delta z \lesssim 0.4$ nm. This indicates that the electrode charges are shielded almost completely by RTIL ions in $\Delta z \lesssim 0.4$ nm ("screening zone"), so that the average electric field in $\Delta z \gtrsim 0.5$ nm essentially vanishes. In view of the ideal EDLC behavior outside of the screening zone noted above, we identify this zone as the electric double layer region. We notice that the screening zone around the anode is considerably smaller ($\Delta z \lesssim 0.2$ nm) than that around the cathode ($\Delta z \lesssim 0.4$ nm). This suggests that the anions screen the electrode charge considerably more efficiently than the cations.

Many features of the RTIL supercapacitor results in Fig. 6 (a) and (c) are shared by the organic electrolyte supercapacitor in Fig. 6 (b) and (d). However, there is one major difference, *i.e.*, acetonitrile makes an important contribution to $\Phi(z)$, especially near the electrodes, in the latter. Nevertheless, the fact that $\Phi_{\rm CH_3CN}$ shows a plateau behavior for $\Delta z \gtrsim 0.5$ nm clearly exposes the limitation of charge screening by electrically-neutral acetonitrile.

To gain further insight into screening, we have analyzed the difference of the cation and anion numbers in the electrolyte volume, extending from the position of the anode surface to $z$, and the resulting net charge of the volume (*cf.* eqn (1) and 2)

$$\Delta N(z) = A_0 \int_{-z_0}^{z} dz' (\overline{n}_+(z') - \overline{n}_-(z')); \quad Q_{\rm IL}(z) = A_0 \int_{-z_0}^{z} dz' (\overline{\rho}_+(z') + \overline{\rho}_-(z')). \quad (4)$$

 

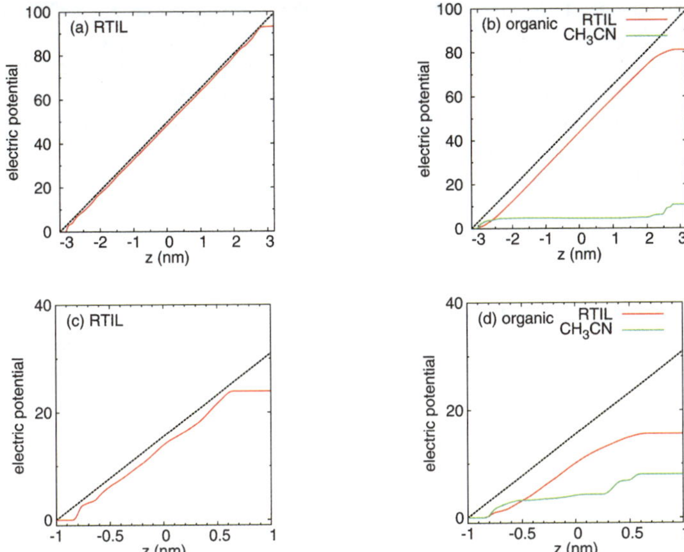


**Fig. 6** Components of the Poisson potential $\Phi(z)$ between the charged electrodes (units: V): $\Phi_{IL}(z)$(red solid line) and $-\Phi_{\sigma}(z)$ (black dotted line). The electrode/electrolyte configurations are the same as in Fig. 5. In (b) and (d), $\Phi_{CH_3CN}$is plotted in green. For convenience, all potentials are set to 0 at the anode located at $-z_0$.

The reader is reminded that the coordinates of the center-of-mass of ions are used for $\bar{n}_{\alpha}(z)$, while the extended atomic charge distributions of ions are employed to calculate $\bar{\rho}_{\pm}(z)$. The results for $\Delta N(z)$ are shown in Fig. 7. Marked oscillations in the charged RTIL supercapacitors C6.4 and C2.0 are the direct consequence of alternating layered structures of counter-ions and co-ions in Fig. 2. The organic electrolyte supercapacitors show much weaker oscillations in $\Delta N(z)$ than the RTIL case because of their low ionic concentration.

Since the total charge of the electrified anode located at $-z_0$ is $10e$, its full screening occurs when $\Delta N(z) = -10$. One can show easily that the complete screening of the cathode of total charge $-10e$ at $z_0$ is given by the same condition $\Delta N(z) = -10$. According to Fig. 7, the difference of cations and anions in the vicinity of the charged anode, *i.e.*, $\Delta z \lesssim 0.4$nm (*cf.* Fig. 2 (b)), is $\Delta N(z) \approx -25$ for the RTIL supercapacitor in both C6.4 and C2.0 configurations. This means that there are much more counter-ions present in the screening zone than needed to fully shield

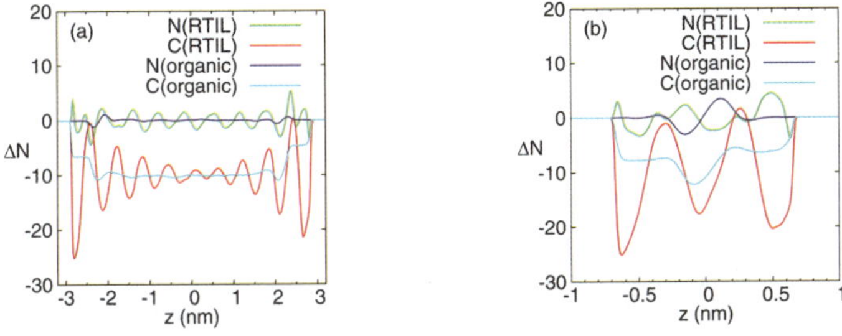


**Fig. 7** $\Delta N(z)$ in eqn (4) in the RTIL and organic electrolyte supercapacitors: (a) $d = 6.4$ nm and (b) 2.0 nm.

the anode charge. This "over-screening"[23] of the electrode charge is well exposed in Fig. 8, where $Q_{IL}(z)$ of the RTIL supercapacitor shows a large negative peak of $-30e$ near the anode irrespective of $d$. Furthermore, oscillations of $Q_{IL}(z)$ about the full screening value, $-10e$, confirms the alternation of the counter-ion rich and co-ion rich regions observed in Fig. 2 (b) and (d) above. $\Delta N(z)$and $Q_{IL}(z)$ near the negatively-charged cathode, though considerably lesser in extent, exhibit a similar over-screening behavior with oscillations. Nevertheless, the difference in $\Delta N(z)$ and $Q_{IL}(z)$ between the two electrodes is a clear indicator of differing screening efficiency between $EMI^+$ and $BF_4^-$ ions.

General characteristics of screening behavior in the charged organic electrolyte supercapacitors in Fig. 8 (b) and (d) (*i.e.*, red lines there) are quite similar to the corresponding RTIL supercapacitor cases in (a) and (c). However, $Q_{CH_3CN}(z)$ is largely responsible for both the over-screening and oscillatory behaviors in the former. Because of low ionic concentration, $Q_{IL}(z)$ does not yield over-screening.

We turn to capacitance of the RTIL and organic electrolyte supercapacitors. Specific capacitance $c^S$ normalized to the electrode surface area is obtained as

$$c^S = \frac{\sigma_S}{\Delta\Delta\Phi^S}; \quad \Delta\Phi^S(\text{charged}) - \Delta\Phi^S(\text{discharged}), \tag{5}$$

where $\Delta\Phi^S$ is the potential drop associated with the electrode with surface charge density $\sigma_S$ in eqn (5) and $\Delta\Delta\Phi^S$ is the corresponding difference with respect to the potential drop at PZC (potential at zero charge). The results for $\Delta\Delta\Phi^S$ and $c^S$ are compiled in Table 1. The total cell capacitance $c^{tot}$ determined as

$$\frac{1}{c^{tot}} = \frac{1}{c^{(+)}} + \frac{1}{c^{(-)}} \tag{6}$$

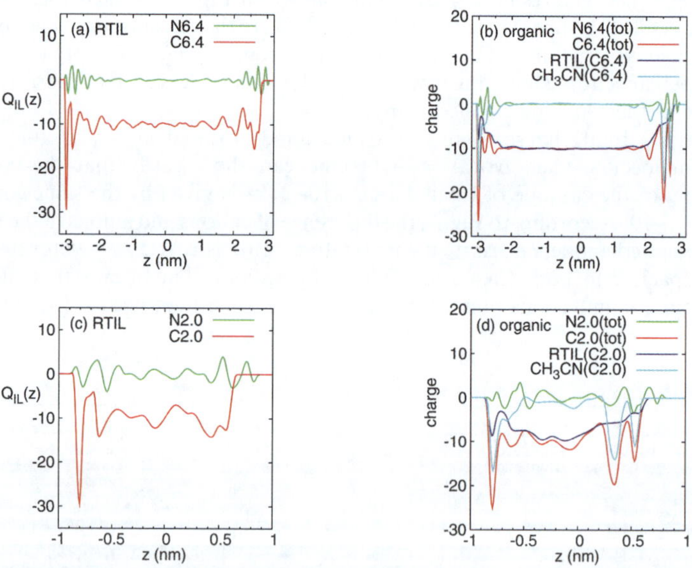


**Fig. 8** Total charges contained in the electrolyte volume from $-z_0$ to $z$ and their RTIL and acetonitrile components. For the RTIL supercapacitors in (a) and (c), only $Q_{IL}(z)$ is shown because it is the same as the total charge. In (b) and (d), the RTIL and acetonitrile contributions to the total charge for the charged organic electrolyte supercapacitor as well as the total charges are presented. In each panel, two vertical lines of the frame, which define the $y$ range of the plot, represent the positions of two electrodes.

 

is also presented there. There are several noteworthy features. First, the anode capacitance $c^{(+)}$ is considerably larger than the cathode capacitance $c^{(-)}$ for both the RTIL and organic electrolyte supercapacitors regardless of their configurations, *i.e.*, $\sigma_S$ and $d$. This disparity is attributed to the difference in size and shape between the cations and anions above.[37] Smaller and simpler $BF_4^-$ anions screen the anode more efficiently than larger and more complicated $EMI^+$ cations shield the cathode despite the π-stacking capability of the latter. This is in good accord with prior MD studies on various RTIL supercapacitors.[16,20,23] Second, since $c^{(+)}$ is larger than $c^{(-)}$ by a factor of 3–4, the total capacitance $c^{tot}$ is primarily governed by the smaller $c^{(-)}$. Therefore the screening efficiency of cations at the cathode controls the overall capacitance of the supercapacitor systems considered here.

Another interesting result is that the specific capacitance depends on the electrode surface charge density. To be specific, for given $d$, both $c^{(+)}$ and $c^{(-)}$ tend to decrease with increasing $|\sigma_S|$. This is attributed to the expansion of the electric double layer region due to the screening zone saturation. As $|\sigma_S|$ increases, $|\Delta N(z)|$ in the screening zone, *i.e.*, electric double layer region, would grow in proportion to $|\sigma_S|$. If the double layer region were to remain unchanged, the electric field strength in that region would increase linearly with $|\sigma_S|$. Because of finite size of ions, however, the original screening zone would not be sufficient to accommodate the increase in counter-ions due to shortage of free space. (In a lattice-gas theory description of ionic solutions where each lattice site can be occupied by only one ion, this situation of screening zone saturation is referred to as lattice saturation.)[38] This means that the screening zone would need to expand to make room for additional counter-ions. Since the potential drop is given by the integration of the electric field across the double layer, the resulting $|\Delta\Phi^S|$ in the expanded double layer region would increase more rapidly than a linear behavior in $|\sigma_S|$. As a consequence, $c^S$ would decrease with increasing $|\sigma_S|$. We note that this $c^S$ trend is in line with earlier simulations with RTIL electrolytes although the numerical values of $|\Delta\Phi^S|$ obtained for a high surface charge density exceed somewhat an electrochemical window allowed for the realistic RTIL supercapacitor.[20,37] Our MD results indicate that the screening zone saturation is not limited to the RTIL supercapacitors; the organic electrolyte supercapacitors are also characterized by decreasing $c^S$ with increasing $|\sigma_S|$. We also observe in Table 1 that $c^S$ varies with the electrode separation $d$. Due to insufficient data (*viz.*, only two different $d$ values), however, we will not attempt to analyze potential correlations between $c^S$ and $d$ in the present study.

For perspective, we make brief contact with our earlier study on single-electrode capacitance, where a single-sheet graphene electrode is inserted in the middle of an electrolyte extending from $z = -6.4$ nm to 6.4 nm with both sides of the electrode exposed to the electrolyte.[23] For simplicity, we will refer to this electrode-electrolyte arrangement as "double-sided" electrode configuration. We note that supercapacitators in this double-sided electrode structure show a couple of interesting differences from those in the parallel plate (*i.e.*, single-sided) H6.4 electrode configuration considered in the present study. First, disparity of $c^S$ between the anode and cathode is much weaker in the former than in the latter configuration. For example, $c^{(+)}$- $c^{(-)}$ difference in the parallel plate electrode structure with $d = 6.4$ nm is about 7 $\mu$F cm$^{-2}$ for both the RTIL and organic electrolyte supercapacitors (Table 1), whereas the corresponding difference in the double-sided electrode configuration is less than 1 $\mu$F cm$^{-2}$.[23] Second, the difference in $c^S$ between the RTIL and organic electrolyte supercapacitors is smaller in the parallel plate electrode configuration than in the double-sided electrode configuration. Their respective differences are ~1 and ~2 $\mu$F cm$^{-2}$. In both configurations, $c^S$ of the RTIL supercapacitor is found to be larger than that of the corresponding organic electrolyte supercapacitor with the exception of $c^{(+)}$ of C2.0 and H2.0 in the parallel plate electrode configuration.

To gain insight into the differences above, average electric potential and electrolyte charge density of the two configurations are compared in Fig. 9 and 10, respectively. The $\Phi(z)$ results show that $\Delta\Phi^{(+)}$ and $|\Delta\Phi^{(-)}|$ in eqn (3), respectively, decreases

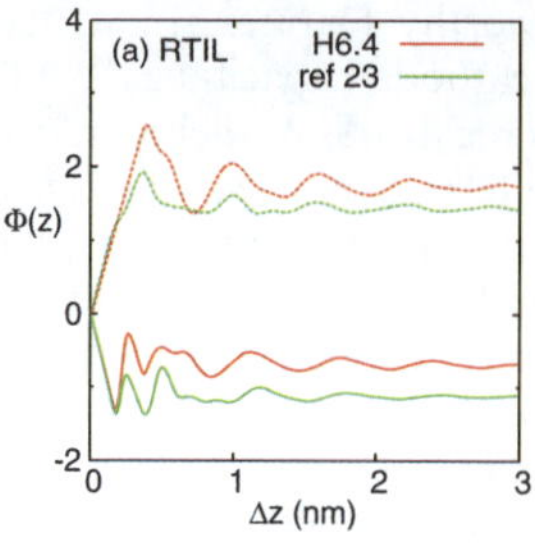


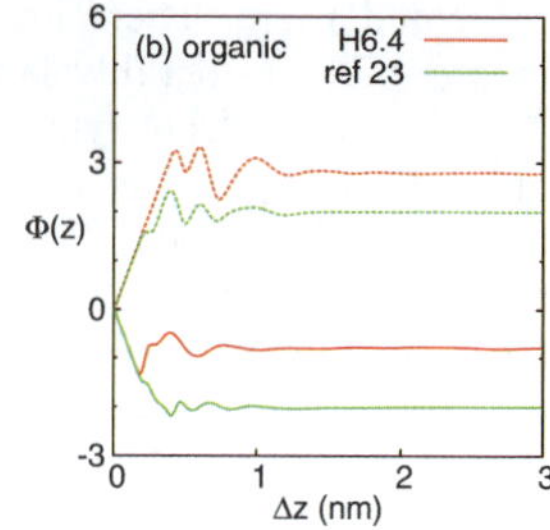


**Fig. 9** Comparison of $\Phi(z)$ in the parallel plate H6.4 configuration (red lines) and in the two-sided electrode configuration of ref. 23 (green lines) around the anode (solid lines) and cathode (dotted lines) of (a) RTIL and (b) organic electrolyte supercapacitors at 350 K. For clear exposition, $\Phi(z)$ at each electrode surface is set to 0. The surface charge density of the single-electrode supercapacitor is $\sigma_S = \pm 0.86e\ \text{nm}^{-2}$ and that of the parallel-electrode supercapacitor is $\sigma_S = \pm 0.43e\ \text{nm}^{-2}$.

and increases in the parallel plate single-sided electrode configuration, compared to the double-sided electrode case of ref. 23. This is due to differing changes in electrolyte charge density in the screening zone around the anode and cathode as shown Fig. 10. Compared to the double-sided electrode configuration, the electrolyte charge density around the anode becomes enhanced in the parallel plate configuration for both the RTIL and organic electrolyte supercapacitors. This would yield better screening and thus larger $c^{(+)}$ in the latter than in the former. By contrast, the screening of the cathode in the parallel plate arrangement occurs further away from the electrode, *i.e.*, larger $\Delta z$, by ~0.2 nm than in the double-sided electrode configuration. This results in poor screening and thus reduced $c^{(-)}$ in the former, compared to the latter.

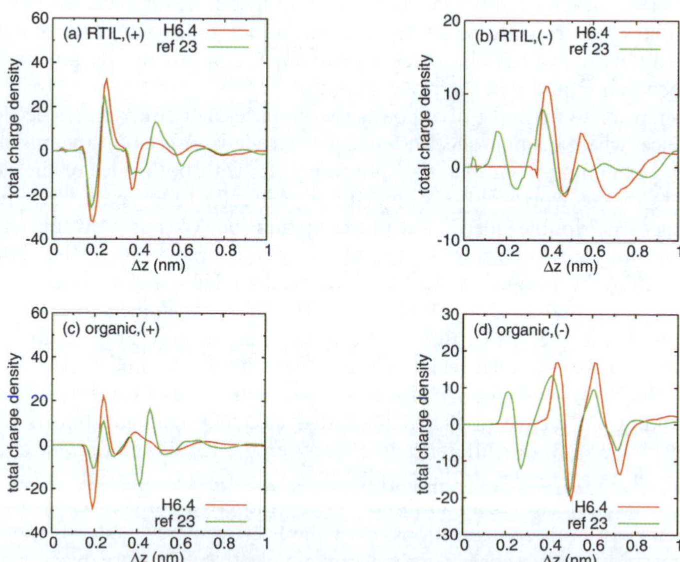


**Fig. 10** Comparison of average local charge density $\bar{\rho}(z)$ (units: $e\ \text{nm}^{-3}$) in the parallel plate H6.4 configuration (red) and in the two-sided electrode configuration[23] (green): (a) anode and (b) cathode of the RTIL electrolyte; (c) anode and (d) cathode of the organic electrolyte supercapacitor.

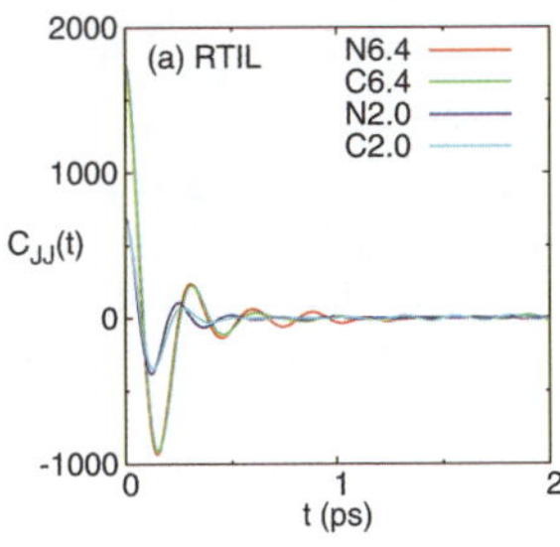

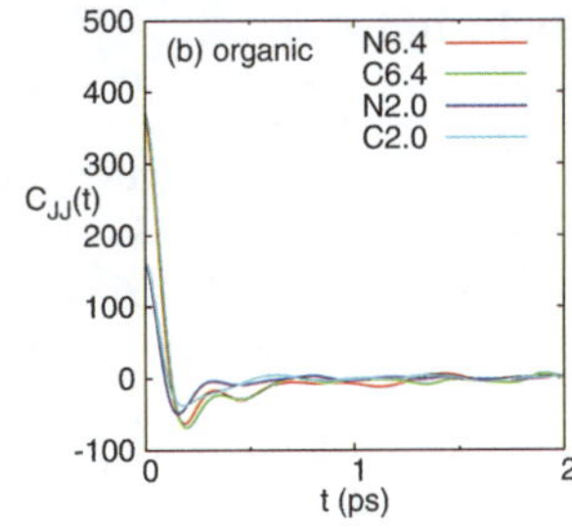


**Fig. 11** Time correlation function of the ionic current normal to the electrode surface in (a) RTIL and (b) organic electrolyte supercapacitors.

## Ion conductivity

To gain insight into power density of the supercapacitors, we briefly examine ion conductivity in the direction normal to the electrode surface using the Green-Kubo (GK) formula as in our previous study.[23] In the GK formulation, the ion conductivity is related to the collective current $J_z(t)$ as

$$\sigma_z^{\mathrm{GK}} = \frac{1}{3Vk_{\mathrm{B}}T}\int_0^\infty dt C_{JJ}(t); \quad C_{JJ}(t) = \langle J_z(0)\cdot J_z(t)\rangle; \quad J_z(t) = \sum_{i=1}^{N} q_i v_{z,i}(t), \tag{7}$$

where $q_i$ and $v_{z,i}$ are the charge and the $z$ component of center-of-mass velocity of $i$ -th ion and $\langle\ldots\rangle$ represents an equilibrium ensemble average.

The results for $C_{JJ}(t)$ are presented in Fig. 11 and those for $\sigma_z^{\mathrm{GK}}$ in Table 2. We note that the structure of $C_{JJ}(t)$ in Fig. 11 (a) is very similar to that of the double-sided electrode case in ref. 23 as well as in bulk ionic liquids.[39] The big discrepancy in $C_{JJ}(0)$ between the $d = 2.0$ nm and 6.4 nm cases arises from the difference in the total number of ions. We notice that $C_{JJ}(t)$ librational motions become faster in the former than in the latter. This indicates that short-time ion relaxation dynamics become accelerated as the confinement size becomes reduced, especially in the RTIL.

We notice in Table 2 that for given $d$, the ion conductivity does not vary much with the electrode surface charge density, consonant with the double-sided electrode case.[23] Interestingly, in the present parallel plate configuration, $\sigma_z^{\mathrm{GK}}$ is larger in the RTIL supercapacitor than in the organic electrolyte supercapacitor by 50–100%. This trend is the opposite of the single-electrode case in ref. 23. Our results here seem to imply that power density of parallel plate capacitors with small electrode separation would be higher with RTILs employed as an electrolyte than with organic electrolytes. However, a detailed analysis including nonequilibrium relaxation dynamics would be needed to address this issue accurately.

**Table 2** Results for ion conductivities (S m$^{-1}$) along the $z$ direction.

| Solvent | $\sigma_S$ (e nm$^{-2}$) | $d$ (nm) | $\sigma_z^{\mathrm{GK}}$ |
|---|---|---|---|
| $EMI^+BF_4^-$ | ±0.86 | 6.4 | 0.27 |
| | 0.0 | 6.4 | 0.26 |
| | ±0.86 | 2.0 | 0.32 |
| | 0.0 | 2.0 | 0.32 |
| $CH_3CN/EMI^+BF_4^-$ | ±0.86 | 6.4 | 0.18 |
| | 0.0 | 6.4 | 0.18 |
| | ±0.86 | 2.0 | 0.16 |
| | 0.0 | 2.0 | 0.13 |

## 4 Conclusion

Supercapacitors consisting of two single-sheet graphene electrodes in the parallel plate configuration were investigated *via* MD simulations. Pure $EMI^+BF_4^-$ and a 1.1 M solution of $EMI^+BF_4^-$ in acetonitrile were considered as a prototypical RTIL and an organic electrolyte. Regardless of the electrode separation we studied, both the RTIL and organic electrolyte supercapacitors show nearly an ideal electric double layer capacitor behavior. We found that the cathode and anode show a significant gap in the specific electrode capacitance; to be specific, the anodic capacitance $c^{(+)}$ ranges 8–12 $\mu F\ cm^{-2}$, while the cathodic value is $c^{(-)} = 2$–$4\ \mu F\ cm{-}^2$. This is attributed to the efficient screening behavior of $BF_4^-$ compared to $EMI^+$. Due to this strong cathode–anode asymmetry of capacitance, the higher $c^{(+)}$ does not play a significant role in the determination of total cell capacitance, which is mainly controlled by the lower cathodic capacitance $c^{(-)}$. Therefore improving the screening efficiency of cations through creative design and modifications of their structures would be important to increasing energy storage capability of supercapacitors based on complex cations, such as 1-ethyl-3-methylimidazolium studied here.

We compared the properties of the parallel plate electrode configuration with those of the double-sided electrode structure we analyzed previously,[23] where both sides of the electrode interfaces with the electrolyte. We found that electrostatic interactions of ions on one side of the electrode with those on the other play an important role in modulating the screening of the electrode charges in the latter configuration. To be specific, these "through the graphene electrode" interactions, respectively, enhance and reduce the screening and thus specific capacitance of the cathode and anode in the double-sided electrode configuration, compared to the single-sided parallel plate electrode configuration, in which these interactions are absent. One important consequence of this is that the cathode–anode asymmetry of capacitance becomes considerably weakened in the double-sided electrode configuration than in the parallel plate arrangement.

For insight into power density, ion conductivity in the direction perpendicular to the electrode surface was examined using the Green-Kubo formula. Interestingly, RTIL supercapacitors are characterized by higher ion conductivity than organic electrolyte supercapacitors, irrespective of the electrode surface charge density or separation. This result is opposite to the trend obtained for the double-sided electrode configuration.[23] It would thus be worthwhile in the future to study charging and discharging processes *via* nonequilibrium MD and examine if conductivity obtained with equilibrium analysis properly addresses power performance of supercapacitors.

## Acknowledgements

This work is supported by the National Research Foundation of Korea grants (2011-0001212, 2011-0003555, 2011-0018038, and 2011-0013299), and by the grants of the KISTI Supercomputing Center (No. 305-20110004). YS acknowledges the financial support from the BK 21 Program of Korea.

## References

1 B. E. Conway, *Electrochemical Supercapacitors: Scientific Fundamentals and Technological Applications*, Plenum, New York, 1999.
2 R. Kötz and M. Carlen, *Electrochim. Acta*, 2000, **45**, 2483.
3 G. Pandolfo and A. F. Hollenkamp, *J. Power Sources*, 2006, **157**, 11.
4 E. Frackowiak, *Phys. Chem. Chem. Phys.*, 2007, **9**, 1774.
5 H. D. Abruña, Y. Kiya and J. C. Henderson, *Phys. Today*, 2008, **61**, 43.
6 K. S. Novoselov, A. K. Geim, S. V. Morozov, D. Jiang, Y. Zhang, S. V. Dubonos, I. V. Grigorieva and A. A. Firsov, *Science*, 2004, **306**, 666.

 

7 J. C. Meyer, A. K. Geim, M. I. Katsnelson, K. S. Novoselov, T. J. Booth and S. Roth, *Nature*, 2007, **446**, 60.
8 C. Gomez-Navarro, R. T. Weitz, A. M. Bittner, M. Scolari, A. Mews, M. Burghard and K. Kern, *Nano Lett.*, 2007, **7**, 3499.
9 M. J. Allen, V. C. Tung and R. B. Kaner, *Chem. Rev.*, 2010, **110**, 132.
10 S. R. C. Vivekchand, C. S. Rout, K. S. Subrahmanyam, A. Govindaraj and C. N. R. Rao, *J. Chem. Sci.*, 2008, **120**, 9.
11 A. Balducci, R. Dugas, P. L. Taberna, P. Simon, P. D. Plee, M. Mastragonstino and S. Passerini, *J. Power Sources*, 2007, **165**, 922.
12 C. Pinilla, M. G. D. Pópolo, J. Kohanoff and R. M. Lynden-Bell, *J. Phys. Chem. B*, 2007, **111**, 4877.
13 L. Yang, B. H. Fishbine, A. Migliori and L. R. Pratt, *J. Am. Chem. Soc.*, 2009, **131**, 12373.
14 S. A. Kislenko, I. S. Samoylov and R. H. Amirov, *Phys. Chem. Chem. Phys.*, 2009, **11**, 5584.
15 G. Feng, J. S. Zhang and R. Qiao, *J. Phys. Chem. C*, 2009, **113**, 4549.
16 Y. Shim and H. J. Kim, *ACS Nano*, 2010, **4**, 2345.
17 Y. Lauw, M. D. Horne, T. Rodopoulos, A. Nelson and F. A. M. Leermakers, *Phys. Rev. Lett.*, 2009, **103**, 117801.
18 Y. Lauw, M. D. Horne, T. Rodopoulos, A. Nelson and F. A. M. Leermakers, *J. Phys. Chem. B*, 2010, **114**, 11149.
19 M. Trulsson, J. Algotsson, J. Forsman and C. E. Woodward, *J. Phys. Chem. Lett.*, 2010, **1**, 1191.
20 J. Vatamanu, O. Borodin and G. D. Smith, *J. Am. Chem. Soc.*, 2010, **132**, 14825.
21 J. Chmiola, G. Yushin, Y. Gogotsi, C. Portet, P. Simon and P. L. Taberna, *Science*, 2006, **313**, 1760.
22 C. Largeot, C. Portet, J. Chmiola, P. L. Taberna, Y. Gogotsi and P. Simon, *J. Am. Chem. Soc.*, 2008, **130**, 2730.
23 Y. Shim, Y. Jung and H. J. Kim, J. Phys. Chem., *submitted.*
24 G. Hummer, J. C. Rasaiah and J. P. Noworyta, *Nature*, 2001, **414**, 188.
25 Y. Shim and H. J. Kim, *ACS Nano*, 2009, **3**, 1693.
26 J. N. C. Lopes, J. Deschamps and A. A. H. Pádua, *J. Phys. Chem. B*, 2004, **108**, 2038.
27 J. N. C. Lopes, J. Deschamps and A. A. H. Pádua, *J. Phys. Chem. B*, 2004, **108**, 11250.
28 J. de Andrade, E. S. Böes and H. Stassen, *J. Phys. Chem. B*, 2002, **106**, 13344.
29 X. Wu, S. Huang and W. Wang, *Phys. Chem. Chem. Phys.*, 2005, **7**, 2771.
30 A. M. Nikitin and A. P. Lyubartsev, *J. Comput. Chem.*, 2007, **28**, 2020.
31 T. Yan, C. J. Burnham, M. G. D. Pópolo and G. A. Voth, *J. Phys. Chem. B*, 2004, **108**, 11877.
32 D. Jeong, Y. Shim, M. Y. Choi and H. J. Kim, *J. Phys. Chem. B*, 2007, **111**, 4920.
33 T. R. Forester and W. Smith, *DL_POLY user manual, CCLRC*, Daresbury Laboratory, Daresbury, Warrington, U.K., 2001.
34 I. Yeh and M. L. Berkowitz, *J. Chem. Phys.*, 1999, **111**, 3155.
35 T. W. Odom, J.-L. Huang, P. Kim and C. M. Lieber, *Nature*, 1998, **391**, 62.
36 C. Pinilla, M. G. D. Pópolo, R. M. Lynden-Bell and J. Kohanoff, *J. Phys. Chem. B*, 2005, **109**, 17922.
37 M. V. Fedorov and A. A. Kornyshev, *J. Phys. Chem. B*, 2008, **112**, 11868.
38 A. A. Kornyshev, *J. Phys. Chem. B*, 2007, **111**, 5545.
39 Y. Shim and H. J. Kim, *J. Phys. Chem. B*, 2008, **112**, 11028.

# Adsorption, absorption and desorption of gases at liquid surfaces: water on $[C_8C_1Im][BF_4]$ and $[C_2C_1Im][Tf_2N]$

**Alexey Deyko and Robert G. Jones***

***Received 8th April 2011, Accepted 9th June 2011***
**DOI: 10.1039/c1fd00062d**

The adsorption, absorption and desorption of water vapour at the liquid, and frozen solid, surfaces of two ionic liquids (ILs) is described. Surface kinetics were measured by sticking probability ($S$), and temperature programmed desorption, using line of sight mass spectrometry in ultra-high vacuum. The two ILs used were 1-octyl-3-methylimidazolium tetrafluoroborate ($[C_8C_1Im][BF_4]$) and 1-ethyl-3-methylimidazolium bis[(trifluoromethyl)sulfonyl]imide, ($C_2C_1Im]$ $[Tf_2N]$. At room temperature, continuous absorption into the bulk occurred ($S \approx 0.12$), which reduced with decreasing temperature, reaching zero when the ILs froze to glassy solids. At still lower temperatures, monolayer, and then multilayer, growth of water ice occurred at the solid surfaces. Adsorption is thought to occur first into a physisorbed state, and then into an ionic underlayer, the activation energy between the physisorbed state and the underlayer state being $48.0 \pm 0.8$ kJ mol$^{-1}$ and $58 \pm 2.0$ kJ mol$^{-1}$ for $[C_2C_1Im][Tf_2N]$ and $[C_8C_1Im][BF_4]$, respectively. From the underlayer state, water diffuses into the bulk rather than desorbing back into the gas phase. Adsorption and desorption kinetics were consistent with the existence of an outermost, hydrophobic layer covering the ionic underlayer. For $[C_8C_1Im][BF_4]$ this consists of outwardly orientated octyl chains, while for $[C_2C_1Im][Tf_2N]$ it is probably outward orientated ethyl and $CF_3$ groups. Adsorption on the hydrophobic glassy surface was unusual, with $S$ increasing for increasing coverage. The barrier to surface diffusion of water on the glassy surface of $[C_8C_1Im][BF_4]$ was $13 \pm 12$ kJ mol$^{-1}$.

## 1 Introduction

In this paper we present kinetic measurements of a gas interacting with two structured liquid surfaces. The gas was water vapour and the liquids were two ionic liquids (ILs), 1-octyl-3-methylimidazolium tetrafluoroborate ($[C_8C_1Im][BF_4]$) and 1-ethyl-3-methylimidazolium bis[(trifluoromethyl)sulfonyl]imide ($[C_2C_1Im][Tf_2N]$), Table 1. The measurements consisted of sticking probability ($S$) and temperature programmed desorption (TPD) studies, and are the first of their kind to be carried out on liquid surfaces in ultra-high vacuum.

There are both academic and commercial reasons for studying the interactions of gases with ionic liquid surfaces. Academically, ionic liquids present a class of material which undergoes surface freezing, where the surface layers are more structured and ordered, than the bulk material. This has been demonstrated both experimentally[1] and in simulations where the surface structure of ILs containing alkyl chains are found to be ordered, such that the alkyl chains are aligned in a manner similar to the formation of self assembled monolayers (SAMs) on gold surfaces.[2] For the IL

*School of Chemistry, University of Nottingham, Nottingham, NG7 2RD, UK. E-mail: robert.g.jones@nottingham.ac.uk; Tel: +44 115 9513468*

**Table 1** Abbreviations and names for ions, and the ionic liquids studied in this work, and their structures

| Ion abbreviation | Structure |
|---|---|
| $[C_nC_mIm]^+$: dialkylimidazolium | 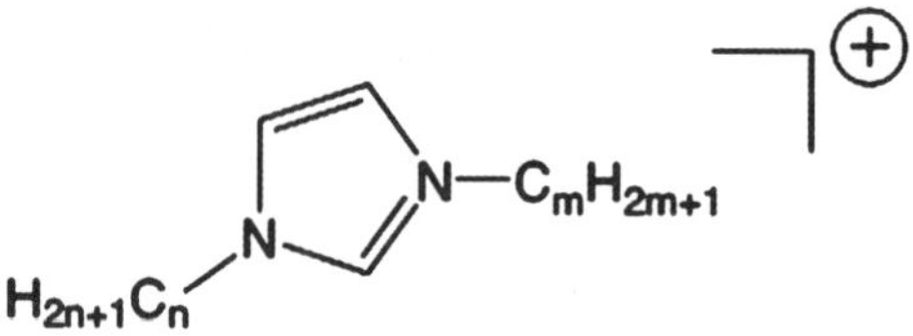 |
| $[C_nC_mPyrr]^+$: dialkylpyrrolidinium | 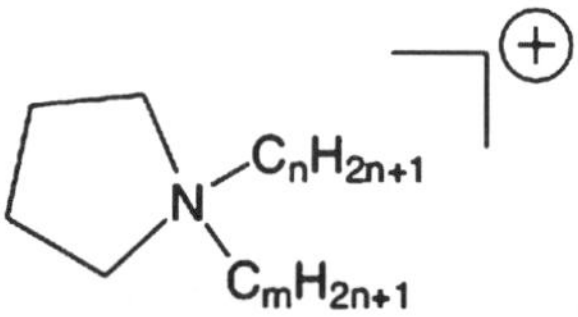 |
| $[TfO]^-$: trifluoro-methanesulfonate | 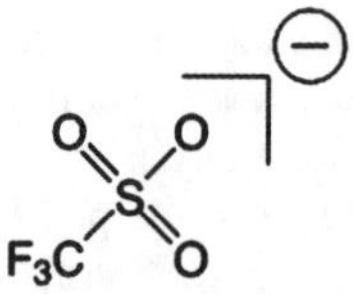 |
| $[Tf_2N]^-$: bis[(trifluoromethyl)-sulfonyl]imide | 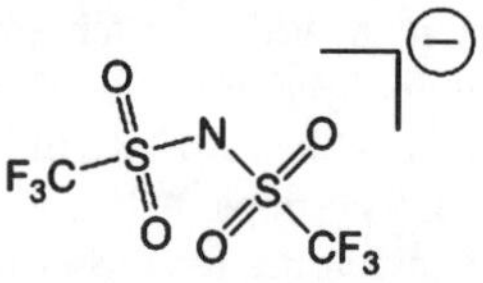 |
| **ILs in this work** | **Structure** |
| $[C_8C_1Im][BF_4]$: 1-octyl-3-methylimidazolium tetrafluoroborate | 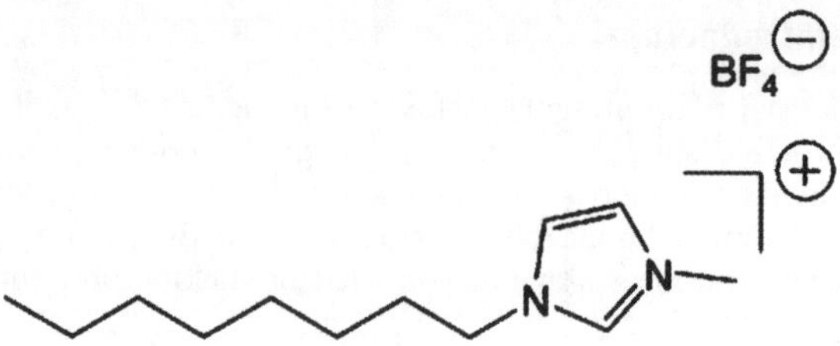 |
| $[C_2C_1Im][Tf_2N]$: 1-ethyl-3-methylimidazolium bis[(trifluoromethyl)-sulfonyl]imide | 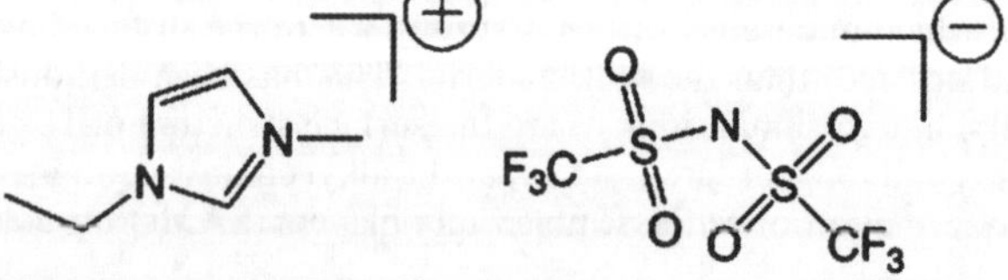 |

 

surfaces, the charged parts of the anion and cation tend to be located at an ionic underlayer below the alkyl chain layer, the entire surface representing a self assembled liquid surface (SALS). Simulations show that for longer alkyl chain lengths (>8 carbon atoms)[3] several layers of interleaved, ordered alkyl chains and ionic layers, can be formed, which become progressively more disordered with depth into the bulk of the liquid, eventually forming the hydrophobic and ionic domain structure of the bulk ionic liquid. SALS offer the possibility of ordering adsorbates either at the hydrophobic, oleaginous layer, or at the ionic underlayer such that directed self-assembly of new species can be envisaged. As the surface is liquid, material transport can occur from the bulk phases below (liquid) and above (gas), as well as within the two dimensional liquid surface itself. By changing the nature of the anions (*e.g.* halides, $[BF_4]^-$, $[PF_6]^-$, triflimides, *etc.*) or cations (*e.g.* imidazoliums, pyrrolidiniums, pyridiniums, phosphoniums, *etc.*) or by functionalising the anions or cations, it is possible to tune the properties of the IL such that it should be possible, eventually, to predetermine the structure and properties of the liquid surface. Such structured liquid surfaces could then be used to direct the synthesis of molecules, through to the fabrication of nanoparticles.

Commercially, the ultra-low vapour pressure of ILs, plus their ability to absorb gases, makes them highly attractive in several areas. Ionic liquids are used in the storage of reactive or toxic gases and due to their selective absorption of gases, ILs are used as gas separation media and are potential carbon capture agents for $CO_2$ (see ref. 4 for a recent review of gas solubility in ionic liquids). In the SILP process, a thin layer of IL containing a catalyst is supported on a high surface area solid; gaseous reactants diffuse into the IL, react, and the products then diffuse out, the process combining the advantages of heterogeneous catalysis with those of homogeneous catalysis.[5] In all these areas, the capture of gas molecules is determined by the interaction of the surface layer of the liquid with the gas at the liquid–vapour interface, any SALS structure at that interface will significantly affect the kinetics of gas penetration through the surface.

Water was chosen for this first study as it can alter the chemistry of ILs. Trace amounts of water will hydrolyse $[BF_4]^-$ and $[PF_6]^-$ anions to produce HF and imidazolium based ILs such as $[C_4C_1Im][BF_4]$ and $[C_4C_1Im][PF_6]$ are hygroscopic and will absorb large amounts of water from the ambient atmosphere. Water is known to reduce the electrochemical window of ILs,[6] decrease their density and viscosity,[7] and decrease their surface tension. The interaction of water with surfaces generally is a large and complex field which was first reviewed by Thiel and Madey in 1982,[8] and then again by Henderson in 2002.[9] In broad terms, water can adsorb molecularly or dissociatively, depending on the reactivity of the substrate. The substrates can be categorised as metals/alloys, oxides, semiconductors, salts or organics. Salts and oxides, possessing ions at their surfaces might be regarded as the closest analogues to ionic liquids. Ordered structure have been reported for water on NaCl(100), MgO(100) and $CaF_2(111)$[9] and Henzler *et al.* found a heat of adsorption of 65 kJ $mol^{-1}$ for water molecularly adsorbed on NaCl(100),[10] the large value being indicative of the water interacting directly with ions at the surface of the salt. By comparison, water adsorption on an organic surface has been studied by adsorption of $D_2O$ on a self assembled monolayer (SAM), hexadecane thiolate on gold.[11] This SAM comprises a close packed layer of alkyl chains and has a very low surface free energy. TPD of water from its surface exhibits zero-order-like behaviour from submonolayer coverages to multilayers, with an activation energy of 35 kJ $mol^{-1}$. This low activation energy, less than the heat of sublimation of water, is consistent with water interacting with the oleaginous, hydrophobic surface. In the multilayer regime, water has a rich adsorption behaviour on all surfaces, due to its ability to cluster and form amorphous or crystalline states which depend upon the substrate, temperature and incident flux.[9]

The SALS structure of ILs has been studied by a number of methods. In an early work, Gannon *et al.*[12] used rare gas recoil spectrometry on a $[C_4C_1Im][PF_6]$ surface to show that both ions are present at the surface and that the imidazolium ring is aligned parallel to the surface normal with the alkyl chains directed away from the surface. Neutron reflectometry of $[C_4C_1Im][BF_4]$ and $[C_8C_1Im][PF_6]$[1] shows multilayer ordering of an alkyl rich top layer, sitting on a region comprising the cationic head groups and anions, which is in turn on top of a further alkyl rich layer. Combined sum frequency generation and surface tension studies on $[C_nC_1Im]$ $[C_mOSO_3]$ (alkyl chain lengths $n$ and $m$ = 1–4)[13] indicate that the alkyl chains of both anions and cations project towards the gas phase, with the charged species located below them. The effect on surface tension was marked, with a monotonic lowering of the surface tension, and hence the surface energy, with increasing alkyl chain length. A particularly graphic demonstration of the SALS effect is provided by Jiang *et al.* who used multiscale coarse graining simulations for $[C_nC_1Im][NO_3]$ ($n$ = 2–12)[3] to show that the longer the alkyl chain the more the outer surface is composed of orientated alkyl chains rather than polar groups. Lovelock *et al.*[14] have studied $[C_nC_1Im][Tf_2N]$ using angle resolved X-ray photoelectron spectroscopy (XPS), and have shown that the surface becomes enriched with alkyl chains with the imidazolium head groups and the anion located at approximately the same depth below the alkyl chain layer. Iwahashi *et al.* have investigated the surface structure of $[C_nC_1Im][A]$ ($n$ = 4,8, and 10 and $[A]^-$ = $[BF_4]^-$, $[TfO]^-$ and $[Tf_2N]^-$) by ultraviolet photoelectron spectroscopy (UPS) and metastable atom electron spectroscopy (MAES).[15] They also found a layered structure with the alkyl chains and $CF_3$ groups orientated outwards at the surface, and the $SO_3$, $SO_2$, and imidazolium polar groups on the bulk side. An extensive angle resolved XPS study of $[C_8C_1Im][A]^-$ over a range of anion, $[A]^-$, sizes[16] showed that the surfaces was enriched in the alkyl chains with the charged ions forming an underlayer, the smaller the anion, the more it tended to form an underlayer.

Sum frequency generation was used by Rivera-Rubero *et al.* to investigate the influence of $H_2O$ on $[C_4C_1Im][BF_4]$.[17] They found that the imidazolium ring was approximately parallel to the surface with the butyl chain projecting into the gas phase for the pure IL, and that this structure did not change appreciably when water was absorbed. Lauw *et al.* used surface X-ray reflectivity to study the effect of water on the surface structure of $[C_4C_1Pyrr][Tf_2N]$.[18] They observed molecular ordering at the surface with a net positively charged cation + water layer at the liquid-gas boundary, below which was a net negatively charged layer due to anions with less water. The alkyl chain on the pyrrolidinium ring was orientated towards the gas phase. Finally we mention an ultrahigh vacuum (UHV) XPS study of water ($D_2O$) adsorbed and absorbed on $[C_8C_1Im][BF_4]$.[30] Using TPD monitored by XPS, a monolayer of water was detected at the surface, which disappeared from the surface with an activation enthalpy of 76 kJ $mol^{-1}$. This was interpreted as being due to water that had adsorbed into an ionic underlayer where it was particularly strongly bound to the either the anions and/or the cations. As the gas phase was not monitored in these experiments, it was not known whether the adsorbed water was lost by desorption into the gas phase or by diffusion into the bulk of the IL.

In this study we have chosen $[C_8C_1Im][BF_4]$ as an ionic liquid with a cation with a long octyl chain paired with an anion which is rather small, which forms a surface consisting of octyl chains covering an ionic underlayer of the imidazolium head groups and $[BF_4]^-$. This underlayer had already been shown to adsorb water strongly. The other ionic liquid, $[C_2C_1Im][Tf_2N]$, was chosen as it has a much smaller cation with a rather short ethyl chain, paired with a rather large, stable anion with $CF_3$ and $SO_2$ groups. It was thought that this IL might expose some charged species at the surface, thus possibly bonding water at the surface, rather than in an ionic underlayer. Our results indicate that both ILs exhibit hydrophobic surfaces.

## 2 Experimental

Line of sight mass spectrometry was used for the sticking probability and TPD experiments.[19–24] In this technique, only species emanating from a specific patch

 

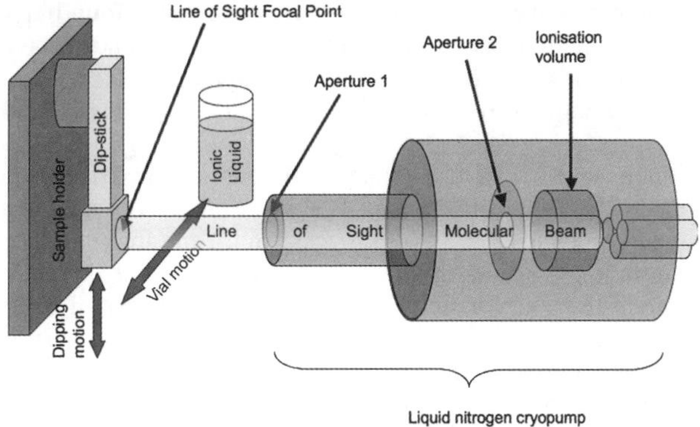


**Fig. 1** Schematic of the experimental arrangement, illustrating the principle of line of sight mass spectrometry (LOSMS) and how a thin film sample of IL is formed on the dip-stick by dipping into a vial of IL in vacuum.

on the sample surface, and traveling by line of sight to the ionisation volume of the mass spectrometer, are detected, Fig.1. Gases from all other sources are pumped by the cryopump surrounding the mass spectrometer. Apertures 1 and 2, Fig. 1, define the line of sight molecular beam from the sample to the ionization volume. This property, of measuring only the signal from the sample surface, is the basis of the sticking probability measurements, as well as providing high quality TPD measurements which are unaffected by desorption from surfaces other than the front face of the sample. At present cryopumping is achieved using a copper shroud which surrounds the mass spectrometer and is cooled by liquid nitrogen to 77 K. This temperature is sufficient to cryopump water vapour to a pressure $\ll 10^{-13}$ mbar, but is insufficient to pump low boiling point species such as $H_2$, $N_2$, or CO, for which a lower temperature would be required.

We define the effective sticking probability, $S(\theta)$, as the ratio of the flux of molecules that stick to the surface, to the flux of molecules that hit the surface. Coverage $\theta$ is defined in the TPD section. If $F_{in}$ and $F_{out}$ are the fluxes (molecules m$^{-2}$ s$^{-1}$) impinging on the surface, and emanating from the surface, then

$$S(\theta) = \frac{F_{in} - F_{out}}{F_{in}}.$$

In LOSMS we measure the intensity ($I$) of an ion, which is proportional to the flux of a species ($H_2O$ in this paper) in the LOS molecular beam. As LOSMS is inherently angular dependent, the following analysis assumes that the angular distribution does not alter during adsorption. Although this has not been proved to be the case for the present system, it is a reasonable hypothesis that there will be a cosine distribution for the disordered liquid, and glassy solid surfaces under study here. For highly ordered, single crystal surfaces, which have been extensively studied using molecular beams, one might expect anisotropic effects leading to non-cosine distributions. With no gas pressure applied to the sample surface $F_{out} = 0$, which is the same as a sticking probability of $S = 1$, so, $I_{S=1}$ (measured) $\propto F_{out} = 0$. When $S = 0$, $I_{S=0}$ (measured) $\propto F_{out} = F_{in}$. So $I_{S=0}$ and $I_{S=1}$ (=0) calibrate the sticking probability scale. Experimentally, there will be a background intensity that is independent of pressure ($I_{Back}$), which has to be subtracted, and the intensities have to be normalized by division by the applied pressure, giving the following equation for $S$

$$S(\theta) = \frac{(I_{S=0} - I_{back})/p - (I - I_{back})/p}{(I_{S=0} - I_{back})/p}.$$

For adsorption experiments on solid surfaces, $I_{S=0}$ can be found at saturation of the adsorbed state, where $S$ has dropped to zero. For ionic liquids, where continuous absorption into the bulk occurs (see below), the liquid has to be frozen to prevent absorption, and a temperature chosen where the adsorbate does not stick ($S = 0$), to give $I_{S=0}$.

The experiments were carried out in a diffusion pumped UHV stainless steel vacuum system with a base pressure of $4 \times 10^{-10}$ mbar. The sample consisted of a polycrystalline silver dip-stick ($6 \times 1.5 \times 25$ mm), the lower part of which was coated in a thin film ($\approx$0.5 mm thick) of the ionic liquid. The dip-stick was affixed to a sample holder designed for single crystal samples, such that it stood forward of the sample holder back plate by a few mm. This allowed the dip-stick to be dipped into a glass vial containing the ionic liquid, mounted on a linear motion drive inside the vacuum chamber, Fig. 1. Two type K thermocouples were used for temperature measurement, one attached to the dip-stick, the other to the sample holder back-plate; The sample could be heated and cooled in the temperature range 100–1000 K with an accuracy and stability of $\approx$ 1 K. A clean IL surface was formed by raising the dip-stick, moving the glass vial beneath it, lowering the dipstick so that the lower half was immersed in the liquid, raising the dip stick (now covered with a fresh film of IL), and then retracting the glass vial, Fig. 1. Heating of the dip-stick was restricted to <353 K to prevent evaporation of the IL.

The LOSMS used a VG SX300 quadrupole mass spectrometer which had been converted to pulse counting, mounted inside a liquid nitrogen cooled cryopump shroud, Fig. 1; see ref. 19–21 for a fuller description. The LOSMS focal spot was $\approx$6.5 mm, equal to the width of the dipstick, such that the mass spectrometer only detected species coming from the IL covered dip-stick surface. Water vapour was admitted to the vacuum chamber *via* a leak valve from a gas handling line such that the vapour flooded the chamber. Pressure measurements were carried out using a Bayard Alpert ionisation gauge set to a sensitivity of 1.0 for nitrogen. The flux of water emanating from the dip-stick surface was monitored using the $[OH]^+$ ion at $m/z = 17$.

Sticking probability measurements were carried out (using $[OH]^+$) while the temperature of the surface, and/or the water vapour pressure above the surface was adjusted for the particular experiment. The sticking probability scale was calibrated as follows. With no water vapour present, the signal at $m/z = 17$ is equal to $I_{S=1}$, corresponding to $S = 1$ (*i.e.* no water vapour emanating from the dip-stick). With water vapour present above the ionic liquid which is at or below its glass transition temperature ($T_g$) but above 150 K, the signal is equal to $I_{S=0}$, corresponding to $S = 0$, because the IL is now a solid, and hence cannot absorb water into the bulk, and the temperature is too high for water to adsorb onto the surface. TPD was carried out by dosing the surface at 100 K with a given exposure of water vapour, and then monitoring the desorption using $[OH]^+$ and a heating rate of 0.1 K $s^{-1}$.

## 3 Results

### *S* as a function of *T*

A cooling curve for $[C_8C_1Im][Tf_2N]$, from 295 K to 213 K, in a fixed pressure of water vapour, $\approx 5 \times 10^{-8}$ mbar, is shown in Fig. 2. When the water vapour was first applied to the surface at 295 K, the detected $[OH]^+$ signal rose to a significant extent and then remained constant, indicating that the sticking probability was of a finite and constant value at room temperature. On cooling the surface, the signal began to rise, indicating a decrease in sticking probability. At $\approx$200 K, the signal flattened and became constant. This flattening was coincident with the glass transition temperature of $[C_8C_1Im][Tf_2N]$ at 192 K.[25] As the formation of a solid, glassy, IL surface must necessarily prevent any possibility of water vapour absorbing into the bulk of the IL, we interpret this constant value of $S$ at low temperature as being

 

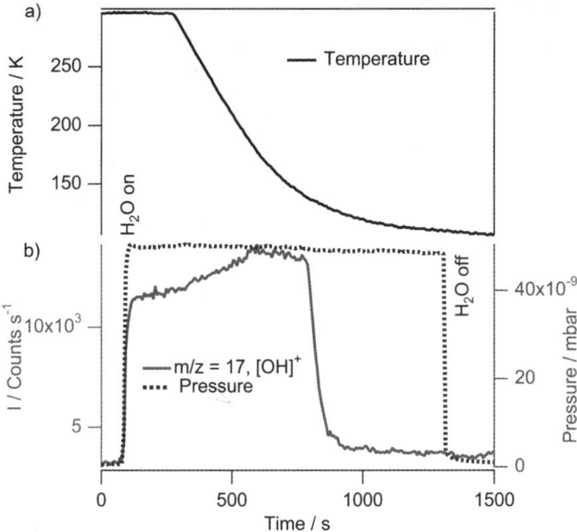


**Fig. 2** Water absorption on $[C_8C_1Im][BF_4]$ during cooling. (a) Temperature *versus* time. (b) Total pressure of water vapour and the detected LOSMS signal for $[OH]^+$ at $m/z = 17$.

due to a value of zero for the water vapour. With this $S = 0$ calibration point, and the $S = 1$ calibration point (signal value for zero water vapour pressure), we are able to define the entire sticking probability scale. On further cooling, the signal rapidly began to drop at $\approx 160$ K, indicating a rapid increase in sticking probability which we interpret as water physisorbing onto the glassy IL surface, first as a monolayer, and then as multilayers. By 140 K the sticking probability had reached 0.9, and by 120 K it had almost reached 1.0. On turning the gas off, there was almost no change in signal, indicating that $S = 1.0 \pm 0.05$.

Fig. 3a shows the data from Fig. 2 plotted as sticking probability, $S$, *versus* temperature, $T$. It can now be seen that at 295 K, $S = 0.12$, decreasing almost linearly with decreasing temperature until it reaches zero at $\approx$ 213 K, slightly ahead of the IL freezing to a glass at 192 K. Further cooling causes a small increase in $S$ for temperatures lower than $\approx$ 177 K, and at 160 K $S$ begins to rise rapidly as water condenses as ice onto the frozen IL surface, reaching a value close to 1.0 by $\approx 110$ K.

Similar cooling experiments were carried out for water sticking to $[C_2C_1Im][Tf_2N]$, again using a pressure of $\approx 5 \times 10^{-8}$ mbar, Fig. 3b, with similar results. At 320 K $S = 0.12$, and drops almost linearly with decreasing temperature to 192 K, where it reaches zero. (The glass transition temperature of $[C_2C_1Im][Tf_2N]$ is 186 K[26]). The sticking probability remains at zero until $\approx 152$ K, where it abruptly begins to increase due to water ice condensation onto the IL glass surface. For temperatures <135 K, $S \approx$ 1.0.

## $S$ as a function of coverage for isothermal adsorption

To elucidate the behaviour of the sticking probability with surface coverage, a series of isothermal adsorption studies were carried out for both ILs, from 113 to 295 K. To understand the behaviour of water at the surface of the liquid, and the glassy solid, two exemplars are described, at 295 K, where the ILs are liquids, and 143 K, where they are glassy solids.

In Fig. 4a we show how the $[OH]^+$ signal behaved when water vapour was applied to the $[C_2C_1Im][Tf_2N]$ surface at 295 K. The profile of the pressure change and the $[OH]^+$ signal with respect to time, are identical within experimental error. This indicates, firstly, that the sticking probability of the water on the liquid surface is constant.

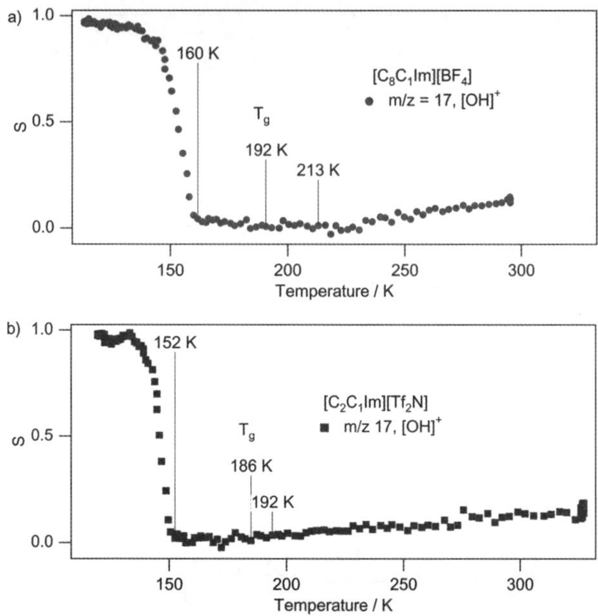


**Fig. 3** *S versus T* for water absorption on the ILs during cooling. (a) $[C_8C_1Im][BF_4]$. (b) $[C_2C_1Im][Tf_2N]$.

Secondly, from Fig. 3b we know that the sticking probability at this temperature is $S = 0.12$, so Fig. 4a shows that the sticking probability remained fixed at this value during the entire adsorption. Fig. 4b shows the sticking probability as a function of exposure to water vapour, using this known value of $S = 0.12$. As we have both the sticking probability ($S = 0.12$, a dimensionless number) and the exposure (in mbar s) the surface coverage (in mbar s) is given simply by $S \times$ exposure. This coverage (or more precisely, amount absorbed) is plotted as a dotted line in Fig. 4b, which shows that the amount absorbed is close to $5 \times 10^{-5}$ mbar s by the end of the experiment. In Fig. 4c the sticking probability is re-plotted as a function of coverage, illustrating that the sticking probability remains constant at $0.12 \pm 0.02$ over the entire range. Such behaviour is not easily explained if the water is accumulating as an adsorbed state at the surface, as it shows no sign of saturating. However, if we assume that the adsorbed water, $H_2O_{ads}$, can pass through the IL surface to absorb into the bulk, forming $H_2O_{abs}$, then $S = 0.12$ is readily understandable as the overall sticking probability of water vapour passing from the gas phase to the bulk phase.

Isothermal adsorption experiments carried out using $[C_8C_1Im][BF_4]$ at 295 K gave very similar results to those for $[C_2C_1Im][Tf_2N]$ above, with a constant sticking probability of $0.12 \pm 0.02$ up to a coverage of $3.5 \times 10^{-6}$ mbar s. The interpretation is the same, that water vapour was absorbing into the bulk of the IL.

For isothermal adsorption at 143 K, both ILs were already frozen to glassy solids, and water was expected to adsorb as multilayers of ice, see Fig. 3. Fig. 5a shows the isothermal adsorption of water vapour on $[C_8C_1Im][BF_4]$ at 143 K. On increasing the pressure, the $[OH]^+$ signal immediately rose to a high value and remained almost constant for a short period. Using the calibration point for $S = 0$ described above, the intensity of the $[OH]^+$ was found to correspond to a sticking probability of zero. With increasing exposure, the signal began to drop, indicating an increasing sticking probability, until after 800 s, when the applied pressure was reduced back to zero, there was a small drop in signal, indicating that $S$ has not quite reached a value of 1.0. Fig. 5b shows the $S$ and coverage *versus* exposure plots, calculated from

 

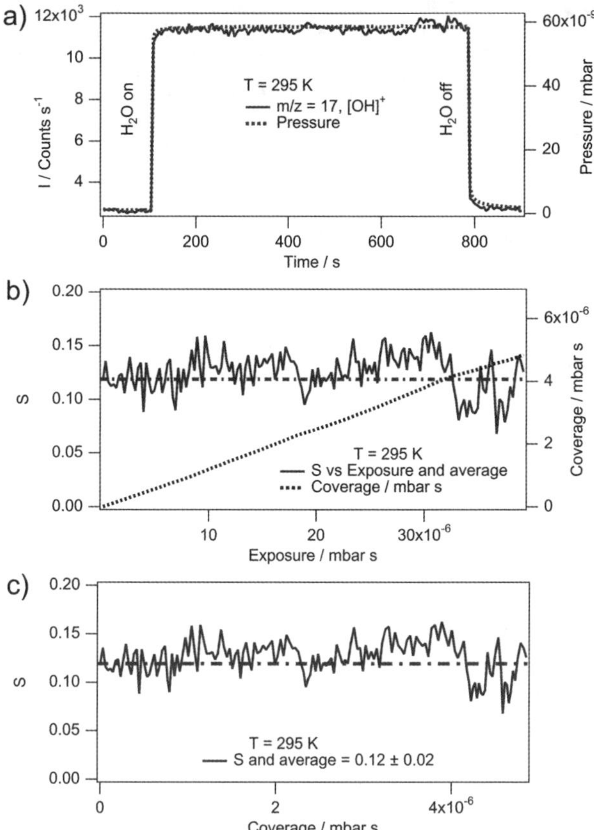


**Fig. 4** Isothermal adsorption of $H_2O$ on [$C_2C_1$Im][$Tf_2$N] at 295 K. (a) Total pressure and the LOSMS signal using $[OH]^+$ *versus* time. (b) *S* and coverage *versus* exposure. (c) *S versus* coverage.

Fig. 5a, while Fig. 5c shows S *versus* coverage extracted from Fig. 5b. The adsorption behaviour in Fig. 5c is unusual in that the sticking probability at zero coverage is zero, but rises rapidly as the coverage increases. Further increase in coverage then leads to further increase in *S*, but at a slowing rate, the value of *S* reaching 0.9 by $2.5 \times 10^{-5}$ mbar s. At this temperature we expect multilayer water ice to occur on the surface of the frozen IL, which accounts for the high coverages in Fig. 5c.

Adsorption systems in which the sticking probability increases for increasing coverage are unusual, but not unknown, in the field of surface science. For instance, nitrogen adsorption on Ru(0001)[27] and noble gases on metal surfaces[28] both exhibit this behaviour, as does Pb on clean and electron irradiated polymethylmethacrylate (PMMA).[29] In the case of Pb on PMMA the sticking probability is zero on the clean surface, but it rapidly increases with increasing coverage when electron irradiation forms nuclei for the growth of large 3D islands. We shall use similar ideas of island growth at nuclei to explain the adsorption behaviour of water on ionic liquid surfaces.

Experiments carried out using [$C_2C_1$Im][$Tf_2$N] at this temperature, showed the same qualitative behaviour, exhibiting an initial sticking probability of zero, which rapidly rose with increasing coverage. If adsorption was interrupted by dropping the dosing pressure to zero after some adsorption had occurred, it was found that on bringing the pressure back to its original value, the sticking probability returned

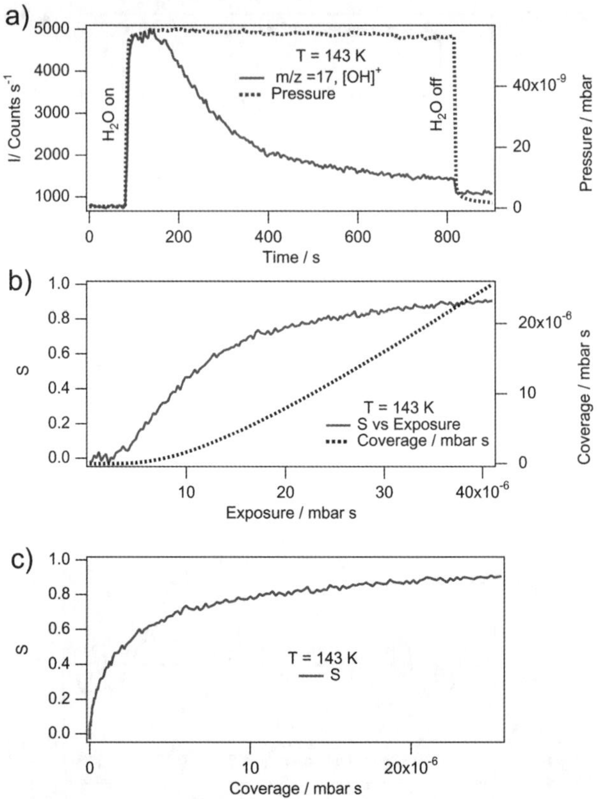


**Fig. 5** Isothermal adsorption of $H_2O$ on $[C_8C_1Im][BF_4]$ at 143 K. (a) Total pressure and the LOSMS signal using $[OH]^+$ *versus* time. (b) *S* and coverage *versus* exposure. (c) *S versus* coverage.

to the value it had immediately before the gas pressure was reduced, and subsequently the sticking probability continued its increase with increasing coverage. For both ILs, *S* tended towards a value of 1.0 for very high coverages.

A set of adsorption isotherms were measured from 113–295 K for each IL, and analysed in the same way as described above for the 295 K and 143 K isotherms. Fig. 6 shows the isotherms for $[C_8C_1Im][BF_4]$, while Fig. 7 shows those for $[C_2C_1Im][Tf_2N]$. For $[C_8C_1Im][BF_4]$, *S* has a constant value of 0.12 at 295 K, the value reducing, but remaining constant with coverage as the adsorption temperature drops to 228 K, Fig. 6b. For adsorption temperatures in the region of the glass transition temperature, 192 K, the value of S becomes zero and constant. At lower temperatures, the initial value of S is zero, but rises as the coverage rises, *e.g.* 146 K. For a continued reduction in adsorption temperature (down to 135 K) the initial value of *S* remains at zero, but the rate of increase of *S* with coverage rises rapidly and tends towards 1.0 for high coverages. At the lowest temperature studied, 116 K, the increase in *S* is so fast that the initial value appears to be 0.98, which then rises to a value of 1.0 for increasing coverage.

These results can be interpreted as follows. For adsorption at temperatures 295 > $T$ > 192 K (the glass transition temperature), $H_2O$ dissolves into the bulk of the IL, but as the temperature drops such dissolution becomes increasingly difficult, so the sticking probability reduces. When the glass transition temperature is reached, dissolution becomes impossible, and the sticking probability becomes zero. Around the

 

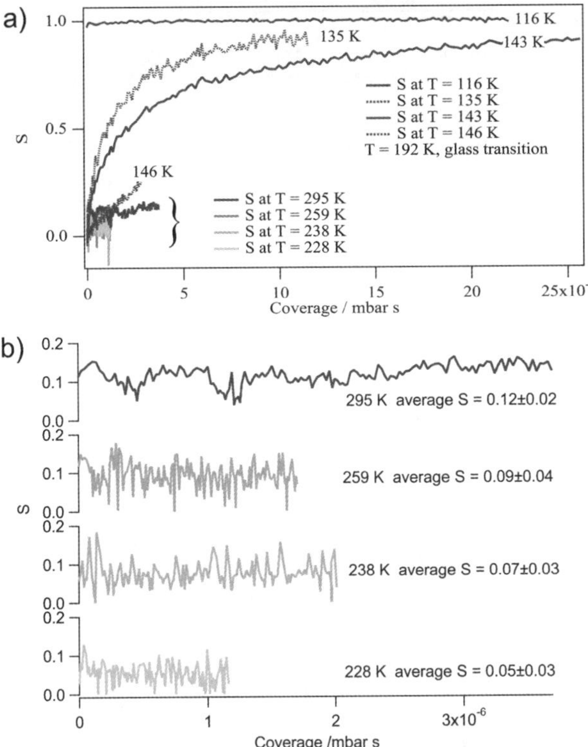


**Fig. 6** (a) *S versus* coverage for $H_2O$ adsorption on $[C_8C_1Im][BF_4]$ in the temperature range 116 K to 295 K. (b) Expanded views of the higher temperature plots in (a).

glass transition temperature, $S = 0$ and no adsorption occurs as the temperature is too high to allow $H_2O$ to populate the physisorbed state on the surface. At still lower temperatures the initial sticking probability is zero, but $S$ then increases as the coverage increases as physisorption occurs. This unusual behaviour is what one would expect from an adsorbate that bonds more strongly to itself, than to the substrate. In other words, the $H_2O\cdots H_2O$ interaction on the surface is stronger than the $H_2O\cdots$IL interaction. This in turn implies that the IL surface is hydrophobic, with very few exposed charged species, which would be expected to bond $H_2O$ rather strongly. As the adsorption temperature falls further, the lifetime of $H_2O$ adsorbed onto the hydrophobic surface increases, allowing it to diffuse across the surface and bond to available pre-adsorbed $H_2O$ nuclei, or islands. This accounts for the rapidly increasing sticking probability with increasing coverage for $T \leq 146$ K.

The adsorption isotherms for water on $[C_2C_1Im][Tf_2N]$ (glass transition temperature of 186 K), Fig. 7, are qualitatively the same as for water adsorption on $[C_8C_1Im][BF_4]$. For adsorption above the glass transition temperature, the similarity between the two might be expected. However, below the glass transition temperature, one might expect the lack of any long alkyl chains in $[C_2C_1Im][Tf_2N]$ to lead to a surface with some exposed charged species at the surface, which water could bond to rather strongly. However, the water behaves as though it were on a hydrophobic surface. This suggests that the $CF_3$ groups of $[Tf_2N]^-$ and the ethyl group of $[C_2C_1Im]^+$ must be orientated outwards at the surface of the IL, presenting, again, a hydrophobic surface to the adsorbing $H_2O$. A quantitative model for adsorption will be presented in the discussion.

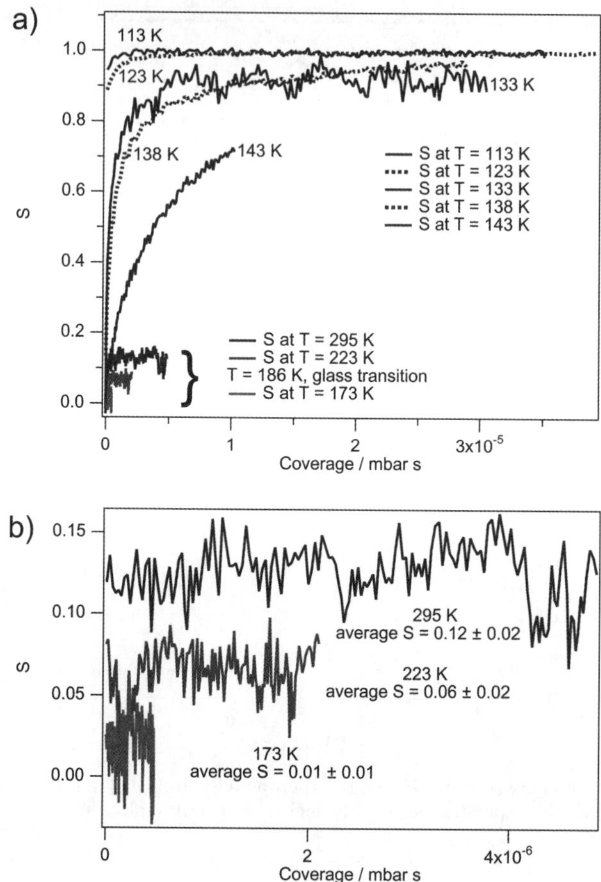


**Fig. 7** (a) *S versus* coverage for $H_2O$ adsorption on $[C_2C_1Im][Tf_2N]$ in the temperature range 113 K to 295 K. (b) Expanded view of the higher temperature plots in (a).

## Temperature programmed desorption

Temperature programmed desorption was carried out using LOSMS to detect the $[OH]^+$ ion to monitor the rate of desorption of water from the IL surface. Dosing was carried out at 100 K, and the initial coverage determined by integrating the area under the desorption peaks.

Desorption was monitored from $\approx$110 K up to 300 K, as previous XPS work[30] had observed the existence of strongly bound adsorbed water on $[C_8C_1Im][BF_4]$ which disappeared from the surface at $\approx$245 K. However during TPD of water using LOSMS, no desorption peak could be detected at this temperature from either the $[C_8C_1Im][BF_4]$ or the $[C_2C_1Im][Tf_2N]$ surfaces. From this lack of any detectable desorption of water into the gas phase, we conclude that the water in this strongly bound state at $\approx$245 K on $[C_8C_1Im][BF_4]$ must be lost from the surface by diffusion into the bulk of the IL, rather than desorption into the gas phase.

Fig. 8 shows the desorption peaks that were observed, in the lower temperature range of 130 K–170 K, where physisorption of $H_2O$ was expected for $[C_8C_1Im][Tf_2N]$. A low initial coverage ($\theta = 0.12$) gave a peak at $\approx$139 K, Fig. 8a. As the initial coverage increased, $0.12 < \theta \leq 1.0$, this desorption peak grew larger, and shifted to higher temperatures, reaching $\approx$150 K after an exposure of $\approx 5 \times 10^{-6}$ mbar s, which we define as monolayer coverage with $\theta = 1.0$. Further increase in

 

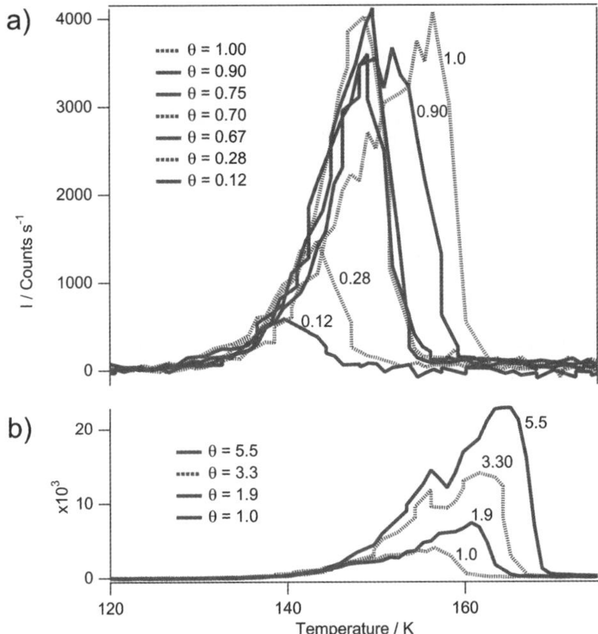


**Fig. 8** (a) Temperature programmed desorption of water from $[C_8C_1Im][BF_4]$ for increasing initial coverages $0 < \theta \leq 1$, using the $[OH]^+$ ion at $m/z = 17$. (b) The $\theta \geq 1$ curves.

the initial coverage then caused the peak to grow further in intensity, with a fractional, or zero order peak shape emerging, Fig. 8b, which we identify as being due to multilayer adsorption of water. (Zero order desorption peaks are characterised by an exponentially increasing signal with increasing temperature, followed by an abrupt drop in signal to zero at high temperature; fractional order peaks are similar but with a less abrupt cut off).

Although the desorption curves in Fig. 8(a) appear to be zero order, they cannot be truly zero order, as the initial coverages are all less than a monolayer. (True zero order desorption can only occur for multilayer desorption, where the desorbing area remains fixed as the multilayers are stripped from the surface, giving rise to an exponentially increasing signal followed by an abrupt cut off, see ref. 23 and 24 for examples). Here we expect first order desorption as we have molecularly adsorbed water desorbing to form molecular water vapour. One way of achieving fractional order desorption kinetics in the monolayer regime is to invoke a model in which water forms islands, surrounded by a dispersed phase, where the desorption kinetics are governed by the length of the island edges which increase as $\sqrt{\theta}$, as the coverage increases. However, this model fails when the coverage moves towards monolayer, as the islands coalesce and hence the length of the island edges will reduce with increasing coverage, becoming zero at monolayer. Clearly the desorption curves in Fig. 8a appear zero-order-like up to initial coverages of a monolayer, so the island perimeter model cannot be used.

A close examination of the curves in Fig. 8(a) shows that they each have the asymmetric shape characteristic of 1st order desorption. Calculated desorption curves have therefore been fitted to the experimental curves for $0 < \theta \leq 1.0$ using first order desorption kinetics and an assumed pre-exponential factor of $10^{13}$ $s^{-1}$, Fig. 9a. The assumed value for the pre-exponential gives an absolute error of $\pm 4$ kJ $mol^{-1}$ in the activation energies for desorption, $E_d$. Rather surprisingly, each curve can be fitted rather well with a first order shape and a fixed activation energy. For the lowest coverage, $\theta = 0.12$, the activation energy was $E_d = 39.9$ kJ $mol^{-1}$. This corresponds

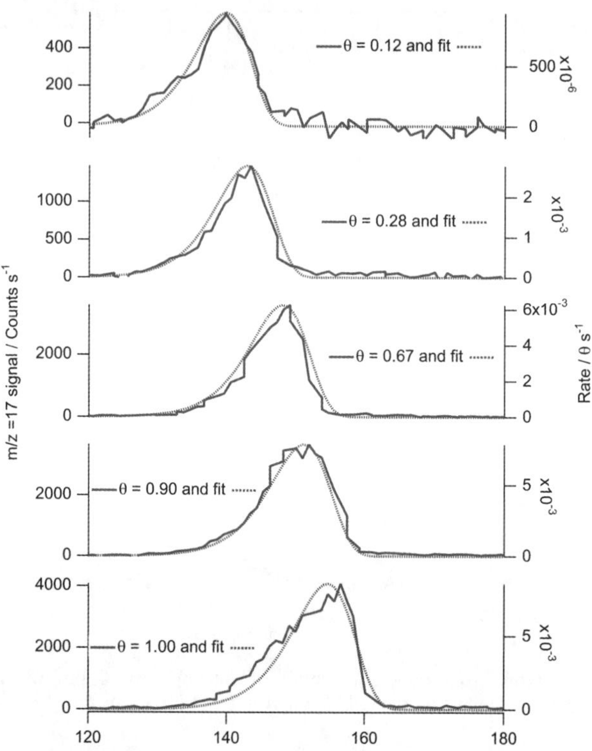


**Fig. 9** TPD peaks for $\theta \leq 1$ from Fig. 8 with fitted first order desorption peak shapes.

to small islands, or possibly individual $H_2O$ molecules, adsorbed on the clean, hydrophobic, IL glass surface. As the initial coverage increased, $E_d$ also increased, reaching 43.4 kJ mol$^{-1}$ for $\theta = 1.0$, see Table 2 and Fig. 11. The entire data set can be fitted with an activation energy which increases with increasing initial coverage, but which does not change during the desorption process itself. This is consistent with attractive interactions existing between the adsorbed $H_2O$ such that as they cluster together on the hydrophobic IL surface, the activation energy for desorption from the water cluster is higher than from a smaller cluster or an individual adsorbed molecule. However, it is not clear why the activation energy for, say, $\theta = 0.9$ remains fixed at 42.4 kJ mol$^{-1}$ throughout the desorption scan, even though the coverage must drop from 0.9 to zero in the process. Such behaviour suggests

**Table 2** Desorption energies, $E_d$, for water from the two ionic liquid surfaces

| $[C_2C_1Im][Tf_2N]$ | | $[C_8C_1Im][BF_4]$ | |
|---|---|---|---|
| $\theta$ | $E_d$/kJ mol$^{-1}$ | $\theta$ | $E_d$/kJ mol$^{-1}$ |
| 0.2 | 41.5 | 0.12 | 39.9 |
| 0.5 | 42.5 | 0.28 | 40.0 |
| 1.0 | 44.4 | 0.67 | 41.5 |
| >1.0 | 46 | 0.7 | 41.6 |
| — | — | 0.75 | 41.8 |
| — | — | 0.9 | 42.4 |
| — | — | 1.0 | 43.4 |

 

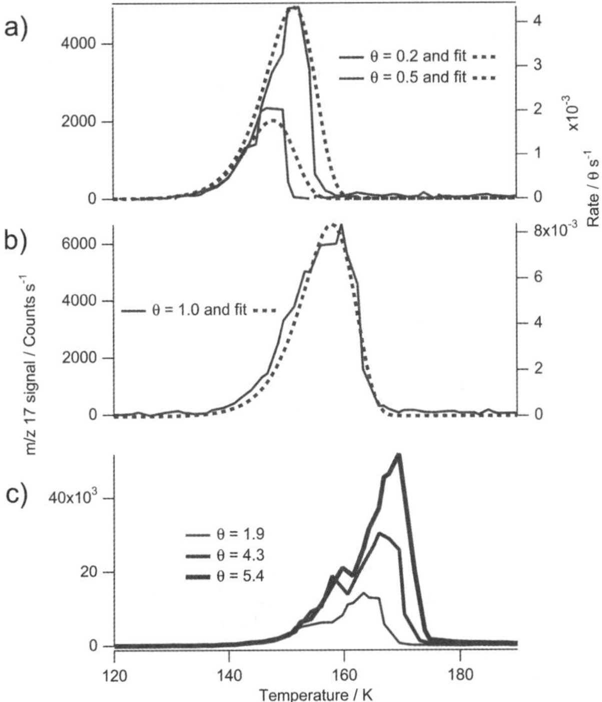


**Fig. 10** Temperature programmed desorption of water from $[C_2C_1Im][Tf_2N]$ for increasing initial coverages ($\theta$), using the $[OH]^+$ ion at $m/z = 17$. (a) and (b) $0 < \theta \leq 1$ with first order fits. (c) $\theta > 1.0$.

a degree of aggregation which is set during initial adsorption, but which then remains fixed during desorption.

Fig. 8b shows multilayer desorption profiles, all of which resemble zero order desorption, with an approximately exponentially increasing leading edge, followed by a precipitate drop at high coverage. For multilayer adsorption, zero order is the expected behaviour, but there is a noticeable, and reproducible, shoulder at 156 K in both the $\theta = 3.3$ and $\theta = 5.5$ curves. It is known that multilayers of water ice deposited at <145 K, grow as a metastable phase and that this metastable phase has a higher vapour pressure than the crystalline phase at higher temperature.[9] During heating, this metastable phase rearranges to the more stable crystalline form, causing a drop in desorption intensity in the TPD. We conclude that the shoulders in the multilayer TPD curves in Fig. 8(b) are due to this rearrangement.

The same desorption experiments were carried out for $[C_2C_1Im][Tf_2N]$, Fig. 10. The behaviour is very similar to that for water on $[C_8C_1Im][BF_4]$. $\theta = 1$ was again chosen at a coverage just lower than that at which multilayer growth became evident from the growth of a fractional order peak shape. For $\theta \leq 1$ the peaks were fitted with first order desorption kinetics, see fits in Fig. 10, with assumed pre-exponentials of $1 \times 10^{13}$ s$^{-1}$ and a heating rate of 0.1 K s$^{-1}$. The activation energy rose from 41.5 kJ mol$^{-1}$ at $\theta = 0.2$ to 44.4 kJ mol$^{-1}$ at $\theta = 1$, see Table 2. Again, a shoulder is observed in the multilayer desorption curves, corresponding to the amorphous to crystalline phase change of the water.

Fig. 11 shows how the activation energy changes for first order desorption in the monolayer regime of $0 < \theta \leq 1$, for both ionic liquids. Although the absolute error in these values is $\pm 4$ kJ mol$^{-1}$, the random errors are much less, at

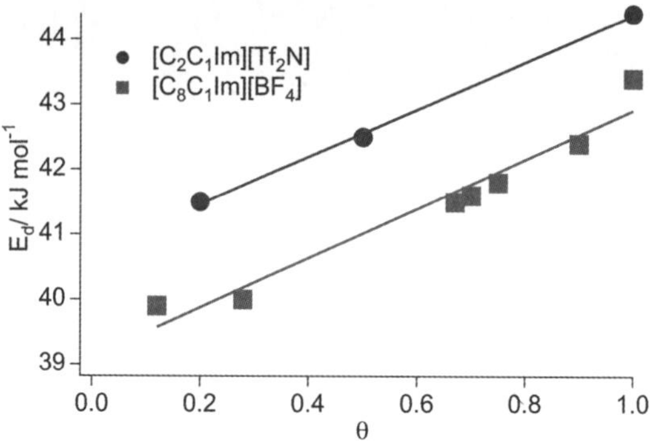


**Fig. 11** $E_d$ *versus* $\theta$ for initial coverages of $\theta < 1$ for water desorbing from $[C_2C_1Im][Tf_2N]$ and $[C_8C_1Im][BF_4]$. Linear regression fits are shown for both ILs.

$\pm$ 0.3 kJ $mol^{-1}$. From Fig. 11 we can see that the values of $E_d$ at zero coverage are different for the two ILs, being 39.1 $\pm$ 0.5 and 40.7 $\pm$ 0.2 kJ $mol^{-1}$ for $[C_8C_1Im][BF_4]$ and $[C_2C_1Im][Tf_2N]$, respectively. The lower value for the $[C_8C_1Im][BF_4]$ surface is probably directly attributable to the existence of the large, hydrophobic, octyl chain at the surface of the IL, while the slightly larger value for $[C_2C_1Im][Tf_2N]$ probably reflects the somewhat less hydrophobic ethyl and $CF_3$ groups. It is also clear that $E_d$ increases at the same rate with respect to coverage for both surfaces, which is consistent with the increase being due to $H_2O \cdots H_2O$ lateral interactions, which are relatively unaffected by the bonding to the substrate.

It is worth noting here that TPD of water adsorbed on a hexadecane thiolate SAM on gold,[11] *i.e.* an organic overlayer consisting of orientated $C_{16}$ alkyl chains, exhibits behaviour which is remarkably similar to the data shown in Fig. 8 and 10 for the two IL surfaces. The activation energy for desorption was found to be 35 kJ $mol^{-1}$, slightly lower than our values. This is consistent with the ILs in this study forming an organic overlayer somewhat similar to that of the SAM, which therefore produces a similarly small heat of adsorption for adsorbed water.

## 4 Discussion

### Ionic underlayer

A recent metastable atom electron spectroscopy measurement[15] has shown that for $[C_8C_1Im][BF_4]$, the ions arrange themselves into an ionic underlayer consisting of the $[BF_4]^-$ and the imidazolium groups, with the long octyl chain forming an outer layer which covers the $[BF_4]^-$ ion and the charged imidazolium ion. The same work also shows that if the $[BF_4]^-$ is replaced by $[Tf_2N]^-$, the $CF_3$ groups of the $[Tf_2N]^-$ are orientated towards the vacuum side, with the $SO_2$ groups pointing towards the bulk. In a recent work using X-ray reflectivity,[18] Lauw *et al.* have shown that a significant amount of water can be preferentially adsorbed at the surface of an IL (1-butyl-1-methylpyrrolidinium trifluoromethylsulfonylimide, $[C_4C_1Pyrr][Tf_2N]$) at room temperature, the water being associated with cations to a depth of some 30 Å. A similar underlayer has also been detected by XPS[30] for water adsorption on $[C_8C_1Im][BF_4]$ at low temperatures. In that study the water disappeared from the surface on heating *via* a first order process with a peak maximum at $\approx$240 K corresponding to an activation energy of 74 $\pm$ 4 kJ $mol^{-1}$ with an assumed pre-exponential of $1 \times 10^{13}$ $s^{-1}$. The water was thought to be adsorbed in an ionic underlayer below a surface consisting of ordered octyl chains from the $[C_8C_1Im]^+$. We have shown above that no desorption peak was observed in the region of 240 K for either $[C_8C_1Im][BF_4]$ (the same IL as used in the XPS experiment) or $[C_2C_1Im][Tf_2N]$. We can therefore say that water adsorbed in the underlayer for $[C_8C_1Im][BF_4]$ must diffuse into the bulk at 240 K and above, rather than desorbing back into the gas phase. This implies that the activation energy for moving from the ionic underlayer to the bulk liquid is lower than from the ionic underlayer to the gas phase (both represent a semi-infinite, empty reservoir for the water).

## Absorption kinetics

We now develop a model for the absorption kinetics of water on the hydrophobic surfaces of both ILs in this study. Water in the gas phase, $H_2O_g$, can physisorb onto the hydrophobic surface (consisting of the long octyl chain in the case of $[C_8C_1Im][BF_4]$, or the ethyl chain on the imidazolium and the $CF_3$ groups on the $[Tf_2N]^-$ for the $[C_2C_1Im][Tf_2N]$), to form $H_2O_p$. Water can also reside in the ionic underlayer, $H_2O_u$, and it can dissolve into the bulk of the ionic liquid, $H_2O_b$. The mechanistic steps, and their rate constants, are therefore

$$H_2O_g \rightarrow H_2O_p \ k_1 \quad (1)$$

$$H_2O_p \rightarrow H_2O_g \ k_2 \quad (2)$$

$$H_2O_p \rightarrow H_2O_u \ k_3 \quad (3)$$

$$H_2O_u \rightarrow H_2O_b \ k_4 \quad (4)$$

As water was not observed to desorb back into the gas phase after being adsorbed into the ionic underlayer, we have omitted the back reaction from underlayer to physisorbed state and can simply consider steps (1) to (3) when deriving an expression for the rate of absorption.

A steady state analysis of (1) to (3), where $H_2O_p$ is the intermediate, gives

$$[H_2O_p] = \frac{k_1[H_2O_g]}{k_2 + k_3}. \quad (5)$$

Our measured sticking probability is the flux that sticks, divided by the flux that hits, which gives

$$S = \frac{k_1[H_2O_g] - k_2[H_2O_p]}{k_1[H_2O_g]}. \quad (6)$$

On substituting in the expression for $[H_2O_p]$, this becomes

$$S = 1 - \frac{k_2}{k_2 + k_3}. \quad (7)$$

If $k_2 = k_3$ (the rates for desorption back into the gas phase, and penetration into the ionic underlayer are equal) then $S = 0.5$ exactly. If $k_2 > k_3$ (desorption predominates) then $S < 0.5$, and if $k_2 < k_3$ (penetration into the ionic underlayer predominates) then $S > 0.5$. For both ionic liquids $S < 0.5$, so $k_2$, the rate of desorption back into the gas phase, is higher than $k_3$, the rate of adsorption into the ionic underlayer.

Replacing $k_2$ and $k_3$ with their Arrhenius expressions

$$k_2 = A_2 \exp\left(-\frac{E_2}{RT}\right) \text{and } k_3 = A_3 \exp\left(-\frac{E_3}{RT}\right) \tag{8}$$

and substituting these into (7), we obtain

$$\ln\left(\frac{S}{1-S}\right) = \ln\left(\frac{A_3}{A_2}\right) - \frac{\Delta E}{RT} \tag{9}$$

where $\Delta E = E_3 - E_2$.

Plots of $\ln(S/(1-S))$ *versus* $1/T$ are shown for both ILs in Fig. 12 from which values of $\Delta E$ can be derived of $7.0 \pm 0.8$ kJ mol$^{-1}$ and $16.8 \pm 2.0$ kJ mol$^{-1}$ for $[C_2C_1Im][Tf_2N]$ and $[C_8C_1Im][BF_4]$, respectively. If we regard water residing in the ionic underlayer as a surface species, then these values of $\Delta E$ both represent activated adsorption, where the activation energy for adsorption from the physisorbed layer, 47 kJ mol$^{-1}$ (=40 + 7) for $[C_2C_1Im][Tf_2N]$ and 57 kJ mol$^{-1}$ (=40 + 17) for $[C_8C_1Im][BF_4]$) is greater than the physisorption energy ($\approx$40 kJ mol$^{-1}$ at zero coverage for both ILs). Also, as the water then continues to move into the liquid once it has reached the ionic underlayer, one could also describe the process as activated absorption. Significantly, the $[C_2C_1Im][Tf_2N]$ surface, which consists of relatively small ethyl and $CF_3$ groups, presents a smaller barrier to water penetration than the $[C_8C_1Im][BF_4]$ surface, which consists of larger octyl chains.

From eqn (9) it can be seen that the intercept of the graphs in Fig. 12 are equal to $\ln(A_3/A_2)$. For $[C_8C_1Im][BF_4]$ $A_3/A_2 = 148$, while for $[C_2C_1Im][Tf_2N]$ $A_3/A_2 = 2.2$.

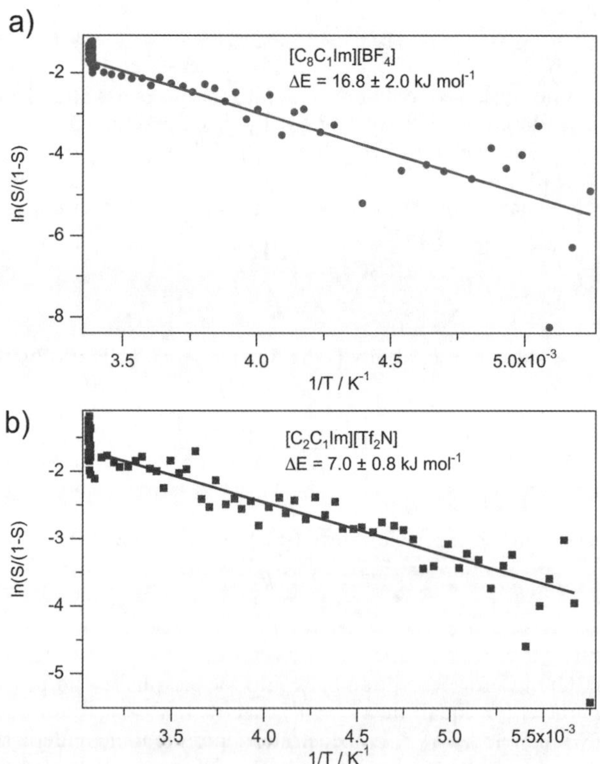


**Fig. 12** $\ln(S/(1-S))$ *versus* $1/T$, best fits, and the corresponding $\Delta E$ values for water adsorption on (a) $[C_8C_1Im][BF_4]$ and (b) $[C_2C_1Im][Tf_2N]$.

 

As $A_2$, the pre-exponential for desorption from the physisorbed state back into the gas phase, might be expected to be the same for both IL surfaces, the difference is due to $A_3$, which is ≈ 68 × larger for penetration through the octyl chains, than through the ethyl and $CF_3$ chains. The reason for this difference is not apparent at present.

For $[C_8C_1Im][BF_4]$ the XPS study[30] provided an activation enthalpy of 76 kJ mol$^{-1}$, while temperature dependent absorption studies of water into $[C_8C_1Im][BF_4]$[31] gave an enthalpy of absorption of 34 kJ mol$^{-1}$. For temperatures in the region of 150 K, R$T$ = 1.25 kJ mol$^{-1}$, so the enthalpy of physisorption (=$E_2$ + R$T$) is 41 kJ mol$^{-1}$ while $\Delta E$ = 17 kJ mol$^{-1}$. These can be combined into a potential energy diagram for water adsorption on $[C_8C_1Im][BF_4]$, Fig. 13A. For $[C_2C_1Im][Tf_2N]$ the enthalpy of activation from $\Delta E$ is 7 kJ mol$^{-1}$ but there are, as yet, no values available for the activation energy from the ionic underlayer to the bulk, nor for the enthalpy of absorption, Fig. 13B.

## Adsorption kinetics

Finally we discuss the adsorption kinetics of water on the glassy IL surfaces at temperatures in the region of 140 K. From the TPD experiments it was possible to determine that an exposure of $\approx 5 \times 10^{-6}$ mbar s at 100 K ($S = 1$) was equivalent to $\theta = 1$ for both of the ILs studied, which allows the coverage in mbar s in Fig. 6 and 7 to be converted to $\theta$ values.

Our adsorption model is the following. We assume that the surface is hydrophobic, but that there is a low concentration of centres ($\theta < 0.01$), due to defects or exposed ionic head groups, where water molecules can adsorb. These centres act as nuclei for $H_2O$ island growth. When water from the gas phase, $H_2O_g$, impinges on one of these nuclei it sticks, or if it impinges on a pre-existing island of water ice, $H_2O_{is}$, it sticks with a sticking probability of 1.0 into the second layer to form multilayer water ice, $H_2O_{ML}$.

$$H_2O_g + H_2O_{is} \rightarrow H_2O_{ML} \tag{10}$$

Water molecules that impinge onto the clean IL surface are adsorbed into a physisorbed state on the hydrophobic surface ($H_2O_p$), as described above, eqn (1), where we assume that the IL liquid surface, and the IL glassy surface are similar in structure.

The physisorbed water has a lifetime, τ, given by

$$\tau = \tau_0 \exp\left(\frac{E_{des}}{RT}\right) \tag{11}$$

where $\tau_0 \approx 10^{-13}$ s$^{-1}$ and $E_{des}$ (=$E_2$ in eqn (8)) is the desorption energy of $H_2O$ from the IL glass at zero coverage. For zero coverage, $E_{des}$ was about 40 kJ mol$^{-1}$ for both ILs, which corresponds to $\tau \approx 20$ s at 146 K. If we assume there are barriers to diffusion across the glassy IL surface, the physisorbed water will execute a random walk from its first point of contact with the surface during its surface lifetime τ.

If a water molecule lands within a certain critical distance of a nucleation site, or an island of water that has grown around the nucleation site, it can diffuse to the nucleation site, or islands edge, where it adsorbs permanently due to the increased stability of either the $H_2O \cdots H_2O$ interaction at the island edge, or the stability conferred by the nucleation site itself. If the water molecule lands further away than this critical distance, then it fails to diffuse to the island edge during the lifetime τ and desorbs back into the gas phase, the net sticking probability being zero. The measured sticking probability is therefore proportional to the perimeter length of the islands, which is proportional to $\sqrt{\theta}$. We can write

$$S = k_0\sqrt{\theta} \tag{12}$$

where $k_0$ is a proportionality constant which depends on the critical distance from the island edge, which itself depends on the square root of the number of diffusion hops, $H$

$$k_0 = C\sqrt{H} \tag{13}$$

where C is another constant. Any water molecules that impinge onto a pre-existing island will have a sticking probability of 1, so the experimentally measured sticking probability becomes

$$S = \theta + k_0\sqrt{\theta}. \tag{14}$$

If the barrier to surface diffusion is $E_{diff}$ then the number of hops is simply the ratio of the lifetime on the surface, $\tau_{surface}$, to the life-time in a site before doing a hop, $\tau_{site}$,

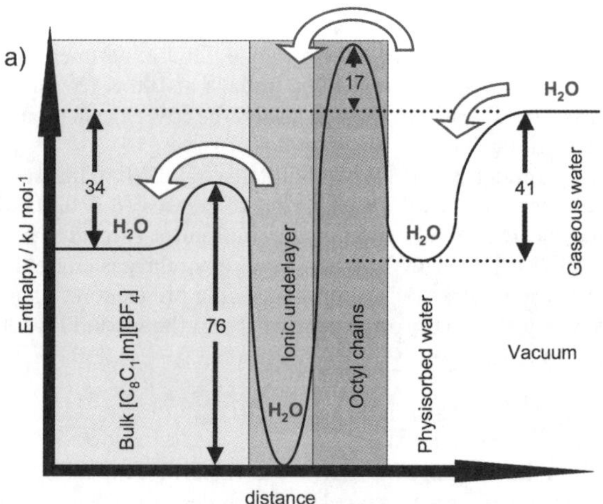


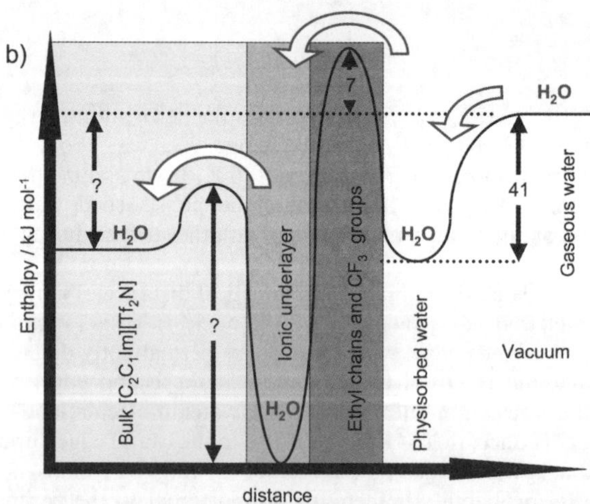


**Fig. 13** Potential energy diagrams of enthalpy *versus* distance along the surface normal for (a) $[C_8C_1Im][BF_4]$ and (b) $[C_2C_1Im][Tf_2N]$. The arrows indicate absorption of water from gas to bulk IL.

 

**Table 3** $k_0$ values derived from $S$ *vs.* $\theta$ plots. Uncertainty is shown in brackets

| $[C_8C_1Im][BF_4]$ | | $[C_2C_1Im][Tf_2N]$ | |
|---|---|---|---|
| $T$/K | $k_0$ | $T$/K | $k_0$ |
| 146 | 0 | 143 | 0.22(2) |
| 144 | 0.56(3) | 138 | 1.91(10) |
| 143 | 0.54(3) | 133 | 2.28(16) |
| 141 | 0.65(4) | — | — |
| 138 | 1.34(5) | — | — |
| 135 | 0.93(7) | — | — |

$$H = \frac{\tau_{\text{surface}}}{\tau_{\text{site}}} = \frac{\tau_0 \exp(E_{\text{des}}/RT)}{\tau_0 \exp(E_{\text{diff}}/RT)}. \quad (15)$$

If $\Delta E_{\text{dd}} = E_{\text{des}} - E_{\text{diff}}$ and we assume the same value for the pre-exponential, $\tau_0$, we can then write

$$k_0 = \text{C}\exp(\Delta E_{\text{dd}}/2RT). \quad (16)$$

Fig. 14a and b shows examples of the fits, using eqn (14), to the experimental $S$ *vs.* $\theta$ data for $[C_8C_1Im][BF_4]$ and $[C_2C_1Im][Tf_2N]$, respectively, while Table 3 lists the $k_0$ values. The fits were carried out for $0 < \theta < 0.05$ because at higher coverages one would expect the islands of water ice to begin to coalesce, and hence the $\sqrt{\theta}$ behaviour would no longer be valid. It can be seen that as the temperature drops, the value of $k_0$ increases. For $[C_8C_1Im][BF_4]$ at 146 K, $k_0$ is effectively zero, as shown by the $S = \theta$ line, which corresponds to a zero width perimeter around the islands from which physisorbed water can successfully diffuse to the island edge to stick. For lower temperatures, the value of $S$ with respect to $\theta$ rises faster than an $S = \theta$ curve, (dotted lines on the $[C_8C_1Im][BF_4]$ plots) indicating a broadening width around each island inside which water molecules can find the island edge and stick.

A plot of ln $k_0$ *vs.* $1/T$, eqn (16), has a gradient of $\Delta E_{\text{dd}}/2R$. For $[C_8C_1Im][BF_4]$ $\Delta E_{\text{dd}} = 26.5 \pm 12.0$ kJ mol$^{-1}$, which for a desorption energy of $E_{\text{des}} \approx 40$ kJ mol$^{-1}$ gives an activation energy for diffusion of $13 \pm 12$ kJ mol$^{-1}$. For a disordered, glassy, surface, this activation energy must be an average over a range of activation energies corresponding to different barriers to diffusion. For $[C_2C_1Im][Tf_2N]$ the value of $\Delta E_{\text{dd}}$ ($73 \pm 38$ kJ mol$^{-1}$) has too large an error to allow any meaningful interpretation, as $\Delta E_{\text{dd}}$ must be smaller than $E_{\text{des}}$. Note that the large errors are a consequence of the very small temperature range over which the data could be measured.

## 5 Conclusion

The interaction of water with the surfaces of both ILs is consistent with the existence of an oleaginous top layer, covering an ionic underlayer. For $[C_8C_1Im][BF_4]$ the surface layer comprises the octyl chains on the imidazolium cation, with the imidazolium head group and the $[BF_4]^-$ cation forming the underlayer. For the $[C_2C_1Im]$ $[Tf_2N]$ the ionic underlayer is shielded from direct interaction with incoming water from the gas phase by, we surmise, the $CF_3$ and ethyl groups on orientated $[Tf_2N]^-$ and $[C_2C_1Im]^+$ ions, respectively, at the surface. The ionic underlayer of $[C_2C_1Im]$ $[Tf_2N]$ then consists of the imidazolium head group and the $SO_2$ parts of the $[Tf_2N]^-$ ion.

Water molecules approaching the surface from the gas phase physisorb onto the oleaginous top layer of both ILs. For $[C_8C_1Im][BF_4]$ the physisorption energy on the

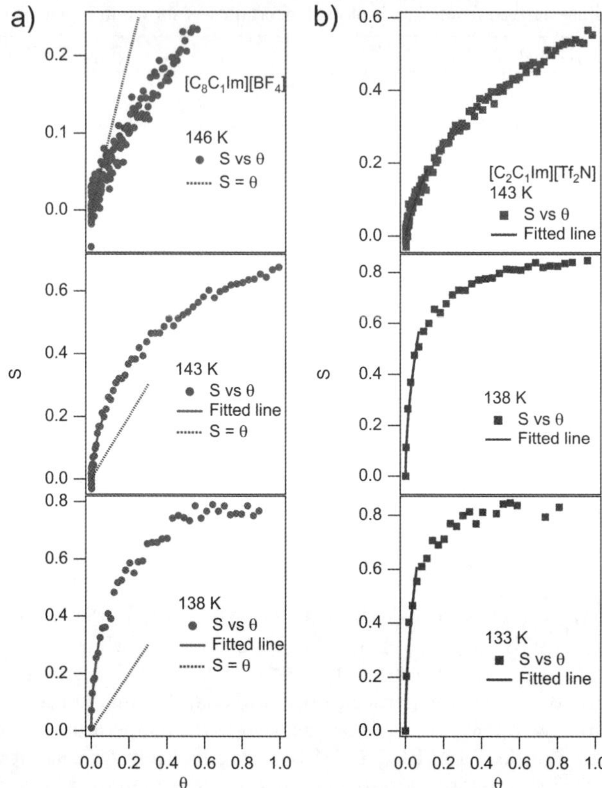


**Fig. 14** Examples of $S$ *vs.* $\theta$ plots at different temperatures for (a) $[C_8C_1Im][BF_4]$ and (b) $[C_2C_1Im][Tf_2N]$, with fits to $S = \theta + k_0\sqrt{\theta}$ for $0 < \theta < 0.05$. Also shown are $S = \theta$ lines for $[C_8C_1Im][BF_4]$.

frozen, glassy surface at zero coverage is 39.1 ± 0.5 kJ mol$^{-1}$, while for $[C_2C_1Im][Tf_2N]$ it is higher at 40.7 ± 0.2 kJ mol$^{-1}$. This binding energy is less than that measured for multilayer desorption of water, 46 kJ mol$^{-1}$, and demonstrates that there is no significant concentration of exposed, charged species at the surface of either IL. The lower physisorption energy for the $[C_8C_1Im][BF_4]$ surface is probably due to the relatively thick layer of hydrophobic octyl chains, compared to the rather thin surface layer afforded by orientated $CF_3$ and ethyl chains for $[C_2C_1Im][Tf_2N]$. For sub monolayer coverages, water on both surfaces exhibited attractive lateral interactions. A sticking probability of zero at zero coverage on the glassy solid surface was found for both ILs. As the coverage increased, the sticking probability increased, the rate of increase of $S$ with $\theta$ becoming more rapid the lower the adsorption temperature. This unusual behaviour is consistent with adsorption of an adsorbate species where the adsorbate···adsorbate interactions exceed the adsorbate··· substrate interactions, *i.e.* water on a hydrophobic surface. The behaviour was modelled using a fixed number of nucleation sites around which water ice islands can grow by addition of water at the island edges, and directly by adsorption from the gas phase onto the top of the islands. The water ice islands are more stable than isolated adsorbed water molecules due to the attractive $H_2O\cdots H_2O$ interactions. Water that lands within diffusion distance of an island edge, for its lifetime at a given temperature, will reach the island edge and stick, while those landing further away desorb back into the gas phase. Using this model, an activation energy of 13 ± 12 kJ mol$^{-1}$ was determined for physisorbed water diffusing across the IL surface.

 

For higher temperatures, above the glass transition temperatures of the ILs, water in the physisorbed state can penetrate through the hydrophobic layer to the ionic underlayer. For both ILs adsorption into the ionic underlayer is an activated processes with an enthalpy of 17 and 7 kJ mol$^{-1}$ for [$C_8C_1$Im][$BF_4$] and [$C_2C_1$Im][$Tf_2N$], respectively, relative to gaseous water, see Fig. 13. Once in the ionic underlayer of the IL, the activation enthalpy for moving into the bulk of the IL is smaller than that necessary to desorb back to the gas phase *via* the physisorbed state. For [$C_8C_1$Im][$BF_4$] the enthalpy of this barrier to moving into the bulk is 76 kJ mol$^{-1}$, determined from TPD-XPS experiments.[30] The consequence of these kinetics is that at room temperature, when the IL is a liquid, water has a small ($S \approx 0.12$) sticking probability. The water physisorbs onto the hydrophobic liquid surface, and a small fraction passes through this oleaginous layer to the underlayer, the remaining physisorbed water desorbing back into the gas phase. The water in the ionic underlayer then diffuses promptly into the semi-infinite volume of the 0.5 mm thick IL sample. As the IL temperature is reduced the sticking probability drops as a smaller fraction of the water penetrates the hydrophobic layer. When the IL freezes to a glass, penetration of the water to the bulk of the IL stops completely and $S = 0$.

## Acknowledgements

We gratefully acknowledge the support of the School of Chemistry at the University of Nottingham, for the award of a studentship (AD) and research funding (RGJ). We also thank Dr Peter Licence and his group at Nottingham for providing the ionic liquids used in this study.

## References

1 J. Bowers, M. C. Vergara-Gutierrez and J. R. P. Webster, *Langmuir*, 2004, **20**, 309–312.
2 F. Schreiber, *J. Phys.: Condens. Matter*, 2004, **16**, R881–R900.
3 T. Jiang, J. Wang, T. Yan and G. A. Voth, *J. Phys. Chem. C*, 2008, **112**, 1132–1139.
4 Y.-F. Hu, Z.-C. Liu, C.-M. Xu and X.-M. Zhang, *Chem. Soc. Rev.*, 2011, **40**, 3802–3823.
5 A. Riisager, R. Fehrmann, M. Haumann and P. Wasserscheid, *Top. Catal.*, 2006, **40**, 91–102.
6 U. Schroder, J. D. Wadhawan, R. G. Compton, F. Marken, P. A. Z. Suarez, C. S. Consorti, R. F. de Souza and J. Dupont, *New J. Chem.*, 2000, **24**, 1009–1015.
7 K. R. Seddon, A. Stark and M. J. Torres, *Pure Appl. Chem.*, 2000, **72**, 2275–2287.
8 P. A. Thiel and T. E. Madey, *Surf. Sci. Rep.*, 1987, **7**, 211–385.
9 M. A. Henderson, *Surf. Sci. Rep.*, 2002, **46**, 5–308.
10 S. Folsch, A. Stock and M. Henzler, *Surf. Sci.*, 1992, **264**, 65–72.
11 L. H. Dubois, B. R. Zegarski and R. G. Nuzzo, *J. Am. Chem. Soc.*, 1990, **112**, 570–579.
12 T. J. Gannon, G. Law, P. R. Watson, A. J. Carmichael and K. R. Seddon, *Langmuir*, 1999, **15**, 8429–8434.
13 C. S. Santos and S. Baldelli, *J. Phys. Chem. B*, 2009, **113**, 923–933.
14 K. R. J. Lovelock, C. Kolbeck, T. Cremer, N. Paape, P. S. Schulz, P. Wasserscheid, F. Maier and H. P. Steinrück, *J. Phys. Chem. B*, 2009, **113**, 2854–2864.
15 T. Iwahashi, T. Nishi, H. Yamane, T. Miyamae, K. Kanai, K. Seki, D. Kim and Y. Ouchi, *J. Phys. Chem. C*, 2009, **113**, 19237–19243.
16 C. Kolbeck, T. Cremer, K. R. J. Lovelock, N. Paape, P. S. Schulz, P. Wasserscheid, F. Maier and H. P. Steinrück, *J. Phys. Chem. B*, 2009, **113**, 8682–8688.
17 S. Rivera-Rubero and S. Baldelli, *J. Phys. Chem. B*, 2006, **110**, 15499–15505.
18 Y. Lauw, M. D. Horne, T. Rodopoulos, N. A. S. Webster, B. Minofar and A. Nelson, *Phys. Chem. Chem. Phys.*, 2009, **11**, 11507–11514.
19 R. G. Jones and C. J. Fisher, *Surf. Sci.*, 1999, **424**, 127–138.
20 R. G. Jones and C. A. Clifford, *Phys. Chem. Chem. Phys.*, 1999, **1**, 5223–5228.
21 A. S. Y. Chan, M. P. Skegg and R. G. Jones, *J. Vac. Sci. Technol., A*, 2001, **19**, 2007–2012.
22 M. G. Roper and R. G. Jones, *Langmuir*, 2005, **21**, 11684–11689.
23 J. P. Armstrong, C. Hurst, R. G. Jones, P. Licence, K. R. J. Lovelock, C. J. Satterley and I. J. Villar-Garcia, *Phys. Chem. Chem. Phys.*, 2007, **9**, 982–990.
24 A. Deyko, K. R. J. Lovelock, J.-A. Corfield, A. Taylor, P. N. Gooden, I. J. Villar-Garcia, P. Licence, R. G. Jones, V. G. Krasovskiy, E. A. Chernikova and L. M. Kustov, *Phys. Chem. Chem. Phys.*, 2009, **11**, 8544–8555.

25 J. M. Crosthwaite, M. J. Muldoon, J. K. Dixon, J. L. Anderson and J. F. Brennecke, *J. Chem. Thermodyn.*, 2005, **37**, 559–568.
26 H. Tokuda, K. Hayamizu, K. Ishii, M. Susan and M. Watanabe, *J. Phys. Chem. B*, 2005, **109**, 6103–6110.
27 P. Feulner and D. Menzel, *Phys. Rev. B*, 1982, **25**, 4295–4297.
28 C. R. Arumainayagam and R. J. Madix, *Prog. Surf. Sci.*, 1991, **38**, 1–102.
29 S. F. Diaz, J. F. Zhu, J. J. W. Harris, P. Goetsch, L. R. Merte and C. T. Campbell, *Surface Sci.*, 2005, **598**, 22.
30 K. R. J. Lovelock, E. F. Smith, A. Deyko, I. J. Villar-Garcia, P. Licence and R. G. Jones, *Chem. Commun.*, 2007, 4866–4868.
31 J. L. Anthony, E. J. Maginn and J. F. Brennecke, *J. Phys. Chem. B*, 2001, **105**, 10942–10949.

 

# Nonlinear vibrational spectroscopic studies on water/ionic liquid([$C_n$mim]TFSA: $n = 4, 8$) interfaces

**Takashi Iwahashi,[a] Yasunari Sakai,[a] Doseok Kim,[b] Tatsuya Ishiyama,[c] Akihiro Morita[c] and Yukio Ouchi*[a]**

***Received 8th April 2011, Accepted 1st June 2011***
**DOI: 10.1039/c1fd00061f**

The interfaces of water/room temperature ionic liquids (RTIL) (1-alkyl-3-methyl imidazolium bis(trifluoromethylsulfonyl)amide ([$C_n$mim]TFSA): $n = 4, 8$) are investigated by infrared-visible sum frequency generation (IV-SFG) vibrational spectroscopy and molecular dynamics (MD) simulation. SFG spectra taken within the SO stretch region drastically differ between air/RTIL and water/RTIL interfaces. When a RTIL surface is in contact with water, a broadened and blue-shifted $SO_2$-ss mode peak is observed in the SFG spectra, indicating an inhomogeneous intermolecular interaction due to hydrogen bonding of the [TFSA]$^-$ anions and water molecules at the water/[$C_n$mim]TFSA interface. MD simulations show the $SO_2$ groups of the anion are preferentially orientated toward the water phase, which is consistent with the SFG spectral features. Polar orientation of the [TFSA]$^-$ anion originates from the ordered structure of the alkyl chains of [$C_n$mim]$^+$ cations.

## Introduction

Molecular ordering at "buried interfaces" of liquids, including liquid/solid and liquid/liquid interfaces, are crucial for various applications such as batteries, capacitors, and phase separable catalysis because the properties of buried interfaces are closely related to performance.[1–3] Electrolyte/electrode interfaces, for example, play a key role in electrochemical applications because the electrode surface provides the electrochemical reaction field.[2,3] Liquid/liquid interfaces also play a vital role in many chemical processes such as separation, extraction, and reaction of materials. In such systems, careful and proper selection of the liquids is necessary for optimal performance of the target processes.[4] As has been reported to date, liquid/liquid interfaces assume various microscopic structures; for instance, unique interfacial environments should form due to orientational layering of carbon tetrachloride molecules and polar orientation of chloroform at oil/water interfaces[5] in which the highly asymmetric and flexible nature of liquid/liquid interfaces should contribute. However, due to difficulties in probing a buried interface, the understanding of a liquid/liquid interface is quite limited.

Recently, room-temperature ionic liquids (RTILs), which are salts in the liquid phase at room temperature,[6] have attracted much interest as totally new fluids due to their unique nature, such as nonvolatility, nonflamability, high ion

[a]*Department of Chemistry, Graduate School of Science, Nagoya University, Nagoya, 464-8602, Japan. E-mail: ohuchi@chem.nagoya-u.ac.jp; Fax: +81 52 789 2944; Tel: +81 52 789 2485*
[b]*Department of Physics, Sogang University, Seoul, 121-742, Republic of Korea*
[c]*Department of Chemistry, Graduate School of Science, Tohoku University, Sendai, 980-8574, Japan*

transportation, and wide electrochemical window.[6,7] RTILs are considered polar solvents, and their polarities have been reported to be similar to alcohols.[8] In general, the miscibility of a solvent depends on its polarity, and fluids with similar polarities are miscible. Interestingly, some RTILs can form liquid/liquid interfaces with both polar and nonpolar solvents. Hence, RTILs increase the variety of liquid/liquid interfaces. Unlike conventional molecular liquids,[9–17] RTILs are known to form unique microscopic structures in the bulk and at the surface. Therefore, the liquid/liquid interfaces of RTILs and conventional molecular liquids may also possess anomalous microscopic structural formation.[18,19] Consequently, the discovery and elucidation of novel liquid/liquid interfacial structures should trigger the development of new reaction systems.

Herein, we employ infrared-visible sum frequency generation (IV-SFG) vibrational spectroscopy and a molecular dynamics (MD) simulation to elucidate the water/RTIL interfacial structure. Because IV-SFG is a powerful nonlinear vibrational spectroscopic technique to investigate the surface and the interface of various media, structural and environmental information can be separately obtained for RTIL cations, anions, and molecular liquids.[14,20,21] Furthermore, combining IV-SFG and MD simulations provides a powerful opportunity to advance the understanding of molecular processes at RTIL interfaces. Thus, we selected 1-alkyl-3-methyl imidazolium bis(trifluoromethylsulfonyl)amide ([$C_n$mim]TFSA ($n = 4,8$); shown in Fig. 1) as a representative RTIL and water as a molecular liquid. The alkyl chain length dependence of the water/[$C_n$mim]TFSA interfaces were examined because the alkyl chain length of the cation may strongly affect the microscopic structural formation at RTIL interfaces[15,20] as well as in the bulk.[10,11,13]

The IV-SFG spectra of the water/[$C_n$mim]TFSA taken in the SO stretch region exhibit the $SO_2$ symmetric stretch ($SO_2$-ss) mode of the $[TFSA]^-$ anion. The peak positions are blue-shifted and the widths are broadened relative to those observed in both the IV-SFG spectra of air/[$C_n$mim]TFSA interface and the IR spectra of [$C_n$mim]TFSA bulk. These observations clearly indicate the presence of an inhomogeneous intermolecular interaction of the $SO_2$ groups of the $[TFSA]^-$ anion, originating from hydrogen bonding of the water molecules and the $[TFSA]^-$ anions. Additionally, the $SO_2$-ss peak amplitude strongly depends on the alkyl chain length of the $[C_n\text{mim}]^+$ cation; the amplitude is much larger for a chain length of $n = 8$ than that of $n = 4$. Our results suggest that the alkyl chain length of the $[C_n\text{mim}]^+$ cation strongly affects molecular ordering of both the $[C_n\text{mim}]^+$ cation and the $[TFSA]^-$ anion at the water/[$C_n$mim]TFSA interface.

## Theoretical

The basic theory of IV-SFG spectroscopy has been described previously;[20,21] here we briefly sketch the essentials needed for the later discussion. We consider a SFG process where the visible beam at frequency $\omega_{vis}$ and the tunable IR beam at $\omega_{ir}$ generate the sum frequency $\omega_{sf} = \omega_{vis} + \omega_{ir}$. The SF signal intensity is proportional

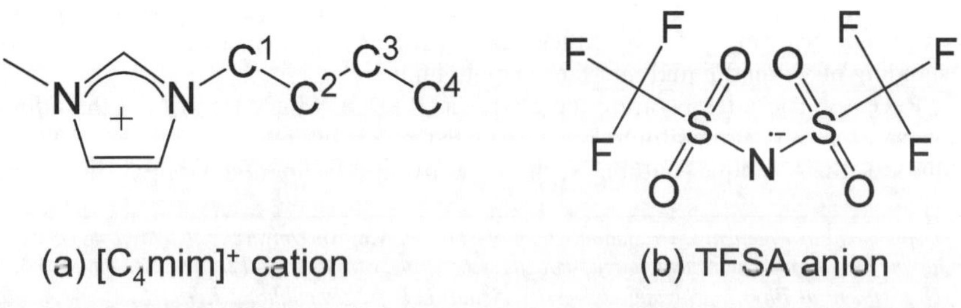

**Fig. 1** Structures of the ionic liquid components (a) 1-butyl-3-methyl imidazolium ($[C_4\text{mim}]^+$) cation and (b) bis(trifluoromethylsulfonyl) amide ($[TFSA]^-$) anion.

 

to the square modulus of an effective surface nonlinear susceptibility, $\chi_{\text{eff}}$, which depends on the incident beam geometry in the form of

$$\chi_{\text{eff}} = [\mathbf{L}(\omega_{\text{sf}})\cdot\hat{\mathbf{e}}(\omega_{\text{sf}})].\ \chi\text{: } [\mathbf{L}(\omega_{\text{vis}})\cdot\hat{\mathbf{e}}(\omega_{\text{vis}})][\mathbf{L}(\omega_{\text{ir}})\cdot\hat{\mathbf{e}}(\omega_{\text{ir}})] \tag{1}$$

where $\hat{\mathbf{e}}(\omega)$ and $\boldsymbol{L}(\omega)$ are the unit polarization vector and the Fresnel factor of the field at frequency $\omega$, respectively. Here we calculate the components of $L(\omega)$ using the expressions

$$L_{XX}(\omega) = \frac{2n_1(\omega)\cos\phi_2}{n_1(\omega)\cos\phi_2 + n_2(\omega)\cos\phi_1} \tag{2}$$

$$L_{YY}(\omega) = \frac{2n_1(\omega)\cos\phi_1}{n_1(\omega)\cos\phi_1 + n_2(\omega)\cos\phi_2} \tag{3}$$

$$L_{ZZ}(\omega) = \frac{2n_1^2(\omega)n_2(\omega)\cos\phi_1}{n_1(\omega)\cos\phi_2 + n_2(\omega)\cos\phi_1}\left(\frac{n_1(\omega)}{n'(\omega)}\right)^2 \tag{4}$$

where $n_1(\omega)$ and $n_2(\omega)$ are the refractive indices of media 1 and 2, respectively. $\phi_1$ and $\phi_2$ are the incident and refracted angles of the light, respectively, and $n'(\omega)$ is the empirical refractive index of the interfacial layer.[19,21] In the case of an azimuthally isotropic interface, there are only four independent nonvanishing components of $\chi$s: e $\chi_{XXZ} = \chi_{YYZ}$, $\chi_{XZX} = \chi_{YZY}$, $\chi_{ZXX} = \chi_{ZYY}$, and $\chi_{ZZZ}$ in space-fixed coordinates. These components are chosen such that $Z$ is along the surface normal and $X$ is in the incidence plane. The ssp (denoting s-polarized sum frequency, s-polarized visible, and p-polarized IR fields, respectively) measurement of an azimuthally isotropic surface probes only the nonlinear susceptibility component $\chi_{YYZ}$. Only vibrations bearing an IR transition moment normal to the interface contribute, and the ppp spectra originate from all the nonvanishing components $\chi_{XXZ} = \chi_{YYZ}$, $\chi_{XZX} = \chi_{YZY}$, $\chi_{XZZ} = \chi_{ZYY}$, and $\chi_{ZZZ}$. Amplitudes $A$ in the SFG spectra taken with ssp and ppp polarization combinations are expressed as

$$\chi_{\text{eff, ssp}} = L_{YY}(\omega_{sf})L_{YY}(\omega_{vis})L_{ZZ}(\omega_{ir})\sin\beta_{ir}\chi_{YYZ} \tag{5}$$

$$\begin{aligned}\chi_{\text{eff, ppp}} = \; & -L_{YY}(\omega_{sf})L_{YY}(\omega_{vis})L_{ZZ}(\omega_{ir})\cos\beta_{sf}\cos\beta_{vis}\sin\beta_{ir}\chi_{YYZ} \\ & -L_{YY}(\omega_{sf})L_{zz}(\omega_{vis})L_{YY}(\omega_{ir})\cos\beta_{sf}\sin\beta_{vis}\cos\beta_{ir}\chi_{YZY} \\ & +L_{ZZ}(\omega_{sf})L_{YY}(\omega_{vis})L_{YY}(\omega_{ir})\sin\beta_{sf}\cos\beta_{vis}\cos\beta_{ir}\chi_{ZYY} \\ & +L_{ZZ}(\omega_{sf})L_{ZZ}(\omega_{vis})L_{ZZ}(\omega_{ir})\sin\beta_{sf}\sin\beta_{vis}\sin\beta_{ir}\chi_{ZZZ}\end{aligned} \tag{6}$$

where $\beta_i$ is the incidence angle of the fields $\boldsymbol{E}(\omega_i)$. The SF signal intensity $I_{\text{sf}}$ is given by

$$I_{\text{sf}}(\omega_{\text{ir}}) \propto \left|\chi_{\text{NR}} + \sum_q \frac{A_q}{\omega_{\text{ir}} - \omega_q + i\Gamma_q}\right|^2 \tag{7}$$

where $\chi_{\text{NR}}$ refers to the nonresonant susceptibility, and $A_q$, $\omega_q$, $\Gamma_q$, are the amplitude, frequency, and damping constant of the $q$th vibrational mode, respectively. $\chi$ is related to molecular hyperpolarizability $\alpha$ by

$$\chi = N_{\text{S}}\langle\alpha\rangle \tag{8}$$

where $N_S$ is the surface number density of molecules and the angular bracket represents the average over the molecular orientational distribution.

## Experimental

### Material preparation

An ultrapure [$C_4$mim]TFSA sample was purchased from Kanto Chemical Co. and used as received. The [$C_8$mim]TFSA sample was prepared according to the literature with a slight modification.[14,20] [$C_8$mim]Br salt was prepared by alkylation of 1-methylimidazole (Aldrich, purity 99+%, used as received) with a slight molar excess of 1-bromooctane (TCI, purity >98%), which was distilled under low pressure. The reaction mixture was stirred at room temperature for 2–3 days. [$C_8$mim]Br salt, which was obtained as a viscous liquid, was washed several times with ethyl acetate. The anion exchange reaction of [$C_8$mim]Br was carried out by adding a slight molar excess of LiTFSA (Aldrich, >99.95%, used as received). The obtained [$C_8$mim]TFSA salt was washed repeatedly with ultrapure water and subsequent evaporation under low pressure to remove the residual water.

The [$C_8$mim]TFSA salt was further degassed by repeatedly heating and cooling in a vacuum for several hours to remove water and volatile solvent. We verified the purity using $^1$H NMR spectroscopy; the spectrum lacked traces of the starting materials, namely 1-methylimidazole, 1-bromooctane, and the [$C_8$mim]Br salt, and solvents. The water content of the [$C_8$mim]TFSA salt, which was determined by Karl Fischer titration, was less than 60 ppm.

### Optical setup and measurement

Details of the experimental setup are described elsewhere.[14,20] A mode-locked picosecond Nd:YAG laser (Ekspla, PL-2143B, 20 ps pulsewidth, 10 Hz) with a difference frequency generation (DFG) unit (Ekspla, DFG 401, mounting LBO and $AgGaS_2$ crystals) was used to generate a visible beam at 532 nm and a tunable IR beam with a wavelength ranging from 2.5 to 10 μm. The wavenumber of the IR beam was calibrated using literature values for the absorption bands of a standard polystyrene film.[22] Typical input intensities were ~350 μJ/pulse for the visible beam and ~100 μJ/pulse for the IR beam. The polarization combinations in the experiments were ssp and ppp.

The sample geometry for the SFG measurements of the air/[$C_n$mim]TFSA interfaces is described elsewhere.[14,20] The visible and IR beams overlapped at the surface of the RTILs in a glass vessel with incidence angles of 70° and 50°, respectively.

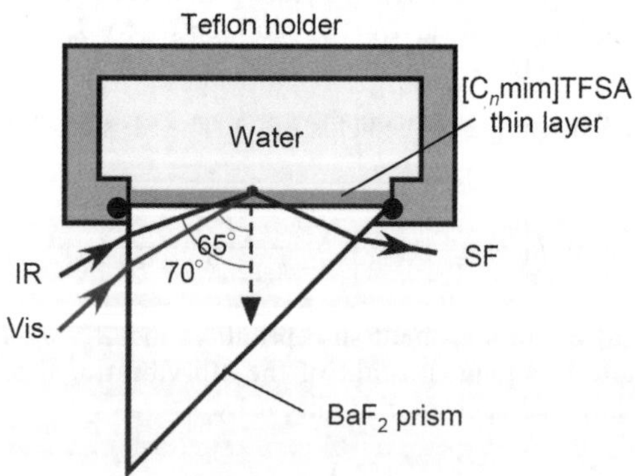


**Fig. 2** Schematic illustration of the sample holder for the SFG measurements of the water/RTIL interfaces.

 

Fig. 2 schematically illustrates the sample geometry for the SFG measurement of water/[$C_n$mim]TFSA interfaces. We used a slight modification of the sample arrangements of C. D. Bain *et al.* and G. L. Richmond *et al.*[5,19,23] A thin layer of [$C_n$mim]TFSA was prepared between a $BaF_2$ rectangular prism and water. IR transmission indicated the thickness of the [bmim]TFSA layer was a couple of microns on average. The visible and IR beams irradiated the water/[$C_n$mim]TFSA interfaces from the [$C_n$mim]TFSA layer side with incidence angles of 65° and 70°, respectively. It should be noted that unlike ref. 5 and 23, our experimental conditions are not the total internal reflection. We confirmed that the SF signal from our [$C_n$mim]TFSA/$BaF_2$ interface was negligible compared to that from the water/[$C_n$mim]TFSA interface.

### Molecular dynamics (MD) simulations

MD simulations were performed to investigate the interfaces of water/[$C_4$mim]TFSA and water/[$C_8$mim]TFSA. The force fields for the constituent species are Price–Brooks' modified TIP3P water model[24] for $H_2O$ and the models by Canongia Lopes and coworkers[25–27] for $[TFSA]^-$ and $[C_n mim]^+$. The force fields in the present simulations incorporated the electronic polarization implicitly in the effective partial charges. These non-polarizable force fields well described the structural properties[28] such as the orientational structure at the interfaces, but explicitly polarizable models may be necessary to accurately describe dynamical and transport properties.[28–30]

The MD simulations were executed in a rectangular simulation cell with 3D periodic boundary conditions. A simulation cell contained 2400 $H_2O$ molecules and 300 pairs of $[C_4mim]^+$ and $[TFSA]^-$ for the water/[$C_4$mim]TFSA, and 250 pairs of $[C_8mim]^+$ and $[TFSA]^-$ for the water/[$C_8$mim]TFSA. The periodic lengths were $L_x = L_y = 34.8$ Å and $L_z = 174.3$ Å for the former system and $L_x = L_y = 35.1$ Å and $L_z = 175.6$ Å for the latter system. These dimensions were optimized to pressure $P = 1$ atm and temperature $T = 300$ K *via* a constant temperature and pressure (NPT) simulation for the systems. Subsequent investigations on the interface structure were carried out at a constant volume and temperature (NVT) simulation for each system using the periodic cell of $L_x \times L_y \times L_z$ and the temperature $T = 300$ K.

For each system, four different atomic configurations were prepared and then equilibrated during a 10 ns MD simulation. Subsequently, the structures of interfaces were sampled over 50 ns for each trajectory, which amounted to a total of 200 ns of sampling for each interface system. The equations of motion were solved with a time step of 2 fs where all the bond lengths related to hydrogen atoms were fixed by SHAKE.[31] The particle mesh Ewald method was used to treat the long-range Coulombic interaction. The cutoff radius of the van der Waals and the real space Coulombic interactions was set to 9 Å. The MD trajectories were calculated with the Amber 10 program package.[32]

## Results and discussion

### SF activity of the $[TFSA]^-$ anion

Fig. 3a shows the SFG spectra of an air/[$C_4$mim]TFSA interface taken within the SO stretch region. The solid lines in the SFG spectra show the fitting results by multi-peak Lorentzian functions using eqn (7). Fig. 3b shows the IR absorption spectrum of [$C_4$mim]TFSA in the same wavenumber region as the SFG spectra. The two peaks at 1140 $cm^{-1}$ and 1240 $cm^{-1}$ in the SFG spectra are assigned to the $SO_2$ symmetric stretch ($SO_2$-ss) mode and the $CF_3$ symmetric stretch ($CF_3$-ss) mode, respectively.[33,34] Similar spectral features are also observed for the air/[$C_8$mim]TFSA interface (not shown here). On the other hand, the IR spectrum possesses an additional prominent peak centered at 1200 $cm^{-1}$, which is derived from the $CF_3$ asymmetric stretch ($CF_3$-as) modes.[33,34] This $CF_3$-as peak is weak in the Raman spectra of the TFSA salts.[33] It should be noted that only modes active for both IR and Raman

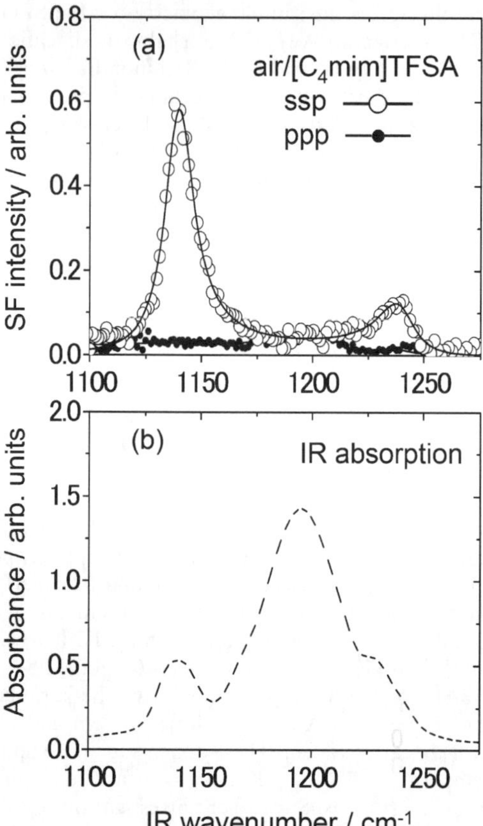


**Fig. 3** (a) SFG spectra of the air/[$C_4$mim]TFSA interface and (b) IR absorption spectrum and [$C_4$mim]TFSA. Solid lines in the SFG spectra show fitting results.

are SF active.[35] Therefore, a very weak SF signal for the $CF_3$-as peak at the air/[$C_n$mim]TFSA interface is due to the low Raman activity of $CF_3$-as modes for the [TFSA]$^-$ anion.

It is noteworthy that the [TFSA]$^-$ anion has two stable conformations, the $C_1$ and $C_2$ conformers.[36] Fig. 4 shows model structures of the two conformers. The SF activity of the normal mode strongly depends on molecular structural symmetry,[20,37] and conformers with different symmetries may exhibit very different specific SF activities. To roughly estimate the SF activity for each conformer, we performed *ab initio* calculations for the [TFSA]$^-$ anion in $C_1$ and $C_2$ to obtain the IR intensity (in units of km mol$^{-1}$) and Raman activity (in units of Å$^4$ amu$^{-1}$) for each normal mode. The product of the IR intensity and Raman activity can be used as an

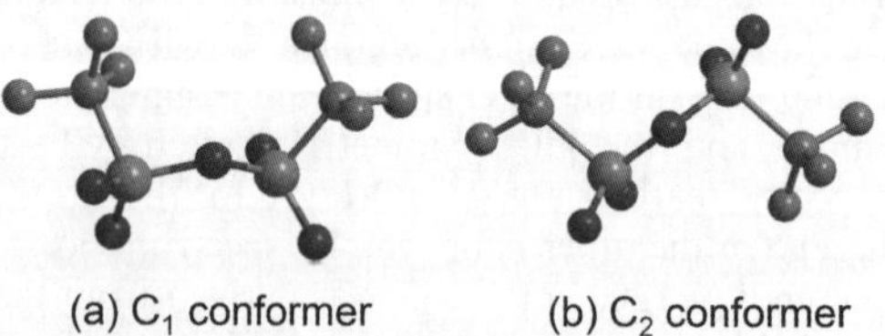


**Fig. 4** Schematic illustrations of (a) $C_1$ and (b) $C_2$ conformers of the [TFSA]$^-$ anion.

 

indicator of the SF activity for each mode. Calculations were performed using the Gaussian 03 package[38] with the B3LYP functional[39] and 6-311+G* basis set. Fig. 5 shows the calculated results for each normal mode along with the values of the IR intensity, Raman activity, and their products for the two conformers in the SO and CF stretch mode regions. As is clearly seen in Fig. 5c, the product values, particularly those for the $SO_2$-ss modes at ~1130 cm$^{-1}$, which are dominant in the observed SFG spectra, are high for the $C_1$ (black bar) conformer but negligible for the $C_2$ (white bar) conformer. These differences should result from symmetry differences of the $C_1$ and $C_2$ conformers. Thus, we can safely conclude that the observed SF signal from the [TFSA]$^-$ anion is dominantly derived from the $C_1$ conformer of [TFSA]$^-$ anion at the interfaces. The $CF_3$-as modes within the region of 1180–1220 cm$^{-1}$ have a strong IR intensity, although their Raman and SF activities are low for both conformers. This result agrees well with the observed IR,

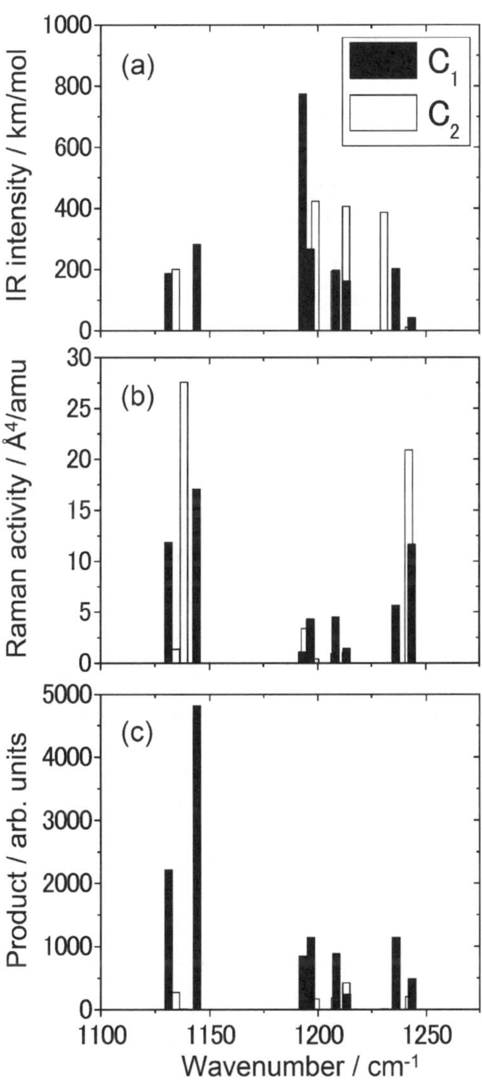


**Fig. 5** Calculated (a) IR intensity, (b) Raman activity, and (c) their product for $C_1$ (black bar) and $C_2$ (white bar) conformers of the [TFSA]$^-$ anion.

Raman, and SFG spectra of the TFSA salts, and thus, confirms the reliability of the calculation results.

### $[TFSA]^-$ anion configuration at the air interface

A recent RTIL surface structural study with high-resolution Rutherford backscattering spectroscopy[40] as well as our metastable atom electron spectroscopy (MAES) study[15] have shown that the $[TFSA]^-$ anion with $[C_4mim]^+$ and trimethyl-propylammonium cations exhibit a polar arrangement at the surface where the $CF_3$ groups point toward the vacuum side while the $SO_2$ groups face the RTIL side at the surface. Such a polar orientation is due to the fact that the $CF_3$ group reduces the surface energy compared to the $SO_2$ group.[41] On the other hand, the $C_2$ conformer is dominant in the bulk liquid because the $C_2$ conformer is energetically stable compared to $C_1$ conformer.[36] Combining all these results, we can safely conclude that the $[TFSA]^-$ anion preferentially takes the $C_1$ conformer at air/RTIL interfaces, differing from the bulk liquid, with the $CF_3$ and $SO_2$ groups pointing away from and toward the $[C_n mim]$TFSA phase, respectively.

### $[C_n mim]^+$ cation configuration at the air interface

The $[C_n mim]^+$ cation configuration has been vastly investigated using various experimental techniques such as neutron reflectometry,[44] X-ray reflectivity measurement,[16,45] nonlinear surface spectroscopy,[16,20,41–43] electron spectroscopy,[15,46–49] and MD simulations.[17,50–53] Recently, our MAES study has demonstrated that the $[C_n mim]^+$ cation preferentially directs the alkyl chain away from the RTIL bulk.[16] This structural configuration is consistent with the results obtained by the aforementioned techniques.

### $[TFSA]^-$ anion configuration at the water interface

Fig. 6 shows the SFG spectra of water/$[C_n mim]$TFSA ($n = 4$ and 8) interfaces taken within the SO stretch region. The SFG spectral features of the water interface change drastically compared to those of the air interface. Considering that the SF activities of the $CF_3$-as and $CF_3$-ss modes are much lower than that of the $SO_2$-ss mode,[33,34] we conclude that the broad peak observed in the SFG spectra of the water/$[C_n mim]$TFSA interfaces is assigned to the $SO_2$-ss mode peak. The width and position of the $SO_2$-ss mode peak at the water/$[C_n mim]$TFSA interfaces are much broader and blue-shifted relative to that at the air/$[C_n mim]$TFSA interface, indicating that the $SO_2$ groups of the $[TFSA]^-$ anion undergo a strong intermolecular interaction in an inhomogeneous microscopic environment.[14,18,19] The $SO_2$ group is a polar moiety; hence such drastic changes in the microscopic interaction are due to the presence of the hydrogen bonding of the $[TFSA]^-$ anion with water molecules. Furthermore, the $[TFSA]^-$ anion observed in the SFG spectra assumes the $C_1$ conformer, as shown in Fig. 4a. The $[TFSA]^-$ anion, which exhibits a broad and blue-shifted peak of the $SO_2$-ss mode, should orient so that the two $SO_2$ groups in $C_1$ preferentially point toward the water phase to form hydrogen bonding at the water/$[C_n mim]$TFSA interfaces.

The peak amplitude $A_q$ of the $SO_2$-ss mode strongly depends on the alkyl chain length. The amplitude is much larger for $n = 8$ than that for $n = 4$, indicating that the $[TFSA]^-$ anion configuration at the interface is strongly affected by the alkyl chain of the $[C_n mim]^+$ cation. Moreover, the orientational change of the target functional group should substantially change the ratio of the SF signal amplitude of its normal mode in the ssp and ppp spectra ($A_{ssp}/A_{ppp}$).[14,20,21] However, the fitting results of the SFG spectra of the water/$[C_n mim]$TFSA interfaces exhibit a similar $SO_2$-ss mode amplitude ratio of $A_{ssp}/A_{ppp}$: $\sim 0.2$ for $n = 4$ and $\sim 0.4$ for $n = 8$. Therefore, the orientational change in the $[TFSA]^-$ anion might be one of the reasons, but

 

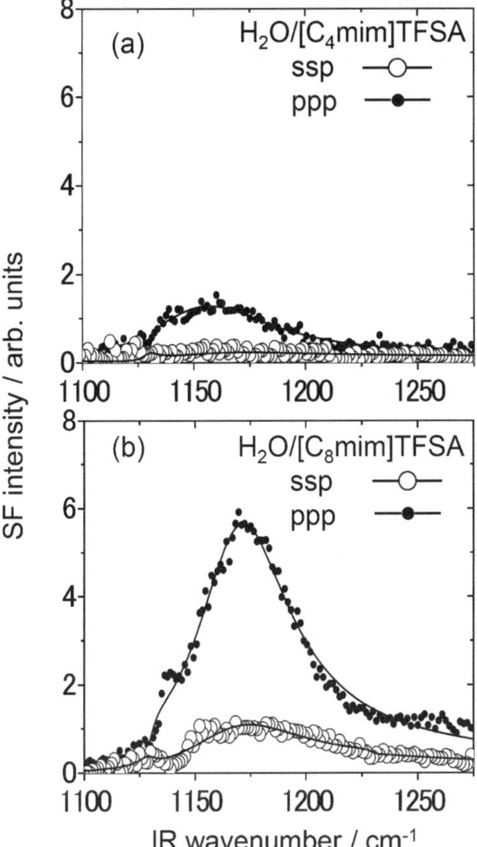


**Fig. 6** SFG spectra of (a) water/[$C_4$mim]TFSA and (b) water/[$C_8$mim]TFSA interfaces.

does not sufficiently explain such large differences in the $SO_2$-ss mode peak amplitude.

Here we propose a plausible scenario for the alkyl chain length dependence of the $SO_2$-ss peak amplitude of the [TFSA]$^-$ anion. The larger $SO_2$-ss mode amplitude for $n = 8$ than $n = 4$ is due to more of the $C_1$ conformers of the [TFSA]$^-$ anion at the water/[$C_8$mim]TFSA interface. Because the spectral features of the SFG spectra in both cases ($n = 4$ and 8) are similar except for peak amplitude, both cases should have a similar SF emission mechanism. Thus, the difference in amplitude $A_q$ is attributable mainly to the density of the locally aligned $C_1$ conformers at the interface. As the alkyl chain length increases, the [TFSA]$^-$ anion changes its conformation from the SF-inactive $C_2$ conformer into the SF-active $C_1$ conformer.

### [$C_n$mim]$^+$ cation configuration at the water interface

Unlike the SFG spectra of the [TFSA]$^-$ anion, a SF signal is not present in the CH stretch region of the [$C_n$mim]$^+$ cation at the water/[$C_n$mim]TFSA interfaces for both $n = 4$ and 8 cases (not shown in this paper).[18] Because the alkyl chain length dependence on the SFG spectral features of the [TFSA]$^-$ anion is clearly observed, we expect that the interfacial structures with respect to the alkyl chains differ for both cases. It is noteworthy that the random orientation of the molecules leads to a negligible SF signal intensity,[21,37] and the SF signal of the functional group submerged into its own phase also tends to be cancelled due to the specific out of phase SF contribution of the bulk molecules.[14,54] Although it is difficult to distinguish these two cases by SFG spectroscopy, they may explain the water/[$C_n$mim]TFSA interfacial structure of $n = 4$ and 8.[53]

## Insights from MD simulations

Fig. 7 shows the results of MD simulations. Fig. 7a represents the density profile $\rho(z)$ of water and $[TFSA]^-$ anions for $n = 4$ and 8 at the interfaces as a function of surface normal coordinate $z$. The thickness of the interface ($\delta_{10-90}$) defined by a 10–90% density of $H_2O$ molecules is 8.5 Å for $n = 4$ and 6.3 Å for $n = 8$, indicating [$C_8$mim]TFSA forms a sharper interface with water compared to [$C_4$mim]TFSA. Fig. 7b depicts the orientation of the $[TFSA]^-$ anion by $\rho\langle\cos\theta\rangle$ where $\theta$ is defined as the angle between the surface normal $z$ and the vector from the sulfur atom S to the adjacent oxygen atom O of $SO_2$ group. A positive $\rho\langle\cos\theta\rangle$ indicates the anion $[TFSA]^-$ orientates its $SO_2$ groups toward the water phase. The amplitude of $\rho\langle\cos\theta\rangle$ for $n = 8$ is larger than that for $n = 4$, strongly suggesting that the $[TFSA]^-$

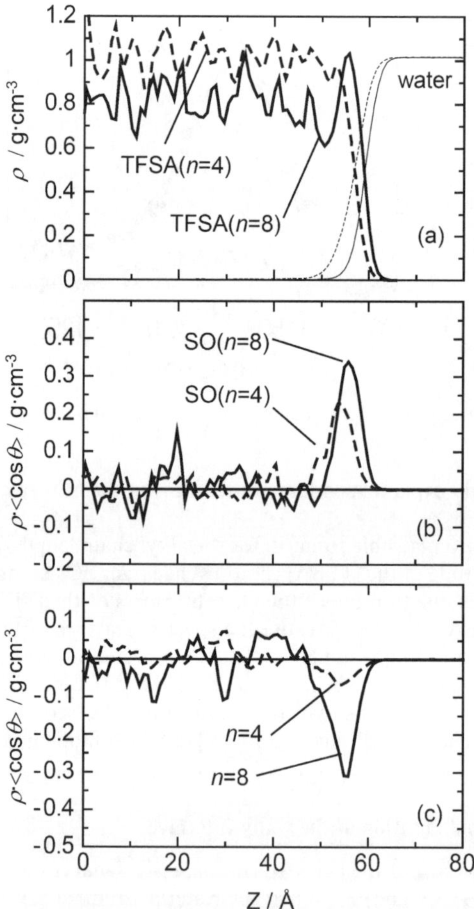


**Fig. 7** MD simulation results for water/[$C_n$mim]TFSA ($n = 4$, 8) interfaces. Dashed and solid lines are results for $n = 4$ and 8, respectively. (a) Density profile $\rho$ of water (thin line), $[TFSA]^-$ anions (thick line). (b) Orientational structure $\rho\langle\cos\theta\rangle$ of SO of $[TFSA]^-$ anion. Positive $\rho\langle\cos\theta\rangle$ indicates an anion orientation where $[TFSA]^-$ directs its $SO_2$ groups toward the water phase. (c) Orientational structure $\rho\langle\cos\theta\rangle$ of alkyl chain of $[C_n\text{mim}]^+$ cation. Negative $\rho\langle\cos\theta\rangle$ indicates a cation orientation where $[C_n\text{mim}]^+$ directs its alkyl chain toward the RTIL side.

 

anion polar orients more for $n = 8$ than for $n = 4$ at the interfaces. As the illustration of the $[TFSA]^-$ anion clearly shows in Fig. 4, the SF-active $C_1$ conformer can point both $SO_2$ groups toward the water phase, whereas the SF-inactive $C_2$ conformer points one $SO_2$ group toward the RTIL phase and the other $SO_2$ group in the opposite direction (the water phase). Thus, the results of MD simulations indicate that the SF-active $C_1$ conformer polar orientates with two $SO_2$ groups facing the water phase more for $n = 8$ than for $n = 4$ at the water/$[C_n\text{mim}]$TFSA interface. This model structure agrees well with the interpretation deduced from the results of SFG spectra.

The orientational profile of the $[C_n\text{mim}]^+$ cation shown in Fig. 7c reveals a negative $\rho\langle\cos\theta\rangle$ region at the interface where $\theta$ is defined as the angle between the surface normal $z$ and the vector from the $C^3$ to $C^4$ carbon (Fig. 1) of the alkyl chain of the $[C_n\text{mim}]^+$ cation. The negative $\rho\langle\cos\theta\rangle$ indicates a cation orientation where $[C_n\text{mim}]^+$ directs its alkyl chain toward the RTIL side, which contrasts the case of the air/RTIL interfaces.[14–20,42,43] Here we should note that the degree of orientational ordering at the water/RTIL interface depends strongly on the alkyl chain length $n$ of the $[C_n\text{mim}]^+$ cation; $n = 8$ shows a more ordered structure than $n = 4$.[53] Such a structural ordering effect for the longer alkyl chain reflects a sharper interface for $n = 8$, as shown in Fig. 7a. The SF-active and polar $C_1$ conformers may align themselves at the interface to direct both $CF_3$ groups toward the RTIL phase and $SO_2$ groups to the water phase, respectively.

Actually, it has been reported that a polar conformer is stabilized as it approaches a medium with a higher dielectric constant. A typical example is 1,2-dichloroethane (DCE), and the population of its polar *gauche* conformer increases at the interface of DCE and water because water has a much larger dielectric constant than DCE.[55] Consequently, the sharp boundary between high (water) and low ($[C_n\text{mim}]$TFSA) dielectrics[56,57] caused by a structural ordering effect for longer alkyl chain of $n = 8$ promotes the formation of the SF-active and polar $C_1$ conformer more than $n = 4$. Furthermore, the well-ordered hydrogen bonding configuration of the $[TFSA]^-$ anions with the water molecules may enhance the orientational ordering of the $[TFSA]^-$ anions. As indicated by eqn (8), a highly-ordered system with a narrow molecular orientational distribution exhibits a stronger SF signal relative to a randomly-ordered system with a wide molecular orientational distribution.[37] Therefore, a well-ordered structural formation and an increase in the SF-active $C_1$ conformer of the $[TFSA]^-$ anion contribute to the stronger $SO_2$-ss mode peak amplitude for the alkyl chain length of $n = 8$ relative to that for $n = 4$.

## Conclusions

IV-SFG vibrational spectroscopy and MD simulations were applied to elucidate the molecular liquid/ionic liquid interfaces of $[C_n\text{mim}]$TFSA ($n = 4$ and 8) and water. The SFG spectra of the air/$[C_n\text{mim}]$TFSA interface and the *ab initio* calculation as well as the previously reported MAES results[15] suggest that the $[TFSA]^-$ anion preferentially takes the SF-active $C_1$ conformation with its $CF_3$ and $SO_2$ groups pointing away from and toward the $[C_n\text{mim}]$TFSA phase, respectively, while the cation points its alkyl chain toward the air. When a RTIL surface is in contact with water, a broadened and blue-shifted $SO_2$-ss mode peak is observed, indicating an inhomogeneous intermolecular interaction due to hydrogen bonding of the $[TFSA]^-$ anions and the water molecules at the water/$[C_n\text{mim}]$TFSA interfaces.

The apparent alkyl chain length dependence of the SF signal amplitude of $SO_2$-ss mode occurs at the water/$[C_n\text{mim}]$TFSA interface, and the $SO_2$-ss mode amplitude is much stronger for $n = 8$ relative to that for $n = 4$. The MD simulation study supported the SFG spectroscopic results; more $SO_2$ groups point toward the water phase for $n = 8$ than for $n = 4$. Moreover, the octyl chain of the $[C_8\text{mim}]^+$ cation preferentially points toward the RTIL phase although the butyl chain of the $[C_4\text{mim}]^+$ cation orients weakly or rather randomly at the water/$[C_n\text{mim}]$TFSA interfaces. Therefore, such a structural ordering effect for the longer alkyl chain of the water/[$C_n$mim]TFSA interfaces might result in the formation of a well-ordered hydrogen bonding configuration of the $[TFSA]^-$ anions with the SF-active $C_1$ conformation and the water molecules, and hence increase the SF signal amplitude of the $SO_2$-ss mode of the $[TFSA]^-$ anion.

## Acknowledgements

This work is supported in part by a Grant-in-Aid for Scientific Research in Priority Areas "Science of Ionic Liquids", Scientific Research (S) (Grant No. 19105005), and Global COE program on Molecular Functions at Nagoya University from the Ministry of Education, Culture, Sports, Science and Technology (MEXT) of Japan. Numerical computations were carried out with the support from the Next-Generation Supercomputer Project, MEXT, Japan. D. K. acknowledges Korean Government (MEST) grant No. 2010-0027660 and the World-Class University program of the Republic of Korea.

## Notes and references

1 L. A. Blanchard, D. Hancu, E. J. Beckman and J. F. Brennecke, *Nature*, 1999, **399**, 28.
2 H. Matsumoto, H. Sakaebe and K. Tatsumi, *J. Power Sources*, 2005, **146**, 45.
3 (*a*) H. Sakaebe and H. Matsumoto, *Electrochem. Commun.*, 2003, **5**, 594; (*b*) B. Garcia, S. Lavallée, G. Perron, C. Michot and M. Armand, *Electrochim. Acta*, 2004, **49**, 4583.
4 J. H. Sinfelt and H. G. Drickamer, *J. Chem. Phys.*, 1955, **23**, 1095.
5 F. G. Moore and G. L. Richmond, *Acc. Chem. Res.*, 2008, **41**, 739.
6 J. D. Holbrey and K. R. Seddon, *J. Chem. Soc., Dalton Trans.*, 1999, 2133.
7 T. Welton, *Chem. Rev.*, 1999, **99**, 2071.
8 A. J. Carmichael and K. R. Seddon, *J. Phys. Org. Chem.*, 2000, **13**, 591.
9 H. Katayanagi, S. Hayashi, H. Hamaguchi and K. Nishikawa, *Chem. Phys. Lett.*, 2004, **392**, 460.
10 S. Shigeto and H. Hamaguchi, *Chem. Phys. Lett.*, 2006, **427**, 329.
11 K. Iwata, H. Okajima, S. Saha and H. Hamaguchi, *Acc. Chem. Res.*, 2007, **40**, 1174.
12 C. Hardacre, J. D. Holbrey, S. E. J. McMath, D. T. Bowron and A. K. Soper, *J. Chem. Phys.*, 2003, **118**, 273.
13 J. N. A. Canongia Lopes and A. A. H. Pádua, *J. Phys. Chem. B*, 2006, **110**, 3330.
14 T. Iwahashi, T. Miyamae, K. Kanai, K. Seki, D. Kim and Y. Ouchi, *J. Phys. Chem. B*, 2008, **112**, 11936.
15 T. Iwahashi, T. Nishi, H. Yamane, T. Miyamae, K. Kanai, K. Seki, D. Kim and Y. Ouchi, *J. Phys. Chem. C*, 2009, **113**, 19237.
16 Y. Jeon, J. Sung, W. Bu, D. Vaknin, Y. Ouchi and D. Kim, *J. Phys. Chem. C*, 2008, **112**, 19649.
17 R. M. Lynden-Bell, M. G. Del Pópolo, T. G. A. Youngs, J. Kohanoff, C. G. Hanke, J. B. Harper and C. C. Pinilla, *Acc. Chem. Res.*, 2007, **40**, 1138.
18 T. Iwahashi, T. Miyamae, K. Kanai, K. Seki, D. Kim and Y. Ouchi, ACS Symposium Series 1030 "*Ionic Liquids: From Knowledge to Application*", ed. N. V. Plechkova, R. D. Rogers, K. R. Seddon (American Chemical Society, Washington, DC, 2009) p. 305.
19 T. Iwahashi, Y. Sakai, K. Kanai, D. Kim and Y. Ouchi, *Phys. Chem. Chem. Phys.*, 2010, **12**, 12943.
20 T. Iimori, T. Iwahashi, K. Kanai, K. Seki, J. Sung, D. Kim, H. Hamaguchi and Y. Ouchi, *J. Phys. Chem. B*, 2007, **111**, 4860.
21 X. Zhuang, P. B. Miranda, D. Kim and Y. R. Shen, *Phys. Rev. B: Condens. Matter*, 1999, **59**, 12632.
22 International Union of Pure and Applied Chemistry. *Commission on Molecular Structure and Spectroscopy* Tables of Wavenumbers for the Calibration of Infra-red Spectrometers; Butterworths: London, 1961.
23 M. M. Knock, G. R. Bell, E. K. Hill, H. J. Turner and C. D. Bain, *J. Phys. Chem. B*, 2003, **107**, 10801.
24 D. J. Price and C. L. Brooks, *J. Chem. Phys.*, 2004, **121**, 10096.
25 J. N. Canongia Lopes and A. A. H. Pádua, *J. Phys. Chem. B*, 2004, **108**, 16893.
26 J. N. Canongia Lopes, J. Deschamps and A. A. H. Pádua, *J. Phys. Chem. B*, 2004, **108**, 2038.
27 J. N. Canongia Lopes, J. Deschamps and A. A. H. Pádua, *J. Phys. Chem. B*, 2004, **108**, 11250.

 

28 H. Nakano, T. Yamamoto and S. Kato, *J. Chem. Phys.*, 2010, **132**, 044106.
29 T. Yan, C. J. Burnham, M. G. Del Pópolo and G. A. Voth, *J. Phys. Chem. B*, 2004, **108**, 11877.
30 T. M. Chang and L. X. Dang, *J. Phys. Chem. A*, 2009, **113**, 2127.
31 M. P. Allen and D. J. Tildesley, *Computer simulation of liquids* (Clarendon Press, Oxford, 1987).
32 D. A. Case, T. A. Darden, T. E. Cheatham, III, C. L. Simmerling, J. Wang, R. E. Duke, R. Luo, M. Crowley, R. C. Walker, W. Zhang, K. M. Merz, B. Wang, S. Hayik, A. Roitberg, G. Seabra, I. Kolossváry, K. F. Wong, F. Paesani, J. Vanicek, X. Wu, S. R. Brozell, T. Steinbrecher, H. Gohlke, L. Yang, C. Tan, J. Mongan, V. Hornak, G. Cui, D. H. Mathews, M. G. Seetin, C. Sagui, V. Babin and P. A. Kollman, *AMBER 10*, University of California, San Francisco, 2008.
33 I. Rey, J. C. Lassègues, J. Grondin and L. Servant, *Electrochim. Acta*, 1998, **43**, 1505.
34 J. C. Lassègues, J. Grondin, R. Holomb and P. Johansson, *J. Raman Spectrosc.*, 2007, **38**, 551.
35 P. Guyot-Sionnest, J. H. Hunt and Y. R. Shen, *Phys. Rev. Lett.*, 1987, **59**, 1597.
36 K. Fujii, T. Fujimori, T. Takamuku, R. Kanzaki, Y. Umebayashi and S. Ishiguro, *J. Phys. Chem. B*, 2006, **110**, 8179.
37 C. D. Stanners, Q. Du, R. P. Chin, P. Cremer, G. A. Somorjai and Y. R. Shen, *Chem. Phys. Lett.*, 1995, **232**, 407.
38 M. J. Frisch, G. W. Trucks, H. B. Schlegel, G. E. Scuseria, M. A. Robb, J. R. Cheeseman, J. A. Montgomery, Jr., T. Vreven, K. N. Kudin, J. C. Burant, J. M. Millam, S. S. Iyengar, J. Tomasi, V. Barone, B. Mennucci, M. Cossi, G. Scalmani, N. Rega, G. A. Petersson, H. Nakatsuji, M. Hada, M. Ehara, K. Toyota, R. Fukuda, J. Hasegawa, M. Ishida, T. Nakajima, Y. Honda, O. Kitao, H. Nakai, M. Klene, X. Li, J. E. Knox, H. P. Hratchian, J. B. Cross, V. Bakken, C. Adamo, J. Jaramillo, R. Gomperts, R. E. Stratmann, O. Yazyev, A. J. Austin, R. Cammi, C. Pomelli, J. Ochterski, P. Y. Ayala, K. Morokuma, G. A. Voth, P. Salvador, J. J. Dannenberg, V. G. Zakrzewski, S. Dapprich, A. D. Daniels, M. C. Strain, O. Farkas, D. K. Malick, A. D. Rabuck, K. Raghavachari, J. B. Foresman, J. V. Ortiz, Q. Cui, A. G. Baboul, S. Clifford, J. Cioslowski, B. B. Stefanov, G. Liu, A. Liashenko, P. Piskorz, I. Komaromi, R. L. Martin, D. J. Fox, T. Keith, M. A. Al-Laham, C. Y. Peng, A. Nanayakkara, M. Challacombe, P. M. W. Gill, B. G. Johnson, W. Chen, M. W. Wong, C. Gonzalez and J. A. Pople, *GAUSSIAN 03 (Revision D.01)*, Gaussian, Inc., Wallingford, CT, 2004.
39 A. D. Becke, *J. Chem. Phys.*, 1993, **98**, 5648.
40 K. Nakajima, A. Ohno, M. Suzuki and K. Kimura, *Nucl. Instrum. Methods Phys. Res., Sect. B*, 2009, **267**, 605.
41 C. S. Santos, S. Rivera-Rubero, S. Dibrov and S. Baldelli, *J. Phys. Chem. C*, 2007, **111**, 7682.
42 T. Iimori, T. Iwahashi, H. Ishii, K. Seki, Y. Ouchi, R. Ozawa, H. Hamaguchi and D. Kim, *Chem. Phys. Lett.*, 2004, **389**, 321.
43 S. Baldelli, *J. Phys. Chem. B*, 2003, **107**, 6148.
44 J. Bowers, M. C. Vergara-Gutierrez and J. R. P. Webster, *Langmuir*, 2004, **20**, 309.
45 E. Sloutskin, B. M. Ocko, L. Tamam, I. Kuzmenko, T. Gog and M. Deutsch, *J. Am. Chem. Soc.*, 2005, **127**, 7796.
46 D. Yoshimura, T. Yokoyama, T. Nishi, H. Ishii, R. Ozawa, H. Hamaguchi and K. Seki, *J. Electron Spectrosc. Relat. Phenom.*, 2005, **144–147**, 319.
47 T. Nishi, T. Iwahashi, H. Yamane, Y. Ouchi, K. Kanai and K. Seki, *Chem. Phys. Lett.*, 2008, **455**, 213.
48 K. Kanai, T. Nishi, T. Iwahashi, Y. Ouchi, K. Seki, Y. Harada and S. Shin, *J. Chem. Phys.*, 2008, **129**, 224507.
49 S. Krischok, M. Eremtchenko, M. Himmerlich, P. Lorenz, J. Uhlig, A. Neumann, R. Öttking, W. J. D. Beenken, O. Höfft, S. Bahr, V. Kempter and J. A. Schaefer, *J. Phys. Chem. B*, 2007, **111**, 4801.
50 B. L. Bhargava and S. Balasubramanian, *J. Am. Chem. Soc.*, 2006, **128**, 10073.
51 R. M. Lynden-Bell and M. G. Del Pópolo, *Phys. Chem. Chem. Phys.*, 2006, **8**, 949.
52 T. Yan, S. Li, W. Jiang, X. Gao, B. Xiang and G. A. Voth, *J. Phys. Chem. B*, 2006, **110**, 1800.
53 A. Chaumont, R. Schurhammer and G. Wipff, *J. Phys. Chem. B*, 2005, **109**, 18964.
54 J. Sung, K. Park and D. Kim, *J. Phys. Chem. B*, 2005, **109**, 18507.
55 I. Benjamin, *J. Chem. Phys.*, 1992, **97**, 1432.
56 C. Wakai, A. Oleinikova, M. Ott and H. Weingartner, *J. Phys. Chem. B*, 2005, **109**, 17028.
57 T. Singh and A. Kumar, *J. Phys. Chem. B*, 2008, **112**, 12968.

# Slow and fast capacitive process taking place at the ionic liquid/electrode interface

**Bernhard Roling,* Marcel Drüschler and Benedikt Huber**

*Received 3rd May 2011, Accepted 19th May 2011*
**DOI: 10.1039/c1fd00088h**

Electrochemical impedance spectroscopy was used to characterise the interface between the ultrapure room temperature ionic liquid 1-butyl-1-methylpyrrolidinium tris(pentafluoroethyl)trifluorophosphate and a Au(111) working electrode at electrode potentials more positive than the open circuit potential (−0.14 V *vs.* Pt pseudo-reference). Plots of the potential-dependent data in the complex capacitance plane reveal the existence of a fast and a slow capacitive process. In order to derive the contribution of both processes to the overall capacitance, the complex capacitance data were fitted using an empirical Cole-Cole equation. The differential capacitance of the fast process is almost constant between −0.14 V and +0.2 V (*vs.* Pt pseudo-reference) and decreases at more positive potentials, while the differential capacitance of the slower process exhibits a maximum at +0.2 V. This maximum leads to a maximum in the overall differential capacitance. We attribute the slow process to charge redistributions in the innermost ion layer, which require an activation energy in excess of that for ion transport in the room temperature ionic liquid. The differential capacitance maximum of the slow process at +0.2 V is most likely caused by reorientations of the 1-butyl-1-methylpyrrolidinium cations in the innermost layer with the positively charged ring moving away from the Au(111) surface and leaving behind voids which are then occupied by anions. In a recent Monte Carlo simulation by Federov, Georgi and Kornyshev (*Electrochem. Commun.* 2010, **12**, 296), such a process was identified as the origin of a differential capacitance maximum in the anodic regime. Our results suggest that the *time scales* of capacitive processes at the ionic liquid/metal interface are an important piece of information and should be considered in more detail in future experimental and theoretical studies.

## Introduction

The investigation of the interfacial properties of room temperature ionic liquids (RTILs) in contact with different electrode materials has become a challenging task in recent years, since RTILs are attractive candidates as liquid electrolytes in a broad range of electrochemical applications. For instance, they can be used in electrochemical supercapacitors,[1] field-effect double-layer transistors[2] and dye sensitised solar cells.[3] Due to the broad electrochemical window of RTILs,[4] their usage in supercapacitors offers new perspectives for improved energy storage. However, the limited understanding of the structure and dynamics of the RTIL/electrode interface retards further developments in the field of RTIL-based supercapacitors.

A fundamental property of the interface related to the charge distribution is the differential interfacial capacitance:

*Department of Chemistry, Philipps-University of Marburg, Hans-Meerwein-Strasse, 35032 Marburg, Germany. E-mail: roling@staff.uni-marburg.de*

 

$$C_{\mathrm{Int}} = \left.\frac{\partial q}{\partial(\Delta\varphi)}\right|_{\mu,T,p} \tag{1}$$

Here, $q$ denotes the electrode charge, while $\Delta\varphi$ is the electrode potential, *i.e.* the electric potential difference between electrode and bulk of the RTIL. When measuring $C_{\mathrm{Int}}$, the chemical potential of the bulk RTIL $\mu$, the temperature $T$, and the pressure $p$ are kept constant. In the classical Gouy-Chapman model for double layers in diluted electrolytes, ions are treated as point charges, and the interactions between individual ions as well as the interactions between ions and metal atoms are treated at the mean-field level. Due to the neglect of the finite ion volume, there is an enormous accumulation of counter ions at high electrode potentials resulting in an exponential increase of $C_{\mathrm{Int}}$ with increasing potential. In 2007, Kornyshev proposed a mean-field lattice gas model as a simple method for taking into account the finite ion volume in dense ionic systems.[5] The outcome of this model for the potential-dependent differential capacitance is clearly distinct from the classical models for diluted electrolytes. A global maximum of $C_{\mathrm{Int}}$ at the potential of zero charge (bell-shaped curve) or two distinct maxima very close to the potential of zero charge (camel-shaped curve) are obtained depending on the ion density. The shape of the $C_{\mathrm{Int}}(\Delta\varphi)$ curves is governed by the entropy penalty for the crowding of counter ions close to the charged electrode surface. At very high electrode potentials, the mean-field lattice gas model predicts a capacitance decay with $C_{\mathrm{Int}} \sim \Delta\varphi^{-1/2}$, which reflects the growth of the thickness of the counter ion layers with increasing electrode potential (lattice saturation).

Of course, Kornyshev's lattice gas model can only be considered a first step for understanding the interface between electrodes and dense ionic systems, since several types of important interactions are treated at the mean-field level or are even completely ignored. For instance, individual ion-electrode and ion-ion interactions can lead to overscreening, *i.e.* the innermost ion layer at the charged electrode surface contains more counter-charges than required for screening the electrode charge. Consequently, the next layer contains an excess of co-charges, followed by another layer with an excess of counter-charges. The resulting charge oscillations decay with increasing distance from the electrode surface. The existence of distinct ion layers in RTIL close to charged surfaces was confirmed experimentally by means of X-ray reflectivity studies[6] and atomic force microscopy (AFM).[7] Charge density oscillations were also found in molecular dynamics (MD) simulations of the interface between charged surfaces and RTILs.[8–20] However in a recently published phenomenological theory,[21] Bazant *et al.* predicted that overscreening is only relevant within a narrow potential range close to the potential of zero charge (pzc), while it is suppressed in favor of lattice saturation at relatively high electrode potentials (about 2.5 V). Therefore, overscreening has only little influence on the general characteristics of potential-dependent differential capacitance curves. These theoretical predictions were critically discussed in a recent publication by Vatamanu *et al.*[20] In MD simulations of a RTIL/graphite interface, they found that the $C_{\mathrm{Int}}(\Delta\varphi)$ curves are mainly determined by the charge distribution in the first two ion layers adjacent to the charged electrode. Furthermore, the transition from overscreening to lattice saturation extends over a much broader potential range (about 15 V) than in the theory proposed by Bazant *et al.* (2.5 V).[21] This raises the question as to whether or not lattice saturation effects are relevant within the electrochemical stability window of RTIL.

If potential-dependent differential capacitance curves are mainly determined by charge distribution in the first two or three layers adjacent to the electrode, they should be very sensitive to conformational changes of the molecular cations and anions within these layers. A strong dependence of the ion conformations on the electrode potential was shown by means of surface-enhanced Raman (SERS) scattering,[22,23] sum frequency generation (SFG) vibrational spectroscopy,[24–28] AFM,[29,30] and MD simulation.[11,13,17–20] An interesting observation was made by Fedorov *et al.* in a recent MC simulation.[13] Long alkyl chains of the RTIL cation can act as 'latent voids', which allow for anions to enter the innermost layer when the alkyl chains change from an orientation parallel to the surface to a tilted or perpendicular orientation moving the positively charged head group away from the positively charged electrode surface.[13,18,20] This process may lead to a maximum of the differential capacitance in the anodic regime, *i.e.* at potentials more positive than the pzc.

It is important to note that in most theoretical studies and simulations, the time scale of the interfacial processes was not explicitly considered. On the other hand, a recent impedance spectroscopic study of the interface between 1-ethyl-3-methylimidazolium tris(pentafluoroethyl)trifluorophosphate ([EMIm]FAP) and Au(111) carried out by our group revealed the existence of two distinct capacitive processes taking place on different time scales.[31] While the time scale of the faster process is determined by the bulk resistance of the RTIL, the slower process seems to require an activation energy in excess of that for ion transport in the RTIL.

These results are confirmed by the present study on the interface between 1-butyl-1-methylpyrrolidinium tris(pentafluoroethyl)trifluorophosphate ([BMPyrr]FAP) and Au(111) in the anodic regime. At any electrode potential, we detect a fast and a slow capacitive process. The overall differential capacitance exhibits a maximum around +0.2 V *vs.* Pt pseudo-reference, which is caused by a maximum in the differential capacitance of the slower process. We suggest that the slow capacitive process reflects charge redistributions in the innermost ion layer where the ions are strongly bound to the electrode surface. This interpretation is in line with a recent impedance spectroscopic study by Pajkossy and Kolb on the interface between Au(100) and 1-butyl-3-methylimidazolium hexafluorophosphate. They found a low-frequency arc in the complex capacitance plane which was attributed to the slow kinetics of anion/cation replacements in the innermost ion layer.[32]

In the case of the [BMPyrr]FAP/Au(111) interface studied here, the capacitance maximum around +0.2 V is most likely caused by a reorientation of the cations as found in the MC simulations in Ref. 13. Since for this reorientation to take place, bonds between the alkyl groups and the gold surface have to be broken, the process is slow. Overall, our results suggest that the time scale of capacitive processes contains important information about the structure and dynamics of the RTIL/electrode interface and should therefore be considered in more detail in future experimental and theoretical studies.

## Experimental

Custom-made [BMPyrr]FAP of the highest available purity (water, alkali metal and halide content all below 10 ppm) was purchased from Merck KGaA. In order to further reduce the $H_2O$ contents below 1 ppm, the IL was dried for several hours under vacuum ($10^{-3}$ mbar) at 100 °C by the Endres group at Clausthal University of Technology. The IL was then sent to our group in sealed ampoules and stored in a glovebox with a high-purity $N_2$ atmosphere (LABstar, MBRAUN GmbH; $H_2O$ and $O_2$ contents of the box < 1ppm). For the electrochemical measurements, a home-made cell (rhd instruments) was used which was placed inside the glove box. The measurements were carried out at 30 ± 0.1 °C in a three-electrode configuration with a polycrystalline Pt wire acting as a pseudo-reference electrode (RE), a Pt crucible (Pt/Rh 80/20, 150 μL volume, purchased from Mettler Toledo GmbH) acting as a container for the RTIL and as the counter electrode (CE), and a Au single crystal (MaTeck GmbH) acting as the working electrode (WE). The Au single crystal exhibited a cylindrical shape with a diameter of 3 mm and a height of 5 mm. On top of the cylinder the Au(111) surface was orientated to values better than 1°. A gold wire soldered to the upper, non-orientated end of the Au cylinder allowed for a comfortable handling during preparation and for an easier connection to the measurement system. Prior to the electrochemical measurements, the single crystal was cleaned and annealed in a propane gas flame to light red glow for 30 s.[29,31,33–35] Subsequently, it was cooled to ambient temperature under Ar flow. The contact between the Au(111) surface and the RTIL was established by using the dipping technique, thus avoiding a contact with the non-orientated side walls of the single crystal.[36] In order to achieve the same conditions for all measurements, the Pt wire acting as RE was polished, carefully rinsed with acetone, dried under vacuum and then annealed in a propane gas flame for a few minutes. This procedure leads to the formation of a thin oxide film on the Pt surface which establishes the potential of the RE in contact with the RTIL.[37] By using this method, the potential shift of the RE during the measurements could be reduced to values less than 100 mV. The electrochemical cell was connected to a Novocontrol modular measurement system which consists of an Alpha-AK high-resolution impedance analyzer and a POT/GAL 15 V/10 A electrochemical interface. Prior to an EIS measurement, a cyclic voltammogram with a scan rate of 10 mV $s^{-1}$ was taken in order to determine the electrochemical window suitable for interfacial capacitance measurements. The EIS measurements were carried out at different WE dc potentials superimposed by a small ac voltage signal (10 mV rms). The frequency range extended from 10 mHz to 1 MHz. After a dc potential step, a waiting time of 10 min was chosen before recording an impedance spectrum. Measured impedance data were fitted using the WinFit software (Novocontrol Technologies).

## Results and discussion

Fig. 1 shows the cyclic voltammogram (CV) of [BMPyrr]FAP in a potential range extending from −1.5 V to +1.0 V *versus* a Pt pseudo-reference electrode (scan rate 10 mV $s^{-1}$). In the following, we use the notation $E$ for the potential difference between WE and RE, *i.e.* $E = \Delta\varphi_{WE} - \Delta\varphi_{RE}$. The scan was started at $E = 0.0$ V, then the potential was swept to −1.5 V, then to +1.0 V, and finally back to 0.0 V. The resulting CV reveals six distinct current peaks $C_1$, $C_2$, $C_3$, $A_2$, $A_4$ and $A_4^*$. A discussion of the peaks $C_1$, $C_2$, $C_3$ and $A_2$ detected at potentials more negative than the ocp (−0.14 V) is beyond the scope of the present paper and can be found in a recent publication.[29] The anodic peaks $A_4$ and $A_4^*$ are related to each other and are most likely caused by an interaction between the $FAP^-$–anion and the Au(111) surface.[31]

The EIS spectra of the [BMPyrr]FAP/Au(111) interface were recorded in a potential range from $E = -0.14$ V (ocp) to $E = +1.0$ V using a dc potential step width of

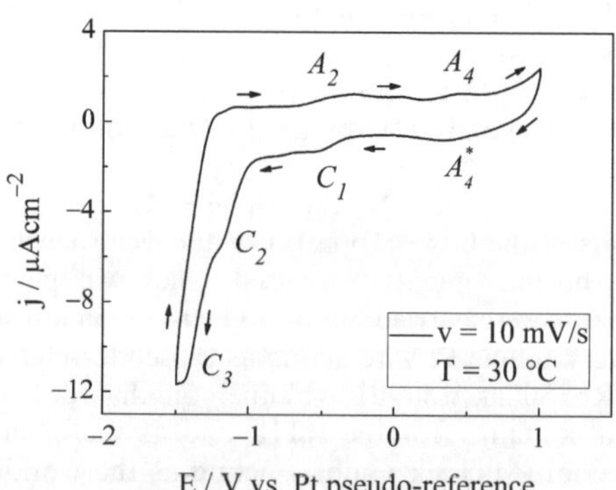


**Fig. 1** Cyclic voltammogram of the [BMPyrr]FAP/Au(111) interface recorded at a scan rate of 10 mV $s^{-1}$.

 

20 mV. In previous papers, we showed that a plot of the impedance data in the complex capacitance plane (CCP) is well suited for resolving different capacitive and non-capacitive interfacial processes.[29,31] Consequently, this type of representation will also be used in the present paper. In Fig. 2, exemplary spectra for the [BMPyrr]FAP/Au(111) interface are shown in the CCP.

At all electrode potentials, two capacitive processes were clearly detectable. At very low frequencies, typically below 0.1 Hz, the onset of a third process was observed. This third process seems to become Faradaic in its low-frequency limit with the phase angle approaching 0°, see Fig. 3. The analysis and interpretations of this ultraslow process is beyond the scope of this paper, but we focus on the two capacitive processes.

The high-frequency semicircle observed in a frequency range from MHz down to typically 20 Hz is only slightly suppressed. The time scale of the underlying fast capacitive process is determined by the bulk resistance of the IL. Typical values for the relaxation time $\tau_{fast} = R_{bulk}C_{fast}$ are in the range of 0.5 ms. Thus, the fast process should encompass all interfacial charge redistributions which do not require an activation energy in excess of that for charge transport in the RTIL. The second capacitance semicircle in a frequency from typically 20 Hz down to typically 0.1 Hz is more strongly suppressed than the high-frequency semicircle. In a potential range

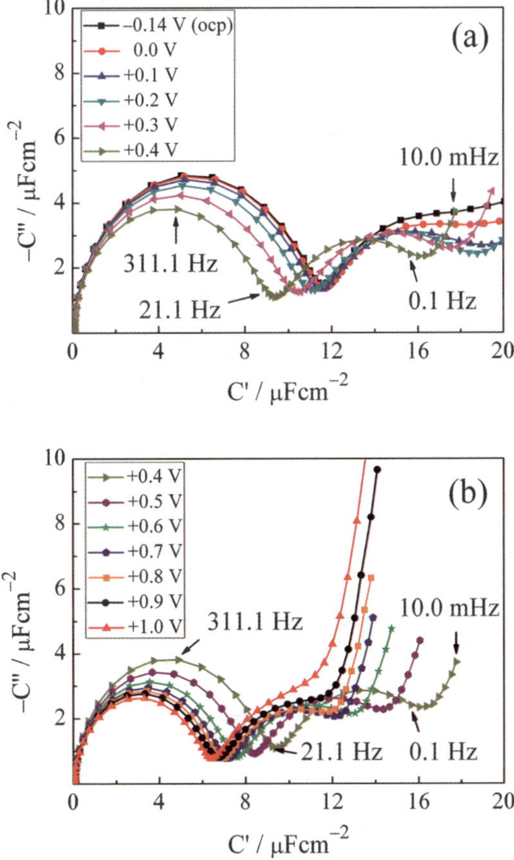


**Fig. 2** (a) Complex capacitance plane plots for the [BMPyrr]FAP/Au(111) interface in a potential range from −0.14 V (ocp) to +0.4 V *vs.* Pt pseudo-reference, (b) and in a potential range from + 0.4 V and +1.0 V *vs.* Pt pseudo-reference.

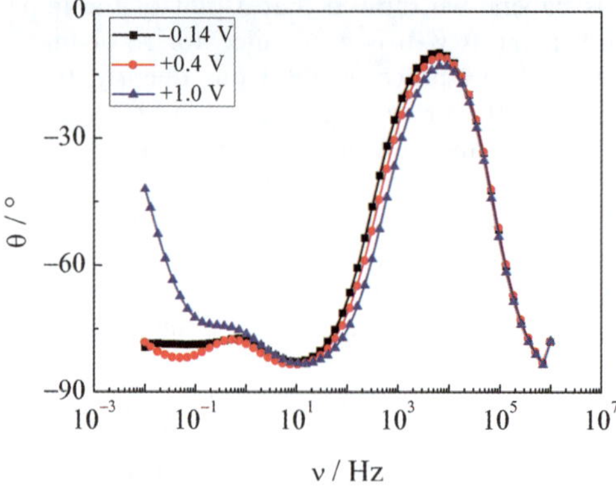


**Fig. 3** Plot of the impedance phase angle $\theta$ *versus* the measurement frequency $\nu$ at −0.14 V (ocp), +0.4 V, and +1.0 V.

from −0.14 V to +0.8 V, the second semicircle is well resolved, while at potentials above +0.8 V, it is superimposed by the capacitance contribution of the ultraslow Faradaic process. The strong influence of the ultraslow process on the low-frequency capacitance data is clearly revealed in the phase angle plot in Fig. 3.

The time scale of the slow capacitive process is of the order of seconds. This suggests that this process requires an activation energy in excess of that for ion transport in the RTIL.

In order to analyse the contribution of both capacitive processes to the overall differential capacitance, the complex capacitance data were fitted to an empirical Cole-Cole type expression:

$$\hat{C}(\nu) = \sum_{i=1}^{n} \frac{\Delta C_i}{1 + (j \cdot 2\pi\nu\tau_i)^{\alpha_i}} \tag{2}$$

where $\Delta C_i$ and $\tau_i$ denote the capacitance relaxation strength and the relaxation time of process *i*.[29,31] In this case, we identify $\Delta C_1 = C_{fast}$ and $\Delta C_2 = C_{slow}$. In Fig. 4, the differential capacitances $C_{fast}$ and $C_{slow}$ as well as the overall differential capacitance of the interface $C_{Int} = C_{fast} + C_{slow}$ are plotted *versus* the electrode potential.

Between −0.14 V and +0.20 V, the differential capacitance of the fast process $C_{fast}$ exhibits a plateau regime with a value of 9.3 μF cm$^{-2}$. With increasingly positive

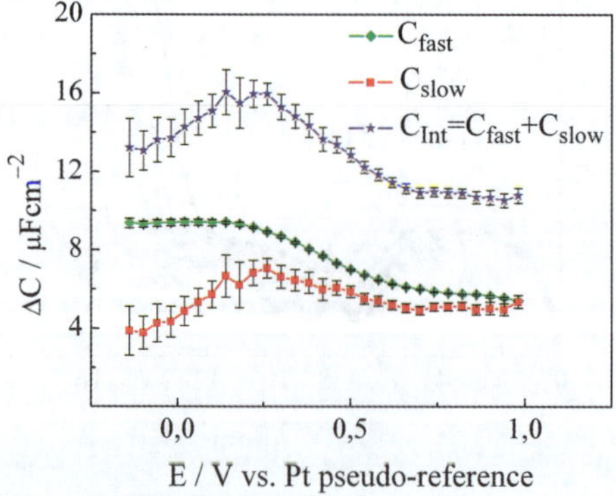


**Fig. 4** Potential-dependent differential capacitance of the fast and of the slow capacitive process as well as of the overall differential capacitance.

 

potential, $C_{fast}$ decreases monotonically and reaches a value of 5.4 μF cm$^{-2}$ at +1.0 V. The differential capacitance of the slow process $C_{slow}$ exhibits a maximum around +0.2 V. This maximum leads to a maximum of the overall interfacial capacitance $C_{Int}$, see Fig. 4.

We note that the data presented here for the anodic regime match well with previous data in the cathodic regime.[29] However, the match is not perfect, which is most likely caused by potential drifts of the Pt pseudo-reference electrodes. In future studies, we will reduce this unwanted drift by using novel micro REs of the second kind recently developed by Huber *et al.*[37]

When we compare our present results to recent theoretical work, it is evident that the experimental features in the $C(\Delta\varphi)$ curves are generally broader than predicted by Kornyshev,[5] Oldham[38] and Bazant *et al.*[21] For instance, in the mean-field lattice gas model by Kornyshev, the chemical potential of the ions at the electrode/RTIL interface is governed by an entropy penalty term for ion accumulation. This term leads to charge redistributions in a relatively narrow electrode potential range and thus to comparatively sharp features in the potential-dependent differential capacitance curves. We conclude that such entropy penalty terms are not sufficient for describing the RTIL/Au(111) interface. In contrast, the potential-dependent features seen in recent MD simulations are considerably broader and bear a stronger resemblance to experimental results.[17,20] Unfortunately, these simulations do not distinguish between capacitive processes taking place on different time scales. Only in Ref. 20, the authors present time-dependent electrode charge relaxation curves after a sudden change of the electrode potential. These curves give indication for a faster and a slower relaxation process, however, the authors do not analyse the contributions of these processes to the overall differential capacitance.

We propose that the slow process is due to charge redistribution in the innermost ion layer. The ions in this layer interact strongly with the electrode surface, in particular *via* image charge interactions and van der Waals interactions. If an ion leaves the innermost layer or performs a reorientational motion within the innermost layer, bonds between the ion and the electrode surface have to be broken. Thus, such a process should require a considerable amount of activation energy. We note that this picture bears a strong analogy to specific ion adsorption in aqueous electrode solutions.

In contrast, the fast capacitive process should comprise all charge redistribution in the layers next the innermost layer. In these layers, the ions are less strongly bound, so that charge redistribution should *not* require an activation energy in excess of that for ion transport in the RTIL.

The differential capacitance of the slow process, $C_{slow}$, exhibits a maximum around +0.2 V. *In situ* AFM studies indicate that the innermost layer is enriched with cations at potentials more negative than the ocp (−0.14 V) and enriched with anions at potentials more positive than +1.0 V.[29,30] Therefore it is reasonable to assume that at the ocp, the $[BMPyrr]^+$ cations in the innermost layer are bound to the Au(111) surface *via* their neutral alkyl chains in a slightly tilted orientation, so that the ring with the positively charged nitrogen atom is not in direct contact to the surface. With increasingly positive potential, the cations change their orientation in a way that the alkyl chains are perpendicular to the Au(111) surface. The resulting 'voids' on the electrode surface can then be occupied by $[FAP]^-$ anions.

We note that in the framework of this picture, it is also possible to explain results for the RTIL/electrode interface in the cathodic regime.[29] Force–distance curves obtained by *in situ* AFM measurements revealed that at any electrode potential, there is a multi-layer structure next to the Au(111) surface. Increasing the potential in the cathodic direction leads to an increase of the number of detectable layers and to an increase of the push-through forces indicating that the layers become more strongly bound to the charged surface. Furthermore, the thickness of the innermost layer decreases with increasingly negative potential. This can be explained by a reorientation of the $[BMPyrr]^+$-cations with the charged ring moving into an orientation parallel to the electrode surface. Furthermore, STM measurements showed that between −0.5 V and −1.2 V, the Au(111) surface undergoes a reconstruction resulting in a herringbone Au(111)(22 x $\sqrt{3}$) superstructure. Both cation reorientation and surface reconstruction should give contributions to the slow capacitive process. Therefore, this process is detectable over a broad range of cathodic potentials.[29]

## Conclusion

We have presented EIS results for the interface between [BMPyrr]FAP and Au(111) at potentials ranging from −0.14 V (ocp) to +1.0 V. At all potentials, we detected two semicircles in the complex capacitance plane. The high-frequency semicircle was observed in a frequency range from 1 MHz to about 20 Hz. The underlying fast capacitive process was attributed to charge redistributions at the interface which do *not* require an activation energy in excess of that for ion transport in the RTIL. The low-frequency semicircle was observed in a frequency range from about 20 Hz to about 0.1 Hz. The underlying slow capacitive process was attributed to charge redistributions in the innermost ion layer. In this layer, the ions are strongly bound to the electrode, and thus charge redistributions require a higher amount of activation energy.

In order to extract the contributions of the fast and the slow process to the overall differential capacitance of the interface, we fitted our data to an empirical Cole-Cole type expression. We found that the overall differential capacitance exhibits a maximum around +0.2 V, which is caused by a maximum of the differential capacitance of the slow process. We suggested that this capacitance maximum is caused by reorientational movements of the $[BMPyrr]^+$-cations with the positively charged ring moving away from the electrode surface and leaving behind 'voids' which are then occupied by anions.

## Acknowledgements

This work was financially supported by the Deutsche Forschungsgemeinschaft (DFG) within the Priority Program SPP 1191 (Ionic Liquids), and by the Fonds der Chemischen Industrie (Ph.D stipend for M.D.). We are very thankful to Prof. Endres and Dr Borisenko (Clausthal University of Technology) for providing [BMPyrr]FAP in ultrapure quality.

## References

1 C. Largeot, C. Portet, J. Chmiola, P. L. Taberna, Y. Gogotsi and P. Simon, *J. Am. Chem. Soc.*, 2008, **130**, 2730.
2 H. Yuan, H. Shimotani, A. Tsukazaki, A. Ohtomo, M. Kawasaki and Y. Iwasa, *Adv. Funct. Mater.*, 2009, **19**, 1046.
3 N. Papageorgiou, Y. Athanassov, M. Armand, P. Bonhôte, H. Pettersson, A. Azam and M. Grätzel, *J. Electrochem. Soc.*, 1996, **143**, 3099.
4 M. Galinski, A. Lewandowski and I. Stepniak, *Electrochim. Acta*, 2006, **51**, 5567.
5 A. A. Kornyshev, *J. Phys. Chem. B*, 2007, **111**, 5545.
6 M. Mezger, H. Schröder, H. Reichert, S. Schramm, J. S. Okasinski, S. Schöder, V. Honkimäki, M. Deutsch, B. M. Ocko, J. Ralston, M. Rohwerder, M. Stratmann and H. Dosch, *Science*, 2008, **322**, 424.
7 R. Atkin and G. G. Warr, *J. Phys. Chem. C*, 2007, **111**, 5162.
8 M. V. Fedorov and A. A. Kornyshev, *Electrochim. Acta*, 2008, **53**, 6835.
9 M. V. Fedorov and A. A. Kornyshev, *J. Phys. Chem. B Lett*, 2008, **112**, 11868.
10 G. Feng, J. S. Zhang and R. Qiao, *J. Phys. Chem. C*, 2009, **113**, 4549.
11 S. A. Kislenko, I. S. Samoylov and R. H. Amirov, *Phys. Chem. Chem. Phys.*, 2009, **11**, 5584.
12 Y. Lauw, M. D. Horne, T. Rodopoulos and F. A. M. Leermakers, *Phys. Rev. Lett.*, 2009, **103**, 117801.
13 M. V. Fedorov, N. Georgi and A. A. Kornyshev, *Electrochem. Commun.*, 2010, **12**, 296.
14 M. Trulsson, J. Algotsson, J. Forsman and C. E. Woodward, *J. Phys. Chem. Lett.*, 2010, **1**, 1191.

 

15 S. Tazi, M. Salanne, C. Simon, P. Turq, M. Pounds and P. A. Madden, *J. Phys. Chem. B*, 2010, **114**, 8453.
16 Y. Lauw, M. D. Horne, T. Rodopoulos, A. Nelson and F. A. M. Leermakers, *J. Phys. Chem. B*, 2010, **114**, 11149.
17 J. Vatamanu, O. Borodin and G. D. Smith, *J. Am. Chem. Soc.*, 2010, **132**, 14825.
18 N. Georgi, A. A. Kornyshev and M. V. Fedorov, *J. Electroanal. Chem.*, 2010, **649**, 261.
19 G. Feng, R. Qiao, J. Huang, S. Dai, B. G. Sumpter and V. Meunier, *Phys. Chem. Chem. Phys.*, 2011, **13**, 1152.
20 J. Vatamanu, O. Borodin and G. D. Smith, *J. Phys. Chem. B*, 2011, **115**, 3073.
21 M. Z. Bazant, B. D. Storey and A. A. Kornyshev, *Phys. Rev. Lett.*, 2011, **106**, 046102.
22 V. O. Santos Jr., M. B. Alves, M. S. Carvalho, P. A. Z. Suarez and J. C. Rubim, *J. Phys. Chem. B*, 2006, **110**, 20379.
23 Y. X. Yuan, T. C. Niu, M. M. Xu, J. L. Yao and R. A. Gu, *J. Raman Spectrosc.*, 2009, **41**, 516.
24 S. Rivera-Rubero and S. Baldelli, *J. Phys. Chem. B*, 2004, **108**, 15133.
25 S. Baldelli, *J. Phys. Chem. B Lett.*, 2005, **109**, 13049.
26 C. Aliaga and S. Baldelli, *J. Phys. Chem. B*, 2006, **110**, 18481.
27 S. Baldelli, *Acc. Chem. Res.*, 2008, **41**, 421.
28 W. Zhou, S. Inoue, T. Iwahashi, K. Kanai, K. Seki, T. Miyamae, D. Kim, Y. Katayama and Y. Ouchi, *Electrochem. Commun.*, 2010, **12**, 672.
29 R. Atkin, N. Borisenko, M. Drüschler, S. Z. El Abedin, F. Endres, R. Hayes, B. Huber and B. Roling, *Phys. Chem. Chem. Phys.*, 2011, **13**, 6849.
30 R. Hayes, N. Borisenko, M. K. Tam, P. C. Howlett, F. Endres and R. Atkin, *J. Phys. Chem. C*, 2011, **115**, 6855.
31 M. Drüschler, B. Huber and B. Roling, *J. Phys. Chem. C*, 2011, **115**, 6802.
32 T. Pajkossy and D. M. Kolb, *Electrochem. Commun.*, 2011, **13**, 284.
33 R. G. Musket, W. McLean, C. A. Colmenares, D. M. Makowiecki and W. J. Siekhaus, *Appl. Surf. Sci.*, 1982, **10**, 143.
34 N. Batina, A. S. Dakkouri and D. M. Kolb, *J. Electroanal. Chem.*, 1994, **370**, 87.
35 M. H. Hölzle, T. Wandlowski and D. M. Kolb, *J. Electroanal. Chem.*, 1995, **394**, 271.
36 D. Dickertmann, F. D. Koppitz and J. W. Schultze, *Electrochim. Acta*, 1976, **21**, 967.
37 B. Huber and B. Roling, *Electrochim. Acta*, 2011, DOI: 10.1016/j.electacta.2011.02.055.
38 K. B. Oldham, *J. Electroanal. Chem.*, 2008, **613**, 131.

# General discussion

**Mr Frolov** opened the discussion of the paper by Professor Endres: Which structures do you actually explore by your STM measurements: the innermost layer or layers of the ionic liquid at the gold surface or/and the gold surface itself?

**Professor Endres** responded: This is a good but difficult to answer question. Bearing in mind the ultra-high vacuum STM experiments we think (without being absolutely sure) that in the *in situ* STM experiment the innermost layer is probed. Under certain circumstances (herringbone structure) it seems that the gold surface is probed through the innermost ionic liquid layer. This is difficult to distinguish.

**Professor Castner** remarked: Your diagram indicating the possibility of charge ordering at the Au(111) surface is most intriguing. Can you create an experimental geometry in which you can apply a bias voltage sufficient to induce a phase transition from the charge-layered structure to the bulk charge-alternant structure, or *vice versa*?

**Professor Endres** replied: We consider an UHV STM experiment where we intend to do electrochemistry in the few monolayer regime. This would be one option to shed light on the raised question. I cannot say more at the moment.

**Professor Jung** remarked: Why does the addition of LiCl cause such a dramatic change in the tip-surface interaction, from repulsive to attractive? Do you think such a change will occur gradually as you increase the salt concentration or will it occur suddenly at a critical concentration? Have you seen the same behavior with other salts?

**Professor Endres** replied: First of all we should declare the real concentration of LiCl, which was recently determined in more detail by Rob Atkin: whilst a 10 wt % solution of LiCl in $[Py_{1,4}]$FAP was prepared, only a very small quantity of the salt was dissolved in the IL. The saturation solubulity of LiCl in $[Py_{1,4}]$FAP near room temperature is 0.083 wt%, after sonication for ~12 h. This is the concentration used in the experiments. Thus, the AFM force curve measurement is for a saturated solution of 0.083 wt% LiCl in $[Py_{1,4}]$FAP.

The same general effect of an attractive, stepwise force profile on AFM tip approach was observed whenever LiCl was dissolved in either $[Py_{1,4}]$FAP or [Emim]FAP. We are now conducting AFM experiments at LiCl concentrations lower than the saturation value to determine the effect of LiCl concentration on its own and as a function of potential. These experiments should reveal the origin of the switch from repulsive to attractive forces, which could stem from a variety of sources.

Thus, while we cannot yet comment why LiCl has such a dramatic effect, but we can already say that the STM results are somewhat different to the results in the pure ionic liquid. A preliminary interpretation would be that the interfacial structure is disturbed by the addition of the salt leading — somehow — to attractive forces. These studies are ongoing both our lab and that of Rob Atkin. We cannot yet comment if other salts have the same effect, but it is likely that there is an effect, too.

**Dr Taylor** asked: In Figure 11 you show the remarkable effect that addition of 10 wt% LiCl has on the AFM force/distance profile of Au(111)/$[Py_{1,4}]$FAP. In the pure ionic liquid, the profile is repulsive, whereas with addition of LiCl it becomes attractive. This is clearly an interesting observation, given the use of ionic liquids in lithium ion batteries and also the prevalence of $Li^{+}$ and $Cl^{-}$ as impurities. I am aware that further work is ongoing, but have you yet observed a LiCl concentration dependence on the force/distance profile? Does the switch from a repulsive to attractive profile occur incrementally with increasing LiCl content or is the attractive profile seen even at very low concentrations, i.e. <100 ppm— a typical level for an impurity in reagent grade ILs?

**Professor Endres** replied: These are ongoing studies and we indeed intend to check if there is a threshold concentration. So far we can already say that the AFM signals are less clear if lower quality liquids are used. Therefore we intended to set a reference of 2 extreme concentrations: one IL with all impurities of below 10 ppm and one saturated with the Lithium salt. Also, the STM results are affected by Lithium salts.

**Professor Ouchi** asked: In the cyclic voltammogram shown in Figures 1 and 2, how can peaks C1 and C2 be assigned? Is there any experimental evidence or a clue on the adsorption structure of the IL layers?

**Professor Endres** replied: These peaks have been discussed in a recent paper.[1]

1 R. Atkin, N. Borisenko, M. Drüschler, S. Zein El Abedin, F. Endres, R. Hayes, B. Huber and B. Roling, *Phys. Chem. Chem. Phys.*, 2011, **13**, 6849–6857.

**Professor Dr Roling** said: Regarding the IL thin film on Au: Can the amount of IL be quantified? If yes, shouldn't it be possible to distinguish between the monolayer and the bilayer model?

**Professor Endres** answered: Yes, it is possible to quantify the amount of IL and it should also be possible to distinguish between the monolayer and the bilayer model. These are experiments for the future, as mentioned in the discussion.

**Professor Pádua** opened the discussion of the paper by Professor Fedorov†: Carbon nanotubes are conducting materials, and their interactions with charged entities are affected by contributions from polarization of the conducting surface. But your simulations were carried out with a force field model of the Lennard-Jones type plus fixed point charges, therefore ignoring any polarization effects explicitly.

If the surface is polarized, then interactions of any charged moieties of the ionic liquid would be enhanced (attractively) compared to what you obtained, therefore affecting the orientations of the ions with respect to the surface.

Unless you can show by comparison with experimental evidence that your results are accurate, then you should explicitly include polarization of the carbon nanotube in your simulations.

**Mr Frolov** answered: In general, carbon nanotubes can be both metallic as well as semiconducting depending on their chirality.[1] Although in our work we, indeed, study a conducting carbon nanotube with (6,6) chirality, we believe that the preferred orientations of molecular ions are mostly determined by the strong Coulomb ion-ion and ion-charged surface interactions and the polarization of the metallic surface has no determinant effect on the ion orientations. We think that an explicit incorporation of the surface polarization would not qualitatively change the main observed trends in our study.

We also note that any classical molecular dynamics simulation is an approximation and the results should be taken as a hypothesis rather than the final answer. Thus, the regularities observed in our simulations should facilitate future

† Professor Fedorov's paper was presented by Mr Frolov, Max Planck Institute for Mathematics in the Sciences, Leipzig, Germany.

experimental and more advanced simulation studies to refine the results. In addition, we would like to note that Iori and Corni showed that the inclusion of the surface polarization affected the preferred orientation of a strong dipole at the neutral gold surface–water interface.[2] This observation suggests that simple speculations on whether the polarization of the surface is important or not, and to which extent it is important, are not sufficient. There should be performed a detailed study on how the inclusion of the surface polarization affects, firstly, capacitance profile; secondly, preferred orientations of ions at the neutral/charged surfaces for the carbon electrode/ionic liquid system. Such a study should make some answers on whether or not the polarization of nanocarbon surfaces significantly contributes to the mechanisms of the electrical double layer formation at the nanocarbon–RTIL interface.

1 R. H. Baughman, A. A. Zakhidov and W. A. de Heer, *Science*, 2002, **297**, 787–792.
2 F. Iori and S. Corni, *J. Comput. Chem.*, 2008, **29**, 1656–1666.

**Professor Canongia Lopes** said: In my opinion your structural analysis of the ionic liquid in the nanotube vicinity could be extended in two ways:

a) The anion bistriflamide has two important conformations in the liquid (bulk) phase.[1] Does the corresponding conformer distribution change when going from the bulk to the vicinity of the nanotube?

b) Whereas the anion can be modeled as a prolate spheroid (with only one director vector), the imidazolium cations have a long longitudinal axis aligned with its side chains but also the characteristic plane of the ring. They should be modeled as scalene ellipsoids with two director vectors. This means that in the case of the cations the analysis presented in the paper could be improved to address the issue of the relative positions of the ring plane and the nanotube surface.

1 J. N. Canongia Lopes, K. Shimizu, A. A. H. Pádua, Y. Umebayashi, S. Fukuda, K. Fujii and S. Ishiguro, *J. Phys. Chem. B*, 2008, **112**, 1465–1472.

**Mr Frolov** answered: We absolutely agree with these comments. Indeed, the work could be extended in these two mentioned ways to make a more detailed analysis of ion orientations. However, as far as it concerns the fundamental aspects of RTIL ineractions with the nanotube surface we were interested in, we think that the current analysis of the molecular ion orientations/conformations provides satisfactory results for our purposes:

a) Firstly, we performed our orientation analysis in a view of the hypothesis proposed in ref. 1. In this paper it was shown that alkyl tails of molecular ions can "play a role of space fillers, or latent voids [in the electric double layer] that can be replaced by charged groups, following rotations and translations of ions. This feature allows an electric field-induced increase of the counter charge density without a substantial compression of the liquid." Since the possible changes of the preferred conformation of the bistriflamide anion would not cause considerable change in the excluded volume of the species, for the first approximation we may neglect this conformational degree of freedom and consider the anion as "a prolate spheroid" or a dumbbell.

b) Secondly, we agree that representing the 1-alkyl-3-methylimidazolium cations by one directional vector is not sufficient to fully describe their orientation degrees of freedom. However, we considered only the longitudinal axis of the cations (the vector connecting the alpha carbon atoms of the alkyl and methyl groups), because the rotation of the cation coherent with this axis would lead to the maximum liberation of the volume in the double layer occupied by the cation non-polar alkyl tails (the discussed "latent voids"). We note that, the preferred orientation of the imidazolium ring plane will be considered in our further studies, as, indeed, changes in these orientations can also contribute (not that pronounced though) to the volume changes in the double layer associated to the rotation of the imidazolium rings.

c) The question of whether the alkyl chains of the imidazolium cations are aligned with the longitudinal axis of the cations is not trivial. For short alkyl chains, e.g. ethyl and butyl chains this is apparently true, because the chains do not have much flexibility. However, the internal flexibility of the longer octyl chain complicates the analysis of the preferred orientation of the cations. In general, we can say that all structural parts of the 1-octyl-3-methylimidazolium cations (OMIm) tend to be adsorbed on the CNT surface (as it is suggested by the first peak on the OMIm center of mass radial density profile at the CNT surface; see Fig. 1 of our paper, third row. More detailed studies on the non-polar tail conformations can be considered in the future studies.

1 N. Georgi, A. A. Kornyshev, M. V. Fedorov, *J. Electroanal. Chem.* 2010, **649**, 261–267.

**Professor Castner** commented: Your simulation results on the solvation structure of ionic liquids in contact with carbon nanotubes are fascinating. I am particularly interested in the fact that you have considered both neat ionic liquids and those thinned with $CH_3CN$ for lower viscosities. At the June 15–18 COIL-4 meeting, Oleg Borodin presented his results for ionic liquids inside 9.7 Å diameter single-walled carbon nanotubes, where the nanotubes were filled with alternant cation–anion–cation–anion spacing for a zero bias, but switched to all-cation or all-anion filling for applied bias voltages of less than 2.0 V. Have you considered the possible filling of open carbon nanotubes? Specifically, how will the combination of the ionic liquid and $CH_3CN$ co-solvent determine the solvent structure inside the nanotube?

**Mr Frolov** answered: Thank you very much for the comments and questions. We have not considered filling of the carbon nanotubes in this study because of the following reasons:

Firstly, as prepared raw nanotubes are usually capped, and this does not allow the electrolyte molecular species to enter inside the inner volume of the carbon nanotubes. Secondly, the STM images of the carbon nanotubes bundles suggest that the carbon nanotubes are bent and twisted, which should also complicate ion diffusion inside the pores.[1] Thus, as far as it concerns inside filling of the chemically unfunctionalized pristine carbon nanotubes, these effects seem to be not very significant. However, we note that some chemical modification can be used to significantly increase the solvent accessible surface area of the carbon nanotubes, which is attributed to the formation of holes on the carbon nanotube surface.[2] The authors of this study[2] reported an increase of the capacitance up to 1.6 times in the case of chemically modified carbon nanotubes compared to as prepared carbon nanotubes. Overall, we think that the questions you have mentioned are indeed very interesting both from theoretical and application points of view, but they should be answered in a separate complementary study on the electrochemical performance of the chemically-modified carbon nanotubes.

1 Hata *et al.*, *Science*, 2004, **306**, 1362–1364.
2 Yamada *et al.*, Energy Fuels 2010, **24**, 3373–3377.

**Professor Dr Roling** said: Did you calculate differential capacitance curves for the IL–CNT interface which can be compared to experimental data?

**Mr Frolov** responded: We have not yet completed this analysis, but this is a topic of our intensive ongoing work.

**Professor MacFarlane** said: What potential would be required in a real system to generate the charge densities you have used ($\pm 0.5$ e nm$^{-2}$)?

**Mr Frolov** answered: From our preliminary estimations we can say that the magnitude of the potentials would correspond to about 1.3–1.5 V. The potential is within the electrochemical window of the investigated electrolytes.

 

**Professor Pádua** opened the discussion of the paper by Professor Jung: As in the previous paper presented by A. Frolov, you have simulated systems containing ions and a conducting material (graphene sheets) without including any surface polarization effects. An important term is missing in your representation of the molecular interactions. Unless you can show that your results are accurate by comparing with experimental evidence, my view is that your model is probably not physically sound.

**Professor Jung** responded: While there is clearly room for improvement to make the model description more realistic, we strongly believe that our non-polarizable model correctly captures important features of RTIL and organic electrolyte supercapacitors. Furthermore, according to a recent study by Madden and coworkers[1] on the IL/graphite system, the inclusion of electrode polarizability does not affect the qualitative aspects of the results much. Nonetheless, we are currently developing a polarizable electrode description that can be easily implemented *via* molecular dynamics.

1 C. Merlet, M. Salanne, B. Rotenberg and P. A. Madden, *J. Phys. Chem. C*, 2011, **115**, 16613.

**Professor Dr Roling** remarked: The Green–Kubo conductivity characterizes the linear response at low electric fields. Does it make sense to apply it to double layers with very high fields?

**Professor Jung** replied: While we agree that care should be taken in applying linear response theory, we nonetheless believe that it provides useful insight into conductivity. Furthermore, under equilibrium conditions, we have few alternatives other than using approaches based on linear response. As briefly mentioned during my presentation at this meeting, we are currently conducting non-equilibrium simulations to analyze solvent response to capacitor charging and discharging processes. This ongoing study should help us to obtain a better and more accurate understanding of conductivity and related power density.

**Mr Frolov** said: In your simulations you consider high surface charge densities $\pm 0.86$ e nm$^{-2}$ which correspond to large potential drops at the graphene electrodes (up to 6 V as it is shown in the paper). The calculated voltage seems to be well beyond the stability region for the electrolytes under investigation in this case (*e.g.* the electrochemical window of the EMIm-$BF_4$ ionic liquid is 3.5 V on the carbon nanotube and 4.5 V on the glassy carbon electrodes).[1] What is the motivation to consider such large surface charge densities in your study?

1 J. Zheng, S. S. Moganty, P. C. Goonetilleke, R. E. Baltus and Dipankar Roy, *J. Phys. Chem. C*, 2011, **115**, 7527.

**Professor Jung** responded: It is sometimes useful to consider conditions that are not exactly realistic, e.g., elevated temperature, because they can help to facilitate theoretical analysis and gain detailed insight. We considered high surface charge density, together with the moderate charge density case, to enable us to obtain a better understanding of, for example, RTIL screening behaviors and saturation of the screening zone.

**Mr Frolov** commented: I think that the simulations performed for $\pm 0.43$ e nm$^{-2}$ surface charge densities are more relevant for comparison with the experimental data than the simulations for $\pm 0.86$ e nm$^{-2}$ surface charge densities. The electrostatic potentials look more realistic in the former case: *e.g.* the overall potential drop of the supercapacitor cell for the neat EMIm-$BF_4$ ionic liquid at $\pm 0.43$ e nm$^{-2}$ is 2.4 V (this is within the electrochemical window for this ionic liquid), while at $\pm 0.86$ e nm$^{-2}$ the electrostatic potential drops seems to be well beyond the electrochemical window of this ionic liquid.

**Professor Jung** responded: I agree. However, I think our response to your previous question answers this.

**Mr Frolov** asked: Did you estimate the electrostatic potential across the simulation box in the case of the pure acetonitrile solvent between the moderately charged ($\sigma = \pm 0.43$ e nm$^{-2}$) electrodes? Does the potential curve come to a plateau at the middle of the simulation box?

**Professor Jung** replied: No, we did not. Based on our results in Fig. 5c of our paper, however, I do not expect that the electric potential will show a plateau behavior anywhere inside acetonitrile in equilibrium with electrodes even if their charge densities are lowered to $\pm 0.43$ nm$^{-2}$.

**Professor Maroncelli** remarked: The preliminary data you have on the solvent response to capacitor discharge is very interesting. As you know these dynamics have much in common with both the dynamics of solute solvation and with dielectric relaxation. Your coworkers Shim and Kim have already studied both of these latter phenomena using the same ionic liquid model you employ here. Have you or are you planning to explore the inter-relationships among these various dynamics?

**Professor Jung** responded: Though not presented in our paper, preliminary results we have obtained thus far indeed show that dynamic response of EMI-$BF_4$ to charging and discharging of electrodes bear interesting similarities to solvation dynamics. Specifically, the initial relaxation dynamics, which account for about 50% of entire RTIL relaxation in response to capacitor charge/discharge, occur on a sub-picosecond time scale, while the ensuing slow decay is completed in ~10 ps. According to several previous MD studies (most notably by Shim and Kim as you pointed out), solvation dynamics of small solutes in RTILs exhibit similar biphasic relaxation behaviors. Despite this similarity, it appears that the time scale of long-time relaxation differs markedly between solvation in bulk RTILs and response to charging/discharging in parallel plate capacitors. Our preliminary results indicate that slow relaxation of the latter is faster than that of the former by at least a couple of orders of magnitude. To gain insight into this issue as well as short-time dynamics, we are continuing our non-equilibrium simulations to improve MD statistics and investigate how various factors, such as confinement and different multipole character of solute and electrode charge distributions, influence RTIL response and its underlying molecular motions. We hope to complete and report on this ongoing study in the near future.

**Professor Dr Roling** said: In your simulations, the anodic and cathodic capacitances differ by a factor of 4–5. Do you know any experimental capacitance data showing such a big difference?

**Professor Jung** responded: I have a few comments on this issue. First, our results for the relative difference between cathodic and anodic capacitances of supercapacitors are probably an overestimation due to the neglect of both the electrode and solvent electronic polarizability in our model description. Second, despite this limitation, our analyses reported here and elsewhere indicate that the anode–cathode asymmetry is quite sensitive to electrode geometry and electrode–solvent configuration. For instance, if both sides of a graphene electrode are exposed to electrolytes, the relative difference of anodic and cathodic capacitances decreases dramatically, compared with the parallel plate case (where only one side of the electrode interacts with electrolytes). This work on double-sided electrodes published in *Journal of*

*Physical Chemistry C*. Furthermore, according to a recent MD study based on the same model description, the ratio of the anodic and cathodic capacitances of microporous carbon electrodes varies with the pore size and ranges between 1 and 1.5.[2] These values are considerably lower than the ratio 3-4 we obtained for the parallel plate case. While further studies with the inclusion of polarizability effects are needed, the observations here seem to suggest that one should pay attention to the details about the electrodes and electrolytes when comparing the anode-cathode asymmetry with measurements. As regards your question, we are not aware of any experimental studies where the differences of cathodic and anodic capacitances of graphene-based electrodes are examined.

1 Y. Jung , H. J. Kim and Y. Shim, *Journal of Physical Chemistry C*, 2011, DOI: 10.1021/jp203458b.
2 Y. Shim and H. J. Kim, *ACS Nano*, 2010, **4**, 2345.

**Professor MacFarlane** remarked: It is surprising that the calculated conductivity (Table 2 of your paper) of the acetonitrile based electrolyte should be lower than the pure IL. How does that compare with measured data?

**Professor Jung** replied: This unexpected result for conductivity is attributed to the system confinement along the $z$ direction. Specifically, the presence of two electrodes in the parallel plate configuration exerts a strong influence on ion transport and conductivity along $z$ and thus alters their behaviors compared to bulk. It is worthwhile to note that the conductivity of the present supercapacitor system along the (unconfined) $x$ and $y$ directions in acetonitrile solution (1.1 S $m^{-1}$) is higher than that in pure EMI-$BF_4$ (0.59 S $m^{-1}$) by a factor of 2. MD yields a similar difference between the two liquid systems in the bulk phase. Experimentally, the conductivity of a 1 M solution of EMI-$BF_4$ in acetonitrile is higher than that of pure EMI-$BF_4$ by a factor of $\sim$3.5 at 298 K.[1] Considering the difference in temperature between experiments (298 K) and MD (350 K), we think that the agreement between the two is very reasonable.

1 A. Stoppa, J. Hunger and R. Buchner, *J. Chem. Eng. Data*, 2009, **54**, 472.

**Professor Dr Roling** remarked: Why do you sample exclusively over $z$ components of the ion velocities?

**Professor Jung** answered: Only the current between the positively- and negatively-charged electrodes contributes to power density of the supercapacitors. For the parallel plate configuration considered here, this corresponds to the $z$ component of the ionic current.

**Dr Mele** opened the discussion of the paper by Professor Jones: How do you make sure that the system is at equilibrium?

**Professor Jones** responded: The results described in the paper are kinetic measurements. We describe water absorbing into the bulk of the IL, and onto the surface of the glassy IL, from the gas phase. We also describe desorption of water from the physisorbed layer on the glassy IL surface back into the gas phase. The system is not at equilibrium.

**Dr Mele** commented: Which ion are you monitoring to follow the process?

**Professor Jones** answered: The $[OH]^+$ ion at $m/z = 17$ was used to monitor the flux of water emanating from the dip-stick surface, as described in the experimental section.

**Professor Hardacre** remarked: In Fig. 13 you have illustrated two activated processes. Could you explain the origins of the activation barriers? In particular, is the second step an activated process as this process is presumably driven by entropy?

**Professor Jones** answered: Both ionic liquids have an activated process for water moving from the physisorbed state to the ionic underlayer, which is 17 kJ mol$^{-1}$ higher than the enthalpy of water in the gas phase for $[C_8C_1Im][BF_4]$, and 7 kJ mol$^{-1}$ for $[C_2C_1Im][Tf_2N]$. These energies were determined experimentally. For $[C_8C_1Im][BF_4]$, the octyl chains are expected to form a relatively compact layer at the surface, and we think the activation energy corresponds to the water squeezing between the alkyl chains as it moves from the physisorbed state to the ionic underlayer. However, for $[C_2C_1Im][Tf_2N]$ the rather small ethyl group on the $[C_2C_1Im]^+$ cation is unlikely to produce a hydrophobic adlayer itself, but if the $CF_3$ groups of the $[Tf_2N]^-$ anion were orientated towards the vacuum at the surface, then there may be enough hydrophobic material (ethyl and $CF_3$ groups) to provide a small barrier to penetration of the water through to the ionic underlayer. Although this molecular argument for a hydrophobic layer on $[C_2C_1Im][Tf_2N]$ is rather weak, there is no evidence that ionic species are exposed to the vacuum, as the measured physisorption energy on $[C_2C_1Im][Tf_2N]$ is essentially the same as on $[C_8C_1Im][BF_4]$, *i.e.* adsorption on a hydrophobic surface.

The second activation barrier, between water in the ionic underlayer and water in the bulk of the ionic liquid, is speculation at this stage. It may not exist, the activation energy being the same as the thermodynamic difference between water in the underlayer, and water in the bulk. If the second activation barrier does exist, we assume that it would be due to the formation of a second, subsurface, layer of alkyl chains, which inhibit the movement of water into the bulk. Water movement from the ionic underlayer to the bulk of the IL is driven by entropy. The bulk of the ionic liquid is empty of absorbed water, so water in the ionic underlayer will continue to move into the bulk until equilibrium is attained.

**Professor Castner** commented: Your models for the structure of the wet/dry ionic liquid interface with vacuum are going to be most significant for helping to develop our understanding of water in ionic liquids for many important applications. Do you have any angle-resolved XPS data for surface depth profiling of the thickness of the "ionic underlayer" shown in Fig. 13 a) and b) of your paper?

**Professor Jones** responded: We have not measured angle resolved XPS data ourselves for the systems in this study. However, such measurements have been carried out by several groups for both the ionic liquid/vacuum interface (see for example ref. 1 and 2) and the ionic liquid/solid interface.[3] In the future we plan to carry out angle resolved XPS in combination with surface X-ray reflectivity, to determine the thicknesses of the various layers at the ionic liquid/vacuum interface.


1 V. Lockett, R. Sedev, S. Harmer, J. Ralston, M. Horne and T. Rodopoulos, *Phys. Chem. Chem. Phys.*, 2010, **12**, 13816–13827.

2 C. Kolbeck, T. Cremer, K. R. J. Lovelock, N. Paape, P. S. Schulz, P. Wasserscheid, F. Maier and H. P. Steinrück, *J. Phys. Chem. B*, 2009, **113**, 8682-8688.

3 T. Cremer, M. Stark, A. Deyko, H. P. Steinruck and F. Maier, *Langmuir*, 2011, **27**, 3662–3671.


**Professor Canongia Lopes** addressed Professor Jones and Professor Pádua : In my opinion after water reaches the surface polar layer (at the surface if the ionic liquid has very short alkyl side chains; beneath a thin non-polar layer if the IL has longer alkyl side chains; *c.f.* Ref. 1 and Fig. 1) it can diffuse into the bulk using paths provided by the polar network (thus avoiding the non-polar domains of the ionic liquid). This means that the barriers and energy levels shown in the left side of both images of figure 13 of paper 15 have probably lower values than those depicted in the image, calculated from "average" bulk phase data.

1 A. S. Pensado, M. F. Costa Gomes, J. N. Canongia Lopes, P. Malfreyt and A. A. H. Pádua, *Phys. Chem. Chem. Phys.*, 2011, **13**, 13518–13526.

**Professor Jones** answered: The potential energy diagrams for $[C_8C_1Im][Tf_2N]$ and $[C_2C_1Im][BF_4]$ are drawn to show hypothetical barriers between water absorbed in the bulk, and water adsorbed in the ionic underlayer. The sizes of the barriers are not known for either IL, so the heights in Figs 14A and 14B (in my paper) are simply to indicate that a barrier might exist. For $[C_8C_1Im][Tf_2N]$, the activation energy for dissolution of water from the ionic underlayer to the bulk was determined by XPS[1] to be 76 kJ $mol^{-1}$, while the enthalpy of solution[2] was determined from Henry's constants to be 34 kJ $mol^{-1}$. These two energies are measured from the bottom of the ionic underlayer adsorption well, and gaseous water, respectively, so they cannot be related to each other to give the barrier height. Consequently, it is not known whether the value of 76 kJ $mol^{-1}$ corresponds to the thermodynamic enthalpy difference between the two states, or to a smaller thermodynamic difference, plus a barrier height.

The simulations indicate that there are contiguous ionic pathways between the bulk ionic domain structure, and the surface ionic underlayer, that a water molecule can follow as it moves from the ionic underlayer to the bulk, or *vice versa*. Such paths remove the necessity for the water to pass through a hydrophobic region between the bulk and the ionic underlayer, which is where an activation barrier would arise. Hence, from the simulations, there would appear to be no reason for an activation energy on passing from bulk to underlayer, the pathway simply proceeding "down-hill" from the bulk absorption enthalpy to the adsorption enthalpy in the ionic underlayer. If this is true, with a an activation barrier of

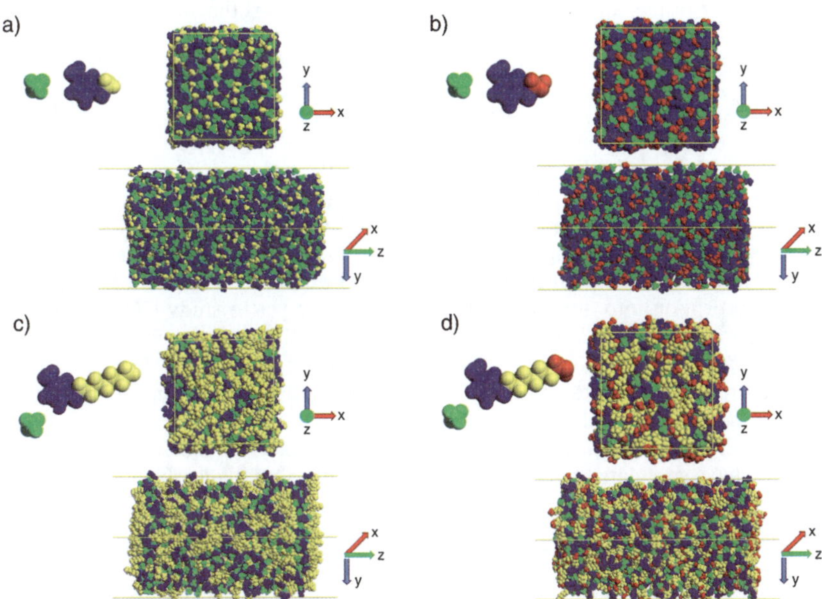


**Fig. 1** Snapshots of the simulation boxes: (a) 1-ethyl-3-methylimidazolium tetrafluoroborate; (b) 1-(2-hydroxyethyl)-3-methylimidazolium tetrafluoroborate; (c) 1-octyl-3-methylimidazolium tetrafluoroborate; (d) 1-(8-hydroxyoctyl)-3-methylimidazolium tetrafluoroborate.

zero, then for $[C_8C_1Im][Tf_2N]$ we can now relate the enthalpy of solution ($-34$ kJ mol$^{-1}$) to the enthalpy of adsorption in the ionic underlayer ($-34 + -76 = -110$ kJ mol$^{-1}$). Clearly the simulations make this very attractive, and probable, but experimental evidence is also required to prove it to be the case.

1 K. R. J. Lovelock, E. F. Smith, A. Deyko, I. J. Villar-Garcia, P. Licence and R. G. Jones, *Chem. Commun.*, 2007, 4866–4868.
2 J. L. Anthony, E. J. Maginn and J. F. Brennecke, *J. Phys. Chem.* B, 2001, **105**, 10942–10949.

**Professor Maginn** asked: Can you study gas mixtures with your approach?

**Professor Jones** replied: Yes, but we have not done so yet. When absorbing from a gas mixture it is important that the composition of the gas impinging on the surface is known. This is most easily achieved by leaking gas into the vacuum chamber from a mixture of known composition. However difficulties may arise if the components of the mixture are pumped at different speeds by the vacuum system. For instance, in a mixture of methane and water, the water would be rapidly pumped by a titanium sublimation pump, whereas the methane would not be pumped at all, leading to a lowering of the mole fraction of water impinging on the surface, relative to that at the inlet valve. Alternatively, two or more gases could be mixed within the vacuum chamber itself and a second (non line of sight) mass spectrometer used to measure the partial pressures of the gases. In this case appropriate calibrations of the mass peak intensities to the component gas pressures would have to be carried out. The second mass spectrometer would also need to be positioned close to the sample, so that it sensed the gas composition impinging on the sample, rather than some other composition (caused by pressure gradients within the vacuum system) at a distance from the sample.

**Professor Maginn** remarked: We have performed simulations of $CO_2$ and water in contact with $[C_6mim][NTf_2]$ and observe that the $CO_2$ adsorbs to the outer layer but the water localizes just below the outer region in an area of high charge density. These latter results appear to be consistent with your experimental observations.[1]

1 M. E. Perez-Blanco and E. J. Maginn, *J. Phys. Chem. B*, 2011, **115**, 10488–10499.

**Professor Jones** answered: $[C_6mim][NTf_2]$ in our nomenclature is $[C_6C_1Im][Tf_2N]$, so the cation in this ionic liquid only differs by a $C_2H_4$ group from the cation in $[C_8C_1Im][BF_4]$, while the anion is the same as the anion in $[C_2C_1Im][Tf_2N]$, the two ILs used in the experimental study. The system in the simulation is therefore highly relevant to interpreting the experimental results of the present study. It is pleasing to see that the simulations place the water in a high charge density region just below the surface, which is what we found experimentally and have referred to as absorption into an ionic underlayer. We have yet to study $CO_2$ adsorption on IL surfaces, as the vapour pressure of $CO_2$ at 80 K (our cryopumping temperature) is too high to allow line of sight mass spectrometry (LOSMS) to operate. However, we are now designing a LOSMS system with a cryopumping temperatures of $<40$ K, which will allow $CO_2$ adsorption studies on surfaces to be carried out. $CO_2$, although lacking a permanent dipole, does have an inherent quadrupole, and we have speculated that this quadrupole might cause it adsorb within the ionic underlayer of an IL surface. The simulations, however, suggest that this will not be the case.

**Professor Ouchi** asked: According to the MAES spectroscopic studies of ref. 15 in your manuscript, the top surface structures of long and short alkyl chain ILs are very different. However, as shown in Figure 13, a depth of physisorption of $H_2O$ molecule is almost the same for $[C_8C_1Im][BF_4]$ and $[C_2C_1Im][Tf_2N]$. Why?

 

**Professor Jones** answered: I discussed the physisorption energy of water in the answer to both Professor Hardacre's and Prof. Canongia Lopes' questions. Briefly, we were expecting a higher physisorption energy for water on $[C_2C_1Im][Tf_2N]$, due to exposed charged head groups at the surface, than for water on $[C_8C_1Im][BF_4]$, where we expect the vacuum interface to consist of octyl chains. The experimentally determined physisorption energy of water on $[C_2C_1Im][Tf_2N]$ (40.7 ± 0.2 kJ mol$^{-1}$ at zero coverage) was found to be slightly higher than for water on $[C_8C_1Im][BF_4]$ (39.1±0.5 kJ mol$^{-1}$ at zero coverage), but the difference is small, and both are lower than the physisorption energy of water on water ice. This was a surprise and suggests there are no charged species at the vacuum interface for either ionic liquid. The result is readily understandable for $[C_8C_1Im][BF_4]$, but not for $[C_2C_1Im][Tf_2N]$.

**Professor Lynden-Bell** said: Please can you explain in the context of the enthalpy profiles of Fig. 13 why the sticking probability is a low as 0.1 at low temperatures when the well depth of the physisorbed layer is 41 kJ mol$^{-1}$.

**Professor Jones** replied: Experimentally we measure the flux hitting the surface, $F$, and the flux leaving the surface, $F_{out}$. These fluxes are used in the paper to define the sticking probability, $S$, as the ratio of the flux of gas that disappears at the surface ($F_{in}-F_{out}$) divided by the flux hitting the surface, $F_{in}$. In other words the sticking probability is the experimental or effective sticking probability and takes into account the gas adsorbing (we assume with unit probability, see below) into the physisorbed state, equation (1), desorption from the physisorbed state back into the gas phase, equation (2) and penetration of the gas into the ionic underlayer, equation (3). The steady state kinetic analysis is given in the paper, equations (5) and(6) which provides the relationship between $S$ and the rate constants for these three processes, equation (7). Assuming the pre-exponentials for desorption and penetration (absorption) to be the same, and also assuming the activation energies for these two processes are the same, e.g. 41 kJ mol$^{-1}$, then physisorbed water has equal probability of desorbing or absorbing, leading to an effective sticking probability of 0.5. If the barrier to penetration (absorption) were higher than that for desorption (as is the case for both ILs here), then water desorbs in preference to absorbing, and the sticking probability drops below 0.5 (in our case to about 0.12). If the barrier to penetration is lower than that for desorption then the sticking probability rises above 0.5. The value of $S$ is therefore directly related to the barrier height difference for the two competing processes, desorption and absorption. As the derivation relies on the steady state approximation, it is assumed that the steady state concentration of physisorbed water is attained in a time much shorter than the duration of the measurements. This is true down to the lowest of the two glass transition temperatures (186 K), where the lifetime of a water molecule in the physisorbed state (41 kJ mol$^{-1}$) is still only 33 ms (assuming a pre-exponential of $10^{-13}$ s for the lifetime calculation), which is much shorter than the tens of seconds taken for each measurement of $S$.

It is worth mentioning that there is an alternative definition of sticking probability, defined as the probability that an incident gas molecule will enter into the physisorbed state instead of being reflected back into the gas phase, on collision with the surface. (This definition then requires a definition of what constitutes the molecule as residing in the physisorbed state, e.g. some minimum time such as a vibrational period on the surface.) For water vapour (at 300 K) impinging on our ionic liquid surfaces (100–300 K) we would expect a highly facile energy exchange to occur between the incoming water molecules and the IL surfaces, such that all the water molecules lose the physisorption energy, 41 kJ mol$^{-1}$, to become adsorbed into the physisorbed state. This would lead to a sticking probability of 1.0 using this alternative definition. So by this alternative definition of $S$, a value of 0.1 at low temperature would indeed appear unusual, but this is not the definition of S we use in our analysis (but we have implicitly assumed a value of 1 for this alternative definition of sticking probability within our analysis).

**Professor Kim** asked: About the physisorption well, is it determined by the attractive Coulomb interaction between the water dipole and its image in the bulk ionic liquid? Then the dielectric constants of two ionic liquids would make the difference.

**Professor Jones** replied: The adsorption of water has been studied on a huge range of surfaces, see reference 8 and 9 in the paper. Generally the adsorption energy is larger on ionic surfaces (65 kJ $mol^{-1}$ on NaCl), and smallest on oleaginous materials (35 kJ $mol^{-1}$ on hexadecane thiolate on gold), with adsorption on metals being in the middle. Metals are very polarisable, the electrons in the metal being able to move such as to form an oppositely charged image of any charge distribution above the surface. For a dipole at a given distance above the metal surface, the image dipole corresponds to the same charge distribution an equal distance below the surface but with the opposite polarity. The dipole and image dipole therefore show an attractive, or bonding interaction. For water adsorption on metal surfaces this image force contributes to the physisorption energy. For water adsorbed above an ionic liquid surface, the formation of such an image dipole would require the subsurface ions to redistribute themselves in a similar manner to the electrons in metals. Given the size and reduced mobility of the ions in ionic liquids, relative to the electrons in metals, we feel the ions in ionic liquids would be unable to form such an image dipole, or only capable of forming a weak image dipole. We therefore feel that differences in dielectric constants for different ionic liquids are unlikely to influence the physisorption energy via a dipole/image-dipole effect. However, this is a qualitative argument and computational studies are needed to show whether it is correct or not.

**Dr Fitzpatrick** asked: At high temperature the [$Tf_2N$] ionic liquid is hydrophilic, if you reduce the temperature the ionic liquid becomes hydrophobic (i.e. is a phase switching ionic liquid) what effect will this have on activation energies of the physical processes described?

**Professor Jones** answered: An interesting question. We would speculate as follows.

Hydrophilicity/hydrophobicity is a bulk property of the ionic liquid, so a switch from hydrophilic to hydrophobic with reduction in temperature, would correspond to a change in the domain structure within the bulk of the ionic liquid. Simulations of the liquid/vacuum interface indicate that layering at the surface, with formation of an ionic underlayer covered by alkyl chains, is a surface manifestation of the formation of domain structures within the bulk. So it would seem likely that a change in the bulk structure (as indicated by a switch from hydrophilic to hydrophobic) would also cause a change in the surface structure, thus affecting the kinetics of water vapour transport through the liquid/vacuum interface. At elevated temperatures, it could be that the hydrophobic, alkyl chain covered, surface is disrupted, allowing a greater density of ions at the liquid/vacuum interface and hence increasing the physisorption energy of water, and decreasing the absorption energy for water in an underlayer. A quantitative analysis of the kinetics would be required to determine whether this would increase the effective sticking probability on the hydrophilic ionic liquid, relative to what it would be on the surface of the hydrophobic ionic liquid.

**Professor Costa Gomes** asked: Can you measure, using your experimental technique, the solubility of water vapour in the ionic liquid? How long would it take to reach thermodynamic equilibrium?

**Professor Jones** replied: Our experiments use a relatively thick (0.5 mm) layer of ionic liquid, formed by dipping the solid support (silver) into a vial of the liquid in ultra-high vacuum. As described in a separate answer, this thick layer of ionic liquid represents a semi-infinite reservoir for adsorbed gases for the experimental pressures ($10^{-7}$ mbar) used. Therefore our experiments did not reach equilibrium.

 

To estimate how long it would take an absorbing gas to reach equilibrium, we have to estimate the absorption capacity of the ionic liquid, which depends upon the enthalpy of absorption (as determined from Henry's constants) and the applied pressure. We suggest that the enthalpy of absorption of gases into the bulk of ionic liquids is likely to be pressure dependent for low pressures (< 1 bar) where Henry's constants have yet to be measured. Ionic liquids aggregate into domains, with alkyl chains in one domain, and the ionic head groups in another domain. This may have a profound effect on the absorption energy of gases at low pressures. For a polar molecule such as water, one would expect a distribution of absorption enthalpies for water residing in different local geometries of ions. Under equilibrium conditions, as the applied water vapour pressure is reduced, water from geometries with shallower absorption energies will be depopulated, while those in geometries with deeper absorption energies will remain. One would then expect, for equilibrium adsorption at low pressures, that the enthalpy of absorption will rise for decreasing pressure, eventually reaching the value due to the ionic underlayer (if this is the deepest absorption energy well in the liquid), or possibly even higher if there are bulk geometries with even deeper energies. For higher pressures (1 bar) where the bulk concentration of the water in the ionic liquid is relatively high, the relatively small population of deep absorption wells will be saturated, with the majority of the absorbed water residing in the higher population, shallower absorption energies. This would give a constant, lower, enthalpy of absorption at higher pressures. This argument does not suggest that the deep absorption energy geometries are static, merely that there will be a statistical distribution of them at a given temperature, which will be populated with absorbed gas at equilibrium. Both simulations and experimental measurements are required to determine whether such high absorption enthalpies do exist at low pressures.

**Professor Costa Gomes** said: Is the ionic liquid layer thick enough to have the same properties as the liquid bulk?

**Professor Jones** answered: The liquid layers used were about 0.5 mm thick, so we assume that they exhibit the same properties as the bulk liquid; including a surface which is characteristic of the liquid/vacuum interface.

**Professor Hardacre** remarked: Is there any advantage of making the measurements using a supersonic molecular beam of water to increase the rate of the processes and the overall concentration of water in the ionic liquid, for example?

**Professor Jones** answered: This is a good idea. By using a supersonic beam of water vapour (*i.e.* accelerated using He), one could bring the water molecules onto the ionic liquid surface with a kinetic energy that exceeds the barrier (17 kJ $mol^{-1}$ for $[C_8C_1Im][BF_4]$) for water absorption into the ionic underlayer. Experimentally, we would expect the effective sticking probability to increase for increasing incident kinetic energy. The effective sticking probability may reach 1.0 when the incident energy equals the activation energy, but most likely the incident energy will have to exceed the activation barrier before a sticking probability of 1.0 is achieved. This is due to energy losses which occur from the water molecule, to the surface, as it traverses the physisorption well and the alkyl chain layer on its inward trajectory. Even though the initial energy might suggest the water molecule can reach the underlayer, in fact it loses too much energy to penetrate through the alkyl chains to the underlayer. Such experiments, though difficult, would provide enormously detailed information on the nature of the ionic liquid/vacuum surface, and its response to adsorbing and absorbing species.

**Professor Endres** remarked: Would your experiment also work if instead of a bulk film thin layers of ionic liquids would be applied to the surface?

**Professor Jones** answered: In the present experiment a film of IL of about 0.5 mm thickness was used. Assuming a monolayer of IL is about 0.5 nm thick (depends upon the IL and the geometry of the monolayer), this corresponds to $10^5$ layers, which is a semi-infinite reservoir for water molecules under ultra-high vacuum conditions. At the beginning of each experiment, this ionic liquid reservoir was empty, due to the low partial pressure of water in the vacuum chamber ($<10^{-10}$ mbar). During the course of an absorption experiment only a few monolayers of water were absorbed into the IL, leaving the IL reservoir below the IL surface still close to empty of water. Hence for all stages of the experiments there was a very low concentration of water dissolved in the IL, allowing transport from the gas phase to the absorbed phase to occur.

If we were to use thin layers of ionic liquid (e.g. 10–20 monolayers thick), the capacity of the IL to absorb water would be greatly reduced. For water adsorption into such a thin layer, we would expect, initially, a measurable sticking probability, followed by a decrease in sticking probability to zero as the thin layer saturated with water, as it achieved equilibrium with the applied water vapour pressure (about $10^{-7}$ mbar). This is a very different situation to adsorption on 0.5 mm thick IL layers. It is then interesting to speculate on the exact location of the water within the thin film, which will presumably consist of a layered structure at the liquid/vacuum and solid/liquid interfaces, with bulk IL between them. Water is most likely to reside where the ions are densest, in the ionic underlayer(s) at the liquid/vacuum interface, but also possibly in any ionic layers formed at the solid/liquid interface, as well as within the bulk-like IL between them. Depending on the solid substrate used, water may also diffuse through the IL and preferentially stick to the solid surface below the IL layer. Our previous work,[1] reference 30 in the paper, for water absorption onto $\approx$ 0.5 mm thick $[C_8C_1Im][BF_4]$, shows a temperature $<$ 240 K was necessary to populate the ionic underlayer at the liquid/vacuum interface with water for XPS analysis. We therefore expect similar low temperatures would be necessary to increase the water concentration in thin IL layers, such that surface sensitive techniques such as XPS could detect them.

1 K. R. J. Lovelock, E. F. Smith, A. Deyko, I. J. Villar-Garcia, P. Licence and R. G. Jones, *Chem. Commun.*, 2007, 4866–4868.

**Professor Endres** remarked: Have you an idea of what happens at the interface support/ionic liquid during the experiment?

**Professor Jones** responded: For the experiments described in the paper we used polycrystalline silver to support a thick ($\approx$ 0.5 mm) layer of ionic liquid. The IL layer was never removed to expose the underlying silver surface for analysis, so we cannot say what happens at the polycrystalline silver/ionic liquid interface. However, at present we are carrying out surface studies of thin layers (0–10 monolayers) of ionic liquids on gold (Au(110) and Au(111)) and copper (Cu(111)) single crystal surfaces, using ultra-violet photoelectron spectroscopy (UPS), X-ray photoelectron spectroscopy (XPS), temperature programmed desorption, low energy electron diffraction (LEED), and scanning tunnelling microscopy (STM). The ionic liquid adsorbs molecularly and strongly to the metal surfaces, forming specific interface states, as observed by UPS, and ordered structures as observed by LEED and STM. The first part of this work, for adsorption of $[C_2C_1Im][Tf_2N]$ on Au(110), will be submitted for publication shortly.

**Professor Canongia Lopes** opened the discussion of the paper by Professor Ouchi: Water and bistriflamide-based ionic liquids are immiscible but not completely immiscible. Some water can dissolve in the ionic liquid. Was this taken into account during the equilibration of the simulation boxes? How does the water density profile change along the simulation box? Is it compatible to the known values of the solubility of water in these ionic liquids?

 

**Professor Ouchi** replied: According to the literature,[1] IL concentration in the $H_2O$ phase and $H_2O$ concentration in the IL phase was reported to be 0.032 mol% and 27 mol%, respectively. Our simulation shows 0.00874 mol% for IL concentration and 6.5 mol% for $H_2O$ concentration after 280 ns of production run, in which the $H_2O$ density in the IL phase is almost the same along the simulation box within one digit of log scale.

1 J. M. Crosthwaite, S. N. V. K. Aki, E. J. Maginn, and J. F. Brennecke, *J. Phys. Chem. B*, 2004, **108**, 5113–5119.

**Professor Canongia Lopes** remarked: Even in the presence of the limiting boundary conditions, liquid–liquid interfaces tend to deviate (sometimes strongly) from a planar arrangement. This causes problems in terms of the definition and calculation of the density and charge profiles normal to the surface. How was this problem tackled in the present work?

**Professor Ouchi** replied: At the interface, instantaneous deviation from a planar arrangement may happen time to time. Our density profile is simply calculated by $\rho(z) = \langle m_i \delta(z_i - z) \rangle$, where the slab thickness is set at 0.7 Å along the $z$ axis for practical calculation, and is averaged over more than 200 ns. No special methods or treatments are incorporated in our calculation.

**Professor Canongia Lopes** asked: The solubilities of water in the $C_4$mim- and $C_8$mim-based ionic liquids is low in both cases but quite different between them. Is this fact apparent in the distribution of water molecules in the boundary region? Can this difference explain the differences found in the structural features between the two ionic liquids?

**Professor Ouchi** replied: Maybe yes, but we are not sure yet. We have checked the density profile at the interface for the judgment of its equilibrium and indeed ~200 ns is reasonably long enough to obtain the variables related to the interface. Though our calculated (bulk) solubility seems acceptable as shown above, it will be necessary to conduct a longer scale calculation for more reliable prediction of the bulk solubility and its comparison with surface properties.

**Professor Johansson** asked: The way you use the product of the calculated IR and Raman intensities for SF activity: a) Is this a well-known strategy? b) How about the sensitivity to the often large errors in calculated DFT Raman intensities? c) Do you take into account the Boltzmann distribution of the two C1 and C2 conformers in the SFG spectra?[1]

1 M. Herstedt, M. Smirnov, P. Johansson, M. Chami, J. Grondin, L. Servant and J. C. Lassègues, *J. Raman Spectrosc.*, 2005, **36**, 762–770.

**Professor Ouchi** responded: a) The way we used in the paper is a very rough estimation of the relative magnitude of nonlinear susceptibility of C1 and C2. As for more detailed calculation of non-linear susceptibility, please refer to our paper (ref. 14) and its electronic supporting information. When the molecular structure is simple and straight-forward, it is not difficult to calculate each component of nonlinear susceptibility. b) Our crude estimation was to support a qualitative discussion of SF activity of C1 and C2. Obviously, the level of the approximation is very low but a possible error is minor for our qualitative argument. c) We have not assumed any types of distribution functions on the analysis of SF spectra.

**Professor Castner** asked: Your combined infrared-visible sum frequency generation (IV-SFG) spectra and molecular dynamics (MD) simulations of ionic liquid–

vacuum and ionic liquid–water interfaces provide major insights into the nature of the 1-butyl-3-methylimidazolium TFSA and 1-octyl-3-methylimidazolium TFSA ionic liquids. Do you have any preliminary information from either the IV-SFG or MD methods for ionic liquids where the cation has either a terminal alcohol group or an ether chain replacing the alkyl chain?

**Professor Ouchi** answered: Sorry to say, no, we do not. However, our results clearly show possibilities of controlling interface structures by careful selection of molecular liquid and ionic liquid, your suggested derivatives are one of the materials for further study.

**Professor Maginn** asked: Can you say anything from the experiments about how the imidazolium ring orients at the interface?

**Professor Ouchi** replied: Experimentally, it is difficult to answer. We have not obtained a rather small peak of C–H vibration mode of the imidazolium ring. Since we have not observed even larger C–H vibration mode peaks of the alkyl chain of the cation, it is suggested that the rings at the interface may also hold the inversion symmetry.

**Professor Maginn** asked: Are you able to compute the orientation of the rings from your simulations? If so, how do the rings organize at the interface?

**Professor Ouchi** responded: According to our simulation, the orientation of the ring is found to be rather flat at the interface. Since our simulation successfully provides consistent pictures with those obtained from SFG spectra, the calculated results of flat orientation of the imidazolium ring strongly suggest that a very small C–H vibration mode of the ring is hardly observable even though it may lose the inversion symmetry by any means.

**Professor Endres** said: The interfaces CCl4–IL–$H_2O$ should show a dependence on temperature. Maybe it is worth doing temperature dependent experiments.

**Professor Ouchi** replied: Thank you for your suggestion. Yes, we are now planning to study a temperature dependence of molecular liquid–ionic liquid interfaces.

**Mr Frolov** said: Can the infrared-visible sum frequency generation (IV-SFG) vibrational spectroscopy be used to probe interfaces between liquids and solid surfaces?

**Professor Ouchi** replied: Yes. IV-SFG has also been used to study liquid–solid interfaces as well as air–liquid and air–solid interfaces.

**Mr Frolov** remarked: For which metal surfaces do you have the IV-SFG spectroscopy data?

**Professor Ouchi** answered: We have studied ILs–Pt, Au interfaces so far. As for the results on the IL–Pt interface, please refer to our paper.[1]

1 W. Zhou *et al.*, *Electrochem. Commun.*, 2010, **12**, 672.

**Professor Endres** asked: This is an elegant experiment. Are there, as in the studies on electrode surfaces, also hysteresis effects?

**Professor Ouchi** answered: It is a very interesting question. However, we have not done such experiments yet.

 

**Professor Dr Roling** commented: What could be the origin of slow processes at the ionic liquid–molecular liquid interface? For instance, interdiffusion processes over the length scale of some nm or some 10 nm should not take 20 min to 1 h.

**Professor Ouchi** answered: At this moment, we do not have a clear answer yet. Hopefully, we will be able to report it soon.

**Professor Endres** asked: It would be tempting to shake the 3 layers. If you do so, how do your interfacial structures at the phase boundaries change ?

**Professor Ouchi** replied: After three or four hours, the original three-layer-structure comes back again.

**Professor Canongia Lopes** said: The peculiar character of ionic liquids as solvents can be appreciated in Fig. 2 where an ionic liquid phase (yellow) sits between aqueous (blue, due to the presence of copper sulfate) and an organic (transparent) phases without mixing with neither of them. The idea can be explored even further by adding a fourth immiscible liquid phase composed by a perfluorinated alkane.

**Professor Ouchi** replied: It is an interesting demonstration for further studies of liquid–liquid interfaces.

**Mr Frolov** remarked: Do the 1-alkyl-3-methyl imidazolium ions lay on the solid (metal) surfaces, or do they stick with their tails against the solid surfaces?

**Professor Ouchi** responded: Please refer to the ref. 1 and references there in.

1 W. Zhou *et al.*, *Electrochem. Commun.*, 2010, **12**, 672.

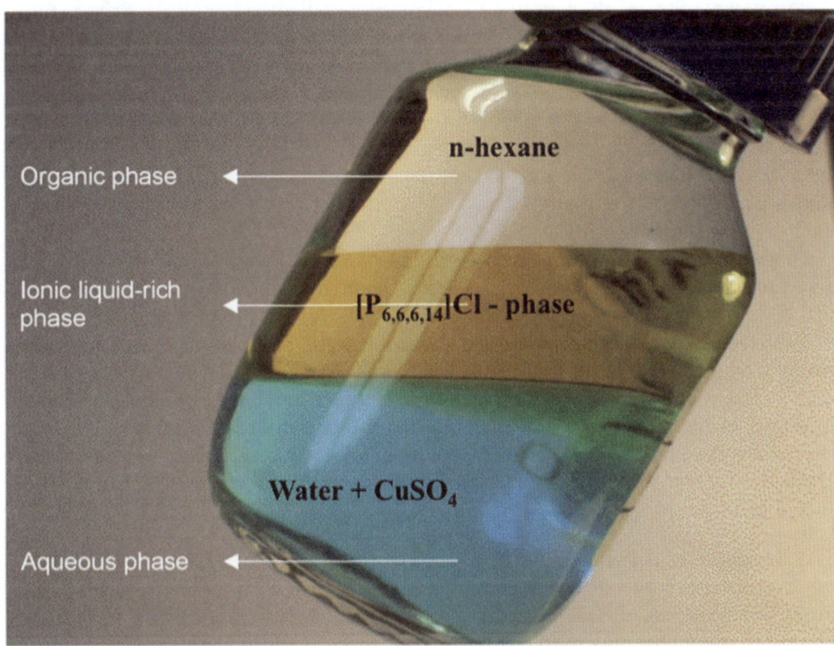


**Fig. 2**

**Mr Frolov** remarked: Based on the infrared-visible sum frequency generation (IV-SFG) vibrational spectroscopy measurements could you compare how the preferential orientation of molecular ions of the 1-alkyl-3-methylimidazolium bis(trifluoromethylsulfonyl)amide ionic liquids (ILs) changes between vacuum–IL, water–IL and metal–IL interfaces: 1. Does the preferential orientation of the same 1-alkyl-3-methylimidazolium cation change from interface to interface? 2. Does the length of the alkyl tail of the imidazolium-based cation affect the preferential orientation of the molecular cations at the interface? By the preferential orientation we consider parallel and perpendicular to the interface orientations of the alkyl non-polar tails and the imidazolium rings of the molecular cations.

**Professor Ouchi** responded: 1. Even within our limited experimental results, yes, it does. Please compare with those of our previous papers (ref. 18 and 19). 2. Yes, it does. As we have shown in Fig. 7(c) of our paper, $[C_4mim]^+$ and $[C_8mim]^+$ cations orient at the interface differently.

**Professor Endres** asked: A further maybe interesting experiment would be to put a rod of an inert metal and to test how the phase boundaries change.

**Professor Ouchi** replied: Thank you for your suggestion. We are willing to try.

**Professor Pádua** remarked: It is not clear to me from how the sum-frequency cancellation can be so complete given that at the interface there are significant breakages of symmetry due to spatial, orientational and conformational arrangements of the ions near the surface.

**Professor Ouchi** responded: As has been discussed in ref. 18 of our paper, and contrary to our present case, the $CCl_4$–$[C_4mim]PF_6$ interface exhibits very clear C–H vibrational mode peaks of the alkyl chain of the cation. Thus the symmetry argument is not that simple and straightforward. “Local symmetry breaking” is a convenient phrase often used to conclude overall discussions. However, more specifically, we need to define a physical parameter in each case to be discussed and analyze how the interface is formed. Our paper (ref.19) has discussed (d)-alcohol–$[C_4mim]PF_6$ interfaces, where a $CD_3$ symmetric stretching mode ($CD_{3ss}$) of (d)-alcohol is completely cancelled while a $CD_3$ asymmetric stretching mode ($CD_{3as}$) of (d)-alcohol is strongly enhanced. This is probably because the local interface structure is so constructed that the interface is symmetric in terms of $CD_{3ss}$, and is asymmetric in terms of $CD_{3as}$.

**Dr Delle Site** opened the discussion of the paper by Professor Dr Roling: Could you provide a concrete example of the time scale involved?

**Professor Dr Roling** responded: For both ultrapure ionic liquids we have studied up to now, namely [EMIm]FAP (see ref. 1) and [BMPyrr]FAP (this Faraday discussion paper) we find that the time scale of the fast process is in the millisecond regime and that of the slow process in the second regime. A millisecond time scale is expected for the charging of double layers when the working electrode/counter electrode distance is of the order of mm and when the diffusion coefficient of the ions in the IL in the range of $10^{-7}$–$10^{-6}$ cm$^2$ s$^{-1}$.

1 M. Drüschler, B. Huber, and B. Roling, *J. Phys. Chem. C*, 2011, **115**, 6802–6808.

**Dr Delle Site** asked: Given the time scale involved, current atomistic simulations cannot be used. One may need a coarse-grained model which would imply the loss of chemical specificity.

 

**Professor Dr Roling** replied: I do not agree with respect to the fast process. A millisecond time scale is expected for electrode distances in the millimeter range. However in simulations, the distance between the electrodes may be some 10 nm. Thus, the double layer charging should be around $10^5$ times faster than in experiment, *i.e.* in the range of 10 ns. Therefore, double layer formation in ILs at room temperature can be studied in MD simulations.

On the other hand, we have not yet studied the influence of the electrode separation distance on the slow process. Therefore, we cannot estimate the time scale of the slow process at electrode distances of some 10 nm. It is possible that the slow process cannot be observed in MD simulations, so that coarse-grained models are needed.

**Professor Angell** inquired: It seems that a study of the temperature dependence of the characteristic time for the slow process could be helpful in pinning down the origin of this relaxation. Yet you have not reported temperature effects. Is there any special reason for that?

**Professor Dr Roling** responded: I agree that the temperature dependence of the slow process should provide information about the origin of the process. Such temperature-dependent measurements are currently under way in our lab.

**Professor Endres** remarked: There are very slow processes at the interface IL–gold which occur on the time scale of hours and which might be correlated with defect formation in the IL layer.

**Professor Dr Roling** answered: We also detect very slow processes in our broadband capacitance spectra with time scales clearly exceeding minutes. These processes appear to become Faradaic in the limit of low frequencies (see our Faraday discussion paper) with the impedance phase angle approaching zero degrees. An important point is that the charge flow related to these ultraslow processes is much larger than the charge storable in IL double layers. Therefore, we think that defect formation in the IL layer might be one sub-process, but other ultraslow sub-processes with much more charge flow take place in addition.

**Professor Ouchi** asked: Dear Prof. Roling, As for a mechanism of the slow process, alkyl chain interaction with electrode has been emphasized. If so, an octyl chain case should have even slower process. Do you have any preliminary experimental information to support your conclusion?

**Professor Dr Roling** answered: No, we have not yet done experiments on ultrapure liquids with an octyl chain at the cation. But this is a very good suggestion.

**Dr Taylor** asked: You have proposed that the fast and slow capacitive processes observed for [BMPyrr][FAP] on Au (111) are caused by facile charge redistribution at the interface and the slower reorientation of the pyrollidinium cation to allow the approach of further anions to the gold surface, respectively. Simulations and experiments suggest that in response to a charged surface, ionic liquids form strong solvation layers that oscillate in charge moving away from the interface.[1] Furthermore, the innermost layer of counterions may completely or even overscreen the electrode charge.[2,3] Presumably, if such a layer of $[FAP]^-$ anions forms on short-time scales at the positively charged electrode, its screening effect would discourage the approach of further anions. Consequently, to approach and adsorb at the interface, any anion would have to have enough energy to overcome the repulsive energy barrier caused by the already adsorbed anion layer. Would it be possible that this could cause or contribute to the second, slower capacitive process?

1 F. Endres, O. Höfft, N. Borisenko, L. H. Gasparotto, A. Prowald, R. Al-Salman, T. Carstens, R. Atkin, A. Bund and S. Z. E. Abedin, *Phys. Chem. Chem. Phys.*, 2010, **12**, 1724–1732.
2 R. M. Lynden-Bell and M. V. Fedorov, in *Faraday Discuss.*, RSC, Belfast, 2011, Poster 40.
3 Y. Shim, H. J. Kim and Y. Jung, *Faraday Discuss.*, 2011, DOI: 10.1039/C1FD00086A.

**Professor Dr Roling** answered: This is an interesting suggestion. I could imagine that overscreening leads to a slow process. On the other hand, I do not think that overscreening is the only origin of the slow process. The slow process is observed over broad electrode potential ranges (see our paper on [EMIm]FAP/Au(111)),[1] while overscreening is most relevant at small electrode charges, *i.e.* in a narrow potential range around the potential of zero charge.

1 M. Drüschler, B. Huber, and B. Roling, *J. Phys. Chem. C*, 2011, **115**, 6802–6808.

**Dr Taylor** asked: In Paper 12, Frank Endres outlined the dramatic differences in AFM force/distance profiling and STM imaging at the Au(111) ionic liquid interface between samples of varying purity. Like Professor Endres, you have also employed >99.5% pure commercially available ionic liquids as supplied from Merck. Have you observed similar capacitive behaviour if your experiments are repeated with synthesis grade samples? I ask this question as I am aware that the prohibitive cost of ionic liquids is a barrier to their commercialisation, *e.g.* in energy storage devices, especially if more expensive high purity ILs are needed.

**Professor Dr Roling** replied: We have not yet compared the interfacial capacitance behavior of our ultrapure FAP ionic liquids to synthesis grade samples. I would expect that dramatic differences in AFM and STM measurements also manifest in capacitance measurements, since our capacitance measurements probe essentially the charge distribution in the 2–3 innermost layers. What we have studied systematically is the influence of water on the electrode–IL interface. We have found that above 200 ppm water, the electrical window shrinks considerably and the interfacial capacitance increases.

**Dr Harris** commented: There are two types of diffusion coefficients in ionic liquids where there is a reaction at an electrode. The electrochemical diffusion coefficient describes the diffusion of the electroactive species in the concentration gradient at the electrode *i.e.* the diffusion current. The *intra*-diffusion coefficients of the ions *etc.* in the bulk describe the thermal motion of the species concerned. The electrochemical diffusion coefficient of the electroactive species is only equal to its *intra*-diffusion coefficient at infinite dilution (*i.e.* tracer-diffusion), as in a supporting electrolyte. Bearman[1] discussed the difference in the 1960s in terms of phenomenological (Onsager) friction coefficients, but the problem has yet to be recast in terms of Green–Kubo velocity correlation functions. Recent work by Bond *et al.*[2] at Monash University suggests the differences are probably small in dilute solutions, but little is known about concentrated ionic liquid electrolytes as might be used in batteries *etc.*

1 R. J. Bearman, *J. Phys. Chem.*, 1962, **66**, 2072.
2 A. Solangi, A. M. Bond, I. Burgar, A. F. Hollenkamp, M. D. Horne, T. Rüther and C. Zhao, *J. Phys. Chem. B*, 2011, **115**, 6843.

**Professor Dr Roling** replied: I agree that very little is known about chemical diffusion coefficients in ILs. I think, however, that it is reasonable to assume that self diffusion and chemical diffusion coefficients of the ions in ILs are of the same order of magnitude. In the case of double layer formation, the double layer charging time can be expressed as: $\tau = R_{bulk} \times$ CDL, where $R_{bulk}$ is the bulk resistance of the IL and $C_{DL}$ is the double layer capacitance. $R_{bulk}$ is proportional to $L / D$, with $L$ and $D$ denoting the distance between the electrodes and an average diffusion coefficient of the ions. When we take typical values for the self diffusion coefficients for ions in

 

ILs, which are in the range of $10^{-7}$–$10^{-6}$ cm$^2$ s$^{-1}$, and when we assume that $L$ is of the order of 1 mm, then we obtain $\tau$ values in the range of milliseconds. This is exactly the time scale of our fast process.

**Professor Johansson** commented to Professor Dr Roling and Professor Endres: If we for a moment disregard the need of ultrapurity to perform the AFM, I wonder if you have any thought on if the schematic on potential dependent dynamics, more specifically the hysteresis of cation/anion displacement at the surface would be affect the opposite way if the cation would be twice the anion in size or disappear if they would be of the same size? You are right now only probing an IL with a twice as large anion as cation.

**Professor Dr Roling** answered: First of all, we would like to emphasize that a high purity is not only important for the AFM and STM measurements, but also for impedance/capacitance measurements. The capacitance semicircles shown in our paper reflect essentially the charge distribution in the 2–3 innermost layers directly at the electrode. Any impurities influencing this charge distribution, for instance water, do definitely have a strong influence on the measured interfacial capacitance. Slow processes (e.g. de- and reconstruction of the electrode surface, specific ion adsorption) and the resulting hysteresis effects appear to be a quite universal phenomenon at electrode–electrolyte interfaces and have even been found in the case of diluted aqueous electrolytes.[1,2] Therefore, hysteresis should be always observable whatever is the ion size ratio. Of course, the ion size ratio should influence the shape of the capacitance / voltage curves as shown in theoretical papers.[3,4] For instance, a small size of the anion compared to the cation should result in higher differential capacitances in the anodic regime as compared to the cathodic regime.

1 D. Eberhardt, E. Santos and W. Schmickler, *J. Electroanal. Chem.*, 1996, **419**, 23–31.
2 T. Pajkossy, T. Wandlowski and D.M. Kolb, *J. Electroanal. Chem.*, 1996, **414**, 209–220.
3 M. V. Fedorov, N. Georgi and A. A. Kornyshev, *Electrochem. Commun.*, 2010, **12**, 296–299.
4 Y. Lauw, M. D. Horne, T. Rodopoulos, A. Nelson, F. A. M. Leermakers, *J. Phys. Chem. B*, 2010, **114**, 11149–11154.

**Professor Endres** said: We use ultimate quality ionic liquids with the lowest possible impurity level. It is not that simple to vary liquids as the purity has to be guaranteed. Therefore we have to do everything step by step.

# Protic pharmaceutical ionic liquids and solids: Aspects of protonics†

**Jelena Stoimenovski, Pamela M. Dean, Ekaterina I. Izgorodina and Douglas R. MacFarlane***

*Received 19th April 2011, Accepted 23rd May 2011*
**DOI: 10.1039/c1fd00071c**

A series of new protic compounds based on active pharmaceutical ingredients have been synthesised and characterised. Some of the salts synthesised produced ionic liquids, while others that were associated with rigid molecular structures tended to produce high melting points. The "protonic" behaviour of these compounds was found to be a major determinant of their properties. Indicator studies, FTIR-ATR and transport properties (Walden plot) were used to probe the extent of proton transfer and ion association in these ionic liquids. While proton transfer was shown to have taken place in all cases, the Walden plot indicated strong ion association in the primary amine based examples due to hydrogen bonding. This was further explored *via* crystal structures of related compounds, which showed that extended hydrogen bonded clusters tend to form in these salts. These clusters may dictate membrane transport properties of these compounds *in vivo*.

## 1. Introduction

Many drug substances on the market are organic acids or bases that have an ability to form salts.[1] Converting a drug from its acid or base form to a salt form can often allow a useful manipulation or optimisation of the drug's properties, such as its absorption, pharmacodynamics or pharmacokinetics.[2,3] In addition, by keeping an active ion constant and changing the counterion of the drug salt, either by combining with another active, or other biocompatible ion, one can further alter its chemical and biological properties without changing the structure of the active ingredient. Different salt forms of an active are now accepted to be different drugs, with individual chemical and biological profiles, resulting in varying clinical efficacies and safety.[1] Turning the pharmaceutical salt into an ionic liquid (IL) is a further conceptual step that has been discussed recently in this regard.[4] ILs are salts with melting points below 100 °C and are typically comprised of cations and anions that have either relatively low symmetry or delocalised charge; these structural properties tend to provide a low, or sometimes nonexistent melting point.[5,6] A wide range of different physical and chemical properties can be designed into the liquid by simply varying the cations and anions of which it is comprised.[6] From solvents for chemical reactions,[7] to electrolytes for electrochemical cells,[8] ILs are now of intense interest in a range of applications in different areas. Recently, recognising that more than half of pharmaceuticals currently on the market are organic salts,

*School of Chemistry, Monash University, Clayton, Victoria, 3800, Australia. E-mail: Douglas.MacFarlane@monash.edu; Tel: +61 3 9905 4540*

† Electronic supplementary information (ESI) available. CCDC reference numbers 802624–802630. For ESI and crystallographic data in CIF or other electronic format see DOI: 10.1039/c1fd00071c

ILs have also been proposed as a novel approach to drug design and optimisation in the pharmaceutical world.[4,9–11] In 2007, Rogers and co-workers, in conjunction with our group, filed a series of patents in which the preparation of ILs from known active pharmaceutical ingredients (APIs) was described.[12] The concept involved the replacement of what was usually a "ballast" counterion, for example sodium or bromide, with an alternative counterion that lowered the melting point of the salt such that it was liquid at ambient temperature, thus offering a range of different delivery modalities. If the "counterion" itself has active pharmaceutical properties then the compound can be considered as dual active, potentially presenting a synergistic combination of effects in a single compound. Other groups have been investigating ILs for related applications including anti-microbial[13] and anti-bacterial activity.[14–16]

However, unlike the broader ILs field where the design or choice of cation and anion is, to an extent, flexible such that the desired low melting point can be achieved, this is not a case with pharmaceutical ILs. In this field the identity of at least one of the ions is specified, hence lowering the melting point of its salts becomes a challenging goal. This challenge has been explored over the last few years using an "anti-crystal engineering" approach,[17] aided by a greater understanding of the lattice energy of crystalline organic salts as revealed by the Madelung constant.[18] It was found, not surprisingly, that the extent of specific intermolecular interactions, such as hydrogen bonding, directly influences the phase behaviour of the product. Hence the advantageous combinations of active ion and counterion that can produce the low melting point required to form an IL will typically be those that avoid or disrupt all of these identifiable interactions.

## Protic ionic liquids – the proton transfer conundrum

Protic Ionic Liquids, PILs, are a subclass of ILs formed by reacting a Brønsted acid with a Brønsted base. Unlike *aprotic* ILs, PILs exhibit Brønsted acidity due to the exchangeable proton, and can hence be used as solvents for acid catalysed reactions;[19–21] for the same reason, they are also found to be suitable proton conducting electrolytes for fuel cells.[22] Given their ease of preparation, being simply the product of an acid (HA) - base (B) neutralisation reaction (Scheme 1), this IL family is of significant potential in broad, multifaceted pharmaceutical applications. Many pharmaceuticals are in fact delivered as simple salts of organic bases, or metal ion salts of organic acids. However, the properties manifested by PILs depend predominantly on the degree of proton transfer from the acid to the base as well as the hydrogen bonding in which the proton is involved; the possible outcomes are summarised in Fig. 1 and all of these outcomes have been observed in recent examples. For the PIL to be considered a "pure" salt and not a complex mixture of acid, base and salt, the extent of proton transfer should be at least 99%.[23,24] Despite the fact that in aqueous solution the proton transfer between the acids and bases usually involved would be strongly complete, the situation is more complex in the "neat" salt for complex reasons. Yoshizawa *et al.*[25] originally revealed the problem by correlating observed boiling points with differences in $pK_a^{aq}$ values ($\Delta pK_a^{aq}$) between the acid and the amine base. On the basis of a comparison of Walden products (explained further below) it appeared that $\Delta pK_a^{aq} > 10$ was needed for the full proton transfer to be observed (*i.e.* to achieve the situation in Fig. 1c rather than the acid–base mixture of Fig. 1a).[25]

To further investigate this phenomenon we explored a series of carefully chosen primary and tertiary amines of very similar $pK_a^{aq}$ values.[26] We found that restricted

$$HA + B \rightleftharpoons BH^+ + A^-$$

**Scheme 1** Neutralisation reaction between an acid and a base.

 

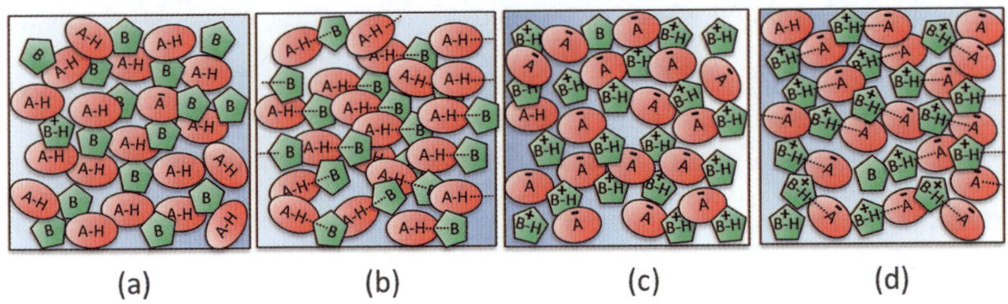


**Fig. 1** The four possible outcomes in PILs **(a)** mostly un-ionised acid and base **(b)** hydrogen bonded acid–base pairs (no proton transfer) **(c)** mostly ionised PIL (proton transfer occurred, independent ions), **(d)** associated PIL (proton transfer occurred but ions associated through hydrogen bonding).

proton transfer was clearly exhibited by the tertiary amine base systems, and $\Delta pK_a^{aq} > 10$ was indeed required to produce substantial proton transfer. On the other hand, the primary amine systems tended to be fully ionised even at $\Delta pK_a^{aq} > 5$. In other words, the primary amine based ILs tend to behave much more like aqueous systems. The strong difference in behaviour appears to be related to the hydrogen bond donor ability of the protonated base cation. In the tertiary amine case, only the single, transferred proton is available to hydrogen bond with the anion, whereas the primary base cation can offer multiple hydrogen bond sites. Free energy of solvation calculations from COSMO-RS confirmed this difference in the solvation environment in the two cases.[26] However, clearly, additional intermolecular interactions stemming from long alkyl chains, $\pi\cdots\pi$ interactions and other hydrogen bonding sites in either ion, for example –OH functionality, can easily disturb these trends. Hence, introduction of such characteristics into tertiary amine based ILs could induce a stronger degree of proton transfer. These phenomena are crucial to attempts to form PILs based on APIs, since the lack of complete proton transfer may render the formulation nothing more than an acid/base/salt mixture. On the other hand, as we will show, the H-bonding introduced, involving both the transferred proton and also the residual conjugate base anion, can also produce strong ion-association in the PIL, potentially producing a further dimension of property variation in the product. We attempt to describe this complex set of interconnected phenomena surrounding the labile proton, its energetics and dynamics, in terms of a single unified concept – protonics. Exploring these phenomena in the context of APIs is the main theme of this paper.

## Pharmaceutically active protic ionic liquids

A number of groups including us have investigated the use of PILs of various compositions as an approach to pharmaceutically active ILs.[27,28] For example, Bica *et al.*[27] described a series of salts of aspirin, *e.g.* lidocainium acetylsalicylate, $T_g = -14$ °C, where lidocaine is a topical anaesthetic and aspirin is a broad analgesic. In the present work we have focused on salt formation and the issue of proton transfer in a number of families of pharmaceutically active acids and bases. For the purposes of a systematic study, a series of acids and bases (Fig. 2) were selected with slight structural variations and physico-chemical properties, with the objective of understanding the factors controlling melting point, H-bonding and proton transfer. A library of nine compounds and four oligomeric PILs was synthesised and characterised. Benzoic, salicylic and gentisic acids were chosen for this study, as these are frequently encountered in pharmaceutical compounds/formulations. Benzoic acid, BzH, a common preservative and pharmaceutical aid (antifungal), is used in a range of products on the market including ointments, mouthwashes and cosmetics. A derivative of benzoic acid, salicylic acid (SalH), is a widely recognized and used

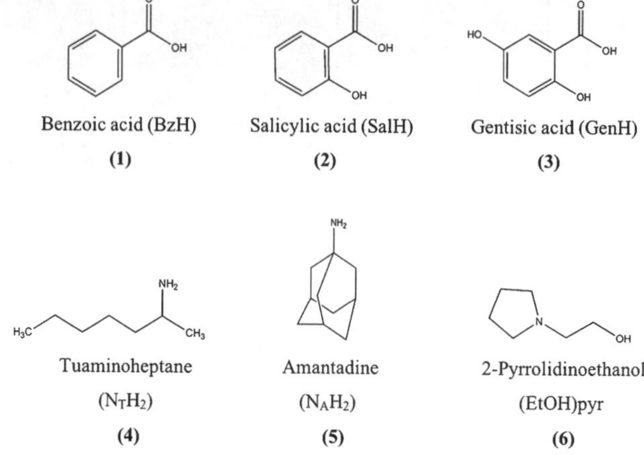


**Fig. 2** Acids and bases used in this study.

keratolytic, also possessing anti-inflammatory, analgesic and antipyretic properties as a sodium salt. Gentisic acid, GenH, another benzoic acid derivative, is used in the pharmaceutical industry as an analgesic and anti-inflammatory.[29]

The three pharmaceutical bases were chosen on the basis of their $pK_a^{aq}$ values and their structure (Fig. 2). Tuaminoheptane, $N_TH_2$, a primary amine base with a $pK_a^{aq}$ value of 10.50, is used as a nasal decongestant. Amantadine, $N_AH_2$, also a primary amine base with a $pK_a^{aq}$ value of 10.10, is an antiviral/anti-Parkinsonian drug. 2-Pyrrolidinoethanol, (EtOH)pyr, a "generally recognised as safe compound", is a tertiary amine with a $pK_a{}^{aq}$ value of 9.44.[29]

## 2. Experimental

### 2.1 General

Tuaminoheptane ($N_TH_2$), amantadine ($N_AH_2$), 2-pyrrolidinoethanol ((EtOH)pyr) and gentisic acid (GenH) were purchased from Aldrich Chemical Company. Benzoic acid (BzH) was purchased from Sigma-Aldrich, while salicylic acid (SalH) was supplied by Chem Supply.

The synthesis procedure and the characterisation of the resulting compounds can be found in the Electronic Supplementary Information.†

### 2.2 Instrumental

Nuclear Magnetic Resonance (NMR) data were recorded on either a Bruker Avance 400 (9.4 Tesla magnet) with a 5mm broadband autotunable probe with Z-gradients and BACS 60 tube autosampler or on a Bruker DPX-200 spectrometer. All samples were measured as solutions in deuterated dimethylsulfoxide supplied by Cambridge Isotope Laboratories. Multiplicities are reported according to their relative splittings, as a singlet (s), doublet (d), doublet of doublets (d(d)), triplet (t), pentet (p), sextet, multiplet (m) or a broad signal (b).

Electrospray Mass Spectrometry (ESI) was carried out on the Micromass Platform II API QMS Electrospray Mass Spectrometer with a cone voltage of 25 V or 35 V, using methanol as the mobile phase. Analyses were conducted in both positive ($ESI^+$) and negative ($ESI^-$) modes.

Density data was collected using an Anton Paar DMA 5000 density meter. The density of the compounds was determined using the "oscillating U-tube principle". The viscosity measurements were performed using a falling ball technique on an Anton Paar AMVn viscosity meter. Conductivity measurements were carried out on a locally designed dip cell probe containing two platinum wires covered in glass. 0.01M KCl solution at 25 °C was used to determine the cell constant. The conductivities were obtained by measuring the complex impedance spectra between 1 MHz and 0.01 Hz on a Solatron SI 1296 dielectric interface.

Where crystals were obtained, the reflection intensity data were measured on a Bruker X8 APEX KAPPA CCD single crystal X-ray diffractometer ((EtOH)pyr$^+$ Gen$^-$; $N_AH_3^+$ Gen$^-$; $N_AH_3^+$ Sal$^-$; $N_TH_3^+$ Gen$^-$; (EtOH)pyr$^+$ Bz$^-$; (EtOH)pyr$^+$ Sal$^-$) and a Nonius KAPPA CCD single crystal X-ray diffractometer ($N_AH_3^+$ Bz$^-$), using graphite-monochromated Mo KR radiation ($\lambda$) 0.71073 Å. Crystals were coated with Paratone N oil (Exxon Chemical Co.,TX, USA) immediately after isolation and cooled in a stream of nitrogen vapor on the diffractometer. Structures were solved by direct methods using the program SHELXS-97 and refined by full matrix least-squares refinement on F2 using SHELXL-97. All non-hydrogen atoms were revealed in the E-map and subsequent difference electron density maps and thus placed and refined anisotropically. All hydrogen atoms were observed in difference syntheses and were either refined isotropically or placed in geometrically idealized positions and constrained to ride on their parent atoms with C–H distances in the range 0.95–1.00 Å and Uiso(H)) $x$Ueq(C), where $x = 1.5$ for methyl and 1.2 for all other C atoms.[30] Specific details of refinement are contained in the relevant CIF files. The crystal structures of most of the synthesised salts are described here for the first time (CCDC Refcodes: 802629 for $N_TH_3^+$ Gen$^-$, 802627 for $N_AH_3^+$ Gen$^-$, 802625 for $N_AH_3^+$ Sal$^-$, 802628 for (EtOH)pyrH$^+$ Bz$^-$, 802630 for (EtOH)pyrH$^+$ Sal$^-$ and 802624 for (EtOH)pyrH$^+$ Gen$^-$). The crystal structures for $N_AH_3^+$ Bz$^-$ has been reported before and can be found in ref. 31. However, we have solved the structure of this compound and submitted the report as part of the supplementary information.

Differential scanning calorimetry (DSC) scans were carried out at a heating/cooling rate of 10 °C min$^{-1}$ in the range of −150 °C to 200 °C using a T.A. Instruments Q100 differential scanning calorimeter. Thermal scans below room temperature were calibrated with the cyclohexane solid-solid transition and melting point at −87.0 °C and 6.5 °C respectively. Thermal scans above room temperature were calibrated with the indium, tin, lead and zinc. Transition temperatures are reported using the peak maximum of the thermal transition.

The FTIR-ATR spectra for all the samples were obtained using Bruker IFS Equinox FTIR system, fitted with a Golden GateTM single bounce diamond. The FTIR spectrometer was connected to a computer equipped with OPUS 6.0 software, which was used for the analysis of spectra.

## 3. Results and discussion

### 3.1 Protic pharmaceutical salts

Table 1 summarises the protic salts made in this work. Of the thirteen compounds synthesised, nine satisfy the definition of an IL by having melting points below 100 °C. Five of these are room temperature ILs. Salts containing the $N_AH_2$ base all possess high melting points. ILs containing the (EtOH)pyrH$^+$ ion all have melting points below 100 °C, while those based on the $N_TH_3^+$ cation have no obvious trend amongst their melting points.

When reacted with the same acid, the two *primary* amines, $N_TH_2$ and $N_AH_2$, give rise to compounds with significantly different melting points. The compounds based on the $N_TH_3^+$ cation, a seven carbon molecule, have melting points of 121 ± 2 °C and below, while those based on the $N_AH_3^+$ cation, a nine carbon molecule, possess melting points of 216 ± 2 °C and above. These differences can be understood from a thermodynamic point of view. Given that the change in Gibbs free energy of fusion ($\Delta G_f$) at the melting point is zero by definition, the enthalpy ($\Delta H_f$) and the entropy ($\Delta S_f$) of fusion of the compound determine $T_m$ (eqn (1)).

**Table 1** The protic pharmaceutical compounds prepared in this study including a number that are ILs

| | Tuaminoheptane ($N_TH_2$) - *Nasal decongestant*<br>(liquid, b.p. 143 °C, p$K_a$ 10.50, *water insoluble*) | | Amantadine ($N_AH_2$) - *Antiviral/antiparkinson*<br>(solid, m.p. 206-208 °C, p$K_a$ 10.10, *water insoluble*) | 2-Pyrrolidinoethanol ((EtOH)pyr) - *GRAS*<br>(liquid, b.p. 80 °C, p$K_a$ 9.44, *water soluble*) | |
|---|---|---|---|---|---|
| **Benzoic acid (BzH)** - *Preservative*<br>(solid, m.p. 122 °C, p$K_a$ 4.19, *soluble in hot water*) | **$N_TH_3^+$ $Bz^-$ 1:1**<br>Δp$K_a$ 6.31<br>solid<br>DSC: $T_m$ 91±2°C<br>*Water soluble* | **$N_TH_3^+$ $(Bz\text{-}Bz)^-$ 1:2***<br>liquid<br>DSC: $T_g$ -43±2°C<br>*Water insoluble* | **$N_AH_3^+$ $Bz^-$ 1:1**<br>Δp$K_a$ 5.91<br>solid<br>DSC: $T_m$ 256±2°C<br>*Poorly water soluble* | **(EtOH)pyrH$^+$ $Bz^-$ 1:1**<br>Δp$K_a$ 5.25<br>solid, DSC: $T_g$ -40±2°C, $T_r$ -24±2°C, $T_m$ 28±2°C<br>*Water soluble* | **(EtOH)pyrH$^+$ $(Bz\text{-}Bz)^-$ 1:2***<br>liquid<br>DSC: $T_g$ -49±2°C<br>*Water insoluble* |
| **Salicylic acid (SalH)** - *Keratolytic*<br>(solid, m.p. 159 °C, p$K_a$ 2.98, *poorly soluble in water*) | **$N_TH_3^+$ $Sal^-$ 1:1**<br>Δp$K_a$ 7.52<br>liquid<br>DSC: $T_g$ -40±2°C<br>*Water soluble* | **$N_TH_3^+$ $(Bz\text{-}Sal)^-$ 1:2***<br>liquid<br>DSC: $T_g$ -40±2°C<br>*Water insoluble* | **$N_AH_3^+$ $Sal^-$ 1:1**<br>Δp$K_a$ 7.12<br>solid<br>DSC: $T_m$ 216±2°C<br>*Water insoluble* | **(EtOH)pyrH$^+$ $Sal^-$ 1:1**<br>Δp$K_a$ 6.46<br>solid<br>DSC: $T_g$ -44±2°C, $T_r$ 6±2°C, $T_m$ 49±2°C<br>*Water soluble* | **(EtOH)pyrH$^+$ $(Bz\text{-}Sal)^-$ 1:2***<br>liquid<br>DSC: $T_g$ -46±2°C<br>*Water insoluble* |
| **Gentisic acid (GenH)** - *Analgesic*<br>(solid, m.p. 200 °C, p$K_a$ 2.93, *soluble in hot water*) | **$N_TH_3^+$ $Gen^-$ 1:1**<br>Δp$K_a$ 7.57<br>solid<br>DSC: $T_m$ 121±2°C<br>*Water soluble* | | **$N_AH_3^+$ $Gen^-$ 1:1**<br>Δp$K_a$ 7.17<br>solid<br>DSC: $T_m$ 250±2°C<br>*Water insoluble* | **(EtOH)pyrH$^+$ $Gen^-$ 1:1**<br>Δp$K_a$ 6.51<br>solid<br>DSC: $T_m$ 97±2°C<br>*Water soluble* | |

[a] Represents compounds with two moles of acid, in some cases two different acids, to one mole of base.

$$\Delta G_f = 0 = \Delta H_f - T_m \Delta S_f \Rightarrow T_m = \Delta H_f / \Delta S_f \quad \text{(Eqn 1)}$$

Being a long-chain molecule, the $N_TH_3^+$ cation has more degrees of rotational freedom that can become active on melting and hence exhibits a larger $\Delta S_f$, this in turn leads to compounds with a lower melting point. On the other hand, $N_AH_3^+$, a rigid multi-ringed structure, has fewer rotational degrees of freedom that can become active on melting, resulting in a smaller $\Delta S_f$ and hence higher melting point of the resulting compounds. Thus, in order to synthesise low melting PILs, the more flexible anion/cation structures are desirable.

### 3.2 Oligomeric compositions

Another domain of PILs that has been explored recently are the protic oligomers, whereby several equivalents of acid were reacted with one equivalent of base to form H-bonded acid-oligomer based ILs (Scheme 2).[32] For example, in the case of 2HA + B, the anionic species present appears to be $[A\text{–}H\text{–}A]^-$. Proton transfer at this stoichiometry can be stronger than at the 1 : 1 stoichiometry.

Bica *et al.* have reported pharmaceutical ILs based on such oligomeric ions; these compositions allow one to alter the dose of one active ingredient over the other active ingredient in the case of dual protic pharmaceutical ILs.[33] Recognising that such 2 : 1 mixtures may also represent "buffer" type properties,[34] they may also have a useful property in controlling pH. A similar approach was investigated in Table 1 in which a second mole of acid, in some cases a different acid, was incorporated into the formulation, in principle forming the "dimer" acid/anion species. In the case of the formulations involving two different acids $HA_1$ and $HA_2$, there is no intrinsic reason why the dimeric species should be uniquely $A_1\text{–}H\text{–}A_2$; the real situation will be a dynamic equilibrium between all of the possible species, with this overall composition.

### 3.3 Physical properties, proton transfer and ion association

As discussed above, certain important physicochemical properties of pharmaceutical acids/bases can be altered by converting them into salts of an appropriate counter ion. For example, in this study the salts of the originally water insoluble/slightly soluble acids and bases have been rendered water soluble in most cases. In the case of the dimer-acid based ILs, an additional acid molecule decreases the water solubility of the compounds. This in turn demonstrates the ease of water solubility manipulation, an important factor in drug delivery.

The effects of proton transfer and ion association on physical properties are illustrated by the two salicylate compounds, one based on a *primary* amine, $N_TH_3^+$ $Sal^-$, and the other on a *tertiary* amine, $(EtOH)pyrH^+$ $Sal^-$. Both compounds have low melting points. Indicator studies, the basis of which was described in detail in our previous paper,[26] were used to show that both the primary amine, $N_TH_2$, as well as the tertiary amine, (EtOH)pyr, have the capability to completely deprotonate the polyprotic indicator-acid, phenol red. This is shown in Fig. 3 where the amines have both turned a fuchsia colour in the presence of the indicator acid, which originally exists as an orange coloured solution. One would observe a yellow colour if only a partial deprotonation occurred. The proton transfer behaviour of the (EtOH)pyr, a tertiary amine, is therefore different to the behaviour of the simple tertiary amine cases studied previously,[26] presumably as a result of the presence of the hydroxyl group.

$$(A\text{-}H)_x + B \rightleftharpoons H_{x-1}(A^-)_x + B^+H$$

**Scheme 2** Anion oligomer based PIL formation.[32]

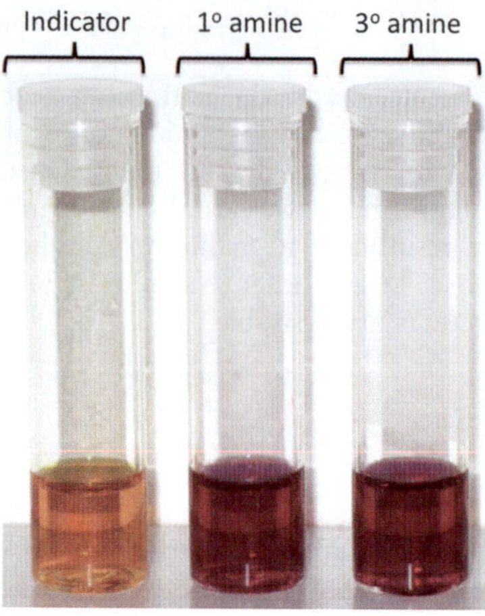


**Fig. 3** The primary amine, $N_TH_2$, and tertiary amine, (EtOH)pyr, with indicator phenol red.

Further evidence of proton transfer in this case is seen in the FTIR-ATR spectra of (EtOH)pyr$H^+$ $Sal^-$, $N_TH_3^+$ $Sal^-$ and the starting materials (EtOH)pyr, $N_TH_2$ and SalH displayed in Fig. 4. The spectra of the PILs (shown in red) are notably different to those of the starting materials. A peak at 1655 $cm^{-1}$ in the SalH spectrum (shown in blue), corresponding to the C=O stretch of the acid, is absent in both PIL spectra, which is indicative of complete proton transfer in the ILs. Also, the disappearance of, or shift in, other peaks when compared to the original starting materials are all consistent with strong proton transfer in these PILs.

The effect of ion association is subtly revealed by the bulk properties in these systems. The density data (Fig. 5) reveals two distinctly different groups of liquids, where the tertiary amine based PILs (including the 2 : 1 mixtures) appear to be denser than the primary amine based examples. This in turn implies that ions in the tertiary amine based PILs pack more tightly than in the primary amine based

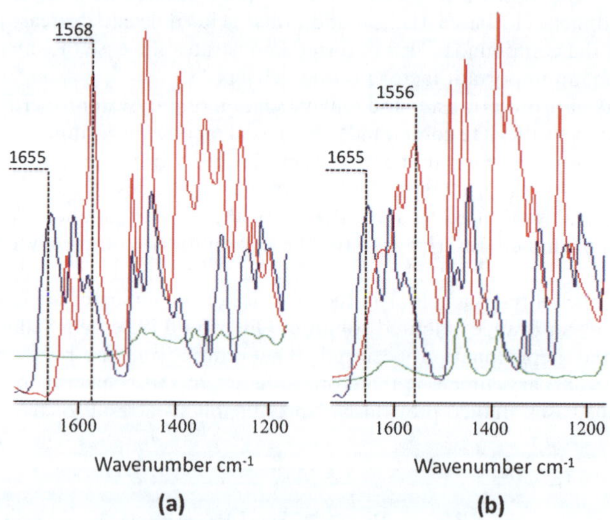


**Fig. 4** FTIR-ATR spectra of **(a)** (EtOH)pyr$H^+$ $Sal^-$ (red), (EtOH)pyr (green) and SalH (blue) and **(b)** $N_TH_3^+$ $Sal^-$ (red), $N_TH_2$ (green) and SalH (blue).

 

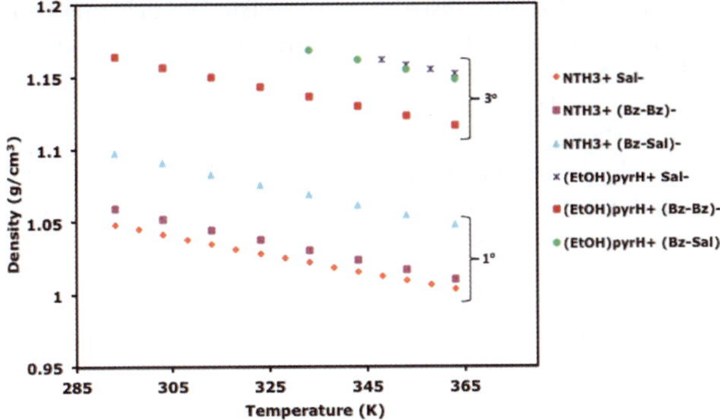


**Fig. 5** Density data of the synthesised PILs.

PILs. As will be shown below, the latter have a tendency to form open hydrogen bonded structures that produce lower densities.

The ionic conductivity (Fig. 6) also shows two distinctly different groups of compounds; the tertiary amine based PILs have a conductivity approximately one order of magnitude greater than the primary amine based ILs, despite the fact that both families appear to be substantially proton transferred. On the other hand, their viscosities (Fig. 6) show no distinct differences. Of all the compounds analysed in this study, the primary amine based PIL, $N_TH_3^+$ $Sal^-$, displayed the highest viscosity followed by the tertiary amine based PIL, (EtOH)pyr$H^+$ $Sal^-$.

The transport property and density data can be combined into the Walden plot, to provide a revealing perspective on proton transfer and ion association. The Walden plot[25,35,36] is a method of comparing transport property data using the well-known Walden rule that states:

$$\Lambda\eta = \mathrm{k} \qquad \text{(Eqn 2)}$$

where $\Lambda$ is the molar conductivity and $\eta$ is the viscosity. An estimate of the constant (k) can be obtained from data for a 0.1M aqueous solution of KCl, which is thought

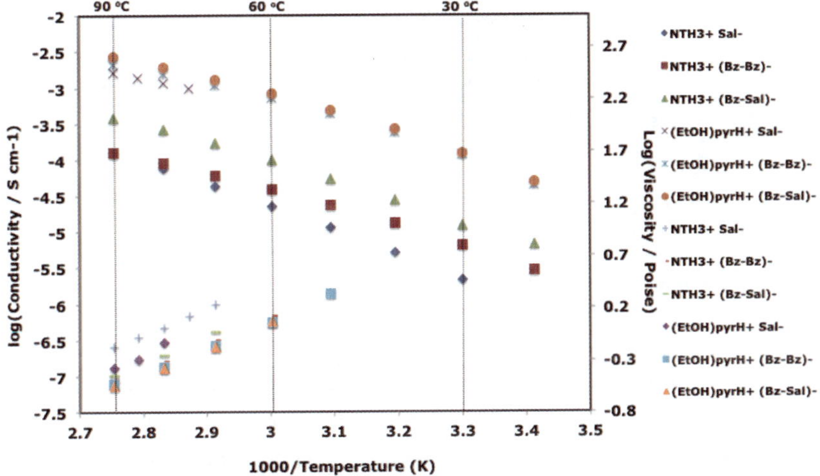


**Fig. 6** Conductivity (full symbols) and viscosity data (hollow symbols) of the synthesised PILs.

to be a good example of a fully dissociated salt. The Walden rule then appears as a straight line on a plot of log ($\Lambda$) *versus* log ($\eta^{-1}$). Previous discussion of this in the literature[35] has shown that many ILs fall quite close to the Walden rule line, as do many molten salts. ILs that fall well below the line are thought of as exhibiting signs of ion association in a way that inhibits conduction, but not viscosity. Additionally, in the protic systems incomplete proton transfer can also produce transport properties that fall well below the line. As a reminder of the impact of the logarithmic scale of Fig. 7, the dashed line indicates the situation where the molar conductivity is only 10% of that expected from the Walden rule. Such systems are clearly showing a low degree of proton transfer and/or strong signs of ion-association.

In the Walden plot in Fig. 7 the PILs based on the $N_TH_3^+$ cation display a significant deviation from the ideal line, whereas those containing the (EtOH)pyrH$^+$ cation show better concordance with the Walden rule. The latter is consistent with the infra-red evidence, which indicates strong proton transfer and might be considered a normal example of a PIL. However, the $N_TH_3^+$ salts appear surprisingly low on the Walden plot, given that the infra-red spectra indicated strong proton transfer. This perhaps indicates that these salts, although proton transferred, are highly associated in the liquid state. Ion pairs and larger clusters that have zero net charge are unable to conduct and therefore cannot contribute to the molar conductivity of the material. The $N_TH_3$ Sal case in Fig. 7 is exhibiting only 1% of the conductivity that would be expected on the basis of the Walden rule; this indicates that the overwhelming majority of the ions in this case are "tied up" in ion pairs and clusters. We will return to the nature of association in these salts later, however, we will now present some data showing how association is of some importance in the membrane transport properties of these pharmaceutically active PILs.

### 3.4 Membrane transport properties

One of the benefits of synthesising IL forms of known actives is the potential to modulate the membrane permeation properties of an active. Tuning and controlling membrane transport is the key to a variety of novel delivery methods including slow, continuous-release transdermal approaches. Recent work by us[38] and others[39] provides some insight into the membrane transport properties of these

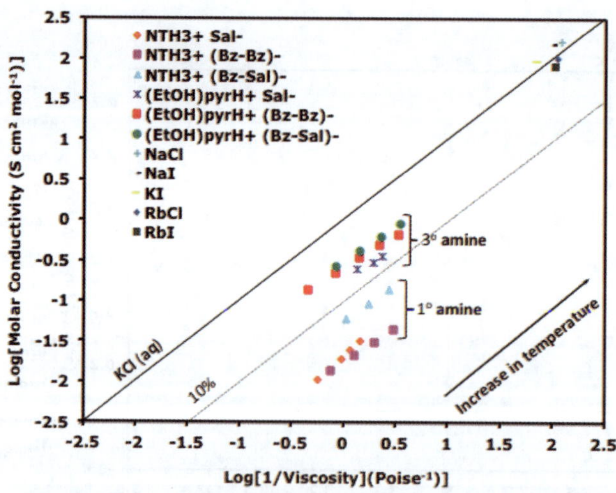


**Fig. 7** Walden plot of transport property data for the synthesised PILs. Each data set represents a temperature series for the compound above its melting point. Also included are some data points for a number of simple inorganic salts at temperatures just above their respective melting points.[37]

 

pharmaceutically active IL compounds. It is a well established fact that salts *do not* readily cross the lipid bilayer unless there are active transport mechanisms available. Hence, it is not immediately obvious that IL forms of an active will be readily membrane permeable. On the other hand, if the IL has a propensity to form neutral species, ion-pairs or clusters then membrane transport may be enhanced as compared to an inorganic salt of the active. To further investigate this hypothesis in respect of these protic examples, we have carried out model membrane transport studies following the method of Tantishaiyakul *et al.*[39] A full publication describing these results will be presented elsewhere;[38] we present here an overview of the trends observed. Typical data, for example for $N_TH_3^+$ $Sal^-$ and its parent acid and base, as well as its simple salt forms that are used commercially, show that while the component acid is rapidly transported through the membrane, the sodium salt is not, nor is the sulfate salt (the form normally used pharmaceutically). On the other hand, the neat IL form, despite being a salt, is found to rapidly transport through the membrane over a period of one or two hours. Therefore the IL form is clearly more membrane permeable than the active ingredients that were the starting materials. Interestingly, the IL is *not* readily transported when dissolved in propylene glycol; this indicates that the situation is not simply one of providing an alternate, but independent, counter-ion. Indeed, Tantishaiyakul *et al.* have speculated that ion-pairing can play a significant role in the transport of these types of compounds.[39] It is interesting to note that $N_TH_3^+$ $Sal^-$ shows signs of strong ion pairing/aggregation by sitting low in the Walden plot and hence this aspect of the behaviour of these PILs may be of importance in their membrane transport properties. The nature of this ion pairing/aggregation is further explored in the next sections. It is also possible that reverse proton transfer to form the neutral acid and base at the point of dissolution into the membrane is the basis of transport – partially the result of a relatively smaller p$K_a$ gap than the inorganic salts. At this stage it is not possible to discriminate between these hypotheses, however doing so will allow us to understand how modulating the counter acid/base in these systems can be used to control their interaction with membranes.

### 3.5 Ion association - what can we learn from crystal structures?

Crystal structures can also provide some insight into the existence of proton transfer and ion association, and thereby serve as a guide to the existence of these phenomena in related liquid phases. Crystal structures for several of the compounds synthesised here have been solved and can be found in the supplementary information.† The two ion-pairs shown in Fig. 8 are the asymmetric units of the crystal structures of the tertiary amine based PIL, (EtOH)pyrH$^+$ Sal$^-$, and of the primary amine based PIL, $N_TH_3^+$ Gen$^-$. As can be seen from the displayed structures, the proton transfer has occurred in both cases. However, we should note that proton transfer in the solid state is likely to be stronger since it incurs a smaller entropic

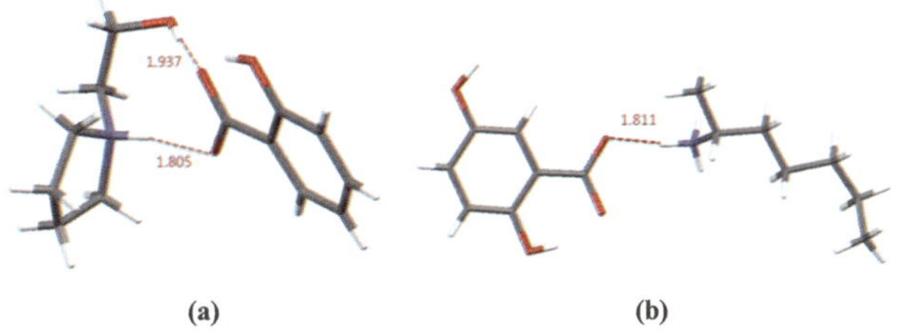


**Fig. 8** Crystal structures **(a)** of the tertiary amine based IL, (EtOH)pyrH$^+$ Sal$^-$ (m.p. 49 ± 2 °C); N–H···O and O–H···O distances are shown in red; **(b)** of the primary amine based salt, $N_TH_3^+$ Gen$^-$ (m.p. 121 ± 2 °C) (N–H···O distance is shown in red).

penalty than in the liquid and therefore produces a larger free energy change. This occurs because proton transfer in the neat liquid case causes a sharp loss of the configurational degrees of freedom that are characteristic of the liquid state and have a substantially negative $\Delta S$.

Fig. 8a also shows that the hydroxyl group of the (EtOH)pyrH$^+$ cation stabilises a resonating carbonyl group in the Sal$^-$ anion, through hydrogen bonding, thus supporting the proton transfer. The O···H distances observed in this pair are 1.805 and 1.937 Å, comparable to the hydrogen bond distances in water.

The asymmetric unit of $N_TH_3^+$ Gen$^-$, shown in Fig. 8b, displays relatively short interionic N–H···O contacts of 1.811 Å, indicating strong hydrogen bonding between the ions. Additional O–H···O and C–H···O hydrogen bonding is also observed between the cations.

Although the proton transfer is clearly complete in both cases, in accord with the FTIR-ATR data, a large difference between their positions in the Walden plot (>1 orders of magnitude) demands explanation - an explanation that cannot be based on lack of proton transfer. Fig. 9 provides a possible answer; $N_TH_3^+$ Gen$^-$ organises itself in the crystal in a cyclic cluster-like complex consisting of two cations and two anions.

If a cyclic arrangement such as this, with its overall zero charge, is reasonably long lived in the liquid state then the molar conductivity will be lower than expected, despite the strong degree of proton transfer. Part of the origin of this cluster-like complex may lie in the inter-ring interactions as seen in Figure S2.† $\pi \cdots \pi$ stacking between phenyl rings in the anions imposes a certain arrangement of ions in the crystal structure that results in a strong hydrogen-bonded lattice.

Another example of this phenomenon was found in the analysis of the $N_AH_3^+$ Gen$^-$ crystal structure (m.p. = 250 ± 2 °C) containing the $N_AH_3^+$ ion (Fig. 10). In this case, a cyclic cluster-like complex is formed with both oxygens of the carboxylic group partaking in hydrogen bonding. Due to spatial constraints, one of the H···O bonds is longer than the other by 0.173 Å. A closer inspection of the packing diagram reveals that the cyclic arrangement extends beyond the cluster-like complex, with oxygens of the COO$^-$ group and the hydrogens of the $NH_3^+$ group forming distinct channels of hydrogen-bonded interactions (Figure S3†). Extended associated structures would also have the effect of placing these compounds low on the Walden plot, despite there being a high degree of proton transfer. Associations such as these would appear to also be crucial to the rapid membrane transport of these salts.

Recently we have developed a generalised approach for the calculation of Madelung constants based on crystal structures.[18] The Madelung constant, M, represents a measure of the net electrostatic energy in ionic solids and thus determines the stability of the ionic lattice with respect to an isolated ion pair. For stable lattices M is expected to be more than 1, with larger values indicating greater stability of

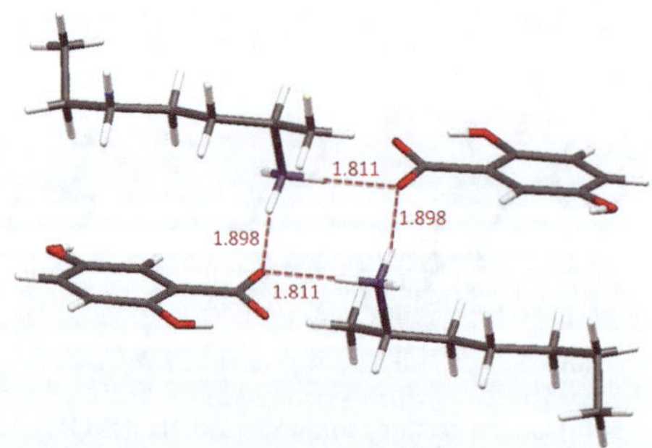


**Fig. 9** Cluster-like complex conformation of the primary amine based IL, $N_TH_3^+$ Gen$^-$.

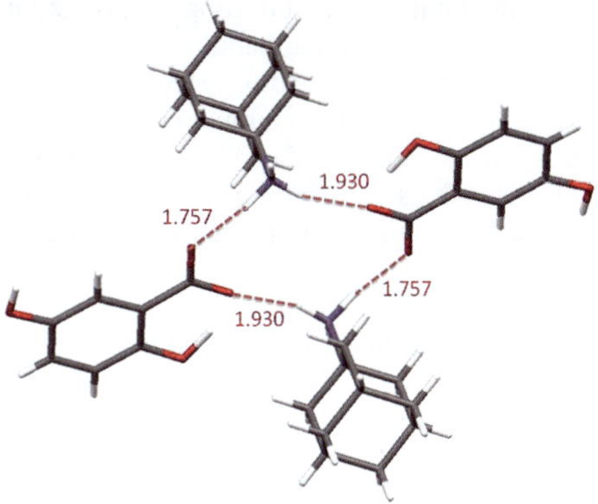


**Fig. 10** Cluster-like complex conformation of the primary amine based IL, $N_AH_3^+$ $Gen^-$.

**Table 2** Madelung constants, electrostatic lattice energies and melting points of protic pharmaceutical salts prepared in this work

| | M | Electrostatic Lattice energy, kJ mol$^{-1}$ | $T_m$, °C |
|---|---|---|---|
| $N_TH_3^+$ $Gen^-$ | 1.2307 | −551 | 121 |
| $N_AH_3^+$ $Bz^-$ | 1.2320 | −624 | 256 |
| $N_AH_3^+$ $Sal^-$ | 1.2566 | −625 | 216 |
| $N_AH_3^+$ $Gen^-$ | 1.2358 | −616 | 250 |
| (EtOH)pyrH $Bz^-$ | 1.1731 | −546 | 28 |
| (EtOH)pyrH $Sal^-$ | 1.0748 | −493 | 49 |
| (EtOH)pyrH $Gen^-$ | 1.0831 | −487 | 97 |

the lattice. The Madelung constant can also be viewed as a geometric parameter; it provides us with a static perspective on cation-anion arrangements that might persist to an extent in the liquid state. Values of the Madelung constant closer to 1 represent ionic solids dominated by ion-pair interactions, whereas larger values indicate an extended lattice of inter-ionic interactions. In this study Madelung constants and electrostatic lattice energies (Table 2) were calculated for the PILs whose crystal structures were resolved in this work. Inspection of Table 2 reveals values that are relatively low; by comparison, a simple symmetrical organic salt such as $[NMe_4]$ $[BF_4]$ has a Madelung constant close to that of NaCl (1.75). Ion association in the solid state is therefore lowering M considerably in all cases. Values for the $(EtOH)pyrH^+$ based salts are closer to 1, indicating that the ion pairs are somewhat more isolated in this case.

The electrostatic lattice energies also seem to generally follow the trend in measured melting points, with $N_AH_3^+$ based salts having the largest lattice energies and highest melting points. As discussed elsewhere,[40] this indicates that the electrostatic part of the lattice energy plays a dominant role in the overall energetics of melting of these compounds, despite the obvious influence of H-bonding and ion association in other properties.

### 3.6 Perspectives on protonics

In a number of fields of application, including ionic liquids and solids for fuel cell electrolytes,[41] protein stabilising ionic liquids[34] and water electrolysis[42] the focus is on understanding the energetics and dynamics of protons and how they impact on the properties of the substance. Just as an understanding of the electronic structure of a molecule is vital to understanding its bonding and reactivity, we have found it useful to adopt a holistic view of protonic behaviour in these contexts in order to fully understand its subtleties. Much of the discussion above has concerned the location of the proton in these protic pharmaceutical systems, the consequences of its transfer from acid to base and the impact of its H-bonding, or lack of, in determining properties. The neat IL provides a rather unique environment in which to observe proton transfer, since there is an almost complete lack of a dielectric solvent to shield the charges from one another and the liquid is so concentrated that the charges are never more than one ion-pair diameter apart. Unlike water and other solvents, in which the protonic properties are heavily dominated by the properties of the *solvent*, in PILs the IL species themselves are the only acid and base present. The proton energy levels involved are therefore more directly revealed and manipulated, but they are also strongly impacted by the molecular structure of the species involved.

The distinction between the H-bonded acid–base complex and the H-bonded, proton-transferred ion-pair becomes very slight in a PIL. In fact, it amounts to simply the matter of where the proton sits on a potential energy surface between the two heavy atoms involved (usually N and O). This is illustrated in Fig. 11 for the case of acetic acid and methylpyrrolidine in the gas phase, where the proton transfer amounts to only a ~0.4 A° shift in an otherwise H-bonded complex. Charge is usually a significant distinction between these two states in a solvent based system, but becomes less so in the PIL, because H-bonding in the neutral complex involves some generation of a partial charge on each atom, while, on the other hand, H-bonding in the ion-pair acts to diminish the unit charge that notionally exists on the heavy atoms. In the case of acetic acid and methylpyrrolidine these minor changes in position and charge do not produce a sufficiently significant free energy change to create a substantial population of the proton transferred species, in either the gas phase or the ionic liquid.[26] As we will discuss in more detail below, further solvation interactions are required to push this proton transfer along.

Angell and coworkers[44,45] have discussed the use of a proton free-energy scale, first introduced by Gurney,[46] to assist in understanding the energetics in these materials. In this perspective, the proton site in an acid moiety is considered as one of high proton chemical potential, and the available site on the base moiety as being one of low proton chemical potential. The energy scale involved bears many features

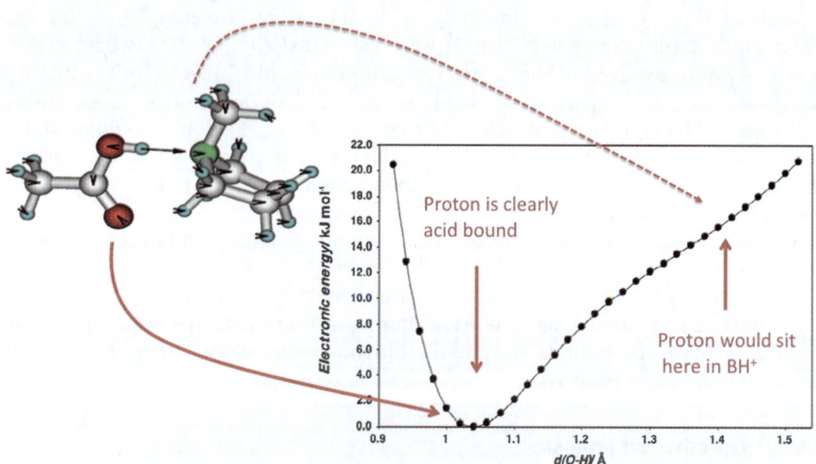


**Fig. 11** *Ab initio* calculations for acetic acid and methylpyrrolidine $\Delta pK_a^{aq} = 5$. The potential energy surface was generated as a relaxed scan of the O–H bond stretch at the MP2/6-31G+(d,p) level of theory. The optimised points on the PES were then improved at the MP2/aug-cc-pVTZ level of theory. All calculations were performed using GAUSSIAN 03.[43]

 

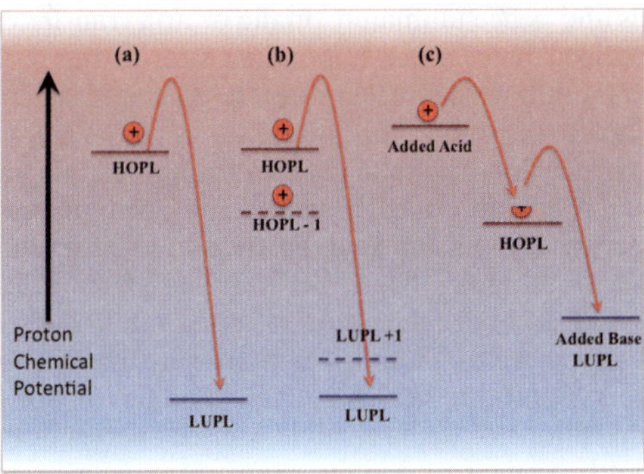


**Fig. 12** Proton Energy levels in **(a)** a simple, solvent-free acid–base mixture **(b)** a more complex case in which either the acid or base is multi-protic and **(c)** a buffer case where the HOPL is ~50% occupied and hence can accept or donate protons to an added strong acid or base to create a buffer action.[34]

in common with electrochemical potential scales; in particular the chemical potential of the proton is the difference in free energies between that of the acid HA and its conjugate base $A^-$. The highest occupied proton level (HOPL) in the system thus corresponds to the species having the highest free energy change available from a proton donation process $HA \rightleftharpoons A^- + H^+$. The lowest unoccupied proton level (LUPL) similarly corresponds to the site offering the largest free energy change from a proton addition process. A number of scenarios are illustrated in Fig. 12. Note that the LUPL of the base, B, is also the HOPL of the corresponding protonated species $BH^+$.

The impact of hydrogen bonding and other solvation interactions on these proton energy levels is the key aspect we focus upon here, being dominated by the absence of, or limited presence of, H-bond donating species in the acid or base. Recognizing that the proton energy levels in Fig. 12 actually represent a free energy change in a process that involves both the protonated and the unprotonated species, it is necessary to further deconstruct this diagram to explicitly identify the free energy of each species and then separately consider the effect of solvation on each. Fig. 13 shows such an energy level diagram. The diagram begins by considering the isolated gas phase species and, as discussed above, in some cases there may be no proton transfer at all, as illustrated. Then, as shown in Fig. 13, providing a source of solvation or additional hydrogen bonding, for example by a primary ammonium cation, impacts

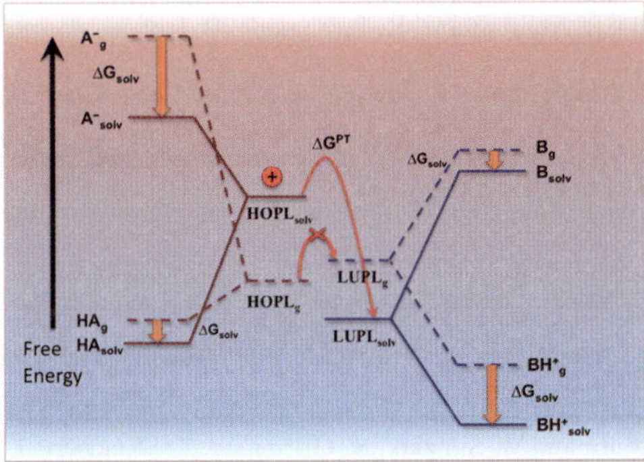


**Fig. 13** Free energy diagram showing the component molecular free energies that make up the proton energy levels in the gas phase and the effect of solvation on these.

on the energy levels and hence ultimately on the proton transfer.[26] In the case of the protic systems described here, other H-bonding sites in the molecules have additional solvating effects and hence further alter the respective positions of the proton energy levels available.

Where proton transfer does not take place in the liquid state, despite the expectation from aqueous p$K_a$ data that it should, another phenomenon related to H-bonding can occur: un-ionised acid–base complex formation, as shown in Fig. 1b. A number of examples from the lidocaine : fatty acid family of liquids have been described recently in the context of pharmaceutically active compounds where the $\Delta pK_a^{aq}$ values were ~3 and it was shown that proton transfer did not take place.[47] In this situation the lack of H-bonding or solvation interactions strongly alters the positions of the individual ionised species in Fig. 13 and thereby the HOPL and LUPL levels, to the point that proton transfer is not energetically favoured. However, a complete suppression of the melting point indicated formation, instead, of a hydrogen bonded acid–base complex. Such un-ionised complexes can be thought of as a liquid analogue of the "co-crystals" that are much studied in the pharmaceutical field. In the liquid co-crystal case the association of the two components produces a stabilising interaction in the liquid state that results in a sharp lowering of the melting point.

## 4. Conclusions

A number of protic ionic liquids and salts based on pharmaceutical acids and bases have been synthesised and characterised. FTIR-ATR and Walden plot studies were used to assess the proton transfer and determine the degree of ionicity in the synthesised compounds. It was found that additional hydrogen bonding functional groups in the tertiary amine case can promote proton transfer. In the case of primary amines, formation of cluster-like species containing protonated base/deprotonated acid pairs can result in salts possessing higher melting points and decreased ionicity. These findings are supported by crystal structures and Madelung constant calculations. Nonetheless, for application purposes, PILs with lower ionicity are more desirable for drug delivery as they may cross the membrane barrier more easily than the more ionised species. Membrane studies are currently being carried out in order to further investigate this phenomenon and are showing promising results.

## 5. References


1 P. Heinrich Stahl, C. G. Wermuth and Editors, *Handbook of Pharmaceutical Salts; Properties, Selection, and Use*, VHCA and Wiley-VCH, 2008.
2 S. M. Berge, L. D. Bighley and D. C. Monkhouse, *J. Pharm. Sci.*, 1977, **66**, 1–19.
3 G. Davies, *The Pharmaceutical Journal*, 2001, **266**, 322–323.
4 J. Stoimenovski, D. R. MacFarlane, K. Bica and R. D. Rogers, *Pharm. Res.*, 2010, **27**, 521–526.
5 P. M. Dean, J. M. Pringle and D. R. MacFarlane, *Phys. Chem. Chem. Phys.*, 2010, **12**, 9144–9153.
6 R. D. Rogers and K. R. Seddon, *Ionic Liquids: Industrial Applications to Green Chemistry*, American Chemical Society, 2003.
7 T. Welton, *Chem. Rev.*, 1999, **99**, 2071–2083.
8 F. Endres, D. MacFarlane, A. Abbott and Editors, *Electrodeposition from Ionic Liquids*, Wiley-VCH, 2008.
9 W. L. Hough, M. Smiglak, H. Rodriguez, R. P. Swatloski, S. K. Spear, D. T. Daly, J. Pernak, J. E. Grisel, R. D. Carliss, M. D. Soutullo, J. J. H. Davis and R. D. Rogers, *New J. Chem.*, 2007, **31**, 1429–1436.
10 W. L. Hough and R. D. Rogers, *Bull. Chem. Soc. Jpn.*, 2007, **80**, 2262–2269.
11 W. L. Hough-Troutman, M. Smiglak, S. Griffin, W. M. Reichert, I. Mirska, J. Jodynis-Liebert, T. Adamska, J. Nawrot, M. Stasiewicz, R. D. Rogers and J. Pernak, *New J. Chem.*, 2009, **33**, 26–33.
12 *Application*: *WO Pat.*, 2006–US39454 2007044693, 2007.


 

13 A. Cieniecka-Roslonkiewicz, J. Pernak, J. Kubis-Feder, A. Ramani, A. J. Robertson and K. R. Seddon, *Green Chem.*, 2005, **7**, 855–862.
14 J. Pernak and J. Feder-Kubis, *Chem.–Eur. J.*, 2005, **11**, 4441–4449.
15 J. Pernak, K. Sobaszkiewicz and J. Foksowicz-Flaczyk, *Chem.–Eur. J.*, 2004, **10**, 3479–3485.
16 J. Pernak, I. Goc and I. Mirska, *Green Chem.*, 2004, **6**, 323–329.
17 P. M. Dean, J. Turanjanin, M. Yoshizawa-Fujita, D. R. MacFarlane and J. L. Scott, *Cryst. Growth Des.*, 2009, **9**, 1137–1145.
18 E. I. Izgorodina, U. L. Bernard, P. M. Dean, J. M. Pringle and D. R. MacFarlane, *Cryst. Growth Des.*, 2009, **9**, 4834–4839.
19 E. Janus, I. Goc-Maciejewska, M. Lozynski and J. Pernak, *Tetrahedron Lett.*, 2006, **47**, 4079–4083.
20 Y. Du and F. Tian, *J. Chem. Res.*, 2006, 486–489.
21 Z. Duan, Y. Gu, J. Zhang, L. Zhu and Y. Deng, *J. Mol. Catal. A: Chem.*, 2006, **250**, 163–168.
22 M. A. B. H. Susan, A. Noda, S. Mitsushima and M. Watanabe, *Chem. Commun.*, 2003, 938–939.
23 B. Nuthakki, T. L. Greaves, I. Krodkiewska, A. Weerawardena, M. I. Burgar, R. J. Mulder and C. J. Drummond, *Aust. J. Chem.*, 2007, **60**, 21–28.
24 D. R. MacFarlane and K. R. Seddon, *Aust. J. Chem.*, 2007, **60**, 3–5.
25 M. Yoshizawa, W. Xu and C. A. Angell, *J. Am. Chem. Soc.*, 2003, **125**, 15411–15419.
26 J. Stoimenovski, E. I. Izgorodina and D. R. MacFarlane, *Phys. Chem. Chem. Phys.*, 2010, **12**, 10341–10347.
27 K. Bica, C. Rijksen, M. Nieuwenhuyzen and R. D. Rogers, *Phys. Chem. Chem. Phys.*, 2010, **12**, 2011–2017.
28 K. Bica and R. D. Rogers, *Chem. Commun.*, 2010, **46**, 1215–1217.
29 R. E. Buntrock, *The Merck Index: An Encyclopedia of Chemicals, Drugs, and Biologicals*, Fourteenth Edition, edited by Maryadele J. O'Neil, Patricia E. Heckelman, Cherie B. Koch, and Kristin J. Roman, 2007.
30 G. M. Sheldrick, *Acta Crystallogr., Sect. A: Found. Crystallogr.*, 2008, **A64**, 112–122.
31 W. N. Zheng and B. Wang, *Acta Crystallogr., Sect. E: Struct. Rep. Online*, 2009, **E65**, o2769.
32 K. M. Johansson, E. I. Izgorodina, M. Forsyth, D. R. MacFarlane and K. R. Seddon, *Phys. Chem. Chem. Phys.*, 2008, **10**, 2972–2978.
33 K. Bica and R. D. Rogers, Chem. Commun. (Cambridge, U.K.), 46, pp. 1215–1217.
34 D. R. MacFarlane, R. Vijayaraghavan, H. N. Ha, A. Izgorodin, K. D. Weaver and G. D. Elliott, *Chem. Commun.*, 2010, **46**, 7703–7705.
35 D. R. MacFarlane, M. Forsyth, E. I. Izgorodina, A. P. Abbott, G. Annat and K. Fraser, *Phys. Chem. Chem. Phys.*, 2009, **11**, 4962–4967.
36 K. J. Fraser, E. I. Izgorodina, M. Forsyth, J. L. Scott and D. R. MacFarlane, *Chem. Commun.*, 2007, 3817–3819.
37 S. I. Smedley, *The Interpretation of Ionic Conductivity in Liquids*, Plenum Press, New York, 1980.
38 J. Stoimenovski and D. R. MacFarlane, *Chem. Commun.*, 2011, DOI: 10.1039/c1cc14314j.
39 V. Tantishaiyakul, N. Phadoongsombut, W. Wongpuwarak, J. Thungtiwachgul, D. Faroongsarng, K. Wiwattanawongsa and Y. Rojanasakul, *Int. J. Pharm.*, 2004, **283**, 111–116.
40 U. L. Bernard, E. I. Izgorodina and D. R. MacFarlane, *J. Phys. Chem. C*, 2010, **114**, 20472–20478.
41 U. Rana, *et.al.*, 2011, PhD Thesis, Monash University.
42 B. Winther-Jensen, K. Fraser, C. Ong, M. Forsyth and D. R. MacFarlane, *Adv. Mater.*, 2010, **22**, 1727–1730.
43 M. J. Frisch, G. W. Trucks, H. B. Schlegel, G. E. Scuseria, M. A. Robb, J. R. Cheeseman, J. A. Montgomery, Jr., T. Vreven, K. N. Kudin, J. C. Burant, J. M. Millam, S. S. Iyengar, J. Tomasi, V. Barone, B. Mennucci, M. Cossi, G. Scalmani, N. Rega, G. A. Petersson, H. Nakatsuji, M. Hada, M. Ehara, K. Toyota, R. Fukuda, J. Hasegawa, M. Ishida, T. Nakajima, Y. Honda, O. Kitao, H. Nakai, M. Klene, X. Li, J. E. Knox, H. P. Hratchian, J. B. Cross, V. Bakken, C. Adamo, J. Jaramillo, R. Gomperts, R. E. Stratmann, O. Yazyev, A. J. Austin, R. Cammi, C. Pomelli, J. Ochterski, P. Y. Ayala, K. Morokuma, G. A. Voth, P. Salvador, J. J. Dannenberg, V. G. Zakrzewski, S. Dapprich, A. D. Daniels, M. C. Strain, O. Farkas, D. K. Malick, A. D. Rabuck, K. Raghavachari, J. B. Foresman, J. V. Ortiz, Q. Cui, A. G. Baboul, S. Clifford, J. Cioslowski, B. B. Stefanov, G. Liu, A. Liashenko, P. Piskorz, I. Komaromi, R. L. Martin, D. J. Fox, T. Keith, M. A. Al-Laham, C. Y. Peng, A. Nanayakkara, M. Challacombe, P. M. W. Gill, B. G. Johnson, W. Chen,

M. W. Wong, C. Gonzalez and J. A. Pople, *GAUSSIAN 03 (Revision C.02)*, Gaussian, Inc., Wallingford, CT, 2004.
44 J. A. Bautista-Martinez, L. Tang, J. P. Belieres, R. Zeller, C. A. Angell and C. Friesen, *J. Phys. Chem. C*, 2009, **113**, 12586–12593.
45 J.-P. Belieres and C. A. Angell, *J. Phys. Chem. B*, 2007, **111**, 4926–4937.
46 R. W. Gurney, *Ionic Processes in Solution*, 1953.
47 K. Bica, J. Shamshina, W. L. Hough, D. R. MacFarlane and R. D. Rogers, *Chem. Commun.*, 2011, **47**, 2267–2269.

 

# Electron solvation dynamics and reactivity in ionic liquids observed by picosecond radiolysis techniques†

James F. Wishart,* Alison M. Funston,‡ Tomasz Szreder,§ Andrew R. Cook and Masao Gohdo

*Received 13th April 2011, Accepted 1st June 2011*
**DOI: 10.1039/c1fd00065a**

On time scales of a nanosecond or less, radiolytically-generated excess electrons in ionic liquids undergo solvation processes and reactions that determine all subsequent chemistry and the accumulation of radiolytic damage. Using picosecond pulse radiolysis detection methods, we observed and quantified the solvation response of the electron in 1-methyl-1-butyl-pyrrolidinium bis (trifluoromethylsulfonyl)amide and used it to understand electron scavenging by a typical solute, duroquinone.

## Introduction

Over the past decade ionic liquids (ILs) have enjoyed ever-expanding attention for their potential uses as media for advanced devices and chemical processes,[1] as well as a great deal of scientific curiosity into their unusual structures and properties.[2] In contrast to molecular solvents, ionic liquids are composed solely of oppositely-charged ions that, due to structural irregularities and a complex balance of interactions, pack together poorly into low-melting ($\leq$100 °C) solids or, in some cases, never crystallize. Although ionic liquids are such a diverse class of materials that exceptions can be found to every generality used to describe them, they usually possess a wide liquidus range, good conductivity, moderate to high viscosity, low volatility, good combustion resistance, the ability to dissolve a wide range of materials including biopolymers, and oftentimes very wide electrochemical windows. In addition, the ability to substitute anions and cations, and the addition of heteroatom functional groups or perfluorination to the alkyl chains, provides a tremendous capacity to tune the properties of the ionic liquid or IL mixture for a given purpose.

Many useful applications of ionic liquids involving charge transfer reactions, such as the electrochemical conversion of solar energy, fuel cells, batteries, supercapacitors and sensors, take advantage of the properties mentioned above that can make ILs more durable compared to conventional solvents. Other important potential uses of ionic liquids, for example in the electrorefining and plating of metals, and

*Chemistry Department, Brookhaven National Laboratory, Upton, NY, 11973, USA. E-mail: wishart@bnl.gov; acook@bnl.gov; mgohdo@bnl.gov; Fax: +1 631 344-5815; Tel: +1 631 344-4327*

† Electronic supplementary information (ESI) available. See DOI: 10.1039/c1fd00065a

‡ Present address: School of Chemistry, Monash University, Clayton, Victoria 3800, Australia. Fax: +61 3 9905 4597; Tel: +61 3 9905 6292; E-mail: Alison.Funston@monash.edu

§ Present address: Department of Radiation Chemistry and Technology, Institute of Nuclear Chemistry and Technology, Dorodna 16, 03-195 Warsaw, Poland. Fax: +48 22 811 1532; Tel: +48 22 504 1204; E-mail: t.szreder@ichtj.waw.pl

in the recycling of spent nuclear fuel, necessitate an understanding of their chemistry in contact with extremely reactive species.

Radiation chemistry and pulse radiolysis have been very useful in the study of charge transfer reactions and in characterizing highly reactive species in conventional solvents, so it stands to reason that they would be valuable techniques for studying the same problems in ionic liquids. It is also important to know what will happen when ionic liquids are exposed to ionizing radiation. However, early in the investigations of ionic liquid radiation chemistry it became clear that the unusual properties of ILs have profound impacts that differentiate their radiation-induced chemistry from that of conventional molecular solvents. The largest factor contributing to the difference is the much slower dynamical response of ionic liquids to the motion of charge, due to a number of factors that have been and currently remain the subjects of intense study in the experimental and theoretical physical chemistry communities. Relative to conventional solvents, the relative dynamical slowness of ionic liquids directly alters the physical behavior of a key primary radiolysis product, the excess electron, as depicted in Fig. 1.

An excess electron is created when it is ejected from a molecule during the primary radiolytic event, leaving behind a vacancy, or "hole". The electron thus liberated is "excess" with respect to the solvent, in that it is in addition to the (local) solvent's normal complement of electrons. When it comes to kinetic rest it is called a "dry" or "pre-solvated" electron, because the solvent still has to reorganize in multiple dynamical steps (curved arrows) in response to the sudden appearance of the electron, eventually reaching a "solvated" state. The complete solvation of the excess electron in molecular solvents is extremely rapid at room temperature, occurring in less than a picosecond in water, and on the order of several picoseconds in alcohols, for example, 6.9 ps in ethanol as measured using photoionisation techniques.[4]

In contrast, the dynamical slowness of ILs means that it can take 100 to 1000 times longer for the ionic liquid medium to reach a completely solvated configuration in response to charge redistribution.[5,6] As a result, the excess electrons persist in energetic, weakly trapped, highly mobile and reactive pre-solvated states for a much longer period of time than in conventional solvents, during which time they may react to give products and product distributions quite unlike those resulting from solvated electrons. For example, in Fig. 1 the initially-produced pre-solvated electron has sufficient potential energy to react with scavenger S, but more relaxed, partially or completely solvated electrons would not.

Being highly mobile and reactive, pre-solvated electrons are efficiently scavenged by relatively low concentrations of scavengers that would not be sufficient to scavenge many solvated electrons. This reactivity is significantly more important in ionic liquids because of the much longer lifetimes of pre-solvated electrons in them. Unless taken into account, these facts can lead to unexpected results, as in the case of a pulse radiolysis study on H-atom reactions with pyrene, phenanthrene and other scavengers in the IL methyltributylammonium bis(trifluoromethylsulfonyl)amide,

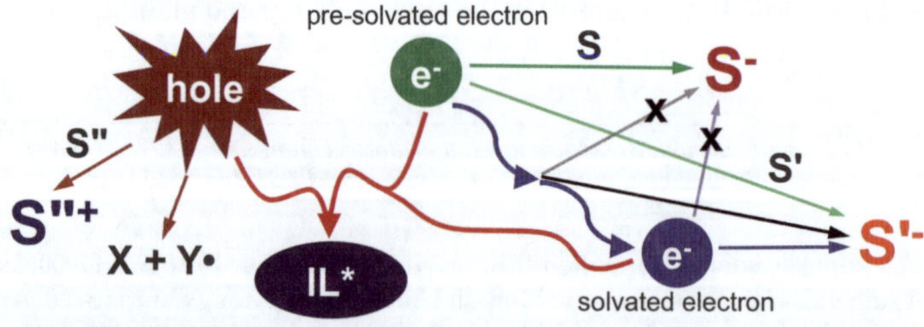


**Fig. 1** Reactions of primary radiation-induced species in ionic liquids. S and S′ represent electron scavengers with different reactivity profiles towards solvated and pre-solvated electrons. Adapted with permission from ref. 3. Copyright 2010 American Chemical Society.

$[N_{1444}][NTf_2]$.[7] The experiments were performed by adding acid to the ionic liquid to react with the electrons to form H-atoms, which then react with the arenes to form H-adducts with characteristic spectra (Reactions 1 and 2). The initial work was done with equipment with insufficient time resolution to directly observe the reactions; instead, competition kinetics were used to infer relative rate constant ratios from the product yields.

$$e^-_{solv} + H^+ \rightarrow H^{\cdot} \tag{1}$$

$$H^{\cdot} + \text{pyrene} \rightarrow \text{H-pyrene}^{\cdot} \tag{2}$$

When the reactions were directly observed on faster equipment, the rate constants obtained did not agree with the ratios obtained *via* the competition kinetics. The reason for the product ratio–observed rate constant disparity was the existence of a second pathway to the same H-pyrene˙ product operating through the direct scavenging of pre-solvated electrons by pyrene, followed by protonation of the pyrene anion (Reactions 3 and 4).

$$e^-_{pre} + \text{pyrene} \rightarrow \text{pyrene}^{\cdot-} \tag{3}$$

$$H^+ + \text{pyrene}^{\cdot-} \rightarrow \text{H-pyrene}^{\cdot} \tag{4}$$

At the pyrene and acid concentrations used, the solvated electron analogue of reaction 3 cannot compete with reaction 1. Therefore, pre-solvated electron reactivity resulted in an unanticipated product distribution in this case.[7] Effects such as this have significant implications for predicting reactivity when IL solutions are exposed to ionizing radiation, as they would be in the recycling of spent nuclear fuel, when undergoing photoionization upon UV exposure, or are subjected to extreme electrochemical conditions as in plasma electrolysis to form metal nanoparticles.[8]

Pre-solvated electron reactivity is usually studied by observing the "missing" fraction of solvated electrons at time zero as the scavenger concentration is increased. An example of this effect is shown in Fig. 2 for increasing concentrations of pyrene in $[N_{1444}][NTf_2]$.

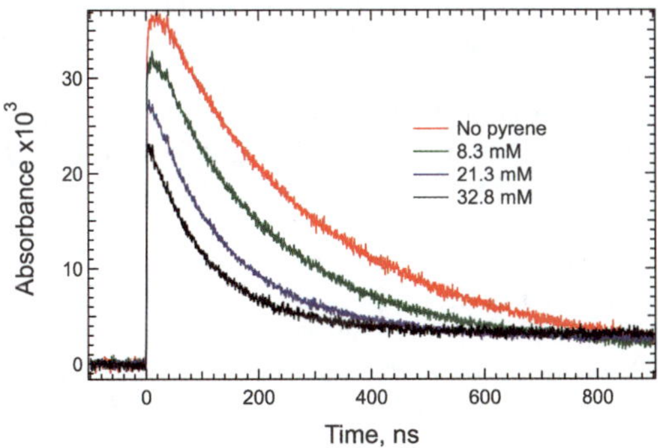


**Fig. 2** Absorbance traces of solvated electron decay at 1030 nm for various concentrations of pyrene in $[N_{1444}][NTf_2]$. Adapted with permission from ref. 9. Copyright 2003 American Chemical Society.

The relative yield (absorbance) of solvated electron for a given scavenger concentration is obtained by fitting the observed solvated electron decay kinetics and extrapolating them to time zero. As the pyrene concentration increases, the "initial" concentration of $e^-_{solv}$ drops, which is attributed to scavenging of the solvated electron's precursor $e^-_{pre}$. This effect is quantified using the relationship $G_c/G_0 = \exp(-c/C_{37})$, where $G_c$ is the yield of solvated electrons at a given scavenger concentration $c$, $G_0$ is the yield in the absence of scavenger, and $C_{37}$ is the concentration where only 1/e (37%) of the electrons survive to be solvated.[10] The lower a given scavenger's $C_{37}$, the more efficient it is at scavenging pre-solvated electrons, and values below 100 mM are considered fairly efficient. Many organic and inorganic scavengers have been rated for their efficiency using this method.[10,11]

However, studying a process by using its effect to make things disappear (such as solvated electrons) can be frustratingly indirect. With our current accelerator technology[12,13] and the larger time window afforded by sluggish ionic liquid dynamics, it is possible to directly observe pre-solvated electrons in ionic liquids and watch them react with solutes. We report here the solvation dynamics of the electron in the ionic liquid 1-methyl-1-butyl-pyrrolidinium bis(trifluoromethylsulfonyl)amide, $[C_4mpyr][NTf_2]$, and observe the reactivity of multiple species of excess electrons with a representative scavenger, duroquinone.

## Experimental

### Synthesis and materials

The synthesis of the ionic liquids used for the pulse-probe radiolysis studies is reported in the supporting information.† Additional $[C_4mpyr][NTf_2]$ used for the OFSS studies (see below) was purchased from IoLiTec (Tuscaloosa, Alabama, USA).

### Instrumentation

Ultraviolet/visible spectra were recorded on an HP 8452A diode array spectrophotometer. $^1H$ and $^{13}C$ NMR spectra were measured using a Bruker AVANCE400 spectrometer. The water contents of the samples were measured after each experiment using a Mettler Toledo DL39 Coulometric Karl Fisher Titrator. Microwave syntheses of quaternary pyrrolidinium salts were carried out using a CEM Discover synthesizer.

### Pulse radiolysis

Radiolysis experiments were carried out at the Brookhaven National Laboratory Laser-Electron Accelerator Facility (LEAF).[12] Two methods were used to measure transient absorption kinetics on the picosecond time scale. The kinetics of excess electron solvation were followed in the visible and near-infrared using a pulse-probe method, analogous to the laser pump–probe technique, for times up to 10 ns and using fast detectors coupled to a transient digitizer for times longer than 2 ns. The electron pulse width was 9–14 ps in the pulse-probe experiment and <60 ps in the transient digitizer experiment.

In pulse-probe measurements the transient absorbance is determined at a single time delay relative to each accelerator pulse using a dual beam (sample and reference) optical setup, requiring a few thousand pulses to obtain a kinetic trace, using a typical average of 20–50 absorption measurements per delay time and 100–140 delay times in a scan. The detection system for the picosecond pulse-probe measurements is described elsewhere.[12] For these experiments, 30–50 mL of argon-saturated ionic liquid was recirculated through a 10 mm path length cell. The circulation system was kept under a positive pressure of argon to minimize infiltration of air and moisture. Optical probe pulses at wavelengths of 600–1700 nm were generated using a Quantronix Topas Optical Parametric Amplifier (OPA). Silicon photodiodes (PIN-10D) were used for detection from 400 to 1000 nm and InGsAs photodiodes (PDA400) from 1000 to 1700 nm. Signals were digitized using a LeCroy 9354 oscilloscope in segment mode, triggered at the repetition rate of the laser. The radiolytic dose for each accelerator shot was measured using a Faraday cup and the results used to correct the measured absorbance data point-by-point to the same dose, and used to normalize the absorbance traces for the purpose of spectral analysis. As is standard for such measurements, the delay time interval between data points was varied in steps between 2 ps around time zero to 200 ps after $t = 3$ ns. For the purpose of singular value decomposition (SVD) analysis using Wavemetrics Igor software, absorbance traces at uniform time intervals of 2 ps were created by interpolation of the original data traces with irregular time intervals and then assembled into a matrix of absorbance values *vs.* wavelength and time for the SVD procedures.

Subsequently, an "Optical Fiber Single-Shot" (OFSS) detection method was developed at the LEAF facility for picosecond time-resolved pulse radiolysis experiments on samples that cannot be flowed or are limited in quantity.[13] The OFSS system uses a bundle of optical fibers, each of different length, to divide a laser probe pulse into a multiplexed spatial array of 143 varied delay times over a span of 5 nanoseconds, with intervals of ~5 ps around time zero. The image of the fiber bundle output is split and transported along a reference arm to a CCD camera and through the irradiated region of the sample to another CCD camera. The integrated light intensities for each fiber are used to calculate an absorbance for the time corresponding to the calibrated delay of the fiber, thus creating an absorbance *vs.* time trace in a single shot, although the data is usually averaged for 25–64 shots for a better signal-to-noise ratio. The OFSS method is superior to the pulse-probe method in terms of damage incurred by the sample (~100-fold less) and acquisition time, but presently it is only available for wavelengths <1000 nm because we have yet to acquire NIR-sensitive area detectors. Consequently, OFSS is not yet available in the region best suited to observe pre-solvated electrons. Samples for OFSS experiments were placed in 5 mm path length Suprasil spectrophotometer cuvettes fitted with valves or septum screw caps and prepared as described below for nanosecond measurements.

The excellent mutual agreement of the pulse-probe and OFSS radiolysis detection techniques is shown in supporting information Fig. 1S,† where kinetic data on the radiolysis of neat, argon-saturated [$C_4$mpyr][$NTf_2$] measured at 800 and 900 nm by both techniques are superimposed. The data were collected on IL samples of different provenance.

For nanosecond experiments the electron pulse was focused into a spectrophotometer cell containing the solution of interest. The monitoring light source was a 75 W xenon arc lamp pulsed to a few hundred times its normal intensity, propagating colinearly with the electron beam but in the opposite direction so as to minimize collection of Cerenkov radiation emitted within the sample by the passage of the electron bunch. Wavelengths were selected using either 40 nm or 10 nm bandpass optical interference filters. Transient absorption signals were detected with either an FND-100Q silicon diode (≤1000 nm) or a GAP-500L InGaAs diode (1100–1700 nm) and digitized using Tektronix TDS-680B, Tektronix TDS-694C or LeCroy 8620A oscilloscopes. Absorption spectra were corrected for the secondary response distortion from the NIR detector[14] by normalization of the signal at time *t* to the signal from the FND100Q silicon diode at the same time and wavelength. The total dose per pulse was determined before each series of experiments using an $N_2O$-saturated solution of aqueous 10 mM KSCN where $G\varepsilon = 4.87 \times 10^4$ ions (100 eV)$^{-1}$ M$^{-1}$ cm$^{-1}$ for the $(SCN)_2^-$ radical at 472 nm. Radiolytic doses of 15–35 Gy, corrected for electron density, were used. The absorption *vs.* time data were analysed with Igor Pro software (Wavemetrics).

The ionic liquids were dried overnight in a vacuum oven at 40 °C immediately prior to use. For nanosecond measurements, the liquids were placed in septum-capped, self-masking semi-micro Suprasil spectrophotometer cuvettes (optical path length 10 mm) and purged by bubbling with argon for at least twenty minutes. During the irradiation experiments the samples were exposed to as little UV light as possible (*via* the use of UV cutoff filters) to avoid photodecomposition; no evidence of this occurring was found within the time frames monitored. Unless stated otherwise, all measurements were performed at 21 ± 1 °C.

## Results and discussion

### Observation of pre-solvated electrons and their solvation process

Fig. 3 depicts the transient absorption pulse-probe traces for the radiolysis of neat $[C_4mpyr][NTf_2]$, recorded at 100 nm intervals from 600 nm to 1600 nm. Unfortunately, data could not be collected at 1100 nm due to weak OPA output and at 1200 nm due to strong sample background absorption. The traces taken at 1400–1600 nm show a quick absorbance decrease over the first few hundred picoseconds, while the traces from 800–1000 nm show an increase over the same interval. On this time scale there is clearly a blue shift of the absorption feature assigned to the radiolytically-generated excess electrons.

The blue shift is more apparent if the kinetic traces are sliced into spectra at selected time intervals as shown in Fig. 4. To guide the eye, the discrete absorption data for each spectral slice have been fitted with a spline curve. The spectra can be compared with the spectrum of the solvated electron at 50 ns after the radiolysis pulse, obtained using the standard transient digitizer pulse radiolysis technique, also included in Fig. 4. The species responsible for this absorption on the longer time scale (≥2 ns) has been unambiguously assigned as the solvated electron based on its spectrum and reactivity in this and other ionic liquids[9,15] and by analogy to red- and NIR-absorbing solvated electrons observed in conventional solvents.[16] The prominent absorption feature on the red side of the early-time spectra is thus reasonably attributed to excess electrons that have not yet reached their equilibrium solvation state, a population of pre-solvated electrons. However, due to the limited time resolution of the pulse-probe radiolysis experiment it is likely that the observed NIR absorption does not correspond to the "driest" form of the pre-solvated electron (green circle in Fig. 1), which would absorb even further in the NIR, but rather to an intermediate, probably penultimate, state of solvation.

It can be seen from the long-wavelength decay kinetics in Fig. 3 and the spectra in Fig. 4 that the relaxation of the excess electron to its equilibrated solvation state is

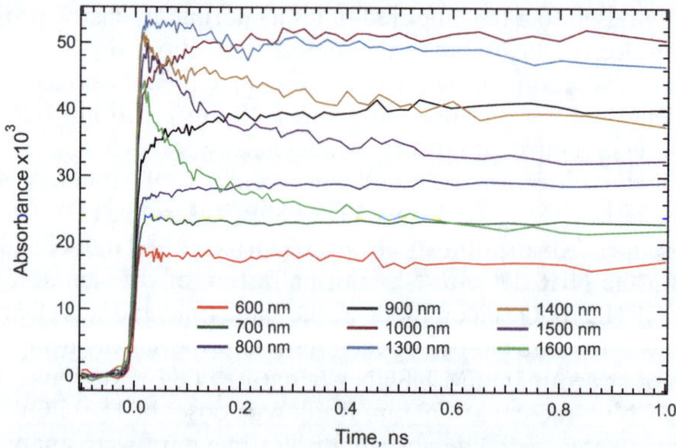


**Fig. 3** The first nanosecond of the time-resolved pulse-probe pulse radiolysis transient absorption traces for neat $[C_4mpyr][NTf_2]$ under argon atmosphere at the wavelengths indicated in the legend. The entire 8.2 ns range collected is shown in supporting information Fig. 2S.†

 

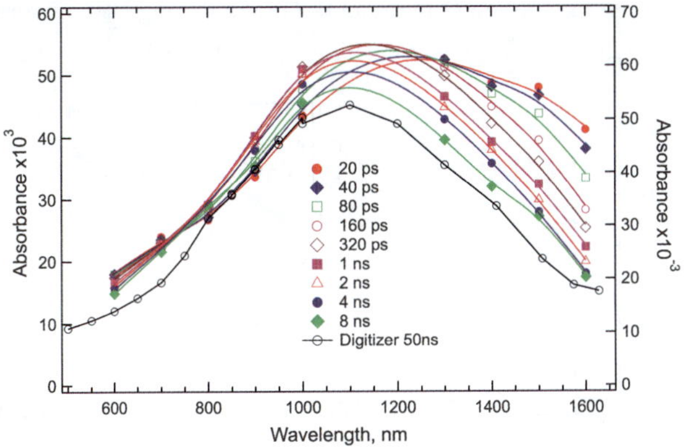


**Fig. 4** Transient absorption spectra constructed by taking slices at indicated delay times from the pulse-probe radiolysis data in Fig. 3 for [$C_4$mpyr][$NTf_2$] under argon atmosphere (left axis). For comparison, the spectrum of the solvated electron in [$C_4$mpyr][$NTf_2$] at 50 ns, constructed from digitized transients using an oscilloscope and Si and InGaAs photodiodes, is included and scaled for visibility (right axis).

complete within one nanosecond. Except for the shifted peak wavelengths of the electron spectra, the results resemble in shape and time scale those obtained by Zhang and Jonah in cold (−30 °C) propanol using a stroboscopic radiolysis technique,[17] and even earlier work by Baxendale and Wardman.[18] The 50-degree difference in working temperatures attests to the effect of slower dynamics in ILs.

In order to quantify the observed spectral shift related to the solvation process, singular-value decomposition (SVD) was applied to the absorbance-time data over all wavelengths. Using SVD, the matrix of absorbance values *vs.* $m$ time points and $n$ wavelengths, $A_{mn}$, can be broken down into the product of three matrices—an orthogonal matrix $U_{\rm mn}$ containing signal amplitude *vs.* time vectors, a diagonal matrix $S_{nn}$ containing the singular values, and the transpose of an orthogonal matrix $V^T_{nn}$ containing the absorbance spectral vectors:

$$A_{mn} = U_{mn} \, S_{nn} \, V^T_{nn} \tag{5}$$

When SVD is applied to the data in Fig. 3 (over the entire 8.2 ns time range), the diagonals of the singular value matrix $S$ are (6.332, 0.250, 0.075, 0.054, 0.043, 0.037, 0.032, 0.026, 0.014). Considering the distribution of the singular values, it is reasonable to consider only the first two vectors of the $U$ and $V^T$ matrices as meaningful and all the rest as dominated by noise. To test this assumption, the absorbance data matrix $A$ was reconstructed using only the first two singular values and plotted against the original data (see supporting information Fig. 3S†). The reconstructed traces coincide well with the original data except for slightly under-predicting the absorbance at 700 and 800 nm at times below 150 ps. This discrepancy is not significant for the analysis.

As shown in Fig. 5, the elements that dominate the SVD analysis of the pulse-probe data consist of the $n = 0$ spectrum and slow decay kinetics (in green) that correspond to the solvated electron, as evidenced by the excellent agreement between the $n = 0$ spectrum and that of the solvated electron at 50 ns (blue dotted line) in the inset. The general shape of the solvated electron decay represented by $U_{m0}$ is consistent with geminate recombination processes typically seen in pulse radiolysis. The amplitude of the $n = 0$ signal corresponding to the solvated electron at time zero (with respect to the ~15 ps resolution of our experiment) indicates that a large component of the relaxation of the initially dry and delocalized excess electrons occurs in less than 15 picoseconds. A dual channel mechanism, invoking direct

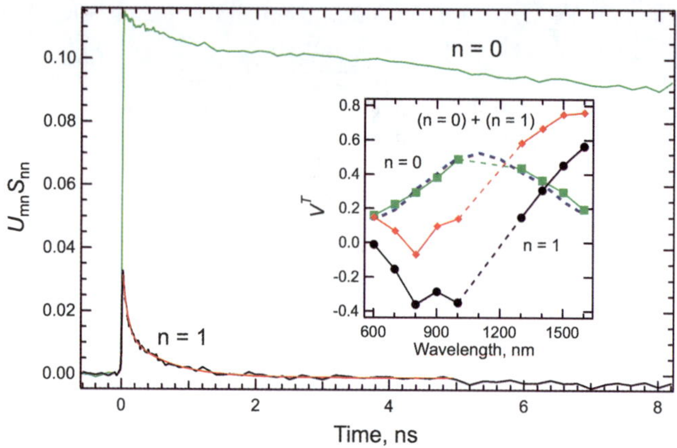


**Fig. 5** SVD results for the pulse probe radiolysis of neat [$C_4$mpyr][$NTf_2$]. In the larger graph, the signal amplitudes *vs.* time $U_{m(n=0,1)}$ are multiplied by their corresponding singular values $S_{00}$ and $S_{11}$. A double-exponential fit to the $n = 1$ vector for the first 5 ns is shown in red. The inset shows the spectral vectors corresponding to the first two singular values and their sum (red trace). The dotted blue line is the measured absorption spectrum of the solvated electron at 50 ns (as in Fig. 4) scaled to coincide with the zero-order spectral vector.

relaxation of delocalized excess electrons into solvated states in addition to the stepwise process, has been postulated in the radiolysis of alcohols.[19,20] In Fig. 1, this corresponds to a direct path from the green electron at the top to the blue one at the bottom. It is also known that in spite of the general slow relaxation of ionic liquids there are dynamical components, often characterized as inertial, that occur on femtosecond and sub-10-picosecond time scales that could facilitate this direct relaxation process.[5,21]

Despite the prevalence of the direct or fast pathways for solvated electron formation, the $n = 1$ vectors correspond to the solvation process of a penultimate form of pre-solvated electron. Lewis and Jonah[19] called this species the trapped electron, $e^-_{tr}$. If the $n = 1$ spectrum corresponds to the difference between the pre-solvated and solvated electron spectra, the sum of the two vectors, depicted in red in the inset and apparently peaking around 1500–1600 nm, must correspond to the spectrum of $e^-_{tr}$. The size and position of this band is consistent with similar spectra obtained in methanol by Keszei and coworkers[20] when allowance is made for the difference in solvated electron peak position in methanol *vs.* this ionic liquid. In both cases the peak of $e^-_{tr}$ is shifted about 2500 $cm^{-1}$ to the red compared to $e^-_{solv}$.

Turning our attention to the kinetics of the $e^-_{tr}$ solvation process represented by the $n = 1$ signal vector, the average solvation time obtained by integrating the transient curve is 270 ps. The response fits well to a double exponential with time constants of 70.4 and 574 ps and amplitudes of 60% and 40%, respectively, as indicated by the red curve in Fig. 5. Both exponential terms are associated with the same spectral change. They represent yet another case of distributed, stretched- or multi-exponential kinetics observed in ionic liquids often attributed to their distribution of heterogeneous local environments.[5,22] However, distributed kinetics can also occur in homogeneous environments, particularly when relaxation phenomena are involved.[23]

It is reasonable to wonder how the solvation dynamics of $e^-_{tr}$ in [$C_4$mpyr][$NTf_2$] compare with those of the popular solvatochromic fluorescent coumarin dye C153 in the same solvent, which has been investigated by several groups. Because many different methods are used to characterize the correlation functions that are typically constructed from fluorescence emission peak frequencies, it is easiest to compare the reported average C153 solvation times with the value of 270 ps obtained here for the electron at 21 °C. Jin *et al.*[6] obtained 380 ps at 25 °C ($\eta = 70$ cP), Funston *et al.*[24]

 

obtained 346 ps at 20 °C ($\eta = 95$ cP) and Mandal and Samanta[25] obtained 500 ps at 25 °C ($\eta = 70$ cP). It must be noted that the operative assumption in the treatment of fluorescence Stokes shifts as solvation dynamics probes is that the shift is continuous, while we have treated the electron solvation as a state-to-state transition using the SVD approach. For comparison, we constructed a correlation function for the electron spectral shift as a continuous process using the same methods employed in ref. 24 by fitting spectra at time intervals and plotting the peak frequencies. The results are presented in the supporting information (Fig. 4S†). The correlation function thus obtained indicates a substantially faster response than the SVD model, with an average solvation time of 113 ps and fitting to a double exponential function with time constants of 35 (62%) and 200 ps (38%). Recently, Bonin *et al.*[26] examined electron solvation in hot glycerol (60 °C, $\eta = 76$ cP) and found that the dynamics could be modelled well either by a stepwise mechanism involving two species or a continuous relaxation one with biexponential functions, although the time constants for glycerol were over ten times shorter than observed here for an IL of similar bulk viscosity.

One might also ask what similarities to expect between the ionic liquid relaxation processes for solvation of $e^-_{tr}$ and dipole reorientation in the case of solvatochromic dyes. It is likely that the two processes interrogate a different profile of response dynamics, but concrete answers must await detailed quantum MD simulations on electron solvation in ILs before comparisons can be made. When available, these simulations may also provide guidance as to whether the continuous or state-to-state transition model is most appropriate.

## Pre-solvated electron scavenging by duroquinone in [C$_4$mpyr][NTf$_2$]

One of the primary motivations for studying electron solvation processes in ionic liquids is to understand how they control pre-solvated electron scavenging reactions, which are of extreme relevance to many important IL applications as laid out in the introduction. We will examine the case of duroquinone (2,3,5,6-tetramethyl-1,4-benzoquinone, DQ) as an electron scavenger in [C$_4$mpyr][NTf$_2$] because it illustrates some interesting features relating to the electron solvation process.

As shown in Fig. 6, when the concentration of duroquinone is increased the initial rise in absorbance at 800 nm due to the solvation process disappears and is replaced by an increasingly large fast decay component. (On longer timescales, $e^-_{solv}$ is scavenged by DQ with a rate constant of $(8.6 \pm 0.3) \times 10^8$ M$^{-1}$ s$^{-1}$.) If the absorbance decays at each DQ concentration are fitted to two exponentials, the rate constant

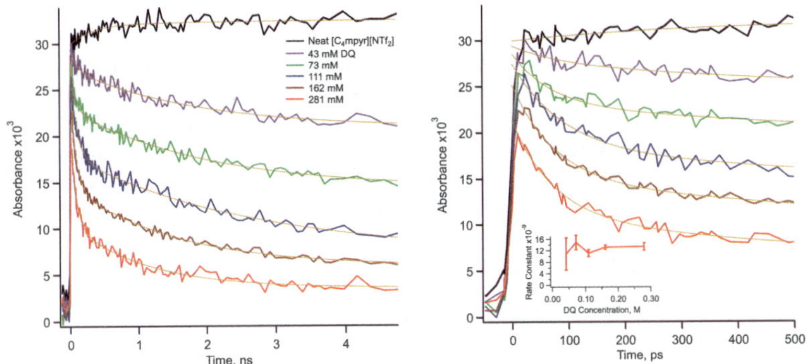


**Fig. 6** OFSS transient absorption data at 800 nm in argon-purged [C$_4$mpyr][NTf$_2$] with several concentrations of duroquinone (DQ) as indicated in the figure legend (left) and expanded view of the first 500 ps (right). DQ$^{\cdot-}$ anion does not absorb appreciably at this wavelength. Inset on right: the faster of the two rate constants used to fit the absorbance decay plotted against DQ concentration, with error bars representing one standard deviation.

of the fast step does not show any concentration dependence. The corresponding time constant is 76 ps at the highest DQ concentration, effectively the same as the 70 ps fast component of the solvation response function obtained from SVD. Fig. 5S† of the supporting information plots the kinetics of the electron solvation process, electron scavenging at 800 nm by 162 and 281 mM DQ, and the raw pulse-probe absorbance data taken at 1600 nm together on one graph to emphasize that the scavenging decay process coincides with the solvation process.

At first glance it appears odd that the rate of the electron absorbance decay due to scavenging remains constant with increasing scavenger concentration even as the decay increases in amplitude, but it is important to recall that very little diffusion occurs on this time scale in [$C_4$mpyr][$NTf_2$]. One explanation for these observations is that true to its name, $e^-_{tr}$ is trapped and its diffusion is restricted, but being in a more energetic state than a solvated electron it is more likely to tunnel to a quencher if one is nearby,[27] and increasing the DQ concentration increases the density of quenchers, resulting in a higher proportion of quenching (absorbance decay at 800 nm) at the expense of solvated electron production.

Compared to the time-zero distribution between $e^-_{tr}$ and $e^-_{solv}$ implied by Fig. 5 it appears that there is a large fraction of $e^-_{solv}$ missing in Fig. 6 at higher [DQ]. This is due to the reaction of DQ with the energetic electron precursor state that produces $e^-_{solv}$ directly without passing through $e^-_{tr}$. It is also worthwhile to point out that according to the standard $C_{37}$ approach to quantifying pre-solvated electron reactivity, the fast process shown in Fig. 6 would not even be observed and the $C_{37}$ would have been calculated on the basis of the absorbances measured at 4 nanoseconds. The existence of earlier processes would be unknown.

DQ shows one type of reactivity pattern with respect to the various types of excess electron species, but there are many others. Nitrate anion, for example, efficiently targets the production of $e^-_{tr}$ by reacting with its precursor, without strongly affecting the direct production of $e^-_{solv}$ from higher states. Because of the importance of pre-solvated electron scavenging to radiation chemistry in general, and in particular the use of kinetic clocks to explore early events in radiolysis (by increasing scavenger concentrations [S] to push the scavenging power $k_S$[S] to very short times), a lot of effort has gone into detailing the reactivity profiles of many quenchers.[11,28] Due to their slower dynamical responses, ionic liquids afford luxurious amounts of time (*i.e.* nanoseconds) in which to probe the mechanistic details of scavenging processes. As shown in the example just above, this allows us to conveniently see a wealth of mechanistic detail heretofore much more difficult to observe.

## Conclusions

The experiments described here confirm, as has been demonstrated in other cases, that dynamical processes and reactions proceed by similar mechanisms in ionic liquids than they do in conventional solvents, but also that the unusual properties of ionic liquids provide additional wrinkles that enable new insights and opportunities for new science and new technologies. In the case of ionic liquid radiation chemistry, understanding very early electron solvation and scavenging processes will enable us to control the chemistry of primary radiolytic species and direct that reactivity towards intended products and away from undesirable ones. In practical terms, this can improve the radiolytic synthesis of nanoparticles, control photo- and electrochemical degradation of solar cells, display devices and batteries, and minimize the deleterious effects of radiation on spent nuclear fuel recycling processes.

## Acknowledgements

This work, and use of the LEAF Facility of the BNL Accelerator Center for Energy Research, was supported by the US Department of Energy, Office of Basic Energy

 

Sciences, Division of Chemical Sciences, Geosciences, and Biosciences under contract # DE-AC02-98CH10886.

## Notes and references


1 J. F. Wishart, *Energy Environ. Sci.*, 2009, **2**, 956–961; N. V. Plechkova and K. R. Seddon, *Chem. Soc. Rev.*, 2008, **37**, 123–150; W. L. Hough and R. D. Rogers, *Bull. Chem. Soc. Jpn.*, 2007, **80**, 2262–2269.
2 J. F. Wishart and E. W. Castner, *J. Phys. Chem. B*, 2007, **111**, 4639–4640; R. D. Rogers and G. A. Voth, *Acc. Chem. Res.*, 2007, **40**, 1077–1078; F. Endres, *Phys. Chem. Chem. Phys.*, 2010, **12**, 1648–1648; D. R. MacFarlane and K. R. Seddon, *Aust. J. Chem.*, 2007, **60**, 3–5; H. Weingartner, *Angew. Chem., Int. Ed.*, 2008, **47**, 654–670.
3 J. F. Wishart, *J. Phys. Chem. Lett.*, 2010, **1**, 3225–3231.
4 P. K. Walhout and P. F. Barbara, in *Ultrafast Processes in Chemistry and Photobiology*, ed. M. A. El-Sayed, I. Tanaka and Y. Molin, Blackwell Scientific, Oxford (England), Cambridge, Massachusetts, 1995, pp. 83–104.
5 A. Samanta, *J. Phys. Chem. Lett.*, 2010, **1**, 1557–1562.
6 H. Jin, G. A. Baker, S. Arzhantsev, J. Dong and M. Maroncelli, *J. Phys. Chem. B*, 2007, **111**, 7291–7302.
7 J. Grodkowski, P. Neta and J. F. Wishart, *J. Phys. Chem. A*, 2003, **107**, 9794–9799.
8 A. Keppler, M. Himmerlich, T. Ikari, M. Marschewski, E. Pachomow, O. Hofft, W. Maus-Friedrichs, F. Endres and S. Krischok, *Phys. Chem. Chem. Phys.*, 2011, **13**, 1174–1181; N. Kulbe, O. Hofft, A. Ulbrich, S. Z. El Abedin, S. Krischok, J. Janek, M. Polleth and F. Endres, *Plasma Process. Polym.*, 2011, **8**, 32–37.
9 J. F. Wishart and P. Neta, *J. Phys. Chem. B*, 2003, **107**, 7261–7267.
10 R. K. Wolff, M. J. Bronskill and J. W. Hunt, *J. Chem. Phys.*, 1970, **53**, 4211–4215; J. E. Aldrich, M. J. Bronskill, R. K. Wolff and J. W. Hunt, *J. Chem. Phys.*, 1971, **55**, 530–539; K. Y. Lam and J. W. Hunt, *Int. J. Radiat. Phys. Chem.*, 1975, **7**, 317–338.
11 C. D. Jonah, J. R. Miller and M. S. Matheson, *J. Phys. Chem.*, 1977, **81**, 1618–1622.
12 J. F. Wishart, in *Radiation Chemistry: Present Status and Future Trends*, ed. C. D. Jonah and B. S. M. Rao, Elsevier Science, Amsterdam, 2001, pp. 21–35; J. F. Wishart, A. R. Cook and J. R. Miller, *Rev. Sci. Instrum.*, 2004, **75**, 4359–4366.
13 A. R. Cook and Y. Z. Shen, *Rev. Sci. Instrum.*, 2009, **80**, 073106.
14 J. A. Cline, C. D. Jonah and D. M. Bartels, *Rev. Sci. Instrum.*, 2002, **73**, 3908–3915.
15 A. M. Funston and J. F. Wishart, in *Ionic Liquids IIIA: Fundamentals, Progress, Challenges and Opportunities*, ed. R. D. Rogers and K. R. Seddon, American Chemical Society, New York, 2005, pp. 102–116; K. Takahashi, T. Sato, Y. Katsumura, J. Yang, T. Kondoh, Y. Yoshida and R. Katoh, *Radiat. Phys. Chem.*, 2008, **77**, 1239–1243.
16 L. M. Dorfman and J. F. Galvas, in *Radiation Research. Biomedical, Chemical and Physical Perspectives*, ed. O. F. Nygaard, H. J. Adler and W. K. Sinclair, Academic Press, New York, 1975, pp. 326–332.
17 X. J. Zhang and C. D. Jonah, *Chem. Phys. Lett.*, 1996, **262**, 649–655.
18 J. H. Baxendale and P. Wardman, *J. Chem. Soc., Faraday Trans. 1*, 1973, **69**, 584–594.
19 M. A. Lewis and C. D. Jonah, *J. Phys. Chem.*, 1986, **90**, 5367–5372.
20 P. Holpar, T. Megyes and E. Keszei, *Radiat. Phys. Chem.*, 1999, **55**, 573–577.
21 S. Arzhantsev, H. Jin, G. A. Baker and M. Maroncelli, *J. Phys. Chem. B*, 2007, **111**, 4978–4989; S. Arzhantsev, H. Jin, N. Ito and M. Maroncelli, *Chem. Phys. Lett.*, 2006, **417**, 524–529.
22 Z. H. Hu and C. J. Margulis, *Proc. Natl. Acad. Sci. U. S. A.*, 2006, **103**, 831–836.
23 C. Khurmi and M. A. Berg, *J. Phys. Chem. Lett.*, 2010, **1**, 161–164.
24 A. M. Funston, T. A. Fadeeva, J. F. Wishart and E. W. Castner, *J. Phys. Chem. B*, 2007, **111**, 4963–4977.
25 P. K. Mandal and A. Samanta, *J. Phys. Chem. B*, 2005, **109**, 15172–15177.
26 J. Bonin, I. Lampre, P. Pernot and M. Mostafavi, *J. Phys. Chem. A*, 2008, **112**, 1880–1886.
27 J. R. Miller, *J. Chem. Phys.*, 1972, **56**, 5173–5183.
28 B. Pastina and J. A. LaVerne, *J. Phys. Chem. A*, 1998, **103**, 209–212; S. M. Pimblott and J. A. LaVerne, *J. Phys. Chem. A*, 1998, **102**, 2967–2975.


 

# Investigating the origin of entropy-derived rate accelerations in ionic liquids†

Hon Man Yau,[ab] Anna K. Croft[b] and Jason B. Harper*[a]

*Received 8th April 2012, Accepted 10th June 2012*
**DOI: 10.1039/c1fd00060h**

The effects on the rate and activation parameters of a series of Menschutkin processes on changing from a molecular solvent to an ionic liquid were investigated. The removal of delocalised π-systems from the reagents does not affect the change in activation parameters on changing solvent. In each of the cases investigated, rate accelerations observed on moving to the ionic liquid could be attributed to an increase in reaction entropy. This suggests a specific interaction of the ionic liquid with the nucleophilic centre, rather than the delocalised π-systems of either the electrophile or the nucleophile.

## Introduction

Organic reactivity in ionic liquids is often reported to differ from that observed in molecular solvents.,[1]‡ Ionic liquids, which are typically made up of a bulky organic cation and a charge diffuse anion,[2–6] have numerous advantages over molecular solvents, not least very low vapour pressures[2,7] and the ability to change solvent properties through modification of the components.[5–10] However, the widespread application of these room temperature molten salts for synthetic purposes is currently restricted by the limited understanding of the underlying principals that govern reactivity in these solvents.[1]

In an effort to develop a framework for the understanding of reaction outcomes in ionic liquids, we have previously examined a series of substitution[11–15] and cycloaddition processes[16,17] in both molecular solvents and ionic liquids, focusing on the ionic liquid 1-butyl-3-methylimidazolium ($[Bmim]^+$) bis(trifluoromethanesulfonyl) imide ($[N(CF_3SO_2)_2]^-$). Particularly, we recently noted the rate acceleration in the Menschutkin reactions between the benzyl bromides **1** and pyridine **2** (Scheme 1) on changing solvent to an ionic liquid. This acceleration was found to be due to an increase in the entropy of activation of the process. Given the increasing charge development in the transition state, this entropy difference is unlikely to be the effect of a simple reduction in organisation around the transition state complex; rather, it suggests increased organisation about either one or both of the starting materials.

[a]*School of Chemistry, University of New South Wales, Sydney, NSW, 2052, Australia. E-mail: j.harper@unsw.edu.au; Fax: +61 2 9385 6141; Tel: +61 2 9385 4692*
[b]*School of Chemistry, University of Wales Bangor, Bangor, Gwynedd, LL57 2UW, United Kingdom*

† Electronic supplementary information (ESI) available: Full details for the preparation of the imine **7**, description of the computational methods used, three-dimensional coordination profile for each of the bromides **1d** and **4**, and each of the amines **2** and **7** in the ionic liquid $[Bmim][N(CF_3SO_2)_2]$, activation parameters and temperature-dependent rate constants from which Table 1 is derived, ESP from GAMESS and Cartesian geometries for rigid bodies for compounds **1d**, **2**, **4** and **7**. See DOI: 10.1039/c1fd00060h

‡ In many cases, components of the ionic liquid participate directly in the reaction; this is exemplified by protic cations[34,35] and basic anions.[36,37] Such cases do not rely on fundamental properties of the ionic liquid and will not be referred to further here.

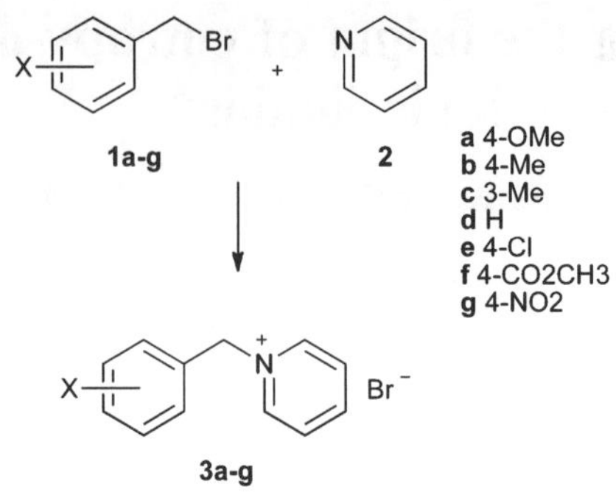

1a-g 2

a 4-OMe
b 4-Me
c 3-Me
d H
e 4-Cl
f 4-CO2CH3
g 4-NO2

3a-g

**Scheme 1** The Menschutkin reaction between the benzyl bromides **1a–g** and pyridine **2** carried out in either acetonitrile or the ionic liquid [Bmim][N($CF_3SO_2$)$_2$].[13]

The work described herein aims to deconvolute the origins of this organisation. Specifically, the relative importance of interactions between the components of the ionic liquid with either the nitrogen centre of the nucleophile or the delocalised π-systems of the starting materials are considered. The effect of changing solvent on the reactions of either the amine **5** or the imine **7**, with either the benzyl bromide **1d** or the alkyl bromide **4** is considered (Scheme 2); that is, both nucleophiles and electrophiles with and without a delocalised π-system have been examined. The presence of a methyl group in the imine **7** is required to prevent the trimerisation in solution that occurs for the corresponding non-methylated analogue.[18] Hence, the picoline **5** is used (rather than pyridine **2** previously) for consistency.

## Experimental

The bromides **1d** and **4** were commercially available and used without further purification. [Bmim][N($CF_3SO_2$)$_2$] was prepared from the corresponding chloride[6] and dried to constant weight under reduced pressure (< 5 Torr) at *ca.* 60 °C. Picoline **5** was purified using a literature method,[19] whilst the imine **7** was prepared through modification of literature procedures (for full details, see ESI†).[20–23] Analytical grade acetonitrile was used as received. $^1$H nuclear magnetic resonance (NMR) and $^1$H-$^{13}$C heteronuclear single quantum coherence (HSQC) spectra for the reaction between the bromide **4** and the imine **7** were recorded on a Bruker Avance 700 (700 MHz) spectrometer. For all other reactions, $^1$H NMR spectra were recorded either on a Bruker Avance 600 spectrometer (600 MHz), a Bruker Avance 500 spectrometer

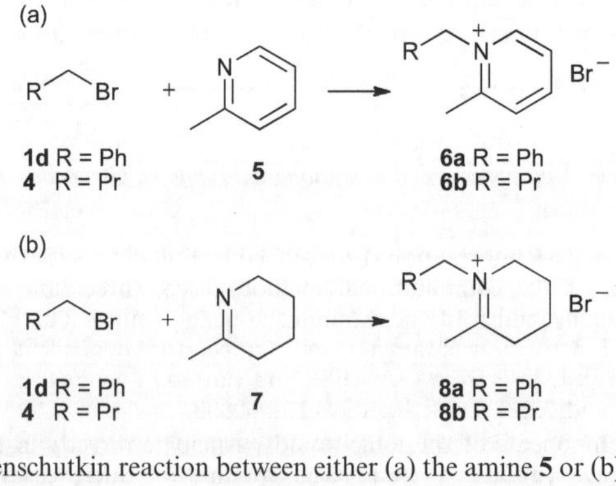

(a)

1d R = Ph
4 R = Pr

5

6a R = Ph
6b R = Pr

(b)

1d R = Ph
4 R = Pr

7

8a R = Ph
8b R = Pr

**Scheme 2** The Menschutkin reaction between either (a) the amine **5** or (b) the imine **7** and the bromides **1d** and **4** carried out in either acetonitrile or the ionic liquid [Bmim][N($CF_3SO_2$)$_2$].

 

(500 MHz) or on a Bruker Avance 400 spectrometer (400 MHz). In all cases, 500 μL of reaction mixture (details below) were examined in a 5 mm NMR tube. Replicate experiments carried out on different instruments gave the same outcome, demonstrating that the results were independent of the instrument used.

Kinetic analyses of the Menschutkin reactions were carried out in solutions containing the appropriate bromide (*ca.* 0.05 mol $L^{-1}$) and nucleophile (*ca.* 10 equiv.), either inside an NMR machine for the reactions involving benzyl bromide **1d** or using a temperature-controlled water bath for the reactions involving *n*-bromobutane **4**. The reactions were examined over a range of temperatures between 268 K and 335 K, with the temperature measured using a thermocouple inside an NMR tube, containing either ethanol or ethylene glycol, fitted to an OMEGA HH23 microprocessor-based thermometer. For the reaction between benzyl bromide **1d** and 2-methylpyridine **5**, the reaction was followed using $^{1}H$ NMR spectroscopy until more than 95% of the starting material was consumed. All other reactions were followed up to *ca.* one half-life, using either $^{1}H$ NMR or $^{1}H$-$^{13}C$ HSQC spectroscopy. In all cases spectra were taken at regular intervals during the reaction. At least twenty spectra were obtained for each kinetic run involving the bromide **1d** and the nucleophile **5**, and at least five for all other reactions.

For reactions monitored using $^{1}H$ NMR spectroscopy, the extent of reaction was deduced by integration of the signal corresponding to the protons on the bromide-bearing carbon atom in the starting materials (for details refer to ESI†). For reactions monitored using $^{1}H$-$^{13}C$ HSQC spectroscopy, the extent of reaction was deduced by integration of the signal corresponding to the carbon-proton cross coupling of the bromide-bearing carbon atom.

From this information, the pseudo-first order rate constants for the reactions under these conditions were calculated and, from the concentration dependence of these values, the second order rate constants at each temperature were subsequently calculated. The activation parameters were then determined using the Eyring equation.[24]

## Results and discussion

Our previous report[13] highlighted the importance of organisation about the starting materials of the substitution reaction presented in Scheme 1; this is in contrast to the simple systems studied previously.[11,12] However, it does not determine the origin of such an effect.

Each of the starting materials **1** and **2** have delocalised π-systems and molecular dynamics simulations show significant ordering of the ionic liquid [Bmim][N$(CF_3SO_2)_2$] about each of the aromatic systems **1d** and **2** (Figures S1 and S2†). In addition, there is a key interaction of the cation of the ionic liquid with the nucleophilic nitrogen on the pyridine **2**. Such organisation is consistent with that observed about similar aromatic systems[15,25–27] and nitrogen nucleophiles.[28] In order to determine the contributions of interactions with the delocalised π-systems to the effect of ionic liquids on the Menschutkin reactions, reagents which lack such a π-system are required.

The bromide **4** was chosen because it represents a non-aromatic analogue of the benzylic system **1d**, and, importantly for kinetic studies, it is readily available, practical to use and reacts through a bimolecular substitution mechanism at a reasonable rate.§ Molecular dynamics simulations confirm the appropriateness of this choice, with little organisation observed in the three-dimensional profile at the same probability cut-offs as was used for the solutes **1d** and **2** (Figure S3†); this is consistent with our earlier studies of an alkyl chloride.[12]

§ Whilst other saturated analogues of the bromide **1d** might be considered (such as cyclohexylmethyl bromide), they are either not readily available, are impractical to use or react either too fast or too slow to be practically observed.

Similar to above, the imine **7** was chosen as it retains the $sp^2$ hybridised nitrogen centre of pyridine **2**, but without the aromatic $\pi$-system. As mentioned before, as the imine **9** trimerises in solution,[18] substitution at the 2-position to give the analogue **7** is necessary for kinetic studies and, therefore, will be compared directly with the picoline **5**. Once again, molecular dynamics simulations of the organisational profile about the amine **7** are crucial to show the appropriateness of this nucleophile. The organisation is shown to be dominated by the key interaction between the cation of the solvent and the nucleophilic nitrogen centre; little organisation about the isolated double bond is observed (Figure S4†).

Since these molecular dynamics simulations suggest that there is significant organisation about both the starting materials **1d** and **2**, due to the aromatic electron density, and that this can be ameliorated by removing the aromaticity, temperature-dependent kinetic studies using the electrophiles **1d** and **4** and the nucleophiles **5** and **7** were undertaken in order to determine activation parameters for the processes in either acetonitrile or the ionic liquid [Bmim][$N(CF_3SO_2)_2$] for deconvolution of the effect of the delocalised $\pi$-systems. The effect of moving to the ionic liquid solvent on the activation parameters is shown in Table 1.

The first thing to note from this data is the negligible effect of adding a methyl group to the nucleophile **2** on the values of $\Delta(\Delta H^{\ddagger})$ and $\Delta(\Delta S^{\ddagger})$ when changing the solvent to the ionic liquid from acetonitrile. This suggests that the methyl group either has a limited effect on organisation of the solvent about the starting material and the transition state complex, or that any such effect is similar in both solvents. It is worth noting that absolute values for the activation entropies in each solvent (see ESI†) remain similar on going from nucleophile **2** to nucleophile **5**. The additional steric bulk introduced by the methyl substituent is reflected largely in the enthalpy of activation, as is consistent with a decrease in the ability of the nucleophilic lone-pair to interact on going from nucleophile **2** to nucleophile **5**.

The effect of removing the aromaticity from the electrophile is observed by contrasting the effect on activation parameters on changing solvent between the reactions of compounds **1d** and **4** with the nucleophile **5**. As expected, the absolute value of the activation enthalpies was larger in the butyl bromide **4** case (see ESI†) due to the lack of electronic stabilisation in the transition state. On the other hand, the activation entropies are similar between bromide **4** and bromide **1d**, indicating that aromaticity on the nucleophile has little influence on solvent reorganisation. Irrespective, the effect on the activation parameters of solvent change is virtually identical, irrespective of the combination of electrophile and nucleophile.

**Table 1** Changes in the activation parameters for the Menschutkin reaction of the reagents shown on going from acetonitrile to [Bmim][$N(CF_3SO_2)_2$] as the solvent

| Reagents | $\Delta(\Delta H^{\ddagger})$/kJ $mol^{-1}$ [a] | $\Delta(\Delta S^{\ddagger})$/J $K^{-1}$ $mol^{-1}$ [b] |
|---|---|---|
| **1d** and **2** | $6.6 \pm 1.2$ [b] | $29 \pm 4$ [b] |
| **1d** and **5** | $7.3 \pm 1.2$ | $32 \pm 4$ |
| **4** and **5** | $6.7 \pm 0.3$ | $28 \pm 1$ |
| **1d** and **7** | $4.4 \pm 1.6$ | $23 \pm 6$ |
| **4** and **7** | $26.6 \pm 10.5$ | $94 \pm 33$ |

[a] Uncertainties quoted are standard errors from linear regression. [b] Reproduced from Yau *et al.*[13]

 

Thus, initial organisation of the $[bmim]^+$ cation about the aromatic electrophile **1d** has a negligible effect and is not responsible for the observed solvent effects.

In a similar fashion to above, by considering the reactions of the electrophile **1d** with each of the nucleophiles **5** and **7**, the effect of delocalised π-electrons in the nucleophile can be determined. Whilst the absolute activation parameters decrease on moving to the imine **7** (see ESI†), presumably because of a higher degree of freedom as facilitated by the relatively more flexible backbone, there is no difference in the effect of changing solvent from acetonitrile to the ionic liquid. This demonstrates that the interaction of the solvent with the delocalised π-system of the nucleophile is also not responsible for changes in reaction rates.

Whilst the above results are conclusive, it was also considered useful to examine the case where neither of the reagents have a delocalised π-system (the reaction between compounds **4** and **7**). This proved distinctly more problematic than previous cases with signal overlap in the $^1H$ NMR spectra requiring either high field strengths (acetonitrile case) or 2D NMR techniques (ionic liquid case) to obtain kinetic data. Long reaction times also meant that competing decomposition of the nucleophile **7** was observed. The data presented in Table 1 highlights issues with reproducibility, as the uncertainties are much larger than observed with the other cases.

Having said that, the significant enhancement in the rates of reaction facilitated by the ionic liquid, over the range of temperatures examined, mirror those observed in all other combinations of nucleophiles and electrophiles. Qualitatively, increases in both activation parameters were also observed on going from acetonitrile to [Bmim] $[N(CF_3SO_2)_2]$; this is consistent with what was observed in all other cases. This change in activation parameters indicates that the underlying interactions, which give rise to the solvent effects observed in the reaction between the bromide **4** and the nucleophile **7**, are likely the same as for all the other cases discussed above. Note that the origins of the larger effects in this case are not clear; the effects may be due to either decreased interactions between solvent ions and the transition state or increased coordination of the starting material, but which it is cannot be determined from these data.

The key outcome of the complete set of results is that removing the aromatic component of either the electrophile or the nucleophile cannot account for the general changes in activation parameters on moving from acetonitrile to [Bmim] $[N(CF_3SO_2)_2]$. Whilst there are distinct interactions between the aromatic portions of the reagents and the components of the ionic liquids (as observed through molecular dynamics simulations), either there is little change in the magnitude of these interactions on going to the transition state complex or any changes in one solvent-ion interaction are offset in the other.¶ This suggests that the key solvent–solute interactions in this series of reactions (Scheme 2) are likely to be those with the lone pair on the nucleophilic nitrogen centre; these interactions must be overcome to achieve the bond forming process.

Previous work has shown similar entropy driven rate enhancements for $S_N2$ processes with $sp^3$ hybridised amine nucleophiles.[28] These enhancements were attributed to the interaction of the anion of the ionic liquid with the nucleophile through hydrogen-bonding, and interaction of the cation with the nucleophile was highlighted as unfavourable. Clearly this is not the case for $sp^2$ hybridised amine nucleophiles with no possibility of acting as hydrogen bond donors and the arguments presented above more cogently explain the observed rate increases in the reactions of benzyl bromide with either methylimidazole[29] or dimethylimidazole.[30]

It is of interest to consider the nature of the interaction between the cation and the nucleophile in these cases. While the obvious interaction is likely to be with the C-2

¶ Given the nature of the transition state is non-trivial and varies with the nature of the reagents and solvents used,[13] no examination of the interactions of the solvent with the transition state was undertaken.

proton of the imidazolium cation, which is clearly the most acidic based on carbene formation,[31] neutron diffraction studies demonstrate organisation of anions about all of the ring hydrogens along with above and below the plane of the ring.[32,33] Kinetic investigations with modified ionic liquids are underway to determine the principle site(s) of interaction of the cation of the ionic liquid with the nitrogen nucleophile.

## Conclusions

The presence of a delocalised π-system does not change the effects of an ionic liquid on the activation parameters of a bimolecular substitution process. In each case, the rate enhancement observed over the range of temperatures examined is as a result of an increase in the entropy of activation, which is partially offset by a small increase in the enthalpy of activation. The interactions with the aromatic portions of some reagents, demonstrated by molecular dynamics simulations, must either remain almost unchanged on going to the transition state or any changes in one ion-solvent interaction are offset in the other. The origin of the entropy driven rate increase is interaction of the ionic liquid cation with the nucleophilic centre, which demonstrates that such interactions can be favourable for reaction outcome.

## Acknowledgements

HMY acknowledges the support of the Graduate Research School, UNSW, Postgraduate Research Student Support Scheme along with the Australian government through the receipt of an Australian Postgraduate Award. AKC is grateful to the Royal Society for the award of a Travel Grant. JBH acknowledges financial support from the UNSW Faculty Research Grants Programme and the receipt of a UNSW Goldstar Award. The authors also wish to acknowledge the use of the UK National Grid Service in carrying out this work.

## References

1 J. B. Harper and M. N. Kobrak, *Mini-Rev. Org. Chem.*, 2006, **3**, 253–259.
2 C. L. Hussey, *Pure Appl. Chem.*, 1988, **60**, 1763–1772.
3 R. H. Dubois, M. J. Zaworotko and P. S. White, *Inorg. Chem.*, 1989, **28**, 2019–2020.
4 J. S. Wilkes and M. J. Zaworotko, *J. Chem. Soc., Chem. Commun.*, 1992, 965–967.
5 P. Bonhôte, A. Das, N. Papageorgiou, K. Kalanasundram and M. Grätzel, *Inorg. Chem.*, 1996, **35**, 1168–1178.
6 J. G. Huddleston, H. D. Willauer, R. P. Swatloski, A. E. Visser and R. D. Rogers, *Chem. Commun.*, 1998, 1765–1766.
7 K. R. Seddon, *Kinet. Catal. Engl. Transl.*, 1996, **37**, 693–697.
8 H. Stegemann, A. Rhode, A. Reiche, A. Schnittke and H. Füllbier, *Electrochim. Acta*, 1992, **37**, 379–383.
9 A. Elaiwi, P. B. Hitchcock, K. R. Seddon, N. Srinivasan, Y.-M. Tan and T. Welton, *J. Chem. Soc., Dalton Trans.*, 1995, 3467–3472.
10 K. R. Seddon, *J. Chem. Technol. Biotechnol.*, 1997, **68**, 351–368.
11 B. Y. W. Man, J. M. Hook and J. B. Harper, *Tetrahedron Lett.*, 2005, **46**, 7641–7645.
12 H. M. Yau, S. A. Barnes, J. M. Hook, T. G. A. Youngs, A. K. Croft and J. B. Harper, *Chem. Commun.*, 2008, 3576–3578.
13 H. M. Yau, A. G. Howe, J. M. Hook, A. K. Croft and J. B. Harper, *Org. Biomol. Chem.*, 2009, **7**, 3572–3575.
14 H. M. Yau, S. J. Chan, S. R. D. George, J. M. Hook, A. K. Croft and J. B. Harper, *Molecules*, 2009, **14**, 2521–2534.
15 S. G. Jones, H. M. Yau, E. Davies, J. M. Hook, T. G. A. Youngs, J. B. Harper and A. K. Croft, *Phys. Chem. Chem. Phys.*, 2010, **12**, 1873–1878.
16 C. E. Rosella and J. B. Harper, *Tetrahedron Lett.*, 2009, **50**, 992–994.
17 S. R. D. George, G. L. Edwards and J. B. Harper, *Org. Biomol. Chem.*, 2010, **8**, 5354–5358.
18 H. Kessler, H. Möhrl and G. Zimmerman, *J. Org. Chem.*, 1977, **42**, 66.
19 W. L. F. Armarego and C. Chai, *Purification of Laboratory Chemicals*, 5th Edition, Butterworth-Heinemann, Boston, 2003.

 

20 E. P. Anderson, J. V. Crawford and M. L. Sherrill, *J. Am. Chem. Soc.*, 1946, **68**, 1294.
21 Y. Chan, X. Li, C. Zhu, X. Liu, Y. Zhang and H. Leung, *J. Org. Chem.*, 1990, **55**, 5497.
22 M. J. Panigot, M. J. Robarge and R. W. Curley, *AAPS PharmSci*, 2001, **3**, 26.
23 C.-H. Lee, J.-S. Lee, H.-K. Na, D.-W. Yoon, H. Miyaji and W.-S. Cho, *J. Org. Chem.*, 2005, **70**, 2067.
24 H. Eyring, *J. Chem. Phys.*, 1935, **3**, 107–115.
25 C. G. Hanke, A. Johansson, J. B. Harper and R. M. Lynden-Bell, *Chem. Phys. Lett.*, 2003, **374**, 85–90.
26 J. B. Harper and R. M. Lynden-Bell, *Mol. Phys.*, 2004, **102**, 85–94.
27 R. M. Lynden-Bell, M. G. Del Pópolo, T. G. A. Youngs, J. Kohanoff, C. G. Hanke, J. B. Harper and C. C. Pinilla, *Acc. Chem. Res.*, 2007, **40**, 1138–1145.
28 L. Crowhurst, N. L. Lancaster, J. M. P. Arlandis and T. Welton, *J. Am. Chem. Soc.*, 2004, **126**, 11549–11555.
29 R. Bini, C. Chiappe, C. S. Pomellic and B. Parisi, *J. Org. Chem.*, 2009, **74**, 8522–8530.
30 A. Skrzypczak and P. Neta, *Int. J. Chem. Kinet.*, 2004, **36**, 253–258.
31 F. McLachlan, C. J. Mathews, P. J. Smith and T. Welton, *Organometallics*, 2003, **22**, 5350–5357.
32 C. Hardacre, S. E. J. McMath, M. Nieuwenhuyzen, D. T. Bowron and A. K. Soper, *J. Phys.: Condens. Matter*, 2003, **15**, S159–S166.
33 C. Hardacre, J. D. Holbrey, S. E. J. McMath, D. T. Bowron and A. K. Soper, *J. Chem. Phys.*, 2003, **118**, 273–278.
34 S. S. Palimkar, S. A. Siddiqui, T. Daniel, R. J. Lahoti and K. V. Srinivasan, *J. Org. Chem.*, 2003, **68**, 9371–9378.
35 H. Zhang, F. Xu, X. Zhou, G. Zhang and C. Wang, *Green Chem.*, 2007, **9**, 1208–1211.
36 J.-M. Xu, Q. Wu, Q.-Y. Zhang, F. Zhang and X.-F. Lin, *Eur. J. Org. Chem.*, 2007, 1798–1802.
37 B. C. Ranu, R. Jana and S. Sowmiah, *J. Org. Chem.*, 2007, **72**, 3152–3154.

# *Ab initio* simulations of thermal decomposition and of electron transfer reactions in room temperature ionic liquids

**Pietro Ballone*[a] and Robinson Cortes-Huerto[b]**

*Received 10th April 2011, Accepted 5th May 2011*
**DOI: 10.1039/c1fd00064k**

Selected aspects of the *ab initio* modelling of room temperature ionic liquids are discussed in our contribution, focusing on thermal decomposition reactions, and on the determination of the electrochemical stability window of these compounds. In both cases, we emphasise the role of *ab initio* simulation methods, able to deal simultaneously with the ionic and electronic side of the systems and phenomena under investigation.

## 1 Introduction

During the last twenty years, room temperature ionic liquids (RTILs, or ILs for short) have emerged as a vast and fascinating family of compounds.[1] A large number of RTIL applications have been proposed, ranging from their use as low volatility solvents[2] to high performance lubricants at high pressure and relatively high temperature conditions.[3]

The variety of compounds, and the wide range of conditions and of applications in which RTILs play a role, make it more urgent the need for accurate and versatile methods to model these systems, and more challenging the task of predicting their thermodynamic, structural and chemical properties by purely theoretical and/or computational means. Over the years, many different approaches have been used to this aim, from semi-empirical correlations of computed and measured quantities (see, for instance, COSMO[4] and COSMO-RS[5]) to refined (and expensive) quantum chemistry methods.[6]

RTILs are remarkable especially for their stability. In many applications RTILs behave virtually as inert species, and only their peculiar thermodynamic properties are of primary interest. As a result, most computational investigations have been devoted to structural and thermodynamic properties, using computer simulations[7] based on empirical and classical force field models.[8] This approach is justified precisely by the fact that in most cases the chemical identity of the species appearing in any given state are the same. Moreover, as it is often the case in organic chemistry, chemical bonds are well defined entities, whose properties are transferable among large sets of homologous compounds. In many respects, the major ingredient missing in the simplest version of empirical force fields is explicit polarisability†, which, however, has already been accounted for in a few simulations based on polarisable force fields.[9]

[a]*Atomistic Simulation Centre, Queen's University Belfast, Belfast, BT7 1NN, UK. E-mail: p.ballone@qub.ac.uk*
[b]*CNRS, Ctr Interdisciplinaire Nanosci Marseille, Campus de Luminy, 13288 Marseille 9, France*

† To some extent, polarisability effects are implicitly contained in the bonded and especially non-bonded parameters.

 

A few *ab initio* studies have been carried out on RTILs, including static investigations of cohesive and electronic structure properties,[10] as well as *ab initio* molecular dynamics (MD) studies,[11] based on density functional theory[12] (DFT). The interest of these studies is undeniable and difficult to overestimate, but also their high computational cost is worth mentioning. Moreover, and more importantly, both the complexity of the method, and the great effort required to carry out computations often conceals the fact that even the best among *ab initio* methods are still approximate, and sometimes they turn out to be severely inadequate in practice if not in principle.

However, and perhaps more interestingly for our discussion, a smaller, but potentially relevant class of RTIL applications rely on their role in chemical transformations, and thus on their electronic structure. Electrochemistry[13] and catalysis[1,2,14] are representative examples, both expected to provide in the near future many opportunities for RTIL applications of sizeable economic impact. In all these cases, involving the interplay of structural and electronic properties, *ab initio* methods, especially if coupled to molecular dynamics simulations, represent an irreplaceable tool to study ILs.

Our contribution is devoted to selected aspects of the computational modelling of RTIL systems, focusing on cases in which *ab initio* methods, and *ab initio* simulation in particular, represent the method of choice to investigate the phenomena of interest. To exemplify our arguments, (i) we investigate the thermal stability/decomposition of RTILs, and (ii) we discuss the determination of the electrochemical stability window by a density functional approach.

*Ab initio* computations have already been used, at least as an auxiliary tool, to discuss the thermal stability[15,16] of RTIL compounds. To the best of our knowledge, previous computations have been limited to a quasi-static point of view, based on the determination of minimum energy configurations along a pre-defined reaction coordinate. In principle, the method allows mapping of the bottom of a valley on the potential energy surface leading to a reaction, corresponding in this case to a decomposition/formation process. In this way, transition states and energy barriers are determined, and translated into a temperature-dependent decomposition rate using standard relations of chemical kinetics. In our computations we introduce the molecular dynamics simulation aspect, that allows us to include entropy contributions, as well as the effect of a condensed environment made of multiple interacting molecules. In addition, we carried out a limited exploration of the effect on thermal stability of an extra molecule (water) representing a common contaminant of RTILs under ambient condition.

Again to the best of our knowledge, very little has been reported in the literature[17] on the second subject of our present study, the *ab initio* determination of electrochemical properties of RTILs, mainly because of basic limitations in this respect of the density functional approach used in most *ab initio* quantum chemistry studies. We discuss a simple approach to overcome these difficulties, and we provide an estimate of the electrochemical stability window of a few prototypical RTIL molecules, *i.e.*, [bmim][Cl], [bmim][$BH_4$], [bmim][$PF_6$], [bmim][$NTf_2$] chemisorbed on an idealised (structureless) model of an aluminium surface. More refined methods[18] could be used as well, but they require more extensive computations. Moreover, our simple approach provides a more intuitive view of the role of DFT eigenvalues in electrochemical processes.

In both parts of our discussion, we emphasise the role of both *ab initio* methods and of molecular dynamics simulation, that represent the two pillars of a reliable and comprehensive description of RTIL systems undergoing transformations in their chemical identity and/or electronic structure.

## 2 Computational method

Computations were carried out by *ab initio* molecular dynamics in the CPMD implementation.[19]

 

The system is represented as a set of $N_a$ ionic cores, and $N$ valence electrons. The ionic configuration is specified by classical position vectors $\{\mathbf{R}_I, I = 1,\ldots, N_a\}$, while the electronic configuration is described by two continuous functions $\rho_+(\mathbf{r})$ and $\rho_-(\mathbf{r})$ giving the density of spin-up and spin-down electrons, respectively, at each position in space.

Following the standard Kohn–Sham (KS) formalism,[12] the spin-up and spin-down electron densities are expressed as the sum of contributions from single electron orbitals:

$$\rho_+(\mathbf{r}) = \sum_{i=1}^{N_+} |\psi_i(\mathbf{r})|^2 \tag{1}$$

$$\rho_-(\mathbf{r}) = \sum_{i=N_++1}^{N} |\psi_i(\mathbf{r})|^2 \tag{2}$$

where $N_+$ and $N_- = N - N_+$ is the number of spin-up and spin-down electrons, respectively. In what follows, we shall use also the electron density, given by $\rho(\mathbf{r}) = \rho_+(\mathbf{r}) + \rho_-(\mathbf{r})$.

The system potential energy for any given configuration of the ionic cores is obtained by minimising the functional of the electron density:

$$E_{\mathrm{KS}}[\rho|\mathbf{R}_I] = \sum_{i=1}^{N} \left\langle \psi_i \middle| -\frac{1}{2}\nabla^2 \middle| \psi_i \right\rangle + \sum_{i=1}^{N} \left\langle \psi_i \middle| \frac{1}{2}\varphi + \hat{V}_{\mathrm{ps}} \middle| \psi_i \right\rangle + E_{\mathrm{XC}}[\rho] + E_{\mathrm{ions}} \tag{3}$$

under the constraint of ortho-normality for the KS orbitals $\{\psi_i(\mathbf{r}); i = 1,\ldots, N\}$. In eqn (3), $\varphi(\mathbf{r})$ is the electrostatic potential of the electron distribution, the operator $\hat{V}_{\mathrm{ps}}$ represents the electron-ionic cores interaction, $E_{\mathrm{XC}}[\rho]$ is the exchange correlation energy, and $E_{\mathrm{ions}}$ is the classical Coulomb repulsion among the ionic cores. Atomic Hartree units are used throughout the paper.

Only valence electrons are explicitly included in our computation, and electrons interact with the ionic cores *via* norm-conserving, soft pseudopotentials[20] of the non-local, *ab initio* type. The system under consideration is enclosed into a cubic simulation box of side $L$, periodically repeated in space. Consistently with this periodicity, Kohn–Sham orbitals are expanded in plane waves, up to a kinetic energy cutoff of 80 Ry ($E_{\mathrm{cut}} = 320$ Ry for the electron density), which is sufficient to converge the 2s and 2p orbitals of nitrogen and oxygen. The Brillouin zone is sampled at the $\Gamma$ point only. The coefficients $\{c_{\mathbf{G}}^{(i)}\}$ of the plane wave expansion for the KS orbitals, together with the position vector of the ionic cores, represent the dynamical variables in our simulations, evolving in time according to:

$$M_I \ddot{\mathbf{R}}_I = -\frac{\partial E_{\mathrm{KS}}[\rho|\mathbf{R}_I]}{\partial \mathbf{R}_I} \tag{4}$$

$$\mu \ddot{c}_{\mathbf{G}}^{(i)} = -\frac{\partial E_{\mathrm{KS}}[\rho|\mathbf{R}_I]}{\partial \left[c_{\mathbf{G}}^{(i)}\right]^*} + \sum_{j=1}^{N} \Lambda_{ij} c_{\mathbf{G}}^{(j)} \tag{5}$$

where $M_I$ is the mass of the atomic species, and $\mu$ is a parameter whose role is comparable to that of a mass. The $\Lambda_{ij}$ introduced in eqn (5) is a matrix of Lagrange multipliers introduced to ensure ortho-normality of the KS orbitals. Step by step integration of these equations of motion[7] provides a direct view of the dynamical and statistical properties of the system under investigation. Further insight on these equations can be gained from ref. 21.

In our computations we resort to a spin-density formulation, to account for possible spin effects in bond-breaking/bond-forming processes. We used the gradient corrected Perdew–Burke–Ernzerhof (PBE) functional[22] to approximate the exchange–correlation energy. This approximation has known limitations, but still represents the state of the art in the condensed matter computational physics community, and its computational efficiency allows for fairly extensive simulations.

## 3 Thermal stability of prototypical RTIL compounds

The first series of *ab initio* simulations has been devoted to the investigation of thermal stability of RTIL molecules. A fairly large number of experiments has been carried out to systematically explore the thermal stability of different RTIL families, in most cases using thermogravimetric analysis.[15,16,23] The results show that the choice of the anion is the most important factor for thermal stability. RTILs whose anions are a bare halide, for instance, are the least stable, with typical decomposition temperatures of the order of 500–550 K, apparently because of the high affinity of halides for hydrogen atoms in the cation. RTILs whose anions are larger and chemically less reactive typically exhibit enhanced thermal stability. The choice of the cation, of course, is not irrelevant, imidazolium-based RTILs being more stable than pyridinium-based and tetraalkylammonium-based compounds. The stability of common RTILs such as [bmim][$NTf_2$], for instance, extends up to about 700 K. Systems of this kind amply justify the statement found in many papers that RTILs enjoy a relatively high thermal stability.

As already mentioned, *ab initio* approaches have been used in the past to investigate the thermal stability of RTILs, at least as a tool complementing experimental measurements.[15,16] Standard approaches are based on the identification of possible decomposition reactions, and on the determination of transition states, that represent saddle-points on the potential energy surface. Then, simple considerations of chemical thermodynamics allow the computation of decomposition rates.

Methods based on simulation, able to include the effect of thermal motion and of entropy, have been far less used until now, partly because of the difficulty of comparing measured and computed decomposition rates. Experiments, in fact, measure thermal decomposition over macroscopic times, of the order of several minutes. This is far beyond the reach of brute force simulation methods, that, at most, provide suggestions on possible reaction pathways, usually obtained from simulations carried out at unrealistically high temperature. Nevertheless, a number of more sophisticated methods based on molecular dynamics simulations and on free-energy concepts, have been developed over the years.[24,25]

We re-visited both these stages of computational approaches devised to investigate thermal stability by simulation methods. In a first naive exploration, we carried out a series of plain MD simulations on systems of a few RTIL ions, at increasing temperature up to about $T$ = 1000 K. This temperature is significantly higher than the stability limit measured in experiments, but is required (in fact, not even sufficient) to observe bond-breaking events during the short times allowed by simulation, and *ab initio* simulations in particular.

To include at least approximately the effect of a condensed phase, simulations have been carried out for two ion pairs, enclosed in a cubic box of 24 a.u. periodicity. The model, therefore, represents a fairly dense, high-temperature vapour. The computation is still relatively inexpensive, but it includes the effect of temperature, and of the interaction with the environment. Simulations have been carried out for [bmim][Cl], [bmim][$BH_4$], [bmim][$PF_6$], and [bmim][$NTf_2$]. A snapshot of the simulation cell from a high temperature unconstrained simulation of [bmim][$NTf_2$] is shown in Fig. 1.

Trajectories of a few ps have been generated for each system and at a few selected temperatures ($T$ = 300 K, 500 K, 800 K and 1000 K), covering a total of some 40–50 ps for each molecular species. Needless to say, this time scale is very short, but comparable or even longer than that of most other *ab initio* simulations. As could have been expected, analysis of snapshots did not show any change in the bonding,

 

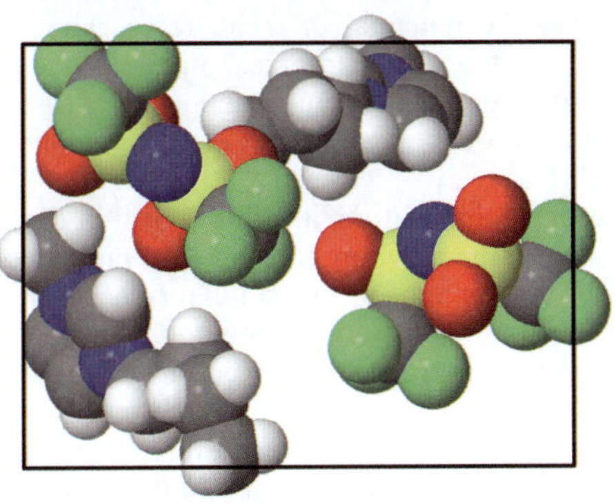

**Fig. 1** Snapshot from an unconstrained *ab initio* molecular dynamics simulation of [bmim][$NTf_2$] at $T = 800$ K.

despite the fast motion and fairly violent molecule-molecule collisions observed at the highest simulation temperatures.

As a further step in this preliminary stage, we investigated by the same brute force approach the effect of the addition of a water molecule on the stability of RTILs at high temperature. Simulations could only provide a useful result in cases where water significantly decreased the thermal stability of the RTIL, bringing the decomposition rate within the short simulation time span, at least at the highest temperatures considered in our simulations. This, however, was not the case, and, in all systems, we did not observe any reaction. Again in all cases, spin did not play any role, despite the spin-unrestricted approach used in our simulations.

As already pointed out, the negative result of our brute force simulations is hardly surprising, taking into account the short simulation times. The simulation, however, provided a test of the thermal stability of RTILs as given by density functional modelling, and produced a library of configurations for the following, more systematic step.

The *ab initio* simulation tools and the computer facilities that are currently available, allow us to sample a large number of different decomposition mechanisms, that in the case of the relatively simple molecules considered in our study, could practically exhaust all possibilities. To illustrate this point, we report here the results of our constrained-MD investigation of thermal stability for [bmim][Cl].

Our choice of [bmim][Cl] as a test case is motivated by its relatively low thermal stability, in line with the general trends concerning halide-RTILs already discussed above. Moreover, the same system has already been studied by quasi-static methods,[15,16] and thus comparison with our finite temperature results allows us to highlight the effect of entropy and thermal motion.

For the sake of simplicity, we probed a number of different reactions taking place along a reaction coordinate $\xi$ corresponding to the distance of a tagged $[Cl]^-$ anion from a C or H atom in the $[bmim]^+$ cation. Starting from a configuration extracted from the plain *ab initio* MD simulations, this distance is progressively reduced in a few discrete (short) steps, until we observe a reaction to take place. More in detail, at each step: (i) the $[Cl]^-$ ion is moved towards the target atom by $\sim$0.2 Å; (ii) the system is equilibrated at *constant* value of the reaction coordinate $\xi$ for about 5 ps, and statistics is accumulated over another 5 ps, during MD runs at constant $T = 300$ K. The constraint of fixed reaction coordinate (*i.e.*, fixed $[Cl]^-$ - target atom distance) is imposed using the method described in ref. 26.

In the case that the reaction coordinate is a single interatomic distance, statistical mechanics arguments[27] show that the average force needed to maintain the constrain (mean force) is minus the derivative of the system free energy as a function of the reaction coordinate. The integral of the mean force along the reaction coordinate provides the free energy variations experienced by the system during the chemical reaction, also known as the potential of mean force. The simulation, therefore, allows us to determine the free energy profile, to identify the transition point, and thus to estimate temperature dependent reaction rates. Changing temperature and/or density allows us to quantify the effect of entropy, and of interactions with the environment.

In what follows, we refer to the atom labelling shown in Fig. 2. In the first of our simulations we considered the reaction coordinate represented by the distance of a $[Cl]^-$ ion (one of the two present in our simulation cell) and the methyl-carbon $C_1$. As already mentioned, our constrained simulations started from a configuration extracted from the unconstrained MD at $T = 300$ K. In the selected configuration, the $[Cl]^-$ -$C_1$ separation turned out to be $\xi = 5.2$ Å, distance at which the mean force averaged over 5 ps virtually vanishes. The mean force as a function of the reaction coordinate $\xi$, computed at $T = 300$ K according to the protocol described above, is shown in Fig. 3. Analysis of snapshots shows that for $\xi \leq 2.35$ Å the system has evolved into the product configuration, with a neutral $CH_3Cl$ molecule separated from the former imidazolium moiety to which the $C_1$ methyl carbon was originally attached.

Integration of the mean force with respect to the reaction coordinate gives us the potential of mean force $\Delta F(\xi)$, representing, in this case the Helmholtz free energy profile along the reaction coordinate (see Fig. 4). From the graph, we obtain the value of the reaction barrier ($\Delta F = 135$ kJ mol$^{-1}$), and the $C_1$-Cl distance at the transition point ($\xi = 2.5$ Å). The product molecules are ~22 kcal mol$^{-1}$ less stable than the original [bmim][Cl] compound. These results are in fair agreement with those of previous studies. Our value for $\Delta F$, in particular, is slightly higher than the one obtained in refs. 15 and 16. Because of the finite temperature of the computational experiment, it is not possible to identify a single geometry to be associated to characteristic points of the reaction, such as the transition point or the product species. Nevertheless, it is possible to extract from the simulation trajectory a wide range of thermally averaged structural parameters that display characteristic changes along the reaction coordinate.

Quenching the atomic structure for every value of the reaction coordinate allows us to approach the quasi-static conditions of previous simulations.[15,16] In this way, at the same time, we can compare with data obtained by other researchers, and we can estimate the temperature dependence of the barrier. The results of this stage of our investigation show that the reaction barrier decreases by roughly 8% as the temperature decreases from $T = 300$ K to $T = 0$ K. The $T = 0$ K estimate turns out to be in fair agreement with the data of refs. 15 and 16 for the same decomposition reaction. As a second result, we find that our estimate for the thermal stability range of [bmim][Cl] is similar but somewhat higher than the one obtained in refs. 15 and 16, and in fair agreement with experimental data. Given the large gap in the time scale probed by simulation and by experiments, this is indeed a remarkable result. The

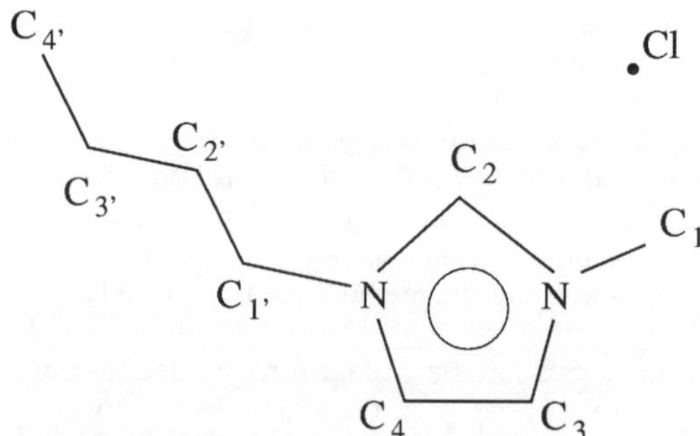


**Fig. 2** Labelling of atoms in the simulation of thermal decomposition of [bmim][Cl].

 

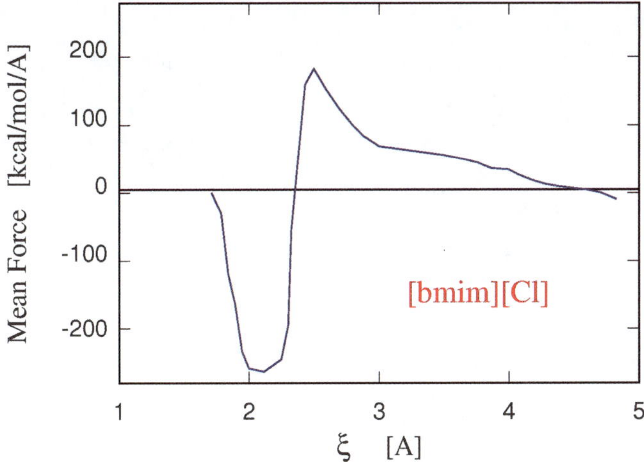


**Fig. 3** Mean force as a function of the reaction coordinate $\xi$, corresponding to the $[Cl]^-$ -$C_1$ separation in [bmim][Cl], from constrained MD simulations at $T = 300$ K.

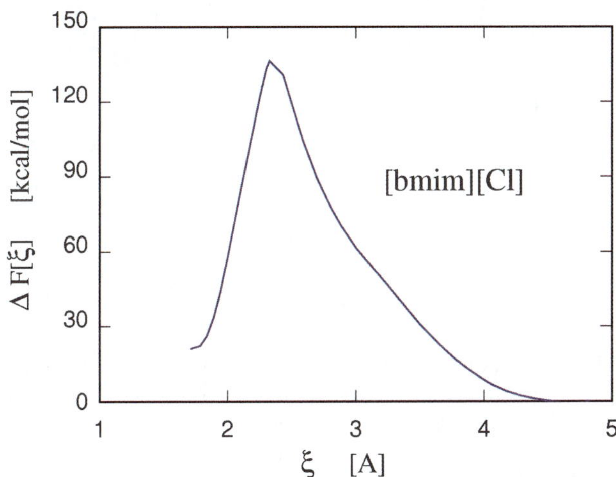


**Fig. 4** Potential of mean force $\Delta F(\xi)$ for the same decomposition reaction of [bmim][Cl] whose mean force is shown in Fig. 3. $\Delta F(\xi)$ is obtained by numerical integration of the mean force with respect to the reaction coordinate.

quantitative discrepancy of the estimate obtained by different methods at equal conditions ($T = 0$ K) may depend on the slightly different exchange–correlation approximation, by the different conditions of our computations (periodic boundary conditions in our case, gas phase configuration and localised basis functions in previous studies), and by the different methods of geometry optimisation.

The non-negligible dependence of the reaction parameters on temperature implies that a fully quantitative description of thermal decomposition would require a few constrained-MD scans of the reaction coordinate, carried out at different temperatures up to the estimated limit of the [bmim][Cl] stability range.

For the quenched simulation, it is possible to plot snapshots illustrating the system evolution along the reaction coordinate, as shown in Fig. 5.

For a more systematic view of possible decomposition reactions of [bmim][Cl], we repeated the same sequence of constrained-MD simulations described above, using,

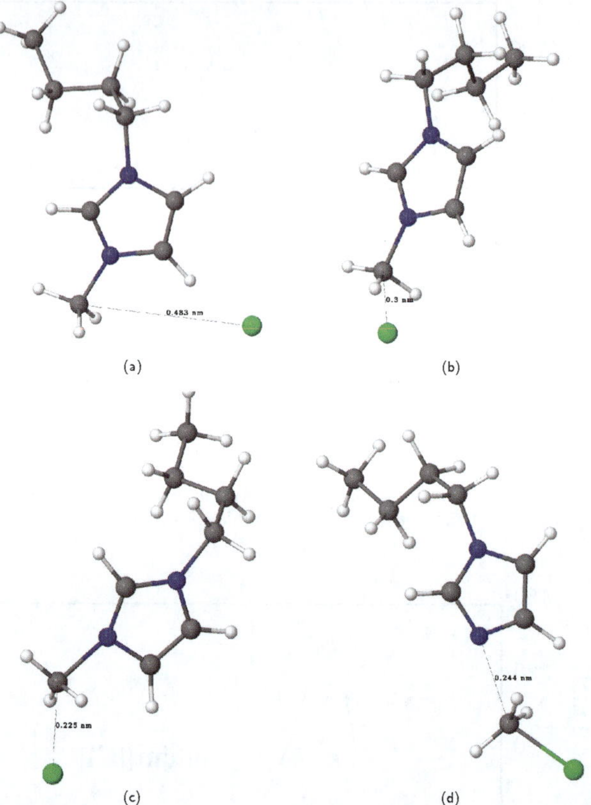


**Fig. 5** Snapshots illustrating the progression (at $T = 0$ K) of [bmim][Cl] along the reaction coordinate discussed in the text. Geometries have been determined by minimising the energy for each value of the reaction coordinate $\xi$.

in turn, $H_2$ (hydrogen bonded to $C_2$), $C_{1'}$, $C_{2'}$, $C_{3'}$ and $C_{4'}$ as target atoms for the approaching [$Cl^-$]. In all cases in which a comparison was possible, our results agree fairly well with those of refs. 15 and 16. Quantitative differences are due to the different approach used in our (room temperature simulation) and in previous (quasi-static search) computational studies.

In the case of [$Cl^-$] attacking $C_{3'}$ we observed a completely unexpected reaction taking place, involving one of the hydrogens originally bonded to $C_{3'}$, ending up being covalently bonded to the $C_4$ carbon in the other imidazolium molecule present in our simulation box. The reaction takes place through the migration of $H_{3'}$ over a fairly long distance, and in our DFT simulation this is facilitated by the interaction with the alkane tails of [bmim]$^+$. This unlikely mechanism might indeed be an artifact of the DFT approximation underlying our simulations, similar to other faulty results seen in the DFT analysis of chemical reactions.[28] In such a case, the observation of the spurious reaction could provide one additional piece of evidence documenting the limitations of present day *ab initio* methods, and would emphasise, once again, that even the results of *ab initio* computations need to be judged with some care.

## 4 *Ab initio* determination of the electrochemical stability window of RTILs

Electrochemistry is one of the most likely applications of ILs, and the one in which ionic and electronic aspects of these molecules are more closely intertwined.

 

We focus on one specific problem, *i.e.*, the determination of the electrochemical window by *ab initio* simulations based on density functional theory.

By definition, the electrochemical stability window is the voltage span over which one molecule is neither oxidised nor reduced. In other words, molecules of the compound under consideration do not take or give away an electron in/from their valence configuration when in contact with an electrode at a potential $\Delta V$ (within the electrochemical window) with respect to a reference electrode, often represented by the normal hydrogen electrode. The role of the reference electrode is less crucial in the case that only the width of the stability window is of interest, irrespective of the anodic and cathodic limits. The result, however, will depend on the choice of the electrode (working electrode) in contact with the molecules under investigation, for reasons that will be discussed in detail below.

Popular electrochemistry textbooks discuss this problem in terms of single-electron energy diagrams, focusing on the highest occupied (HOMO) and lowest unoccupied (LUMO) molecular orbital. Seen from this point of view, the problem is virtually intractable within DFT, at least using standard approaches, since Kohn–Sham eigenvalues are not single electron energies,[12] and it is not clear what replaces them in that role.

The aim of our discussion is to show that a big deal of information on electrochemical processes can still be obtained from DFT computations. Our argument is that it is possible to identify conditions for the charge transfer across an electrochemical interface, although, in principle, these would involve the exact single electron energies for the system of interest, which are not available in DFT.

Let us first discuss oxidation and reduction in terms of the genuine (*i.e.*, not DFT) density of states (DOS) at the metal/molecule interface. Fig. 6 schematically represents the relevant portion of the density of states for (a) a metal electrode and (b) for a (closed shell) molecule, supposed to be isolated from each other. Whenever they are brought into contact, the two sets of levels align, equalising the chemical potential of electrons over the entire system. At large distances, this is achieved by slight distortions (polarisation) of the electron distribution of the molecule and of the metal, mediated by multipolar electrostatic forces. Levels localised on the molecule, and corresponding to its HOMO and LUMO, are symmetrically placed below and above the system Fermi energy. At short distances, comparable to those of chemisorbed molecules, the alignment is due to chemical interactions, *i.e.*, to overlapping orbitals. At this stage, the HOMO and LUMO do not need to be symmetrically placed around the metal Fermi energy‡ (See Fig. 6(c)).

Strictly speaking, no charge transfer can take place for the system of Fig. 6(c), since metal electrons do not have enough energy to sit on the molecular LUMO, and the energy of molecular electrons is below the metal Fermi energy. However, if we apply a voltage $\Delta V$ to the electrode, shifting the metal chemical potential, charge will flow at the voltage $\Delta V$ such that $E_F$ is raised above the LUMO (the molecule is reduced) or $E_F$ is lowered below the HOMO (the molecule is oxidised). Thus, if one could rigidly shift the metal energy levels by the application of a bias potential, leaving unchanged the energy of levels localised on the molecule, the electrochemical stability window would be simply the energy difference between the HOMO and LUMO levels. Needless to say, because of the interaction of the metal and molecular moieties, this is never the case. Nevertheless, seen at this atomistic/molecular detail, the problem is reduced to the computation of the single electron energies as a function of the external bias. An additional difficulty that might arise in some cases is that sometimes it might not be easy to attribute electron levels to one or the other side of the interface, because of the mixing of levels due to the interaction. With relatively un-reacting species such as RTILs, this is a problem of limited impact.

‡ At any non-zero temperature, the chemical potential and the Fermi energy, in general, do not coincide. For normal metals, however, the difference is small, and we shall use the two terms as equivalent.

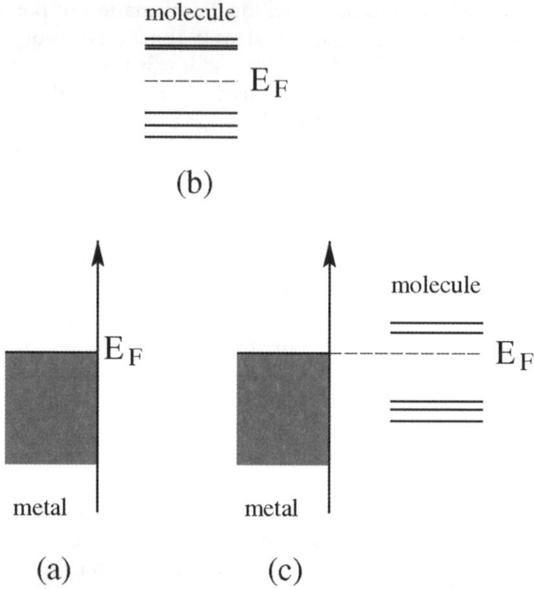


**Fig. 6** Schematic plot of the density of states for a molecule approaching a metal surface. (a) Isolated metal; (b) isolated molecule; (c) molecule adsorbed on the metal surface.

Up to now we assumed that the density of states schematically shown in Fig. 6 reflects the energy distribution of the true single-electron levels, that are in fact quasi-particle states.[29] In practice, when using DFT, the only density of states that is available is the one computed for Kohn–Sham energies. Even though these do not correspond to single electron energies, the condition for transferring charge across the interface is similar: charge will flow, and thus adsorbed molecules will be oxidised or reduced whenever the application of a bias on the metal electrode will cause occupied and empty Kohn–Sham states on the metal and on the molecules to cross each other. This condition is a consequence of a relation known as Janak's theorem,[30] stating that:

$$\frac{\partial E_{\mathrm{KS}}}{\partial f_i} = -\varepsilon_i \tag{6}$$

which is strictly valid in DFT even in the case of approximate XC functionals. In eqn (6), $0 \leq f_i \leq 1$ is the occupation number for the Kohn–Sham level $i$, and $\varepsilon_i$ is the corresponding Kohn–Sham eigenvalue. If we apply this relation to the charge transfer between two distinct Kohn–Sham levels ($i$ and $j$), the first occupied and the second empty, being localised one on the metal and the other on an adsorbed molecule, we obtain:

$$\frac{\partial E}{\partial f_i} - \frac{\partial E}{\partial f_j} = -(\varepsilon_i - \varepsilon_j) \tag{7}$$

implying that charge will be transferred from $i$ to $j$ whenever $\varepsilon_i > \varepsilon_j$.

The true difficulty in interpreting charge transfer reactions in terms of DFT eigenvalues, therefore, is that the dependence of Kohn–Sham energies on the charge state and on any other external perturbation is much stronger then that of genuine single-electron energies (*i.e.*, quasi-particle energies), and thus the simple analysis of ground state eigenvalues gives results drastically different from the available experimental values. To use Kohn–Sham energies to predict electrochemical properties, we need to investigate the dependence of Kohn–Sham energies on the bias applied to the metal, and on the charge added to or taken away from the molecule.

To illustrate the approach that could be used to predict the electrochemical stability window of RTILs, we resort to a very idealised model, in which the metal electrode is replaced by a droplet of $N_m$ valence electrons confined by the electrostatic potential of a spherical distribution of positive charge, representing the ionic cores of the metal. In other words, we resort to a finite jellium model for the electrode, and we investigate the electronic properties of the combined system obtained upon adsorbing molecules on this very idealised electrode.

The computations presented below refer to a jellium sphere made of $N_m = 40$ electrons, interacting with the positive charge through the potential:

$$V_J(\mathbf{r}) = \begin{cases} -\dfrac{1}{2}\dfrac{N_m}{R_b}\left(3 - \dfrac{r^2}{R_b^2}\right) & r < R_b \\ -\dfrac{N_m}{\mathrm{r}} & r \geq R_b \end{cases} \tag{8}$$

In the equation above, $R_b$ is the radius of the positive charge distribution, given by:

$$R_b = \left(\frac{3N_m}{4\pi\rho_+}\right)^{1/3} \tag{9}$$

and $\rho_+$ is the density of the positive charge within the sphere. In our computations $\rho_+$ is taken equal to the valence electron density of aluminium.

As expected, the KS density of states of the model electrode is highly discretised (see Fig. 7(a)), and thus very different from the continuous density of states of an extended metal. Discretisation is due to the finite (and small) size of the metal droplet, and to the bunching of states into degenerate shells of equal angular momentum.[31] This observation alone will prevent a quantitative comparison of our results with experimental data. In fact, many other considerations prevent a quantitative comparison of computed and measured values for the electrochemical stability window. First of all, experimental measurements of the electrochemical window of RTILs are also affected by sizeable uncertainties, and the values measured by different groups on systems that are nominally the same may differ by as much as ~1 V. Moreover, measurements are carried out in the condensed phase, while our computations are performed on a single adsorbed molecule. Finally, the choice of the working electrode is also important, and this is not accounted for by our simplified (jellium) model.

Nevertheless, our study shows that meaningful results on electrochemical properties can be obtained within standard DFT. Moreover, trends and relative values for the electrochemical properties of RTIL are expected to be far less sensitive than absolute values to our approximations. More quantitative results could be obtained with larger conducting spheres and with a more detailed description of the metal side. Larger spheres with up to a few hundred electrons would require a large but still manageable computational effort. Even the small sphere considered in the present computation, however, provide a qualitatively correct model of a metal electrode, because of their fairly low HOMO–LUMO gap, and high density of states at the Fermi energy.

A few prototypical RTIL molecules ([bmim][Cl], [bmim][$BH_4$], [bmim][$PF_6$], [bmim][$NTf_2$]) in contact with the jellium-like electrode have been simulated by *ab initio* MD. To provide a comparison, a water molecule adsorbed on the 40-electron jellium sphere has been considered as well. Fairly long simulations of the combined jellium-RTIL and jellium-$H_2O$ samples enclosed in a cubic cell of 15 Å side confirm that the model is fully stable at room temperature. Progressive annealing to very low

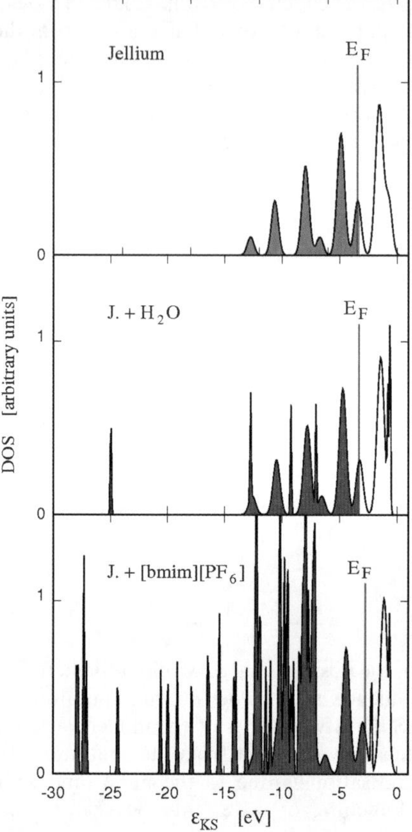


**Fig. 7** Density of Kohn–Sham eigenvalues computed for: (a) a 40-electron jellium sphere at the valence electron density of aluminium; (b) a water molecule adsorbed on the same 40-electron jellium sphere; (c) a [bmim][Cl] molecule (ion pair) adsorbed on the 40-electron jellium sphere. All levels have been convoluted with a normalised Gaussian of 0.05 eV width. Degenerate levels have been displaced by 0.05 eV to display multiplicities. Kohn–Sham eigenvalues are computed by CPMD, using a periodically repeated simulation cell and a plane wave expansion for the orbitals (see text).

temperature (a few K) provide realistic ground state structures for the adsorbed molecule adsorbed on jellium. The electronic properties discussed below have been investigated for these low temperature/low energy structures. KS eigenvalues have been computed with the same periodic geometry used in the simulation. By removing periodic boundary conditions in a few test computations, we verified that periodicity affects the absolute value of the eigenvalues but not their relative position or energy separation. In other words, periodicity introduces a rigid shift of all eigenvalues that does not affect our discussion.

The combined DOS for a water and for a [bmim][$PF_6$] molecule adsorbed on jellium are shown in Fig. 7(b) and (c), respectively. The alignment of levels equalising the Fermi energy over the entire system automatically results from the minimisation of the total energy, that implies a constant chemical potential over the entire system.

The case of water is somewhat simpler, and is used here to illustrate the general picture. Comparison of the isolated water eigenvalues (see Table 1) with the results displayed in Fig. 7 confirms that upon adsorption, the energy of electron levels localised on water shifts nearly rigidly, maintaining on water a gap of more than 6 eV (for an isolated water molecule we compute $E_g = 6.45$ eV). If the metal Fermi

 

**Table 1** Kohn–Sham eigenvalues for an isolated water molecule computed in the periodic super-cell geometry described in the text

| $\varepsilon_i$ | −25.21 | −12.93 | −9.29 | −7.19 (HOMO) | −0.74 (LUMO) |
|---|---|---|---|---|---|

level could be rigidly shifted with respect to those of water by the application of a suitable bias, then the electrochemical stability window would be $E_g = 6.45$ eV, a value that appears greatly overestimated with respect to the experimental measurements. The electrochemical stability window of water is often quoted as 1.27 eV. The evaluation of the dependence of water levels upon the application of the bias greatly reduces the discrepancy.

To simulate the bias, we modify the potential $V$, shifting it by a constant value $\delta V$ within the spherical volume occupied by the positive charge:

$$\tilde{V}_J(\mathbf{r}) = \begin{cases} V_J(\mathbf{r}) + \delta V & r < R_b \\ V_J(\mathbf{r}) & r \geq R_b \end{cases} \tag{10}$$

The bias we apply, therefore, corresponds to a neutral perturbation. The relation with a perturbation that changes the system charge is discussed below. We anticipate that both give rise to the same effect in terms of the electrochemical window.

First of all, we compute the variation of Kohn–Sham eigenvalues for the isolated jellium sphere. The effect of the perturbation is not a rigid shift of the jellium level by the same quantity $\delta V$, *i.e.*, $\delta\varepsilon_i \neq \delta V$. Low energy levels are more affected than high energy ones, because of their different localisation within the sphere of radius $R_b$. To quantify this effect, we computed the derivative of the eigenvalues with respect to the bias. The results are reported in Table 2. It is apparent that for all levels, the shift is much less than the applied perturbation $\delta V$. This is due to the finite volume of the metal sample. In the case of an extended metal, shifting the potential on the entire metal volume would shift the energy levels by the same amount.

More interesting is the result of the same computation for the combined jellium-water system. First of all, we observe that the derivative of eigenvalues for states localised on the jellium sphere (recognised by their energy and by their degeneracy) is nearly the same as in the isolated jellium case. Moreover, we observe that KS states localised on water do change upon the application of the external bias, and their derivative with respect to $\delta V$ tends to be negative. Both observations provide a way to discriminate levels belonging to the metal or to the molecule, even without a detailed analysis of their wave functions. Dubious results are certainly expected to arise for other molecules, especially in the case of surface reactions, and then a more detailed analysis of orbital localisation will be required.

By changing $\delta V$, we verified that the dependence of eigenvalues on the metal and on the molecule is remarkably linear in $\delta V$. Moreover, and more importantly, we carried out computations for $H_2O$/jellium without periodic boundary conditions, and with a net (positive or negative) charge on the system, that changes $E_F$. Once the discretisation of the metal DOS is taken into account, the derivative with respect to $E_F$ of the KS energies for states localised on the molecule turns out to be equal to the one already computed for the neutral perturbation.

The $\delta V$-dependence of the eigenvalue for states localised on the adsorbed molecule is the crucial ingredient needed to interpret electrochemical measurements, going beyond the simple analysis of the ground state Kohn–Sham density of states. To express this dependence, we compute the derivative:

$$\frac{\partial \varepsilon_i}{\partial E_F} = \frac{\partial \varepsilon_i}{\partial \delta V} \frac{\partial \delta V}{\partial E_F} \tag{11}$$

**Table 2** Derivative of the Kohn–Sham energies of a 40-electron jellium sphere with respect to the amplitude of the perturbation $\delta V$ (see text)

| | | | | | | | | |
|---|---|---|---|---|---|---|---|---|
| $\varepsilon_i$ (eV) | −12.75 | −10.62 | −10.62 | −10.62 | −7.96 | −7.96 | −7.96 | −7.96 |
| $f_i$ | 1.0 | 1.0 | 1.0 | 1.0 | 1.0 | 1.0 | 1.0 | 1.0 |
| $\frac{\partial \varepsilon_i}{\partial \delta V}$ | 0.18 | 0.23 | 0.23 | 0.23 | 0.25 | 0.25 | 0.25 | 0.25 |
| $\varepsilon_i$ (eV) | −7.96 | −6.71 | −4.89 | −4.89 | −4.89 | −4.89 | −4.89 | −4.89 |
| $f_i$ | 1.0 | 1.0 | 1.0 | 1.0 | 1.0 | 1.0 | 1.0 | 1.0 |
| $\frac{\partial \varepsilon_i}{\partial \delta V}$ | 0.25 | 0.13 | 0.23 | 0.23 | 0.23 | 0.23 | 0.23 | 0.23 |
| $\varepsilon_i$ (eV) | −4.89 | −3.43 | −3.43 | −3.43 | −1.58 | −1.58 | −1.58 | −1.58 |
| $f_i$ | 1.0 | 1.0 | 1.0 | 1.0 | 0.0 | 0.0 | 0.0 | 0.0 |
| $\frac{\partial \varepsilon_i}{\partial \delta V}$ | 0.23 | 0.05 | 0.05 | 0.05 | 0.17 | 0.17 | 0.17 | 0.18 |

where we identify $E_F$ with the energy of the highest occupied level localised on the electrode. The results for water and for the RTIL molecules considered in our study are reported in Table 3. In the case of water, these results explain why the electrochemical stability is much less that which would be suggested by the Kohn–Sham density of states for the ground state configuration. If we move $E_F$ upwards, water levels move downwards. If we move $E_F$ downwards, water levels move upwards. In both cases, the drift of the levels on water is faster than that of the metal levels close to the Fermi energy.

In the case of RTIL molecules adsorbed on the jellium sphere, the derivative with respect to $\delta V$ of eigenvalues for molecular states can be either positive or negative, and its absolute value tends to be less than in the case of water, probably because of a somewhat weaker chemical interaction of RTIL with the metal surface.

Once the equivalence of neutral and charged perturbations has been established, simple arithmetics shows that the electrochemical stability window (*ESW*) is given by:

$$ESW = \frac{\varepsilon(\mathrm{L}) - E_F}{\left(1 - \frac{\partial \varepsilon(\mathrm{L})}{\partial E_F}\right)} + \frac{E_F - \varepsilon(\mathrm{H})}{\left(1 - \frac{\partial \varepsilon(\mathrm{H})}{\partial E_F}\right)} \tag{12}$$

**Table 3** Derivative of selected Kohn–Sham energies with respect to the variation of $E_F$ due to the perturbation $\delta V$ described in the text, computed for each of the four molecules listed on top, adsorbed on a 40-electron jellium sphere. The electrochemical stability window (ESW) computed by the method discussed in the text is also given. $E_F$ is identified with the KS energy of the highest occupied level localised on jellium. $\varepsilon_H$(mol) and $\varepsilon_L$(mol) are the KS energies of the HOMO and LUMO states localised on the adsorbed molecule, respectively.

| Jellium+ | $H_2O$ | [bmim] [Cl] | [bmim] [$BH_4$] | [bmim] [$PF_6$] | [bmim] [$NTf_2$] |
|---|---|---|---|---|---|
| $E_F$ (eV) | −3.20 | −2.78 | −3.57 | −2.76 | −2.97 |
| $\varepsilon_H$(mol) (eV) | −7.02 | −4.62 | −3.88 | −7.06 | −4.52 |
| $\partial\varepsilon_H(\mathrm{mol})/\partial E_F$ | −1.02 | 0.28 | 0.21 | −0.82 | 0.51 |
| $\varepsilon_L$(mol) (eV) | −0.70 | −2.56 | −1.99 | −2.20 | −2.46 |
| $\partial\varepsilon_L(\mathrm{mol})/\partial E_F$ | −1.52 | 0.45 | 0.22 | 0.50 | 0.16 |
| ESW (V) | 2.88 | 2.96 | 2.42 | 3.48 | 3.77 |

 

where $\varepsilon(H)$ and $\varepsilon(L)$ are the HOMO and LUMO KS eigenvalues. The constant term of 1 in the denominator accounts for the fact that $\partial E_F/\partial E_F = 1$. Apparently, a negative value for $(\partial\varepsilon_i/\partial E_F)$ will tend to reduce $ESW$, while positive values of $(\partial\varepsilon_i/\partial E_F)$ enhance $ESW$.

Replacing the numerical values of Table 3 in eqn (12), in the case of water we obtain $ESW = 2.88$ eV. This is still higher than the $ESW = 1.27$ eV of experiments, but, first of all, our result refers to a metal surface that does not exist in reality, and, second, the DFT-PBE approximation we are using is certainly not exact.

With the support of this result for water, we turn to the RTIL molecules. Using again the values of Table 3 in eqn (12), we compute $ESW = 2.96$ V for [bmim][Cl], $ESW = 2.42$ V for [bmim][$BH_4$], $ESW = 3.48$ V for [bmim][$PF_6$], and $ESW = 3.77$ V for [bmim][$NTf_2$]. Quantitative comparison with experimental data is prevented by the many reason listed above, but we think that the values we obtain represent the first prediction of the electrochemical stability window for RTIL molecules based on DFT, and going beyond the simple (and inadequate) analysis of Kohn–Sham eigenvalues for the ground state molecules. The most uncertain value is the one for [bmim][Cl], because the mixing of molecular and jellium states is higher than in the other cases, apparently because of the high reactivity of $[Cl]^-$.

## 5 Summary and conclusions

Our contribution to the present Faraday discussion is devoted to selected aspects of the *ab initio* simulation of RTILs. We discussed, in particular, the application of *ab initio* molecular dynamics to the investigation of thermal decomposition and to the determination of the electrochemical stability window of prototypical RTIL compounds. These two phenomena involve changes in the configuration of covalent bonds (thermal decomposition), or in the electronic structure (electrochemical oxidation or reduction) of the molecules under investigation. In both cases, the ability of *ab initio* methods to simultaneously describe structural and electronic properties represents a decisive advantage with respect to other empirical or semi-empirical methods.

We carried out an extensive simulation study of thermal decomposition mechanisms in [bmim][Cl], [bmim][$BH_4$], [bmim][$PF_6$], and [bmim][$NTf_2$]. For reasons of length, in the present paper we report only the results for [bmim][Cl]. Our investigation included an exploration of RTIL molecular dynamics up to high temperature, and a detailed analysis of decomposition mechanisms based on a more refined statistical mechanics approach.

Not surprisingly, the brute force approach, covering a time span of pico-seconds, did not produce any decomposition reaction. This negative result, however, confirms that over short times the DFT model is stable, and does not show any spurious feature up to at least 1000 K. Simulations also show that spin does not play any role in common RTIL compounds up to high temperature.

A fairly large number of constrained molecular dynamics simulations have been carried out to probe the potential of mean force along a reaction coordinate corresponding, in each case, to the distance between a pair of atoms on the cation and on the anion. Our results for [bmim][Cl] are in fair agreement with those of previous computations, based on a quasi-static approach. Differences are also apparent, and could be attributed to the temperature dependence of the activation free energy. This effect turns out to be non-negligible already at room temperature. Overall, the computational picture turns out to be consistent with experiments, and could be used to estimate the thermal stability of new compounds, especially in cases such that poisonous species (HF, $H_2S$, ...) could be expected to form.

In one case, an unexpected thermal decomposition mechanism has been observed for [bmim][Cl]. We suspect that this is an artifact of the DFT-PBE approximation, pointing to limitations of the density functional scheme underlying our simulations. This observation would represent a second useful aspect of our study, highlighting a clear case in which the computational methods used by our community could and should be improved.

The section on the determination of the electrochemical stability window of RTILs is focused on the discussion of basic problems of the DFT approach in describing charge transfer processes across an electrochemical interface, and describes the development and validation of a simple method able to overcome these difficulties. Our discussion starts from the well known fact that Kohn–Sham eigenvalues are not single-electron energies. However, their dependence on neutral and charged perturbations is predictable, and can be used to investigate charge transfer processes such as those involved in the oxidation and reduction of molecules adsorbed on a metal electrode. Following these lines, we have explored the reason why the electrochemical stability window is not equal to the energy gap in the ground state molecule, and we have shown how the energy correction can be estimated. The results are not and cannot be in quantitative agreement with experiments, first of all because we used a very idealised model for the metal electrode. Moreover, the electronic structure approach underlying our computations is approximate. Nevertheless, computations using more detailed models and possibly better DFT approximations could indeed provide a tool to model and predict electrochemical properties.

Both in the case of thermal decomposition, and in the case of the electrochemical stability window, the addition of simulation to the basic density functional approach has added a new and important ingredient to our ability to explore molecular systems of the complexity of RTIL compounds.

## Acknowledgements

Computations have been carried out on the HPC facilities of Queen's University Belfast, and of CINaM-CNRS at Aix-Marseille Université.

## References


1 (*a*) T. Welton, *Chem. Rev.*, 1999, **99**, 2071; (*b*) A. Blanchard, D. Hancu, E. J. Beckmann and J. Brennecke, *Nature*, 1999, **399**, 28.

2 (*a*) P. Wasserscheid and T. Welton, *Ionic liquids in Synthesis*, Wiley-VCH, Weinheim, Germany, 2002; (*b*) *Ionic liquids: Industrial applications for green chemistry*, ed. R. D. Rogers and K. R. Seddon, American Chemical Society, Washington, DC, 2002.

3 (*a*) M. F. Fox and M. Priest, *Proc. IMechE*, 2008, **222J**, 291; (*b*) C.-F. Ye, W. M. Liu, Y. X. Chen and L. G. Yu, *Chem. Commun.*, 2001, 2244; (*c*) H. Z. Wang, Q. M. Lu, C.-F. Ye, W. M. Liu and Z. J. Cui, *Wear*, 2004, **256**, 44; (*d*) Z.-G. Mu, W. M. Liu, S. X. Zhang and F. Zhou, *Chem. Lett.*, 2004, **33**, 524.

4 A. Klamt and G. Schüürmann, *J. Chem. Soc., Perkin Trans. 2*, 1993, 799.

5 A. Klamt, V. Jonas, T. Bürger and J. J. W. Lohrenz, *J. Phys. Chem. A*, 1998, **102**, 5074.

6 (*a*) R. Ahlrichs, M. Bär, M. Häser, H. Horn and C. Kölmel, *Chem. Phys. Lett.*, 1988, **162**, 165; (*b*) M. J. Frisch, G. W. Trucks, H. B. Schlegel, G. E. Scuseria and J. A. Pople, 2008, *Gaussian 03*, Gaussian Inc.: Wallingford, CT.

7 D. Frenkel and B. Smit, *Understanding Molecular Simulation*, Academic, San Diego, 1996.

8 (*a*) C. G. Hanke, S. L. Price and R. M. Lynden-Bell, *Mol. Phys.*, 2001, **99**, 801–809; (*b*) J. K. Shah, J. F. Brennecke and E. J. Maginn, *Green Chem.*, 2002, **4**, 112.

9 (*a*) T. Y. Yan, Y. T. Wang and C. Knox, *J. Phys. Chem. B*, 2010, **114**, 6905; (*b*) T. Y. Yan, C. J. Burnham, M. G. D. Popolo and G. A. Voth, *J. Phys. Chem. B*, 2004, **108**, 11877.

10 (*a*) P. A. Hunt and I. R. Gould, *J. Phys. Chem. A*, 2006, **110**, 2269; (*b*) M. G. D. Popolo, C. Pinilla and P. Ballone, *J. Chem. Phys.*, 2007, **126**, 144705.

11 (*a*) E. I. Izgorodina, *Phys. Chem. Chem. Phys.*, 2011, **13**, 4189; (*b*) S. Zahn, J. Thar and B. Kirchner, *J. Chem. Phys.*, 2010, **132**, 124506; (*c*) H. Valencia, M. Koyama, S. Tanaka and H. Matsumoto, *J. Chem. Phys.*, 2009, **131**, 244705; (*d*) S. Bhargava and S. Balasubramanian, *J. Phys. Chem. B*, 2008, **112**, 7566; (*e*) M. Bühl, A. Chaumont, R. Schurhammer and G. Wipff, *J. Phys. Chem. B*, 2005, **109**, 18591; (*f*) G. S. A. Bagno and F. D'amico, *ChemPhysChem*, 2007, **8**, 873; (*g*) B. Kirchner, A. P. Seitsonen and J. Hutter, *J. Phys. Chem. B*, 2006, **110**, 11475; (*h*) M. G. D. Popolo, R. M. Lynden-Bell and J. Kohanoff, *J. Phys. Chem. B*, 2005, **109**, 5895.


 

12 R. M. Martin, *Electronic structure: basic theory and practical methods*, Cambridge University Press, Cambridge, 2004.
13 M. Galinski, A. Lewandowski and I. Stepniak, *Electrochim. Acta*, 2007, **51**, 5567.
14 (*a*) V. I. Parvulescu and C. Hardacre, *Chem. Rev.*, 2007, **107**, 2615; (*b*) M. J. Earle and K. R. Seddon, *Pure Appl. Chem.*, 2000, **72**, 1391; (*c*) H. Olivier-Bourbigou and L. Magna, *J. Mol. Catal. A*, 2002, **182–183**, 419; (*d*) L. Gonsalvi, I. W. C. E. Arends and R. A. Sheldon, *Chem. Commun.*, 2002, 202.
15 Y. Hao, J. Peng, S. Hu, J. Li and M. Zhai, *Thermochim. Acta*, 2010, **501**, 78.
16 M. C. Kroon, W. Buijs, C. J. Peters and G.-J. Witkamp, *Thermochim. Acta*, 2007, **465**, 40.
17 G.-C. Tian, X.-J. Zhou, J. Li and Y.-X. Hua, *Trans. Nonferrous Met. Soc. China*, 2009, **19**, 1639.
18 J. Cheng, M. Sulpizi and M. Sprik, *J. Chem. Phys.*, 2009, **131**, 154504.
19 D. Marx and J. Hutter, in *Modern Methods and Algorithms of Quantum Chemistry*, ed. J. Grotendorst, John von Neumann Institute for Computing, Jülich, 2nd edn, 2000, vol. 3, pp. 329–477.
20 N. Troullier and J. L. Martins, *Phys. Rev. B: Condens. Matter*, 1991, **43**, 1993.
21 G. Galli and A. Pasquarello, in *Computer Simulation in Chemical Physics*, ed. M. P. Allen and D. J. Tildesley, Kluwer, Dordrecht, 1993, p. 159.
22 J. P. Perdew, K. Burke and M. Ernzerhof, *Phys. Rev. Lett.*, 1996, **77**, 3865.
23 (*a*) R. Liang, M. Yang and X.-P. Xuan, *Catalysis, Kinetics and Reactors*, 2010, **18**, 736; (*b*) N. M. Yunus, M. I. A. Mutalib, Z. Man, M. A. Bustam and T. Murugesan, *J. Chem. Thermodyn.*, 2010, **42**, 491; (*c*) C. P. Fredlake, J. M. Crosthwaite, D. G. Hert, S. N. V. K. Aki and J. F. Brennecke, *J. Chem. Eng. Data*, 2004, **49**, 954; (*d*) M. Kosmulski, J. Gustafsson and J. B. Rosenholm, *Thermochim. Acta*, 2004, **412**, 47; (*e*) H. L. Ngo, K. LeCompte, L. Hargens and A. B. McEwen, *Thermochim. Acta*, 2000, **357–358**, 97.
24 E. A. Carter, G. Ciccotti, T. J. Hynes and R. Kapral, *Chem. Phys. Lett.*, 1989, **156**, 472.
25 M. Iannuzzi, A. Laio and M. Parrinello, *Phys. Rev. Lett.*, 2003, **90**, 238302.
26 J. P. Ryckaert, G. Ciccotti and H. J. C. Berendsen, *J. Comput. Phys.*, 1977, **23**, 327.
27 M. Sprik and G. Ciccotti, *J. Chem. Phys.*, 1998, **109**, 7737.
28 R. Vassilev, M. J. Louwerse and E. J. Baerends, *Chem. Phys. Lett.*, 2004, **398**, 212.
29 A. A. Abrikosov, *Fundamentals of the Theory of Metals*, Elsevier, Amsterdam, 1988.
30 J. F. Janak, *Phys. Rev. B*, 1978, **18**, 7165.
31 M. Brack, *Rev. Mod. Phys.*, 1993, **65**, 677.

# The ion speciation of ionic liquids in molecular solvents of low and medium polarity†

Yanping Jiang,[a] Holger Nadolny,[a] Stefan Käshammer,‡[a] Sebastian Weibels,[a] Wolffram Schröer[b] and Hermann Weingärtner*[a]

*Received 19th April 2011, Accepted 5th May 2011*
**DOI: 10.1039/c1fd00075f**

The ion speciation of ionic liquids dissolved in molecular liquids of low and medium polarity is studied experimentally and theoretically. The ion speciation in some representative systems is characterized by dielectric relaxation spectroscopy and electrical conductance measurements. A corresponding-states approach is used to compare the experimental results with predictions for a reference system of charged hard spheres in a dielectric continuum. Topics of special interest are the formation of ion pairs and larger ion clusters in dilute solutions and their fate at high concentrations, where they have to redissociate in the excess ionic liquid to form the charge-ordered structure of the fused salt. In solvents of low polarity this transition leads to electrical conductance minima and liquid–liquid immiscibilities.

## Introduction

Ionic liquids (ILs) have revolutionized many chemical methodologies.[1] In addition to applications as neat solvents, ILs are often used as co-solvents of molecular liquids. Mixtures with molecular liquids also play a key role in many non-solvent applications of ILs, for example as ions in chemical reactions or biotransformations. The huge number of potential IL-solvent combinations in such applications renders the systematization and molecular understanding of their properties mandatory.

In principle, mixtures of ILs with molecular solvents form electrolyte solutions. Studies of simple electrolyte solutions performed over many decades[2] therefore offer important concepts for understanding the properties of dissolved ILs. However, many characteristics of ILs are not encountered in the case of simple salts.[1]

First, ILs enable a transition from typical electrolyte solution behaviour to fused salt behaviour. For simple salts a miscibility up to the fused salt is a very unusual phenomenon, which has only been characterized for a few systems at high temperatures.[3] Moreover, the molecular nature of the ions allow the tuning of their properties for given applications. The huge number of potential cation–anion combinations offers a versatility not encountered in the case of simple salts. Finally, the tunable polarity and hydrophobicity of ILs can be used to enhance miscibility with solvents of medium or low polarity. In this way one can design electrolyte systems with very unusual properties.

[a]*Department of Physical Chemistry 2, Faculty of Chemistry and Biochemistry, Ruhr-University Bochum, D-44780 Bochum, Germany. E-mail: hermann.weingaertner@rub.de; Fax: +0234-32-1429*

[b]*Institut für Anorganische und Physikalische Chemie, Universität Bremen, D-28359 Bremen, Germany*

† Work supported by the Deutsche Forschungsgemeinschaft within the priority program SPP 1191 ("Ionic Liquids").

‡ Present address: BASF SE, CZT-E100, 67056 Ludwigshafen, Germany.

The present study aims at pinpointing the basic characteristics of ILs in solvents of low and medium polarity, for which little is known experimentally and theoretically. The target is the ion speciation and its dependence on the solvent.

In solvents of low dielectric constant the depth of the Coulomb potential is large compared to the thermal energy, which stabilizes cation–anion pairs and larger ion clusters.[4] This feature is not encountered in highly polar solvents. To form the charge-ordered structure of the neat IL these ion clusters have to redissociate in the excess ionic liquid. Understanding the formation of ion pairs and larger ion clusters and their fate in the excess IL concentrations is a key prerequisite for understanding the solution properties of ILs. In solvents of low polarity these phenomena lead to electrical conductance minima and liquid–liquid miscibility gaps.[4,5]

Traditionally, ion pairing is studied by electrical conductance measurements.[2] Here, we combine such experiments with dielectric relaxation spectroscopy (DRS). DRS is a powerful tool for detecting ion pairs because their high electric dipole moments result in distinct modes in dielectric spectra. DRS has been employed to neat ILs,[6–9] and is just beginning to be applied to ILs in molecular solvents.[10,11]

Theories which quantitatively describe electrolytes at the conditions of interest are not available now. A beneficial tool for elucidating properties of electrolytes over wide ranges of concentration, temperature and dielectric constant of the solvent is found in the corresponding-states principle, which can be used to relate the properties of real systems to those of a reference system.[12] Its utility for describing electrolyte solutions results from the fact that the charged hard sphere (CHS) fluid satisfies this principle.[4] The CHS fluid is the molecular model underlying many electrolyte theories, including Debye–Hückel (DH) theory and its descendants and Bjerrum's (Bj) theory of ion pairing. It provides a natural reference system for states beyond the range of applicability of standard electrolyte solution theories.[4]

While the corresponding-states principle implies that the real solution and CHS reference fluid are similar in the Coulombic part of the interionic potential, the molecular nature of the ions may introduce a wide spectrum of additional short-range ion-ion and ion-solvent interactions. Although corresponding-states conditions will therefore hardly apply rigorously for real electrolyte systems, an approximate validity has been observed in many cases.[4,5] Large deviations indicate specific molecular interactions in the particular system.

As a major prerequisite, a corresponding-state approach requires reliable data for the reference fluid. In principle, an "exact" way to obtain such data is provided by Monte-Carlo (MC) or molecular dynamics (MD) simulations. For the CHS fluid such simulations are essentially limited to the characterization of its phase behaviour, which is key to the understanding of some unusual critical phenomena of ionic fluids.[5,13] In practice, one has therefore to resort to analytical equations of state. We adopt here an equation of state derived by Weiss and Schröer (WS),[14] which extends the familiar DHBj theory of ion pairing by terms for dipole–ion (DI) and dipole–dipole (DD) interactions in a system comprising free ions and dipolar ion pairs. This approach is particularly apt for solvents of low polarity, where ion pairing is strong.

The paper is organized as follows: First, we introduce the corresponding-states principle for ionic fluids and discuss the equation of state used for modelling the properties of the CHS reference fluid. Next, we briefly describe the DRS technique for probing the ion speciation. The discussion of our experimental and theoretical results begins with a survey of the global properties of the corresponding-states diagram. We then proceed to examples for ILs in solvents of medium polarity, where the special focus is on the formation of ion pairs and their fate in the excess ionic liquid. Particular features in this regime are minima of the effective degree of dissociation of the IL, which reveal themselves by electrical conductance minima. We then proceed to solvents of lower polarity, where ILs and molecular liquids may become immiscible.

 

## The corresponding-states principle for ionic systems

For a long time corresponding-states ideas have been employed for correlating the equilibrium properties of molecular fluids. This success invites further study by applying this concept to electrolyte systems.

The corresponding-states principle implies that the scaling of the temperature and particle density with regard to their values in an appropriate reference state provides a universal representation of the thermodynamic properties. The requirement for corresponding-states behaviour is the similarity of the interaction potential with regard to energy and distance scaling. Statistical thermodynamics shows that this principle holds rigorously, if the interaction potential is of the form $u(r) = u_0 \cdot \varphi(\sigma/r)$, where $u_0$ and $\sigma$ define the characteristic energy and length scales.[12] A reduced temperature $T^*$ and reduced density $\rho^*$ can then be defined as

$$T^* = k_B T/u_0 \tag{1a}$$

$$\rho^* = \rho\sigma^3 \tag{1b}$$

where $k_B$ is the Boltzmann constant and $\rho$ is the number density of the particles. $T^*$ reflects the ratio of the thermal energy $k_BT$ at temperature $T$ to the depth of the potential. $\rho^*$ is proportional to the excluded volume of the particles.

The CHS fluid satisfies these conditions.[4,5] In the fully symmetrical version used here it models the electrolyte by hard (*i.e.* non-polarizable) cations and anions with the same diameters $a$ and with equal charges $q_+ = |q_-|$, which are immersed in a dielectric continuum with the relative dielectric permittivity (dielectric constant) $\varepsilon$.[4] By setting the energy scale by the depth of the potential of two ions at contact, $q^2/4\pi\varepsilon\varepsilon_0 a$, and the length scale by the hard sphere diameter $a$, one finds

$$T^* = \frac{4\pi\, k_B T \varepsilon\varepsilon_0 a}{q^2} \tag{2a}$$

$$\rho^* = \rho a^3 = (\rho_+ + \rho_-)a^3 \tag{2b}$$

$\rho_i$ represents the number density of the ions, $\varepsilon_0$ is the dielectric permittivity of the vacuum. At fixed temperature $T$ all thermodynamic properties are then universal functions of the dielectric constant $\varepsilon$ of the medium and of the hard sphere diameter $a$ of the ions.

## The equation of state of the charged hard sphere fluid

The development of an equation of state for the CHS fluid is challenging. In solvents of high dielectric constant, such as water, the thermal energy is of the same order of magnitude as the depth of the interionic potential. As a consequence, the Boltzmann factor $\exp[-u(r)/k_BT]$ of the interionic potential can be linearized in $1/T$, which is a key ingredient of most electrolyte theories including the DH theory and its descendants.[2] In such cases stable ion pairs and ion clusters are of little relevance, unless their formation is driven by ion-specific short-range interactions.

In solvents of low dielectric constant the depth of the interionic potential is large compared to the thermal energy. These strong interionic interactions favour the formation of stable cation–anion pairs, which cannot be captured by a linearized Boltzmann factor. The common procedure is to consider these pairs separately from the remainder of the system by treating them as chemically distinct species.[2,4,5]

The degree of dissociation, $\alpha$, and the ion pair association constant $K_A(T)$ are then given by the mass action law

$$\frac{\rho_p}{\rho_+ + \rho_-} = \frac{2(1-\alpha)}{\rho\alpha^2} = K_A(T)\frac{\gamma_+\gamma_-}{\gamma_p} \tag{3}$$

where $\rho_i$ represents the particle densities of the free ions and ion pairs, and $\gamma_i$ represents the corresponding activity coefficients.

The overwhelming number of electrolyte theories presume that the ion pairs behave as thermodynamically ideal species, explicitly or implicitly setting $\gamma_p = 1$. DH theory augmented by Bjerrum's model of ion pairs (DHBj) provides a familiar theory of this type.[4] In sufficiently polar solvents, typically with dielectric constants $\varepsilon_{solv} > 12$, DHBj theory satisfactorily rationalizes the experimental properties of solutions at low and moderate salt concentrations.[2] In such systems, the molar electrical conductance $\Lambda$ strongly decreases with increasing salt concentration because neutral ion pairs do not contribute to the conductance.

Because most neat ILs possess a comparatively high electrical conductivity,[1] the transition from dilute solutions to the neat IL requires the re-appearance of charged species. In other words, the ion pairs have to redissociate in the excess IL. This redissociation is evidenced by minima of the molar conductance $\Lambda$ in solvents of low dielectric constant.[2] For a long time these conductance minima have been related to the formation of charged triple ions.[2,15] While this mechanism is certainly an option in some ranges of concentration, this interpretation cannot rationalize the transition to the dissociated fused salt.[4]

A proper description of these phenomena has to account for the activity coefficient $\gamma_p$ of the ion pairs. The rationale is that ion pairs possess high electric dipole moments, which give rise to strong dipole–ion (DI) and dipole–dipole (DD) interactions with free ions and other pairs. Moreover, strong DD interactions lead to an increase of the dielectric constant of the medium surrounding the ions.[14,16,17] Taken together, these effects are expected to destabilize ion pairs in the excess IL.

Collecting the various possibilities, one can speculate about the following scenario for the state of the ions:

free solvated ions ↔ ion pairs ↔ triple ions ↔ larger ion clusters
↔ redissociation of the ion aggregates in the excess IL
↔ charge-ordered structure of the neat IL.

Except for subtle differences in the last two steps, this scenario corresponds to a model for ILs suggested by Dupont.[18]

Expressing the Helmholtz free energy $A$ in terms of a reduced free energy $\Phi$, one can capture all effects of intermolecular interactions by an excess term, yielding an equation of state of the general form

$$\Phi = A\sigma^3/k_BTV = \Phi^{ideal} + \Phi^{excess} \tag{4}$$

where $\Phi^{ideal}$ is the ideal gas contribution. In this spirit, Weiss and Schröer (WS) have extended the DHBj theory by terms for screened DI and DD interactions (DHBj + DI + DD).[14] For a system comprising free ions and ion pairs, the resulting expression for the excess free energy reads

$$\Phi^{excess} = \Phi^{hc} + \Phi^{ii} + \Phi^{id} + \Phi^{di} + \Phi^{dd} + \rho_p^* \ln K_A(T) \tag{5}$$

The excess term comprises contributions from hard core (hc), ion–ion (ii), ion–dipole (id), dipole–ion (di) and dipole–dipole (dd) interactions $\rho_p^*$ is the reduced density of the ion pairs. By symmetry one has $\Phi^{id} = \Phi^{di}$. Based on the DD interactions WS theory also accounts for a state-dependent dielectric constant of the solution.[14] Prior to their work, the basic concepts were developed in a pioneering study by Levin and Fisher,[19] who have derived an equation of state up to the DI level (DHBj + DI).

 

## Experimental methods

All ILs were purchased from Iolitec (Heilbronn, Germany). Their purity and handling is described in detail elsewhere.[8] Briefly, all ILs were dried at 60 °C under vacuum for up to 48 h. Solutions were made up from molecular solvents of the highest available purity (Baker). The water content of the solutions (typically <200 ppm by Karl-Fisher titration) primarily resulted from water traces in the ILs. As noted elsewhere,[8] effects of chloride impurities at the level quoted by the manufacturer were negligible.

Our broadband DRS equipment is described in detail elsewhere.[7,8] Briefly, we use a coaxial reflection technique to determine the complex dielectric permittivity

$$\tilde{\varepsilon}(\nu) = \varepsilon'(\nu) - i\varepsilon''(\nu) + \kappa/i2\pi\nu\varepsilon_0 \quad (6)$$

of the sample in the frequency range 10 MHz $\leq \nu \leq$ 20 GHz.

In Equ. (6) the real part $\varepsilon'(\nu)$ of the complex permittivity accounts for dielectric dispersion, which essentially reflects the ability of the system to follow the oscillating electric field. Zero-frequency extrapolation of $\varepsilon'(\nu)$ yields the dielectric constant $\varepsilon$ of the solution. For electrically conducting liquids this procedure forms the only way to determine $\varepsilon$.[6–8] The imaginary part, $\varepsilon''(\nu) + \kappa/2\pi\nu\varepsilon_0$, reflects the dielectric loss. Its relaxation part $\varepsilon''(\nu)$ provides the same information as $\varepsilon'(\nu)$. In conducting solutions $\varepsilon''(\nu)$ is superimposed by the so-called Ohmic loss, $\kappa/\omega\varepsilon_0$. The Ohmic loss results from the direct current conductivity $\kappa$ of the solution, thus yielding information on the molar conductance $\Lambda = \kappa/C$ of the solution.

In principle, data fitting of $\varepsilon'(\nu)$ and $\varepsilon''(\nu) + \kappa/2\pi\nu\varepsilon_0$ to a suitable relaxation model yields $\kappa$ and the parameters of the relaxation model. However, in most systems of interest, the intensity of the relaxation contribution $\varepsilon''(\nu)$ is low compared to the Ohmic loss, and we based our fits on the real part $\varepsilon'(\nu)$. We ensured, however, that the results were consistent with those for the imaginary part $\varepsilon''(\nu)$.

## Results

### General features of the corresponding-states diagram of ionic fluids

To pinpoint the most important features of the corresponding-states diagram, we consider in Fig. 1 a typical IL at 25 °C in media with dielectric constants in the range $2.5 \leq \varepsilon \leq 25$. Specifically, we consider an IL with a hard sphere diameter of $a = 5.4$ Å, which corresponds to the distance of the charge centers in [$C_2$-mim] [$NTf_2$],[20] and is typical for many other ILs.

Following Friedman and Larsen,[4] we divide in Fig. 1 the regime of dominant ion pairing from the range, where "free" ions dominate, by the locus of states for which $\alpha = 1/2$. The branch of this locus at low IL concentrations (solid line) is derived from the mass action-law for ion pairing based on the DHBj expression. In this regime DI and DD interactions are not yet quantitatively relevant and the dielectric constant $\varepsilon$ of the medium can be equated with the dielectric constant $\varepsilon_{solv}$ of the neat solvent. In such cases $\gamma_p = 1$ is a reliable approximation. It is seen that a decrease of the dielectric constant of the solvent stabilizes the ion pairs, as expected by simple electrostatic reasonings.

The continuation of this locus at high IL concentrations (dashed line) reflects the reappearance of charged species, which signals the reorganization of the solution structure towards that of an electrically conducting molten salt. As a consequence, the regime, where the ions of the IL are in a dissociated state, shifts toward lower dielectric constants. In principle, a formulation of DHBj theory already predicts this behaviour, if proper account is made for the screening effect of the free ions.[21] However, a more quantitative description has to account for DI and DD interactions between the dipolar pairs and the remainder of the system. The strong

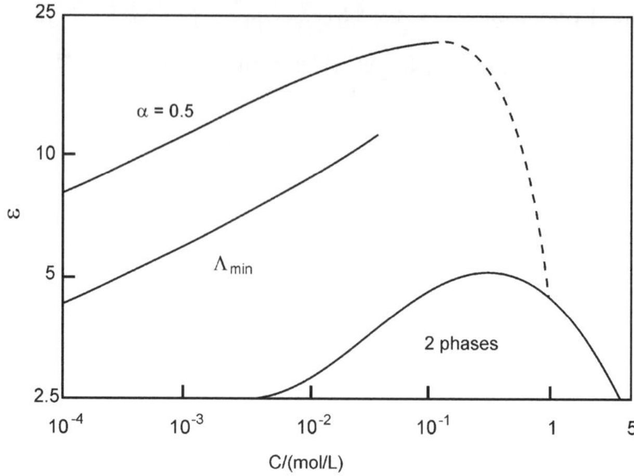


**Fig. 1** Predicted corresponding-states diagram of an IL with $a = 5.4$ Å at 298 K in media with dielectric constants $2.5 \leq \varepsilon \leq 25$ as a function of the molar concentration of the IL. Note the logarithmic scales of both axis.

DD interactions also result in a state-dependent dielectric constant. These effects are accounted for by WS theory, but this theory is not very accurate at high IL concentrations. Thus, the displayed locus of $\alpha = 1/2$ remains somewhat speculative.

The described phenomena imply a minimum of the effective degree of dissociation at intermediate IL concentrations.[21] Experimentally, this minimum of $\alpha$ reveals itself by a minimum of the molar conductance $\Lambda$ of the solutions. A decrease of $\varepsilon_{solv}$ is known to shift the conductance minimum to lower salt concentrations.[2] Fig. 1 shows the locus of this minimum in the corresponding states plot, as estimated from experimental data.[22] It is known that in solvents of high polarity, with $\varepsilon_{solv} > 10$–12 say, the conductance minimum disappears. Thus, the experimental locus of $C_{min}$ has an upper end point.

The third characteristic feature displayed in Fig. 1 is an instability regime at high IL concentrations in an environment of low polarity. In real systems this instability regime corresponds to a liquid–liquid immiscibility of ILs in solvents of low dielectric constant.[23,24] Unfortunately, the observed critical concentrations are much too high to enable an accurate description by WS theory or by any other equation of state. MC simulations for the CHS fluid provide an estimate, but the available simulations show some discrepancies.[5] Fig. 1 provides a schematic sketch of the coexistence curve based on the MC simulations by Yan and de Pablo.[25] Interestingly, the estimated locus of $\alpha = 1/2$ approaches the two-phase coexistence in a regime close to the critical point. Thus, at low dielectric constant of the solvent liquid–liquid phase equilibria may be qualitatively rationalized by a dipolar fluid of ion pairs at coexistence with an expanded molten salt.[4]

## The conductance minimum

Much of the knowledge concerning the ion speciation in electrolyte solutions results from electrical conductance data.[2] To a first approximation, the conductance reflects ion pairing because neutral ion pairs do not contribute to the charge transport. One of the basic conclusions derived from such data is that, despite being a macroscopic quantity, the dielectric constant of the solvent is of crucial importance in understanding these phenomena.[2,22]

Data for solutions of simple salts[2] and ILs[26] show that in polar solvents such as acetonitrile or acetone there is a monotonous decay of $\Lambda$ with increasing solute concentration. In solvents of low dielectric constants, $\varepsilon_{solv} < 12$ say, this decay turns over into a more complex behaviour, resulting in conductance minima and in maxima at high IL concentrations. We focus here on the conductance minima, which reflect major changes in the solution structure. Fig. 2 illustrates this transition by data for 1-ethyl-3-methylimidazolium trifluoromethanesulfonate ([$C_2$-mim][TfO]) dissolved in 1,2-dichloroethane (DCE, $\varepsilon_{solv} = 10.4$ at 25 °C), dichloromethane (DCM, $\varepsilon_{solv} = 8.9$) and tetrahydrofurane (THF, $\varepsilon_{solv} = 7.4$).

By plotting the ratio $\Lambda/\Lambda^0$, where $\Lambda^0$ is the limiting molar conductance at infinite dilution, the comparison between the various solvents is facilitated because the different solvent viscosities are taken into account. The observed conductances are small fractions of $\Lambda^0$, which prevents the direct extrapolation of $\Lambda^0$ from the experimental data. Instead, we resorted to correlations of $\Lambda^0$ with the ion size (Stokes' law) and viscosity of the solvent (Walden's rule) to deduce $\Lambda^0$ from data for [$C_4$-mim][$NTf_2$] and [$C_4$-mmim][$NTf_2$] in DCM.[27] These correlations predict $\Lambda^0 \cong 135$, 120 and 71 S cm$^2$ mol$^{-1}$ for [$C_2$-mim][TfO] in DCM, THF and DCE, respectively. In comparison with highly accurate $\Lambda^0$ values obtained by direct extrapolation of conductance data in more favourable situations, this procedure introduces a considerable uncertainty of $\Lambda^0$, amounting to several percent.[22]

Fig. 2 identifies DCE ($\varepsilon_{solv} \cong 10.4$) as a limiting case, where conductance minima are beginning to develop. The concentrations $C_{min}$ of the minima and the conductance ratios $\Lambda_{min}/\Lambda^0$ decrease with decreasing dielectric constant $\varepsilon_{solv}$. The occurrence of liquid–liquid phase equilibria prevents the observation of the fate of these minima in solutions in solvents with markedly lower dielectric constants than considered in Fig. 2. These liquid–liquid immiscibilities obviously correspond to the instability regime of the CHS fluid displayed in Fig. 1.[4]

By employing highly hydrophobic tetraalkylammonium salts one can achieve complete miscibility with some nonpolar solvents, such as benzene or tetrachloromethane.[28] Fig. 3 shows that in such systems the conductance minimum is shifted to very low concentrations, so that only the branch of increasing conductivity of the conductance curve is accessible by conductance experiments.

The results are consistent with a conjecture by Walden[29] that the location of the conductance minimum satisfies the approximate relation

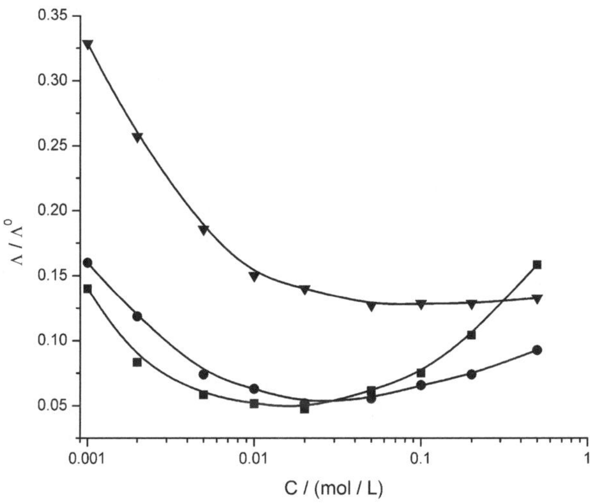


**Fig. 2** Concentration dependence of the molar conductance of [$C_2$-mim][TfO] in DCE (triangles), DCM (circles) and THF (squares) at 25 °C. The data are plotted as ratios $\Lambda/\Lambda^0$ based on $\Lambda^0$ values given in the text.

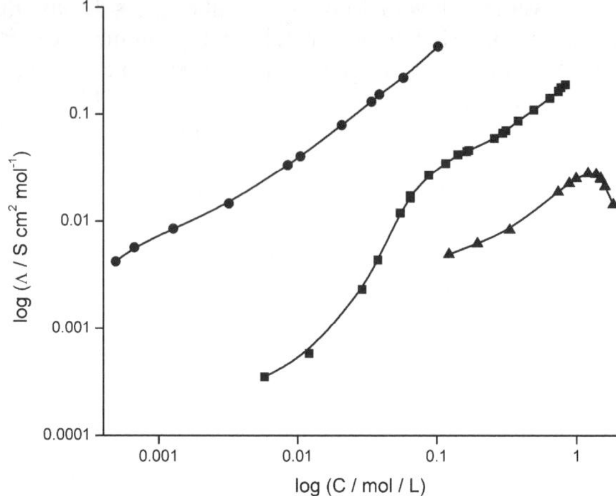


**Fig. 3** Electrical conductance of solutions of tetrabutylammonium bromide (squares) and tetrabutylammonium tetrabutylborate (circles) in benzene at 25 °C. Data reported by us some time ago[28] for tetrabutylammonium bromide in $CCl_4$ are displayed as well (triangles).

$$\varepsilon_{\text{solv}} \cdot C_{\text{min}}^{-1/3} = const \tag{7}$$

Noting that at fixed temperature $T$, conductance minima form corresponding states, an insertion of Equ. (2a) into (2b) predicts Walden's relation (7) to hold rigorously for the CHS fluid. Moreover, the resulting product $\varepsilon_{\text{solv}} \cdot C_{\text{min}}^{-1/3}$ does not depend on the mean diameter $a$ of the ions, as approximately observed in many cases.[2]

## The ion speciation of imidazolium-based ionic liquids in tetrahydrofurane

The origin of the conductance minimum may be further elucidated by DRS, which can be used to characterize the ion speciation in an IL solution.[6,10,11] Due to rather high IL concentrations required for meaningful experiments, the concentration range covered by DRS experiments ($C \geq 0.05$ mol $L^{-1}$) is typically above the concentration of the conductance minimum. Thus, we observe the fate of the ion pairs along the increasing branch of the conductance curves, but by some extrapolation, and in combination with conductance data for low IL concentrations, a detailed molecular picture can be developed.

DRS experiments have been conducted by us for several ILs in THF, DCM and DCE, respectively, yielding essentially the same conclusions concerning the ion speciation. Here, we discuss the results for two ILs in THF. The dielectric constant of THF is typical for many organic solvents of potential use for chemical processes. Its low viscosity ($\eta = 0.46$ mPa·s at 25 °C) renders it an excellent solvent for DRS studies. The corresponding states diagram in Fig. 1 places solutions in THF in a regime slightly above the instability region of the CHS.

In parts, our interest in solutions in THF resulted from a preceding study, in which we have employed DRS to probe the ion speciation driving chirality transfer in the ruthenium-catalyzed hydrogenation of a keto-functionalized prochiral imidazolium cation.[10] In this hydrogenation reaction the chiral camphorsulfonate anion was found to induce chirality at the cation *via* ion pairing. In THF a maximum enantiomeric excess was found at an IL concentration of $C = 0.15$ mol $L^{-1}$, which could be linked to the characteristics of the ion speciation in this regime.[10]

The experiments shown here concern solutions of 1-ethyl-3-methylimidazolium bis(trifluoromethanesulfonyl)imide ([$C_2$-mim][$NTf_2$]) and its homologue based on the 1-ethyl-2,3-dimethylimidazolium cation ([$C_2$-mmim][$NTf_2$]). In the latter case the hydrogen at carbon C-2 of the imidazolium ring is replaced by a methyl group. The effect of this substitution is of interest because the hydrogen at carbon C-2 can form a hydrogen bond to anions. In the case of camphorsulfonate we found evidence for very strong non-electrostatic contributions to ion pair formation, presumably caused by the formation of such a hydrogen bond.[10]

Fig. 4 displays the concentration dependence of the molar conductance $\Lambda$ of the two systems. The correlation based on Stokes' law and Walden's rule yields $\Lambda^0 = 113$ S cm$^2$ mol$^{-1}$ for [C$_2$-mim][NTf$_2$] and $\Lambda^0 = 110$ S cm$^2$ mol$^{-1}$ for [C$_2$-mmim][NTf$_2$]. Clearly, the rapid initial decay of $\Lambda$ reflects the formation of ion pairs, but the data do not cover sufficiently large segments of the low-concentration branch to quantitatively extract the ion pair association constant $K_A$. Moreover, above $C = 0.02$ mol L$^{-1}$ the viscosity, $\eta$, of the solutions begins to deviate from the viscosity $\eta^0$ of the neat solvent. At such conditions, the effective degree of dissociation, $\alpha_{eff}$, is usually approximated by the conductance–viscosity product ("Walden product"), setting $\alpha_{eff} \cong \Lambda\eta/\Lambda^0\eta^0$. Fig. 4 shows the resulting concentration dependence of $\alpha_{eff}$. It is worthwhile to note that the viscosity correction removes the conductance maximum near 1 mol L$^{-1}$, in agreement with the common interpretation that such maxima result from a strong increase of viscosity of the solutions.[22]

Based on the Walden products we roughly estimate $K_A \cong 1.3 \times 10^5$ L mol$^{-1}$ for [C$_2$-mim][NTf$_2$] compared with $K_A \cong 5.0 \times 10^4$ L mol$^{-1}$ for [C$_2$-mmim][NTf$_2$]. The main result is that in [C$_2$-mim][NTf$_2$] ion pairing is slightly stronger than in [C$_2$-mmim][NTf$_2$], supporting the conjecture that a hydrogen bond at carbon C-2 stabilizes ion pairs. The observed values of $K_A$ are somewhat higher than values derived by Katsuta *et al.*[27] for [C$_4$-mim][NTf$_2$] and [C$_4$-mmim][NTf$_2$] in DCM. Such a difference is expected on grounds of the lower dielectric constant of THF, but the quantitative comparison is hampered by lacking knowledge on the exact ion sizes. Note that by treating the hard core diameter *a* as an adjustable parameter

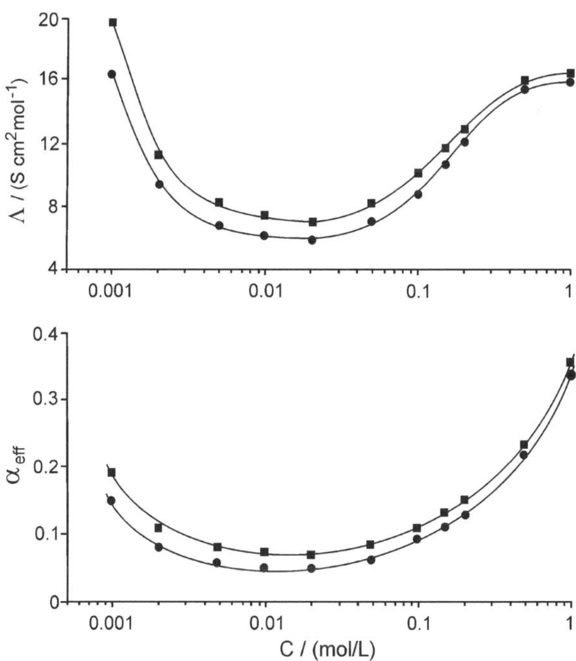


**Fig. 4** Molar electrical conductance $\Lambda$ of [C$_2$-mmim][NTf$_2$] (squares) and [C$_2$-mim][NTf$_2$] (circles) in THF at 25 °C and effective degree of dissociation $\alpha_{eff}$ derived from the Walden product.

in fits to the conductance equation, Katsuta *et al.* observed no clear trends for different salts.[27] For this reason, comparison with theoretical $K_A$ values derived from DHBj theory are also rather inconclusive, except for the observation that the experimental data are of the expected order of magnitude.

Having established the general dissociation behaviour, we focus on the DRS data. Microwave DRS spectra of electrolyte solutions in solvents of low and medium polarity are often bimodal with a strong solvent mode, which decreases upon addition of salt, and a solute mode, which increases with increasing salt concentration.[6] Due to the low viscosity of neat THF, molecular reorientation of the THF dipoles driving relaxation is very fast. As a consequence the solvent mode is centered near 70 GHz,[30] which is beyond the frequency window up to 20 GHz of our equipment. In practice, it yields only a background contribution to the dispersion spectrum below 20 GHz, which is not detectable by visual inspection of the measured DRS spectra and is easily corrected for. Moreover, the mode of THF is not markedly shifted by addition of salts.[30]

Fig. 5 shows as an illustrative example the dielectric dispersion curve $\varepsilon'(\nu)$ of a solution of $[C_2\text{-mim}][NTf_2]$ in THF at $C = 0.5$ mol $L^{-1}$. Except for a minor correction for THF relaxation at the upper edge of the spectrum, the dispersion regime reflects the reorientational dynamics of dipolar solute species such as ion pairs. Note that the free ions also form potential relaxators, but at the concentrations under consideration their contributions can be safely neglected because their dipole moments are small fractions of the dipole moments of the dipolar pairs and the concentrations of the free ions are low.

A more detailed analysis of the spectrum in Fig. 5 shows that the relaxation regime assigned to the solute (henceforth numbered "1") is very broad and exceeds by far the Lorentzian line width caused by relaxation of a single dipolar species in a homogenuous environment. The observed mode may therefore form the envelope of a distribution of overlapping processes. As in similar cases,[10,11] we fitted the contribution $\Delta\tilde{\varepsilon}_1(\nu)$ of the solute mode to the Cole-Cole equation[31]

$$\Delta\tilde{\varepsilon}_1(\nu) = \frac{S_1}{(1 + i2\pi\nu\tau_1)^{1-h}} \tag{8}$$

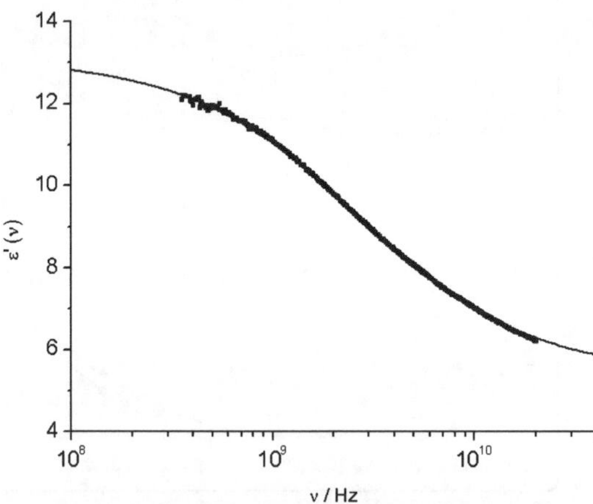


**Fig. 5** Dielectric dispersion curve $\varepsilon'(\nu)$ of $[C_2\text{-mim}][NTf_2]$ in THF at $C = 0.5$ mol $L^{-1}$. The temperature is $(25 \pm 0.1)$ °C. The displayed dispersion regime reflects the solute contribution to the dielectric spectrum. The dispersion regime of THF is beyond the frequency window up to 20 GHz, and provides only a background contribution to the observed dispersion curve. The solid line shows the fit to a Cole-Cole-type ansatz for the spectral distribution.

 

The Cole-Cole representation presumes a symmetrical distribution of relaxation times, characterized by a width parameter $0 \leq h < 1$, which accounts for a symmetrical broadening relative to the Lorentzian line shape caused a single Debye-type relaxation process. $\tau_1$ is a relaxation time describing the co-operative reorientation of the solute dipoles. $S_1$ is the relaxation amplitude which reflects the contribution of the orientational polarization of the solvent dipoles to the static dielectric constant $\varepsilon$ of the solution. For $h = 0$ one obtains the Lorentzian line shape of a single Debye-type relaxation process. Fig. 6 shows the parameterization of the DRS spectra of [C$_2$-mim][NTf$_2$] and [C$_2$-mim][NTf$_2$] in THF.

As a striking feature, at the level of accuracy of the various parameters the results for [C$_2$-mim][NTf$_2$] and [C$_2$-mmim][NTf$_2$] practically coincide. Obviously, the electrical conductance is the only property which shows a marked effect caused by the presumed cation–anion hydrogen bond at carbon C-2. Note that results for the two ILs in DCM and DCE (not shown here) also proved to be insensitive to methyl subsitution at C-2. A possible rationale is that the electrical conductance reflects effects of H-bonded configurations at short (<10 ps) time scales, which do not survive on the time scale (>10 ps) of the molecular reorientation of the pairs. This insensitivity to methyl substitution can, however, not continue up to the neat ILs because neat [C$_2$-mmim][NTf$_2$] shows a much larger viscosity and much longer dielectric relaxation time than [C$_2$-mim][NTf$_2$].[7,8] In the range considered here the viscosities of the solutions of [C$_2$-mmim][NTf$_2$] and [C$_2$-mim][NTf$_2$] do not yet differ markedly.

Considering the results in Fig. 6 in detail, the empirical width parameter $h$ indicates an unusually broad distribution of relaxation times of the solute. At low concentrations $h$ should extrapolate to zero, because one may expect ion pairs to form the only dipolar species. The expected decrease of $h$ sets on at the low-concentration edge of our data, but extrapolation suggests that near the conductance minimum the ion speciation still shows a notable heterogeneity.

Further information on the ion speciation is provided by the relaxation times $\tau_1$. A common procedure relates $\tau_1$ by the hydrodynamic Stokes–Einstein-Debye (SED) relation $\tau_1 = 3V_{\text{eff}}\eta/k_BT$ to the effective volume $V_{\text{eff}}$ of a rotating species in a medium of viscosity $\eta$. Neglecting corrections for collective effects, extrapolation of the relaxation times in Fig. 6 to low IL concentrations yields a "Stokes radius" of about 0.2 nm. Altough SED theory is not an accurate one, this result signals the presence of contact ion pairs, as opposed to expanded solvent-separated structures. At higher concentrations $\tau_1$ increases, indicating the presence of larger dipolar ion clusters, and passes a maximum near 0.2 mol L$^{-1}$ beyond which the cluster size decreases. Because the hydrodynamic approximation is known to fail for neat ILs,[32] the SED analysis may be increasingly inappropriate in concentrated solutions.

Fig. 6 also displays the concentration dependence of the dielectric constant $\varepsilon$ of the solutions and the solute contribution $S_1$ to $\varepsilon$ derived from the solute mode of the dielectric spectrum. When increasing the IL concentration, the magntiude of $\varepsilon$ is increasingly controlled by the solute contribution $S_1$, but at high IL concentrations the concentration dependence of both, $\varepsilon$ and $S_1$, levels off. This phenomenon can be readily understood by noting that the relaxation amplitude $S_1$ provides information on the effective electric dipole moment $\mu$ of the relaxators.

Calculations of $\mu$ by means of the Cavell model[33] reveal a strongly concentration-dependent dipole moment, which is displayed in Fig. 6. At high dilution $\mu$ extrapolates to a value of about 40 D (1 D = $3.30 \times 10^{-30}$ C m) for the ion pair. This very high value points implies a crucial importance of DD interactions between ion pairs for the ion speciation. Perhaps, the most interesting information concerns, however, the decrease of $\mu$ with increasing IL concentration. This decrease signals the formation of ion configurations with low effective dipole moments. These may reflect a preferred antiparallel arrangement of neighbouring ion pairs, which is known to reduce the effective dipole moment, or may even reflect the formation of distinct, long-lived ion clusters with low dipole moments, such as neutral quadruple ions.

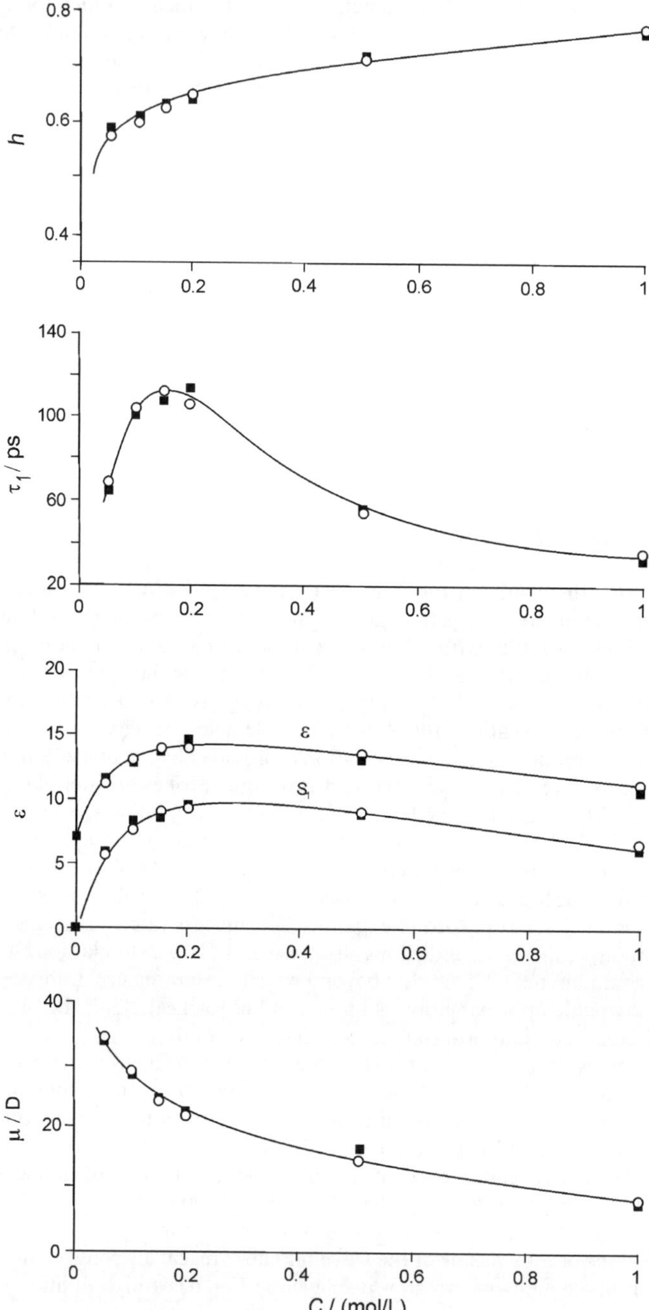


**Fig. 6** Relaxation parameters and molecular parameters extracted from DRS spectra of solutions of [$C_2$-mim][$NTf_2$] (squares) and [$C_2$-mmim][$NTf_2$] (circles) in THF at 25 °C. The lines are a guide to the eyes. The following properties are displayed from top to bottom: width parameter $h$ of the Cole-Cole relaxation time distribution of the solute mode; relaxation time $\tau_1$ of the solute mode; dielectric constant $\varepsilon$ of the solutions and amplitude $S_1$ of the underlying solute contribution; effective dipole moment $\mu$ of the solute.

 

There are procedures to estimate association constants for equilibria between ion pairs and quadruple ions from dielectric constant data.[22]

We have developed a dielectric theory for treating the effect of DD interactions on the dielectric constant of the CHS fluid at different levels of approximation. Fig. 7 compares results of model calculations for an IL with $a = 5.4$ Å in a solvent with $\varepsilon_{solv} = 7.4$ with the measured concentration dependence of the dielectric constant of $[C_2\text{-mmim}][NTf_2]$ in THF.

WS theory forms the simplest level for treating DD interactions in the CHS fluid.[14] WS theory accounts for DD interactions between ion pairs in the same way as Onsager's theory of the reaction field, but includes screening of the DD interactions by the free ions. The dipolar species are assumed to be spherical. Fig. 7 shows that this approach largely overestimates the increase of the dielectric constant with increasing solute concentration and fails to predict $\varepsilon$ to level off at high IL concentrations.

This failure is clearly related to the decrease of the effective dipole moment of the relaxing species shown in Fig. 6. A more sophisticated approach[17] supplements the Onsager model by including dipole–dipole correlations among the ion pairs, corresponding to the well-known Kirkwood-Fröhlich theory of dipolar molecular fluids.[34] Kirkwood-Fröhlich theory represents deviations of the dielectric constant from the Onsager prediction by the Kirkwood factor $g$. We note that the $g$-coefficient is unity in the case of a random distribution of the dipoles, while deviations describe dipolar correlations reflecting preferred antiparallel ($g < 1$) or parallel ($g > 1$) orientations of the dipoles. The Onsager model implies $g = 1$. Phaenomenologically one can determine $g$ from our experimental data by adapting the procedure employed for molecular fluids to solutions of ILs. Then, $g$ is approximated by the relation of the squared effective dipole moment at concentration $C$ to the extrapolation value at $C \to 0$, where interactions with neighbouring dipoles are absent

$$g = \mu^2(C)/\mu^2(C \to 0) \tag{9}$$

The experimental results for $g$ are displayed in Fig. 8. The figure points toward very strong antiparallel dipolar correlations of the ion pairs. Even stronger

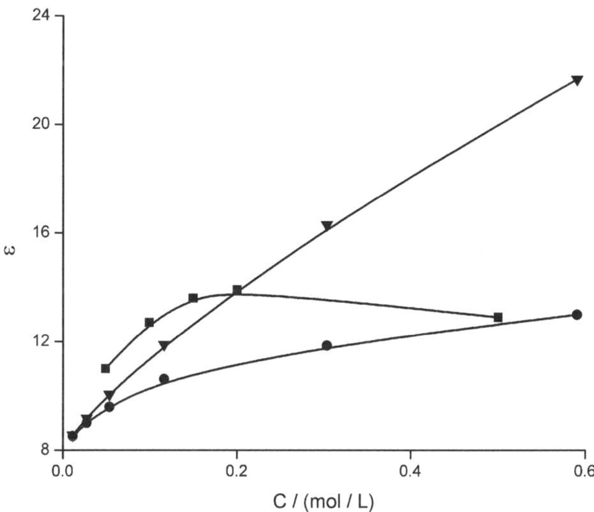


**Fig. 7** Concentration dependence of the dielectric constant of $[C_2\text{-mmim}][NTf_2]$ in THF (squares) compared with theoretical predictions for the CHS fluid based on an Onsager-type approach for dipolar spheres underlying WS theory (triangles) and its extension to include orientational correlations between hard dipolar ellipsoids (circles).

correlations were reported by us for tetrabutylammnium bromide in $CCl_4$, as expected on grounds of the lower dielectric constant of the solvent ($\varepsilon_{solv} = 2.3$). These experimental results are also displayed in Fig. 8.

To elucidate the observed behaviour we have modeled the *g*-coefficient and dielectric constant in THF by a Kirkwood-Fröhlich based theory at the level of the second virial coefficient for a system of dipolar hard ellipsoids,[35] thereby accounting for the screening by the free ions. For dipolar hard ellipsoids estimates of the second dielectric virial coefficient are available in the literature.[36] In the calculations we assumed an aspect ratio of 2 : 1 for the ellipsoidal model of the ion pair. Extrapolating the numerical data to the temperature region of the experiment, we observe a flattening of the concentration dependence of the dielectric constant (Fig. 7) caused by a decrease of the Kirkwood factor (Fig. 8). It is seen that an account for an ellipsoidal shape of the ion pairs considerably improves the theoretical predictions for the concentration dependence of the dielectric constant.

In concluding this section, we comment on the role of charged triple ions for the conductance minimum. This widely used traditional interpretation goes back to the work of Fuoss and Kraus,[15] which ascribes the minimum by a mass action law between neutral ion pairs MX and free ions $M^+$ and $X^-$ to form charged triple ions $[M_2X]^+$ and/or $[MX_2]^-$. This hypothesis has become so popular that conductance minima are often taken as a definite proof of the triple ion concept.[2] While triple ions may play a role in some ranges of concentrations, the appearance of conductance minima (*i.e.* minima in the effective degree of dissociation) are readily described by accounting for strong DI and DD interactions of ion pairs with the surrounding medium and the resulting increase of the dieletric constant of the solution, which drive redissociation. On general grounds, a high stability of triple ions can, however, be doubted, and Kraus himself seems to have rejected the triple ion concept in later work in favour of a redissociation approach.[3c] Thus, the key for modeling the re-appearance of charged species seems to be founded in a proper treatment of the activity coefficient of the ion pairs in the mass action law (Equ. 3) taking account for a realistic model for the effect of DD interactions on the dielectric constant of the solutions.

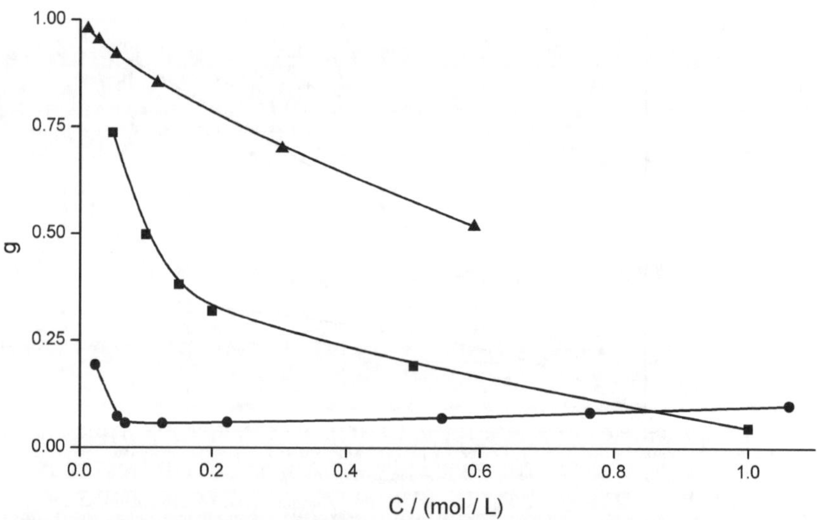


**Fig. 8** Comparison of experimental Kirkwood correlation factors for solutions of $[C_2\text{-mmim}][NTf_2]$ in THF (squares) with theoretical calculations for ion pairs modeled as dipolar hard ellipsoids (triangles). Circles represent data for tetrabutylammonium bromide in $CCl_4$.

 

## Liquid–liquid immiscibilities and critical points

When lowering the dielectric constant of the solvent to values near $\varepsilon_{solv} = 5$, one approaches conditions, where in some range of concentration electrolyte solutions become thermodynamically instable.[4] Experimentally, this instability manifests itself by liquid–liquid immiscibilities.[23,24] Homologous long-chain alcohols enable a stepwise variation of $\varepsilon_{solv}$ which allows to study the development of such immiscibilities with decreasing polarity of the environment.[23,24]

As an illustrative example, Fig. 9 shows the liquid–liquid phase diagram of the solution of $[C_{10}\text{-mim}][NTf_2]$ in dodecanol. The experimental data (upper curve) were obtained by the cloud-point method.[24b] The curve is a fit based on the assumption of Ising critical behavior and on the linear diameter rule, yielding an estimate of the critical temperature of $T_c = 325.9$ K and a critical concentration of $C_c = 1.28$ mol L$^{-1}$. The experimental curve is compared with the prediction of the CHS model (lower curve). The data points are transcribed from the simulation results of Yan and de Pablo[25] for $T^*$ and $\rho^*$ to the real system by using the dielectric permittivity of dodecanol at the upper critical solution temperature ($e_{solv} = 4.9$), and assuming $a = 5.4$ Å, as used earlier. The curve is obtained by fitting the transformed simulation data in the same way as the experimental data. The predictions obtained for the critical data by the CHS model are $T_c = 314.8$ K and $C_c = 0.73$ mol L$^{-1}$, which in view of the simplicity of the model, is in fair agreement with the experimental data.

Fig. 10 shows the electrical conductance isotherms of three IL–solvent systems, which exhibit liquid–liquid consolute points near room temperature, namely ethylammonium nitrate — *n*-octanol ($T_c = 315$ K),[37] tetrabutylammonium picrate — tridecanol ($T_c = 341$ K),[38] and ethyltrimethylammonium bromide – chloroform ($T_c = 294$ K).[39] The conductances were recorded in the one-phase region, typically 5 K away from the critical isotherms, which implies that they refer to corresponding states near the critical isotherm. The conductance curves show minima and are normalized to their values at these minima by plotting $\Lambda/\Lambda_{min}$ *vs.* $C/C_{min}$.

Below the conductance minimum, the data yield an almost perfect master curve, which practically coincides with the prediction of the DHBj model. Larger scatters are observed at concentrations above the minimum. The data suggest that along the low-concentration segment one finds a universal behaviour of the ion distribution, but at higher concentrations the degree of dissociation behaves in a more system-specific manner. The critical concentrations $C_c$ are located in this high-concentration

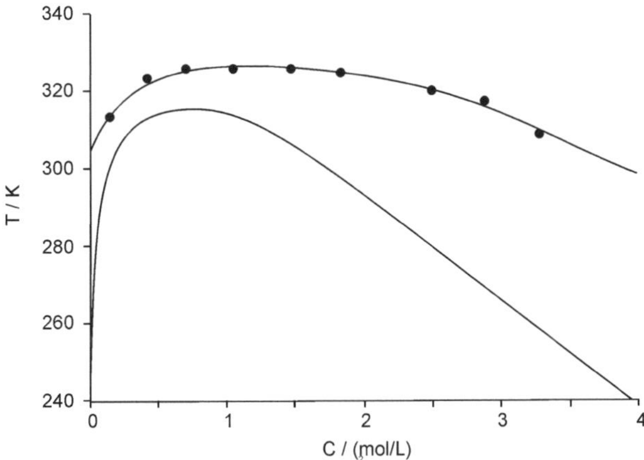


**Fig. 9** Liquid–liquid coexistence curve of $[C_{10}\text{-mim}][NTf_2]$ in dodecanol (circles, upper curve) compared to the prediction based on the instability regime of the CHS fluid (lower curve).

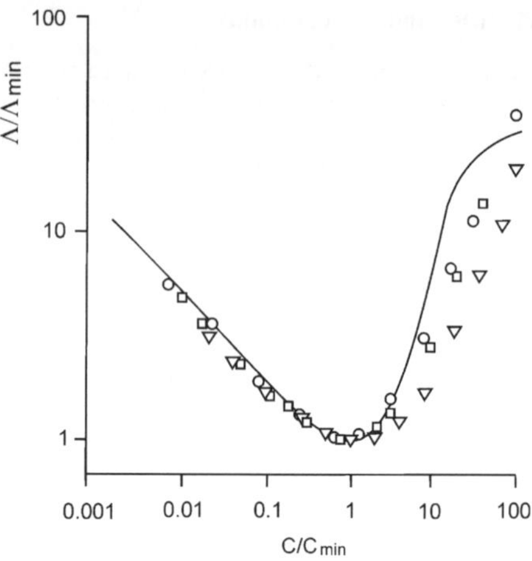


**Fig. 10** Near-critical conductance isotherms in ethylammonium nitrate — *n*-octanol (circles), tetrabutylammonium picrate — tridecanol (squares), and ethyltrimethylammonium bromide — chloroform (triangles) normalized by the concentrations and conductances at the conductance minima. The solid curve shows the prediction by simple DHBj theory.

regime, roughly at $C_{min}/C_c \cong 0.01$, where DI ion and DD interactions play an important role. The DHBj + DI + DD model overestimates, however, the degree of dissociation at the critical point,[21] which can be traced back to the strong concentration dependence of the predicted dielectric constant. It has been shown elsewhere[21] that an overestimate of $\alpha$ causes the predicted coexistence curve to be shifted to lower concentrations than experimentally observed.

## Conclusions

The remarkable properties of solutions of ILs in solvents of low polarity, such as the conductance minimum and the existence of a liquid–liquid miscibility gap can be understood —at least on a semi-quantiative level— by an equation of state which accounts for DI and DD interactions between ion pairs and the remainder of the solution and for the increase of the dielectric constant caused by the DD interactions. DRS in combination with electrical conductance measurements provides a beneficial tool for experimentally characterizing the underlying ion speciation.

## References


1 (*a*) P. Wasserscheid and T. Welton (ed.), *Ionic Liquids in Synthesis*, 2nd ed., Wiley-VCH, Weinheim, 2008; (*b*) N. V. Plechkova and K. R. Seddon, *Chem. Soc. Rev.*, 2008, **37**, 123; (*c*) H. Weingärtner, *Angew. Chem., Int. Ed.*, 2008, **47**, 654; (*d*) N. V. Plechkova, R. D. Rogers and K. R. Seddon (ed.), *Ionic Liquids. From Knowledge to Application, ACS Symp. Series*, Vol. 1030, American Chemical Society, Washington D.C., 2009.

2 R. A. Robinson and R. H. Stokes, *Electrolyte Solutions*, Butterworths, London, 1969.

3 (*a*) C. A. Kraus, *J. Phys. Chem.*, 1954, **58**, 673; (*b*) K. S. Pitzer, *J. Am. Chem. Soc.*, 1980, **102**, 2902; (*c*) L. C. Kenausis, E. C. Evers and C. A. Kraus, *Proc. Natl. Acad. Sci. U. S. A.*, 1963, **49**, 141.

4 H. L. Friedman and B. Larsen, *J. Chem. Phys.*, 1979, **70**, 92.

5 W. Schröer and H. Weingärtner, *Structure, Thermodynamics and Critical Properties of Ionic Fluids, in: New Approaches to the Structure and Dynamics of Liquids: Experiments, Theories and Simulations*, ed. J. Samos and V. Durov, Kluwer, Amsterdam, 2004, pp. 503–537.

6 R. Buchner and G. Hefter, *Phys. Chem. Chem. Phys.*, 2009, **11**, 8984; and references cited therein.

7 (*a*) C. Wakai, A. Oleinikova, M. Ott and H. Weingärtner, *J. Phys. Chem. B*, 2005, **109**, 17028; (*b*) C. Daguenet, P. J. Dyson, I. Krossing, A. Oleinikova, J. Slattery, C. Wakai


 

and H. Weingärtner, *J. Phys. Chem. B*, 2006, **110**, 12682; (*c*) M. Krüger, S. Funkner, E. Bründermann, H. Weingärtner and M. Havenith, *J. Chem. Phys.*, 2010, **132**, 101101.
8 M.-M. Huang, Y. Jiang, P. Sasisanker, G. W. Driver and H. Weingärtner, *J. Chem. Eng. Data*, 2011, **56**, 1494.
9 K. Nakamura and T. Shikata, *ChemPhysChem*, 2010, **11**, 285.
10 M.-M. -Huang, K. Schneiders, P. S. Schulz, P. Wasserscheid and H. Weingärtner, *Phys. Chem. Chem. Phys.*, 2011, **13**, 4126.
11 J. Hunger, A. Stoppa, R. Buchner and G. Hefter, *J. Phys. Chem. B*, 2008, **112**, 12013.
12 See for example: E. A. Guggenheim, *J. Chem. Phys.*, 1945, **13**, 253.
13 H. Weingärtner and W. Schröer, *Adv. Chem. Phys.*, 2001, **116**, 1.
14 W. Schröer and V. C. Weiss, *J. Chem. Phys.*, 1998, **108**, 7747.
15 R. M. Fuoss and C. A. Kraus, *J. Am. Chem. Soc.*, 1933, **55**, 476.
16 E. A. S. Cavell and P. C. Knight, *Z. Phys. Chem.*, 1968, **57**, 331.
17 W. Schröer, *J. Mol. Liq.*, 2001, **92**, 67.
18 J. Dupont, *J. Braz. Chem. Soc.*, 2004, **15**, 341.
19 Y. Levin and M. E. Fisher, *Phys. A*, 1995, **225**, 164.
20 B. Qiao, C. Krekeler, R. Berger, L. del Site and C. Holm, *J. Phys. Chem. B*, 2008, **112**, 1743.
21 V. C. Weiss, W. Schröer and H. Weingärtner, *J. Chem. Phys.*, 2000, **113**, 762.
22 T. Sigvartsen, B. Gestblom, E. Noreland and J. Songstad, *Acta Chem. Scand.*, 1989, **43**, 203.
23 H. Weingärtner, T. Merkel, U. Maurer, J.-P. Conzen, H. Glasbrenner and S. Käshammer, *Ber. Bunsenges. Phys. Chem.*, 1991, **95**, 1579.
24 (*a*) W. Schröer and V. R. Vale, *J. Phys.: Condens. Matter*, 2009, **21**, 424119; (*b*) V. R. Vale, B. Rathke, S. Will Stefan and W. Schröer, *J. Chem. Eng. Data*, 2010, **55**, 4195; (*c*) D. Saracsan, C. Rybarsch and W. Schröer, *Z. Phys. Chem.*, 2006, **220**, 1417.
25 Q. L. Yan and J. J. de Pablo, *J. Chem. Phys.*, 1999, **111**, 9509.
26 H. Wang, J. Wang, S. Zhang, Y. Pei and K. Zhuo, *ChemPhysChem*, 2009, **10**, 2516.
27 S. Katsuta, Y. Shiozawa, K. Imai, Y. Kudo and Y. Takeda, *J. Chem. Eng. Data*, 2010, **55**, 1588.
28 H. Weingärtner, H. G. Nadolny and S. Käshammer, *J. Phys. Chem. B*, 1999, **103**, 4738.
29 P. Walden, *Z. Phys. Chem.*, 1902, **30**, 513; 1922, 100, 512.
30 D. Saar, J. Bauner, H. Farber and S. Petrucci, *J. Phys. Chem.*, 1978, **82**, 1943.
31 (*a*) K. S. Cole and R. Cole, *J. Chem. Phys.*, 1941, **9**, 341; (*b*) C. J. Böttcher and P. Bordewijk, *Theory of Dielectric Polarization*, Elsevier, Amsterdam, 1978, Vol. 2.
32 M.-M. Huang, S. Bulut, I. Krossing and H. Weingärtner, *J. Chem. Phys.*, 2010, **133**, 101101.
33 A. S. Cavell, P. C. Knight and M. A. Sheikh, *J. Chem. Soc., Faraday Trans. I*, 1971, **67**, 2225.
34 C. J. Böttcher, *Theory of Dielectric Polarization*, Elsevier, Amsterdam, 1973, Vol. 1.
35 W. Schröer and C. Rybarsch, *Chem. Phys. Lett.*, 1986, **126**, 342.
36 C. G. Joslin, *Mol. Phys.*, 1982, **47**, 771.
37 H. Weingärtner, T. Merkel, S. Käshammer, W. Schröer and S. Wiegand, *Ber. Bunsenges. Phys.Chem.*, 1993, **97**, 970.
38 H. Weingärtner, S. Wiegand and W. Schröer, *J. Chem. Phys.*, 1992, **96**, 848.
39 S. Wiegand, M. Kleemeier, J.-M. Schröder, W. Schröer and H. Weingärtner, *Int. J. Thermophys.*, 1994, **15**, 1045.

# Measurements of the complete solvation response of coumarin 153 in ionic liquids and the accuracy of simple dielectric continuum predictions

**Mark Maroncelli,*[a] Xin-Xing Zhang,[bc] Min Liang,[a] Durba Roy[a] and Nikolaus P. Ernsting[b]**

*Received 2nd April 2011, Accepted 19th May 2011*
**DOI: 10.1039/c1fd00058f**

The complete solvation response of coumarin 153 (C153) has been determined over the range $10^{-13}$–$10^{-8}$ s in a variety of ionic liquids by combining femtosecond broad-band fluorescence upconversion and picosecond time-correlated single photon counting measurements. These data are used together with recently reported dielectric data in eight ionic liquids to test the accuracy of a simple continuum model for predicting solvation dynamics. In most cases the features of the solvation response functions predicted by the dielectric continuum model are similar to the measured dynamics of C153. The predicted dynamics are, however, systematically faster than those observed, on average by a factor of 3–5. Computer simulations of a model solute/ionic liquid system also exhibit the same relationship between dielectric predictions and observed dynamics. The simulations point to spatial dispersion of the polarization response as an important contributor to the over-prediction of solvation rates in ionic liquids.

## 1. Introduction

Solvation dynamics, the response of a solvent to perturbations of a molecular solute, is a fundamental aspect of liquid state behavior and an important determinant of solvent effects on chemical reactions. Solvation dynamics in conventional solvents was vigorously studied in the 1980s and 1990s with the result that we now possess a good, albeit incomplete, understanding of solvation dynamics in dipolar liquids.[1–6] The emergence of room temperature ionic liquids as a new class of solvent has rekindled interest in this phenomenon[7,8] and its relationship to charge transfer processes occurring in these solvents.[9,10] Central themes in this research have been to discover the extent to which our understanding of solvation in dipolar liquids is transferrable to ionic liquids, and to learn what distinctive features of solvation arise in a purely ionic environment.

Experimental characterizations of solvation dynamics in ionic liquids began roughly a decade ago,[11] with some related studies in higher melting salts[12] preceding the current wave of interest in room-temperature ionic liquids. With few exceptions,[13,14] experimental work has entailed dynamic Stokes shift measurements, most often using the time-correlated single photon counting (TCSPC) technique, which affords no better than 20 ps time resolution. In contrast to simple dipolar

[a]*Departments of Chemistry, The Pennsylvania State University, University Park, PA, USA. E-mail: maroncelli@psu.edu*
[b]*Humboldt Universität zu Berlin, Germany*
[c]*College of Physical Science, Nankai University, Tianjin, China*

solvents, where solvation is complete on such timescales, early studies showed that a significant portion of the dynamics occur in the $10^{-11}$–$10^{-8}$ s time-window accessible with such time resolution. The response measured in this way was found to be broadly distributed in time and strongly correlated to bulk viscosity.[8,15] But it was also recognized early on that roughly half of the solvation response in common ionic liquids is faster than can be detected with TCSPC.[16] Measurements with sufficient time resolution to accurately characterize the faster portions of solvation in ionic liquids have been few[13,17–20] and measurements of the complete response, which spans the range $10^{-13}$–$10^{-8}$ s, fewer still.[13,17,18,20]

The purpose of the present work is twofold. First, we report results of new dynamic Stokes shift measurements using the gold-standard probe coumarin 153 (C153, Fig. 1). To capture the entire solvation response, we combine the techniques of broadband fluorescence upconversion spectroscopy (FLUPS) having 80 fs time resolution[21,22] with TCSPC to cover times out to 20 ns. The second purpose of this work is to use these data to examine the accuracy of simple dielectric continuum models for predicting the solvation response of ionic liquids. These models are dynamical extensions of well-known Born–Onsager type expressions for solvation energies which use as their only input the frequency-dependent dielectric response of the medium. Despite the simplicity of such models, they have been shown to provide remarkably accurate predictions of solvation dynamics in a variety of conventional dipolar solvents.[3,23–26] It is natural to ask whether the same success will pertain in the case of ionic liquids, and several prior studies have suggested contrary answers to this question.[18,19,27,28] In the present study we take advantage of the fact that a number of groups have begun to measure the dielectric response of ionic liquids up to the high gigahertz and terahertz frequency regimes needed for rigorously testing these models.[29–36] We find that in a number of cases the solvation response functions predicted on the basis of simple continuum dielectric models are strikingly similar to those measured with C153. In the eight ionic liquids where dielectric data are available for comparison, the predicted and observed solvation times are clearly related; however the predictions are found to systematically overestimate the rate of solvation by a factor of 3–5. To amplify these observations, we also briefly describe computer simulations of solvation and dielectric response which show analogous behavior. These results suggest that, while simple dielectric theories are not quantitatively accurate, they do provide a sensible starting point for understanding solvation dynamics in ionic liquids. We also briefly discuss additional simulations which explore how inclusion of spatial dispersion in the dielectric response might improve the accuracy of dielectric continuum models.

## 2. Methods

### A. Experimental methods

Coumarin 153 (laser grade) was obtained from Lambda Physik and Exciton and used without further purification. Ionic liquids used here (see Fig. 1 for notation)

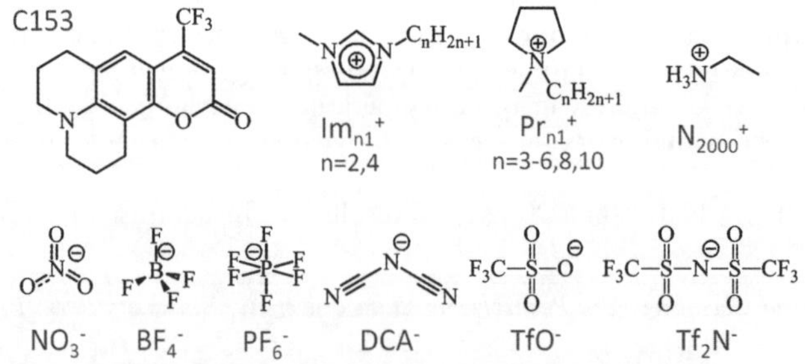

C153 $CF_3$ $C_nH_{2n+1}$ $Im_{n1}^+$ n=2,4 $Pr_{n1}^+$ n=3-6,8,10 $N_{2000}^+$ $NO_3^-$ $BF_4^-$ $PF_6^-$ $DCA^-$ $TfO^-$ $Tf_2N^-$

**Fig. 1** Structures of the probe solute and ionic liquid components studied.

 

were obtained from several sources. The methylimidazolium liquids $[Im_{21}][BF_4]$, $[Im_{41}][BF_4]$, $[Im_{41}][TfO]$, and $[Im_{41}][PF_6]$, as well as ethylammonium nitrate $[N_{2000}][NO_3]$, were obtained from Iolitec. $[Im_{41}][Tf_2N]$ was obtained from Kanto Chemical. The series of pyrrolidnium ionic liquids, $[Pr_{n1}][Tf_2N]$ with $n = 3, 4, 5, 6, 8$, and 10 were prepared by Gary Baker as described in ref. 37 and $[Im_{41}][DCA]$ was provided by Richard Buchner. The liquids were all dried in vacuum to water levels of less than 100 ppm by weight prior to use.

Femtosecond time-resolved emission measurements were made using a home-built broadband fluorescence upconversion spectrograph described in detail in refs. 21 and 22. A Ti : sapphire laser (FEMTOLASERS *sPro*) provides 30 fs, 500 μJ pulses at 800 nm and a 500 Hz repetition rate. This beam is split in a 6 : 1 ratio. The major portion (430 μJ) is used to drive a travelling-wave optical parametric amplifier of superfluorescence (TOPAS, LIGHTCONVERSION) which delivers 60 μJ pulses at 1340 nm to be used for optical gating. For optical pumping, the rest of the fundamental light is frequency doubled to 400 nm pulses having a width of 40 fs (FWHM) after compression. Their polarization is set with a half-wave plate. After traversing a variable delay stage, attenuated pump pulses (~ 1 μJ) are focused by a thin lens (200 mm focal length, fused silica) onto a sample cell to a diameter of about 0.1 mm. The cell is made from a pair of 0.2 mm fused silica windows spaced 0.2–0.3 mm apart. The sample solution is protected from water by enclosure within a sealed flow system consisting of a glass chamber containing $P_2O_5$ desiccant flushed with argon. With this sample arrangement the FLUPS samples typically maintained water levels of < 200 ppm over the course of an experiment. However, water contents near 1000 ppm were found in the cases of $[Im_{21}][BF_4]$ and $[Im_{41}][DCA]$. To check for the influence of water, experiments in $[Im_{21}][BF_4]$ having seven different water contents between 650–5000 ppm were measured. No systematic differences could be detected as a function of water content. The FLUPS experiments were run at room temperature, 20.5 ± 1 °C.

Steady-state and ps-ns fluorescence measurements were performed in the manner described in detail in ref. 15 Time-resolved emission was measured using the time-correlated single photon counting (TCSPC) technique. The excitation was supplied by the doubled output of a cavity-dumped Ti : sapphire laser which delivered ~150 fs pulses at 800 nm and a repetition rate of 2.7 MHz. Samples were contained in sealed 1 cm cuvettes into which a yellow filter glass was inserted to remove reflections near time zero. Emission was collected at magic angle through an ISA H10 monochromator with an emission band-pass of 8 nm. The overall response time of the TCSPC instrument was 25–30 ps (FWHM), as measured with a scattering solution. Emission transients were collected over ~20 wavelengths across the emission spectrum and the broadening effect of the instrumental response partially removed by fitting to a multi-exponential form with an iterative reconvolution algorithm. Time-resolved spectra were then reconstructed from the fits to these data using methods described in ref. 38. Samples for steady-state and TCSPC measurements were maintained at 20.5 ± 0.1 °C with water from a circulating bath.

Spectra covering the time window 100 fs–20 ns were obtained by combining the FLUPS (100 fs–600 ps) and TCSPC (50 ps–20 ns) data. To match the frequencies of the time-evolving spectra measured by these independent experiments, spectra were compared over times between 100 and 600 ps. Based on the average differences in the frequencies observed over this time window a shift, typically on the order of ~200 $cm^{-1}$, was applied to the FLUPS data. This procedure was deemed sensible because the TCSPC spectra are referenced to calibrated steady-state spectra whereas the frequencies of the FLUPS spectra are less certain due to variability of the frequency characteristics of the upconversion gate. Spectra from both experiments were then fit to log-normal lineshape functions and the fitted peak frequencies used to determine the spectral response function over the 100 fs–20 ns range.

### B. Simulation methods

Simulations of the "ILM2" ionic liquid model[39] were performed with a modified version of the DL_POLY program[40] as described in refs. 39 and 41. The ILM2 model is of standard interaction site form, consisting of a rigid 3-site cation and an atomic anion which interact through Lennard-Jones plus Coulomb terms. Potential parameters are provided in ref. 39. Neat ionic liquid simulations consisted of 343 ion pairs in a cubic periodic array. Equations of motion were integrated using a Verlet leapfrog algorithm with a time step of 5 fs. Long-range electrostatic interactions were treated *via* the standard Ewald method. Equilibrium simulations were run in the NPT ensemble at 1 bar and 350 K with Nose-Hoover (Melchionna) algorithms with relaxation times of 0.5 and 2.0 ps for the thermostat and barostat, respectively. The dielectric response of the neat liquid was determined from a 150 ns simulation of the total system dipole moment and electrical current time correlation functions as described in refs. 27, 42 and 43. Simulations of wave-vector dependent charge density correlation functions were carried out in the manner described in ref. 44 from an NVT simulation of 85 ns duration.

Simulations of solvation entailed embedding a single immobile atomic solute into the system of 343 ion pairs and performing equilibrium simulations, again in the NPT ensemble (1 bar, 350 K). The Lennard-Jones parameters of the solute were approximately those of a neutral single-site representation of benzene ($\sigma = 5.07$ Å, $\varepsilon = 3.09$ kJ mol$^{-1}$). The solvation response to charge ($S_q(t)$) and dipolar ($S_\mu(t)$) perturbations were calculated in the linear response limit as time correlation functions of fluctuations in the electrical potential $V$ and electric field $\vec{E}$[45] at the center of the solute:

$$S_q(t) \cong \frac{\langle \delta V \delta V(t) \rangle}{\left\langle (\delta V)^2 \right\rangle} \tag{1}$$

$$S_\mu(t) \cong \frac{\left\langle \vec{E} \cdot \vec{E}(t) \right\rangle}{\left\langle |\vec{E}|^2 \right\rangle} \tag{2}$$

where $\langle x \rangle$ denotes an average of $x$ over an equilibrium trajectory. Data for these correlation functions were collected over trajectories totalling 150 ns.

## 3. Results and discussion

### A. Experimental measurements

Representative time-resolved emission spectra of C153 in the ionic liquid [Im$_{41}$][PF$_6$] are shown in Fig. 2. Blue curves are FLUPS spectra and green points are TCSPC data which have been fit to log-normal functions (green curves). Using the peak frequencies $\nu(t)$ of such spectra the normalized spectral response function

$$S_\nu(t) = \frac{\nu(t) - \nu(\infty)}{\nu(0) - \nu(\infty)} \tag{3}$$

is calculated, using estimates of $\nu(0)$ derived from steady-state spectra[46] and $\nu(\infty)$ from extrapolations of TCSPC data at long times. We assume here that these spectral functions are dominated by solvation and are therefore equivalent to the solvation response functions of interest theoretically. A collection of some of the response functions observed in imidazolium ionic liquids is provided in Fig. 3. As found previously,[17,18,20] these functions consist of a sub-picosecond component followed by a broadly distributed relaxation which extends out to nanosecond times. Consistent with earlier work,[18] we find the sub-picosecond component to correlate with inertial

 

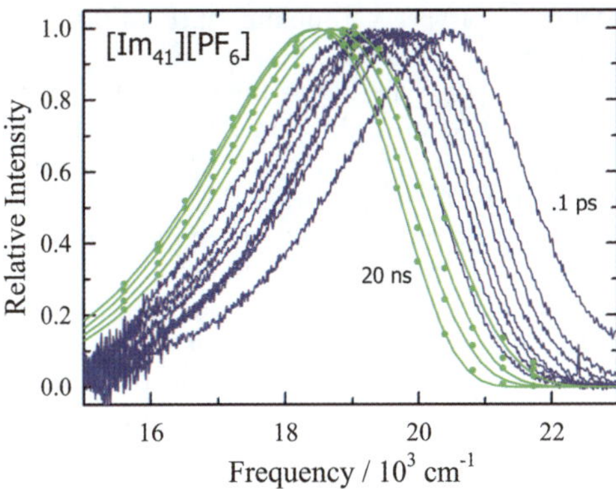


**Fig. 2** Time-resolved emission spectra of C153 in $[Im_{41}][PF_6]$. Blue curves are FLUPS spectra at times of 0.1, 0.5, 2, 10, 50, 100, and 500 ps. Green points are TCSPC data at times of 1, 2, 5, and 20 ns and the smooth green curves are lognormal fits to these data. Spectra, shown as the relative photon flux per unit frequency, are peak normalized to most clearly display the dynamic Stokes shift.

properties of the solvent ions and the remaining relaxation to be reasonably represented by a stretched exponential function having an average time constant proportional to solvent viscosity.[18,20] We note that while the general features of these response functions are similar to those observed with the solutes 4-dimethylamino-4′-cyanostilbene[17,18] and 4′-*N*,*N*-diethylamino-3-hydroxyflavone[20] the amplitudes of the inertial components obtained with C153 here are considerably larger (40–60%) than previously observed. We do not know whether this difference is a real solute effect or the result of the better time resolution in the present experiments.

Two other studies have previously reported fs measurements of C153 in three ionic liquids to which we can make comparison. We find that the relative amplitude of the inertial component of C153 in $[N_{2000}][NO_3]$ reported by Halder *et al.*,[19] ~30% within 300 fs, is consistent with our observations (see Fig. 6 below). A second study, by Lang *et al.*[13] using an experiment with 230 fs time resolution, reported that only 10–15% of the solvent response occurs in the first 10 ps in the ionic liquid $[Im_{31}][BF_4]$. This result contrasts sharply with the data shown in Fig. 3 for the two adjacent homologues $[Im_{21}][BF_4]$ and $[Im_{41}][BF_4]$, where we find $S_\nu(t)$ to have decayed by

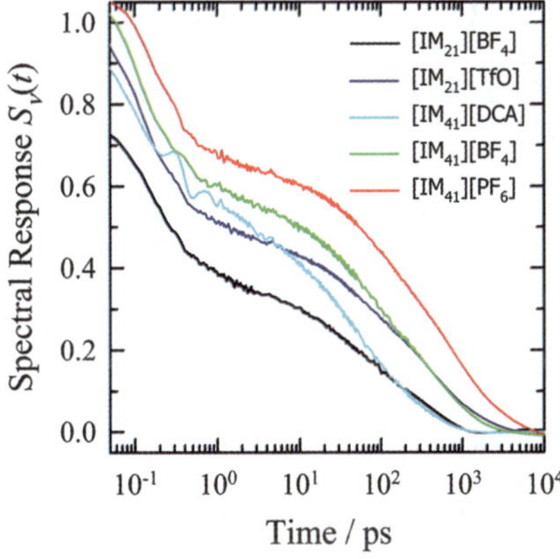


**Fig. 3** Spectral response functions of several imidazolium ionic liquids.

70% and 50% respectively in 10 ps. A troubling aspect of the results of Lang *et al.* is that the total magnitude of the dynamic Stokes shifts they observed is only ~500 $cm^{-1}$. Other experiments, with only 25 ps time resolution report shifts of twice this magnitude for C153 in comparable ionic liquids,[15] and the shifts observed here are more than four times as large. Thus we suspect some sort of systematic error is present in the measurements reported in ref. 13.

We now turn to comparison of the solvation response functions observed here to those predicted from a simple dielectric continuum model of solvation. The model we employ assumes the solute to be a point dipole centered in a spherical cavity of dielectric constant $\varepsilon_u = 2$ surrounded by a solvent having the generalized dielectric response function $\hat{\eta}(\nu)$. The calculations are essentially the same as those described in ref. 18 and previous work.[3] Normalized response functions are determined using the expressions

$$S_{dc}(t) = \frac{L_p^{-1}\left\{\left[\hat{\chi}(\infty) - \hat{\chi}(p)\right]/p\right\}}{\hat{\chi}(\infty) - \hat{\chi}(0)} \tag{4}$$

$$\hat{\chi}(p) = \frac{\hat{\eta}(p) - 1}{\hat{\eta}(p) + \frac{1}{2}\varepsilon_u} \tag{5}$$

where $L_p^{-1}$ denotes an inverse Laplace transform with respect to the variable $p = 2\pi i\nu$. An important distinction between the present work and that of ref. 18 is the use of the generalized function,[47]

$$\hat{\eta}(\nu) = \hat{\varepsilon}(\nu) + \frac{2i\sigma_0}{\nu} \tag{6}$$

This function describes the frequency dependence actually observed in dielectric measurements, whereas $\hat{\varepsilon}(\nu)$ is the function typically reported after removal of limiting conductivity contributions ($\sigma_0 = \hat{\sigma}(\nu \rightarrow 0)$) which cause divergence of the dielectric response at low frequencies.[33,48] In ref. 18 we neglected to add back the $\sigma_0$ term in eqn (6) and the calculations presented in that work are therefore incorrect.

The dielectric relaxation data that can be used to predict solvation dynamics in the ionic liquids studied here are summarized in Table 1. The majority of available data extend only to gigahertz frequencies and do not provide the complete characterization of $\hat{\varepsilon}(\nu)$ needed for accurate calculation of the solvation response. As indicated by the column labeled $\Delta\varepsilon = \varepsilon_\infty - n_D^2$ in Table 1, a significant portion of the dielectric response is not observed in many cases. Although the fraction missed $(\varepsilon_\infty - n_D^2)/(\varepsilon_0 - n_D^2)$ is typically <25%, this missing portion of the dielectric spectrum can have a large impact on the predictions and must be accounted for in some manner. Fig. 4 and 5 illustrate the problem and our method of approximating this missing contribution. In Fig. 4 we show dielectric data reported by three laboratories for the liquid $[Im_{41}][PF_6]$.[33,35,36] In panel (a) the parameterized functions are shown as reported. The frequency coverage of each data set is indicated by the solid portions of these curves. We note that the data of Stoppa *et al.*[33] (black) and Nakamura and Shikata[35] (green) are in good agreement over their region of overlap, whereas the data of Mizoshiri *et al.*[36] is rather different between 100 MHz–20 GHz. Of these three data sets, only the data of Stoppa *et al.*[33] extends to the THz range necessary to reduce $\varepsilon'(\nu)$ to close to its value $n_D^2$ at optical frequencies.

The solvation response functions computed directly from the three dielectric parameterizations of Fig. 4(a) are shown in Fig. 5(a). The short-time ($t$ < 10 ps) dynamics predicted by these three dielectric representations differ widely, as would be expected. Moreover, the transformations in eqn (4) and 5 are such that even at times of 100 ps the calculated $S(t)$ functions differ significantly. Inclusion of an effective high-frequency component, in the form of a damped harmonic oscillator

 

**Table 1** Dielectric sources and frequency ranges covered[a]

| Ionic liquid | Frequency range | $\varepsilon_0$ | $\varepsilon_\infty$ | $n_D$ | $\Delta\varepsilon$ | Reference |
|---|---|---|---|---|---|---|
| $[Im_{21}][BF_4]$ | 100 MHz–10 THz | 15.9 | 1.94 | 1.391 | 0.0 | T′09[34] |
| | 1 MHz–20 GHz | 13.6 | 6.7 | 1.411 | 4.7 | N′10[35] |
| $[Im_{21}][TfO]$ | 100 MHz–10 THz | 17.7 | 2.12 | 1.433 | 0.1 | B′11[59] |
| $[Im_{41}][BF_4]$ | 100 MHz–3 THz | 12.2 | 1.06 | 1.420 | −1.0 | S′08[33] |
| | 200 MHz-20 GHz | 11.0 | 4.08 | 1.421 | 2.1 | S′07[30] |
| | 1 MHz–20 GHz | 14.1 | 5.45 | 1.421 | 3.4 | N′10[35] |
| $[Im_{41}][PF_6]$ | 100 MHz–3 THz | 11.8 | 2.1 | 1.409 | 0.1 | St′08[33] |
| | 1 MHz–20 GHz | 14.1 | 4.8 | 1.407 | 2.8 | N′10[35] |
| | 40 MHz–40 GHz | 12.4 | 3.26 | 1.409 | 1.3 | M′10[36] |
| $[Im_{41}][DCA]$ | 100 MHz–10 THz | 12.3 | 2.36 | – | | T′09[34] |
| | 100 MHz–3 THz | | 2.13 | – | | St′08[33] |
| $[Im_{41}][Tf_2N]$ | 200 MHz-20 GHz | 11.5 | 3.03 | 1.427 | 1.0 | D′06[29] |
| | 1 MHz–20 GHz | 13.7 | 4.25 | 1.426 | 2.2 | N′10[35] |
| | 40 MHz–40 GHz | 12.7 | 2.7 | 1.426 | 0.7 | M′10[36] |
| $[Pr_{41}][Tf_2N]$ | 200 MHz-20 GHz | 11.7 | 2.42 | 1.423 | 0.4 | W′07[31] |
| $[N_{2000}][NO_3]$ | 10 MHz–1 THz (wet) | 26.4 | 3.2 | 1.452 | 1.1 | K′10[32] |

[a] All dielectric data correspond to a temperature of 25 °C. Mizoshiri *et al.* (M′10)[36] recorded data between 30–70 °C and in this case we extrapolated these data to 25 °C. (a) $\Delta\varepsilon = \varepsilon_\infty - n_D^2$.

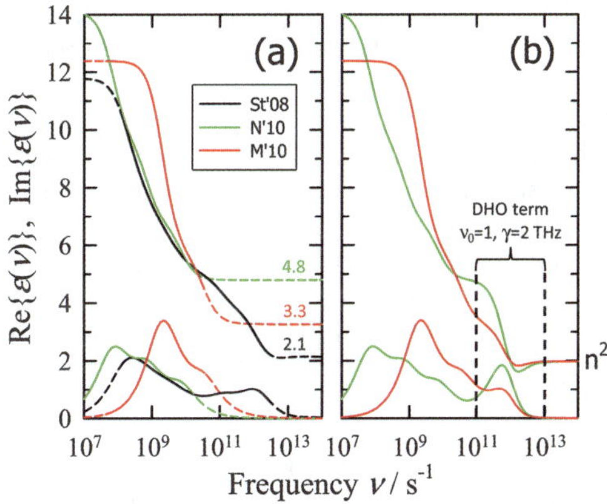


**Fig. 4** Dielectric permittivity ($\varepsilon'(\nu)$, upper curves) and loss ($\varepsilon''(\nu)$, lower curves) functions of $[Im_{41}][PF_6]$ calculated from the parameterizations provided in refs. 33, 35 and 36 (the legend refers to references as coded in Table 1). Panel (a) shows the functions as reported. The solid portions of these curves indicate the frequency region covered by each measurement. Limiting values of $\varepsilon_\infty = \varepsilon'(\nu \to \infty)$ are indicated. Panel (b) shows these same functions modified by the addition of a damped harmonic oscillator of amplitude $\Delta\varepsilon = \varepsilon_\infty - n_D^2$, which is used to account for unobserved high-frequency components of $\hat{\varepsilon}(\nu)$. (The data of Stoppa *et al.*[33] is not modified in this fashion because $\Delta\varepsilon \cong 0$.)

$$\Delta\hat{\varepsilon}(p) = (\varepsilon_\infty - n_D^2)\frac{\nu_0^2}{\nu_0^2 + p(p+\gamma)} \tag{7}$$

to account for the difference between $\varepsilon_\infty$ and $n_D^2$, results in much better agreement among the solvation predictions. Fig. 4(b) shows how this additional term, with parameters $\nu_0 = 1$ THz and $\gamma = 2$ THz, crudely mimics the behavior observed by Stoppa *et al.*[33] in the frequency range between $10^{11}$–$10^{13}$ Hz. Fig. 5(b) shows the

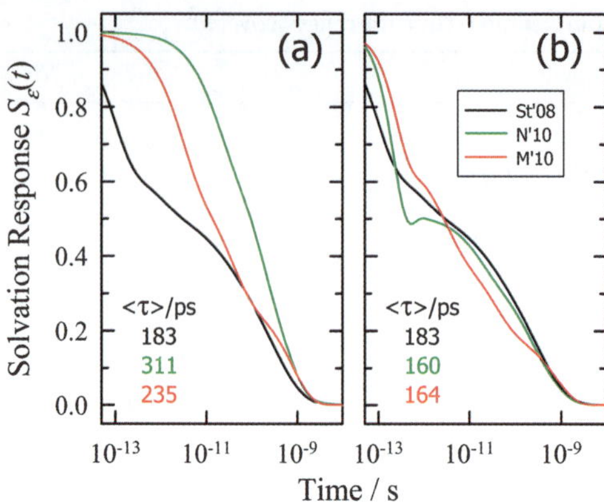


**Fig. 5** Solvation response functions calculated on the basis of the dielectric data in Fig. 4. In panel (a) the dielectric functions as parameterized are used and in panel (b) these functions are modified by the addition of a damped harmonic oscillator (eqn (5)). The integral times <$\tau$> of these response functions are indicated.

effect of this additional component on the $S(t)$ predictions. As illustrated in Fig. 5 (b), the two $\hat{\varepsilon}(\nu)$ representations that were close in agreement over the $10^8$–$10^{10}$ Hz range (St′08 and M′10) agree very well for times longer than 10 ps and are in moderately good agreement at shorter times as well. For the following comparisons we therefore include such an effective high-frequency contribution (with $\nu_0$ and $\gamma$ fixed at 1 and 2 THz) in all cases where the deficit at high frequencies $\Delta\varepsilon$ is greater than 0.2.

Four representative comparisons between dielectric continuum predictions and the solvation response functions observed with C153 are shown in Fig. 6. In most cases the dielectric continuum predictions are similar in shape to the observed $S_\nu(t)$ functions. For example, the amplitudes of the sub-picosecond inertial component as well as the temporal extent (stretching exponent in stretched exponential fits) of the longer time diffusive component are comparable in the predicted and observed functions. In some cases, especially [$Im_{41}$][DCA], the calculated and observed curves are strikingly similar, even to the extent of apparent agreement on the locations of small oscillations in $S_\nu(t)$.

Fig. 7 provides comparisons of times of the predicted and observed $S(t)$ functions for all eight of the ionic liquids listed in Table 1. Plotted here are the times required for $S(t)$ to relax to four different levels, 0.7, 0.5, 0.3, and 0.1. Roughly speaking, the times at which $S(t)$ reaches levels of 0.7 and 0.5 characterize the short time dynamics, whereas the 0.3 and 0.1 times characterize the slower components of the response. As illustrated in Fig. 7, the longer-time portion of the response is observed to be an average of 4.4-fold slower than the continuum predictions. Although there is some spread of predicted times due to differences in reported dielectric data, this spread is smaller than the average deviation between the measured and predicted times and does not obscure the basic result of this comparison. Figures 6 and 7 indicate that, while dielectric continuum calculations are a good starting point for modeling the solvation response, predicted dynamics are systematically faster than those observed.

## B. Molecular dynamics simulations

Computer simulations provide an alternative means to examine the relationship between solvation dynamics and dielectric relaxation in ionic liquids. Simulations are helpful in this regard in that they circumvent problems associated with experimental inaccuracies and provide insight into the reasons for a given experimental

 

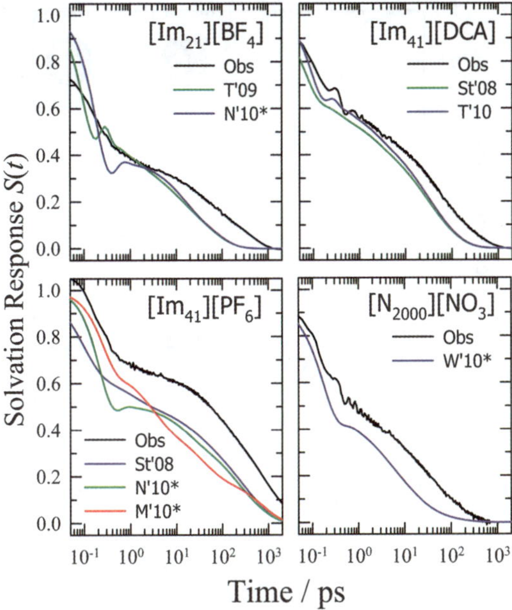


**Fig. 6** Comparison of C153 spectral response functions (black curves) with dielectric continuum calculations (colored curves). The legend to each plot indicates the source of the dielectric data according to the notation in Table 1. Asterisks indicate dielectric data that have been modified by the addition of a damped harmonic oscillator (eqn (7)) to account for unobserved dielectric components.

observation. From an extensive study of solvation in idealized ionic liquid models[49] we select one example designed to test the validity of continuum dielectric predictions under the most favorable circumstances. Fig. 8 shows simulations of the solvation response to two electrostatic perturbations in an atomic solute dissolved in the model ionic liquid "ILM2". ILM2 consists of the 4-site representation shown at the

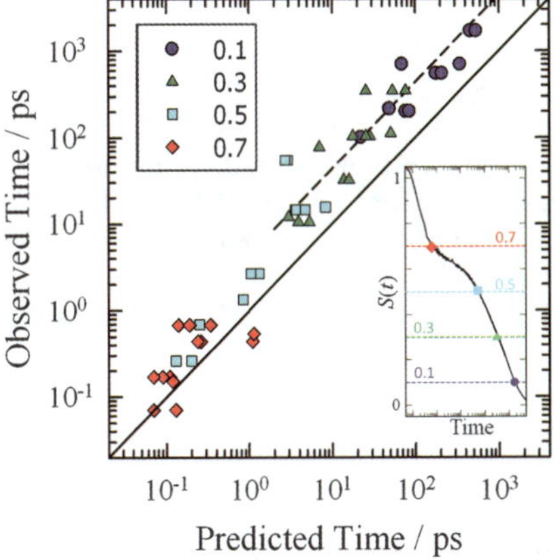


**Fig. 7** Comparison of the times required for observed and predicted response functions to reach levels of 0.7, 0.5, 0.3, and 0.1. The inset illustrates the meaning of these times. The solid line denotes equality between predicted and observed times and the dashed line is $t_{obs} = 4.4\ t_{calc}$. Eight different ionic liquids are represented here; multiple points having the same observed value indicate that multiple sets of dielectric data were used (Table 1).

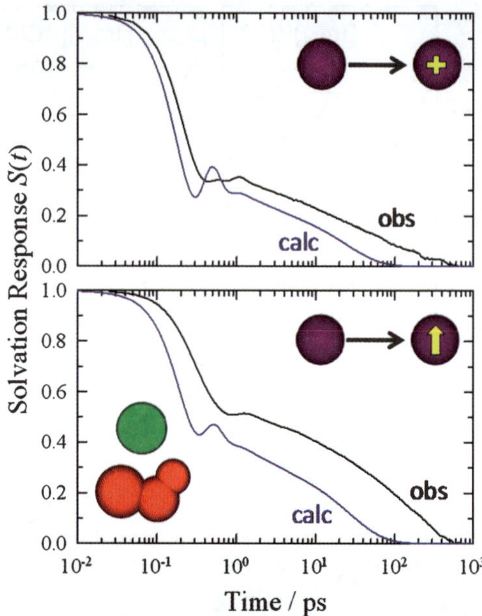


**Fig. 8** Simulated solvation response functions corresponding to ionic (top) and dipolar (bottom) perturbations in an atomic solute interacting with a 4-site model ionic liquid (black curves). The ionic liquid consists of the 3-site cation and single-site anion shown at the lower left. The probe solute is an uncharged immobile atom having Lennard-Jones parameters similar to those of the solvent anion. Solvation response functions are computed from time-correlation functions of the electrical potential and electric field observed in equilibrium simulations assuming a linear solvent response. The blue curves are the dielectric continuum predictions based on the dielectric response function simulated for the neat solvent.

lower left of the figure. It was designed as a generic model for an ionic liquid having properties close to those measured for $[Im_{41}][PF_6]$.[39] The simulated solvation dynamics ("obs") in Fig. 8 are linear response estimates (eqn (1)–(2)) rather than results of non-equilibrium simulations for the reason that the dielectric continuum models used here assume solvent linearity. But linear response estimates are in fact surprisingly accurate in these systems, even for changes of a full electronic charge.[41] The solvation response functions to a dipole perturbation shown in the bottom panel of Fig. 8 are quite similar to those observed in experiment, except that the extent of the inertial response in the atomic solute is slightly larger (50%) than observed experimentally with C153 (30–40%). In the case of the charge perturbation the inertial component is more prominent and the overall response is faster.

The blue curves in Fig. 8 are the results of applying the same dielectric continuum calculations described by eqn (4)–(6) with the dielectric response function simulated for the neat ILM2 liquid. ($\varepsilon_c = \varepsilon_\infty = 1$ for these nonpolarizable models.) The same sort of differences between the dielectric continuum predictions and the solvation response functions found experimentally are observed in these simulations. Whereas the shapes of the predicted and observed response functions are similar, the actual response is considerably slower — by factors of 4 and 7 at long times in the ionic and dipolar perturbations, respectively. Thus these simulations support the experimental conclusion that the speed of solvation is overestimated by dielectric continuum models.

It should be noted that the only other simulation comparison of this sort reported behavior opposite to that shown in Fig. 8. In simulations of a diatomic solute in a model of $[Im_{21}][PF_6]$, Shim and Kim[27] observed large departures from linearity when transitions were made between a non-dipolar (neutral pair or NP) solute and an "ion-pair" solute (IP) having $\pm 1e$ charges on the atoms. The NP $\rightarrow$ IP and IP $\rightarrow$ NP perturbations exhibited markedly different dynamics at long times, and response to both perturbations were found to occur significantly faster than dielectric continuum predictions.[27] We have examined the NP and IP solutes in the ILM2 solvent, and find that the different behavior in the two simulations primarily reflects differences in solute mobilities. The solutes in Fig. 8 are fixed in space whereas the solutes in the simulations of Ref. 27 are mobile. In the case of a dipolar perturbation, solute rotation speeds the solvation response significantly,[41] and small diatomic solutes like NP, undergo rapid jump-like reorientations[50] which can relax much of the solvation energy independently of solvent motion. The equilibrium correlation function representing the NP $\rightarrow$ IP perturbation to an immobile NP solute in ILM2 falls between the two simulated response functions of atomic solutes shown in Fig. 8. Although the effects observed by Shim and Kim are likely to correctly reflect the behavior of small solutes, simulations of mobile *versus* immobile models of C153 in ILM2,[49] as well as the large difference in solvation *versus* rotation times measured in ionic liquids,[15] suggest that the results in Fig. 8 are more representative of the situation pertaining to experimental measurements on fluorescent solutes.

We conclude this work by discussing preliminary simulations designed to reveal some of the reasons why solvation dynamics of even simple solutes such as those in Fig. 8 depart from dielectric continuum predictions. One reason is that dielectric experiments measure the neat solvent's response to time-varying electric fields that are constant in space, whereas a molecular solute interacts with solvent molecules through electric fields that vary significantly over molecular length scales. This idea was the starting point for numerous molecular theories of solvation dynamics in dipolar solvents,[2,5,6,51–53] most of which focus on the fact that the dielectric function needed for understanding molecular solvation is not $\eta(\nu)$ (or $\varepsilon(\nu)$ in non-conducting cases) but its wave-vector dependent generalization, $\eta_L(\nu, k)$ (or $\varepsilon_L(\nu, k)$). These latter functions describe the solvent response to sinusoidal electric fields which vary not only in time but also in space. Although this $k$-dependence cannot be measured experimentally, it is available through simulation, and an obvious starting point for understanding departures from simple continuum models is to examine how dynamics vary as a function of $k$. In both dipolar and ionic liquids represented by interaction site models, the $k$-dependence is most usefully described in terms of the charge density correlation function[54]

$$C_{qq}(k,t) = \frac{1}{V}\left\langle \rho_k(\vec{k},0)\rho_k(-\vec{k},t)\right\rangle \quad \rho_q(\vec{k}) = \sum_j q_j \exp(i\vec{k}\cdot\vec{r_j}) \tag{8}$$

where $V$ is the system volume, and $q_j$ is the charge at solvent site $j$. $C_{qq}(k,t)$ monitors fluctuations in charge density on different length scales in the neat solvent, *i.e.* in the absence of any solute. ($\eta_L(\nu, k)$ is simply related to the Laplace transform of $C_{qq}(k,t)$,[55] but the relation is not needed here.)

Fig. 9 shows some characteristics of $C_{qq}(k,t)$ in the ionic liquid ILM2. Panel (a) shows normalized time-correlation functions $\Phi_{qq}(k,t) = C_{qq}(k,t)/C_{qq}(k,0)$ at a few select wavevectors $k$. At the smallest wavevector accessible in these simulations, $k = 0.1$ Å$^{-1}$, the decay of $\Phi_{qq}(t)$ is approximately the same as the dielectric continuum prediction for a charge perturbation. At larger wavevectors, it decays progressively more slowly until reaching $k \sim 1$ Å$^{-1}$, which corresponds to polarization fluctuations with wavelengths ($2\pi/k$) close to the nearest neighbor separation in the liquid. Fig. 9(c) shows that the correlation time $\tau_{qq}$ is sharply peaked at this particular wavevector. At the maximum, the charge density fluctuations decay roughly 30 times more slowly than at the continuum limit, $k = 0$. The amplitude of the charge density fluctuations ($S_{qq}(k)$ in Fig. 9(b)) is similarly sharply peaked near $k = 1$ Å$^{-1}$.

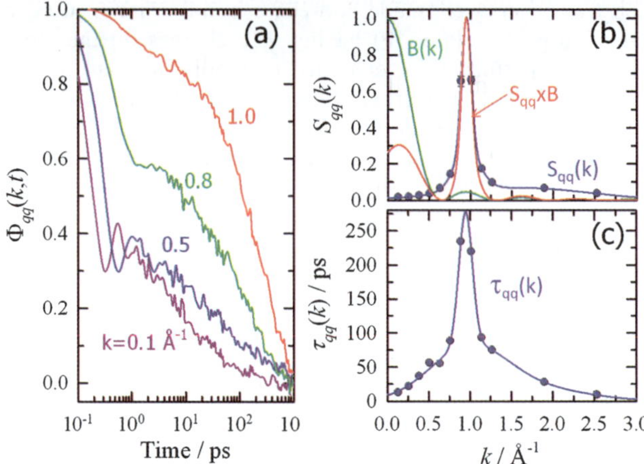


**Fig. 9** Characteristics of the charge density correlation function $C_{qq}$(k,t) of the ILM2 liquid. Panel (a) shows normalized time-correlation functions $\Phi_{qq}(k,t) = C_{qq}(k,t)/C_{qq}(k,0)$ at several wavevectors $k$. Panel (b) shows how the relative amplitudes $S_{qq}(k) = C_{qq}(k,0)$ (blue points & curve) and panel (c) the (integral) correlation times $\tau_{qq}$ of $C_{qq}$ vary with $k$. Also shown in panel (b) is the coupling function B($k$) appropriate for an ionic solute of radius $R = 2.5$ Å, the radius of the solute examined in Fig. 8. The red curve is the product $S_{qq}B$ which weights the contribution of different response times in the overall response.

The solvation response can be approximately expressed in terms of $C_{qq}(k,t)$ using what has been dubbed the non-local dielectric cavity theory ("DCT") by Raineri and coworkers.[56] The relations are:

$$\Delta U_{\rm el} = \langle U_{\rm el}(0)\rangle - \langle U_{\rm el}(\infty)\rangle \cong \frac{\beta}{(2\pi)^3}\int d^3\vec{k}S_{qq}(k)B(k) \tag{9}$$

$$S(t) = \frac{\langle U_{\rm el}(t)\rangle - \langle U_{\rm el}(\infty)\rangle}{\langle U_{\rm el}(0)\rangle - \langle U_{\rm el}(\infty)\rangle} \cong \frac{\int d^3\vec{k}C_{qq}(k,t)B(k)}{\int d^3\vec{k}S_{qq}(k)B(k)} \tag{10}$$

where $U_{\rm el}(t)$ is the electrostatic solute–solvent interaction energy at time $t$, $\beta = 1/k_{\rm B}T$, and $S_{qq}(k) = C_{qq}(k,0)$. The function $B(k)$ describes the wave vector dependence of the solute electrical potential. In the case of a spherical ion of radius $R$ whose charge $Q$ is switched off at $t = 0$ this function is given by[54]

$$B(k) = \left(\frac{4\pi Q}{k^2}\frac{\sin(kR)}{kR}\right)^2 \tag{11}$$

Equations (9) and (10) provide a convenient description of solvation in terms of a coupling function $B(k)$ which dictates how different portions of the solvent's $k$-dependent response are sensed by a solute. Such a description is not exact even in the linear response limit because no account is made for the effect of the solute on the solvent's structure and dynamics. But it nevertheless provides a useful framework for including some essential effects of solvent molecularity upon solvation. For example, within this description it is obvious why the solvation response functions in Fig. 8 should be slower than simple dielectric continuum predictions. The $k = 0$, continuum limit is only the slowest of a spectrum of relevant polarization modes. As shown in Fig. 9(b), an ionic solute of the size shown in Fig. 8 samples a range of slower polarization modes of the liquid and one therefore expects significantly slower relaxation than the continuum prediction.

 

Unfortunately, even for the atomic solutes examined here, the above DCT description is too simplistic to provide quantitatively accurate results. This fact is illustrated in Fig. 10 where we compare the predictions of equations (9)–(11) for solvation energies $\Delta U_{el}$ and integral solvation times $<\tau_{solv}>$ (solid curves) to charge jumps $S^+ \rightarrow S^0$ in atomic solutes of varying size. Although the model explains in a qualitative way how the solvation energy (*via* the dependence of $S_{qq}$ on $k$) and the solvation time (*via* the similar dependence of $\tau_{qq}$ on $k$) should both be greater than simple continuum estimates (dashed curves) the predictions are not more accurate. The oscillatory dependence of the solvation energies and especially the solvation times predicted by this approximation are unrealistic. This failure of the DCT approximation is disappointing but also predictable. In the case of dipolar solvents, Raineri and coworkers have already shown that the solute's influence on the solvent plays an important role in determining the solvation response.[6,56] They further showed how a "surrogate Hamiltonian" approach enables one to incorporate key features of solvation structure into the coupling function $B(k)$ while maintaining $C_{qq}(k,t)$ of the neat solvent as the only source of dynamical information.[56] We intend to use this surrogate Hamiltonian method to further investigate the connection between dielectric and solvation dynamics in future work in order to better understand the similarities and differences between solvation in dipolar and ionic solvents. For now, the results in Fig. 9 and 10 serve to partially answer the question of why systematic differences are observed between the experimental measurements of solvation dynamics and predictions based on dielectric continuum models.

## 4. Summary and conclusions

In this paper we have reported new measurements of the solvation response of coumarin 153 in room-temperature ionic liquids. The combination of broadband

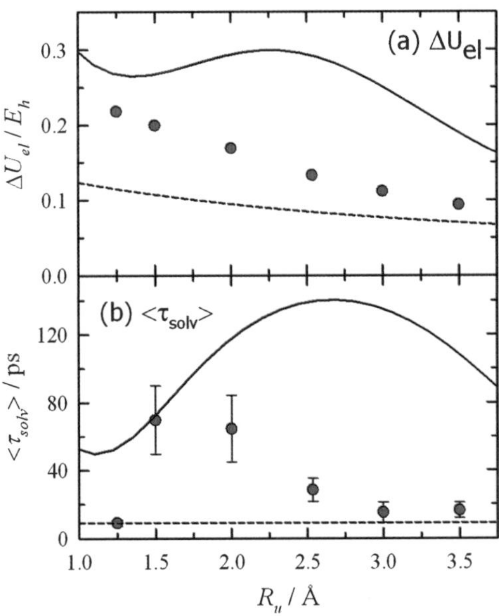


**Fig. 10** (a) Solvation energy changes and (b) solvation times corresponding to the processes $S^+ \rightarrow S^0$ in a series of atomic solutes in ILM2. The points are linear response estimates based on simulations of immobile solutes. The solid curves are the predictions of eqn (9)–(11) using the $C_{qq}(k,t)$ data from Fig. 9, and the dashed curves are simple continuum ($k = 0$) predictions.

fluorescence upconversion and time-correlated single photon counting provided temporal coverage over times of 100 fs to 20 ns, enabling observation of the complete solvation response in a variety of ionic liquids. These data were used to examine the applicability of simple dielectric continuum models of solvation to ionic liquids. The central result of this work was to show that while dielectric continuum models are likely to provide a good starting point for predicting solvation dynamics in ionic liquids, they do not offer the same accuracy as they do in conventional dipolar solvents. The eight experimental comparisons available here, as well as examples from idealized simulations, showed that dielectric continuum models predict solvation response functions which are too fast by factors on the order of 3–5. Computer simulations suggested that one reason for this systematic shortcoming of continuum models is related to neglect of spatial dispersion in the polarization response of ionic liquids. In future work we hope to investigate more sophisticated models shown to accurately incorporate this spatial dispersion in the case of dipolar solvents[56] and also compare the experimental data to two recent theories designed specifically for modeling solvation in ionic liquids.[28,57,58]

## Acknowledgements

The authors would like to thank Gary Baker for preparation of the pyrrolidinium ionic liquids studied here. We also acknowledge the help of Mohsen Sajadi in performing the FLUPS experiments and Richard Buchner and Thomas Sonnleitner for communicating the results of some dielectric measurements prior to publication. Funding for the Penn State portion of this research was provided by the Division of Chemical Sciences, Geosciences, and Biosciences, Office of Basic Energy Sciences of the U.S. Department of Energy through Grant DE-FG02-89ER14020. The Humboldt University work was supported by the German Research Foundation (DFG) priority program SPP 1191. X-XZ acknowledges support from the China Scholarship Council. MM also gratefully acknowledges a Research Award from the Alexander von Humboldt Foundation which enabled an extended stay at the Humboldt University during which time much of this work was performed.

## Notes and References

1 P. F. Barbara and W. Jarzeba, Ultrafast Photochemical Intramolecular Charge Transfer and Exited State Solvation, *Adv. Photochem.*, 1990, **15**, 1–68.
2 M. Maroncelli, The Dynamics of Solvation in Polar Liquids, *J. Mol. Liq.*, 1993, **57**, 1–37.
3 M. L. Horng, J. A. Gardecki, A. Papazyan and M. Maroncelli, Sub-Picosecond Measurements of Polar Solvation Dynamics: Coumarin 153 Revisited, *J. Phys. Chem.*, 1995, **99**, 17311–17337.
4 G. R. Fleming and M. Cho, Chromophore-Solvent Dynamics, *Annu. Rev. Phys. Chem.*, 1996, **47**, 109–134.
5 B. Bagchi and R. Biswas, Polar and nonpolar solvation dynamics, ion diffusion, and vibrational relaxation: role of biphasic solvent response in chemical dynamics, *Adv. Chem. Phys.*, 1999, **109**, 207–433.
6 F. O. Raineri and H. L. Friedman, Solvent Control of Electron Transfer Reactions, *Adv. Chem. Phys.*, 1999, **107**, 81–189.
7 A. Samanta, Dynamic Stokes Shift and Excitation Wavelength Dependent Fluorescence of Dipolar Molecules in Room Temperature Ionic Liquids, *J. Phys. Chem. B*, 2006, **110**, 13704–13716.
8 A. Samanta, Solvation Dynamics in Ionic Liquids: What We Have Learned from the Dynamic Fluorescence Stokes Shift Studies, *J. Phys. Chem. Lett.*, 2010, **1**, 1557–1562.
9 X. Li, M. Liang, A. Chakraborty, M. Kondo and M. Maroncelli, Solvent-Controlled Intramolecular Electron Transfer in Ionic Liquids, *J. Phys. Chem. B*, 2011, in press.
10 E. W. Castner Jr., C. J. Margulis, M. Maroncelli and F. Wishart James, Ionic Liquids: Structure and Photochemical Reactivity, *Annu. Rev. Phys. Chem.*, 2011, **62**, 85–105.
11 R. Karmakar and A. Samanta, Solvation Dynamics of Coumarin-153 in a Room-Temperature Ionic Liquid, *J. Phys. Chem. A*, 2002, **106**, 4447–4452.

 

12 E. Bart, A. Meltsin and D. Huppert, Static and Dynamic Solvation Measurements of Excited Large Dipole in Molten Salt, *Chem. Phys. Lett.*, 1992, **200**, 592–596.
13 B. Lang, G. Angulo and E. Vauthey, Ultrafast Solvation Dynamics of Coumarin 153 in Imidazolium-Based Ionic Liquids, *J. Phys. Chem. A*, 2006, **110**, 7028–7034.
14 M. Muramatsu, Y. Nagasawa and H. Miyasaka, Ultrafast Solvation Dynamics in Room Temperature Ionic Liquids Observed by Three-Pulse Photon Echo Peak Shift Measurements, *J. Phys. Chem. A*, 2011, **115**, 3886–3894.
15 H. Jin, G. A. Baker, S. Arzhantsev, J. Dong and M. Maroncelli, Solvation and Rotational Dynamics of Coumarin 153 in Ionic Liquids: Comparisons to Conventional Solvents, *J. Phys. Chem. B*, 2007, **117**, 7291–7302.
16 J. A. Ingram, R. S. Moog, N. Ito, R. Biswas and M. Maroncelli, Solute Rotation and Solvation Dynamics in a Room-Temperature Ionic Liquid, *J. Phys. Chem. B*, 2003, **107**, 5926–5932.
17 S. Arzhantsev, H. Jin, N. Ito and M. Maroncelli, Observing the Complete Solvation Response of DCS in Imidazolium Ionic Liquids, from the Femtosecond to Nanosecond Regimes, *Chem. Phys. Lett.*, 2006, **417**, 524–529.
18 S. Arzhantsev, H. Jin, G. A. Baker and M. Maroncelli, Measurements of the Complete Solvation Response in Ionic Liquids, *J. Phys. Chem. B*, 2007, **111**, 4978–4989.
19 M. Halder, L. S. Headley, P. Mukherjee, X. Song and J. W. Petrich, Experimental and Theoretical Investigations of Solvation Dynamics of Ionic Fluids: Appropriateness of Dielectric Theory and the Role of DC Conductivity, *J. Phys. Chem. A*, 2006, **110**, 8623–8626.
20 Y. Kimura, M. Fukuda, K. Suda and M. Terazima, Excited State Intramolecular Proton Transfer Reaction of 4'-*N*,*N*-Diethylamino-3-hydroxyflavone and Solvation Dynamics in Room Temperature Ionic Liquids Studied by Optical Kerr Gate Fluorescence Measurement, *J. Phys. Chem. B*, 2010, **114**, 11847–11858.
21 L. Zhao, J. Luis Perez Lustres, V. Farztdinov and N. P. Ernsting, Femtosecond fluorescence spectroscopy by upconversion with tilted gate pulses, *Phys. Chem. Chem. Phys.*, 2005, **7**, 1716–1725.
22 X.-X. Zhang, C. Wurth, L. Zhao, U. Resch-Genger, N. P. Ernsting and M. Sajadi, Femtosecond Broadband Fluorescence Upconversion Spectroscopy: Improved Setup and Photometric Correction, *Rev. Sci. Instrum.*, 2011, **82**, 063108.
23 M. Maroncelli, Continuum Estimates of Rotational Dielectric Friction and Polar Solvation, *J. Chem. Phys.*, 1997, **106**, 1545–1555.
24 X. Song and D. Chandler, Dielectric Solvation Dynamics of Molecules of Arbitrary Shape and Charge Distribution, *J. Chem. Phys.*, 1998, **108**, 2594–2600.
25 J. Ruthmann, S. A. Kovalenko, N. P. Ernsting and D. Ouw, Femtosecond relaxation of 2-amino-7-nitrofluorene in acetonitrile: Observation of the oscillatory contribution to the solvent response, *J. Chem. Phys.*, 1998, **109**, 5466–5468.
26 M. Sajadi, T. Obernhuber, S. A. Kovalenko, M. Mosquera, B. Dick and N. P. Ernsting, Dynamic Polar Solvation Is Reported by Fluorescing 4-Aminophthalimide Faithfully Despite H-Bonding, *J. Phys. Chem. A*, 2009, **113**, 44–55.
27 Y. Shim and H. J. Kim, Dielectric Relaxation, Ion Conductivity, Solvent Rotation, and Solvation Dynamics in a Room-Temperature Ionic Liquid, *J. Phys. Chem. B*, 2008, **112**, 11028–11038.
28 X. Song, Solvation dynamics in ionic fluids: an extended Debye–Hückel dielectric continuum model, *J. Chem. Phys.*, 2009, **131**, 044503.
29 C. Daguenet, P. J. Dyson, I. Krossing, A. Oleinikova, J. Slattery, C. Wakai and H. Weingärtner, Dielectric Response of Imidazolium-Based Room-Temperature Ionic Liquids, *J. Phys. Chem. B*, 2006, **110**, 12682–12688.
30 C. Schröder, C. Wakai, H. Weingärtner and O. Steinhauser, Collective Rotational Dynamics in Ionic Liquids: A Computational and Experimental Study of 1-Butyl-3-Methyl-imidazolium Tetrafluoroborate, *J. Chem. Phys.*, 2007, **126**, 84511.
31 H. Weingartner, P. Sasisanker, C. Daguenet, P. J. Dyson, I. Krossing, J. M. Slattery and T. Schubert, The Dielectric Response of Room-Temperature Ionic Liquids: Effect of Cation Variation, *J. Phys. Chem. B*, 2007, **111**, 4775–4780.
32 M. Krueger, E. Bruendermann, S. Funkner, H. Weingaertner and M. Havenith, Polarity fluctuations of the protic ionic liquid ethylammonium nitrate in the terahertz regime, *J. Chem. Phys.*, 2010, **132**, 101101.
33 A. Stoppa, J. Hunger, R. Buchner, G. Hefter, A. Thoman and H. Helm, Interactions and Dynamics in Ionic Liquids, *J. Phys. Chem. B*, 2008, **112**, 4854–4858.
34 D. A. Turton, J. Hunger, A. Stoppa, G. Hefter, A. Thoman, M. Walther, R. Buchner and K. Wynne, Dynamics of Imidazolium Ionic Liquids from a Combined Dielectric Relaxation and Optical Kerr Effect Study: Evidence for Mesoscopic Aggregation, *J. Am. Chem. Soc.*, 2009, **131**, 11140–11146.

35 K. Nakamura and T. Shikata, Systematic Dielectric and NMR Study of the Ionic Liquid 1-Alkyl-3-Methyl Imidazolium, *ChemPhysChem*, 2010, **11**, 285–294.
36 M. Mizoshiri, T. Nagao, Y. Mizoguchi and M. Yao, Dielectric permittivity of room temperature ionic liquids: A relation to the polar and nonpolar domain structures, *J. Chem. Phys.*, 2010, **132**, 164510.
37 H. Jin, B. O'Hare, J. Dong, S. Arzhantsev, G. A. Baker, J. F. Wishart, A. Benesi and M. Maroncelli, Physical Properties of Ionic Liquids Consisting of the 1-Butyl-3-Methylimidazolium Cation with Various Anions and the Bis(trifluoromethylsulfonyl) imide Anion with Various Cations, *J. Phys. Chem. B*, 2008, **112**, 81–92.
38 M. Maroncelli and G. R. Fleming, Picosecond Solvation Dynamics of Coumarin 153: The Importance of Molecular Aspects of Solvation, *J. Chem. Phys.*, 1987, **86**, 6221–6239.
39 D. Roy and M. Maroncelli, An Improved Four-Site Ionic Liquid Model, *J. Phys. Chem. B*, 2010, **114**, 12629–12631.
40 *DL_POLY_2.13*, edited by W. Smith and T. R. Forester (CCLRC Daresbury Laboratory, Daresbury, UK, 2001).
41 D. Roy, N. Patel, S. Conte and M. Maroncelli, Dynamics in an Idealized Ionic Liquid Model, *J. Phys. Chem. B*, 2010, **114**, 8410–8424.
42 C. Schroeder, M. Haberler and O. Steinhauser, On the computation and contribution of conductivity in molecular ionic liquids, *J. Chem. Phys.*, 2008, **128**, 134501.
43 C. Schroeder and O. Steinhauser, Using fit functions in computational dielectric spectroscopy, *J. Chem. Phys.*, 2010, **132**, 244109.
44 B. Perng and B. Ladanyi, Longitudinal dielectric properties of molecular liquids: Molecular dynamics simulation studies of $CH_3CN$, $C_6H_6$, and $CO_2$, *J. Chem. Phys.*, 1999, **110**, 6389–6405.
45 M. Maroncelli and G. R. Fleming, Computer Simulation of the Dynamics of Aqueous Solvation, *J. Chem. Phys.*, 1988, **89**, 5044–5069.
46 R. S. Fee and M. Maroncelli, Estimating the Time-Zero Spectrum in Time-Resolved Emission Measurements of Solvation Dynamics, *Chem. Phys.*, 1994, **183**, 235–247.
47 Note that the sign of the conductivity correction here depends upon the choice of sign in the dielectric function $\hat{\varepsilon}(\nu) = \varepsilon'(\nu) \pm i\varepsilon''(\nu)$. Equation 6 applies to the positive sign convention.
48 R. Buchner, G. T. Hefter and P. M. May, Dielectric Relaxation of Aqueous NaCl Solutions, *J. Phys. Chem. A*, 1999, **103**, 1–9.
49 D. Roy and M. Maroncelli, "Solvation in an Idealized Ionic Liquid Model," manuscript in preparation (2011).
50 Y. Shim, D. Jeong, M. Y. Choi and H. J. Kim, Rotational dynamics of a diatomic solute in the room-temperature ionic liquid 1-ethyl-3-methylimidazolium hexafluorophosphate, *J. Chem. Phys.*, 2006, **125**, 061102.
51 V. Kapko and S. A. Egorov, Polar solvation dynamics in supercritical fluids: a mode-coupling treatment, *J. Chem. Phys.*, 2004, **121**, 11145–11155.
52 D. V. Matyushov, On the microscopic theory of polar solvation dynamics, *J. Chem. Phys.*, 2005, **122**, 044502-044501-044511.
53 K. Nishiyama, T. Yamaguchi and F. Hirata, Solvation Dynamics in Polar Solvents Studied by Means of RISM/Mode-Coupling Theory, *J. Phys. Chem. B*, 2009, **113**, 2800–2804.
54 F. O. Raineri, Y. Zhou and H. L. Friedman, Ion Solvation Dynamics in an Interaction-Site Model Solvent, *Chem. Phys.*, 1991, **152**, 201–220.
55 F. O. Raineri and H. L. Friedman, Static Transverse Dielectric Function of Model Molecular Fluids, *J. Chem. Phys.*, 1992, **93**, 1–24.
56 F. O. Raineri, H. Resat, B.-C. Perng, F. Hirata and H. L. Friedman, A Molecular Theory of Solvation Dynamics, *J. Chem. Phys.*, 1994, **100**, 1477–1491.
57 H. K. Kashyap and R. Biswas, Dipolar Solvation Dynamics in Room Temperature Ionic Liquids: An Effective Medium Calculation Using Dielectric Relaxation Data, *J. Phys. Chem. B*, 2008, **112**, 12431–12438.
58 H. K. Kashyap and R. Biswas, Solvation Dynamics of Dipolar Probes in Dipolar Room Temperature Ionic Liquids: Separation of Ion-Dipole and Dipole–Dipole Interaction Contributions, *J. Phys. Chem. B*, 2010, **114**, 254–268.
59 R. Buchner, private communication (2011).


 

# High pressure studies of the transport properties of ionic liquids

**Kenneth R. Harris*[a] and Mitsuhiro Kanakubo[b]**

*Received 29th April 2011, Accepted 31st May 2011*
**DOI: 10.1039/c1fd00085c**

High pressure measurements have been made of viscosities, ion self-diffusion coefficients and electrical conductivities of ionic liquids, mainly of imidazolium salts. We review how these properties have been analysed in terms of the empirical Stokes–Einstein, Walden and Nernst–Einstein equations, and examine trends revealed by the phenomenological approach of velocity correlation coefficients and the more general theory of density scaling. Finally we examine the possibility of dynamic crossover in the transport properties of ionic liquids.

## 1. Introduction

In recent years we have made a number of measurements of the transport properties of pure ionic liquids, using high-pressure techniques wherever possible.[1] The use of high pressure allows the separation of the joint effects of temperature and density, something not possible when temperature is the only experimental variable. The analysis of the data has been made through several parallel, but inter-related routes. Here we first outline these methods and then examine the trends found for a number of ionic liquid groups.

Approaches such as the Stokes–Einstein equation and the related Walden relation are empirical, but are extremely useful for testing the consistency of diffusion, conductivity and viscosity data for a given material. The Nernst–Einstein relation is also empirical in its application to molten salts: it is of course only exact at infinite dilution of a salt in an electrolyte solution. Deviations from this relation are general for molten salts as the movement of an ion causes a distortion of the surroundings of a given ion (asymmetric in the case of nearly all ionic liquids). Berne and Rice showed long ago[2] that this distortion results in an anti-parallel internal electric field being added to any applied external field. Hence the charge mobility is less than the diffusive mobility, resulting in deviations from the Nernst–Einstein relation. This has been confirmed in many molecular dynamics simulations,[3] as well as experiment. It is not necessarily due to ion association.

A phenomenological analysis is provided by the use of velocity cross-correlation coefficients (VCCs),[3b,3c,4] which are time and ensemble averages of integrals of Kubo velocity cross-correlation functions, just as self-diffusion coefficients ($D_i$) are time and ensemble averages of integrals of velocity auto-correlation functions. VCCs for ionic liquids show consistent regularities when plotted against the viscosity,[1c,1g,1k,5] and this is explored further below. It can be shown that the

[a]*School of Physical, Environmental and Mathematical Sciences, University College, University of New South Wales, Australian Defence Force Academy, PO Box 7916, Canberra, BC, ACT 2610, Australia. E-mail: k.harris@adfa.edu.au; Fax: +61-2-6268 8017*

[b]*National Institute of Advanced Industrial Science and Technology (AIST), 4-2-1 Nigatake, Miyagino-ku, Sendai, 983-8551, Japan. E-mail: m-kanakubo@aist.go.jp; Fax: +81-22-232-7002; Tel: +81-22-237-2016*

Nernst–Einstein deviation parameter, Δ, is a function of differences between the VCCs ($f_{ij}$).[1c,1g,3b,5] Thus, for a 1 : 1 salt,

$$\Delta = \frac{c(2f_{+-} - f_{++} - f_{--})}{(D_{+} + D_{-})} \tag{1}$$

where $c$ is the molarity. Our work suggests that Δ is, if not constant for a given ionic liquid, only a very weak function of temperature and pressure.

A third approach is that of thermodynamic or density scaling. Originally applied to the relaxation times and viscosities of glassy and super-cooled complex liquids,[6] it has been extended to ionic liquids.[7] In essence, the transport properties are scaled as functions of the group of the group ($TV^{\gamma}$) where $\gamma$ is a constant for a given liquid. For a model inverse power law (IPL) fluid, the scaling is exact, with $\gamma = n/3$, $n$ being the power of the repulsive intermolecular potential. For model fluids with attractive terms, such as the Lennard-Jones, and for real liquids, the values of $\gamma$ obtained from scaling of the transport properties also depend on the attractive forces, that is, the scaling is to an *effective* IPL. Schrøder *et al.*[8] argue that scaling should not necessarily apply tò molten salts as the interaction potential includes Coulombic terms, not just Van der Waals repulsions and attractions. However, scaling for ionic liquids does appear to work just as well as for molecular liquids, though with generally much smaller values of $\gamma$, similar to those found for certain glassy liquids.[7]

Very recent work in our laboratories has revealed that some 1-ethyl-3-methylimidazolium salts appear to exhibit "crossover" effects, that is the temperature dependence of the transport properties changes over a narrow temperature range. The effect is a weak one, but seems to be similar to that shown by certain molecular liquids such as alkyl-substituted benzenes with chains longer than ethyl, di-methyl phthalate,[9] di-isobutyl phthalate,[9,10] and propylene carbonate.[11] Surprisingly little work has been done on diffusion in such simple liquids, except by Tyrrell and Watkiss,[12] with most effort having been concentrated on the dielectric relaxation times and the viscosity of fragile glass-formers,[13] though there are exceptions.[14] In two cases we find a crossover in both the viscosity and conductivity, but not the ion self-diffusion coefficients. This is similar to the result of Tyrrell and Watkiss for tracer diffusion in 2-methylpentane-2,4-diol.[12] A third example, taken from literature data, also seems to show a viscosity crossover. Our new measurements show that propylene carbonate has a self-diffusion crossover in addition to the known viscosity crossover.[11] Here we summarise our results for these systems: the experimental work, and a comparison with other 1-ethyl-3-methylimidazolium salts that do not show crossover behaviour, will be given in more detail elsewhere.[15]

## 2. Analysis of ionic liquid high pressure transport properties

### 2.1 The Stokes–Einstein and Walden relations

It has been shown previously that the hard sphere[16] and Lennard-Jones and non-hydrogen bonded molecular fluids[17] conform to fractional forms of the Stokes–Einstein relation for the self-diffusion coefficients and the viscosity over a wide range of thermodynamic states. For ionic liquids, the fractional form of the Walden relation between the molar conductivity (Λ) and the viscosity ($\eta$) also applies

$$\Lambda \propto \frac{D}{T} \propto \left(\frac{1}{\eta}\right)^{t} \tag{2}$$

with $t \leq 1$. In each case data for high-pressure isotherms and the atmospheric pressure isobar scale onto single lines. Fig. 1 shows such plots for [BMIM][$Tf_2N$]. Table 1 summarises our results for 1-methyl-3-alkyl imidazolium salts, $Pyr_{14}Tf_2N$ and $P_{2225}Tf_2N$, together with the Walden exponent for [$N(C_4H_9)_4$][$B(C_4H_9)_4$], calculated from high-pressure literature data.[19]

 

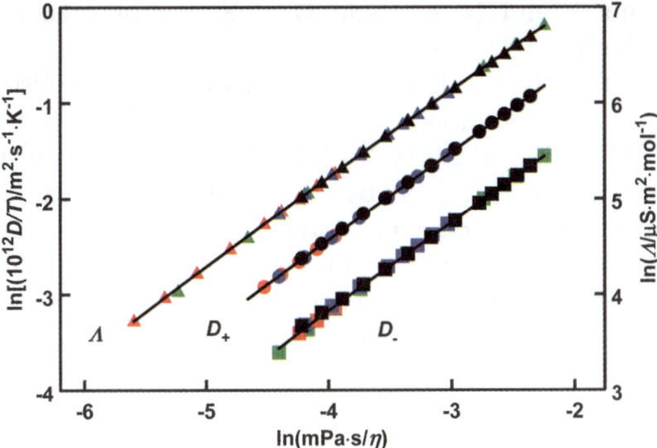


**Fig. 1** Stokes–Einstein and Walden plots for [BMIM][$Tf_2N$]. Triangles, molar conductivity; circles cation $D$; squares, anion $D$; green symbols, atmospheric $p$ isobaric points; red, 25 °C isotherm; blue, 50 °C isotherm; black, 75 °C isotherm.

**Table 1** Stokes–Einstein and Walden exponents, $t$ (see eqn (2))

| Substance[a] | $t(D_+)$ | $t(D_-)$ | $t(\Lambda)$ | $\ln(\Lambda T)$ *vs.* $\ln(D_+ + D_-)$ | Data refs |
|---|---|---|---|---|---|
| [BMIM][$BF_4$] | $0.85_2$ | $0.89_8$ | $0.87_8$ | $1.00_1$ | 1*f*,1*g* |
| [OMIM][$BF_4$] | $0.88_8$ | $0.88_8$ | $0.93_6$ | $1.01_9$ | 1*b*,1*d*,1*g* |
| [BMIM][$PF_6$] | $0.88_3$ | $0.92_0$ | $0.91_3$ | $0.98_6$ | 1*a*,1*c* |
| [HMIM][$PF_6$] | $0.84_5$ | $0.86_8$ | $0.91_2$ | $1.00_9$ | 1*e*,1*g* |
| [OMIM][$PF_6$] | $0.86_0$ | $0.94_8$ | $0.96_2$ | $0.98_6$ | 1*b*,1*d*,1*g* |
| [BMIM][$Tf_2N$] | $0.92_5$ | $0.93_3$ | $0.91_4$ | $0.97_8$ | 1*e*,15 |
| [$Pyr_{14}$][$Tf_2N$] | $0.93_7$ | $0.93_3$ | $0.92_5$[b] | $0.99_2$[b] | 1*i* |
| [$P_{2225}$][$Tf_2N$] | $0.96_3$[b] | $0.94_1$ | $0.97_3$[b] | $0.94_2$[b] | 18 |
| [$N(C_4H_9)_4$][$B(C_4H_9)_4$] | — | — | 0.79 | — | 19 |

[a] [BMIM]$^+$ = 1-n-butyl-3-methylimidazolium, H = n-hexyl, O = n-octyl, [$Pyr_{14}$]$^+$ = 1-n-butyl-1-methyl pyrrolidinium, [$P_{2225}$]$^+$= triethyl-n-pentylphosphonium, [$Tf_2N$]$^-$ = bis(trifluoromethylsulfonyl)amide. [b] Conductivities have been measured only at atm $p$ for these substances, thus far. [c] Standard devn $t$: ± 0.005, except [$N(C_4H_9)_4$][$B(C_4H_9)_4$], ± 0.015.

## 2.2 Velocity cross-correlation coefficients

The velocity cross-correlation coefficients are given in terms of the experimental transport properties by the relations of Schönert:[4b]

$$f_{++} \equiv \frac{N_A V}{3}\int_0^\infty \langle v_{+\alpha}(0)v_{+\beta}(t)\rangle dt = RT\kappa\left(\frac{M_-}{z_-FcM}\right)^2 - \frac{D_+}{\nu_+ c} \tag{3}$$

$$f_{--} \equiv \frac{N_A V}{3}\int_0^\infty \langle v_{-\alpha}(0)v_{-\beta}(t)\rangle dt = RT\kappa\left(\frac{M_+}{z_+FcM}\right)^2 - \frac{D_-}{\nu_- c} \tag{4}$$

and

$$f_{+-} \equiv \frac{N_A V}{3}\int_0^\infty \langle v_{+\alpha}(0)v_{-\beta}(t)\rangle dt = RT\kappa\frac{M_+M_-}{z_+z_-(FcM)^2} \tag{5}$$

where $N_A$ is the Avogadro constant, $V$ the volume of the ensemble, $\kappa$ is the conductivity, $c$ the amount concentration (molarity) of salt, $z_i$ ion charge numbers, $\nu_i$ ion stoichiometric numbers, and $M$, $M_+$, and $M_-$ are the molar masses of salt, cation, and anion, respectively.

The $f_{ij}$ have been found to scale with the viscosity, and indeed the values for the imidazolium salts with common anions scale almost identically, *e.g.* $f_{++}$ for [BMIM][$PF_6$], [HMIM][$PF_6$] and [OMIM][$PF_6$] fall on common curves when plotted against the fluidity, or reciprocal viscosity, $\phi$.[1g] This can be explained as follows. From eqn (5), the group $(z_+z_-cf_{+-}/T)$ is proportional to the molar conductivity $\Lambda$ ($= \kappa/c$). Hence, a graph of $\ln(z_+z_-cf_{+-}/T)$ *versus* $\ln(\eta)$ is really a Walden plot. The similarity of the fractional Stokes–Einstein equation parameters, $t$, for the ion-self-diffusion coefficients and that for the Walden plots (Table 1) therefore results from the similarity of the viscosity scaling of the three velocity cross-correlation functions, $f_{ij}$.[5] In other words, both velocity auto-correlation functions (self-diffusion) and velocity cross-correlation coefficients scale similarly with the viscosity.

Fig. 2 illustrates this for three examples, [BMIM][$BF_4$], [BMIM][$PF_6$] and [BMIM][$Tf_2N$]. The $[BF_4]^-$ and $[PF_6]^-$ salts, including those of the larger cations, $[HMIM]^+$ and $[OMIM]^+$, have $f_{++} < f_{--}$ (*n.b.* the $f_{ij}$ are negative quantities, so $f_{++}$ is larger in magnitude than $f_{--}$), whereas the opposite is true for [BMIM][$Tf_2N$] and the other $[Tf_2N]^-$ salts we have examined, namely, those of $[HMIM]^+$, $[OMIM]^+$ and $[Pyr_{14}]^+$. ($f_{+-}$ is always smaller in magnitude than the other $f_{ij}$). It would be interesting to see whether there are corresponding differences in liquid structure. These effects seem unrelated to the relative magnitudes of $D_+$ and $D_-$ in a given ionic liquid, which follow more complex patterns (Table 2). As is well known, in many cases anions diffuse more slowly than the larger cations, though the ratio $D_+/D_-$ generally decreases with increasing temperature and our data show it can increase or decrease with increasing pressure, depending on the salt (Fig. 3), and the temperature. At this stage, the regularities shown by the VCCs suggest that they might be as profitably simulated by computer calculations as the self-diffusion coefficients.

### 2.3 Nernst–Einstein equation

The Nernst–Einstein equation relating ion self-diffusion coefficients and the salt molar conductivity is

$$\Lambda = \frac{F^2}{RT}(\nu_+ z_+^2 D_+ + \nu_- z_-^2 D_-)(1 - \Delta) \tag{6}$$

As mentioned above, the Nernst–Einstein deviation parameter $\Delta$ is related to differences between the cross-correlation functions, as given, for a 1 : 1 salt, by eqn (1). (An analogous equation, involving cross-correlations between the velocities of distinct molecular pairs, exists for the mutual diffusion coefficient in a binary solution when written in terms of the two self-diffusion coefficients in the Hartley-Crank or Darken equation.[22]) $\Delta \leq 1$, except in the case of "super-ionicity".[3b,3i,5] $\Delta$ cannot be considered to be simply due to ion-association: this is well evidenced in molecular dynamics simulations.[3] We find that, within the ranges of temperature and pressure available, $\Delta$ is very nearly independent of temperature and pressure for a given ionic liquid. Values are given in Table 3. The trends seem to follow the cation, with $[OMIM]^+ > [HMIM]^+ > [BMIM]^+ > [Pyr_{14}]^+$.

### 2.4 Density scaling

As described in the Introduction, thermodynamic or density scaling can be applied to ionic liquids. In the first application of such scaling, to high-pressure viscosities,[7a] isotherms were superposed as functions of the group $(TV^\gamma)$ where $\gamma$ is the scaling parameter. $\gamma$ values of 2.25, 2.4 and 2.9 were obtained for [OMIM][$BF_4$], [OMIM][$PF_6$] and [BMIM][$PF_6$], respectively. Later, Pensado *et al.*[7b] made use of the theoretical expression of Casalini, Mohanty and Roland,[24] derived from the Avramov

 

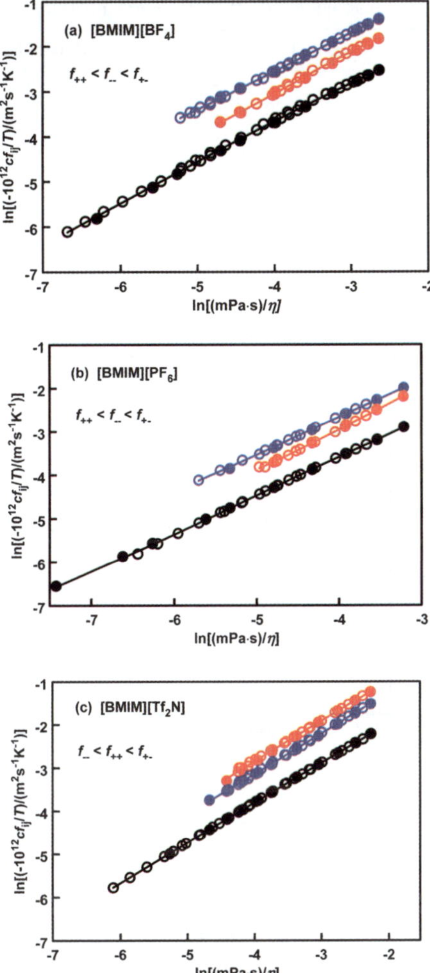


**Fig. 2** Plot of $\ln[(-c_i \cdot f_{ij}/T)/(10^{-15}\ \mathrm{m^2\ s^{-1}\ K^{-1}})]$ *versus* $\ln[(\mathrm{mPa \cdot s})/\eta]$ for [BMIM][$BF_4$], [BMIM][$PF_6$] and [BMIM][$Tf_2N$]. $\eta$ is the viscosity, the $f_{ij}$ are VCC and the ($cf_{ij}$) are the distinct diffusion coefficients of Freidman and Mills.[20] The slopes, $t$, are (in the order, $f_{++}, f_{--}$, and $f_{+-}$), for [BMIM][$BF_4$], $0.84_9$, $0.91_6$ and $0.88_9$; for [BMIM][$PF_6$], $0.86_3$, $0.95_4$ and $0.88_1$; and for [BMIM][$Tf_2N$], $0.93_6$, $0.93_5$ and $0.92_2$, all $\pm$ 0.005. Note the reversal of the order of the $f_{ij}$ for cation-cation and anion-anion interactions for [BMIM][$Tf_2N$]. Blue symbols, $f_{++}$; red symbols, $f_{--}$; black symbols, $f_{+-}$; solid circles, atmospheric $p$ isobar; open circles, high pressure isotherms at 25, 50 and 75 °C, (70 °C for [BMIM][$PF_6$]).

entropic model for glass-transition dynamics,[25] again for viscosities. This is a thermal activation model, not one based on kinetic theory. For the viscosity, the expression, which contains two fitted parameters, $\gamma$ and $\phi$, is:

$$\eta\,(T, V) = \eta_0 \exp\left[\left(\frac{A_\eta}{T{V_m}^{\gamma_\eta}}\right)^{\phi_\eta}\right] \tag{7}$$

where $V_m$ is the molar volume. $\gamma$ and $\phi$ for a given property are inter-related, and can be empirically fitted to an equation of the form

$$\phi = \frac{Z}{1 + \gamma} \tag{8}$$

**Table 2** The ratio $D_+/D_-$ for ionic liquids at atmospheric pressure

| | $[BF_4]^-$ | $[PF_6]^-$ | $[Tf_2N]^-$ | $[CF_3SO_3]^-$ | Data ref. |
|---|---|---|---|---|---|
| $[EMIM]^+$ | — | — | >1 | >1 | 15 |
| $[BMIM]^+$ | ~1 | >1 | >1 | — | 1*c*,1*g* |
| $[HMIM]^+$ | <1 | >1 | ~1 | — | 1*c*,1*g* |
| $[OMIM]^+$ | <1 | <1 | ~1 | — | 1*c*,1*g* |
| $[Pyr_{14}]^+$ | — | — | >1 | — | 1*i* |
| $[P_{2225}]^+$ | — | — | <1 | — | 18 |
| $[3\text{-ABN}_{13}]^{+a}$ | — | — | <1 | — | 1*i* |
| $[\text{choline}]^{+b}$ | — | — | >1 | — | Unpublished[21] |

[a] *N*-methyl-*N*-propyl-3-azabicyclo[3.2.2]nonaneium. [b] 2-hydroxy-*N*,*N*,*N*-trimethylethanaminium.

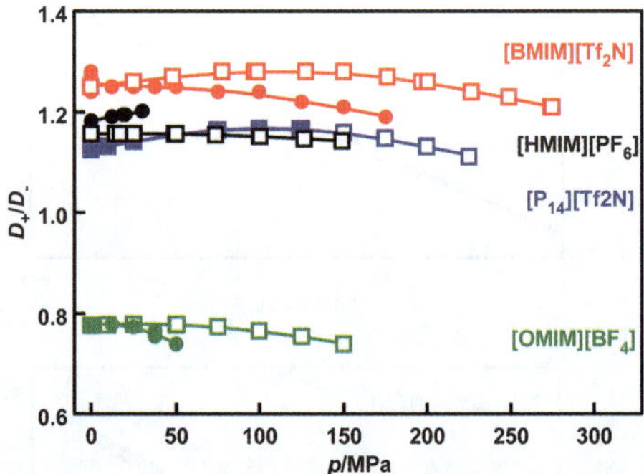


**Fig. 3** Pressure dependence of $D_+/D_-$ for ionic liquids. Red symbols, $[BMIM][Tf_2N]$; black, $[HMIM][PF_6]$; blue, $[Pyr_{14}][Tf_2N]$; green, $[OMIM][BF_4]$; solid circles, 50 °C (except $[OMIM][BF_4]$, 60 °C; open squares, 75 °C.

**Table 3** Nernst–Einstein deviation parameter, Δ

| | $[BF_4]^-$ | $[PF_6]^-$ | $[Tf_2N]^-$ | Data ref. |
|---|---|---|---|---|
| $[BMIM]^+$ | 0.41 ± 0.01 | 0.38 ± 0.01 | 0.37 ± 0.01 | 1*g*,1*c*,15 |
| $[HMIM]^+$ | — | 0.44 ± 0.02 | 0.40 ± 0.01 | 1*g*,21,23 |
| $[OMIM]^+$ | 0.51 ± 0.01 | 0.47 ± 0.01 | 0.49 ± 0.01[a] | 1*g*,21 |
| $[Pyr_{14}]^+$ | — | — | 0.31 ± 0.01[a] | 1*i* |

[a] atm *p*.

where $Z$ depends somewhat on the type of liquid considered, *e.g.* simple molecular liquid, ionic liquid, or glassy polymer, *etc.* This model has also been applied to the high-pressure dielectric relaxation times of 1-butyl-1-methylpyrrolidinium bis(oxalato)borate.[26] $\gamma$ can be linked[24,25,27] to the thermodynamic Grüneisen parameter, [$\gamma_G = (C_p/C_V - 1)/(\alpha_p T)$], which contains the heat capacity ratio and the coefficient of thermal expansion, and a further scaling parameter derived from the equation of state.[26]

In further work, López *et al.*[7c] extended the treatment to ionic liquid self-diffusion coefficients and conductivities. Using the analogous equations

$$D(T,V) = D_0 \exp\left[-\left(\frac{A_D}{TV_m^{\gamma_D}}\right)^{\phi_D}\right] \tag{9}$$

 

and

$$T\Lambda(T,V) = (T\Lambda)_0 \exp\left[-\left(\frac{A_\Lambda}{TV_m{}^{\gamma_\Lambda}}\right)^{\phi_\Lambda}\right] \tag{10}$$

(together with alternate forms), they assessed fits to 10 ionic liquids and 5 tetraalkylammonium salts. The $\gamma$ parameters derived from the viscosities and the molar conductivities tend to be slightly larger than those calculated for the self-diffusion coefficients, though the agreement can be improved by fitting ($D/T$) instead of $D$, by analogy with the Stokes–Einstein relation. The $\gamma$ values lie between 1 and 3.5, being lowest for the conductivities of the tetraalkylammonium tetrafluoroborates. All are much lower than those generally found for simple molecular liquids, *e.g.* toluene, for which $\gamma_D = 4.9$ and $\gamma_\eta = 7.5$. Fig. 4 shows a plot for $D_+$ for [BMIM][$BF_4$], fitted to eqn (9) with $\gamma = 2.33$ and $\phi = 1.91$.

Schrøder *et al.*[8] and Fragiadakis and Roland[6d] have argued that it is reduced forms of the transport properties that should be scaled. For the viscosity and self-diffusion coefficient the appropriate expressions are:

$$\tilde{\eta} = \eta v^{2/3}/\sqrt{mkT} \tag{11}$$

$$\tilde{D} = Dv^{-1/3}\sqrt{m/kT} \tag{12}$$

where $m$ is the molecular mass and $v$ the molecular volume ($V_m/N_A$). Fragiadakis and Roland[6d] showed that this procedure gives values of $\gamma_D$ and $\gamma_\eta$ much closer to one another, the former being increased and the latter decreased, relative to the fits for the unreduced properties. This was more extensively tested by López *et al.*[7c] (using least squares fits to simple polynomials in ($TV^\gamma$) as eqn (8) and (9) cannot apply to the reduced quantities), confirming the Fragiadakis-Roland conclusion for both molecular (*e.g.* toluene, $\gamma_D = 5.95$ and $\gamma_\eta = 5.62$) and ionic liquids, though it remains the case that the $\gamma$ are usually slightly different in value for the two properties. Note that for the ionic liquids, the "molecular" volume is taken to be that of the salt ion-pair.

## 3. Dynamic crossover effect

The term "dynamic crossover effect" refers to change in the temperature dependence of a transport property—generally the viscosity—over a short temperature range. Usually this is observed in fitting data to the Vogel-Fulcher-Tammann (VFT)

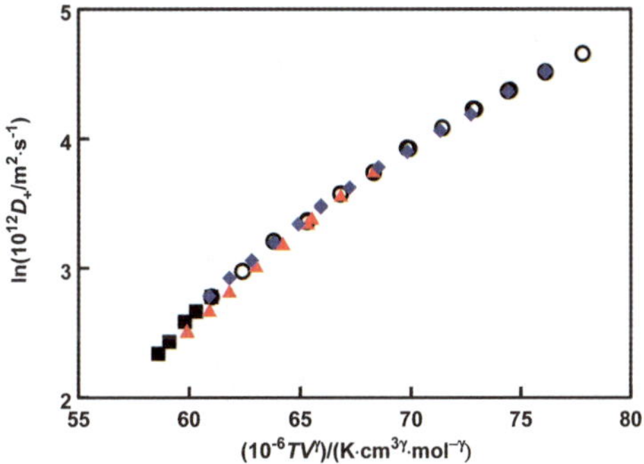


**Fig. 4** Density scaling: fit to eqn (9) for $D_+$ for the ionic liquid [BMIM][$BF_4$]. Black open circles, atm $p$, 25–80 °C; black squares, 25 °C, (0.1–50 MPa); blue diamonds, 50 °C, (0.1–150 MPa); red triangles, 75 °C, (0.1–300 MPa).

equation, with different parameters being required for different ranges of temperature. Early examples appear in the work of Barlow *et al.*[9]—propyl and butyl substituted benzenes, isobutyl halides, and dimethyl and di-isobutyl phthalates. The transition temperatures are close to the freezing points, but sometimes above and sometimes below; like ionic liquids, all of these materials are readily supercooled, and measurements could be made well below the normal freezing points. The effect is subtle, as ethyl and n-hexylbenzene, and di-n-butyl phthalate show no such transition.[9]

More recent work has concentrated on the viscosities and dielectric relaxation times as a function of pressure.[28] An important result is that the transitions occur at critical values of the property concerned, independent of the state parameters: that is, the crossover point is governed by the time-scale of the relaxation.

As mentioned above in the Introduction, there appear to be very few examples of a transition for a diffusion process, particularly for a molecular liquid (water is excluded from this discussion due its special nature[29]). Indeed Watkiss and Tyrrell, while observing a transition in the viscosity of 2-methylpentane-2,4-diol at 20 °C, did not find any corresponding change in tracer diffusion coefficients, which showed a smooth temperature dependence.[12]

Here we report a summary of transport property results for two ionic liquids that seem to show some of these effects (Table 4), both 1-ethyl-3-methylimidazolium salts, the tetracyanoborate, [EMIM][TCB], and the methanesulfonate, [EMIM][$CH_3SO_3$]. We could find no evidence for crossover effects for the self-diffusion coefficients of [EMIM][$Tf_2N$], for any of the transport properties of the trifluoromethanesulfonate, [EMIM][$CF_3SO_3$], nor for the viscosity and cation self-diffusion coefficient of the tris(pentafluoroethyl)trifluorophosphate, [EMIM][FAP]. ($D_-$ was not measured as the $[FAP]^-$ anion has too short a $T_2$ relaxation time.) The detailed experimental measurements will be described in a separate publication.[15] The techniques used were falling body (Canberra)[1a] ($\eta$, expanded uncertainty $\pm$ 2%) and rotating cylinder Stabinger (Sendai) viscometers ($\eta$, expanded uncertainty $\pm$ 2%),[1i] an impedance analyser (Bio-Logic SP-150) for the conductivity measurements (Sendai) ($\Lambda$, expanded uncertainty $\pm$ 2%),[1h] and a constant gradient spin-echo NMR diffusiometer (Canberra) ($D_i$, expanded uncertainty $\pm$ 2%).[1c] Densities were measured with Anton-Paar DMA5000 vibrating-tube densimeters,[1a,1b] (expanded uncertainty $\pm$ 0.0001 g $cm^{-3}$) and water contents were determined by Karl Fischer titration (Metrohm 831 KF Coulometer),[1i] in both laboratories. Differential scanning calorimetry (DSC) was performed with a Bruker DSC3200 unit (Sendai).

We first found evidence for the cross-over effect for the viscosity of [EMIM][TCB]. Usually we fit data to both the VFT and Litovitz equations.

$$\eta = A\exp(B/(T - T_0)) \tag{13}$$

$$\eta = A'\exp(B'/T^3) \tag{14}$$

**Table 4** 1-ethyl-3-methylimidazolium salts showing occurrence of dynamic crossover effects

| Anion | $\eta$ | $\Lambda$ | $D_+$ | $D_-$ |
|---|---|---|---|---|
| $[TCB]^{-a}$ | Yes, 25 °C | Yes, 24 °C | No | — |
| $[CH_3SO_3]^-$ | Yes, 33 °C | Yes, 29 °C | No[c] | No[c] |
| $[CF_3SO_3]^-$ | No | No | No | No |
| $[FAP]^{-b}$ | No | — | No | — |
| $[Tf_2N]^-$ | — | — | No | No |

[a] Tetracyanoborate. [b] Tris(pentafluoroethyl)trifluorophosphate. [c] Not separately measurable by spin-echo NMR at 20 MHz: measurement of average $D$ shows no occurrence.

 

In this case, the 3-parameter VFT equation fits well enough, but, unusually, the Litovitz equation did not, showing an apparent transition at about 25 °C. Barlow *et al.*[9] also detected transitions in the same way, with either Litovitz or Barlow-Lamb plots (where $T^3$ in eqn (14) is replaced by $T^4$), depending on the substance examined. They then fitted their data above and below the transition temperatures with VFT equations with different parameters. Fig. 5 shows Litovitz plots for the viscosity, molar conductivity, and self-diffusion coefficients of [EMIM][TCB], [EMIM][$CH_3SO_3$] and [EMIM][$CF_3SO_3$]. (Note that due to the presence of protons on both ions of [EMIM][$CH_3SO_3$], the diffusion coefficient observed is a value averaged over the relative signal strengths). Our results for the viscosity of [EMIM][$CH_3SO_3$] are in close agreement with the data sets of Hasse *et al.*[30] (0–60 °C) and of Blesic *et al.*[31] (22–86 °C), both of which pass through the transition point. There appear to be weak transitions for the viscosities and molar conductivities of [EMIM][TCB] at about 25 °C, and of [EMIM][$CH_3SO_3$] at about 29–33 °C, but not for the self-diffusion coefficients. As a check, the viscosity data have been fitted to a variant of eqn (14) with the temperature exponent as a fitted coefficient (as has been found necessary for the quite viscous hydrocarbon, squalane[32]), but the change in slope was still apparent.

Literature viscosity data for [EMIM][$MeOHPO_2$] (1-ethyl-3-methylimidazolium methylphosphonate)[30] also appear to show a very weak transition at about 25 °C. We have re-examined our previously published viscosity and conductivity data[1a,b,d–g] for 1-alkyl-3-methylimidazolium salts in the light of these results, but it is unclear why dynamic crossover is not observed for those imidazolium salts with longer alkyl (butyl, hexyl and octyl) chains on the cations.

The transitions for [EMIM][$CH_3SO_3$], which is readily super-cooled, occur in the region of the normal freezing point. DSC scans showed a glass transition at approximately −72 °C, a broad exothermic peak around −6 °C, due to crystallization of the glass during heating, and a broad endothermic melting peak from 19 to 38 °C, turning at 34 °C. Such broad peaks are typical for ionic liquids.[33] A melting peak for [EMIM][TCB] was observed from 0 to 14 °C, turning at 9 °C, but this is somewhat below the transition temperature, 25 °C. Interestingly, the fractional Walden plots are smooth, showing no discontinuity, with exponents of $0.91_5$ and $0.87_7$ for [EMIM][TCB] and [EMIM][$CH_3SO_3$], respectively. So the crossover behaviours for the two properties seem to be complementary.

A much studied molecular liquid showing a high-temperature dynamic crossover is the cyclic molecule propylene carbonate (PC), or 4-methyl-1,3-dioxolan-2-one.[11] (PC also shows a different, low-temperature, crossover at −81 °C: its glass temperature is −117 °C[34]). Fig. 6 shows viscosities due to Barthel *et al.*[35] and self-diffusion coefficients measured for this work. This substance does not fit the Litovitz equations, (as is usual for molecular liquids other than diols or triols, for which it was originally proposed[36]) but does show two distinct linear regions in both a Barlow-Lamb plot and the VFT plot as illustrated, for both the viscosity and self-diffusion coefficient. Here the transition is more sharply defined, at 30.35 °C for the viscosity and 31.6 °C for the self-diffusion coefficient. One can also estimate a value of 35 °C from the Stickel analysis of the same viscosity data by Casalini and Bair[11] (their Fig. 5). An earlier estimate, due to Stickel *et al.*[13a] is 27 °C, based on the dc conductivities of ionic tracers and dielectric loss peak frequencies. Again, it is interesting that the fractional Stokes–Einstein plot shows no discontinuity, with an exponent of $0.95_1$. This contrasts with the case of the two ionic liquids above, where the viscosity and diffusion coefficients behave differently with regard to their temperature dependences.

The VFT equation, with three coefficients, is quite flexible and fits the transport data of [EMIM][TCB], [EMIM][$CH_3SO_3$] and [EMIM][$MeOHPO_2$] quite well, with single fits for each property, including the transition region. Thus the apparent crossover in the ionic liquids is much weaker than for propylene carbonate, where the data could be resolved into two regions with different VFT coefficients. In

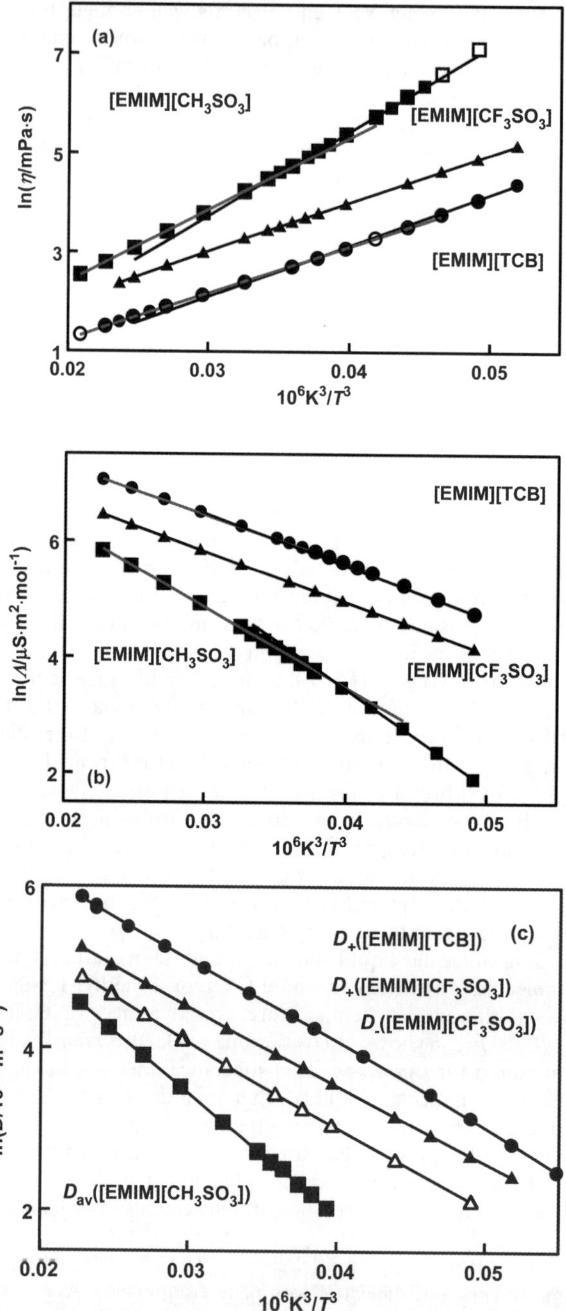


**Fig. 5** **(a)** Litovitz plots for the viscosities of [EMIM][TCB], [EMIM][$CF_3SO_3$] and [EMIM][$CH_3SO_3$]. The blue lines represent the higher temperature data; the open symbols represent the Stabinger viscometer data and the closed ones falling body results. Litovitz eqn fits: [EMIM][TCB], −10–20 °C, ln($\eta$/mPa·s) = −1.1791 + 107.49 × $10^6$ $K^3/T^3$; 23–90 °C, ln($\eta$/mPa·s) = −0.6324 + 93.76 × $10^6$ $K^3/T^3$; [EMIM][$CF_3SO_3$], −5–75 °C, ln($\eta$/mPa·s) = 0.04575 + 99.170 × $10^6$ $K^3/T^3$; [EMIM][$CH_3SO_3$], 0–30 °C, ln($\eta$/mPa·s) = −1.5835 + 176.05 × $10^6$ $K^3/T^3$; 40–90 °C, ln($\eta$/mPa·s) = −0.4553 + 143.52 × $10^6$ $K^3/T^3$. **(b)** Litovitz plots for the molar conductivities of [EMIM][TCB], [EMIM][$CF_3SO_3$] and [EMIM][$CH_3SO_3$]. The blue lines represent the higher temperature data. Litovitz eqn fits: [EMIM][TCB], 0–22.5 °C,

 

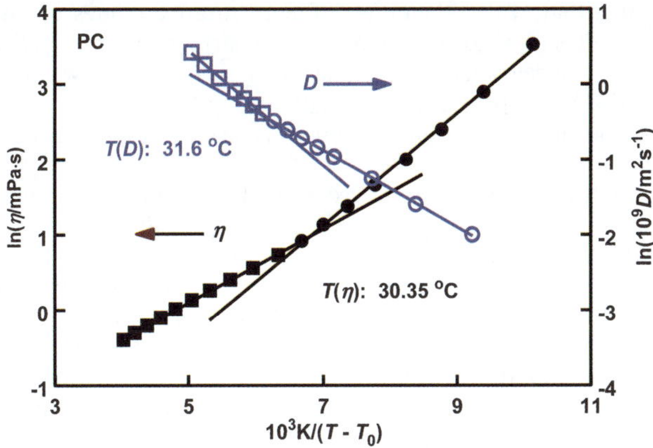


**Fig. 6** Dynamic crossover for propylene carbonate. Viscosity, data of Barthel *et al.*,[35] left $y$-axis, black symbols. Self-diffusion coefficient, this work, right $y$-axis, blue symbols. VFT eqn fits: viscosity, −45–25 °C, $\ln(\eta/\text{mPa·s}) = -1.9714 + 428.39/(T/\text{K} - 150.20)$; 35–125 °C, $\ln(\eta/\text{mPa·s}) = -2.7780 + 673.04/(T/\text{K} - 116.59)$; −21.2–25 °C, $\ln(10^9 D/\text{m}^2\,\text{s}^{-1}) = 2.6930 - 525.49/(T/\text{K} - 141.57)$; 30–80 °C, $\ln(10^9 D/\text{m}^2\,\text{s}^{-1}) = 3.6511 - 796.20/(T/\text{K} - 112.64)$. The origin of the $x$-axis has been adjusted for each of the low temperature lines ($D$: 0.6830; $\eta$: 1.173) to bring the low and high temperature lines into coincidence at the transition points.

addition, no crossover was observed in the ion self-diffusion coefficients for [EMIM][TCB] and [EMIM][$CH_3SO_3$]. Consequently, there may be more than one type of behaviour present, as the Tyrrell and Watkiss work[12] also presents an example of a crossover in the viscosity but not the (solute tracer) diffusion coefficient. The questions of why these ionic liquids seem to show weak crossover effects and whether this can be linked to their liquid structure are not yet answerable. Further work is required to determine whether true cross-over behaviour occurs in ionic liquids, and structural, relaxation and spectroscopic studies could be very useful in giving insight into these transport property measurements.

## 4. Conclusions

High pressure measurements have been made of viscosities, ion self-diffusion coefficients and electrical conductivities of ionic liquids, mainly of imidazolium salts, but including examples of substituted phosphonium and pyrrolidinium salts. These properties have been analysed in terms of the empirical Stokes–Einstein and Walden equations. The similarity of the fractional Stokes–Einstein equation parameters, $t$, for the ion-self-diffusion coefficients and that for the Walden plots is shown to result from the similarity of the viscosity scaling of the three velocity cross-correlation functions, $f_{ij}$. That is, both velocity auto-correlation functions (self-diffusion) and velocity cross-correlation coefficients scale similarly with the viscosity.

$\ln(\Lambda/\mu\text{S m}^2\,\text{mol}^{-1}) = 9.4109 - 94.643 \times 10^6\,\text{K}^3/T^3$; (25–80) °C, $\ln(\Lambda/\mu\text{S m}^2\,\text{mol}^{-1}) = 8.9401 - 82.399 \times 10^6\,\text{K}^3/T^3$; [EMIM][$CF_3SO_3$], 0–80 °C, $\ln(\Lambda/\mu\text{S m}^2\,\text{mol}^{-1}) = 8.4253 - 86.947 \times 10^6\,\text{K}^3/T^3$; [EMIM][$CH_3SO_3$], 0–30 °C, $\ln(\Lambda/\mu\text{S m}^2\,\text{mol}^{-1}) = 9.8831 - 161.69 \times 10^6\,\text{K}^3/T^3$; 40–90 °C, $\ln(\Lambda/\mu\text{S m}^2\,\text{mol}^{-1}) = 8.9466 - 136.28 \times 10^6\,\text{K}^3/T^3$. **(c)** Litovitz plots for the ion self-diffusion coefficients for [EMIM][TCB], [EMIM][$CF_3SO_3$] and [EMIM][$CH_3SO_3$. There appear to be weak transitions for the viscosities and molar conductivities of [EMIM][TCB] and [EMIM][$CH_3SO_3$], but not for the self-diffusion coefficients. Litovitz eqn fits: [EMIM][TCB], −10–80 °C, $\ln(10^{12}D_+/\text{m}^2\,\text{s}^{-1}) = 8.0029 - 157.51 \times 10^6\,\text{K}^3/T^3$; [EMIM][$CF_3SO_3$], −5–80 °C, $\ln(10^{12}D_+/\text{m}^2\,\text{s}^{-1}) = 7.4546 - 97.667 \times 10^6\,\text{K}^3/T^3$; 0–80 °C, $\ln(10^{12}D_-/\text{m}^2\,\text{s}^{-1}) = 8.0029 - 151.51 \times 10^6\,\text{K}^3/T^3$; [EMIM][$CH_3SO_3$], 20–80 °C, $\ln(10^{12}D_{av}/\text{m}^2\,\text{s}^{-1}) = 7.9573 - 150.29 \times 10^6\,\text{K}^3/T^3$.

We have found that, within the ranges of temperature and pressure available, the Nernst–Einstein deviation parameter $\Delta$ is very nearly independent of temperature and pressure for a given ionic liquid. The trends seem to follow the cation, with $[OMIM]^+ > [HMIM]^+ > [BMIM]^+ > [Pyr_{14}]^+$.

Examples are given of the usefulness of density scaling of the transport properties of ionic liquids.

There appear to be weak dynamic crossover transitions for the viscosities and molar conductivities of [EMIM][TCB] and [EMIM][$CH_3SO_3$], but not for the self-diffusion coefficients. Literature viscosity data for [EMIM][$MeOHPO_2$] also seem to show this effect. Other [EMIM] salts were examined: $[FAP]^-$, $[Tf_2N]^-$ and $[CF_3SO_3]^-$, but they were not found to exhibit dynamic crossover. The transitions for [EMIM][$CH_3SO_3$], occur just below the normal freezing point. The fractional Walden plots are smooth, showing no discontinuity, with exponents $0.91_5$ and $0.87_7$ for [EMIM][TCB] and [EMIM][$CH_3SO_3$], respectively. A comparison was made with propylene carbonate, a molecular liquid showing crossover behaviour in both the viscosity and self-diffusion coefficient, using new measurements for the latter, and with literature results for 2-methylpentane-2,4-diol, which also shows a crossover, but only for the viscosity. The fractional Stokes–Einstein plot for PC shows no discontinuity, in contrast to those of [EMIM][TCB] and [EMIM][$CH_3SO_3$].

Several inter-related transport properties have been collected for a series of ionic liquids over a wide range of pressure as well as temperature. They can be generalized in terms of empirical and phenomenological relationships with parameters nearly independent of temperature and pressure for a given ionic liquid. This suggests that the strong interactions existing in ionic liquids are predominant in determining such properties under the conditions examined. However, these general relationships break down in certain ionic liquids at temperatures near the normal freezing point, as is shown by the dynamic crossover effect. The questions of why these particular ionic liquids seem to show such weak crossover effects and whether this can be linked to their liquid structure are not yet answerable. Further work is required to determine whether true cross-over behaviour occurs in ionic liquids, and structural, relaxation and spectroscopic studies could be very useful in interpreting the transport property measurements.

## Acknowledgements

We have been greatly assisted by our collaborations with J. Fernández and her colleagues, D. Kodama and his colleagues, K. Ibuki and his colleagues, T. Rüther and, particularly, L. A. Woolf. It is a pleasure to thank Miss Eriko Niitsuma for her careful experimental work, and the following UNSW students in the Chief of Defence Force Student Programme at ADFA for their assistance with the measurements for [EMIM]TCB as part of their BSc studies: density—MIDN Kate Gallaway (RAN), diffusion—OFCDT Nathan Segal (RAAF) and viscosity—OFCDT Wilson Wee (RSAF). This work was supported, in part, by grants from the Research Fund of the Royal Society of Chemistry to KRH and the Industrial Technology Research Grant Program from the New Energy and Industrial Technology Development Organization (NEDO) of Japan in 2007 to MK.

## References

1 (*a*) K. R. Harris, L. A. Woolf and M. Kanakubo, *J. Chem. Eng. Data*, 2005, **50**, 1777; (*b*) K. R. Harris, L. A. Woolf and M. Kanakubo, *J. Chem. Eng. Data*, 2006, **51**, 1161; (*c*) M. Kanakubo, K. R. Harris, N. Tsuchihashi, K. Ibuki and M. Ueno, *J. Phys. Chem. B*, 2007, **111**, 2062; M. Kanakubo, K. R. Harris, N. Tsuchihashi, K. Ibuki and M. Ueno, *J. Phys. Chem. B*, 2007, **111**, 13867; (*d*) M. Kanakubo, K. R. Harris, N. Tsuchihashi, K. Ibuki and M. Ueno, *Fluid Phase Equilib.*, 2007, **261**, 414; (*e*) K. R. Harris, L. A. Woolf and M. Kanakubo, *J. Chem. Eng. Data*, 2007, **52**, 1080; (*f*) K. R. Harris,

 

L. A. Woolf and M. Kanakubo, *J. Chem. Eng. Data*, 2007, **52**, 2425; K. R. Harris, L. A. Woolf and M. Kanakubo, *J. Chem. Eng. Data*, 2008, **53**, 1230; (*g*) K. R. Harris, M. Kanakubo, N. Tsuchihashi, K. Ibuki and M. Ueno, *J. Phys. Chem. B*, 2008, **112**, 9830; (*h*) M. Kanakubo, H. Nanjo, T. Nishida and J. Takano, *Fluid Phase Equilib.*, 2011, **302**, 10; (*i*) K. R. Harris, L. A. Woolf, M. Kanakubo and T. Rüther, *J. Chem. Eng. Data*, 2011, DOI: 10.1021/je2006049.

2 B. Berne and S. A. Rice, *J. Chem. Phys.*, 1964, **40**, 1347.

3 (*a*) S. I. Smedley and L. V. Woodcock, *J. Chem. Soc., Faraday Trans. 2*, 1974, **70**, 955; (*b*) J. A. Padró, J. Trullàs and G. Sesé, *Mol. Phys.*, 1991, **72**, 1035; (*c*) J. A. Padró and J. Trullàs, *Phys. Rev. B: Condens. Matter*, 1997, **55**, 12210; (*d*) R. Sharma and K. Tankeshwar, *J. Chem. Phys.*, 1998, **108**, 2601; (*e*) C. Rey-Castro, A. L. Torma and L. F. Vega, *Fluid Phase Equilib.*, 2007, **256**, 62; (*f*) J. Picálek and J. Kolafa, *J. Mol. Liq.*, 2007, **134**, 29; (*g*) H. V. Spohr and G. N. Patey, *J. Chem. Phys.*, 2008, **129**, 064517; (*h*) W. Zhao, F. Leroy, B. Heggen, S. Zahn, B. Kirchner, S. Balasubramanian and F. Müller-Plathe, *J. Am. Chem. Soc.*, 2009, **131**, 15825; (*i*) T. Yamaguchi and S. Koda, *J. Chem. Phys.*, 2010, **132**, 114502; (*j*) J. A. Armstrong and P. Ballone, *J. Phys. Chem. B*, 2011, **115**, 4927.

4 (*a*) H.-G. Hertz, *Ber. Bunsenges. phys. Chem.*, 1977, **81**, 656; (*b*) H. Schönert, *J. Phys. Chem.*, 1984, **88**, 3359; (*c*) F. O. Raineri and H. L. Friedman, *J. Chem. Phys.*, 1989, **91**, 5642.

5 K. R. Harris, *J. Phys. Chem. B*, 2010, **114**, 9572.

6 (*a*) C. Alba-Simionesco, A. Cailliaux, A. Alegría and G. Tarjus, *Europhys. Lett.*, 2004, **68**, 58; (*b*) R. Casalini and C. M. Roland, *Phys. Rev. E: Stat., Nonlinear, Soft Matter Phys.*, 2004, **69**, 062501; (*c*) R. Casalini and C. M. Roland, *J. Non-Cryst. Solids*, 2007, **353**, 3936; (*d*) D. Fragiadakis and C. M. Roland, *J. Chem. Phys.*, 2011, **134**, 044504.

7 (*a*) C. M. Roland, S. Bair and R. Casalini, *J. Chem. Phys.*, 2006, **125**, 124508; (*b*) A. S. Pensado, A. A. H. Pádua, M. J. P. Comuñas and J. Fernández, *J. Phys. Chem. B*, 2008, **112**, 5563; (*c*) E. R. López, A. S. Pensado, M. J. P. Comuñas, A. A. H. Pádua, J. Fernández and K. R. Harris, *J. Chem. Phys.*, 2011, **134**, 144507.

8 T. Schrøder, N. P. Bailey, U. R. Pedersen, N. Gnan and J. C. Dyre, *J. Chem. Phys.*, 2009, **131**, 23503.

9 A. J. Barlow, J. Lamb and A. J. Matheson, *Proc. R. Soc. London, Ser. A*, 1966, **292**, 322.

10 H. J. V. Tyrrell and P. J. Watkiss, *J. Chem. Soc., Faraday Trans. 1*, 1979, **75**, 1417.

11 R. Casalini and S. Bair, *J. Chem. Phys.*, 2008, **128**, 084511.

12 H. J. V. Tyrrell and P. J. Watkiss, *J. Chem. Soc., Faraday Trans. 1*, 1984, **80**, 1279.

13 For example, (*a*) F. Stickel, E. W. Fischer and R. Richert, *J. Chem. Phys.*, 1996, **104**, 2043; (*b*) S. Pawlus, R. Casalini, C. M. Roland, M. Paluch, S. J. Rozka and J. Ziolo, *Phys. Rev. E: Stat., Nonlinear, Soft Matter Phys.*, 2004, **70**, 061501.

14 (*a*) R. Richert and K. Samwer, *New J. Phys.*, 2007, **9**, 36; (*b*) J. R. Rajian and E. L. Quitevis, *J. Chem. Phys.*, 2007, **126**, 224506.

15 K. R. Harris and M. Kanakubo, paper in preparation.

16 D. M. Heyes, M. J. Cass, J. G. Powles and W. A. B. Evans, *J. Phys. Chem. B*, 2007, **111**, 1455.

17 K. R. Harris, *J. Chem. Phys.*, 2009, **131**, 054503.

18 M. Kanakubo, D. Kodama and K. R. Harris, paper in preparation.

19 (*a*) G. Morrison and J. E. Lind, Jr, *J. Chem. Phys.*, 1968, **49**, 5310; (*b*) R. J. Speedy, *J. Chem. Soc., Faraday Trans. 1*, 1977, **73**, 471.

20 H. L. Friedman and R. Mills, *J. Solution Chem.*, 1981, **10**, 395.

21 M. Kanakubo and K. R. Harris, unpublished data.

22 D. W. McCall and D. C. Douglass, *J. Phys. Chem.*, 1967, **71**, 987.

23 J. A. Widegren, E. M. Saurer, K. N. Marsh and J. W. Magee, *J. Chem. Thermodyn.*, 2005, **37**, 569.

24 R. Casalini, U. Mohanty and C. M. Roland, *J. Chem. Phys.*, 2006, **125**, 014505.

25 (*a*) I. Avramov and A. Milchev, *J. Non-Cryst. Solids*, 1988, **104**, 253; (*b*) I. Avramov, *J. Chem. Phys.*, 1991, **95**, 4439; (*c*) I. Avramov, *J. Non-Cryst. Solids*, 2000, **262**, 258.

26 M. Paluch, S. Haracz, A. Grzybowski, M. Mierzwa, J. Pionteck, A. Rivera-Calzada and C. Leon, *J. Phys. Chem. Lett.*, 2010, **1**, 987.

27 D. Fragiadakis and C. M. Roland, *Phys. Rev. E: Stat., Nonlinear, Soft Matter Phys.*, 2011, **83**, 031504.

28 (*a*) R. Casalini, M. Paluch and C. M. Roland, *J. Chem. Phys.*, 2003, **118**, 5701; (*b*) R. Casalini and C. M. Roland, *Phys. Rev. B: Condens. Matter Mater. Phys.*, 2005, **71**, 014210; (*c*) C. M. Roland, *Soft Matter*, 2008, **4**, 2316.

29 K. R. Harris, *J. Chem. Phys.*, 2010, **132**, 231103.

30 B. Hasse, J. Lehmann, D. Assenbaum, P. Wasserscheid, A. Leipertz and A. P. Fröba, *J. Chem. Eng. Data*, 2009, **54**, 2576.

31 M. Blesic, M. Swadźba-Kwaśny, T. Belhocine, H. Q. Nimal Gunaratne, J. N. Canongia Lopes, M. F. Costa Gomes, A. A. H. Pádua, K. R. Seddon and L. P. N. Rebelo, *Phys. Chem. Chem. Phys.*, 2009, **11**, 8939.

32 K. R. Harris, *J. Chem. Eng. Data*, 2009, **54**, 2729.
33 K. Nishikawa, S. Wang, T. Endo and K. Tozaki, *Bull. Chem. Soc. Jpn.*, 2009, **82**, 806.
34 H. Fujimori and M. Oguni, *J. Chem. Thermodyn.*, 1994, **26**, 367.
35 (*a*) J. Barthel, M. Utz, K. Groβ and H. J. Gores, *J. Solution Chem.*, 1995, **24**, 1109; (*b*) J. Barthel, R. Neueder and H. Roch, *J. Chem. Eng. Data*, 2000, **45**, 1007.
36 T. A. Litovitz, *J. Chem. Phys.*, 1952, **20**, 1088.

 

# General discussion

**Dr Driver** opened the discussion of the paper by Professor MacFarlane: Is it surprising that primary ammonium systems, or, onium systems in general, with all their different flavours, might appear so poorly on the Walden plot, considering the Walden rule fails to account for hydrogen bonding effects, as well as those effects due to local electric fields produced from the asymmetric distribution of dipole density, that would oppose ionic motion? Is it not possible to encounter highly ionised onium systems that just exhibit poor transport properties?

**Professor MacFarlane** answered: The primary ammonium systems are those that we often observe as more strongly adhering to the Walden rule.[1] The examples in the present paper, we believe, are exceptions to the rule because of special structuring effects. At a very basic level, H-bonding affects both viscosity and molar conductivity, unless its effect is to create strong ion-pairs—the impact of these is much stronger in the molar conductivity. It certainly is possible to encounter onium systems which exhibit poor transport properties because of H-bonding, but to deviate strongly from the "ideal" line on the Walden plot the conductivity needs to be much more strongly affected than the viscosity. Anything that creates neutral pairs or aggregates has that effect.

1 J. Stoimenovski, E. I. Izgorodina and D. R. MacFarlane, *Phys. Chem. Chem. Phys.*, 2010, **12**, 10341.

**Professor Angell** commented: In trying to get insight into the contributions being made by ion pairs we point out that an ion pair relaxation peak can be observed in electrical relaxation studies. Particularly if one converts dielectric susceptibility data into electrical modulus data so that the loss due to ionic motion can be rendered as a peak in the modulus spectrum, then if there are ion pairs present in abundance, a second peak at higher frequencies can be observed. This was seen by Richert and coworkers and contrasted with the case of a typical aprotic IL $EMI^+$ $PF_6^-$ which is a "good" ionic liquid. We want to emphasize that for "good" ionic liquids of high conductivity, there will still be a deficit in the measured *vs* Nernst–Einstein conductivity, because of cation–anion friction, the subject of a theoretical paper by Berne and Rice, in the 60's, which it seems no one remembers, or at least, wants to deal with. To get good agreement with the Nernst–Einstein equation the conductivity should be kept rather low.

1 N. Ito, W. Huang and R. Richert, *J. Phys. Chem. B*, 2006, **110**, 4371.
2 N. Ito and R. Richert, *J. Pys. Chem. B*, 2007, **111**, 5016.

**Professor Canongia Lopes** said: Speciation of protic ionic liquids in the liquid phase is a problem of paramount importance. Recently we have explored this issue in the context of reactive azeotropy, thus establishing a link between vapour pressure and boiling point determinations (liquid vapour *p-x-y* or *T-x-y* phase diagrams) and the speciation of the liquid, cf ref. 1. In my opinion it will be extremely difficult to know the amount of ionization of the protic ionic liquid using solely vapour pressure data, although such approach can be useful when combined with other techniques.

1 J. N. Canongia Lopes and L. P. N. Rebelo, *PhLys. Chem. Chem. Phys.*, 2010, **12**, 1948–1952.

**Professor MacFarlane** answered: I think this is a very important approach to this problem of speciation in protic ionic liquids, because it is one of the relatively few direct thermodynamic measurements that links to this state of equilibrium and therefore thermodynamic models are at least approximately useful. On the other hand, the transport properties which are the basis of the Walden plot are, of course, only in-direct indicators of the position of the proton transfer equilibrium and, though easily measured, are therefore of lower fundamental value.

**Professor Quitevis** asked: Watanabe and coworkers[1] have used the molar conductivity ratio, $\Lambda_{imp}/\Lambda_{NMR}$, where $\Lambda_{imp}$ is the molar conductivity obtained from impedance measurements and $\Lambda_{NMR}$ is the molar conductivity calculated from ionic self diffusion constants obtained from pulse-field-gradient spin-echo NMR measurements, as a quantitative measure of the ionicity of ILs. They showed that $\Lambda_{imp}/\Lambda_{NMR}$ approach is able to reveal smaller differences in the ionic nature of ILs than Walden plots. Have you considered using the $\Lambda_{imp}/\Lambda_{NMR}$ approach to quantify the extent of proton transfer and the degree of ionicity in these pharmaceutically active ILs?

1 K. Ueno, H. Tokuda, and M. Watanabe, *Phys. Chem. Chem. Phys.*, 2010, **12**, 1649.

**Professor MacFarlane** replied: The NMR approach, which we have also used in respect of aprotic ionic liquids,[1] is a little more difficult to apply in the present context because (i) it is more difficult to unambiguously separate the NMR responses for nuclei on the cation from those on the anion and (ii) there is a contribution to conductivity from the labile proton that needs to be identified in the NMR response and quantified in order to use the Nernst–Einstein equation correctly. We are currently investigating a number of simpler protic systems in which this will be unambiguously possible to confirm the technique.

1 D. R. MacFarlane, M. Forsyth, E. I. Izgorodina, A. P. Abbott, G. Annat and K. Fraser, *Phys. Chem. Chem. Phys.*, 2009, **11**, 4962.

**Dr Harris** commented: The position of the high temperature molten salts on the Walden plot (Fig. 7 of your paper) is interesting. Though these lie on the "ideal" line, the Nernst–Einstein deviation parameter, $\Delta$, as shown by early MD studies, such as those of Hansen and McDonald[1] and Padro *et al.*,[2] is about 0.9 or less, though there can be no ion-pairing in such simple salts.

Again, it is possible to show (see paper 24) that it is the difference between the cation-anion velocity cross-correlation coefficient (VCC) and the mean of the cation-cation and anion-anion velocity cross-correlation coefficients that governs the magnitude of $\Delta$, so it is not necessary to invoke ion-pairing when $\Delta < 1$. One needs evidence from other techniques, more sophisticated than transport property measurements, to determine whether ion-pairing is present, and on what time-scale.

1 J.-P. Hansen and I. R. McDonald, *J. Phys. C: Solid State Phys.*, 1974, **7**, L384.
2 J. A. Padró, J. Trullàs and G. Sesé, *Mol. Phys.*, 1991, **72**, 1035.

**Professor MacFarlane** responded: I totally agree with these points about the molten salt data. We discussed some of this in our 2009 PCCP paper "On the Concept of Ionicity in Ionic Liquids" where we discussed this in terms of a Walden-like plot that tested only the Nernst–Einstein Equation and which we called an Ionicity Plot.[1] This typical deviation parameter of around 0.9 arises, as you say, from what one might call first order fundamental effects in the motion of ions in an electric field that are *not* accounted for by the Nernst–Einstein equation. We have therefore defined a line on the Walden and Ionicity plots about 0.1 log units below the usual line as describing an "Ideal Ionic Liquid" (a concept, of course, which more closely describes simple molten salts than real ionic liquids—nonetheless it is a useful limiting behaviour in the same way as is the ideal gas).

 

(Note to other readers: this $\Delta$ parameter = 1 when the Nernst–Einstein equation is completely followed and decreases as ion-pairing *etc.* creeps in. Other parameters that we and others have used for the same phenomenon = 0 under these conditions and increase as ion-pairing creeps in).

1 D. R. MacFarlane, M. Forsyth, E. I. Izgorodina, A. P. Abbott, G. Annat and K. Fraser, *Phys. Chem. Chem. Phys.*, 2009, **11**, 4962.

**Professor Ballone** remarked: The deviation of molar conductivity from "ideal" behaviour, as expressed by the Walden or Nernst–Einstein rules, is not an unambiguous sign of ion pairing. As stressed in the contribution to this discussion by K. R. Harris and M. Kanakubo, deviations from ideality can be expressed in terms of velocity cross correlations. The effect of these correlations on conductivity, however, does not have a definite sign. For instance, enhancement of conductivity with respect to the Nernst–Einstein value (*i.e.*, the opposite of the decrease of conductivity often attributed to ion pairing) has been observed several times in computer simulations (see, for instance, ref. 1–3, or, more recently, ref. 4.

In principle, even an "ideal" conductivity in a strongly correlated ionic fluid could result from the compensation of contributions from pairs of ++, +− and −− charge. It is apparent that the wide variety of behaviours cannot be covered by the simple interpretation of conductivity anomalies as due to ion pairing.

1 J. Trullàs and J. A. Padró, *Phys. Rev. B*, 1997, **55**, 12210.
2 Ç. Tasseven, J. Trullàs, O. Alcaraz, M. Silbert and A. Giró, *J. Chem. Phys.*, 1997, **106**, 7286.
3 T. Yamaguchi, A. Nagao, T. Matsuoka and S. Koda, *J. Chem. Phys.*, 2003, **119**, 11306.
4 J. A. Armstrong and P. Ballone, *J. Phys. Chem. B*, **115**, 2011, 4927.

**Professor MacFarlane** replied: We totally agree regards the origin of these effects in the cross-correlation functions, and clearly it is possible for these terms in principle to produce an enhancement rather than a reduction in the total conductivity. Interestingly, relatively few electrolyte materials have ever been observed which strongly exceed the Nernst–Einstein (or Stokes–Einstein) "ideal". In these cases there is typically a special mechanism, such as the Grotthus mechanism of conduction of protons or iodide ions, that is the origin of the extra conduction. It is important to note that we do not ascribe conductivity anomalies as solely due to ion pairing. In fact we believe that "Liquid Ion Pairs" are a relatively rare phenomenon that turns up when the ion sizes are substantially different, as in the famous P6,6,6,14 Cl. Extended H-bonding between ions that creates highly correlated motions (large cross-terms in the velocity and current correlation functions) is a common observation—this is very much the subject of the present paper. Normally these are an "annoyance" in terms of achieving high conductivity, but in the present case of ionic liquids that are rapidly mobile across membranes, these strong cross correlations in the transport properties appear to be very usefully important.[1]

1 J. Stoimenovski and D. R. MacFarlane, *Chem. Commun.* 2011, **47**, 11429

**Dr Driver** commented: To prepare a Walden plot, we make use of the molar conductivity, $\Lambda$ (S cm$^2$ mol$^{-1}$), a quantity taken from the specific conductivity, $\sigma$ (S cm$^{-1}$) normalised for ion concetration (via $V_m$, in cm$^3$ mol$^{-1}$). We also make use of the reciprocal shear viscosity, or, the fluidity, ($\eta^{-1} = \Phi$, in P$^{-1}$), where concentration effects are not considered. In your opinion, would there be any utility in defining a new quantity, such as the *molar fluidity*, $\xi$ (mol cm$^{-3}$ P$^{-1}$), where a log $\Lambda$ versus log $\xi$ plot would then encompass the proportionality between two *molar* quantities? In this way, we may take concentration effects into account for both transport properties.

**Professor MacFarlane** responded: Very interesting perspective. The lack of an explicit concentration effect in viscosity is one of the main distinguishing features between these two properties. On the other hand if we look at electrolyte solution properties, which are the historical, and not always useful, origins of these quantities, it is clear that molar conductivity tends to 0 as ion concentration increases while viscosity approaches that of the pure solvent. In polymer solution phenomena it is common to use (viscosity- viscosity of the pure solvent) as a key property, and this quantity, when divided by concentration, is often used as a "molar" quantity. I would be very interested to see whether processing the data for ionic liquids in this fashion would provide any new insights.

**Dr Harris** asked: You have described scaling in terms of the Walden and Stokes–Einstein relations. There is a relatively new alternative, but one which needs data over a significant range of density (pressure) to apply it. Work by Roland, Casalini and Bair on the density scaling of ionic liquid viscosities[1] and by Josefa Fernandez at Santiago de Compostela and coworkers on IL viscosities, self-diffusion coefficients and conductivities[2,3] has shown that this method of scaling appears to work reasonably well. The data are scaled with the group $(TV^{\gamma})$, a method that is exact for liquids with an inverse power law intermolecular potential when reduced units are used. The scaling coefficients are, however, much lower than those of typical molecular liquids.

1 C. M. Roland, S. Bair and R. Casalini, *J. Chem. Phys.*, 2006, **125**, 124508.
2 A. S. Pensado, A. A. H. Pádua, M. J. P. Comuñas and J. Fernández, *J. Phys. Chem. B*, 2008, **112**, 5563.
3 E. R. López, A. S. Pensado, M. J. P. Comuñas, A. A. H. Pádua, J. Fernández and K. R. Harris, *J. Chem. Phys.*, 2011, **134**, 144507.

**Professor MacFarlane** replied: The scaling methods of Roland *et al.* and Fernandez *et al.* do indeed work impressively well. However the ionic liquids tested thus far are all those that would sit close to the "ideal" line on the Walden plot, *i.e.* adhering moderately well to the simple forms of the Stokes–Einstein and Nernst–Einstein equations (or scaling relations). The examples highlighted in my paper deviate quite strongly (by an order of magnitude or more) from these relations and it would therefore be very interesting to see how such systems would appear on the scaling plots of Roland *et al.* and Fernandez *et al.* and which parameters would reflect the differences. Of course a wider range of data is needed to do that and those researchers that have the appropriate gear are encouraged to look into these interesting systems —either the aprotic systems which show this phenomena (see ref. 1) or some of the protic systems described in the present paper.

1 Fraser *et al.*, *Chem. Commun.*, 2007, 3817.

**Dr Taylor** opened the discussion of the paper by Dr Wishart: You discuss the mechanism of electron solvation in ILs, in particular the slow solvation dynamics of ILs causing electrons to persist in "weakly trapped, highly mobile and reactive pre-solvated states for a much longer period of time than in conventional solvents." Would a similar solvation mechanism apply to radical species generated in ILs and similarly affect their reactivity?

**Dr Wishart** replied: It could, bearing in mind that part of the relaxation of molecular radicals is intramolecular, so that the influence of solvation is not the only factor. Also, a distinction should be drawn between the cases where neutral radicals may be created through homolytic bond cleavage and those reactions involving the creation of charged radicals from neutral molecules and neutral radicals from ions. In the former case, solvent dynamics may affect the response of the system to large structural rearrangements, as in the case of the neutral lophyl radical.[1] However, I believe your question focuses on the latter case where radicals are formed by addition or removal of charge. The best example I know involves the highly solvatochromic benzophenone radical anion, whose solvation dynamics have been extensively studied in alcohols at low-temperature by following the blue shift of the strong absorption band from ~800 nm to ~650 nm.[2] The range of that spectral shift works well with our OFSS detection system, and we have performed similar experiments in two ionic liquids where we also observe such shifts. This will be the basis of a subsequent publication. The solvation of a nascent molecular anion is a much easier process to model by molecular dynamics simulations than the process of electron solvation, so we hope that our experimental work will be complemented by such simulations.

1. V. Strehmel, J. F. Wishart, D. E. Polyansky and B. Strehmel, *ChemPhysChem*, 2009, **10**, 3112–3118.
2. J. L. Marignier and B. Hickel, *Chem. Phys. Lett.*, 1982, **86**, 85–90; J. L. Marignier and B. Hickel, *J. Phys. Chem.*, 1984, **88**, 5375–5379; T. Ichikawa, Y. Ishikawa and H. Yoshida, *J. Phys. Chem.*, 1988, **92**, 508–511; Y. Lin and C. D. Jonah, *J. Phys. Chem.*, 1992, **96**, 10119–10124; Y. Lin and C. D. Jonah, *J. Phys. Chem.*, 1993, **97**, 295-302; Z. Xujia and C. D. Jonah, *Chem. Phys. Lett.*, 1995, **245**, 421-426; X. J. Zhang and C. D. Jonah, *J. Phys. Chem.*, 1996, **100**, 7042–7049; X. J. Zhang, Y. Lin and C. D. Jonah, *Radiat. Phys. Chem.*, 1999, **54**, 433–440.

**Professor Quitevis** remarked: With respect to the state of the solvated electron in ionic liquids, is it localized or delocalized?

**Dr Wishart** answered: The question of whether the solvated electron is localized or delocalized in ionic liquids is a matter of degree and of definition. Even in molecular solvents there is a wide range of behavior, ranging from the fairly localized hydrated electron in water and solvated electron in alcohols where the electron absorption peaks appear around 650–700 nm, to the more diffuse, weakly trapped electrons in solvents like tetrahydrofuran where the peak occurs at 2200 nm. In certain hydrocarbons, excess electrons can even thermally access the conduction band. Based on their solvated electron absorption spectra peaking in the range of 1100–1500 nm, ionic liquids would seem to be halfway between those cases. In ionic solids, excess electrons are localized in F-centers, which are anion vacancies forming voids surrounded by cations, and it is reasonable to expect similar behavior in ILs but experimental evidence is lacking at this time. Finally, an operational argument in favor of solvated electron localization in ionic liquids is the fact that their diffusion-limited rate constants for attachment to various aromatic scavengers are slower than the corresponding H-atom reactions, indicating that their diffusivity is more like that of a molecular anion than a quantum particle, suggesting a high degree of localization while not precluding tunneling between sites as the likely mechanism of diffusion.

**Professor Wynne** commented: The dielectric spectrum of an ionic liquid (or any liquid) has many contributions (librational, cage rattling, low-frequency torsional modes, translational diffusion, and rotational diffusion) on many different timescales from 100s of femtoseconds to nanoseconds and slower.[1,2] The radiolysis technique use in your presentation has a resolution of 5 ps. Does this mean that you could miss certain fast processes? Could the electron solvation process involve multiple timescale? Which molecular motions are likely to be most strongly coupled to solvation considering the final location of the solvated electron?

1 D. A. Turton, J. Hunger, A. Stoppa, G. Hefter, A. Thoman, M. Walther, R. Buchner and K. Wynne, *J. Am. Chem. Soc.*, 2009, **131**, 11140–11146.
2 D. A. Turton, J. Hunger, A. Stoppa, A. Thoman, M. Candelaresi, G. Hefter, M. Walther, R. Buchner and K. Wynne, *J. Mol. Liq.*, 2011, **159**, 2–8.

**Dr Wishart** replied: Yes, there is no doubt that there are processes of electron solvation, and possibly migration between sites, that occur before the time resolution of our experiments. Solvation dynamics experiments on solvatochromic dyes in ILs, such as the elegant contribution to this Discussion by Prof. Maroncelli,[1] provides evidence for these fast processes, and electron solvation should also be extended over multiple time scales. On the time scales we observe the solvation processes are coupled to translational and rotational diffusion, with the translation of the ions likely making the stronger contribution. Using different electron scavengers, we can detect differences in the reactivities of the precursor electron states that we cannot observe directly. Laser photoionization experiments would be a useful complement to radiolysis studies but not a direct replacement, since there are differences between the two techniques in terms of the energetics of the generated excess electrons and in the electron-hole pair distance distributions.

1 M. Maroncelli, X. X. Zhang, M. Liang, D. Roy and N. P. Ernsting, *Faraday Disc.*, 2011, **154**, DOI: 10.1039/C1FD00058F.

**Professor Castner** asked Professor Wynne: Given that all atomic and molecular liquids will have interatomic/intermolecular librational and vibrational dynamics on timescales from tens to hundreds of femtoseconds, shouldn't the question be, "Which modes couple to solvation and reaction coordinates for electrons or molecules?" There has been speculation that aromatic cations with delocalized charges such as 1,3-dialkylimidazoliums may have librational motions analogous to phenyl-ring librations that do couple strongly with electron solvation, whereas for tetralkyl ammonium/phosphonium (including pyrrolidinium) cations, the charge is much less delocalized, and closer to the nitrogen or phosphorus center, thus leading to greatly reduced coupling to electron solvation or reaction coordinates.

**Professor Wynne** replied: Yes, you are absolutely correct. Thus, it would be incorrect to just use the dielectric spectrum (or just the low-frequency Raman spectrum) in order to predict solvation of the electron. Rather, you would somehow have to figure out the coupling strength of each "mode".

**Professor Kim** addressed the room: The weak, but measurable absorption beyond the absorption edge and the fluorescence in the visible range that follows the excitation wavelength have been observed in ionic liquids, especially from the ones having imidazolium cations. Can you comment on the possible relation of the above unusual absorption and fluorescence phenomena to the solvated electrons in ionic liquids?

**Professor Maroncelli** responded: We have examined the fluorescence of neat ionic liquids in some detail. At one point I had thought that this fluorescence might be due to the presence of electrons trapped in some special type of ionic environment. But the bulk of evidence now points to emission being due to the presence of very low (micromolar) concentrations an oligomeric impurity that is impossible to remove completely.

**Professor Quitevis** asked: Has anyone observed ionic liquid solvated electrons in the gas phase?

**Dr Wishart** answered: I do not believe that the experimental conditions required have been attained yet. I suggest that the species you are referring to would be the ionic liquid analogues of water cluster anions, which have been extensively studied experimentally and theoretically by numerous groups. I would imagine that to generate them, you would need a source of ion pair clusters $(C^+A^-)_n$, onto which you would spray electrons. Both $C^+$ and $A^-$ would have to be relatively inert towards excess electrons. Once prepared, one would attempt to determine which values of $n$ yield stable clusters and how strongly the electron is bound by photoelectron spectroscopy. I think it would be very challenging.

**Professor Maroncelli** commented: In light of the discussion of electron solvation in ionic crystals and liquids, I would like to ask for you to comment on how this process compares to the situation in conventional liquids. My understanding is that in conventional liquids, after injection and loss of sufficient kinetic energy, electrons first localize into pre-existing defect sites or electrostatically favorable voids. After these localization sites are selected the surroundings reorganize to better solvate the electron and it is the latter process that can be observed experimentally. Is this the correct way to view electron solvation in ionic liquids? Also, please comment on the similarities and differences you find between the equilibrated spectrum of the solvated electron in ionic liquids and conventional solvents.

**Dr Wishart** responded: The picture you have described for conventional liquids generally applies as well to ionic liquids, with a few additional considerations due to the charged-ordered and locally heterogeneous nature of ILs. It is possible in some cases that the excess electron could transiently localize on a site occupied by an anion (and thus surrounded by cations) on the way to becoming solvated. The interfaces between the polar and nonpolar regions provide other possible sites for the initial localization steps. As you stated, we experimentally observe the later stages of the reorganization process leading to the solvation of localized excess electrons.

The spectra of solvated electrons in ionic liquids conform well to the trends observed in conventional solvents (see reference 15 for IL examples and 16 in my article for a wider collection of conventional solvents) in terms of intensity *vs.* peak energy, implying similar oscillator strength. (Refer also to the answer to Professor Quitevis' question.) In addition, solvated electrons in ILs with alcohol functionalities have peak absorptions typical of those found in molecular alcohols (see article reference 15 and 1 below).

1. J. F. Wishart, S. I. Lall-Ramnarine, R. Raju, A. Scumpia, S. Bellevue, R. Ragbir and R. Engel, *Radiat. Phys. Chem.*, 2005, **72**, 99–104.

**Professor Ouchi** asked: How high can the density of pre-solvated and/or solvated electrons be in ILs at a maximum? When considering practical applications such as described in conclusions, these numbers will important and necessary.

**Dr Wishart** replied: It is difficult to say what the maximum density (concentration) of excess electrons in ILs could be. If the electrons are produced by radiolytic or photolytic ionization the answer is not very high, because second-order recombination with holes would limit the electron lifetime and the maximum concentration obtainable under steady-state excitation. To my knowledge, contacting ionic liquids with highly reducing metals such as Li or Na has not been shown to generate solutions of solvated electrons, as it does in ammonia. This does not mean that it might not be possible, but it has not been reported.

**Professor Castner** asked: Can you say something about why one would choose a particular ionic liquid as a good *vs.* a bad solvent for using solvated electrons as reducing agents in ionic liquids?

**Dr Wishart** replied: Neglecting the effects of impurities, solvated electrons are "stable" as transient species in ILs with fairly negative cathodic electrochemical windows such as tetraalkylammonium, dialkylpyrrolidinium and tetraalkylphosphonium salts of irreducible anions such as $NTf_2^-$, $OTf^-$, beti$^-$, $PF_6^-$, and dicyanamide. Ultimately, however, the lifetime of radiolytically-produced solvated electrons in ILs is limited to the microsecond time scale by their eventual recombination with the products of the holes they originally came from, if they are not inadvertently scavenged first.

However, in many other ionic liquids solvated electrons are not stable or even observable because they are reactive towards several of the ions typically used in ILs, including aromatic cations such as imidazoliums and pyridiniums, protic species such as ammonium cations and acidic anions, and other electron acceptors such as nitrate and bis(oxalato)borate anions.

Because of their reactivity and consequent short lifetimes, solvated electrons are problematic as intentional reducing agents in ILs, just as they often are in normal solvents. Their high reducing power makes them useful for production of very reducing species, for example biphenyl anion, and for studying reactions that occur on short time scales. However, in some cases where investigators hoped to use solvated electrons for reductions in ILs, their reactant concentrations were too low to compete for the electrons before they reacted away. For most work it is better to capture the electron with a redox mediator in order to stabilize the reducing equivalent and control its reactivity. Radiation chemists have previously worked out schemes of controlling radiolytically-induced redox chemistry in many conventional solvents.

**Professor Dr Roling** opened the discussion of the paper by Dr Harper: Can you explain why $\Delta(\Delta S^{\ddagger})$ is positive?

**Dr Harper** replied: Whilst the absolute values of $\Delta S^{\ddagger}$ are negative, due to the associative nature of the process, it is the differences between the values (ionic liquid minus molecular solvent) which are positive. Given that there is charge development in the transition state, this entropy difference is not likely to be a decrease in organisation around the transition state. We argue this suggests organisation about the nucleophile.

**Professor MacFarlane** remarked: Is this form of data presentation the best way to understand the effects? Should we be in fact comparing $\Delta H^{\ddagger}$ and $T\Delta S^{\ddagger}$ values?

**Dr Harper** answered: Whilst $T\Delta S^{\ddagger}$ values would certainly convey the same information, the fact that the uncertainties in the $\Delta(\Delta X^{\ddagger})$ values are comparatively large (compounded by subtraction) the principle utility—comparing what dominates in $\Delta(\Delta G^{\ddagger})$—is somewhat lost (and is explicitly available in rate information in the ESI of my paper). Rather, the fact that both $\Delta(\Delta S^{\ddagger})$ and $\Delta(\Delta H^{\ddagger})$ values are positive demonstrates an entropic benefit and an enthalpic cost on going from a molecular solvent to an ionic liquid for this reaction. Hence we would argue that comparing values as has been done is appropriate.

**Professor Maroncelli** commented: With respect to Professor MacFarlane's question concerning the comparing $\Delta H^{\ddagger}$ and $T\Delta S^{\ddagger}$, it seems to me that these two contributions to $\Delta G^{\ddagger}$ are of comparable magnitude and largely cancel one another. Is the reason that you focus on $\Delta S^{\ddagger}$ for interpretation simply that $T\Delta S^{\ddagger}$ is larger (at 25 °C) and so it is $\Delta S^{\ddagger}$ that determines the sign of $\Delta G^{\ddagger}$?

**Dr Harper** answered: Yes, the $\Delta H^{\ddagger}$ and $T\Delta S^{\ddagger}$ values are comparable in magnitude and do cancel one another out to a significant extent. We focus on changes in $\Delta S^{\ddagger}$ to explain observed rate increases (over the $T$ values studied), as both $\Delta(\Delta S^{\ddagger})$ and $\Delta(\Delta H^{\ddagger})$ values are positive, implying the entropic benefit dominates. That is, it determines the sign of $\Delta(\Delta G^{\ddagger})$.

 

**Dr Driver** commented: This work seems to complement previous determinations of partial molar excess thermodynamic quantities of infinitely dilute solutes in various ionic liquid systems. Excess heats of mixing, $H^E$, of varieties of acid–base accepting-donating solutes, as deduced from the temperature dependence of their activity coefficients at infinite dilution, $\gamma^\infty$, (where the maximum non-ideality of the mixture becomes apparent due to an absence of solute-solute interactions), in various ionic liquids, generally indicate the main contribution to excess Gibbs energies of mixing, $G^E$, to be strongly entropic rather than enthalpic in nature. This indicates that solvent–solute contacts at hand are often neither more nor less (*i.e.* enthalpic) than would be expected statistically for purely entropic interactions. This thermodynamic behaviour would seem relevant in this work where entropy driven reaction rate increases have been observed. In this way, it would seem possible, that knowledge of existing $\gamma^\infty$ values, across varieties of ILs, may be utilised as an additional criterion towards reaching the goal of identifying a particluar ionic liquid to perform a specific synthetic transformation.

**Dr Harper** replied: This is an excellent suggestion. Thank you.

**Professor Lynden-Bell** asked: In Table 1 it seems that only by changing both aromatic to aliphatic and the amine 5 to the imine 7 do you get a big change in the activation parameters. How do you reconcile this with the assertion that the important interaction is the solvation of the lone pair on the nitrogen?

**Dr Harper** responded: We argue that qualitatively the results mirror the other systems (that is, a systematic removal of delocalised pi electron density still results in an entropy driven rate increase) and that the remaining interaction (with the nucleophilic centre) is key. The origin of the increase in magnitude of these values is not clear and hindered by the measurement issues with these systems.

**Professor Maroncelli** said: Please elaborate on why the data in Table 1, and the lack of delocalization effect they imply, lead you to conclude that the rate differences observed between ionic liquids and acetonitrile are due to some difference in how these solvents interact with the nucleophilic center.

**Dr Harper** replied: By removing systematically the delocalised pi systems from the electrophile and the nucleophile, the data in Table 1 suggest that the interaction with such systems in ionic liquids is not significant in the entropy-derived rate increases observed. The one remaining key interaction, that with the nucleophilic nitrogen centre (which is displayed well in both Fig. S2 and S4), is therefore reasoned to be responsible for the observed entropy-derived rate increase.

**Professor Hardacre** asked: From your data you have assigned the effect to a change in the entropy of activation associated with an interaction of the IL cation with the nucleophilic centre. Have you tried any less coordinating cations, *e.g.* phosphonium or ammonium systems? Obviously, other effects, such as viscosity *etc.*, will have an influence but comparative changes should provide some interesting information.

In addition have you considered performing nOe experiments to clarify the nature of the interactions postulated?

**Dr Harper** responded: Not yet. Preliminary data with a substituted imidazolium cation ionic liquid gives identical results to those presented. The next step is likely to be the use of an ionic liquid based on the pyrrolidinium cation.

Initial nOe investigations met with little or no success. We hope to try and optimise these in the future.

**Professor Canongia Lopes** asked: The spatial distribution functions (SDFs) of ionic liquid ions around individual reactant molecules (shown during the 5 min presentation but not present in paper 20) are very similar to the ones we have published in a systematic study of the influence of the dipole and quadrupole moment of a molecule (fluorinated derivatives of benzene) on the distribution of ionic liquid ions around it, *cf.* ref. 1. In the case of pyridine such distribution is caused by the interaction between the aromatic quadrupole moment and the dipole moment induced by the nitrogen heteroatom. Perhaps a similar SDF for the non-aromatic pyridine analog with just one double bond (or SDFs for methylated pyridine and its non-aromatic analog) can help to rationalize the effect of knocking-out the quadrupole moment of the molecule on the reactions entropic barriers.

1 K. Shimizu, M. F. Costa Gomes, A. A. H. Pádua, L. P. N. Rebelo and J. N. Canongia Lopes, *J. Phys. Chem. B*, 2009, **113**, 9894.

**Dr Harper** responded: These figures are present as Fig. S1–S4 in the supplementary information.

**Professor Castner** opened the discussion of the paper by Professor Ballone: How can we calculate the electrochemical stability window for a new class of ionic liquids? Could you please provide a very brief outline of how experimentalists might proceed to do this, including comments on how large a jellium surface might be appropriate, how long the *ab initio* molecular dynamics simulation might need to be run for, and how many cations/anions must be included within the simulation box?

**Professor Ballone** responded: The quantitative prediction of the electrochemical stability window of a new IL following the method we propose in our manuscript requires the computation of the density of states (DOS) and its derivative with respect to the external bias, using ab initio simulations with a realistic description of the (planar) metal electrode. The model system will consist of a slab representing the electrode, and of a suitable number of ions representing the ionic liquid. The metal could be described in atomistic detail, without resorting to the jellium model. The simulation would require a sizable computation, even though not as big as several other ab initio simulations that have already been done. To give an idea of the system size, we expect that a realistic estimate could be obtained using a metal slab consisting of 50–60 atoms, and about 100 ion pairs, with a total of some 1500–1800 valence electrons. More in detail, the DOS could be computed by averaging over a given number (about one hundred) of instantaneous configurations generated during the simulation. Moreover, the DOS needs to be projected on the metal and on the IL side of the interface, and dubious cases arising from surface reactions will need to be identified. This part of the computation might take some initial tuning of the method, but in our opinion it can be done at an acceptable computational cost.

Then, to estimate the derivative of the metal-DOS and of the IL-DOS, the computation has to be repeated upon the application of a small external bias (a fraction of an eV, such as 0.1 eV), much smaller than the voltage required to transfer a charge from or to the electrode.

Excluding the case of chemical reactions at the surface, these two computations should allow the estimation of the electrochemical stability window.

The case of chemical reactions at the surface, which is certainly very relevant in practice, would need to be analysed on a case by case basis.

Two specific points: (1) in our computation we applied a bias consisting of a shift of the external potential over the region of space occupied by the jellium background. This will probe the short range (chemical) part of the system response to an external potential, but it will not give rise to the long range electric fields that in principle arise from polarising a metal electrode in contact with a charge reservoir.

This part of the computation can be improved by changing the form of the bias, resorting, for instance to a piece-wise (zig-zag) linear profile commensurate with the system periodicity in the direction perpendicular to the electrode. In this respect, however, the major limitation might be represented by the periodic boundary conditions used in plane wave computations. (2) A further improvement could be obtained by adopting a hybrid QM/MM (quantum mechanics / molecular mechanics) approach, in which part of the system is treated at the *ab initio* level, and a larger portion is described by classical force fields. This, in principle, would allow to include the effect of remote charges. It is important to consider, however, that the screening by the IL will effectively reduce the range of Coulomb interactions down to the nm scale.

To summarise, we think that a first principle determination of basic electrochemical properties of IL could be carried out at an acceptable cost, with computational tools that are already available. The basic ingredient is represented by Kohn–Sham eigenvalues, and their derivative with respect to the external bias.

**Professor Johansson** commented: When the electrochemical stability calculations are considered I did not understand why DFT methods were dismissed so fast:

a) There are plenty of studies made on ionization potentials where quite reasonable values are obtained for transfer from anion to anion radical (oxidation) for anions used in ILs (Ue, Johansson, Tasaki) —by using both ground state and excited state energies and calculating the difference according to the Franck–Condon principle. How do your values compare to those studies and/or that method?

b) The fundamental nature of the DFT Kohn–Sham orbitals and eigenvalues and how they in fact can be used is often dismissed too rapidly, as I note on pages 9 and 15 in your paper —please see the seminal paper by Stowasser and Hoffmann,[1] do you have any comment?

c) In Table 3: Are there no experimental values to compare with your calculated for the ESW? How then can we be assured the validity of the presented approach?

Stowasser and Hoffmann, *J. Am. Chem. Soc.*, 1999, **121**, 3414.

**Professor Ballone** replied: We are well aware of other schemes to compute electrochemical properties based on density functional theory, and we certainly did not intend to dismiss/neglect these previous studies. A few references to methods of this kind are given in the manuscript as an example.

We are also aware that ionisation energies can be computed fairly accurately as total energy differences, even though the application of this approach might be difficult in the case that (small) anions are involved. This statement concerning ionisation potentials is widely accepted by the DFT community, and its validity does not depend much on the choice of the exchange–correlation energy functional, even though some schemes might perform slightly better than others.

What is also generally accepted, however, is that ionisation potentials, and, more in general, charge transfer properties, cannot be quantitatively predicted by looking only at Kohn–Sham eigenvalues. Again, some schemes such as the self-interaction corrected approach (but is it *bona fide* DFT?) might be better than others, but the basic difficulty is that Kohn–Sham eigenvalues, by definition, are not meant to be single electron energies. This statement is not contradicted by the Stowasser and Hoffmann paper, that we briefly discuss at the end of our reply.

Nevertheless, we do not dismiss completely the idea of looking at electrochemical properties on the basis of Kohn–Sham eigenvalues. In fact, we simply point out that what is missing in the eigenvalues to provide the correct answer is a quantity that can be computed fairly easily, representing the derivative of the eigenvalues with respect to the external bias. To summarise: Kohn–Sham eigenvalues alone are not enough, while eigenvalues together with their linear dependence on the external bias do provide a quantitative description of charge transfer properties.

Similar arguments provide the basis for a related approach, proposed by one of us to predict ionisation potentials of atoms and molecules based on Kohn–Sham eigenvalues and their derivatives with respect to the occupation number.[1] These derivatives are computed very easily, and there is no reason to neglect their contribution to the ionisation potential. This result provides a simple interpretation of the linear relation between Kohn–Sham eigenvalues and ionisation potentials presented without explanation in the paper by Stowasser and Hoffmann. Finally, concerning point 3 of the question: electrochemical stability windows of IL's have been measured, and reported in several (but not many) papers. Unfortunately, published values are rather scattered, apparently reflecting the difficulty of these measurements. More importantly, we know that the terms we include in our scheme, *i.e.*, the derivative of the eigenvalues with respect to the external bias, depend on the matrix elements of the Kohn–Sham Hamiltonian with respect to a state on the ions and a state on the electrode. These matrix elements depend on the electronic structure of the electrode, and we cannot simply compare the results obtained using jellium to experimental data measured on platinum, gold, *etc.* However, realistic simulations are feasible, even though they would represent fairly substantial computations.

1 M. Harris and P. Ballone, *Chem. Phys. Lett.*, 1999, **303**, 420.

**Professor Ouchi** inquired: In your *ab initio* calculations of ion pairs, how large is the HOMO–LUMO gap calculated? How do these HOMO–LUMO characters of ion pairs change when they are calculated together with a 40-electron jellium sphere as in Figure 7?

**Professor Ballone** answered: The HOMO–LUMO gaps computed using the Kohn–Sham eigenvalues of the gas-phase single ion pair are as follows.

[bmim][Cl]: $E_{gap} = 2.93$ eV, [bmim][PF6]: $E_{gap} = 4.91$ eV, [bmim][NTf2]: $E_{gap} = 3.48$ eV and [bmim][BH4]: $E_{gap} = 2.48$ eV. Computations have been carried out using the PBE gradient corrected approximation for exchange and correlation. The CPMD program we are using is based on plane waves and pseudopotentials.

Needless to say, the $E_{gap}$ values computed in this way are affected by the well known limitations of simple DFT approximations, and are expected to underestimate the true gap.

Both the HOMO and LUMO levels located on the molecular ions do change by a different amount upon adsorption on the jellium sphere. We computed the energy gap in the density of states "localised" on the adsorbed molecule, and we found variations of up to 1 eV with respect to the gas-phase result. Variations are not systematic, and depend on the IL molecule, on the valence density of the jellium sphere, and on the adsorption geometry, in the case that several isomers are present. Changes in the Kohn–Sham eigenvalues depend on the matrix elements of the Kohn–Sham Hamiltonian, and these can be understood in terms of bonding/anti-bonding concepts, but certainly are not easy to predict for systems with many valence electrons. An explicit computation is needed to determine values and trends.

**Professor Ouchi** remarked: Madelung potential is one of the important parameters of a large scale ionic system, and should affect many properties of ILs as well as the energy levels of ILs (relative position of cation and anion energy levels). I am wondering how the *ab initio* results on a limited number of ion pairs can be realistic for the prediction of large scale physico-chemical IL properties, and how these many-body-type complexities are overcome in a small scale *ab initio* calculation.

**Professor Ballone** answered: Our computations concern samples enclosed in a cubic box periodically repeated in space. As such, they automatically include Madelung energy and forces. To be precise, in a few computations we removed the periodicity, using open boundary conditions implemented in the CPMD code that we used for our test studies, but this was the exception and not the rule. Admittedly, the system sizes that we used in our computations with periodic boundary conditions are very small, and the description of the Madelung energy cannot be fully realistic. This, however, is only due to practical considerations, and not a fundamental limitation. Computations can certainly be done for much larger systems at the experimental charge density.

On the other hand, the investigation of highly inhomogeneous systems such as those consisting of an electrolyte in contact with a conducting electrode, might require the accurate modelling of electric fields arising from remote sources. In these cases, an hybrid QM/MM (quantum mechanics/molecular mechanics) approach might represent the method of choice.

**Dr Wishart** commented: In practice, it is difficult to compare the electrochemical potential windows that you calculate with the observable experimental ones because the latter are often defined by chemical reactivity (EC mechanisms, *etc.*) rather than reversible couples. It may be more relevant to compare your calculations with the potentials of energetic transient species generated by photolysis or radiolysis. At present there is very limited information about the energetics of such species in ionic liquids, but sufficient information exists for such comparisons in some molecular solvents, including water. Looking at it the other way, your calculations may help predict the energetics of reactive transient species in ionic liquids.

**Professor Ballone** responded: Computing the energetics of reactive transient species in ionic liquids is an interesting suggestion that we intend to explore.

On the other hand, DFT-based simulation methods could also be used to investigate systems whose electrochemical stability window is defined by chemical reactivity. We avoided such cases for the sake of simplicity, but they are not outside the reach of the method. In fact, *ab initio* simulations could provide useful insight on these systems and on their electrochemical reactions, even though, great care and possibly extensive calibrations are required for these challenging simulations.

**Professor Lynden-Bell** asked: How sensitive are calculated values of the electrochemical stability window to the electrode material? How sensitive are the experimental measurements to the electrode material?

**Professor Ballone** responded: The dependence of the electrochemical stability window on the electrode material arises primarily from the variation of the Kohn–Sham eigenvalues of the adsorbate upon changing the bias on the electrode. This variation, in turn, depends on the matrix elements connecting states on the electrode and states on the adsorbate. We would expect some chemical regularity for the results obtained for different materials.

The experimental values for the electrochemical stability window of IL that we found in the literature all referred to fairly similar electrode materials, typically noble and quasi-noble metals. Perhaps for this reason, for each IL the values measured for different electrode were similar. Moreover, and unfortunately, differences were comparable to those found by comparing the value measured by different teams on nominally the same system, implying that experimental uncertainties are fairly large.

In our manuscript, we pointed out the dependence of the electrochemical stability window on the electrode material to emphasise the fact that values obtained using jellium, *i.e.*, a simple-metal model, are not directly comparable to experimental data obtained for noble or transition metal electrodes.

However, computations for more realistic electrode models are certainly feasible.

**Professor Maginn** asked: You showed that the "brute force" approach to simulating decomposition fails even at 1000 K. Would there be any merit in increasing the temperature any further, say to 2000 K or even 3000 K? Would one learn anything about the decomposition process at these temperatures, or is it just too unphysical?

**Professor Ballone** replied: As we wrote in our manuscript, "brute force" simulations have been carried out at temperatures up to 1000 K. We did not try even higher temperatures, such as 2000 K or 3000 K. It is known that, in general, the dominant decomposition mechanism can change upon changing temperature and/or pressure. Since the decomposition reactions that might affect IL applications take place at temperatures of about 700 K, there are reasons to doubt that any decomposition reaction that is observed at 2000 K or 3000 K is also important in the realistic temperature range. Nevertheless, high temperature simulations could provide useful suggestions, possibly highlighting unexpected decompositions paths. It is essential, however, to verify the relevance of such reactions at lower temperature by the methods discussed in our contribution.

**Professor Maginn** remarked: You examined certain reaction coordinates in your decomposition studies, but it is not obvious to me how one goes about defining these reaction coordinates *a priori*. Do you have some heuristics that can be used to help identify reasonable reaction coordinates?

**Professor Ballone** answered: As we wrote in our manuscript, we made the simplest choice of reaction coordinates, in each case using a single distance between two atoms to monitor the system progression along the reaction path. In this way, it is easy to visualise the system evolution, and the statistical mechanics framework needed to transform the quantities measured during the simulation into free energy differences is trivial. Moreover, the IL's we considered are in fact relatively simple molecules, and our choice seems to be adequate for a fairly exhaustive characterisation of their thermal decomposition.

However, the definition of coordinates suitable to describe more complex reactions is a very active research field, and efficient approaches have been proposed for many important cases. Without any claim of completeness, we mention two examples, *i.e.*, the Blue-Moon approach[1] and the meta-dynamics approach.[2]

To further complement our reply we consider the following question: does an unconstrained MD simulation provide any indication of what will be the decomposition path? In principle the answer is yes. It would be possible to carry out a few equilibrium simulations at different temperatures, bracketing the expected decomposition temperature. The thermal dependence of average bond lengths and of inter-molecular distances might indeed suggest possible mechanisms for the decomposition reaction, without resorting to the extreme temperatures at which decomposition takes place within the short times allowed by ab initio simulations.

1 G. Ciccotti, R. Kapral and E. Vanden-Eijnden, *ChemPhysChem*, 2005, **6**, 1809.
2 A. Laio and M. Parrinello, *Proc. Natl. Acad. Sci. USA*, 2002, **99**, 12562.

**Dr Salanne** asked: Did you try other reaction coordinates, and if yes what changes did you observe?

**Professor Ballone** answered: In our computations we tested several different reaction paths for each IL molecule, corresponding to different choices for the pair of atoms whose distance is used as a reaction coordinate. These different choices give fairly different estimates for the barrier, and it is usually possible to determine which reaction is the first to occur. However, we did not try more complex (and least constrained) choices of reaction coordinates, possibly involving more than one coordinate. We think that IL's (at least those that we considered, and listed in the manuscript) are still relatively simple, and a more general choice of reaction coordinates might not be needed.

**Professor Jung** remarked: It seems that you have a chosen a very simple minded reaction coordinate in the simulation, that is, an inter-ionic distance of an ion pair. Given the complex nature of reactions occurring in a liquid, however, it may well be the case that the true reaction coordinate may involve collective coordinates of many ions. Have you ever considered such a possibility?

**Professor Ballone** replied: We certainly made a simple choice for the reaction coordinate, perhaps not really simple minded. As stated in the replies to some of the previous questions, we think that IL molecules are still fairly simple, and our choice of a single interatomic distance as a reaction coordinate might be adequate.

**Professor Jung** asked: One way to know if a chosen reaction coordinate is a good one or not is to calculate the transmission coefficient from the free energy profile constructed along the chosen reaction coordinate. Have you ever done it?

**Professor Ballone** responded: We did not compute absolute reaction rates for the thermal decomposition of IL's. More precisely, if we assume that the reaction follows a simple Arrhenius behaviour, we did compute the corresponding activation energy, but we did not compute the pre-exponential factor. We did not carry out this last part of the computation mainly because of the scarcity of experimental data available for a quantitative comparison with the computational results. First of all, the number of papers discussing the thermal decomposition of IL's is relatively low. Moreover, in many cases, only the "onset" temperature is given, together with the result of a chemical analysis of the products. Nevertheless, in the few cases in which a comparison is possible, our computations are remarkably close to the experimental data. In the case of [bmim][Cl], for instance, the activation energy determined by thermogravimetric analysis[1] is 129 kJ $mol^{-1}$, to be compared with the 135 $\pm$ 3 kJ $mol^{-1}$ determined in our computation.

1 Y. Hao, J. Peng, S. Hu, J. Li and M. Zhai, *Thermochim. Acta*, 2010, **501**, 78.

**Dr Taylor** remarked: As a research group we have been investigating the evaporation and vapour phase of ionic liquids for several years. The temperatures required to evaporate and distill ionic liquids make understanding their thermal decomposition critical. We have observed in our bulk distillation experiments ($T > 525$ K, $p \approx 10^{-5}$ mbar) that 1-alkyl-3-methylimidazolium tetrafluoroborate ionic liquids, $[C_nC_1Im][BF_4]$ react to form neutral carbene-borane adducts[1] (see Fig. 1).

More recent work has shown that the hexafluorophosphate analogues react in a similar manner.[2] These results were something of a surprise as (a) we expected distillation of the ionic liquid itself to occur and (b) previous *ab initio* calculations had predicted that decomposition of $[C_4C_1Im][BF_4]$ and $[C_4C_1Im][PF_6]$ occurred

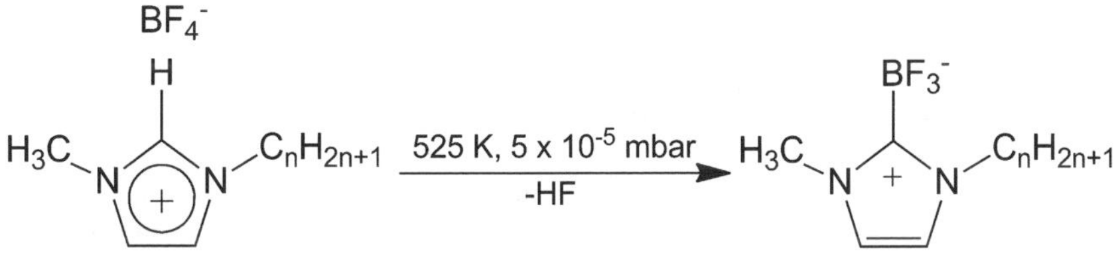

$$[C_nC_1Im][BF_4] \xrightarrow[-HF]{525\ K,\ 5 \times 10^{-5}\ mbar} C_nC_1Im\text{-}BF_3$$

**Fig. 1**

*via* demethylation of the cation to give 1-butylimidazole, methylfluoride and $BF_3/PF_5$.[3] In your paper, you mention that you have carried out simulations on $[C_4C_1Im][BH_4]$ and $[C_4C_1Im][PF_6]$ but only the results for $[C_4C_1Im]Cl$ are described. What do your systematic *ab initio* simulations suggest is the most likely thermal decomposition pathway for $[C_4C_1Im][PF_6]$ and is the result consistent with our experiments? Have you observed the formation of carbene-adducts, as seen in our work, in any of the simulations performed and, if so, are you able to deduce a possible reaction mechanism?

1 A. W. Taylor, K. R. J. Lovelock, R. G. Jones and P. Licence, *Dalton Trans.*, 2011, **40**, 1463–1470.
2 C. Clarke, S. Puttick, T. Sanderson, W. Lewis, A. W. Taylor and P. Licence, *In preparation*, 2011.
3 M. C. Kroon, W. Buijs, C. J. Peters and G.-J. Witkamp, *Thermochim. Acta*, 2007, **465**, 40–47.

**Professor Ballone** answered: In our computations we considered [bmim][BH4] instead of the (more relevant) [bmim][BF4] because the former has 24 valence electrons less than the latter. Moreover, we were not aware of the very recent experimental data discussed in ref. 1. For this reason, right now we do not have data to compare directly to the experimental results given in ref. 1.

In the case of [bmim][BH4], the reaction of BH4 with the cation C2 atom giving methylimidazolium-trifluoroborane and $H_2$ does not seem to be relevant. The situation is different for [bmim][PF6]. Also in this case we analysed a few different decomposition reactions, including both the analogue of the reaction proposed in ref. 1 for [bmim][BF4], and the dealkylation of the methyl group. At variance from the computational results given in ref. 3, we find that the barrier for the two reactions is comparable. To be fully quantitative in our estimation of the two reaction barriers, we need to re-analyse the data, and perhaps refine a couple of points along the reaction coordinate. This will take longer than a few days, but we intend to do it, given the interest of comparing our computational results with the experimental data of ref. 2.

1 A. W. Taylor, K. R. J. Lovelock, R. G. Jones and P. Licence, *Dalton Trans.*, 2011, **40**, 1463–1470.
2 C. Clarke, S. Puttick, T. Sanderson, W. Lewis, A. W. Taylor and P. Licence, *In preparation*, 2011.
3 M. C. Kroon, W. Buijs, C. J. Peters and G.-J. Witkamp, *Thermochim. Acta*, 2007, **465**, 40–47.

**Professor MacFarlane** opened the discussion of the paper by Professor Weingartner: Regarding the lack of observation in your work of distinct ion-pairs in the neat ionic liquids, is it possible that the lifetime of each ion pair could be somewhat shorter than the timescale of the measurement, but that, nonetheless, each ion spends a significant fraction of time engaged in ion pairing?

**Professor Weingaertner** replied: Our spectra extend to 20 GHz, which roughly corresponds to a relaxation time of 8 ps. Assuming that a meaningful definition of ion pairs is possible on a time scale which is long compared to the time scale of intermolecular vibrations, ions may indeed spend some fraction of time in ion pairing, without being detected by our experiments.

**Professor Maroncelli** commented: You stated that dielectric data are clear in denying the presence of ion pairs in neat ionic liquids even for times as short as a few picoseconds. Is it correct to interpret this statement to mean that the dielectric data show no evidence for the presence of dipole moments of the magnitude expected for independent ion pairs even on such short time scales?

 

**Professor Weingaertner** answered: Dielectric relaxation spectroscopy (DRS) probes fluctuations of molecular electric dipole moments. Because ion pairs have very high electric dipole moments DRS should be highly sensitive to such pairs. However, DRS is non-specific with regard to the nature of the relaxing species. Spectral assignment has therefore to resort to plausible expectations. Specifically designed molecular dynamics simulations of dielectric spectra are also of help. In this sense our statement that ion pairs are absent on the time scale of our experiments indeed mean that the spectra show no modes of the magnitude expected for independent ion pairs, or at least none that could not be more easily rationalized in terms of other mechanisms.

**Professor Wynne** asked: In your presentation, you speak of ion pairs in molten salts but what is the meaning of an "ion pair" when the density is high? It could be that an anion and a cation remain close together for a considerable time such that they could give rise to relative motions leading to a peak in the dielectric spectrum. However, in a simple molten salt (like NaCl) pairing will always be somewhat arbitrary. Therefore, should an ion-pair band in the dielectric spectrum not correlate with the transverse optical phonon bands in the crystalline state? Is the probability of ion pairing directly related to directional forces between the ions or particular patterns of stacking resulting in asymmetric packing?

**Professor Weingaertner** responded: In a dense ionic system ions are necessarily in direct contact. I fully agree with Prof. Wynne that in a simple molten salt such as NaCl ion pairing will therefore ever be somewhat arbitrary. However, ionic liquids are not simple molten salts. Variations of the cations and anions may cause drastic changes in the magnitude and balance of the multi-faceted intermolecular interactions. Watanabe and coworkers have shown that the ionicity of ionic liquids deduced from the conductance correlates with their ionic structure and have demonstrated that many other physico-chemical properties reflect this ionicity.[1] In particular, there are salt families which represent poor ionic liquids with conductivities which are much lower than expected. Poor ionic liquids must therefore have a high degree of correlation in the motion of cations and anions. Although this does not say anything on the time scale of these processes, these may well result in peaks in the dielectric spectrum on time scales which are long compared to intermolecular vibrations.

1 K. Ueno, H. Tokuda and M. Watanabe, *Phys. Chem. Chem. Phys.*, 2010, **12**, 1649.

**Dr Buchner** addressed Professor Weingaertner and Prof. Angell: I fully agree that most ionic liquids, like for instance imidazolium salts, permanently experience ion-ion contacts but do not form long-lived ion pairs.[1] However, there are exceptions. Recently we studied the sodium salt of 2,5,8,11-tetraoxatridecan-13-oate ([Na][TOTO]), an IL-forming oligoether carboxylate.[2] In this IL, which has a conductivity of only 0.5 μS/cm at 25 °C, the sodium cations are strongly bound to the carboxylate groups, forming ion pairs which rotate on the micro-second time scale at room temperature. The interaction of the ether groups with $Na^+$ is much weaker but apparently sufficient to cross-link the ion pairs to a three-dimensional network.

1 J. Sangoro *et al.*, *J. Chem. Phys.*, 2008, **128**, 214509.
2 O. Zech *et al.*, *Phys. Chem. Chem. Phys.*, 2010, **12**, 14341.

**Prof. Angell** replied: The carboxylates are famous for hanging on to their alkali cations. The first attempts to make cation conducting polymer electrolytes were made with alkali neutralized polycarboxlic acid and they were quite hopeless.

**Professor Maroncelli** asked: Given the very low values of the Walden product for trihexyl(tetradecyl)phosphonium chloride one might expect to find dielectric evidence for ion pairing in this case. Do you know of data on this liquid?

**Professor Weingaertner** answered: We have recently attempted to measure the dielectric spectrum of trihexyl(tetradecyl)phosphonium chloride. Due to a very low intensity of the signal we could not obtain high-quality data. Qualitatively, the spectrum does not differ from those of other ionic liquids.

**Dr Driver** remarked: There appears to be some confusion as to what constitutes the existence of an ion pair in the pure ionic liquid, given the ion concentrations involved and their possible interference effects, as well as in their solutions with organic solvents, and therefore whether or not one should even look for them, given the various experimental time frames. Is it possible to define an ion pair more succinctly, in terms of their rotational motions? A definition applied for translational motions which may be of some service by analogy, is that an ion pair is classified as such if the two ions comprising the pair reorient together over distances several magnitudes of their paired diameter, regardless of time frame. Another possibility may be to consider an ion pair as a complex with a lifetime that is in the minimum several orders of magnitude longer than the time of a single vibration.[1]

1 J. O'M. Bockris, A. K. N. Reddy, *Modern Electrochemistry 1*: *Ionics*, 2nd Edition, Klewer, New York, 2002.

**Professor Weingaertner** replied: I agree with Dr Driver that for many applications it will be sufficient to define ion pairs without definite reference to the time frame by the requirement that pairs to diffuse and rotate together on a time scale that is several orders of magnitude longer than the time scale of vibrations. However, many experiments are more specific with regard to the life time of these pairs (or to upper bounds for the life time). Thus, this convention sometimes loses important information.

**Professor Maroncelli** asked: As an explanation of the fact that [$C_2$-mim][$NTf_2$] and [$C_2$-mmim][$NTf_2$] differ significantly in their conductance but not in their dielectric relaxation you stated "a possible rationale is that the electrical conductance reflects effects of H-bonded configurations at short (<10 ps) time scales, which do not survive on the time scale (>10 ps) of molecular reorientation of the pairs." I am puzzled by this explanation. Doesn't the DC conductivity probe the zero-frequency or long time limit of molecular motion?

**Professor Weingaertner** answered: The DC conductivity is given by the time intergal over correlation function of the total electric current. As for example shown by Harris and Kanakubo in their contribution to this conference, this integral can be decomposed into contributions from the self-diffusion coefficients of the ions and from velocity cross-correlations of different ionic species. A low DC conductance implies the existence of correlated motions of cations and anions in the same direction, but does not say something about the time scale of these processes. However, a long-time ago, molecular dynamics simulations by Hansen and McDonald have shown that in the case of simple molten salts such as NaCl the relevant contributions occur on comparatively short time scales (see for example reference 1). In this sense such processes can indeed reduce the DC conductance without contributing to dielectric spectra in the microwave regime.

1 J. P. Hansen and I. R. McDonald, *Phys. Rev. A.*, 1975, **11**, 2111.

 

**Professor Castner** commented: What do you propose would be the best framework for us to unify the terminology for describing ion motions detected from NMR self-diffusion, conductivity, and broadband dielectric relaxation experiments? Is the "ionicity" terminology introduced in the earlier papers from Prof. M. Watanabe's group (in which they compared NMR self-diffusion coefficients with conductivity results) the way to proceed, or do you have another alternative?

**Professor Weingaertner** responded: Among the many criteria suggested in the literature the "ionicity" concept developed by Watanabe and coworkers offers a measure for ion pairing which is well founded in statistical mechanics. This scale is defined by comparison of the DC conductivity of the solution with predictions from the self diffusion coefficients of the ions based on the Nernst—Einstein equation. Any conductance reduction compared to this prediction implies that over some time cations and anions undergo correlated translational motions. A low conductance therefore implies that cations and anions move together over a considerable fraction of time, as expected in the case of ion pairing. The resulting scale correlates with several other methodologies to asses the ionic nature of ILs and also reveals interesting correlations with the ionic structure, see for example reference 1. For many solvent applications this scale is therefore of considerable predictive value. In more general, it does, however, not say anything about the time scale of the correlated motions. On the other hand, the multi-faceted interionic interactions probed by different experiments render it difficult to develop a simple unified terminology for ion pairing. Probably, molecular dynamics simulations may offer the most appropriate way for characterizing ion pair formation by probing residence times of ions at contact with counter ions.

1 K. Ueno, H. Tokuda and M. Watanabe, *Phys. Chem. Chem. Phys.*, 2010. **12**, 1649.

**Dr Harris** commented: Though the self-diffusion coefficient is given in terms of velocity auto-correlation functions in the Green–Kubo formalism, and is identified as a single species property, it still depends on the collective interactions in the system. In this sense it is no different from the conductivity which is written in terms of the cation-anion cross-correlation functions.

**Professor Maginn** asked: Is there some type of correlation function or some other property one could compute from an MD simulation that would enable direct comparison with what you measure in your experiments?

**Professor Weingaertner** answered: The frequency-dependent complex dielectric permittivity is related to the Fourier-Laplace transform of the time correlation function of the total electric dipole moment of the sample. The simulation of this collective correlation function is highly challenging. In a series of papers Steinhauser, Schröder and coworkers at the Univerity of Vienna have condutced such simulations for some prototypical ionic liquids, thus guiding the assignment of experimental dielectric spectra.[1,2]

1 C. Schröder, C. Wakai, H. Weingärtner and O. Steinhauser, *J. Chem. Phys.*, 2007, **126**, 084511.
2 C. Schröder and O. Steinhauser, *J. Chem. Phys.*, 2009, **131**, 114504.

**Professor Weingaertner** opened the discussion of the paper by Professor Maroncelli: Several authors have strongly questioned predictions by dielectric continuum models for describing properties of ionic liquids, see for example ref. 1. Do you consider your results to be unexpected?

1 M. Kobrak and H. L. Li, *Phys. Chem. Chem. Phys.*, 2010, **8**, 1922–1932.

**Professor Maroncelli** responded: I too was originally of the opinion that dielectric continuum models would be of little use in understanding the behavior of ionic liquids. But I now think that this opinion was based on an inappropriate neglect of the conductivity portion of the dielectric response. As you recognized some time ago, there is no obvious relationship between the dielectric constants of ionic liquids and solvatochromic measures of solvation energy (see Fig. 5 of the Kobrak article).[1] On solvatochromic scales, ionic liquids tend to register comparably to the most polar of conventional solvents yet their dielectric constants tend to be much smaller, often in the range of 10–15.[2] But these dielectric constants are values obtained by extrapolating MHz dielectric data to zero frequency after subtracting the diverging conductivity contribution. A more appropriate and physically sensible way of applying dielectric continuum models for equilibrium properties is to use a value of infinity for the (generalized) dielectric constant ($\eta$ in Eq. 6) in the zero-frequency limit. In such an approach, dielectric models predict solvation energies in ionic liquids to depend only upon the solute cavity size, which is in turn a function of the sizes of the ions involved. Both simulations[3,4] and solvatochromic experiments[5] show that such a description of electrostatic solvation energies is reasonable. The results we present here likewise suggest that, as long as conductivity contributions are included, a dielectric continuum description is a good starting place for understanding the time-dependence of solvation.

1 M. Kobrak and H. L. Li, *Phys. Chem. Chem. Phys.*, 2010, **8**, 1922–1932.
2 M.-M. Huang *et al.*, *J. Chem. Eng. Data*, 2011, **56**, 1494–1499.
3 R. Lynden-Bell, *J. Chem. Phys.*, **129**, 2008, 204503.
4 Roy and Maroncelli, unpublished work.
5 H. Jin *et al.*, *J. Phys. Chem. B* , 2007, **111**, 7291.

**Dr Wishart** commented: Regarding the apparent temporal offset between the spectral responses for various ionic liquids measured using C153 and those predicted from the dielectric data (as shown in Figs. 6 and 7), have you tried other probe molecules and if so, how do they compare, particularly regarding the difference in offset?

**Professor Maroncelli** answered: You raise an important point. Because C153 has been shown to probe primarily non-specific electrostatic interactions, our inclination is to view the dynamics being reported here to be intrinsic to the solvent and little influenced by the solute. When the dynamics involve changes to specific interactions such as solute–solvent hydrogen bonding, which could be the case with other probes, one does expect to observe solute-specific dynamics. Even in the absence of specific interactions the nature of the electrostatic perturbation produced by the electronic transition, as well as the rate of probe rotation can have some influence over the solvation response. (Mohsen Sajadi and Niko Ernsting have some interesting results concerning the latter effect in conventional solvents.) At this point we do not know how much the dynamics being measured in ionic liquids are solute specific. An early study of ours[1] measured the dynamics in $[Im_{41}][PF_6]$ with six varied solutes and found less than a factor of 2 variation in the slower components of solvation. But these experiments missed half of the solvation response and so are not conclusive. We intend to examine one or two other solutes using the approach described here in order to answer this question more definitively, but we have not done so yet.

1 Ito *et al.*, *Chem. Phys. Lett.*, 2004, **396**, 83.

**Professor Castner** commented: Your body of prior work on solvation dynamics in neutral organic solvents and water has shown that the observed solvation responses depend only very weakly on the choice of fluorescent probe molecule. Given the intrinsic binary (or more complex) nature of ionic liquids, would you expect that specific solvation effects will be more likely to change the observed solvation correlation functions for ionic liquids than for neutral solvents?

**Professor Maroncelli** responded: I agree that given the greater complexity of ionic liquids there is no guarantee that the relative insensitivity to solute observed in conventional solvents will pertain to these solvents. But my guess is that, apart from ionic liquids in which the domain structure is significantly larger than the probe solute, we will find only modest probe dependence when using non-specific solvation probes like C153.

**Professor Castner** asked: But if we consider two of the prototypical electron-transfer molecules, $[Fe(II)CN_6]^{-4}$ and $[Ru(II)bpy_3]^{2+}$, shouldn't we expect rather different solvation structures of the ionic liquid cations and anions, thus leading to rather different observed dynamics?

**Professor Maroncelli** replied: For such charged solutes one does expect rather different solvation structure, with first coordination shells primarily composed of cations in the first case and anions in the second case. So one might expect significant differences in solvation dynamics. But it should be kept in mind that the solvation response is a highly collective phenomenon. Because it involves the coupled dynamics of a large number of solvent ions the response could be much less sensitive to solute and solvation structure than one might at first guess. Evidence for this insensitivity comes from simulations of the non-equilibrium response to four different charge perturbations $0 \leftrightarrow +1e$ and $0 \leftrightarrow -1e$ in the model system described in Fig. 8. Apart from modest differences in the inertial dynamics, which come largely from the first solvation shell ions, we find the responses to these four perturbations to be surprisingly similar.

**Professor Wynne** asked: In your presentation, the influence of the choice of ionic liquid on the solvation dynamics as measured using the dye coumarin 153 seems relatively weak. It appears that the technique is not sensitive to (or is an argument against) mesoscopic structure observed by many authors in these liquids.[1–7] The dielectric constant is likely to vary from high in the ionic phase to low in the non-polar phase, so an effect on the dynamic Stokes shift would be expected. Would it be possible to use a non-polar probe such as pyrene to probe the dynamics of the non-polar phase separately? Are there probes that could be used to measure the dynamics in the ionic phase separately?


1 O. Russina, A. Triolo, L. Gontrani, R. Caminiti, D. Xiao, L. G. J. Hines, R. A. Bartsch, E. L. Quitevis, N. Pleckhova and K. R. Seddon, *J. Phys.: Condens. Matter*, 2009, 21, 424121.
2 D. A. Turton, J. Hunger, A. Stoppa, G. Hefter, A. Thoman, M. Walther, R. Buchner and K. Wynne, *J. Am. Chem. Soc.*, 2009, **131**, 11140–11146.
3 C. Hardacre, J. D. Holbrey, C. L. Mullan, T. G. A. Youngs and D. T. Bowron, *J. Chem. Phys.*, 2010, **133**, 074510.
4 W. Zheng, A. Mohammed, L. G. J. Hines, D. Xiao, O. J. Martinez, R. A. Bartsch, S. L. Simon, O. Russina, A. Triolo and E. L. Quitevis, *J. Phys. Chem. B*, 2011, **115**, 6572–6584.
5 D. A. Turton, J. Hunger, A. Stoppa, A. Thoman, M. Candelaresi, G. Hefter, M. Walther, R. Buchner and K. Wynne, *J. Mol. Liq.*, 2011, **159**, 2–8.
6 T. L. Greaves, D. F. Kennedy, A. Weerawardena, N. M. K. Tse, N. Kirby and C. J. Drummond, *J. Phys. Chem. B*, 2011, **115**, 2055–2066.
7 B. Aoun, A. Goldbach, M. A. González, S. Kohara, D. L. Price and M.-L. Saboungi, J. Chem. Phys., 2011, 134, 104509.


**Professor Maroncelli** answered: In the present study, as well as in our prior work with coumarin 153[1], we have found little evidence for the effect of polar/nonpolar domain structure on equilibrium solvation energies or solvation dynamics in most series of ionic liquids. Lack of an observable effect could be due to the fact that C153 is not small compared to the domain sizes expected in most ionic liquids and/or to the fact that the solute could alter the structure present in the neat liquid. But I would add that the results of Costa Gomez on solvation free energies of ethane in imidazolium ionic liquids suggest that even relatively small and presumably weakly perturbing solutes can be insensitive to the putative domain structure. As you suggest, other solutes such as the nonpolar pyrene, might provide different windows on domain structure. Unfortunately, such probes are only weakly sensitive to environmental polarity and do not provide the same sort of handle on solvation dynamics as do coumarins or other highly solvatochromic molecules.

I would like to point out that we have observed atypical solvation dynamics of C153 in one set of ionic liquids. These are liquids based on the highly alkylated cation trihexyl(tetradecyl)phosphonium. For the ionic liquids composed of this cation and 5 different anions we observed a very different relationship between solvation time and solvent viscosity than in other ionic liquids, (See Fig. 10 of ref. 1). We conjecture that this distinctive behavior might result from the solvation response entailing a translocation of the solute between different domains or perhaps the solute inducing significant domain restructuring between its $S_0$ and $S_1$ states.

1 H. Jin *et al.*, *J. Phys. Chem. B* , 2007, **111**, 7291.

**Professor Canongia Lopes** asked: Perhaps the whale-in-the-swimming pool effect can help to explain the differences found between the solvation response experiments and the dielectic data? After all coumarin 153 is a large molecule when compared to the size of the structural heterogeneities found in ionic liquids.

**Professor Maroncelli** replied: Even if the only structural change produced by a solute is removal of some solvent molecules, the solute does impact solvation dynamics.[1] So I do expect that the presence of the chromophore will have some effect on the dynamics even if it doesn't significantly perturb the structural heterogeneity of the liquid. The spatial dispersion of the solvent response described in the text is a different aspect of the problem, one which does not depend on the solute causing a change in solvation structure. It is related to the variation of the relaxation times of the neat liquid to charge fluctuations occurring on different length scales (see Fig. 9). My current opinion is that this latter aspect of molecular solvation is the more important cause of the systematic deviation of solvation times from dielectric continuum predictions. But, as illustrated in Fig. 10, accounting for the $k$ dependence of $C_{qq}$ without accounting for solvation structure does not provide an accurate description. A more complete theoretical treatment is therefore needed before we can come to any definite conclusions.

1 Song *et al.*, *J. Phys. Chem.*, **100**, 1996, 11954.

**Professor Weingaertner** said: In the case of charged dipolar particles both, dielectric and solvation spectroscopy, reflect translational as well as rotational motions. Do you expect them to be differently weighted in the two types of experiments?

**Professor Maroncelli** answered: Yes. I would expect the weighting of translational and rotational motions to be different in the solvation and dielectric experiments. My guess is that translational motions might become more prominent in the high-$k$ components of $C_{qq}(k,t)$ and therefore play a larger role in solvation than in the ($k = 0$) dielectric experiment.

**Professor Weingaertner** opened the discussion of the paper by Dr Harris: There are several sets of coefficients for decomposing diffusion and conductance data into contributions from different interactions. In particular, Friedman and Mills have repeatedly pointed out that a well-defined transcription may convert the velocity correlation coefficients used in your paper into collective diffusion coefficients which the named "joint diffusion coefficients" or "distinct diffusion coefficients", see *e.g.* ref. 1. What are the advantages of the description used in your paper over other possible representations?

1 H. L. Friedman and R. Mills, *J. Solution Chem.*, 1981, **10**, 395.

**Dr Harris** responded: For the limiting case of a one-component salt, the distinct diffusion coefficients can be simply obtained from the VCC as you say. Fig 2 is actually a Stokes–Einstein plot of the Friedman–Mills distinct *D*. One can see the regularity. It is interesting that a) salts with the same anion can be superposed on a such a plot, and b) the order of the distinct *D*s appears to depend on the nature or perhaps, the size, of the anion. At least, with the limited number of systems studied to date, $Tf_2N^-$ salts appear to behave differently to $PF_6^-$ and $BF_4^-$ salts. This information does not obtain from traditional Stokes–Einstein or Walden plots.

**Professor Weingaertner** remarked: Let me come back to the distinct diffusion coefficients. The difference between the velocity correlation coefficients and distinct diffusion coefficients is essentially founded in a different normalisation of the two sets with regard to the molar volunes of the species. Have you tried to elucidate the role of molar volumes in your data anaylsis?

**Dr Harris** answered: Not specifically for the VCC or the distinct *D*, though we did examine the pressure dependence. We have of course used density scaling of the transport properties as described in Section 2.4.

**Professor Maroncelli** said: Could you say something about what sort of physical model predicts density scaling of the sort embodied in Eqns. 7–10?

**Dr Harris** responded: The theory of thermodynamic or density scaling of transport properties is complex, and there are several approaches. The central idea is that mass and momentum transport are dominated by the repulsive part of the pair potential, and work on supercooled liquids and glassy polymers (ref. 6 of the paper) showed that properties such as dielectric relaxation times and viscosities collapsed on to common curves when scaled as a function of $(TV^{\gamma})$. Molecular dynamics simulations show that for an inverse power potential the scaling is exact with the repulsive exponent, $n = \gamma/3$. One can then fit data obtained over a range of thermodynamic states (which requires measurements under high pressure) with $\gamma$ as a fitting parameter. For Lennard–Jones fluids (see ref. 6d of the paper, for example) larger values of $\gamma$ result due to the effect of the attractive terms in the potential and, in fact, empirical fits give a wide range of values for $\gamma$ (ref. 7 of the paper) for real liquids.

Casalini *et al.* (ref. 23 of the paper) introduced the Avramov model which assumes thermally activated molecular jumps, (as in the old Eyring theory of transport processes), with a distribution of barrier heights. This leads to an expression for the relaxation time in terms of entropy, which in turn, with some assumptions, yields an analogue of eqn (7) for this property. The constants $\gamma$ and $\phi$ can be expressed in terms of thermodynamic properties, with $\gamma$ equal to the Gruneisen constant, $(V\alpha_p/(C_V\kappa_T)$, where $\alpha_p$ is the expansivity, $C_V$ the constant volume heat capacity and $\kappa_T$ is the compressibility, and with the second parameter $\phi$ equal to $C_V/(ZR)$, where $Z$ is the degeneracy of the system, roughly equal to the coordination number of the molecule or ion in the liquid.

In practice $\gamma$ and $\phi$ are obtained from empirical fits, again described in ref. 7, and, with some goodwill, one can correlate the dependence of $\phi$ on $\gamma$ with the class of liquid being examined.

As mentioned in the paper, Fragiadakis and Roland (ref. 26 of the paper) and the Dyre group (ref. 8 of the paper) prefer the use of reduced transport properties, using only $\gamma$ as a fitted parameter. The theory of Dyre's group is based on so-called isomorphs defined in terms of the scaling of the configurational energies of state points of the liquid. The use of reduced variables, rather than unreduced ones, leads to $\gamma$ values very much closer together for the diffusion coefficient and the viscosity. However I suspect the general nature of the fractional form of the Stokes–Einstein relation, which means that the self-diffusion coefficient and the viscosity have slightly different dependence on $T$ and $V$, must result in slightly different values of $\gamma$ for the two properties for a given liquid.

**Professor Maroncelli** asked: Is the fact that volume exponents ($\gamma$) needed for density scaling in ionic liquids are much smaller than those used in most conventional liquids merely related to the smaller compressibilities of ionic liquids?

**Dr Harris** answered: No. We assume it is related to the superpositioning of coulombic interactions on the Van der Waal interactions, but further work is needed to clarify the significance of the scaling parameters in ILs. The model is exact for inverse power liquids, with $\gamma = n/3$, where $n$ is the repulsive exponent, and molecular liquids tend to give $\gamma$ values near 6–9, if reduced transport properties are scaled. (See reference 26 of the paper by Roland and Fragiadakis). Interestingly the parameters for self-diffusion and viscosity in molecular liquids (for those few examples, like toluene, where data are available over a wide range of thermodynamic states) differ slightly: this is probably a reflection of the fractional Stokes–Einstein dependence. That is, the self-diffusion coefficient and the viscosity scale slightly differently with density. Water and short-chain alcohols give poor density scaling fits, very likely due to the strong H-bonding.

**Professor Wynne** remarked: In Fig. 6 of your paper, one can see a discontinuity in the diffusion coefficient and viscosity as a function of inverse temperature. Do you believe that this effect is due to a liquid–liquid phase transition as discussed by Prof. Austen Angell[1] and others?

1 C. A. Angell, *Science*, 2008, **319**, 582–587.

**Dr Harris** answered: We did differential scanning calorimetry experiments with [EMIM][TCB] and [EMIM][$CH_3SO_3$]. As described in the article, there was nothing to indicate any transition for the former, but the transition for the latter is close to the melting point. This is similar to the pattern established by Barlow *et al.* (ref. 9 of the paper) for molecular liquids, where some substances showed viscosity transitions near the melting point, but not all. Figure 6 shows separate VFT plots for the PC viscosities and self-diffusion coefficients, both above and below the apparent transition points, so there are really two scales on the $x$-axis for each property, corresponding to separate $T_o$.

**Dr Harris** remarked: One can use a Stickel analysis to detect dynamic crossover - essentially taking the derivative of an Andrade or Arrhenius plot of the appropriate property. However one has to either differentiate numerically, which requires closely spaced data, or assume a particular dependence. An Arrhenius dependence gives a constant Stickel function, and a VTF dependence a sloping line. Many authors unfortunately do not publish primary numerical data, merely a graph, often small scale, and often do not say how the data were processed. This makes it very difficult for other workers to re-examine data, particularly when one wishes to compare cross-over points determined from different types of data, *e.g.* dielectric relaxation times and viscosities or in our case diffusion coefficients.

 

**Dr Turton** commented: With respect to the dynamic crossover, the VFT function has proved a valuable tool for characterising liquids (over limited temperature ranges) but it has no real physical basis; isn't it dangerous to deduce a change in physical properties from a model that has no physical basis? *i.e.*, since there is no other indication of a dynamic crossover, wouldn't it be simpler to assume that the VFT function is an inadequate model?

**Dr Harris** answered: I think it is fair to say that all the equations used in the analysis of the transport properties, Andrade (Arrhenius), Litovitz, Barlow-Lamb or VFT are empirical. Generally the high-temperature low-pressure dynamic crossover and the low-temperature high-pressure crossover are detected using plots of these forms. In high-pressure studies the latter also shows as an inflexion in $(\mathrm{dln}\eta/\mathrm{dp})$, as first experimentally demonstrated by Bridgman. Stickel plots of the parameter $[\mathrm{dln}(\text{transport property})/\mathrm{d}(1/T)]^{1/2}$ against $(1/T)$ can be used for the same purpose, but this really requires closely spaced data to be able to perform a satisfactory numerical differentiation, something usually very tedious to achieve with say, viscosity, or self-diffusion measurements. What we have found is that this small group of three ionic liquids —and there may well be more— do not fit a single Litovitz equation for the viscosity and molar conductivity, whereas the vast majority of ionic liquids so far examined do. This is what we find to be strange and unusual and, we think, is worthy of investigation by techniques other than simple transport property measurements. If it is a real effect, it should show in high-pressure studies of these same materials, as for the larger crossover found in propylene carbonate and other molecular liquids. This is something for future work.

**Professor Maroncelli** asked: Paul and Samanta reported apparent discontinuities in the temperature dependence of the fluorescence quantum yields of a fluidity probe in two ionic liquids.[1] This behavior resembles the dynamic crossover effect you report for viscosities and conductivities in some ionic liquids. It would be interesting to pursue similar fluorescence probe measurements in the liquids you have studied.

1 A. Paul and A. Samanta, *J. Phys. Chem. B*, 2008, **112**, 16626.

**Dr Harris** answered: I suspect the mechanism may be different. As yet we have found the viscosity crossover for only the three EMIM salts mentioned in the paper. We do not observe any crossover in two of the systems studied by Paul and Samanta, namely [BMIM][$BF_4$] or [BMIM][$PF_6$], though we do have results in the region examined in their study. It is possible their effect is related to a change in the interaction of the fluorescence probe molecule with the ionic liquid in some way.

**Professor Maginn** remarked to Professor Ribeiro: I believe you were the first to show a possible mechanism for the observation that imidazolium cations often have higher self-diffusivities than the smaller anions they are paired with. Is it simply a geometric effect or is there more to it?

**Dr Ribeiro** responded: In the MD simulations of imidazolium ionic liquids,[1], the projections of ionic displacement and velocity along different directions of the cation structure showed significant anisotropy of cations motion that seemed to account for the finding that diffusion is faster for cations than anions. Soon after that work, we simulated an ammonium ionic liquid, thus a more spherical moiety,[2] and now the ionic displacements were essentially isotropic. In this simulation of ammonium ionic liquid we could observe the cation self-diffusion coefficient $D_+$ changing from higher to smaller than $D_-$ as the temperature was increased. Of course, although D is calculated by a single particle time correlation function, it is the result of coupling to other single particle dynamics, such as reorientational motion, and collective dynamical modes of the liquid. For instance, keeping the same geometry of the ammonium cation, but replacing $CH_2$ by oxygen atom in the long chain, further MD simulations,[3] suggested higher ionic mobility for the ether derivative proper to more flexibility of the ether functionalized chain within the domains of chain segregation. Thus, anisotropy due to geometric effect of cation structure is involved in early stages of translational dynamics, but more than that is acting at longer times.

1 S. M. Urahata and M. C. C. Ribeiro, *J. Chem. Phys.*, **122**, 2005, 024511.
2 L. J. A. Siqueira and M. C. C. Ribeiro, *J. Phys. Chem. B*, **111**, 2007, 11776.
3 L. J. A. Siqueira and M. C. C. Ribeiro, *J. Phys. Chem. B*, **113**, 2009, 1074.

**Professor Maginn** asked: Can your analysis help shed light on why sometimes cations have greater self-diffusivities than anions while for some systems, the trend is reversed? Is there a physical explanation?

**Dr Harris** replied: The work of Schröder is interesting in this regard: he believes collective effects are important where $D_+/D_-$ is greater than 1.[1]

1 C. Schröder, *J. Chem. Phys.*, 2011, **135**, 024502.

**Professor Maginn** asked: The issue of whether the self-diffusivity is greater for the cation or anion is an interesting one. We found several years ago in a joint NMR/MD paper that it is possible with pyridinium-based ionic liquids to observe a cross-over where sometimes the cation has a greater self-diffusivity and sometimes the anion does. I am wondering if you have any insight into the physical mechanisms for such behavior.

**Dr Harris** replied: That is intriguing. Was the effect observed experimentally or only in simulations?

**Professor Maginn** responded: The phenomenon was observed both in the simulations and in the experiments.

 

# Concluding remarks

R. M. Lynden-Bell*

*Received 31st August 2011, Accepted 26th October 2011*
**DOI: 10.1039/c1fd00108f**

Since the beginning of this century there has been an enormous increase in interest in ionic liquids, that is in molten salts that happen to be liquid at ambient conditions or at least below 100 °C. The contributions to this Discussion gave a good overview of the knowledge of the physical chemical aspects of these materials in 2011. Many experimental and computational results were reported and questions for future examination posed.

## 1 Introduction

In his introductory lecture[1] Austen Angell gave us an interesting account of the history of these liquids in addition to telling us about recent developments. Overall this Faraday Discussion has provided an opportunity to assess our progress in measuring and understanding physical chemical properties of these materials.

One striking fact is the range of techniques that have been used to study these liquids and the way in which information from different techniques can often be complementary. An example of this is the work presented by Turton *et al.*[2] in which they used optical Kerr spectroscopy, dielectric relaxation spectroscopy and terahertz time domain spectroscopy to study dynamics in protic ionic liquids, and there are many other papers with similar ranges of techniques. As a simulator I was particularly glad to see several papers combining experiments and simulations. One example of this is the paper by Castner and co-workers[3] which reports results from X-ray scattering and simulations of the same system.

## 2 Properties of ionic liquids

A surprising property of many ionic liquids was the discovery of spatial heterogeneity and microsegregation. The existence of spatial inhomogeneities was first found in simulations,[4,5] and the existence of a 'pre-peak' in the static structure factor $S(Q)$ found in simulations[6] and scattering experiments. We have had two papers investigating this.[3,7] Russina and Triolo[7] show that the size of the domains varies linearly for $[C_nMIM][Cl]$ from $n = 6$ to 10, but interestingly a 1 : 1 mixture of $[C_6MIM][Cl]$ and $[C_{10}MIM][Cl]$ has larger domains (about 2.4 nm) than neat $[C_8MIM][Cl]$ (about 2.25 nm). They interpret this in terms of mixing of alkyl chains in hydrophobic domains. The interpretation of the existence of a 'pre-peak' as caused by the formation of such domains is controversial. The paper by Castner and co-workers[3] presents an alternative, more local interpretation. They use simulations combined with X-ray scattering to show that in the phosphonium ionic liquid $[P_{14,666}][NTf_2]$ the principal contribution to the prepeak arises from anion-anion correlations. An important result presented by Russina and Triolo is that the prepeak (at 4 nm$^{-1}$) disappears when the alkyl side chain in $[C_6MIM][NTf_2]$ is replaced by a diether side chain of the same length. They conclude that it is the contrast between hydrophobic and charged parts of the cation that leads to

*Department of Chemistry, University of Cambridge, Lensfield Road, Cambridge, CB2 1EW, UK. E-mail: rmlb@cam.ac.uk; Tel: +44 1223 336535*

domain formation. The work by Costa Gomes and co-workers[8] is indirectly relevant to this. They measured the solubility of ethane and butane in imidazolium ionic liquids with different chain lengths. Although it could have been expected that the solubility of small alkanes would increase rapidly once the liquid became heterogeneous, the trend of increased solubility seems to be approximately linear with the alkyl chain length. However, it was pointed out in the discussion that any solute perturbs the solvent structure and butane may be creating local alkane-rich regions. Certainly there seems to be good evidence of preferential solvation by the side chains of the imidazolium ions which is mainly driven by a favourable solvation entropy.

Another possibly controversial topic is the existence (or not) of ion pairs in ionic liquids. I think that most people who do simulations would agree that in neat ionic liquids cations interact with a number of anions and vice versa. There are certainly strong correlations between the dynamics of cations and anions which leads to deviations from the Nernst–Einstein relation relating conductivity to diffusion constants.

Weingartner and co-workers[9] present some illuminating results on the speciation of ionic liquids as a function of concentration in solvents with dielectric constants lower than 10. In these solvents $[C_2MIM][NTf_2]$ shows a minimum conductivity with respect to concentration at about 1 mol $L^{-1}$. Although the initial decrease is traditionally interpreted as the onset of formation of ion pairs, and the minimum to the onset of formation of ion triples, the authors argue from dielectric relaxation measurements and theoretical grounds that in fact clusters of ions rather than just ion triples form. At higher concentrations these redissociate, and finally form the charge-ordered structure of a neat ionic liquid. I found a similarity between these results on speciation in solvents with low dielectric constants and the results found by Rai and Maginn[10] using Monte Carlo simulation for the state of $[C_nMIM][NTf_2]$ in the vapour phase. In the latter case the number of charged clusters is very small, possibly highlighting the difference between a solvent with $3 < \varepsilon < 10$ and no solvent with $\varepsilon = 1$. This study is mainly concerned with the thermodynamics of vapour-liquid coexistence.

Ionic liquids are frequently used as solvents and there is much empirical knowledge about the contrasts between polar solvents and ionic liquids. Harper and co-workers[11] contribute to this and show that the increased rate of the Menshutkin reaction in $[C_4MIM][N(CF_3SO_2)_2]$ is due to an increase of activation entropy, but that previous explanations for such an increase are not applicable in this case. Obviously, there is more work to be done here.

I was intrigued by various examples of the effects of $Li^+$ ions. As these are the smallest cations available they have a high field strength. Lithium salts dissolve readily in ionic liquids, albeit with viscosity increases. Matic and co-workers[12] study the effects of $Li^+$ on properties such as conductivity and the glass transition in detail. Another example of the effects of $Li^+$ ions was found by Endres *et al.*[13] in their AFM study of the ionic liquid/electrode interface where the addition of 10 wt% LiCl changes the interaction of an AFM probe with the first layer from being repulsive to being attractive.

## 2.1 Novel ionic liquids

Several examples of novel (at least to me) ionic liquids have been investigated by various authors. Dicationic ionic liquids in which two imidazolium head groups are tethered by a $C_{10}$ alkyl chains were used by Matic and co-workers[12] in their study of lithium doping. These seem to offer favourable alternatives to monocationic liquids as their conductivity and glass transition temperatures change less with $Li^+$ concentration.

The ideas presented by MacFarlane and co-workers[14] of using protic ionic liquids based on pharmaceutical compounds struck me as being very promising. The challenge seems to be to obtain salts of drugs with low enough melting points and the best degree of ionicity for trans-membrane transport.

In a different direction the nanoparticle ionic liquids investigated by Panagiotopoulos and co-workers[15] are at least half way to being colloids. Nanosized cores of an inorganic material are grafted with charged groups and dissolved in the counterions, which often have long tails and provide liquidity.

# 3 Ionic liquids near surfaces

## 3.1 Solid–liquid surfaces

The characteristic of any ionic liquid near a solid surface is that it has alternating layers of dense and less dense material. If the surface is charged then successive layers also show charge alternation. Endres and co-workers[13] combine AFM and STM measurements with cyclic voltammograms of a pyrrolidinium ionic liquid on a single crystal gold (111) surface. The AFM measurements show the existence of about 5 successive layers. They found that the position of the first layer varied with voltage indicating structural changes. He showed beautiful STM images obtained *in situ* which give some indication of the changes in the gold structure as ions are adsorbed. Pádua and co-workers[16] presented results for simulations of a hydroxylated imidazolium ionic liquid on a number of glasses. As the glass surfaces were neither exactly planar nor chemically homogeneous the layer structure in the ionic liquid is somewhat diminished. The distribution of ions in the first layer is determined by local interactions with the glass and shows some regions with more anions and other regions with more cations. These patches propagate into the second layer, but by the third layer the lateral structure is almost homogeneous.

Carbon nanotubes are currently very popular as a research area and the ionic liquid community is no exception. Indeed ionic liquids and carbon nanotubes are possible candidates for supercapacitors. In their contribution Frolov *et al.*[17] examined the local environment around nanotubes in a number of ionic liquids. In a similar way to the behaviour near flat surfaces, successive cylindrical layers of different density form adjacent to the nanotubes and the ions in the first layer are preferentially oriented. The authors investigated the effects of changing the surface charge density on the layer structure and the orientations of the ions.

In a somewhat related study, Shim *et al.*[18] used simulation to investigate the behaviour of the ionic liquid [$C_2$MIM][$BF_4$] confined between two graphene sheets. They were particularly interested in the behaviour of the neat ionic liquid compared with that of a solution in an organic solvent in the context of supercapacitors. They find very efficient shielding in the liquid, especially by the $BF_4^-$ ion. The total capacitance was dominated by the cathodic contribution which led them to suggest that consideration of the choice of cations should be important in the design of supercapacitors.

## 3.2 Liquid–liquid and liquid–vapour surfaces

Deyko and Jones[19] have investigated the adsorption and absorption of water on an ionic liquid/vacuum interface using line of sight mass spectrometry with temperature programmed desorption. They were able to interpret their results in terms of enthalpy profiles for water entering the ionic liquid. These showed a physisorbed minimum at the surface, with a second minimum for water in the first ionic layer. To make the transition between these two the water must penetrate the oily layer formed by alkyl chains which gives quite a high barrier. Not surprisingly the height of this barrier increased from roughly 50 kJ mol$^{-1}$ for [$C_2$MIM][$NTf_2$] to about 60 kJ mol$^{-1}$ for [$C_8$MIM][$BF_4$] with a thicker oily layer.

Sum frequency spectroscopy provides a sensitive way of examining interfaces as the signal is generated in the anisotropic environment of the interface. Ouchi and co-workers[20] used this spectroscopy to look at the orientation of $SO_2$ groups near the interfaces of [$C_4$MIM][TFSA] and [$C_8$MIM][TFSA] with water. They found that the $SO_2$ groups tend to point towards the water, and the shift in the symmetric stretch frequency shows that they hydrogen-bond to water, while although the alkyl tails are somewhat disordered in the butyl case, they definitely point towards the ionic liquid in the octyl case.

## 4 Dynamics

In the opening lecture Austen Angell[1] pointed out the importance of bulk transport properties, in particular viscosity and conductivity. He showed that the Walden plot, in which the logarithm of the molar conductivity is plotted against the logarithm of the inverse viscosity, gives a useful classification of ionic liquids. Solutions of KCl fall on a straight line with unit slope. The plots for many ionic liquids lie quite close to this line but generally below it. There are a few examples of super-ionic liquids with higher than expected conductivity, but also some with significantly lower conductivity. In protic ionic liquids this decrease in ionicity can be attributed to equilibria between the ionised and non-ionised forms, but in other cases it must be due to the strong correlations between the motion of cations and anions. Harris and Kanakubo also addressed the question of transport properties.[21] If the motion of ions were uncorrelated then the Nernst–Einstein equation for a 1 : 1 salt

$$\Lambda_{\mathrm{NE}} = \frac{F^2}{RT}\left(D^+ + D^-\right) \tag{1}$$

would hold. In this equation $\Lambda_{\mathrm{NE}}$ is the predicted molar conductivity, $D^+$ and $D^-$ are the self diffusion constants of the cation and anion, respectively, and the other constants have their usual meaning. One can define a Nernst–Einstein deviation parameter $\Delta$ from the difference between the observed ($\Lambda_{\mathrm{obs}}$) and predicted ($\Lambda_{\mathrm{NE}}$) conductivities by

$$\Lambda_{\mathrm{obs}} = (1 - \Delta)\Lambda_{\mathrm{NE}} \tag{2}$$

and $\Delta$ can be shown to depend on the difference between the velocity correlation functions of unlike and like ions. Harris and Kanakubo found that $\Delta$ is not sensitive to pressure and that the values of $\Delta$ lie between 0.3 and 0.5 for a range of neat ionic liquids.

Turton and coworkers[2] used OKE (optical Kerr effect spectroscopy) and dielectric spectroscopy to span a large frequency range for ethyl and propyl ammonium nitrates. The interpretation of dielectric properties and OKE spectra is difficult as the responses are collective in nature; while they were able to analyse the response into various components and compare the contribution of different components in the two spectroscopies, not all the components can be identified. They conclude that the presence of long lived aggregates or mesoscopic structures in alkyl ammonium nitrate liquids at low temperatures can be ruled out. Maroncelli and coworkers[22] also managed to make measurements over a remarkably wide time range. They investigated the solvation response to the dye coumarin 153 in a variety of ionic liquids. They compared the measured response to that predicted from the dielectric continuum model using experimentally determined frequency-dependent permittivities, $\varepsilon(\nu)$. Although there is a qualitative similarity in shape, the dye response is always slower. This difference persists in data generated from simulations. They argue that the difference must be attributed to the difference in the length scales of the experiments and suggest that in the response to the dye the distance dependence of the dielectric function should be included in addition to its frequency dependence. $\varepsilon(\nu)$ must be replaced by $\varepsilon(\nu, q)$. This conclusion is supported to some extent by simulation data, although it is still not clear whether a continuum description can ever be quantitative.

Roling and his coworkers[23] used electric impedance spectroscopy to investigate the dynamics of the response of an ionic liquid to changing electrical potentials. They observed two processes with very different time scales, one milliseconds and one seconds. Very fast processes could not be observed in their apparatus. They attributed the fast (millisecond) process to the time scale for ion diffusion and deduce that the slower process must have a high activation barrier and is probably associated with movement of ions that are strongly bound to the surface. Certainly such slow processes could not be probed directly in simulations.

Wishart and co-workers[24] give an example where the viscosity of ionic liquids is a help rather than a hindrance. They used picosecond pulse radiolysis to study the solvation of electrons in a pyrrolidinium ionic liquid. The slower motions in ionic liquids compared to normal liquids gives the electron more time to become solvated rather than scavenged so that the solvation process and the properties of the solvated electron can be studied more easily.

## 5 Insights from simulation

Many papers used simulations to complement and interpret experiments, and this is most encouraging. There were also some pure simulation studies that give important insights and pointers for future simulations. Ballone and Cortes-Huerto[25] show how *ab initio* simulations can now be used to study properties of ionic liquids involving breaking bonds or electronic changes. As the time scales and system sizes that can be used in such simulations are restricted, 'brute force' simulations are not useful. Rather, one has to use carefully chosen conditions. These authors use guided simulations to investigate various paths for the thermal breakdown of $[C_4MIM][Ntf_2]$ and $[C_4MIM][Cl]$. They also discuss the technical problems in computing the oxidation and reduction of an ionic liquid in contact with a charged electrode and propose a simple method to find this from DFT calculations.

Most classical calculations use non-polarisable force fields as they are cheaper and easier to use. We all know that polarisation effects are likely to be important, but hope that they will not be too important. Salanne *et al.*[26] investigate speciation in $[C_2MIM][AlCl_4]$ using a polarisable model for the chloride ion. Only the tetrahedral $AlCl_4$ ion is found in this material, and excellent agreement with experimental values of viscosities and conductivities is found. It is well known that viscosities tend to be much too high and mobilities too low in simulations using classical force fields with full ionic charges. Several force fields have been proposed with reduced charges on the ions and it has not been clear whether this just compensates for the omission of polarisation or whether it is in some sense a real effect. The paper by Delle Site and co-workers[27] addresses this question using results from *ab initio* simulations to construct the charge distributions for classical simulations. The resulting charges are 60–80% of the expected value of $\pm 1e$. They suggest that this reduction represents the effects of the electronic polarisabilty of the material and that the scaling is related to the high frequency dielectric constant by $q_{\rm eff} = q/\sqrt{\varepsilon_{\rm el}} \approx 0.7q$. It should be noted that their preferred method of measuring charges (Blöchl charges) from their *ab initio* calculations is carried out in reciprocal space and so includes both nearby and distant parts of the electric field. Reducing charges reduces dynamical timescales. However, they note that the ab initio calculations show broad distributions for ion dipole moments suggesting that polarisability should really be included as well. Whether that would make the liquids too mobile or not remains to be seen.

Agilio Pádua pointed out a couple of times during the discussion that it is important (but rarely done) to include polarisability of the surface when simulating an ionic liquid in contact with a conducting support such as a metal electrode or a carbon nanotube. This can be done using image charges or polarisable surface sites.[28,29] I must plead guilty myself for making this omission.

## 6 Where next?

A number of interesting suggestions for future work arose during the discussions. One idea was to measure the time dependence of the different peaks in $S(Q)$. A complete investigation of $S(Q, \omega)$ will surely give more insight into the structure and dynamics of ionic liquids and in particular the origin of the low $q$ 'pre-peak'. We now have information about the nature of ionic liquid vapours,[10] and it would be interesting to examine the mechanism of boiling using simulation. We shall probably see more studies of sub-ionic liquids and here it was suggested that engineering specific ionic liquids would be useful, for example with pairs of ions that would preferentially form ion pairs because of steric hindrance. Similarly studies of ions with branched chains could be compared to those with straight chains to help elucidate the formation of hydrophobic regions in ionic liquids.

## Acknowledgements

I should like to thank the organising committee of this conference and in particular Chris Hardacre and Mario del Pópolo for arranging such a good conference. Not only were the papers of high quality, but the discussion was informative and good-humoured.

## References

1 A. Angell, *Faraday Discuss.*, 2011, **154**, DOI: 10.1039/C1FD00112D.
2 D. A. Turton, T. Sonnleitner, A. Ortner, M. Walther, G. Hefter, K. R. Seddon, S. Stana, N. V. Plechkova, R. Buchner and K. Wynne, *Faraday Discuss.*, 2012, **154**, DOI: 10.1039/C1FD00054C.
3 H. K. Kashyap, C. S. Santos, H. V. R. Annapureddy, N. S. Murthy, C. J. Margulis and E. W. Castner, Jr, *Faraday Discuss.*, 2012, **154**, DOI: 10.1039/C1FD00059D.
4 Y. T. Wang and G. A. Voth, *J. Am. Chem. Soc.*, 2005, **127**, 12192.
5 J. N. C. Lopes and A. A. H. Pádua, *J. Phys. Chem. B*, 2006, **110**, 3330.
6 S. M. Urahata and M. C. C. Ribeiro, *J. Chem. Phys.*, 2004, **120**, 1855–1863.
7 O. Russina and A. Triolo, *Faraday Discuss.*, 2012, **154**, DOI: 10.1039/C1FD00073J.
8 M. F. Costa Gomes, L. Pison, A. S. Pensado and A. A. H. Pádua, *Faraday Discuss.*, 2012, **154**, DOI: 10.1039/C1FD00074H.
9 Y. Jiang, H. Nadolny, S. Käshammer, S. Weibels, W. Schröer and H. Weingärtner, *Faraday Discuss.*, 2012, **154**, DOI: 10.1039/C1FD00075F.
10 N. Rai and E. J. Maginn, *Faraday Discuss.*, 2012, **154**, DOI: 10.1039/C1FD00090J.
11 H. M. Yau, A. K. Croft and J. B. Harper, *Faraday Discuss.*, 2012, **154**, DOI: 10.1039/C1FD00060H.
12 J. Pitawala, J.-K. Kim, P. Jacobsson, V. Koch, F. Croce and A. Matic, *Faraday Discuss.*, 2012, **154**, DOI: 10.1039/C1FD00056J.
13 F. Endres, N. Borisenko, S. Z. El Abedin, R. Hayes and R. Atkin, *Faraday Discuss.*, 2012, **154**, DOI: 10.1039/C1FD00050K.
14 J. Stoimenovski, P. M. Dean, E. I. Izgorodina and D. R. MacFarlane, *Faraday Discuss.*, 2012, **154**, DOI: 10.1039/C1FD00071C.
15 B. Hong, A. Chremos and A. Z. Panagiotopoulos, *Faraday Discuss.*, 2012, **154**, DOI: 10.1039/C1FD00076D.
16 K. Shimizu, A. Pensado, P. Malfreyt, A. A. H. Pádua and J. N. Canongia Lopes, *Faraday Discuss.*, 2012, **154**, DOI: 10.1039/C1FD00043H.
17 A. I. Frolov, K. Kirchner, T. Kirchner and M. V. Fedorov, *Faraday Discuss.*, 2012, **154**, DOI: 10.1039/C1FD00080B.
18 Y. Shim, H. J. Kim and Y. Jung, *Faraday Discuss.*, 2012, **154**, DOI: 10.1039/C1FD00086A.
19 A. Deyko and R. G. Jones, *Faraday Discuss.*, 2012, **154**, DOI: 10.1039/C1FD00062D.
20 T. Iwahashi, Y. Sakai, D. Kim, T. Ishiyama, A. Moritac and Y. Ouchi, *Faraday Discuss.*, 2012, **154**, DOI: 10.1039/C1FD00061F.
21 K. R. Harris and M. Kanakubo, *Faraday Discuss.*, 2012, **154**, DOI: 10.1039/C1FD00085C.
22 M. Maroncelli, X.-X. Zhang, M. Liang, D. Roy and N. P. Ernsting, *Faraday Discuss.*, 2012, **154**, DOI: 10.1039/C1FD00058F.
23 B. Roling, M. Drüschler and B. Huber, *Faraday Discuss.*, 2012, **154**, DOI: 10.1039/C1FD00088H.
24 J. F. Wishart, A. M. Funston, T. Szreder, A. R. Cook and M. Gohdo, *Faraday Discuss.*, 2012, **154**, DOI: 10.1039/C1FD00065A.
25 P. Ballone and R. Cortes-Huerto, *Faraday Discuss.*, 2012, **154**, DOI: 10.1039/C1FD00064K.

 

26 M. Salanne, L. J. A. Siqueira, A. P. Seitsonen, P. A. Madden and B. Kirchner, *Faraday Discuss.*, 2012, **154**, DOI: 10.1039/C1FD00053E.
27 K. Wendler, F. Dommert, Y. Y. Zhao, R. Berger, C. Holm and L. Delle Site, *Faraday Discuss.*, 2012, **154**, DOI: 10.1039/C1FD00051A.
28 A. S. Pensado and A. Pádua, *Angew. Chem. Int. Ed.*, 2011, **50**, 8683–8687.
29 F. Iori and S. Corni, *J. Comput. Chem.*, 2008, **29**, 1656–1666.

# Poster titles

Vapour structure of room temperature ionic liquids, **M Bier,** *Max Planck Institute for Metals Research, Germany*

On the interactions between ionic liquids and amino acids in aqueous solutions: molecular dynamics simulation studies, **L I N Tomé, M Jorge, J R B Gomes and J A P Coutinho,** *Universidade de Aveiro, Portugal*

Molecular dynamics simulations of imidazolium based ionic liquids: effect of the cation structure on the thermodynamic properties, **L I. N Tomé, L M N B F Santos, C Herdes and J A P Coutinho,** *Universidade de Aveiro, Portugal*

XPS study of electrochemical processes at the ionic liquid/electrode and ionic liquid/ ultra high vacuum interface, **A Foelske-Schmitz, D Weingarth, R Kötz and A Wokaun,** *Paul Scherrer Institut, Switzerland*

Spectroscopic and thermal-gravimetric investigation of two new rhenium (V) ionic liquids complexes, **N S AlHokbany,** *King Saud University, Saudi Arabia*

Dynamic Change of Ionic liquids under an External Electric Field, **Y Wang and R Shi,** *Chinese Academy of Sciences, China*

Structure and short-time dynamics of ionic liquids: A molecular dynamics simulation and Raman spectroscopy study, **S M Urahata, L J A Siqueira, L T Costa, B G Nicolau, T C Penna and M C C Ribeiro,** *Universidade de São Paulo, Brazil*

Broadband dielectric and optical Kerr-effect spectroscopy of the dynamics of aprotic ionic liquids, **T Sonnleitner, D Turton, A Ortner, M Walther, K Wynne, and R Buchner,** *Universität Regensburg, Germany*

Physicochemical characterization of new pyridinium ionic liquids, **P Verdía, M Hernaiz, N Urangab and E Tojo,** *University of Vigo, Spain*

Signatures of structural and dynamical heterogeneity in the ultrafast and ultraslow dynamics of ionic liquids, **E L Quitevis,** *Texas Tech University, USA*

Insights on capacitive and structural properties of [BMI][PF6] confined between graphite electrodes from molecular dynamics, **C Merlet, M Salanne, B Rotenberg and P A Madden,** *UPMC Univ Paris 06, France*

Molecular dynamics simulation of polarizable ionic liquids, **C Schröder and O Steinhauser,** *University of Vienna, Austria*

Slow and fast capacitive processes taking place at the interface Ionic liquid/Au(111), **M Drüschler, B Huber, T Jänsch, J Wallaner, N Borisenko, F Endres and B Roling,** *University of Marburg, Germany*

Cation adsorption at Au(111) in imidazolium-based ionic liquids, **Y-Z Su, J-W Yan and B-W Mao,** *Xiamen University, China*

Physicochemical properties and activity coefficients at infinite dilution for organic solutes and water in the ionic liquid 4-(2-methoxyethyl)-4-methylmorpholinium bis(trifluoromethylsulfonyl)-amide, **A Marciniak and M Wlazło,** *Warsaw University of Technology, Poland*

Separation of aliphatic hydrocarbons from aromatic hydrocarbons using [BMPYR][TCB] ionic liquid, **A Marciniak, M Królikowski, U Domańska and W E Acree,** *Warsaw University of Technology, Poland*

Phase behavior of the binary systems {quinolinium, or isoquinolinium – based ionic liquid + hydrocarbon, or + an alcohol}, **M Zawadzki and U Domańska,** *Warsaw University of Technology, Poland*

Measurements of activity coefficients at infinite dilution of organic compounds and water in isoquinolinium-based ionic liquid [$C_8$iQuin][$NTf_2$], **M Zawadzki, U Domańska, M Królikowska, M M Tsibangu, D Ramjugernath and T M Letcher,** *Warsaw University of Technology, Poland*

Unique fluorescence behaviour in ionic liquids having imidazolium cations, **S Cha, T Shim, Doseok Kim and Y Ouchi,** *Sogabg University, South Korea*

Multinuclear NMR studies on translational and rotational motions for four ILs composed of BF4 anion, **K Hayamizu and S Seki,** *National Institute of Advanced Industrial Science and Technology, Japan*

Hydrogen Bonding in Ionic Liquids?, **P Johansson, J Grondin, J-C Lassègues, D Cavagnat, T Buffeteau, K Angenendt and R Holom,** *Chalmers University of Technology, Sweden*

Ionic liquids binary mixtures: the influence of the composition on their properties and catalytic ability, **F D'Anna, S Marullo, P Vitale and R Noto,** *Viale delle Scienze-Parco d'Orleans II, Italy*

Novel ionic liquid processes: Large scale recovery of valuable material and synthesis and recovery, **P Fitzpatrick, R James and Y Wharton,** *C-Tech Innovation Ltd, UK*

Ionic liquids in lithium salts: Characterization of solid electrolytes with high LiTFSI content, **M Marczewski and P Johansson,** *Chalmers University of Technology, Sweden*

Molecular substructure of a free ionic liquid surface: A combined resonant soft and hard x-ray reflectivity study, **M Mezger, B M Ocko, H Reichert and M Deutsch,** *Max Planck Institute for Polymer Research, Germany*

Integrating experiments and computations to model the conductivity of ionic liquids, **B Ahlström, M Marczewski and P Johansson,** *Chalmers University of Technology, Sweden*

Investigation of morphological and mechanical properties of [bmim][Tf2N] thin layers on solid surfaces by Scanning Probe Microscopy, **S Bovio, M Galluzzi, A Podestà, C Lenardi and P Milani,** *Università degli Studi di Milano, Italy*

Characterization of electrical properties of thin films of [bmim][Tf2N] using Atomic Force Microscopy, **M Galluzzi, A Podestà, S Bovio and P Milani,** *Università degli Studi di Milano, Italy*

 

Structure and fluidity of ionic liquids confined to thin films, **K R J Lovelock, L Crowhurst, H Niedermeyer, T Welton, A M Smith, N N Gosvami and S Perkin,** *University College London, UK*

Production of polyetheretherketone (PEEK™) in ionic liquid media, **C R Langrick, A V Puga, K R Seddon and K Whiston,** *INVISTA Textiles (UK) Ltd, UK*

New ionic liquids from azepane and 3-methylpiperidine exhibiting wide electrochemical windows, **T Belhocine, S A Forsyth, H Q N Gunaratne, A V Puga, K R Seddon, G Srinivasan and K Whiston,** *INVISTA Textiles (UK) Ltd, UK*

Penicillin ionic mixtures: "salting" antibiotic properties, **D Garcia, A Rocha, E J Cabrita, I Sá-Nogueira, R Frade and P Vidinha,** *REQUIMTE, Portugal*

Synthesis and applications of ionic and polyionic liquids, **S M McDermott, W D Fogarty, D A Rooney and C B Breslin,** *National University of Ireland Maynooth, Republic of Ireland*

New ionic liquids based on pyrazolium cation: synthesis, physical and structural properties, **D Mendola, F Castiglione, A Famulari, A Mele, C Chiappe, A Sanzone and C Brocchetta,** *Politecnico di Milano, Italy*

Adsorption, absorption and desorption of three gases at an ionic liquid surface, **S G Hessey and R G Jones,** *University of Nottingham, UK*

Intrinsic analysis of the vapour/liquid interface of a room-temperature ionic liquid, **G Hantal, M N D S Cordeiro and M Jorge,** *Universidade do Porto, Portugal*

Investigating ionic liquids for pre-treatment of plant biomass, **R Gammons, J Slattery, S Shimizu, N Bruce and S McQueen-Mason and T James,** *University of York, UK*

Comparing electron transfer in ionic liquids and neutral solvents, **H Y Lee, E W Castner, Jr. and J F Wishart,** *Rutgers University, USA*

Simulation of electrode screening by ionic liquids: implications for electrochemistry, **R M Lynden-Bell and M V Fedorov (presented by A Frolov),** *Max Planck Institute for Mathematics in the Sciences, Germany*

Ion interactions with carbon nanomaterial surfaces in non-aqueous solutions: insights from molecular modeling, **A I Frolov and M V Fedorov,** *Max Planck Institute for Mathematics in the Sciences, Germany*

New strategy for the resolution/separation of sec-alcohols combining ionic acylating agents with supercritical $CO_2$, **S Rebocho, N Lourenço, S Barreiros, C Afonso and A Paiva,** *REQUIMTE/CQFB, Portugal*

Characterization of ion jelly using nmr techniques, **V Augusto, P Vidinha and E J Cabrita.** *REQUIMTE, Portugal*

Viscosity, density and speed of sound of 1-alkyl-3-methylimidazolium bis(trifluoromethylsulfonyl)imide ionic liquids, **A O Figueiras, A P C Ribeiro, F J V Santos, A F Santos, P Goodrich, C Hardacre, M J V Lourenço, I M S Lampreia and C A Nieto de Castro,** *Universidade de Lisboa, Portugal*

Economic impact of good measurements. Are we aware?, **J M P França, A P C Ribeiro and C A Nieto de Castro,** *Universidade de Lisboa, Portugal*

Spin-lattice dynamics in protic ionic liquids: C relaxation for fast motions, **G W Driver and P Ingman,** *Abo Akademi University, Finland*

Physicochemical properties and interactions of alkyl-phosphonate ionic liquids, **J Pitawala, P Jacobsson and A Matic,** *Chalmers University of Technology, Sweden*

Molecular dynamics simulation of ionic liquids at metallic surfaces: structure, thermodynamics and friction, **A C F Mendonça, A S Pensado, P Malfreyt, A A H Pádua,** *Université Blaise Pascal Clermont-Ferrand, France*

Structural organization and phase behaviour of 1-butyl-3-methylimidazolium hexafluorophosphate: An high pressure Raman spectroscopy study, **O Russina, B Fazio, C Schmidt and A Triolo,** *Istituto Struttura Materia, Italy*

The Skinner Prize for the best poster was awarded to Ms Celine Merlet of University Pierre and Marie Curie, France, for her poster on insights on capacitive and structural properties of [BMI][$PF_6$] confined between graphite electrodes from molecular dynamics.

# List of participants

Dr B. Ahlstrom, *Chalmers University of Technology, Sweden*
Dr N. Alhakbany, *King Saud University, Saudi Arabia*
Professor C. Angell, *Arizona State University, USA*
Miss V. Augusto, *Universidade Nova de Lisboa, Portugal*
Professor P. Ballone, *Queen's University, Belfast, United Kingdom*
Dr M. Bier, *Max Planck Institute for Intelligent Systems, Germany*
Mr S. Bovio, *Italy*
Dr R. Buchner, *Universität Regensburg, Germany*
Professor Dr J. Canongia Lopes, *Instituto Superior Técnico, Portugal*
Professor E. Castner, *Rutgers, The State University of New Jersey, USA*
Professor R. Compton, *University of Oxford, United Kingdom*
Professor M. Costa Gomes, *Université Blaise Pascal, France*
Dr A. Croft, *Bangor University, United Kingdom*
Ms K. Crossey, *Queen's University, Belfast, United Kingdom*
Ms Y. Dai, *The Netherlands*
Dr F. D'Anna, *Università di Palermo, Italy*
Dr M. Del Pópolo, *Queen's University, Belfast, United Kingdom*
Dr L. Delle Site, *Max Planck Institute for Polymer Research, Germany*
Mr W. Dennis, *Royal Society of Chemistry, United Kingdom*
Mr F. Dommert, *Institute for Computational Physics, Universität Stuttgart, Germany*
Dr G. Driver, *Umeå University, Sweden*
Mr M. Drüschler, *Universität Marburg, Germany*
Professor F. Endres, *Technical University Clausthal, Germany*
Dr P. Fitzpatrick, *C-Tech Innovation Ltd, United Kingdom*
Dr A. Foelske-Schmitz, *Paul Scherrer Institut, Switzerland*
Mrs N. Forero Martinez, *Queen's University, Belfast, United Kingdom*
Professor M. Fox, *University of Leeds, United Kingdom*
Mr A. Frolov, *Max Planck Institute for Mathematics in the Sciences, Germany*
Mr R. Gammons, *University of York, United Kingdom*
Miss D. Garcia, *Faculdade de Ciências e Tecnologia, Universidade Nova de Lisboa, Portugal*
Dr G. Hantal, *University of Porto, Portugal*
Professor C. Hardacre, *Queen's University, Belfast, United Kingdom*
Dr J. Harper, *University of New South Wales, Australia*
Dr K. Harris, *University of New South Wales, Australia*
Dr K. Hayamizu, *National Institute of Advanced Industrial Science and Technology, Japan*
Mr S. Hessey, *University of Nottingham, United Kingdom*
Miss R. James, *C-Tech Innovation Ltd, United Kingdom*
Professor P. Johansson, *Chalmers University of Technology, Sweden*
Professor R. Jones, *University of Nottingham, United Kingdom*
Professor Y. Jung, *Seoul National University, South Korea*
Professor D. Kim, *Sogang University, South Korea*
Ms H. Lee, *Rutgers, The State University of New Jersey, USA*
Mr Y. Li, *Royal Society of Chemistry, United Kingdom*
Dr K. Lovelock, *University College London, United Kingdom*
Professor R. Lynden-Bell, *University of Cambridge, United Kingdom*
Professor D. MacFarlane, *Monash University, Australia*
Professor E. Maginn, *University of Notre Dame, USA*
Dr A. Marciniak, *Warsaw University of Technology, Poland*

Dr M. Marczewski, *Chalmers University of Technology, Sweden*
Professor M. Maroncelli, *Penn State, USA*
Professor A. Matic, *Chalmers University of Technology, Sweden*
Dr S. McDermott, *National University of Ireland, Republic of Ireland*
Dr A. Mele, *Politecnico di Milano, Italy*
Miss C. Merlet, *Laboratoire PECSA, Université Pierre et Marie Curie, France*
Dr M. Mezger, *Max Planck Institute for Polymer Research, Germany*
Dr V. Nguyen, *CNRS, Université Paris-Sud, France*
Professor R. Noto, *Università di Palermo, Italy*
Professor Y. Ouchi, *Nagoya University, Japan*
Professor A. Padua, *Université Blaise Pascal, Clermont-Ferrand, France*
Professor A. Panagiotopoulos, *Princeton University, USA*
Mr J. Pitawala, *Chalmers University of Technology, Sweden*
Dr A. Puga, *Universitat Rovira i Virgili, Spain*
Dr E. Quitevis, *Texas Tech University, USA*
Miss S. Rebocho, *Universidade Nova de Lisboa, Portugal*
Miss A. Ribeiro, *Universidade de Lisboa, Portugal*
Dr M. Ribeiro, *Universidade de São Paulo, Brazil*
Professor Dr B. Roling, *Universität Marburg, Germany*
Dr M. Salanne, *Université Pierre et Marie Curie, France*
Prof Dr T. Schafer, *University of the Basque Country, Spain*
Dr C. Schröder, *University of Vienna, Austria*
Dr M. Serrano Santos, *DEEFA, Universitat Rovira I Virgili, Spain*
Mr R. Shi, *Chinese Academy of Sciences, China*
Mr T. Sonnleitner, *Universität Regensburg, Germany*
Ms Y. Su, *Xiamen University, China*
Dr A. Taylor, *University of Nottingham, United Kingdom*
Professor Dr E. Tojo, *University of Vigo, Spain*
Dr L. Tome, *University of Aveiro, Portugal*
Dr A. Triolo, *Consiglio Nazionale Delle Ricerche, Italy*
Dr D. Turton, *University of Glasgow, United Kingdom*
Dr P. Vidinha, *Faculdade de Ciências e Tecnologia, Universidade Nova de Lisboa, Portugal*
Ms S. Wang, *Queen's University, Belfast, United Kingdom*
Professor H. Weingärtner, *Ruhr-Universität Bochum, Germany*
Mrs K. Wendler, *Max Planck Institute for Polymer Research, Germany*
Professor J. Wishart, *Brookhaven National Lab, USA*
Professor K. Wynne, *University of Glasgow, United Kingdom*
Mr M. Zawadzki, *Warsaw University of Technology, Poland*

 

# Index of contributors*

* The page numbers in **bold** type indicate papers submitted for discussions.